Lecture Notes in Computer Science 13350

More information about this series at https://link.springer.com/bookseries/558

Derek Groen · Clélia de Mulatier ·
Maciej Paszynski · Valeria V. Krzhizhanovskaya ·
Jack J. Dongarra · Peter M. A. Sloot (Eds.)

Computational Science – ICCS 2022

22nd International Conference
London, UK, June 21–23, 2022
Proceedings, Part I

 Springer

Editors
Derek Groen ⓘ
Brunel University London
London, UK

Clélia de Mulatier ⓘ
University of Amsterdam
Amsterdam, The Netherlands

Maciej Paszynski ⓘ
AGH University of Science and Technology
Krakow, Poland

Valeria V. Krzhizhanovskaya ⓘ
University of Amsterdam
Amsterdam, The Netherlands

Jack J. Dongarra ⓘ
University of Tennessee at Knoxville
Knoxville, TN, USA

Peter M. A. Sloot ⓘ
University of Amsterdam
Amsterdam, The Netherlands

ISSN 0302-9743   ISSN 1611-3349 (electronic)
Lecture Notes in Computer Science
ISBN 978-3-031-08750-9   ISBN 978-3-031-08751-6 (eBook)
https://doi.org/10.1007/978-3-031-08751-6

This Springer imprint is published by the registered company Springer Nature Switzerland AG
The registered company address is: Gewerbestrasse 11, 6330 Cham, Switzerland

Preface

Welcome to the 22nd annual International Conference on Computational Science (ICCS 2022 - https://www.iccs-meeting.org/iccs2022/), held during 21–23 June, 2022, at Brunel University London, UK. After more than two years of a pandemic that has changed so much of our world and daily lives, this edition marks our return to a – partially – in-person event. Those who were not yet able to join us in London had the option to participate online, as all conference sessions were streamed.

Although the challenges of such a hybrid format are manifold, we have tried our best to keep the ICCS community as dynamic, creative, and productive as always. We are proud to present the proceedings you are reading as a result of that.

Standing on the River Thames in southeast England, at the head of a 50-mile (80 km) estuary down to the North Sea, London is the capital and largest city of England and the UK. With a rich history spanning back to Roman times, modern London is one of the world's global cities, having a prominent role in areas ranging from arts and entertainment to commerce, finance, and education. London is the biggest urban economy in Europe and one of the major financial centres in the world. It also features Europe's largest concentration of higher education institutions.

ICCS 2022 was jointly organized by Brunel University London, the University of Amsterdam, NTU Singapore, and the University of Tennessee.

Brunel University London is a public research university located in the Uxbridge area of London. It was founded in 1966 and named after the Victorian engineer Isambard Kingdom Brunel, who managed to design and build a 214m long suspension bridge in Bristol back in 1831. Brunel is well-known for its excellent Engineering and Computer Science Departments, and its campus houses a dedicated conference centre (the Hamilton Centre) which was used to host ICCS. It is also one of the few universities to host a full-length athletics track, which has been used both for practice purposes by athletes such as Usain Bolt for the 2012 Olympics and for graduation ceremonies.

The International Conference on Computational Science is an annual conference that brings together researchers and scientists from mathematics and computer science as basic computing disciplines, as well as researchers from various application areas who are pioneering computational methods in sciences such as physics, chemistry, life sciences, engineering, arts, and humanitarian fields, to discuss problems and solutions in the area, identify new issues, and shape future directions for research.

Since its inception in 2001, ICCS has attracted increasing numbers of attendees and higher-quality papers, and this year – in spite of the ongoing pandemic—was not an exception, with over 300 registered participants. The proceedings series has become a primary intellectual resource for computational science researchers, defining and advancing the state of the art in this field.

The theme for 2022, "The Computational Planet," highlights the role of computational science in tackling the current challenges of the all-important quest for sustainable development. This conference aimed to be a unique event focusing on recent developments in scalable scientific algorithms, advanced software tools, computational grids, advanced numerical methods, and novel application areas. These innovative novel models, algorithms, and tools drive new science through efficient application in physical systems, computational and systems biology, environmental systems, finance, and other areas.

ICCS is well-known for its excellent lineup of keynote speakers. The keynotes for 2022 were as follows:

- Robert Axtell, George Mason University, USA
- Peter Coveney, University College London, UK
- Thomas Engels, Technische Universität Berlin, Germany
- Neil Ferguson, Imperial College London, UK
- Giulia Galli, University of Chicago, USA
- Rebecca Wade, Heidelberg Institute for Theoretical Studies, Germany

This year we had 474 submissions (169 submissions to the main track and 305 to the thematic tracks). In the main track, 55 full papers were accepted (32%), and in the thematic tracks, 120 full papers (39%). A higher acceptance rate in the thematic tracks is explained by the nature of these, where track organizers personally invite many experts in a particular field to participate in their sessions.

ICCS relies strongly on our thematic track organizers' vital contributions to attract high-quality papers in many subject areas. We would like to thank all committee members from the main and thematic tracks for their contribution to ensure a high standard for the accepted papers. We would also like to thank Springer, Elsevier, and Intellegibilis for their support. Finally, we appreciate all the local organizing committee members for their hard work to prepare for this conference.

We are proud to note that ICCS is an A-rank conference in the CORE classification.

We wish you good health in these troubled times and look forward to meeting you at the next conference, whether virtually or in-person.

June 2022

Derek Groen
Clélia de Mulatier
Maciej Paszynski
Valeria V. Krzhizhanovskaya
Jack J. Dongarra
Peter M. A. Sloot

Organization

General Chair

Valeria Krzhizhanovskaya — University of Amsterdam, The Netherlands

Main Track Chair

Clélia de Mulatier — University of Amsterdam, The Netherlands

Thematic Tracks Chair

Maciej Paszynski — AGH University of Science and Technology, Poland

Scientific Chairs

Peter M. A. Sloot — University of Amsterdam, The Netherlands I Complexity Institute NTU, Singapore

Jack Dongarra — University of Tennessee, USA

Local Organizing Committee

Chair

Derek Groen — Brunel University London, UK

Members

Simon Taylor — Brunel University London, UK
Anastasia Anagnostou — Brunel University London, UK
Diana Suleimenova — Brunel University London, UK
Xiaohui Liu — Brunel University London, UK
Zidong Wang — Brunel University London, UK
Steven Sam — Brunel University London, UK
Alireza Jahani — Brunel University London, UK
Yani Xue — Brunel University London, UK
Nadine Aburumman — Brunel University London, UK
Katie Mintram — Brunel University London, UK
Arindam Saha — Brunel University London, UK
Nura Abubakar — Brunel University London, UK

Thematic Tracks and Organizers

Advances in High-Performance Computational Earth Sciences: Applications and Frameworks – IHPCES

Takashi Shimokawabe	University of Tokyo, Japan
Kohei Fujita	University of Tokyo, Japan
Dominik Bartuschat	Friedrich-Alexander-Universität Erlangen-Nürnberg, Germany

Artificial Intelligence and High-Performance Computing for Advanced Simulations – AIHPC4AS

Maciej Paszynski	AGH University of Science and Technology, Poland

Biomedical and Bioinformatics Challenges for Computer Science – BBC

Mario Cannataro	Università Magna Graecia di Catanzaro, Italy
Giuseppe Agapito	Università Magna Graecia di Catanzaro, Italy
Mauro Castelli	Universidade Nova de Lisboa, Portugal
Riccardo Dondi	University of Bergamo, Italy
Rodrigo Weber dos Santos	Universidade Federal de Juiz de Fora, Brazil
Italo Zoppis	Università degli Studi di Milano-Bicocca, Italy

Computational Collective Intelligence – CCI

Marcin Maleszka	Wroclaw University of Science and Technology, Poland
Ngoc Thanh Nguyen	Wroclaw University of Science and Technology, Poland
Dosam Hwang	Yeungnam University, South Korea

Computational Health – CompHealth

Sergey Kovalchuk	ITMO University, Russia
Stefan Thurner	Medical University of Vienna, Austria
Georgiy Bobashev	RTI International, USA
Jude Hemanth	Karunya University, India
Anastasia Angelopoulou	University of Westminster, UK

Computational Optimization, Modelling, and Simulation – COMS

Xin-She Yang	Middlesex University London, UK
Leifur Leifsson	Purdue University, USA
Slawomir Koziel	Reykjavik University, Iceland

Computer Graphics, Image Processing, and Artificial Intelligence – CGIPAI

Andres Iglesias   Universidad de Cantabria, Spain

Machine Learning and Data Assimilation for Dynamical Systems – MLDADS

Rossella Arcucci   Imperial College London, UK

Multiscale Modelling and Simulation – MMS

Derek Groen   Brunel University London, UK
Diana Suleimenova   Brunel University London, UK
Bartosz Bosak   Poznan Supercomputing and Networking Center,
                                      Poland
Gabor Závodszky   University of Amsterdam, The Netherlands
Stefano Casarin   Houston Methodist Research Institute, USA
Ulf D. Schiller   Clemson University, USA
Wouter Edeling   Centrum Wiskunde & Informatica,
                                      The Netherlands

Quantum Computing – QCW

Katarzyna Rycerz   AGH University of Science and Technology,
                                      Poland
Marian Bubak   Sano Centre for Computational Medicine and
                                      AGH University of Science and Technology,
                                      Poland | University of Amsterdam,
                                      The Netherlands

**Simulations of Flow and Transport: Modeling, Algorithms, and Computation –
SOFTMAC**

Shuyu Sun   King Abdullah University of Science and
                                      Technology, Saudi Arabia
Jingfa Li   Beijing Institute of Petrochemical Technology,
                                      China
James Liu   Colorado State University, USA

**Smart Systems: Bringing Together Computer Vision, Sensor Networks,
and Machine Learning – SmartSys**

Pedro Cardoso   University of Algarve, Portugal
João Rodrigues   University of Algarve, Portugal
Jânio Monteiro   University of Algarve, Portugal
Roberto Lam   University of Algarve, Portugal

Software Engineering for Computational Science – SE4Science

Jeffrey Carver	University of Alabama, USA
Caroline Jay	University of Manchester, UK
Yochannah Yehudi	University of Manchester, UK
Neil Chue Hong	University of Edinburgh, UK

Solving Problems with Uncertainty – SPU

Vassil Alexandrov	Hartree Centre - STFC, UK
Aneta Karaivanova	Institute for Parallel Processing, Bulgarian Academy of Sciences, Bulgaria

Teaching Computational Science – WTCS

Angela Shiflet	Wofford College, USA
Nia Alexandrov	Hartree Centre - STFC, UK

Uncertainty Quantification for Computational Models – UNEQUIvOCAL

Wouter Edeling	Centrum Wiskunde & Informatica, The Netherlands
Anna Nikishova	SISSA, Italy

Reviewers

Tesfamariam Mulugeta Abuhay
Jaime Afonso Martins
Giuseppe Agapito
Shahbaz Ahmad
Elisabete Alberdi
Luis Alexandre
Nia Alexandrov
Vassil Alexandrov
Julen Alvarez-Aramberri
Domingos Alves
Sergey Alyaev
Anastasia Anagnostou
Anastasia Angelopoulou
Samuel Aning
Hideo Aochi
Rossella Arcucci
Costin Badica
Bartosz Balis
Daniel Balouek-Thomert
Krzysztof Banaś

Dariusz Barbucha
João Barroso
Valeria Bartsch
Dominik Bartuschat
Pouria Behnodfaur
Jörn Behrens
Adrian Bekasiewicz
Gebrail Bekdas
Mehmet Ali Belen
Stefano Beretta
Benjamin Berkels
Daniel Berrar
Georgiy Bobashev
Marcel Boersma
Tomasz Boiński
Carlos Bordons
Bartosz Bosak
Giuseppe Brandi
Lars Braubach
Marian Bubak

Adam Glos
Ivo Goncalves
Alexandrino Gonçalves
Jorge González-Domínguez
Yuriy Gorbachev
Pawel Gorecki
Markus Götz
Michael Gowanlock
George Gravvanis
Derek Groen
Lutz Gross
Lluis Guasch
Pedro Guerreiro
Tobias Guggemos
Xiaohu Guo
Manish Gupta
Piotr Gurgul
Zulfiqar Habib
Mohamed Hamada
Yue Hao
Habibollah Haron
Ali Hashemian
Carina Haupt
Claire Heaney
Alexander Heinecke
Jude Hemanth
Marcin Hernes
Bogumila Hnatkowska
Maximilian Höb
Jori Hoencamp
Rolf Hoffmann
Wladyslaw Homenda
Tzung-Pei Hong
Muhammad Hussain
Dosam Hwang
Mauro Iacono
David Iclanzan
Andres Iglesias
Mirjana Ivanovic
Takeshi Iwashita
Alireza Jahani
Peter Janků
Jiri Jaros
Agnieszka Jastrzebska
Caroline Jay

Piotr Jedrzejowicz
Gordan Jezic
Zhong Jin
David Johnson
Guido Juckeland
Piotr Kalita
Drona Kandhai
Epaminondas Kapetanios
Aneta Karaivanova
Artur Karczmarczyk
Takahiro Katagiri
Timo Kehrer
Christoph Kessler
Loo Chu Kiong
Harald Koestler
Ivana Kolingerova
Georgy Kopanitsa
Pavankumar Koratikere
Triston Kosloske
Sotiris Kotsiantis
Remous-Aris Koutsiamanis
Sergey Kovalchuk
Slawomir Koziel
Dariusz Krol
Marek Krótkiewicz
Valeria Krzhizhanovskaya
Marek Kubalcík
Sebastian Kuckuk
Eileen Kuehn
Michael Kuhn
Tomasz Kulpa
Julian Martin Kunkel
Krzysztof Kurowski
Marcin Kuta
Panagiotis Kyziropoulos
Roberto Lam
Anna-Lena Lamprecht
Kun-Chan Lan
Rubin Landau
Leon Lang
Johannes Langguth
Leifur Leifsson
Kenneth Leiter
Florin Leon
Vasiliy Leonenko

Jean-Hugues Lestang
Jake Lever
Andrew Lewis
Jingfa Li
Way Soong Lim
Denis Mayr Lima Martins
James Liu
Zhao Liu
Hong Liu
Che Liu
Yen-Chen Liu
Hui Liu
Marcelo Lobosco
Doina Logafatu
Marcin Los
Stephane Louise
Frederic Loulergue
Paul Lu
Stefan Luding
Laura Lyman
Lukasz Madej
Luca Magri
Peyman Mahouti
Marcin Maleszka
Bernadetta Maleszka
Alexander Malyshev
Livia Marcellino
Tomas Margalef
Tiziana Margaria
Svetozar Margenov
Osni Marques
Carmen Marquez
Paula Martins
Pawel Matuszyk
Valerie Maxville
Wagner Meira Jr.
Roderick Melnik
Pedro Mendes Guerreiro
Ivan Merelli
Lyudmila Mihaylova
Marianna Milano
Jaroslaw Miszczak
Janio Monteiro
Fernando Monteiro
Andrew Moore

Eugénia Moreira Bernardino
Anabela Moreira Bernardino
Peter Mueller
Ignacio Muga
Khan Muhammad
Daichi Mukunoki
Vivek Muniraj
Judit Munoz-Matute
Hiromichi Nagao
Jethro Nagawakar
Kengo Nakajima
Grzegorz J. Nalepa
Yves Nanfack
Pratik Nayak
Philipp Neumann
David Chek-Ling Ngo
Ngoc Thanh Nguyen
Nancy Nichols
Sinan Melih Nigdeli
Anna Nikishova
Hitoshi Nishizawa
Algirdas Noreika
Manuel Núñez
Frederike Oetker
Schenk Olaf
Javier Omella
Boon-Yaik Ooi
Eneko Osaba
Aziz Ouaarab
Raymond Padmos
Nikela Papadopoulou
Marcin Paprzycki
David Pardo
Diego Paredesconcha
Anna Paszynska
Maciej Paszynski
Ebo Peerbooms
Sara Perez-Carabaza
Dana Petcu
Serge Petiton
Frank Phillipson
Eugenio Piasini
Juan C. Pichel
Anna Pietrenko-Dabrowska
Laércio L. Pilla

Armando Pinho
Yuri Pirola
Mihail Popov
Cristina Portales
Roland Potthast
Małgorzata Przybyła-Kasperek
Ela Pustulka-Hunt
Vladimir Puzyrev
Rick Quax
Cesar Quilodran-Casas
Enrique S. Quintana-Orti
Issam Rais
Andrianirina Rakotoharisoa
Raul Ramirez
Celia Ramos
Vishwas Rao
Kurunathan Ratnavelu
Lukasz Rauch
Robin Richardson
Miguel Ridao
Heike Riel
Sophie Robert
Joao Rodrigues
Daniel Rodriguez
Albert Romkes
Debraj Roy
Katarzyna Rycerz
Emmanuelle Saillard
Ozlem Salehi
Tarith Samson
Alberto Sanchez
Ayşin Sancı
Gabriele Santin
Vinicius Santos-Silva
Allah Bux Sargano
Robert Schaefer
Ulf D. Schiller
Bertil Schmidt
Martin Schreiber
Gabriela Schütz
Franciszek Seredynski
Marzia Settino
Mostafa Shahriari
Zhendan Shang
Angela Shiflet

Takashi Shimokawabe
Alexander Shukhman
Marcin Sieniek
Nazareen Sikkandar-Basha
Robert Sinkovits
Mateusz Sitko
Haozhen Situ
Leszek Siwik
Renata Słota
Oskar Slowik
Grażyna Ślusarczyk
Sucha Smanchat
Maciej Smołka
Thiago Sobral
Isabel Sofia Brito
Piotr Sowiński
Robert Speck
Christian Spieker
Michał Staniszewski
Robert Staszewski
Steve Stevenson
Tomasz Stopa
Achim Streit
Barbara Strug
Patricia Suarez
Dante Suarez
Diana Suleimenova
Shuyu Sun
Martin Swain
Jerzy Świątek
Piotr Szczepaniak
Edward Szczerbicki
Tadeusz Szuba
Ryszard Tadeusiewicz
Daisuke Takahashi
Osamu Tatebe
Carlos Tavares Calafate
Kasim Tersic
Jannis Teunissen
Mau Luen Tham
Stefan Thurner
Nestor Tiglao
T. O. Ting
Alfredo Tirado-Ramos
Pawel Topa

Contents – Part I

ICCS 2022 Main Track Full Papers

Developing a Scalable Cellular Automaton Model of 3D Tumor Growth   3
  Cyrus Tanade, Sarah Putney, and Amanda Randles

Human-Level Melodic Line Harmonization   17
  Jan Mycka, Adam Żychowski, and Jacek Mańdziuk

Classification Methods Based on Fitting Logistic Regression to Positive
and Unlabeled Data   31
  *Konrad Furmańczyk, Kacper Paczutkowski, Marcin Dudziński,
  and Diana Dziewa-Dawidczyk*

Robust Control of Perishable Inventory with Uncertain Lead Time Using
Neural Networks and Genetic Algorithm   46
  Ewelina Cholodowicz and Przemyslaw Orlowski

Batch QR Factorization on GPUs: Design, Optimization, and Tuning   60
  Ahmad Abdelfattah, Stan Tomov, and Jack Dongarra

ChemTab: A Physics Guided Chemistry Modeling Framework   75
  Amol Salunkhe, Dwyer Deighan, Paul E. DesJardin, and Varun Chandola

Establishing Metrics to Quantify Underlying Structure in Vascular Red
Blood Cell Distributions   89
  Sayan Roychowdhury, Erik W. Draeger, and Amanda Randles

Coevolutionary Approach to Sequential Stackelberg Security Games   103
  Adam Żychowski and Jacek Mańdziuk

FINCH: Domain Specific Language and Code Generation for Finite
Element and Finite Volume in Julia   118
  Eric Heisler, Aadesh Deshmukh, and Hari Sundar

Classifying Anomalous Members in a Collection of Multivariate
Time Series Data Using Large Deviations Principle: An Application
to COVID-19 Data   133
  Sreelekha Guggilam, Varun Chandola, and Abani K. Patra

Adaptive Regularization of B-Spline Models for Scientific Data 150
David Lenz, Raine Yeh, Vijay Mahadevan, Iulian Grindeanu,
and Tom Peterka

Content-Aware Generative Model for Multi-item Outfit Recommendation 164
Valery Volokha and Klavdiya Bochenina

Retrofitting Structural Graph Embeddings with Node Attribute Information 178
Piotr Bielak, Daria Puchalska, and Tomasz Kajdanowicz

Hierarchical Ensemble Based Imbalance Classification 192
Jie Xie, Mingying Zhu, and Kai Hu

Numerical Approximation of the One-Way Helmholtz Equation Using
the Differential Evolution Method 205
Mikhail S. Lytaev

Iterative Solution for the Narrow Passage Problem in Motion Planning 219
Jakub Szkandera and Ivana Kolingerová

A Productive and Scalable Actor-Based Programming System for PGAS
Applications 233
Sri Raj Paul, Akihiro Hayashi, Kun Chen, and Vivek Sarkar

Is Context All You Need? Non-contextual vs Contextual Multiword
Expressions Detection 248
Maciej Piasecki and Kamil Kanclerz

Out-of-Distribution Detection in High-Dimensional Data Using
Mahalanobis Distance - Critical Analysis 262
Henryk Maciejewski, Tomasz Walkowiak, and Kamil Szyc

Multi-contextual Recommender Using 3D Latent Factor Models
and Online Tensor Decomposition 276
Basem Suleiman, Ali Anaissi, Muhammad Johan Alibasa,
and Harrison Truong

Efficient Computational Algorithm for Stress Analysis
in Hydro-Sediment-Morphodynamic Models 291
Alia Al-Ghosoun, Ashraf S. Osman, and Mohammed Seaid

Enhancing Computational Steel Solidification by a Nonlinear Transient
Thermal Model 305
Fatima-Ezzahrae Moutahir, Youssef Belhamadia, Mofdi El-Amrani,
and Mohammed Seaid

Weakly-Supervised Cell Classification for Effective High Content
Screening 318
   Adriana Borowa, Szczepan Kruczek, Jacek Tabor, and Bartosz Zieliński

Which Visual Features Impact the Performance of Target Task
in Self-supervised Learning? 331
   Witold Oleszkiewicz, Dominika Basaj, Tomasz Trzciński,
   and Bartosz Zieliński

DITA-NCG: Detecting Information Theft Attack Based on Node
Communication Graph 345
   Zhenyu Cheng, Xiaochun Yun, Shuhao Li, Jinbu Geng, Rui Qin,
   and Li Fan

Boosted Ensemble Learning Based on Randomized NNs for Time Series
Forecasting 360
   Grzegorz Dudek

Exploring Ductal Carcinoma In-Situ to Invasive Ductal Carcinoma
Transitions Using Energy Minimization Principles 375
   Vivek M. Sheraton and Shijun Ma

Scaling the PageRank Algorithm for Very Large Graphs on the Fugaku
Supercomputer 389
   Maxence Vandromme, Jérôme Gurhem, Miwako Tsuji, Serge Petiton,
   and Mitsuhisa Sato

Ultrafast Focus Detection for Automated Microscopy 403
   Maksim Levental, Ryan Chard, Kyle Chard, Ian Foster,
   and Gregg Wildenberg

Models and Metrics for Mining Meaningful Metadata 417
   Tyler J. Skluzacek, Matthew Chen, Erica Hsu, Kyle Chard, and Ian Foster

HNOP: Attack Traffic Detection Based on Hierarchical Node Hopping
Features of Packets 431
   Jinbu Geng, Zhenyu Cheng, Zhicheng Liu, Shuhao Li, and Rui Qin

TROPHY: Trust Region Optimization Using a Precision Hierarchy 445
   Richard J. Clancy, Matt Menickelly, Jan Hückelheim, Paul Hovland,
   Prani Nalluri, and Rebecca Gjini

Incremental Mining of Frequent Serial Episodes Considering Multiple
Occurrences 460
   Thomas Guyet, Wenbin Zhang, and Albert Bifet

MHGEE: Event Extraction via Multi-granularity Heterogeneous Graph 473
  Mingyu Zhang, Fang Fang, Hao Li, Qingyun Liu, Yangchun Li,
  and Hailong Wang

Consistency Fences for Partial Order Delivery to Reduce Latency 488
  Nooshin Eghbal and Paul Lu

Dynamic Classification of Bank Clients by the Predictability of Their
Transactional Behavior 502
  Alexandra Bezbochina, Elizaveta Stavinova, Anton Kovantsev,
  and Petr Chunaev

Unveiling User Behavior on Summit Login Nodes as a User 516
  Sean R. Wilkinson, Ketan Maheshwari, and Rafael Ferreira da Silva

n-type B-N Co-doping and N Doping in Diamond from First Principles 530
  Delun Zhou, Lin Tang, Jinyu Zhang, Ruifeng Yue, and Yan Wang

Stock Predictor with Graph Laplacian-Based Multi-task Learning 541
  Jiayu He, Nguyen H. Tran, and Matloob Khushi

Identification of MEEK-Based TOR Hidden Service Access Using the Key
Packet Sequence 554
  Xuebin Wang, Zhipeng Chen, Zeyu Li, Wentao Huang, Meiqi Wang,
  Shengli Pan, and Jinqiao Shi

Towards a Scalable Set Similarity Join Using MapReduce and LSH 569
  Sébastien Rivault, Mostafa Bamha, Sébastien Limet, and Sophie Robert

Cyberbullying Detection with Side Information: A Real-World Application
of COVID-19 News Comment in Chinese Language 584
  Jian Xing, Xiaoyu Zhang, Lin Chen, Yu Ding, Yaru Zhang, Wei Hu,
  Zhicheng Jin, Jingya Wang, Yaowei Chen, and Yi Hong

Designing a Training Set for Musical Instruments Identification 599
  Daniel Kostrzewa, Blazej Koza, and Pawel Benecki

Characterizing Wildfire Perimeter Polygons from QUIC-Fire 611
  Li Tan, Raymond A. de Callafon, and Ilkay Altıntaş

On a New Generalised Iteration Method in the PSO-Based Newton-Like
Method 623
  Ireneusz Gościniak and Krzysztof Gdawiec

Peridynamic Damage Model Based on Absolute Bond Elongation 637
Shangyuan Zhang and Yufeng Nie

Privacy Paradox in Social Media: A System Dynamics Analysis 651
Ektor Arzoglou, Yki Kortesniemi, Sampsa Ruutu, and Tommi Elo

GPU Power Capping for Energy-Performance Trade-Offs in Training
of Deep Convolutional Neural Networks for Image Recognition 667
Adam Krzywaniak, Pawel Czarnul, and Jerzy Proficz

Forecasting Bank Default with the Merton Model: The Case of US Banks 682
Kihwan Jo, Gahyun Choi, Jongwook Jeong, and Kwangwon Ahn

Deep Neural Sequence to Sequence Lexical Substitution for the Polish
Language 692
*Michał Pogoda, Karol Gawron, Norbert Ropiak, Michał Swędrowski,
and Jan Kocoń*

Facial Mask Impact on Human Age and Gender Classification 706
Krzysztof Małecki, Adam Nowosielski, and Mateusz Krzak

Simple and Efficient Acceleration of the Smallest Enclosing Ball for Large
Data Sets in E^2: Analysis and Comparative Results 720
Vaclav Skala, Matej Cerny, and Josef Yassin Saleh

Elastic Resource Allocation Based on Dynamic Perception of Operator
Influence Domain in Distributed Stream Processing 734
*Fan Liu, Weilin Zhu, Weimin Mu, Yun Zhang, Mingyang Li, Ziyuan Zhu,
and Weiping Wang*

PRISM: Principal Image Sections Mapping 749
Tomasz Szandała and Henryk Maciejewski

The New UPC++ DepSpawn High Performance Library for Data-Flow
Computing with Hybrid Parallelism 761
Basilio B. Fraguela and Diego Andrade

Author Index 775

ICCS 2022 Main Track Full Papers

Developing a Scalable Cellular Automaton Model of 3D Tumor Growth

Cyrus Tanade$^{(\boxtimes)}$ (iD), Sarah Putney, and Amanda Randles (iD)

Duke University, Durham, NC 27710, USA
{cyrus.tanade,sarah.putney,amanda.randles}@duke.edu

Abstract. Parallel three-dimensional (3D) cellular automaton models of tumor growth can efficiently model tumor morphology over many length and time scales. Here, we extended an existing two-dimensional (2D) model of tumor growth to study how tumor morphology could change over time and verified the 3D model with the initial 2D model on a per-slice level. However, increasing the dimensionality of the model imposes constraints on memory and time-to-solution that could quickly become intractable when simulating long temporal durations. Parallelizing such models would enable larger tumors to be investigated and also pave the way for coupling with treatment models. We parallelized the 3D growth model using N-body and lattice halo exchange schemes and further optimized the implementation to adaptively exchange information based on the state of cell expansion. We demonstrated a factor of $20x$ speedup compared to the serial model when running on 340 cores of Stampede2's Knight's Landing compute nodes. This proof-of-concept study highlighted that parallel 3D models could enable the exploration of large problem and parameter spaces at tractable run times.

Keywords: Cellular automaton · Parallel computing · Tumor growth

1 Introduction

Three-dimensional (3D) models of tumor growth that leverage parallel computing can provide a means to explore 3D multiscale tumor morphology efficiently. Understanding tumor growth dynamics is fundamental in cancer biology, and parallelized 3D models can efficiently capture large-scale dynamics. Mathematical models of cancer biology are increasingly being used to understand tumorigenesis, metastasis, and responses to treatment [1].

Cancer is fundamentally defined by the uninhibited growth of cells. Models that capture cell-cell interactions that span from individual cells to emergent tumors are needed to better understand different features that influence growth dynamics. Cellular automaton models represent cells as dead or alive and interact within a fixed local neighborhood. Agent-based models (ABMs) can have multiple states and incorporate complex interaction networks beyond local neighbors. Game of Life (GoL) constructs offer the capability to use simple rules that are

© The Author(s), under exclusive license to Springer Nature Switzerland AG 2022
D. Groen et al. (Eds.): ICCS 2022, LNCS 13350, pp. 3–16, 2022.
https://doi.org/10.1007/978-3-031-08751-6_1

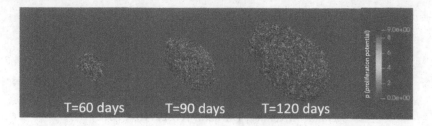

Fig. 1. 3D Tumor Growth Over Time. Tumor growth over 60, 90, and 120 d displaying proliferation potential (0-low, 9-high).

fundamentally rooted in single-cell kinetics to observe emergent dynamics that could span from a few initial seed cells to millions of cells [2,3]. Dissimilar to multi-agent systems - where agents are used to solve a specific problem, GoL and ABM models are used to study emergent behavior. However, implementations of GoL models have conventionally been restricted to 2D due to computational complexity and associated burden [4–6].

The development of tumors resembles Darwinian evolution where cancer cells would need to compete for resources and space. GoL models of cancer cell growth are relevant in the pre-angiogenic phase when tumor growth is driven by cell-cell interactions in small neighborhoods of cells. These GoL models have been shown to model tumor growth well despite simple rules and small parameter spaces [4,7]. However, as these models are expanded from two-dimensional (2D) to three-dimensional (3D) models, the computational space grows exponentially and could become intractable to simulate [6]. Exploring certain applications, such as studying how 3D morphology changes in response to treatment, is only amenable for 3D models and inherently requires more memory than 2D models such that simulations could become limited by the size or duration of tumor growth. Here, we extended a 2D cellular automaton model from Poleszczuk-Enderling [6] to recapitulate 2D tumor dynamics on a slice-level and output 3D growth dynamics for tumor morphology studies. Furthermore, we parallelized the model to realize larger problem sizes.

There are inherent limitations of 2D models. *In vitro* and *in silico* 2D models could reliably mimic *in vivo* tumor dynamics of a range of cancers, and as a result, have been instrumental for understanding cancer cell physiology. However, 2D monolayer representations of tumor growth could fail to recapitulate *in vivo* proliferation, morphology, and cell-cell and cell-matrix interactions. 2D cell culture models lack complex 3D architecture and extracellular-matrix interactions that are pervasive *in vivo*. While there are sophisticated techniques for efficiently capturing the 2D dynamics of tumor growth and proliferation, we found a lack of computationally efficient, scalable methods to capture the 3D morphology. Our study aims to bridge this gap by developing a parallel 3D model of tumor growth. In this study, we developed a proof-of-concept multiscale 3D model that can simulate from a few seed cells to millions of interacting cells (Fig. 1). Such a

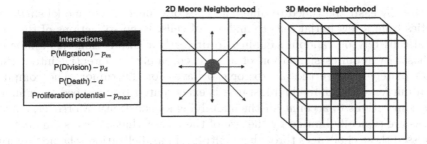

Fig. 2. 3D Model Design. The Poleszczuk-Enderling 2D model used a 2D Moore Neighborhood, where the center cell could interact with it's immediate 8 neighbors. Extending the model to 3D resulted in a 3D Moore Neighborhood with 26 neighbors. Cells have a finite proliferation potential and could interact with the neighborhood via migration, division, or death.

model could serve as a virtual 3D cell culture and enable low-cost investigation of cell pathophysiology and drug discovery.

In this model, cell movement was based on a Monte Carlo implementation. An ensemble of simulations must be completed to buffer randomness in the model, which further highlights the overarching need to reduce time-to-simulation for one individual simulation instance. The Poleszczuk-Enderling 2D tumor growth model took 3–5 h of runtime for 4 months of tumor growth, and running multiple instances to get averaged results could scale quickly.

We demonstrated the ability to capture 3D dynamics and propose techniques for efficient parallelization. As such, we made the following contributions: (1) development of a 3D cellular automaton model, (2) parallelization of a 3D serial model using MPI and optimizations to minimize memory transfer between processes, and (3) performance evaluation of the proof-of-concept parallelization scheme up to 340 tasks (or 5 nodes) on Stampede2 Knight's Landing (KNL) compute nodes. Our results demonstrated that the model could be most useful for simulating dense tumors with a large cell count and long durations.

2   Related Work

3D cellular models can serve as virtual laboratories with fully tunable conditions that enable the investigation of emergent tumor behavior. 3D cellular automaton models have been used to study tumor microenvironments and treatment paradigms, but are usually tuned to one cancer type of interest [1,8,9]. ABMs have been used to explicitly model adhesive, locomotive, drag, and repulsive forces between cells and have been applied to model cellular responses to hypoxia in breast cancer. Such models have been shown to scale to millions of cells [10]. There are also works parallelizing the Poleszczuk-Enderling model, but are limited to thread-level parallelism on local machines [11]. The cellular automaton model in this work is designed to be agile by relying on minimal input parameters and thereby easily be tuned to different types of cancers [6] to model single-cell kinetics from a few initial cells to millions of cells. Even so, actions of the individual cells rely on random sampling to drive their interactions. The inherent stochasticity requires simulation of a large ensemble of potential interactions to adequately capture the macroscopic behavior of the tumor. The computational burden of an individual instance is exacerbated by the need to complete many simulations to account for the underlying stochasticity. Moreover, current models are typically limited by the size of the tumor that can be simulated. We addressed these challenges through a multi-level parallelization scheme targeting 3D tumor models.

3   Methods

3.1   Extension of 2D Cellular Automaton Model to 3D

The Poleszczuk-Enderling model introduced a 2D representation of tumor growth relying on a cellular automaton representation of cancer cells [6]. Tumor growth was captured via cellular interactions, where each cell was modeled as an individual agent. The interactions characterizing cell growth included migration to other discrete lattice points, proliferation via mitotic cell division, and cells could finally die or become quiescent. As proliferation and migration require moving to a different lattice point, communication was needed within a cell's neighborhood to determine if there were empty spaces for interaction. The Poleszczuk-Enderling 2D model used a 2D Moore Neighborhood, where cells could interact with 8 of it's immediate neighbors (Fig. 2). The model included cancer stem cells and non-stem cancer cells. Stem cells were assumed to have infinite proliferation potential, whereas the non-stem cells had a maximum proliferation potential (p_{max}). Cells interacted in the lattice via discrete lattice-based rules governed by probabilities of migration (p_m), proliferation (p_d), and death (α). Each of these traits was kept as trait vectors for each cell such that there were N-body cells. On the other hand, the number of free spots in each neighborhood was stored on the lattice. N-body and lattice components of the simulation were eventually parallelized separately.

   Expansion from 2D to 3D consisted of transitioning the cell lattice, which tracks the number of empty nearest neighbors for a given grid location, from a 2D array to a flattened 3D vector. Helper functions for cell death, proliferation, and migration were adjusted to account for a 3D Moore's Neighborhood of 26 neighbors rather than the initial 2D Moore's neighborhood. We retained input parameter values from the 2D model, but increased the cell division probability (p_d) by 30%. Though this was not a necessary adjustment, it provided the benefit of creating tumors comparable in cross-sectional density to the Poleszczuk-Enderling 2D code and allowed comparison of 2D slices across the center of mass with 2D model outputs. Additionally, this process demonstrated the ease of model tuning.

   The time loop was iterated over a user-specified number of time steps (where each step was a simulation of one hour). Within each time step, every cell in the population was iterated over. Random number generation was used to first check if the cell should spontaneously die and be removed from the population. If the cell did not die, a new random number was generated to determine if the cell should proliferate. Only stem cells or cells that have not reached proliferative exhaustion could divide. If the cell did not proliferate, the random number was checked to determine if the cell should migrate. If the cell did not die, proliferate, or migrate, then nothing would happen and the cell would be added back into the population vector for the next time loop. For each cell, the lattice containing the number of empty nearest neighbors at each grid point was updated based on the cell's actions. After iterating over all the cells, the vector holding the population was refreshed to remove any dead cells and add any new ones.

3.2　N-body and Lattice Parallelization Schemes

Parallelization of the main simulation loop took two forms - an N-body scheme for individual cells and halo exchanges for the underlying lattice. The cell lattice domain was divided into approximately even blocks along the x axis, where each rank was responsible for a local cell lattice of size $lx_{local} * ly * lz$ (where lx_{local} was the length of the x domain local to the rank, ly was the length of the y domain, and lz was the length of the z domain). We were limited to communicating 2D packets of 3000×3000 points because of inherent limits to the size of 2D C++ vectors. A potential future direction would be to optimize these messages further.

Cell movement and cell lattices were parallelized separately. Cells were allowed to move freely into empty neighbors until reaching the edge of their domain, where we implemented a one layer overlap between each rank. This overlap was used as a part of domain decomposition for parallelization and did not influence the mechanics of tumor growth. Cells migrating outwards at the edge of a task's boundary would be transferred to neighboring tasks. This particular overlap allowed cells to move freely into empty spaces, only triggering communication between ranks when a cell entered border regions where a full Moore's neighborhood could not be realized. MPI communication of the cells between ranks occurred in two parts: all the integer properties associated with the moved cells were sent, and then the characters associated with the cells were sent. Communication was implemented using non-blocking sends and receives with the receives posted at the start of each time step and the sends posted at the end of each time step.

The cell lattice was represented as a flattened vector of integers denoting the number of free spaces a cell at a given index had around it. When a cell died, proliferated, or migrated, the lattice values of the surrounding Moore's neighborhood must be updated accordingly. Instead of one layer of overlap between ranks (like that used in the N-body communication), we used an additional layer of communication on each rank that would track a cell's movement into and out of the cell transfer zone. At the end of each time step, before the border cells from neighboring ranks were transferred, the lattice values for these two layers were communicated, compared, and reconciled such that when the data from the

N-body-based communication were added into the rank population, they could access values from an appropriately updated lattice.

Based on our division of ranks along the x $axis$, we placed the initial cell in the simulation in the middle of the median rank. Verification of the parallelized code with the serial code was performed using both odd and even number of ranks.

3.3   Adaptive Communication Scheme to Reduce Overhead

The naïve 3D parallelized model communicated cell and lattice data between each rank in a point-to-point complete manner. Even ranks that contained no cells participated in this communication, which unnecessarily increased memory transfer and associated communication costs. To optimize the communication scheme, before any lattice/cell communication, we gathered across every rank an array of Boolean values indicating the ranks that contained cells and then used this array to determine which ranks needed to communicate. Each rank that had cells would send to both its nearest neighbors, and each rank would receive from any nearest neighbor that contained cells. This optimization meant that ranks still participated in point-to-point communication, but communication of the entire size of the send buffer (up to 2^{22} elements) only occurred when there were actually cells on the rank that necessitated this communication. As a result, the communication expanded adaptively as cells propagated across ranks over time.

3.4   3D Serial and Parallel Verification Protocols

We verified our 3D model by comparing per-slice cell population sizes with the 2D simulation at 60 days of simulation. The morphology was compared using the dice similarity coefficient (DSC), which measures the spatial overlap of the tumor. We verified our parallel code by comparing it to the serial code. To do so, we needed to consider the inherent stochasticity of our simulation model which used multiple randomly generated numbers per cell in the population per time step. While the simulation was capable of being seeded to generate the same tumor population for runs of the same world size, seeding each rank individually would not yield the same results. Consequently, we verified the code by defining 4 variables of interest - cell population size, the proportion of stem cells, the distribution of cell proliferation potential across the population, and the distribution of the number of empty nearest neighbors across the population. We used world sizes of 1, 16, 17, 54, and 55 ranks when completing verification runs and performed 5 different unseeded runs for each case. We calculated the mean and standard error for the 4 variables of interest on the 80th day of simulation.

3.5   Performance Evaluation of the Parallelized 3D Model

The parallelized model was evaluated for performance through strong scaling, throughput, and efficiency of using parallel resources. For all performance evaluation runs, we ran simulations of up to 120 days of tumor growth and up to

2D   **3D**

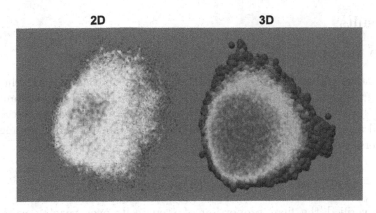

Fig. 3. 3D Model Verification. Per-slice population counts through the center of mass at 60 days resulted in a mean DSC score of 0.94.

340 cores of Intel Xeon Phi cores (clock rate of 1.40 GHz) or 5 nodes on Stampede2 KNL compute nodes with 100 Gb/s Intel Omni-Path network with fat tree topology interconnect. All code were compiled using the -O3 optimization flag. These resources were accessed at TACC via an XSEDE allocation [12]. We simulated up to 120 days because this was a similar time scale used in the Poleszczuk-Enderling 2D model [6].

Strong scaling curves were generated by computing the speedup with an increasing number of ranks for 120 days simulations. However, we found that the serial 3D code was inherently faster for the first 2 3 months of simulation time through some initial testing. This was likely due to the low number of cells in the initial stages of growth and that parallelization only became necessary once the tumor size reached a threshold. To test this observation, we measured strong scaling when neglecting the first 90 days of simulation time to illustrate that the parallel implementation needed to ramp up before eventually outperforming the serial 3D model. Ideal scaling was taken as the number of processors containing cells, averaged over the strong scaling runs due to stochasticity in the model.

We also investigated the throughput of the model by computing the cellular operations per second ($CLOPs$) for different core counts (up to 340 ranks) at different points in time (i.e., 60, 90, and 120 days). Lastly, the efficiency - the number of ranks containing cells relative to the total number of ranks used in the simulation - was measured across all time points and averaged across all the different runs with different number of ranks. This performance measure would indicate the rate of uptake of ranks and the increase in efficiency over time. As the simulations were inherently stochastic, we ran 5 unseeded simulations per data point to compute for means and standard errors.

4   Results

4.1   3D Model Preserves Per-Slice Tumor Morphology

The 3D model agreed with the 2D model on a per-slice basis through the center of mass of the tumor (Fig. 3). The comparison was made at 60 days of simulation. The DSC was 0.94, which demonstrated that the morphology was successfully preserved after increasing the dimensionality of the model and changing input parameters.

4.2   3D Parallel Model Matches 3D Serial Model

We first verified the final proportion of stem cells over varying world sizes (Fig. 4A). The ideal plot would be a horizontal line, indicating that the mean proportion did not change with the number of ranks used. Though the experimental values deviated slightly from a perfect horizontal line, the stem cell proportion values for each world size were within each points' standard error range. There was a similarly horizontal trend for the final cell population size.

Next, we verified the distribution of empty neighbors and cell proliferation potential. Figure 4B demonstrates that the shapes of the distributions for each world size were all similar. The distribution of cell proliferation potential also resulted in similar trends. Therefore, we have verified the parallelized 3D model using the proportion of stem cells, final cell population size, distribution of empty neighbors, and the distribution of cell proliferation potential.

4.3   Strong Scaling Indicates When to Launch the Parallel Model

The parallelized 3D model had relatively modest strong scaling results (Fig. 5). At 120 days of simulation time, there was a factor of 10 speedup compared to the serial model which was roughly equivalent to a parallel efficiency of 20%. Parallel efficiency was taken to be relative to the number of ranks used over the total number of ranks allocated.

Such performance was because the serial model was inherently faster at the beginning of the simulation due to communication overhead. In the early steps of the simulation, the tumor was in the initiation phase going from only a few cells to many and favored the serial code because there were no communication overheads. As the tumor grew and expanded in domain to neighboring ranks, the parallel code would eventually become more efficient than the serial model. The workload demand didn't necessitate use of multiple cores at small tumor sizes. We quantified the speedup when the simulation had surpassed this initiation threshold to test these observations. We assumed *a priori* that 90 days would exceed the initial growth stage. Using this offset, we achieved a factor of $20x$ speedup with a parallel efficiency of 40% at 120 days. This result demonstrated a trade-off between the serial and parallel code, and that there would be an optimum point of switching between the two models. The strong scaling curves

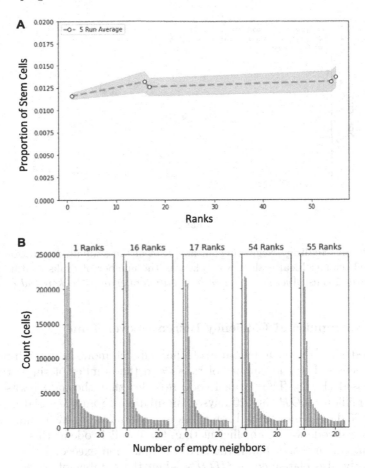

Fig. 4. Verification of Parallelized Model. 80-day simulations of varying world sizes, from 1 to 55 ranks, had consistent (A) stem cell proportion of population and (B) distributions of empty nearest neighbors. Results were averaged over 5 unseeded runs: (A) *average = dotted line, standard error = background area*; (B) *only averages were presented as standard error bounds were too small to visualize.*

had not reached the point of diminishing returns, indicating that running for longer durations could result in better performance.

The serial model was shown to be more efficient for the first few months of simulation until reaching larger cell counts and longer simulation durations, where the parallel model would eventually become more efficient. To quantify a global, world size-invariant cut-off at which this transition occurs, we computed speedup (relative to the serial model) over all time points across all simulations with different world sizes (Fig. 6). The results indicated that the global cut-off occurred at 68 days, which suggested that the parallel code would provide gains for simulations with durations exceeding this cut-off.

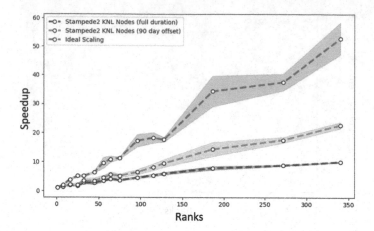

Fig. 5. Strong Scaling. Neglecting initial parts of the simulated duration resulted in improved scaling. Ideal scaling was taken as the number of ranks containing cells, averaged over 5 runs. *Average = dotted line, standard error = background area.*

4.4 Throughput and Efficiency Increases over Time

We defined *CLOPs* as a measure of the parallel model's throughput. The *CLOPs* increased as the number of ranks and the number of simulation days was increased (Fig 7). The parallel code revealed that the model was capable of over 2 million *CLOPs* at 120 days of simulation. From the serial model, the results at 60 days demonstrated some loss in performance, which might indicate that the serial code was more efficient than the parallel code at this point of the simulation. On the other hand, the 90 days simulation exceeded the serial run, but similarly also plateaued in *CLOPs* when the number of ranks increased. The 120 days simulation indicated an increasing trend even at the highest number of ranks, which suggests that the parallel model is more amenable for larger problem sizes over long simulation periods.

We also investigated the efficiency of the number of ranks used relative to all the ranks assigned over time (Fig. 8). In spite of stochasticity at the 120 days time point, the change in percentage rank utilization (or efficiency) was consistent across simulations of different world sizes. At best, the efficiency was just under 20% after 120 days of simulation. The rise in efficiency has not reached the point of diminishing returns and serves as a lower limit of performance.

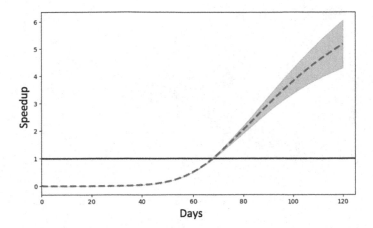

Fig. 6. Cut-off Point for Parallel Model. The serial code was initially faster than the parallel code, where speedup relative to the serial implementation was close to 0 until 50 days. The benefits of the parallel implementation became evident starting at 68 days when speedup crossed unity. The results were aggregated over multiple runs ranging from 8–340 tasks to obtain a global cut-off. *Average = dotted line, standard error = background area, baseline speedup relative to serial code = horizontal solid line.*

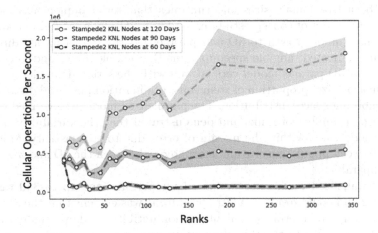

Fig. 7. Scaling cellular operations per second (CLOPs). We defined CLOPs as a measure of throughput. The CLOPs increased with longer simulations and more processors, which may indicate that the model was better suited for larger problem sizes. *Average = dotted line, standard error = background area.*

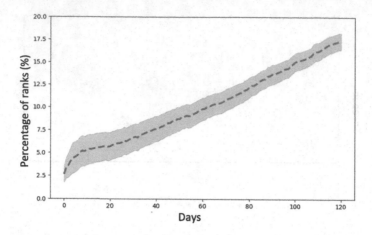

Fig. 8. Efficiency in Rank Utilization. There was a steady increase in efficiency over time. Efficiency was the number of ranks that contained cells relative to the total number of ranks allocated. *Average = dotted line, standard error = background area.*

5   Discussion and Conclusion

3D models have the potential to inform cancer research, however, there is a hard limit to both the domain sizes and runtimes that serial implementations can capture. This proof-of-concept study extended a 2D model of cellular automaton of dense tumor growth to 3D and parallelized the model in one direction. The results indicated tangible speedup gains that enable larger and longer simulations. The 3D parallel model verified well with the serial 3D model regarding cell population size, proportion of stem cells, proliferation potential distribution, and neighboring spaces distribution. The parallelized model was verified with an even and odd number of ranks and demonstrated that the variability between the two models was within the margin of error due to stochasticity. Moreover, the per-slice cell population size and morphology between the 2D and 3D models were comparable.

The parallelized model was evaluated for performance up to 340 tasks on Stampede2 and could result in a $20x$ speedup. It was evident that the 3D model needed to ramp up in terms of rank utilization until it would eventually overtake the serial model. We tested this observation by comparing speedup gains with and without a time delay. We found that the global cut-off point at which the parallel model would provide speedup over the serial model was at 68 days of simulation time. Parallelization increased the lattice size that could fit in memory, from 2^{31} to around 2^{34} elements - a factor of 2^7 increase. Although scaling results have not generally plateaued, we chose to simulate comparable time scales with the Poleszczuk-Enderling model [6] and show a lower limit of strong scaling, throughput, and efficiency results.

There are future paths for additional parallelization and model development. Although we enabled larger domains to be simulated, we only parallelized along the x dimension. Parallelizing over additional axes would decrease the grid sizes allocated per rank and the size of buffers to be communicated between ranks. Though more communication would be required per rank, the size of buffers would be smaller and would provide speedup over the current implementation. In terms of biology, allowing cells to occupy multiple lattice points to model cellular volume expansion could provide a more detailed representation of tumor growth. Speedup could improve with such an implementation as cells may expand over the domain quicker to achieve better rank utilization. This work does not intend to provide an optimal parallel tumor growth model, but offers an evaluation of strategies that warrant further investigation.

The results indicated that the serial model was more efficient for the first 68 days of simulation, but that the model became exponentially slower from that cut-off point onward. This was likely the point at which the cells were starting to expand to a sufficient proportion of ranks that resulted in enough speedup to overcome inherent MPI communication overheads. Therefore, some speedup could be gained by restricting simulations to use the serial code until the cut-off point to avoid unnecessary communication. Checkpointing could be implemented where the serial code could stop at the cut-off and use its outputs to launch parallel tasks.

The performance evaluation also revealed that less than 20% of ranks were actually engaging in the simulation for up to 120 days. This was a clear bottleneck, and ramping up the uptake of ranks faster would likely improve results. Although highly stochastic, the results demonstrated that even across simulations with a different number of tasks, the uptake in ranks were largely consistent. For this particular problem, it could be useful to change the thickness (or proportion of the lattice domain) handled by each rank. Instead of having relatively uniform allocation, it would be useful to allocate as many ranks as possible to the middle 20% of the lattice (i.e., allocate as small of a lattice as possible) and allocate as few ranks as possible (i.e., allocate as much of the lattice as possible) to the edge ranks where cells were unlikely to reach.

Ultimately, the 3D parallelized code verified well with the 3D serial code. The multi-level parallelization scheme of combining N-Body and halo paradigms alongside adaptive communication enabled efficient simulation of 3D tumor growth. This pilot study exemplified that even with parallelization across just one direction, there were clear gains that could enable larger studies and questions to be explored. We have additionally outlined some directions that could provide more significant speedups and allow for even longer simulations. This work could lay the groundwork for future studies of cellular automaton tumor growth models and parallelization methods of computational N-body and lattice-based models.

Acknowledgements. We thank Daniel Puleri, Simbarashe Chidyagwai, Sayan Roychowdhury, and Raveena Kothare for fruitful discussions. This work used an Extreme Science and Engineering Discovery Environment (XSEDE) allocation, which is supported by National Science Foundation grant number ACI-1548562. This work used Stampede2 at TACC through allocation TG-MDE210001. The research of Cyrus

Tanade was supported by the National Science Foundation Graduate Research Fellowship Program under Grant No. NSF GRFP DGE 1644868. The work of Amanda Randles was supported by the National Institutes of Health under award number U01CA253511. The content is solely the responsibility of the authors and does not necessarily represent the official views of the NIH or the NSF.

References

1. Randles, A., et al.: Computational modelling of perivascular-niche dynamics for the optimization of treatment schedules for glioblastoma. Nat. Biomed. Eng. **5**(4), 346–359 (2021)
2. Tang, J., et al.: Irradiation of juvenile, but not adult, mammary gland increases stem cell self-renewal and estrogen receptor negative tumors. Stem Cells (Dayton, Ohio) **32**(3), 649–661 (2014)
3. Gao, X., McDonald, J.T., Hlatky, L., Enderling, H.: Acute and fractionated irradiation differentially modulate glioma stem cell division kinetics. Can. Res. **73**(5), 1481–1490 (2013)
4. Piotrowska, M.J., Angus, S.D.: A quantitative cellular automaton model of in vitro multicellular spheroid tumour growth. J. Theor. Biol. **258**(2), 165–178 (2009)
5. Jiao, Y., Torquato, S.: Emergent behaviors from a cellular automaton model for invasive tumor growth in heterogeneous microenvironments. PLoS Comput. Biol. **7**(12), e1002314 (2011)
6. Poleszczuk, J., Enderling, H.: A high-performance cellular automaton model of tumor growth with dynamically growing domains. Appl. Math. **5**(1), 144–152 (2014)
7. Morton, C.I., Hlatky, L., Hahnfeldt, P., Enderling, H.: Non-stem cancer cell kinetics modulate solid tumor progression. Theor. Biol. Med. Model. **8**(1), 48 (2011)
8. Norton, K.-A., Jin, K., Popel, A.S.: Modeling triple-negative breast cancer heterogeneity: effects of stromal macrophages, fibroblasts and tumor vasculature. J. Theor. Biol. **452**, 56–68 (2018)
9. Alfonso, J.C.L., Jagiella, N., Núñez, L., Herrero, M.A., Drasdo, D.: Estimating dose painting effects in radiotherapy: a mathematical model. PLoS ONE **9**(2), e89380 (2014)
10. Ghaffarizadeh, A., Heiland, R., Friedman, S.H., Mumenthaler, S.M., Macklin, P.: PhysiCell: an open source physics-based cell simulator for 3-D multicellular systems. PLoS Comput. Biol. **14**(2), e1005991 (2018)
11. Salguero, A.G., Capel, M.I., Tomeu, A.J.: Parallel cellular automaton tumor growth model. In: Fdez-Riverola, F., Mohamad, M.S., Rocha, M., De Paz, J.F., González, P. (eds.) PACBB2018 2018. AISC, vol. 803, pp. 175–182. Springer, Cham (2019). https://doi.org/10.1007/978-3-319-98702-6_21
12. Towns, J., et al.: XSEDE: accelerating scientific discovery. Comput. Sci. Eng. **16**(5), 62–74 (2014)

Human-Level Melodic Line Harmonization

Jan Mycka[1] , Adam Żychowski[1](✉) , and Jacek Mańdziuk[1,2]

[1] Warsaw University of Technology, Warsaw, Poland
jan.mycka.dokt@pw.edu.pl, a.zychowski@mini.pw.edu.pl
[2] AGH University of Science and Technology, Krakow, Poland

Abstract. This paper examines potential applicability and efficacy of Artificial Intelligence (AI) methods in automatic music generation. Specifically, we propose an Evolutionary Algorithm (EA) capable of constructing melodic line harmonization with given harmonic functions, based on the rules of music composing which are applied in the fitness function. It is expected that harmonizations constructed in accordance to these rules would be formally correct in terms of music theory and, additionally, would follow less-formalised aesthetic requirements and expectations. The fitness function is composed of several modules, with each module consisting of smaller parts. This design allows for its flexible modification and extension. The way the fitness function is constructed and tuned towards better quality harmonizations is discussed in the context of music theory and technical EA implementation. In particular, we show how could generated harmonizations be modelled by means of adjusting the relevance of particular fitness function components. The proposed method generates solutions which are technically correct (i.e. in line with music harmonization theory) and also "nice to listen to" (i.e. aesthetically plausible) as assessed by an expert - a harmony teacher.

Keywords: Evolutionary algorithm · Harmonization · Music generation

1 Introduction

For centuries, the fine arts have been developed by people and, invariably, a major aspect of this process has been creativity of an artist. For this reason, Artificial Intelligence (AI) researchers attempt to implement *computational creativity* [19] to mimic creative behaviors of AI agents. One of the fields in which AI creativity has been intensively developed is music [9]. AI methods are applied to create new compositions [4] or complement/expand existing pieces [18]. In both tasks they managed to demonstrate human-level performance. Another line of research is imitation of a style of a particular composer, e.g. F. Chopin [10,11]. One of the basic problems in music is the enrichment of a given melodic line by adding chords – the so-called harmonization of the melodic line. Adding new notes is based on the relationship between the notes, both vertically (simultaneous sound) and horizontally (time sequence).

© The Author(s), under exclusive license to Springer Nature Switzerland AG 2022
D. Groen et al. (Eds.): ICCS 2022, LNCS 13350, pp. 17–30, 2022.
https://doi.org/10.1007/978-3-031-08751-6_2

Harmonization is a creative process in which the main role is played by intuition, talent and experience of a musician [14]. The process is constrained by various rules resulting from music theory and centuries of musical practice. The algorithm proposed in this work applies AI methods to create suitable harmonization and is based on harmonization rules defined in music theory. In general, creating music (or other forms of Art) is considered a challenge for AI agents, as mastery in this field requires special gifts that are rare even among humans.

1.1 Contribution

The main contribution of this paper is the following: (1) we propose a novel Evolutionary Algorithm (EA) approach capable of creating melodic line harmonizations that are formally correct, i.e. fulfil music harmonization rules; (2) to this end we design a specific fitness function that corresponds to theoretical rules of harmonization and can be easily extended or tuned to reflect the desired aspects of the resulting harmonizations; (3) the proposed fitness function covers not only harmonic but also melodic aspects (voices leading) of created harmonizations; (4) through modification of the weights of individual components of the fitness function, solutions can be tuned to meet the desired requirements; (5) the resulting harmonizations are evaluated by an (human) expert and assessed as both technically correct and **possessing human-like characteristics**.

2   Problem Definition

Harmonization is a process of creating an accompaniment for a given melody line. Created accompaniment consists of three new melodic lines (voices). Usually, the highest voice (soprano) is the base for a harmonization and the other three voices (alto, tenor and bass) are to be created. Four notes, one from each voice, form a chord. Creating harmonization is mostly based on musician's experience and intuition. However, throughout the years many theoretical rules that harmonization has to fulfill were developed [2,5]. These rules do not specify how harmonization is to be constructed, only whether or not the harmonization is correct.

The aim of the proposed algorithm is to create harmonization fulfilling selected theoretical rules. The harmonization is generated not for the raw melody, but for the melody extended with harmonic functions assigned to each note. The harmonic functions determine which notes (pitches) should constitute a chord. However, they do not specify the number of these notes or the voices in which they should be placed.

3   Related Literature

The melodic line harmonization problem has been addressed in the literature using various AI methods. Popular approaches use neural networks [8] or hidden Markov models [12]. Both papers address the task of harmonization of chorales based on J.S. Bach's style and the resulting harmonizations occasionally do not follow theoretical musical rules.

The algorithm described in this paper relies on a different method, the evolutionary algorithm, in which the required harmonization rules are directly imposed by means of a fitness function. Moreover, unlike neural networks, EA does not require training, which makes this approach independent of the composer's style implicitly present in the training set.

Another approach which uses Markov Decision Processes is presented in [20] and evaluates connections between two consecutive chords. The evaluation rules are based on music theory. Yet another work [6] hybridizes heuristic rules with dynamic programming method.

The use of EAs in the melody harmonization problem has been considered in a few recent works [7,13,15], however each of them addresses a slightly different problem formulation rendering a direct comparison impossible. In [7] multiobjective genetic algorithm is proposed which for a given melody generates a set of suitable harmonic functions, however, without adding new melodic lines. Similarly, in [13], only chords are created, not entire melodic lines. In [15], the algorithm solves a broader problem, i.e. not only adds new melodic lines to a given melody, but also complements the melody with harmonic functions. Furthermore, the method uses a wider range of harmonic functions and fewer theoretical rules than in our approach. The fitness function consists of two parts, the first one evaluates the created harmonic functions and the other one evaluates the melodic lines added.

Each of the above-mentioned works considers a slightly different formulation of the harmonization problem what renders a direct comparison impossible. Hence, the assessment of the resulting melody lines proposed in the paper is two-fold: (1) by means of a numerical fitness value assigned by the algorithm, and (2) by a human expert - a harmony teacher.

4 Proposed Method

To address the harmonization problem we propose the EA that maintains a population of individuals (candidate solutions). In each generation, the current candidate solutions undergo mutation and crossover operations. Subsequent generations are composed of individuals with gradually higher fitness values, i.e. achieve higher evaluations, on average. The core element of the algorithm is the fitness function that evaluates candidate solutions, which is based on theoretical rules of music harmonization.

The algorithm is run for a predefined number of n generations. Afterwards, the best individual in the last population is returned as the final result.

4.1 Problem Search Space - *Admissible* Chords

The input data is a soprano melodic line (the highest voice) with a particular harmonic function assigned to each note. This function indicates at least three and at most five pitches, which have to be used in a chord. If the function indicates only three pitches, one must be doubled, and if five of them, one must be omitted.

In each created chord, the highest note is fixed and derived from the given melodic line (soprano). In addition, an ambitus (the lowest and highest possible pitch of a voice) is defined for each voice, derived from the theory. Thus, for each harmonic function it is possible to define a set of all admissible chords corresponding to this function. The evolutionary operators (mutation, crossover), as well as the construction of the initial population are based on these pre-defined sets of admissible chords associated with particular harmonic functions.

4.2   Population Generation Process

Each initial candidate harmonization is in the form of a sequence of admissible chords. Each chord in a sequence must satisfy the two basic conditions:

(i) the chord corresponds to the function assigned to the completed note,
(ii) the note given in the input voice is located in the chord in the same voice (soprano).

Observe that construction of a solution is performed by manipulating the whole chords, not single notes.

The first population is generated randomly. Every individual consists of randomly chosen chords, which fulfill conditions (i)–(ii). In each subsequent generation, first s_e *elite* (i.e. currently best) individuals are promoted from the previous generation without any adjustments, so as to ensure that the best individuals found in the entire run of the algorithm will not be lost. The rest of the population is generated by means of selection procedure and genetic operators (mutation and crossover), according to Algorithm 1.

```
1  GenerateNewPopulation (P)
2      CalculateFitnessValues(P)  // calculates fitness of each individual
3      P_new ← GetElite(P, s_e)     // population of new individuals
4      while |P_new| < |P| do
5          c_1 ← Selection(P)
6          if rand([0,1]) < p_c then // crossover
7              c_2 ← Selection(P)
8              c_new = Crossover(c_1, c_2)
9          else
10             c_new ← c_1
11         end
12         c_new ← Mutation(c_new)
13         P_new = P_new ∪ {c_new}
14     end
15     return P_new
```

Algorithm 1: Next generation population procedure.

4.3   Selection Method

The selection of individuals is performed in a tournament of size t_s with a roulette element added. In the first step, t_s individuals are drawn uniformly with replacement from the population and their fitness is calculated. Let's define by x_i the i-th individual of the tournament according to the fitness value ranking. Its probability of winning the tournament $p(x_i)$ is calculated according to (1):

$$p(x_i) = \begin{cases} p_s & \text{if} \quad i = 1 \\ (1 - \sum_{j=1}^{i-1} p(x_j)) \cdot p_s & \text{if} \quad 1 < i < t_s \\ (1 - \sum_{j=1}^{i-1} p(x_j)) & \text{if} \quad i = t_s \end{cases} \tag{1}$$

where $p_s \geq 0.5$ is the so-called *selection pressure*.

4.4   Mutation and Crossover

Each individual, before being added to a new generation, undergoes mutation. Each chord in a given harmonization is mutated with probability $\frac{p_m}{l}$, where l is the length of harmonization and p_m is the mutation coefficient (algorithm's parameter). Mutating a chord consists in its replacement by another chord uniformly sampled among those that meet requirements (i)-(ii). An example of mutation is shown in Fig. 1.

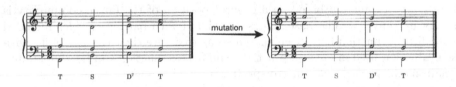

Fig. 1. Example of mutation. Second chord has changed.

The algorithm uses a one-point crossover which happens with probability p_c. In the crossings of the two sampled individuals the whole chords are considered, not the single notes. Let us define by $c[i], i = 1, \ldots, l$ the chord located at the i-th position of harmonization c. One-point crossover combines the initial part of one harmonization with the subsequent part of the other harmonization. The crossover method and its example application are presented in Fig. 2.

4.5   Fitness Function

The fitness function is used to evaluate individuals with respect to fulfilling harmonization rules (referring to chord building) and certain music theory rules (e.g. voice leading or fluidity of the melodic line). The set of considered harmonization rules includes those that are taught at music schools in the first years of

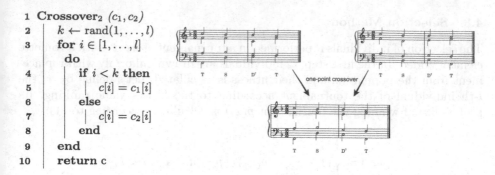

```
1  Crossover₂ (c₁, c₂)
2      k ← rand(1, ..., l)
3      for i ∈ [1, ..., l]
       do
4          if i < k then
5          |   c[i] = c₁[i]
6          else
7          |   c[i] = c₂[i]
8          end
9      end
10     return c
```

Fig. 2. One-point crossover: left - pseudocode, right - example with $k = 2$.

harmonization classes. All of them come from a harmony textbook [17]. Similar rules can be found, for instance, in [3, 16].

As demonstrated in Sect. 5 the selected set of rules allows achieving effective harmonizations, highly graded by the human expert in terms of both formal and aesthetic aspects. Additionally, a modular design of the fitness function allows its easy extension by means of adding new rules, if required.

Each rule used in the construction of the fitness function is assigned a value that affects the final score of the generated harmonization. These values indicate the importance of particular rules. Although in musical practice there is a certain level of subjectivity in evaluating the quality of the constructed harmonization, the essential evaluation based on a general position of the music community is usually unequivocal. The base values associated with particular rules included in the fitness function have been chosen so as to assure their compatibility with the evaluation used in musical practice. The fitness function can be divided into 3 main modules, each referring to specific rules.

1. Strong constraints C_s (high penalty terms) - constraints stemming from the rules, that must absolutely be satisfied in the created harmonization for it to be considered as correct.
2. Weak constraints C_w (lower penalty terms) - constraints derived from rules that do not have to be satisfied in the created harmonization, but their non-fulfillment lowers the harmonization assessment.
3. Added value V_a (reward terms) - the rules that specify chord arrangements or connections between chords that improve the sound of the harmonization.

The fitness function f_t for a given individual X has the following form:

$$f_t(X) = V_a + C_w + (C \cdot t)C_s,$$

$$C_s = \sum_{i=1}^{m_s} \phi_i(X), \quad C_w = \sum_{j=1}^{m_w} \chi_j(X), \quad V_a = \sum_{k=1}^{m_a} \psi_k(X), \qquad (2)$$

where $\phi_i(X) \leq 0$ is the penalty for not fulfilling strong constraint $i, i = 1, \ldots, m_s$, $\chi_j(X) \leq 0$ is the penalty for not fulfilling weak constraint $j, j = 1, \ldots, m_w$,

$\psi_k(X) \geq 0$ is the reward associated with the rule $k, k = 1, \ldots, m_a$, $t \leq n$ is the generation number, C is a constant parameter, $m_s = 9$, $m_w = 9$, $m_a = 4$. Please consult the publicly-available source code [1] for a detailed implementation of the above 22 fitness sub-functions.

Strong Constraints. Strong constraints define harmonization in terms of acceptability. If any of these constraints is not fulfilled then harmonization cannot be considered correct. The following strong constraints are considered (selected examples are presented in Fig. 3).

   i) Doubled prime in the first and last chord – The first and last chord occurring in a harmonization usually is a tonic, so as to emphasize the key in which the harmonization is created.
  ii) Voices are not crossing – The voices in the chords must not cross, that is, the highest note must be in the soprano, the lower one in the alto, yet the lower one in the tenor, and the lowest one in the bass.
 iii) Limited distances between voices – The distance between the three highest voices should not exceed an octave interval. The distance could be up to two octaves between the two lowest voices.
  iv) No quint in the bass on strong downbeat – Downbeats for the meter are given. Chords on the first given downbeat cannot have quint in bass.
   v) Correct notes resolutions – Some rules are specified for note resolution in chords: (a) a sixth must be resolved up by a second, (b) a septim must be resolved down by a second, (c) a ninth must be resolved down by a second, (d) if chord is dominant third must be resolved up by minor second.
  vi) No parallel (or antiparallel) quints, octaves or primes – If there is a fifth interval between two voices in a chord, there cannot be a fifth interval between the same voices in the following chord. An analogous rule applies to octave and prime intervals.
 vii) Voices must move in different directions – The voices, moving from one chord to the next, should move in different directions. The movement of all voices in one direction, up or down between consecutive chords is forbidden.
viii) Penultimate chord bass note – The penultimate chord in harmonization is usually a dominant or subdominant. It is important to emphasize the sound of the chord by doubling the prime (if possible) and not using a fifth in the bass.
  ix) No augmented interval moves – There cannot be an augmented interval between two consecutive notes in one voice.

Weak Constraints. Weak constraints do not have to be strictly satisfied, i.e. violating them does not make a harmonization unacceptable. However, violation of any such constraint lowers the harmonization evaluation. The following strong constraints are considered (selected examples are presented in Fig. 4).

   i) Doubled quint in bass – When the quint is doubled in a chord, one of these quints should be in bass.

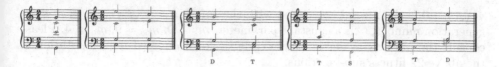

Fig. 3. Examples of strong constraints violation. From left to right: ii) crossed alto and tenor, iii) distance between soprano and alto exceeds an octave, v) incorrect resolution of thirds, vi) antiparallel quints between soprano and bass, ix) augmented jump in bass.

ii) No quint in bass on on-beats – On-beats are sorted by their importance. Quint in bass on on-beat is not preferable.

iii) No tripled prime in tonic function – It is permissible to triple prime in the last chord. However, this is not preferable.

iv) No consecutive chords on quint – In harmonization chords that have a fifth in the bass can occur. However, two such chords should not follow each other directly.

v) Bass movement – The lowest voice (bass) is one of the most significant voices in a harmonization, hence its movement is preferred in chords connections.

vi) Movement of at least two voices – A movement of at least two voices between two following chords is preferred, so that the harmonization does not sound static.

vii) No septim interval – There should not be a septim interval between two consecutive notes in one voice or between three consecutive notes in total in one voice.

viii) Melodic line smoothness – The middle melodic lines, alto and tenor, should be conducted smoothly.

ix) Restricting bass movement – For the bass voice a maximum interval it can take in two consecutive moves is tenth.

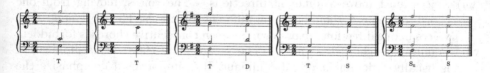

Fig. 4. Examples of weak constraints violation. From left to right: i) Doubled quint but not in bass, iii) A chord in tripled prime, iv) Two consecutive chords with quint in bass, v) No bass movement, vi) Only one voice moved.

Added Value. Certain features of a harmonization improve its quality and, therefore, their occurrence positively contributes to the evaluation score.

i) Parallel sixths – If there is a sixth interval between two voices in a chord and a sixth interval between the same voices in the following chord.

ii) Opposite movement of soprano and bass – Soprano and bass are the two most prominent voices in harmonization. For this reason, the opposite movement of these voices is preferred.

iii) Opposite movement on a perfect interval – Perfect intervals are octaves and quints. The opposite move on such intervals is preferred.

iv) Chord position – Every chord can be in the closed or open position. The preferred position is open.

5   Experimental Setup and Results

The examples used to tune and test the algorithm come from a harmony text-book [17]. Similar examples can be found in other harmony textbooks, as well. 18 examples (melodic lines with harmonic functions) were selected and divided into three groups:

1. long examples (about 20 chords), using only basic functions (7 examples),
2. short examples (about 10 chords), using basic and side functions and added pitches (4 examples),
3. long examples (about 20 chords), using basic and side functions and added pitches (7 examples).

Out of these 18 examples, 3 were used for parameter tuning (one from each group) and 15 in the final tests. All tests were run on a PC with IntelCore i7-9750H (2.6 GHz) processor and 24 GB RAM.

5.1   Parameterization

The algorithm parameters were chosen experimentally based on preliminary tests. For each parameter several values were tested (with the remaining parameters frozen at their basic values) on the 3 examples devoted for parameter tuning. Each test was run five times (with different random seeds) and returned values were averaged.

The following selections/ranges of parameters were tested (the finally selected values are bolded):

- s_p—(*population size*)—[10, 100, 500, **1000**, 1750, 2500, 3500, 5000];
- s_e—(*elite size*)—[0, **3**, 5, 10];
- p_c—(*crossover probability*)—[from 0 to 1 with step 0.1], **0.8**;
- p_m—(*mutation coefficient*)—[from 0 to l with step 1], **1** or **2**;
- p_m (fine tuned)—[from 1 to 2 with step 0.1], **1.1**;
- p_s—(*selection pressure*)—[from 0.5 to 1 with step 0.1], **0.7**;
- n—(*number of generations*)—[1000, 3000, **5000**, 10000];
- t_s—(*tournament size*)—[2, **4**, 8, 10].

5.2   Efficacy of the Algorithm - Formal Aspects

The efficacy of the algorithm was checked on 15 samples which were harmonized by the algorithm. Table 1 shows the time required to find the first correct and the finally returned (best found) solutions, resp. In each run the algorithm found the correct harmonization (satisfying all strong constraints) within the first 86 generations (usually much faster).

The number of generations needed to find the correct solution varies between groups. Shorter problems, with fewer chords, were solved faster (cf. group 1 vs group 2). Likewise easier problems, using fewer functions, turned out to be easier to solve (cf. group 1 vs group 3). Similar relationships can be observed among the finally returned solutions.

Table 1. The number of generations required to find a solution.

Group no.	Example no.	Generation number in which the result was found					
		Correct			Returned		
		Mean	Min	Max	Mean	Min	Max
1	1	16.6	14	22	208.2	86	343
	2	16.8	13	22	268.2	129	744
	3	14.8	12	18	218.4	109	397
	4	16.6	14	19	390	90	803
	5	15.4	14	17	1414.5	145	3914
	6	11.2	7	14	909.6	105	2477
2	7	8.2	6	10	177.6	27	593
	8	3.2	1	5	24	13	36
	9	6.8	5	8	826.2	75	3200
3	10	33.6	19	86	1933.6	96	3576
	11	19.2	17	22	699.8	249	1224
	12	21	18	25	3098.6	2179	3838
	13	20.2	19	21	926.6	73	2507
	14	19.6	17	25	890.8	130	3334
	15	16.2	15	19	1411.6	198	2500

5.3   Human Expert Evaluation Including Aesthetic Aspects

Every created harmonization has its score assigned as a result of the fitness function evaluation. However, this score only indicates how well a harmonization satisfies the *formal* fitness function requirements and does not indicate directly how good is the harmonization in strictly musical terms (*how well does it sound*). For this reason, all 15 generated samples were additionally evaluated by a human expert - a harmony teacher. Harmonizations were evaluated in a 5-point scale, from 1 (the lowest score) to 5 (the highest score).

On the one hand, the expert evaluated theoretical correctness of the constructed solutions, but on the other hand, based on many years of practical experience he/she also evaluated aesthetic and creative elements. In other words, the expert's evaluation was comprehensive and concerned both major harmonization

(a) Long example, from the first group, graded 5.

(b) Short example, from the second group, graded 4.5.

(c) Long example, from the third group, graded 5.

Fig. 5. Example harmonizations created by the algorithm.

aspects: its construction and sound. Figure 5 presents three examples of generated harmonizations rated 5, 4.5 and 5, resp. Out of all created samples, eight were graded 5, six 4.5, and one 4. Two types of problems were distinguished in downgraded harmonizations. One was the lack of adherence to certain theoretical rules. The most common problem was reaching a fifth in the bass of a chord other than a movement by a second. However, **none of the violated rules identified by the experts were included in the fitness function**, which means that their fulfillment was not directly imposed, and the algorithm could only meet them by chance. In contrast, the rules indicated in the fitness function as strong constraints were strictly observed. This means that extending the fitness function with additional rules should potentially solve this problem.

The remaining expert's remarks, formulated in a few cases, were related to minor sound issues, mainly to chords combinations (most often D^7 and T_{VI}). Requirements of this type are hard to be formally expressed in the fitness function which makes their enforcement in the resulting harmonizations difficult.

From the presented examples two were rated 5 (examples Figs. 5a, 5c), and one 4.5 (example Fig. 5b). The indicated place that could be harmonized differently in example Fig. 5b are chords fifth and sixth (D^7, T_{VI}). The created solution is not incorrect, although a better sound would be achieved by placing the prime of D^7 chord in the lowest voice.

To summarize, high grades given by the harmony teacher support the claim that the vast majority of created harmonizations are not only theoretically but also sonically correct. The seven harmonizations contain minor imperfections, some of which should be easily resolved by extending the fitness function. It is also worth mentioning that according to the expert's opinion, **the generated solutions do not expose any features of automatic origin and fully correspond to the products of human harmonization.**

5.4 Running Time

The running times of the algorithm are summarized in Table 2, separately for each group. It can be observed from the table that the running time does not depend on the complexity of the example, but only on its length. Furthermore, a rough comparison of time relationship between groups 1, 2 and 4 suggests its quasi-linear dependence on the harmonization length.

Table 2. The average algorithm's running time in seconds (harmonization time). Group 4 was generated artificially by multiplying 10 times examples from groups 1 and 3.

Group 1			Group 2			Group 3			Group 4		
Mean	Min	Max	Mean	Min	Max	Mean	Min	Max	Mean	Min	Max
531.9	485.5	567.0	225.8	180.7	252.1	536.2	508.7	582.4	5633.2	5126.5	6357.8

5.5 Modeling the Solution

The fitness function consists of 22 smaller functions, each of which addresses and evaluates one particular aspect of harmonization. Each of these evaluations is multiplied by a respective weight (negative for a penalty and positive for a reward). Modifying these weights allows for modeling the solution by increasing/decreasing the relevance of a given aspect with respect to the others.

As an example, Figs. 6a and b present two harmonizations of the same melodic line with different emphasis put on the reward for chords in the open position. In the first case the base fitness function (the one used throughout the paper) was applied and in the second one the respective coefficient was 3 times bigger, so as to reinforce the relevance of this feature. In both figures chords in open position are marked in green. Indeed, the number of chords in the open position in Fig. 6b is clearly greater than in Fig. 6a.

The above example confirms the possibility of modeling harmonization so as to focus on specific aspects. However, it is important to note that too strong reinforcement of specific features may result in others not being met, despite the overall increase of the fitness value.

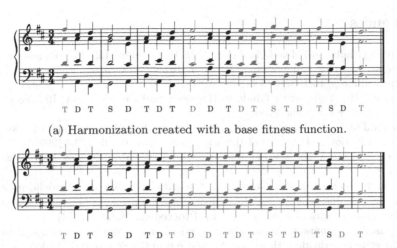

T D T   S D T D T   D D T D T   S T D   T S D   T

(a) Harmonization created with a base fitness function.

T D T   S D T D T   D D T D T   S T D   T S D   T

(b) Harmonization created with enhanced reward for open position.

Fig. 6. Modeling the solution. Chords in open positions are marked in green. (Color figure online)

6   Conclusions and Future Work

The problem of melodic line harmonization considered in this paper is part of the music composition process and as such requires creativity. The outcome (a melody) is generally hard to assess due to its subjective nature. With the above caution, this article points out that it is possible to achieve human-level performance in this task (melody harmonization) using evolutionary computation.

The proposed evolutionary algorithm creates harmonizations by means of carefully designed evolutionary operators and the fitness function that reflects music theory rules. The fitness function is composed of three general terms: (1) the rules that must be fulfilled if harmonization is to be considered correct, (2) the rules that should be fulfilled, otherwise the score of harmonization is lowered, (3) rules whose fulfillment improves harmonization. The design of the fitness function makes it easily extendable with other music rules and allows to emphasize various aspects in the resulting harmonization. Furthermore, the proposed algorithm does not require any training, which makes it independent from the styles/biases implicitly present in the training data.

Harmonizations generated by the algorithm are correct not only in terms of music theory but also sonically. According to the expert's opinion, obtained solutions do not exhibit any features of artificial origin and fully correspond to the products of human harmonization. Additionally, as presented in Table 2, the process of computationally generating harmonizations is relatively fast.

Our future plans involve removing harmonic functions added to the notes and performing harmonization of the melodic line itself, so as to enable creation of choral adaptations in small ensembles or provide harmonization assistance for less advanced musicians.

References

1. https://github.com/MelodicLineHarmonization/melodicLineHarmonization.git
2. Agmon, E.: The Languages of Western Tonality. Springer, Heidelberg (2013). https://doi.org/10.1007/978-3-642-39587-1
3. Benham, H.: A Student's Guide to Harmony and Counterpoint. Rhinegold Publishing Limited, London (2006)
4. Carnovalini, F., Rodà, A.: Computational creativity and music generation systems: an introduction to the state of the art. Front. Artif. Intell. **3**, 14 (2020)
5. Crocker, R.L.: A History of Musical Style. Dover Publications Inc., New York (2018)
6. Evans, B., Fukayama, S., Goto, M., Munekata, N., Ono, T.: Autochoruscreator: four-part chorus generator with musical feature control, using search spaces constructed from rules of music theory. In: Proceedings ICMC (2014)
7. Freitas, A., Guimaraes, F.: Melody harmonization in evolutionary music using multiobjective genetic algorithms. In: Proceedings of the Sound and Music Computing Conference (2011)
8. Hild, H., Feulner, J., Menzel, W.: HarmoNet: a neural net for harmonizing chorales in the style of J.S.Bach. In: NIPS 2091: Proceedings of the 4th International Conference on Neural Information Processing Systems, pp. 267–274 (1991)
9. Kaliakatsos-Papakostas, M., Floros, A., Vrahatis, M.N.: Artificial intelligence methods for music generation: a review and future perspectives. In: Nature-Inspired Computation and Swarm Intelligence, pp. 217–245 (2020)
10. Mańdziuk, J., Goss, M., Woźniczko, A.: Chopin or not? A memetic approach to music composition. In: 2013 IEEE Congress on Evolutionary Computation, pp. 546–553 (2013)
11. Mańdziuk, J., Woźniczko, A., Goss, M.: A neuro-memetic system for music composing. In: Iliadis, L., Maglogiannis, I., Papadopoulos, H. (eds.) AIAI 2014. IAICT, vol. 436, pp. 130–139. Springer, Heidelberg (2014). https://doi.org/10.1007/978-3-662-44654-6_13
12. Moray, A., Williams, C.K.I.: Harmonising chorales by probabilistic inference. Adv. Neural. Inf. Process. Syst. **17**, 25–32 (2005)
13. Olseng, O., Gambäck, B.: Co-evolving melodies and harmonization in evolutionary music composition. In: International Conference on Computational Intelligence in Music, Sound, Art and Design (2018)
14. Pachet, F., Roy, P.: Musical harmonization with constraints: a survey. Constraints **6**(1), 7–19 (2001)
15. Prisco, R.D., Zaccagnino, G., Zaccagnino, R.: Evocomposer: an evolutionary algorithm for 4-voice music compositions. Evol. Comput. **28**(3), 489–530 (2020)
16. Rimsky-Korsakov, N.: Practical Manual of Harmony. C. Fischer, New York (2005)
17. Sikorski, K.: Harmonia cz. 1. PWM (2020)
18. Vechtomova, O., Sahu, G., Kumar, D.: LyricJam: a system for generating lyrics for live instrumental music. In: Proceedings of the 11th International Conference on Computational Creativity (2021)
19. Wiggins, G.A.: Searching for computational creativity. N. Gener. Comput. **24**(3), 209–222 (2006). https://doi.org/10.1007/BF03037332
20. Yi, L., Goldsmith, J.: Automatic generation of four-part harmony. In: Proceedings of the Fifth UAI Bayesian Modeling Applications Workshop (2007)

Classification Methods Based on Fitting Logistic Regression to Positive and Unlabeled Data

Konrad Furmańczyk$^{(\boxtimes)}$ ⓘ, Kacper Paczutkowski ⓘ, Marcin Dudziński ⓘ, and Diana Dziewa-Dawidczyk ⓘ

Institute of Information Technology, Warsaw University of Life Sciences, Warsaw, Poland
{konrad_furmanczyk,kacper_paczutkowski,
marcin_dudzinski,diana_dziewa_dawidczyk}@sggw.edu.pl

Abstract. In our work, we consider the classification methods based on the model of logistic regression for positive and unlabeled data. We examine the following four methods of the posterior probability estimation, where the risk of logistic loss function is optimized, namely: the naive approach, the weighted likelihood approach, as well as the quite recently proposed methods - the joint approach, and the LassoJoint method. The objective of our study is to evaluate the accuracy, the recall, the precision and the F1-score of the considered classification methods. The corresponding assessments have been carried out on 13 machine learning model schemes by conducting some numerical experiments on selected real datasets.

Keywords: Positive unlabeled learning · Logistic regression · Empirical risk minimization · Thresholded lasso

1 Introduction

Learning from positive and unlabeled data, i.e. the so-called PU learning, is an approach where the only information the researcher has consists of positive examples and unlabeled data. In the PU setting, the training data contains positive and unlabeled examples, which means that the true labels $Y \in \{0, 1\}$ are not observed directly in the data and we only observe the surrogate variable $S \in \{0, 1\}$, which indicates whether an example is labeled (and consequently positive, $S = 1$ then) or not ($S = 0$ in this case). The history of PU learning dates back to the early 2000s (see, e.g., [10]) and this idea has gained much attention throughout recent years. The main reason for such a rapid development of the PU learning scheme is that this setting is very useful in numerous important

Supplementary Information The online version contains supplementary material available at https://doi.org/10.1007/978-3-031-08751-6_3.

applications. In particular, the PU learning method can be applied in the case when under-reporting is present in survey data (see [1]). It is quite common while analyzing some records from medical surveys, when we wish to predict the presence of a specific disease. Namely, it often happens that, although some respondents openly admit to suffering from a disease (the surrogate variable $S = 1$ and consequently, the true label $Y = 1$ in this case), there also exists a group of respondents who do not report such a disease (we put $S = 0$ then). This second group includes both the respondents who in fact have an examined disease, but do not admit to it (we have $Y = 1$ and $S = 0$ in this case) and the respondents who actually do not suffer from it (we have $Y = 0$ and $S = 0$ then). Such the under-reporting phenomenon is frequently justified by the fact that individuals suffering from some diseases (e.g. - from HIV or alcoholism) are often negatively perceived and treated by the rest of society. Some other interesting examples where the under-reporting is present may be found in the papers by Bekker and Davis [1] and Teisseyre et al. [15].

Now, suppose that X is a feature vector and, as previously, $Y \in \{0,1\}$ stands for a true class label and $S \in \{0,1\}$ denotes the surrogate variable that indicates, whether an example is labeled ($S = 1$ in this case) or not ($S = 0$ then). We consider a single sample scenario, where it is assumed that, there is a certain unknown distribution P, of (Y, X, S), such that $(Y_i, X_i, S_i), i = 1, \ldots, n$, form an iid sample obtained from this distribution, and that only empirical data $(X_i, S_i), i = 1, \ldots, n$, are observed. Thus, we do not have a traditional sample (X_i, Y_i), which is considered in standard classification problems, and we only observe a sample (X_i, S_i), where S_i are the observations of variable $S \in \{0, 1\}$ (since S is a surrogate of the true label Y, then each S_i depends on (X_i, Y_i)). In the considered concept only positive examples (i.e., examples for which $Y = 1$) may be labeled, which means that $P(S = 1|X, Y = 0) = 0$. It should be emphasized that in the PU design, the true class labels Y are only partially observed, which means that if $S = 1$, then we know that $Y = 1$, but if $S = 0$, then Y may be either 1 or 0.

The following constraint, called the Selected Completely At Random (SCAR) condition, is assumed

$$P(S = 1|Y = 1, X) = P(S = 1|Y = 1).$$

The SCAR assumption implies that X and S are independent given Y, since $P(S = 1|Y = 0, X) = P(S = 1|Y = 0) = 0$. Let $c = P(S = 1|Y = 1)$. The parameter c is called the label frequency and plays a key role in the PU learning scheme.

The main objective of our study is to apply the PU learning concept in order to estimate the posterior probability $f(x) = P(Y = 1|X = x)$, where, as previously, $Y \in \{0, 1\}$ denotes a true class label and X stands for the feature vector. Based on logistic model, three basic methods of this estimation have been proposed so far. They consist in minimizing the empirical risk of logistic loss function and are known as the naive method, the weighted method and the joint method (the latter has been quite recently introduced in Teisseyre et al. [15]).

Now, let us briefly describe the above mentioned methods.

First, we aim to present the naive method. In this case, having the empirical data (X_i, S_i), we minimize the empirical risk of the form

$$\widehat{R_1}(b) = -\frac{1}{n} \sum_{i=1}^{n} \left[S_i log(\sigma(X_i^T b)) + (1 - S_i)log(1 - \sigma(X_i^T b)) \right],$$

where $\sigma(s) = 1/(1+exp(-s))$. Then, the corresponding estimate of the posterior probability $f(x)$ is determined as

$$\widehat{f}_{naive}(x) = c^{-1}\sigma(x^T \widehat{b}_{naive}),$$

where c stands for the label frequency (i.e., $c = P(S = 1|Y - 1)$) and $\widehat{b}_{naive} = argmin_b \widehat{R_1}(b)$.

Using the weighted likelihood method (the weighted method in short, see [1]), we minimize the weighted empirical risk given by

$$\widehat{R_2}(b) = -\frac{1}{n} \sum_{i:S_i=1} \left[c^{-1}log(\sigma(X_i^T b)) + (1 - c^{-1})log(1 - \sigma(X_i^T b)) \right]$$
$$+ \sum_{i:S_i=0} log(1 - \sigma(X_i^T b)).$$

Then, the corresponding estimator of the posterior probability $f(x)$ is expressed as

$$\widehat{f}_{weighted}(x) = \sigma(x^T \widehat{b}_{weighted}),$$

where $\widehat{b}_{weighted} = argmin_b \widehat{R_2}(b)$.

The joint method from Teisseyre et al. [15] consists in minimizing - with respect to both the parameter vector b and the label frequency c - the following empirical risk

$$\widehat{R_3}(b, c) = -\frac{1}{n} \sum_{i=1}^{n} \left[S_i log(c\sigma(X_i^T b)) + (1 - S_i)log(1 - c\sigma(X_i^T b)) \right].$$

Then, the corresponding estimator of the posterior probability $f(x)$ is stated as follows

$$\widehat{f}_{joint}(x) = \sigma(x^T \widehat{b}_{joint}),$$

where $\left\{ \widehat{b}_{joint}, \widehat{c}_{joint} \right\} = argmin_{b,c} \widehat{R_3}(b, c).$

In order to optimize $\widehat{R}_3(b,c)$, the Broyden-Fletcher-Goldfarb-Shanno (BFGS) algorithm has been applied in Teisseyre et al. [15]. This algorithm enables to determine the formula for partial derivatives of $\widehat{R}_3(b,c)$. It is worth mentioning in this place that the Minorisation-Maximisation (MM) algorithm has been considered for the purpose of optimization by Łazęcka et al. [11].

In the most recent time, Furmańczyk et al. [5] proposed the LassoJoint procedure. It derives its name from the fact that it combines the thresholded Lasso procedure with the joint method from Teisseyre et al. [15]. It is a three-step procedure. Namely, in its two first steps, we perform - for some prespecified level - the thresholded Lasso procedure, in order to obtain the support for coefficients of a feature vector X, while in the third step, we apply - on the previously determined support - the joint method proposed by Teisseyre et al. [15]. More precisely, the LassoJoint method may be described as follows:

(1) For available PU dataset (S_i, X_i), $i = 1, \ldots, n$, we perform the ordinary Lasso procedure (see Tibshirani [18]) for some tuning parameter $\lambda > 0$, i.e. we compute the following Lasso estimator of β^*

$$\widehat{\beta}^{(L)} = \arg \min_{\beta \in R^{p+1}} \widehat{R}(\beta) + \lambda \sum_{j=1}^{p} |\beta_j|,$$

where

$$\widehat{R}(\beta) = -\frac{1}{n} \sum_{i=1}^{n} \left[S_i \log \left(\sigma(X_i^T \beta) \right) + (1 - S_i) \log \left(1 - \sigma(X_i^T \beta) \right) \right]$$

and subsequently, we obtain the corresponding support $Supp^{(L)} = \{1 \leq j \leq p : \widehat{\beta}_j^{(L)} \neq 0\}$;

(2) We perform the thresholded Lasso for some prespecified level δ and obtain the support $Supp^{(TL)} = \{1 \leq j \leq p : \left| \widehat{\beta}_j^{(L)} \right| \geq \delta\}$;

(3) We apply the joint method from Teisseyre et al. [15] for the predictors from $Supp^{(TL)}$.

It should be stressed that under some mild regularity conditions, the LassoJoint procedure obeys the screening property (all significant predictors of the model are chosen, with high probability, in the first two steps, see Theorem 1(b) in [5]).

Apart from the works where different learning methods - based on application of the logistic regression model for PU data - have been proposed, there are also some other interesting articles where various machine learning tools are used in the PU learning problems. In this context, it is worthwhile to mention: the papers of Hou [8] and Guo [6] - where the generative adversial networks (GAN) for the PU problem have been employed, the work of Mordelet and Vert [13] - where the bagging Support Vector Machine (SVM) approach for the PU data has been applied. Most relevant methods regarding the learning from PU data may be found in Lee and Liu [10] and Sansone et al. [14].

There are two essential objectives of our research. Its first goal is to verify and compare the accuracy, the recall, the precision and the F1-score of classifications obtained by the so far introduced primary methods of the posterior probability estimation, providing that Y is governed by the logistic regression model and PU data are available. For the corresponding comparisons, we aim to use AdaSampling methods (see [20]). In turn, our second goal is to give a recommendation for the method that seems the most stable and efficient. The details regarding our study have been given in further parts of our work. The remainder of our paper is structured as follows. In Sect. 1, we present the ideas and concepts used in our investigations, especially the methods that enable attaining the set objectives. Furthermore - in Sect. 2 we introduce the applied models, in Sect. 3 we present our numerical experiments together with the obtained results, while Sect. 4 summarizes our study. In order to carry out our simulations, we used the RStudio server module from the ICM UW Topola server[1]. We implemented the following libraries: AdaSampling [21], e1071 [12], glmnet [4], and some additional libraries available from two selected GitHub repositories: PUlogistic [16], PU_class_prior [17].

2   Objectives and Methods

The first goal of our study is to check and compare the accuracy, the recall, the precision and the F1-score of classifications obtained with use of the recently proposed methods - the joint method from Teisseyre et al. [15] and the LassoJoint approach from Furmańczyk et al. [5], as well as with use of the earlier established estimation methods consisting in fitting the logistic model, i.e. by additional application of the naive method, the weighted method and the oracle method for the case when the vector of coefficients is known.

The accuracy, recall, precision and F1-score metrics are defined as follows:

$$Accuracy = \frac{TP + TN}{TP + FP + FN + TN},$$

$$Recall = \frac{TP}{TP + FN},$$

$$Precision = \frac{TP}{TP + FP},$$

$$F1 = \frac{2 \cdot Precision \cdot Recall}{Precision + Recall},$$

where TP, FN, TN and FP stand for: the number of true positives, false negatives, true negatives and false positives, respectively.

The assessments of the mentioned metrics have been gained by conducting some numerical studies on nine real datasets from the UCI Machine Learning

---

[1] This research was carried out with the support of the Interdisciplinary Centre for Mathematical and Computational Modelling (ICM) at the University of Warsaw, under computational allocation No. g88-1185.

Repository [2] and the 'caret' package [9]. The second purpose of our research is to recommend the most reliable and efficient estimation method for the posterior probability $f(x) = P(Y = 1|X = x)$ assessment, where - as in the previous procedures - it is assumed that Y is governed by the logistic model and the PU data are available. Our study was constructed on 13 machine learning (ML) model schemes. We applied the LassoJoint method from [5] by considering the joint method for two scenarios - with the BFGS or the MM algorithm. The LassoJoint approach is a three-step procedure. In its first step, the initial selection of predictors is carried out by employing the Lasso method, for which the tuning parameter λ is either obtained by using the 10-fold cross-validation technique or is fixed. In turn, in the second step, the thresholded Lasso is performed, whereas in the last step, the joint method is employed for the variables selected in the second step. The naive logistic regression approach, the joint method, the Lasso-Joint approach and the weighted method for c estimated from the joint method (for the BFGS or the MM algorithm) have been employed and the corresponding results have been compared with the results obtained by implementing the oracle method when the true label variable Y is known. Moreover, in order to compare the classification methods based on fitting the logistic regression model, the two machine learning methods - namely, the Support Vector Machine (SVM) approach and the k-nearest neighbors algorithm (KNN) have been used - both in the AdaSampling scheme (see Yang et al. [19] and Yang et al. [20]). An application of the AdaSampling design results in constructing an adaptive sampling-based noise reduction method, which enables dealing with noisy data. We have also performed the min-max transformation of our features, which - compared to the original data - greatly improved the accuracy of all of the obtained results.

3 Numerical Experiments

3.1 Datasets

We consider nine datasets from the UCI Machine Learning Repository [2] and the 'caret' package [9]. In Table 1, we present basic characteristics of each dataset (from left to right: the number of features, the number of observations, the number of binary and continuous variables, the number of negative and positive cases, the percentage of positive cases). The values of these characteristics are obtained through fundamental preprocessing, including the one-hot encoding and removing the missing values. In our simulations, we set 1-class as a larger class for each dataset. The selection of datasets was conducted by taking into account various types of potential difficulties that may appear while applying the ML methods. Thus, we tested both a strict low-dimensional datasets ('Banknote') and datasets with many predictors ('Dhfr'). In addition to that, we also considered the sets with only binominal ('Vote') or continuous predictors ('Wdbc', 'Spambase') and mixed instances.

Table 1. Basic characteristics of the datasets. (from left to right: dataset name, no. of features, no. of observations, no. of binary variables, no. of continuous variables, no. of negative instances, no. of positive instances, percentage of positive instances)

Dataset	p	n	nbin_variables	ncon_variables	n_0	n_1	% of 1-class
Banknote	4	1372	0	4	610	762	55.5%
Breastc	9	683	0	9	239	444	65.0%
Credit_a	37	653	31	6	296	357	54.7%
Credit_g	24	1000	12	12	300	700	70.0%
Dhfr	228	325	11	217	122	203	62.5%
Diabetes	8	768	0	8	268	500	65.1%
Spambase	57	4601	0	57	1813	2788	60.6%
Vote	32	435	32	0	168	267	61%
Wdbc	30	569	0	30	212	357	62.7%

The naive logistic regression approach, the joint method, the LassoJoint approach and the weighted method for c estimated with use of the joint method have been employed. The corresponding results have been compared with the results obtained by implementing the oracle method. We deal with the problem of PU data classification. From the above, completely labeled datasets, we randomly select c% of the labeled observations S, for $c = 0.1; 0.3; 0.5; 0.7; 0.9$, and then, we randomly split these datasets into the training sample (80%) and the test sample (20%). By applying the LassoJoint method in its first step, we use the Lasso method with tuning parameters λ, chosen either on the basis of the 10-fold cross-validation scheme - in the first scenario (where lambda.min gives the minimum mean cross-validated error, while lambda.1se stands for the largest value of λ such that an error is within 1 standard error of the cross-validated errors for lambda.min.) or by putting the fixed λ of the form $\lambda = ((\log p)/n)^{1/3}$ - in the second scenario, as in [5]. In the second step, we apply the thresholded Lasso design for $\delta = 0.5\lambda$, with λ selected in the first step. Next, we determine the classification metric by simulating from 100 Monte Carlo replications of our experiment. Subsequently, in order to compare the logistic regression-based classification methods, the tools of machine learning, such as an AdaSampling (see Yang et al. [19] and Yang et al. [20]) together with the Support Vector Machine (SVM) concept and the k-nearest neighbors algorithm (KNN) have been employed.

3.2 Results

We conducted our simulation study on 13 ML model schemes based on the four methods described in Introduction. In our work, we applied four measures based on the confusion matrix: the accuracy, the recall, the precision, the F1-score. All of our metrics are the averages of the obtained values of metrics on a test subset after 100 repetitions. We decided to set a cut-off point at the level of 0.5. This level is typical in cases when the logistic or the logistic-based models are fitted. In the examples from the *AdaSampling* package documentation [21] the level of 0.5 is commonly used. The average values of the accuracy and the recall are given in Fig. 1 and Fig. 2. Additionally, we provide a dedicated visualization for comparison between the joint-wise models with and without the Lasso component (see Fig. 3 and Fig. 4). Tables presenting the precise values of some metrics and the charts depicting the values of the remaining measures are available in our Supplementary Materials[2]. These Supplementary Materials also include all of our codes in R.

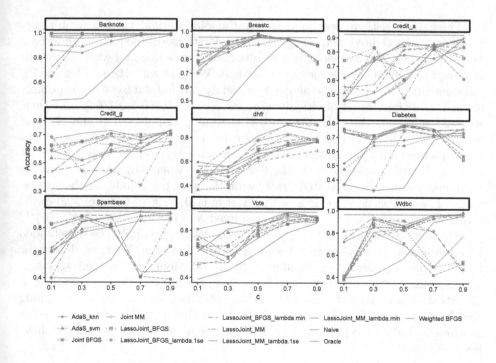

Fig. 1. The accuracy for the test datasets

---

[2] http://github.com/kfurmanczyk/ICCS22.

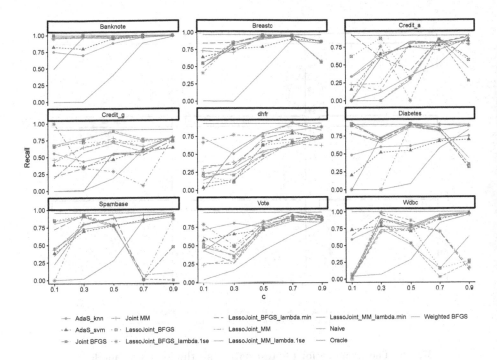

Fig. 2. The recall for the test datasets

It is clear that in the considered scenarios, performance of the oracle method may be perceived as a natural top ('the best') benchmark. On the other hand, in many scenarios the bottom ('the worst') benchmark is connected with performance of the naive method, but it may not always be treated as a strict rule.

Apart from obtaining appropriate metric values, we have also developed, for each value of c, the corresponding ranking methods. The ranking has been obtained on the basis of calculating the average values of ranks in a single scenario (the greater rank value is, the worse a given method is in our ranking). The ranking results are collected in Tables 2, 3, 4 and 5. The best methods (except the oracle approach) are underlined in the columns. Some additional comments and remarks regarding the obtained results are contained in the next section.

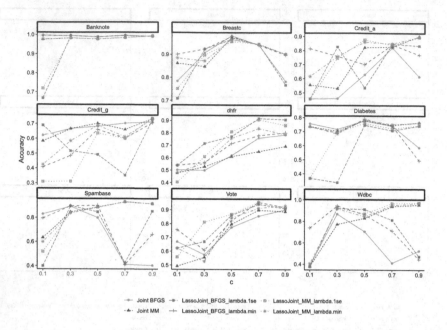

Fig. 3. The accuracy for the test datasets - the joint-wise models

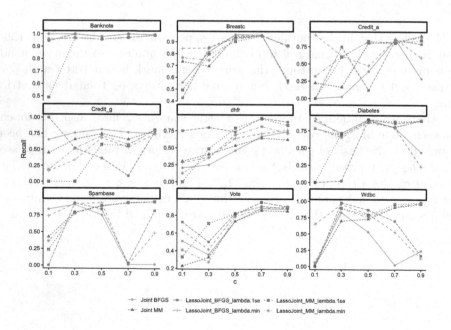

Fig. 4. The recall for the test datasets - the joint-wise models

Table 2. Avg.rank method based on the accuracy

Method	$c = 0.1$	0.3	0.5	0.7	0.9
Oracle	1.00	1.00	2.00	1.11	1.56
LassoJoint_BFGS_lambda.min	4.11	5.33	5.00	7.22	9.11
LassoJoint_MM_lambda.min	7.22	8.22	5.56	6.33	5.22
LassoJoint_MM_lambda.1se	11.22	6.67	5.44	5.56	4.44
LassoJoint_MM	7.67	8.78	6.89	6.33	4.56
LassoJoint_BFGS	4.44	4.89	7.11	8.56	10.00
Joint MM	7.11	8.22	6.56	7.56	5.67
Joint BFGS	5.44	6.00	7.00	7.78	10.22
AdaS_svm	7.78	8.67	8.67	6.56	5.44
LassoJoint_BFGS_lambda.1se	9.00	5.22	6.78	7.56	8.56
AdaS_knn	6.11	8.22	8.78	8.44	8.56
Weighted BFGS	8.11	7.56	9.00	8.11	10.33
Naive	11.78	12.22	12.22	9.89	7.33

Table 3. Avg.rank method based on the recall

Method	$c = 0.1$	0.3	0.5	0.7	0.9
Oracle	1.89	1.44	1.67	1.78	1.00
LassoJoint_BFGS_lambda.min	3.67	5.56	5.22	6.44	8.44
LassoJoint_BFGS	4.78	4.89	6.22	6.67	8.89
LassoJoint_BFGS_lambda.1se	8.11	4.78	5.89	6.56	7.56
LassoJoint_MM_lambda.1se	11.56	6.44	5.33	5.44	5.11
Joint BFGS	5.78	6.33	6.33	5.89	9.89
Joint MM	7.33	8.22	6.56	7.11	6.00
AdaS_knn	5.78	7.44	8.11	8.11	6.89
LassoJoint_MM_lambda.min	7.44	8.44	6.78	7.44	6.33
LassoJoint_MM	7.89	9.33	7.67	7.67	5.67
Weighted BFGS	7.11	6.78	8.67	8.22	10.33
AdaS_svm	7.89	9.11	10.22	8.67	6.11
Naive	11.78	12.22	12.33	11.00	8.78

Table 4. Avg.rank method based on the precision

Method	$c = 0.1$	0.3	0.5	0.7	0.9
AdaS_svm	5.44	_3.89_	_2.89_	_3.22_	_5.00_
Oracle	3.00	5.67	5.67	6.00	5.56
LassoJoint_MM_lambda.min	_3.00_	5.11	6.11	7.22	8.00
LassoJoint_MM	5.33	4.67	7.22	6.56	7.56
LassoJoint_BFGS_lambda.min	6.78	7.67	7.78	6.56	6.11
Naive	10.22	7.78	5.00	4.00	8.11
LassoJoint_BFGS	5.56	7.56	9.00	8.56	4.67
Joint BFGS	8.22	7.67	7.33	7.44	4.78
Joint MM	5.89	6.56	6.89	8.44	8.89
LassoJoint_MM_lambda.1se	10.11	7.33	6.33	8.00	8.22
AdaS_knn	7.22	8.00	6.78	8.78	10.56
LassoJoint_BFGS_lambda.1se	10.22	9.11	9.67	6.78	8.00
Weighted BFGS	10.00	10.00	10.33	9.44	5.56

Table 5. Avg.rank method based on the F1-score

Method	$c = 0.1$	0.3	0.5	0.7	0.9
Oracle	1.00	1.00	2.00	1.11	1.56
LassoJoint_BFGS_lambda.min	_4.56_	5.44	_5.33_	7.56	9.11
LassoJoint_MM_lambda.1se	10.33	6.67	5.78	_5.44_	_4.11_
LassoJoint_BFGS_lambda.1se	7.89	_4.00_	6.00	7.00	8.11
LassoJoint_MM_lambda.min	7.89	8.56	6.22	6.44	5.56
LassoJoint_BFGS	4.22	5.11	7.11	8.56	10.00
Joint BFGS	5.33	6.33	6.00	7.89	10.56
Joint MM	7.22	8.44	7.00	7.78	5.89
LassoJoint_MM	8.33	9.44	7.44	6.78	5.22
AdaS_svm	8.44	8.11	9.22	6.67	5.56
AdaS_knn	6.44	8.33	8.33	8.00	7.78
Weighted BFGS	7.56	7.22	8.22	7.89	9.89
Naive	11.78	12.33	12.33	9.89	7.67

4   Conclusions

The primary purpose of our study was to conduct a comprehensive evaluation of
13 ML model schemes, including the methods from literature and the methods
obtained as a result of some modifications we implemented in some other procedures (such that conducting the MM optimization in the LassoJoint procedure).
We decided to apply a few measures in our research, in order to get a guarantee

of a proper complexity of our assessments. In a typical approach to PU problems, the main attention is focus on calculating the AUC and Accuracy metrics, whereas in our work we provide additional analysis regarding different assessment measures. This extension enables to evaluate fractions of the true '1-class' and fractions of the predicted '1-class', among the real positive instances, which may be very useful in many applications regarding popular PU problems. For instance, in the credit risk management, we want to detect all frauds, even if we label too many observations (equivalently - we agree for a larger type I error). In this case we need to control the recall measure with a greater emphasis. On the other hand, in various marketing campaigns related models (e.g., such as uplifting models), we wish to focus our attention on customers who actually want to buy certain products. In this case we prefer to control the precision measure. The results of our numerical experiments show that if c increases, then the percentage of correct classifications increases as well in most cases. Usually, the LassoJoint procedure helps to improve the classification metrics and prevails over other methods (see Tables 2, 3, 4 and 5 and Tables 1–35 in the Supplementary Materials). The LassoJoint method has been constructed for the high-dimensional cases (i.e., when $p > n$), but it has to be stressed that it may be also so in the low-dimensional cases (i.e., when $p < n$), as we observe that the joint method performance improves while applying the basic metrics on most of the tested datasets, except for the *Credit_g*, *Diabetes*, and *Spambase*. Only in few cases, the method based on the BFGS optimization performs worse for large values of c, but the corresponding accuracy is still acceptable for small values of c. We may also observe that the classifications obtained by applying the LassoJoint method with the MM algorithm result in smaller classification errors (and thus in better classification accuracy) for larger labeling levels c. Moreover, the methods with tuning values λ, obtained by using the cross-validation scheme, display better accuracy than the methods with fixed values. Based on the obtained accuracy, recall and F1-score, we recommend using the LassoJoint method with: (a) the BFGS variant - for small values of c, (b) the MM variant for the values of c above 0.5 (for comparison - see Fig. 3). Furthermore, it is worth mentioning that considering the selected cases with small values of c, we do not observe classification instances from the '1-class'. Most of this cases are connected with the naive method for $c = 0.1, 0.3$, which can be seen in Fig. 2 and therefore, using more complex methods is highly recommended in these cases. However, it is not easy to point out a general winning method by taking into account all of considered measures. For example, an application of AdaSampling with the SVM kernel provides the classification results of the highest precision for almost every dataset scenarios. This high level of precision assures greater certainty that the predicted positives are real positives. On the other hand, the values of the accuracy, the recall and the F1-score are not satisfactory in most cases. In addition to that, the obtained simulations show that the labels noising can boost the precision metrics, since some methods provide better values of precision measures than the oracle approach (see Table 4). It is important to remember that all of the methods based on fitting the logistic regression model assume the celebrated SCAR condition. It is a common approach to impose this assumption in majority of methods dealing with PU learning and only in very few approaches the researchers try to omit this constraint (see [1]). In further investigations, it would be interesting to introduce some new methods which will not require the SCAR assumption. It would also be interesting to check robustness of existing methods under some disturbances of the SCAR condition.

References

1. Bekker, J., Davis, J.: Learning from positive and unlabeled data: a survey. Mach. Learn. **109**(4), 719–760 (2020). https://doi.org/10.1007/s10994-020-05877-5
2. Dua, D., Graff, C.: UCI Machine Learning Repository (2019). [http://archive.ics. uci.edu/ml]. Irvine, CA: University of California, School of Information and Computer Science
3. Friedman, J., Hastie, T., Simon, N., Tibshirani, R.: Glmnet: Lasso and elastic-net regularized generalized linear models. R package version 2.0 (2015)
4. Friedman, J., Hastie, T., Tibshirani, R.: Regularization paths for generalized linear models via coordinate descent. J. Stat. Softw. **33**(1), 1–22, (2010). https://www. jstatsoft.org/v33/i01/
5. Furmańczyk, K., Dudziński, M., Dziewa-Dawidczyk, D.: Some proposal of the high dimensional PU learning classification procedure. In: Paszynski, M., Kranzlmüller, D., Krzhizhanovskaya, V.V., Dongarra, J.J., Sloot, P.M.A. (eds.) ICCS 2021. LNCS, vol. 12744, pp. 18–25. Springer, Cham (2021). https://doi.org/10. 1007/978-3-030-77967-2_2
6. Guo, T., Xu, C., Huang, J., Wang, Y., Shi, B., Xu, C., Tao, D.: On positive-unlabeled classification in GAN. In: CVPR, pp. 8385–8393 (2020)
7. Hastie, T., Fithian, W.: Inference from presence-only data; the ongoing controversy. Ecography **36**, 864–867 (2013)
8. Hou, M., Chaib-draa, B., Li, C., Zhao, Q.: Generative adversarial positive-unlabeled learning. In: Proceedings of the Twenty-Seventh International Joint Conference on Artificial Intelligence (IJCAI-18), pp. 4864–4873 (2018)
9. Kuhn, M. caret: Classification and Regression Training. R package version 6.0-90 (2021). https://CRAN.R-project.org/package=caret
10. Lee, W.S., Liu, B.: Learning with positive and unlabeled examples using weighted logistic regression. In: ICML, Washington, D.C., AAAI Press, pp. 448–455, August 2003
11. Łazęcka, M., Mielniczuk, J., Teisseyre, P.: Estimating the class prior for positive and unlabelled data via logistic regression. Adv. Data Anal. Classif. **15**(4), 1039–1068 (2021). https://doi.org/10.1007/s11634-021-00444-9
12. Meyer, D., Dimitriadou, E., Hornik, K., Weingessel, A., Leisch, F. e1071: Misc functions of the department of statistics, probability theory group (Formerly: E1071), TU Wien. R Pack. Vers. **1**, 7–9 (2021). https://CRAN.R-project.org/ package=e1071
13. Mordelet, F., Vert, J.P.: A bagging SVM to learn from positive and unlabeled examples. Pattern Recogn. Lett. **37**, 201–209 (2013)
14. Sansone, E., De Natale, F. G. B., Zhou, Z.H.: Efficient training for positive unlabeled learning. TPAMI **41**(11), 2584–2598 (2018)

15. Teisseyre, Paweł, Mielniczuk, Jan, Łazęcka, Ma.łgorzata: Different strategies of fitting logistic regression for positive and unlabelled data. In: Krzhizhanovskaya, V.V., et al. (eds.) ICCS 2020. LNCS, vol. 12140, pp. 3–17. Springer, Cham (2020). https://doi.org/10.1007/978-3-030-50423-6_1

16. Teisseyre, P.: Repository from https://github.com/teisseyrep/Pulogistic. Accessed 25 Jan 2022

17. Teisseyre, P.: Repository from. https://github.com/teisseyrep/PU_class_prior. Accessed 25 Jan 2022

18. Tibshirani, R.: Regression shrinkage and selection via the lasso. J. Roy. Stat. Soc. Ser. **58**, 267–288 (1996)

19. Yang, P., Liu, W., Yang. J.: 6. Positive unlabeled learning via wrapper-based adaptive sampling. In: 6 International Joint Conferences on Artificial Intelligence (IJCAI), pp. 3272–3279 (2017)

20. Yang, P., Ormerod, J., Liu, W., Ma, C., Zomaya, A., Yang, J.: 7. AdaSampling for positive-unlabeled and label noise learning with bioinformatics applications. IEEE Trans. Cybern. **49**(5), 1932–1943 (2018) https://doi.org/10.1109/TCYB.2018.2816984

21. Yang, P.: AdaSampling: adaptive sampling for positive unlabeled and label noise learning. R Pack. Vers. **1**, 3 (2019). https://CRAN.R-project.org/package=AdaSampling

Robust Control of Perishable Inventory with Uncertain Lead Time Using Neural Networks and Genetic Algorithm

Ewelina Cholodowicz[✉] [iD] and Przemyslaw Orlowski [iD]

West Pomeranian University of Technology in Szczecin, al. Piastów 17, 70-310 Szczecin, Poland
{ewelina.cholodowicz,przemyslaw.orlowski}@zut.edu.pl

Abstract. The expansion of modern supply chains constantly triggers the need of maintaining resilience and agility for higher profit. There is a need to change the standard methods of inventory control to new approaches that are highly adaptable to uncertainties that emerged as a result of supply chains globalization. In this paper, a novel approach based on neural network, state-space control and robust optimization is proposed to support the perishable inventory replenishment decisions subject to uncertain lead times. We develop an approach based on the Wald criterion to compute optimal robust (i.e. "best of the worst" case) controller parameters. We incorporate lead-time specific perturbations through plausible scenarios using several lead times sets. Based on extensive numerical experiments, the obtained solutions highlight that the approach provides stable and robust solutions even for high lead times.

Keywords: Inventory control · Simulation · Optimization · Uncertain lead time · Neural networks · Genetic algorithm

1 Introduction

Over the last decade, the inventory systems have expanded significantly. Nowadays, they are exposed to the highly changeable environment. Not only the uncertainty of market demand can contribute to rising costs, but also the uncertainty of perishability process, variable lead-time, delays. Nowadays, one of the utmost important goals of modern supply chains with growing uncertainty is to build and maintain agility [1]. Fullfilling orders can be challenging tasks in case of variable environment where customers expect more flexibility than ever. Increasing the efficiency of order management systems in terms of automating many steps that requires manual involvement can enhance the goods flow, increase profitability and prevent shortages.

There is a lot of work with optimal inventory policies dedicated to the system with demand uncertainty while including no uncertainties connected to the production and distribution processes instabilities [2]. For example in [3], the effect of time value of money and inflation on optimal ordering policy is investigated but in the case of zero lead-time. Many policies have been proposed for inventory problems under stochastic demand and constant lead time, for example, the basestock policy (also called "order-up-to" policy) and is widely implemented in industry, but the existed methods are not sufficient to keep the modern supply chain at optimal levels because of constant lead-time assumption. For example, in [4] the optimal basestock levels are calculated in the subject of uncertain demand. Therefore, there is a need to develop methods that cope also with uncertainty in the supply chains in terms of lead times. Lead time in inventory management is the time between placing an order to replenish inventory and order receiving. Lead-time uncertainty is usually concerned with unexpected shipment (or production) delays [2]. Lead time is one of the utmost important factors that affect the stock level at any point in time. The areas which are affected by this kind of uncertainty are the agri-food, electrical, medicine (e.g. blood supply chains) and many more. With a view to the above matters, a lot of practitioners and researchers are active in this area of study. In [5] a model to minimize the total cost of an integrated vendor-buyer supply chain when the lead time is stochastic is proposed with constant demand rate assumption. Another example is study [6] in which an inventory model with the randomly variable lead time is developed also under constant demand assumption. In real supply chains constant demand is not often encountered, hence some more advanced methods based on robust optimization started to be implemented in industry.

Robust optimization is considered as a promising approach to deal with uncertainties [7]. The robust optimization has been widely studied in supply chains problems showing promising computational results for problems under demand uncertainty (e.g., see [4, 8–11]).

In the above papers involving demand uncertainty, the supply-side is assumed to be deterministic and order lead times are assumed to be either zero or fixed. There are a few papers that deal with supply and demand uncertainty. An inventory control model under demand and lead time uncertainty is studied in [12] where the tri-level optimization-based approach is used, but without considering the perishable products. Furthermore, there is a work that includes lead-time uncertainty and uses a robust optimization approach [13] – there is an approach based on Benders' decomposition to calculate optimal robust policy parameters. The work proposes the approach for robust optimization and applies it to the basestock problem. We want to extend this study to the case with perishable products and developing also a controller based on neural networks in the proposed approach, not only an optimization method.

In this paper, we proposed a method to reduce the influence of lead-time uncertainty on the robustness of the inventory system with perishable products. The presented approach for inventory control uses the combination of artificial neural networks and genetic algorithm optimization. For method validation, the nonlinear, discrete-time model of inventory system is implemented in Matlab environment together with neural network and applied to the problem of control the perishable goods flow. The proposed method is tested with the use of the set of initial conditions, different lead times, a variety of lead-times uncertainties, and two fixed shelf-lives. For developed controllers, the testing errors are calculated and the analysis of lead-time uncertainty on testing error, stock level and order quantity is presented.

2 Problem Definition and Assumptions

In this paper, we focus on the problem of inventory system control with lead-time uncertainty and perishable products. The problem considers the calculation of order quantity while balancing two conflicting goals: deliver a sufficient number of products on time and keep inventory levels down. The purpose of such an inventory control system is to determine when and how much to order. This decision should be based on the current inventory state, the expected demand, the lead-time, possible delays, and other cost factors. In this paper, we propose the approach, which includes the solution steps for the problem of inventory system optimization in case of uncertain lead-time. For the offline testing of the control approach the nonlinear, discrete-time perishable inventory with fixed lifetime products, proposed in [14], is implemented in Matlab environment. The considered class of inventory system assumes that stored products have a fixed shelf-life. The following assumptions are used for formulating the model and the investigated problem:

1. The review period is constant and equals one day.
2. The products are sold according to FIFO policy.
3. The inventory contains a single type of product.
4. Lead-time s may be uncertain.
5. Shortages are allowed but are not backlogged. Excess demand is lost.
6. There is one stocking point in each period.
7. Demand is a time-varying function.
8. Deterioration occurs as soon as the items are received into inventory.
9. The shelf-life l is fixed and known a priori. After l days all items from the same batch are expired and became unsellable waste. Lost units are not replaced.

The applied notation of applied inventory model is presented in the Table 1.

Table 1. The model parameters and variables – applied notation

Symbol	Definition
N	Length of the simulation horizon
$k \in \{1,2,\ldots,N\}$	Discrete-time
l	The shelf-life of a product
$i \in \{1,2,\ldots,l\}$	Index of state variables
s	Lead time
s_Δ	Uncertain lead time
s_0	Nominal lead time
Δ	Lead time perturbation
d_{max}	The maximum demand in one period k

(*continued*)

Table 1. (*continued*)

Symbol	Definition
$\mathbf{x}(k)$	Vector of state variables
$y(k)$	Inventory level (on-hand stock)
$u(k)$	Order quantity
$d(k)$	Aggregated demand
$d_i(k)$	Demand for products of age i
$h(k)$	Aggregated amount of sold products
$h_i(k)$	Sold products of age i
n	Number of neurons in the hidden layer
$\mathbf{v}$	The vector of network weights
a_j	Activation function in the first layer
e_j	Activation function in the second layer
c_j	Transformation function in the second layer

In the applied inventory model, the demand is modelled as an unknown a priori, bounded function of discrete-time: $0 \le h(k) \le d(k) \le d_{max}$. There is full demand satisfaction when the number of sold products: $h(k) \in \mathbb{R}_{\ge 0}$ is equal to the current demand $d(k) \in \mathbb{R}_{\ge 0}$, $h(k) = d(k)$. The maximum value of imposed demand for products per k period is constrained by $d_{max} \in \mathbb{R}_{>0}$. The orders are calculated in regular intervals on the basis of the expected demand $d(k)$ and the inventory state $x_i(k) \in \mathbb{R}_{\ge 0}$. The inventory state can be divided into two parts: (a) the on-hand stock per age i $x_{s+1}(k), x_{s+2}(k), \ldots, x_l(k)$; (b) work-in-progress deliveries $x_1(k), x_2(k), \ldots, x_s(k)$. In this model, i represents the age of products, e.g. $d_{s+1}(k)$ is the demand for the freshest products available in the inventory. The total amount of the sold products is given by: $h(k) = \sum_{i=1}^{l} h_i(k)$, where: $h_i(k) \in \mathbb{R}_{\ge 0}$ – sold products of age i.

In general we assume that lead-time s may be not known exactly. In such case the uncertain lead-time is denoted as s_Δ, and takes the following additive form:

$$s_\Delta = s_0 + \Delta \qquad (1)$$

where: s_0 is a nominal value of lead-time and Δ is unknown, but bounded perturbation such that $|\Delta| \le \delta$.

As inventory systems become more complex, representing them with differential equations or state-space models becomes highly advanced. Considering that, for efficient implementation in Matlab, the model is formulated using a state-space approach. State-space representation of this system is given by l equations:

$$\begin{cases} x_1(k+1) = u(k) \\ x_2(k+1) = x_1(k) - h_1(k) \\ \vdots \\ x_l(k+1) = x_{l-1}(k) - h_{l-1}(k) \end{cases} \qquad (2)$$

State variable $x_i(k) \in \mathbb{R}_{\geq 0}$ keeps the information about products quantity of age i. Order quantity $u(k)$ is a nonnegative and real number. A more profound explanation of inventory model fundamentals is presented in [14].

3 Proposed Approach

The main purpose of the proposed approach is to calculate order quantities and their frequency for the inventory system under lead-time uncertainty while obtaining optimal performance in terms of shortage and holding costs minimization. The proposed approach uses two main tools: (a) genetic algorithm (GA) which is used for the learning stage; (b) neural network (NN) which is designed for the goods flow control in the perishable inventory system with lead-time uncertainty. In Table 2, there are main parameters that are assumed in the proposed approach.

We adopted the artificial neural network as a controller to control the flow of perishable products in case of lead-time uncertainty. Furthermore, the proposed approach includes also genetic algorithm application for the learning process of neural networks. A genetic algorithm is used as an optimization tool for calculating neural network weights. The proposed approach can be represented by diagram in Fig. 1.

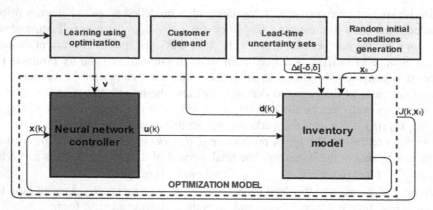

Fig. 1. The diagram of proposed approach.

The proposed approach can be explained as follows: the first step is to generate the random initial conditions of inventory state in range (0, 2). Next step is the optimization process. The goal of the optimization process is to tune the weights of neural networks that minimize the quality cost for the worst-case scenario of lead-time uncertainty. The quality cost is represented as a weighted sum of lost sales and holding cost with the assumption that the cost of lost sales is 3 times higher than the cost of holding cost. Finally, testing process of the obtained results is performed using a set of different initial conditions of inventory and different lead-times uncertainties (from the selected uncertainty range).

Table 2. Parameters of GA and NN.

Approach part	Parameter	Value
GA	The number of variables	$n(l+2)$
	The maximum number of generation	4000
	Population size	2000
	Parallel computing	True
NN	ANN model	Multi-layered perceptron
	The number of neurons in hidden layer	3
	The number of input node	l
	The number of output node	1
	The number of hidden layer	1
	The number of hidden node	3
	Activation function on the hidden layer	Satlin
	Activation function on the output layer	Poslin

The developed neural network controller consists of three layers: input, hidden and output layer. The applied structure of the neural network is depicted in Fig. 2. The input of the neural network controller is the state vector $\mathbf{x}(k) \in \mathbb{R}_{\geq 0}$ which is the number of products on every shelf – shelf represents the age of the product. The products are picked from the shelf to fullfill the current orders. The output of the neural network is the control $u(k) \in \mathbb{R}_{\geq 0}$ which is the order quantity generated in order to satisfy the demand $\mathbf{d}(k) \in \mathbb{R}_{\geq 0}$. The applied structure is a feed-forward network, in which the activation functions: saturating linear transfer function a_j, positive linear transfer function e_j and transformations c_j and $\breve{u}$ occur. The controller on the basis of current stock age and work-in-progress deliveries is able to generate the optimal order quantity for each day k. The weights are the elements of vector $\mathbf{v}$.

The learning process is formulated as an optimization problem of a perishable inventory system with uncertainty with the use of the genetic algorithm. The objective of the considered optimization problem is to establish weights of the neural network (Fig. 2) so that the inventory system may satisfy the customers' needs (3) and minimize the holding cost (4) at the same time. The first criterion is describing the number of lost sales due to stock shortages:

$$J_h = \sum_{k=s+1}^{N} (d(k) - h(k)) \tag{3}$$

As a second criterion for optimization, the surplus of stock over demand is considered:

$$J_y = \sum_{k=s+1}^{N} m(k) \tag{4}$$

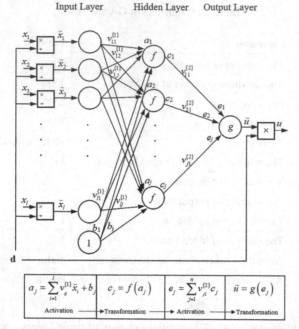

Fig. 2. The applied structure of the neural network controller.

where:

$$m(k) = \begin{cases} y(k) - d(k) \text{ for } y(k) \geq d(k) \wedge \widehat{y}(k) \leq d(k) \\ 0 \quad \text{for} \quad \text{otherwise} \end{cases} \tag{5}$$

The inequalities in above relationship (5) eliminate the penalty for the stock which results only from the initial conditions $\mathbf{x}_0$, where $\widehat{y}(k)$ is free response of the system. Formulated criteria can be written as the weighted cost function:

$$J = 3 \cdot J_h + J_y \tag{6}$$

Formally, the optimization problem may be stated as follows:

$$\min_{\mathbf{v}} \max_{\Delta} J(\mathbf{v}, \Delta)$$

$$\text{s.t.} \quad -\delta \leq \Delta \leq \delta \tag{7}$$

The optimization is performed for assumed set of initial inventory states $\mathbf{x}_0$. As a result of the optimization process, the vector of weights $\mathbf{v}$ is obtained. In this way, the inventory controller can be optimized with a view to uncertain demand, perishability and the state vector $\mathbf{x}(k)$. This approach provides flexibility and resilience, making the inventory system being more robust for uncertain changes of lead-time. In the further part of the work, the proposed approach is called robust neural network controller (in short: RNN controller).

4   Simulation Results

In this section, we apply the proposed approach to control the perishable inventory system with uncertain delay. The simulation research is divided into five parts. The first one is focused on analyzing the performance of the proposed controllers in terms of lead-time influence on testing error. In the second part, the effect of the lead-time perturbation on stock level and cost function is investigated. In order to present the performance of proposed approach, we prepare the numerical example with the data extracted from one of the retail outlet [15]. The data contains the daily demand for milk in one month. In the next part, we extend the simulation research by applying larger lead-time uncertainty. In order to show the numerical example of the performance of RNN controllers, the fourth part contains the time responses of the perishable inventory control system. The fifth part of the simulation study is devoted to the analysis of testing error using different sizes of test sets in case of different lead-time perturbations.

For simulation purposes, the learning set contains 180 different inventory states. The initial conditions of the state vector are generated using random numbers in the range: $(0, 2)$. Single inventory state represents a different level of initial stock level of product of different shelf-life. The general simulation parameters take the following values: the review period is one day, simulation horizon equals month (31 days), shelf-life l is fixed, adopted issuing policy: FIFO. The parameters of the main parts of the approach are listed in Table 2 (previous section).

4.1   Lead-Time Influence on Testing Error

In this subsection, the results of the testing process for the following nominal lead-times $s_0 \in \{2, 3, 4, 5\}$, lead-time perturbation of one day and shelf-life of 9 days are presented. The size of the test set is 1000 different initial inventory states. The obtained results are listed in Table 3.

Table 3. Cost function value and testing error for different lead-times.

s_0 (days)	J	Testing error (%)
2	1.4686	1.69%
3	1.3408	2.23%
4	1.4584	1.98%
5	2.4153	2.66%

The results show that testing error is the smallest for $s_0 = 2$ among considered cases and the biggest for the highest considered lead-time $s_0 = 5$. Nevertheless, the testing error is not exceeded about 3% in all considered cases. The cost function value J takes the highest value for the highest considered nominal lead-time.

4.2 Lead-Time Perturbation Influence on the Learning Process

This subsection is devoted to the investigation of lead-time perturbation and its influence on the learning process. In Table 4 the results of the learning process for the lead-time perturbation bounds: $\delta \in \{0, 1, 2, 3\}$, the nominal lead-time value of 5 days and products with the shelf-life of 12 days are presented. The lowest value corresponds to the no perturbation scenario and the highest to the high lead time uncertainty scenario. The estimated learning time increase with the perturbation bound used for the controller tuning. Learning time was approximately in the range 50–100 min on computer with Ryzen 5950X CPU.

Table 4. Cost function for worst-case scenario for different lead-time perturbation bounds.

δ (days)	J	Cost increase
0	0.8287	0
1	1.2011	0.37
2	1.7234	0.89
3	2.4317	1.60

In the analyzed case, a threefold increase in lead-time perturbation bound leads to about 2 times higher costs in terms of holding space and lost sales. In the assumed scenario, the inventory system without uncertainty in the lead-time is able to generate about 31% less cost J in comparison to the smallest assumed lead-time perturbation bound ($\delta = 1$).

4.3 Robustness of Proposed Approach

For the purpose of robustness analysis the simulation with different lead-time uncertainty is performed. The simulation scenario is prepared as follows: the demand for milk product is extracted from the retail outlet; the simulation scenario starts with a sufficient level of stock in the inventory – it means that inventory initial states are adopted to the lead-time in the analyzed case; the assumed expiration date equals 12 days; the weights of the design RNN are optimized for different perturbation bounds $\delta \in \{0, 1, 2, 3\}$ and the nominal lead-time of 5 days, whereas the simulation is conducted for the perturbated lead-times ranging from 1 to 9 days. In Fig. 3, the surface of the cost function is presented.

Figure 3 visualizes the cost function values of optimized robust neural controllers for the different variants of lead-time uncertainty. It can be seen that the smallest cost function is achieved for perturbations Δ smaller than 0. The most interesting situation is for the $\Delta > 0$. As it can be seen in Fig. 3 the controller, which does not include the uncertainty during the learning process $\delta = 0$, obtains high cost function values for $\Delta > 0$. This is because the controller was not able to be prepared for unknown uncertainties and it causes a lot of shortages in the inventory. On the other hand, the most robust behaviour for the highest lead-time is achieved by the controller of perturbation bound $\delta = 3$. In this case, other controllers ($\delta < 3$) obtain worse control quality.

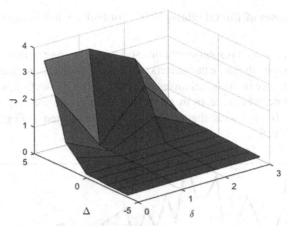

Fig. 3. Cost function values for controllers optimized for different values of perturbation bound δ and simulated using different lead-time perturbations Δ.

Moreover, in order to analyze the effect of lead-time influence on the level of stock the surface with the stock level is also generated (see Fig. 4).

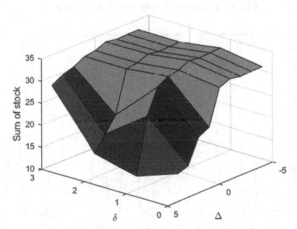

Fig. 4. Sum of stock for controllers optimized for different values of perturbation bound δ and simulated using different lead-time perturbations Δ.

It can be seen that for $\Delta \leq 0$ the stock level is similar for all optimized controllers. The change starts to be visible for $\Delta > 0$ where the stock level decreases. It can be observed that for $\Delta = 4$ the following relationship is satisfied: the higher δ the more stock is stored in the inventory. It means that controller optimized using the highest perturbation bound ($\delta = 3$) is more accurate in determining a sufficient amount of stock to minimize shortages.

4.4 Time Responses of the Obtained NNC Controllers for Long-Lead-Time Scenario

In this subsection, the time responses are investigated. The case with the high lead-time is selected ($s_\Delta = 8$) and the same parameters of the simulation are assumed as in point 4.3 with the only difference in the initial inventory state. In this subsection, we assumed zero initial inventory state, which means that inventory is completely empty at the beginning of the simulation. To start with, the monthly demand is plotted in Fig. 5 and lost sales are illustrated in Fig. 6.

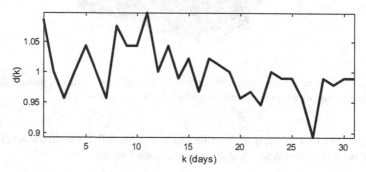

Fig. 5. Demand scenario for milk products.

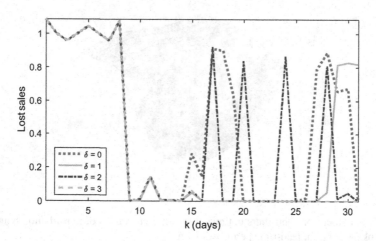

Fig. 6. Lost sales for controllers optimized using different perturbation bounds δ.

It is clearly visible that least lost sales are for $\delta = 3$. This observation implies that variable demand is satisfied with the highest level of service for $\delta = 3$ among considered controllers. It is important to highlight that the significant shortages characterize the non-robust controller ($\delta = 0$). The next time response is the order (Fig. 7).

In Fig. 7, it can be seen that the controller tuned for the perturbation $\delta = 3$ calculates the orders that follows the changes in demand without oscillations and overshoots. The other controllers generate the highly oscillating order quantities which result in higher shortages which can be seen in Fig. 6.

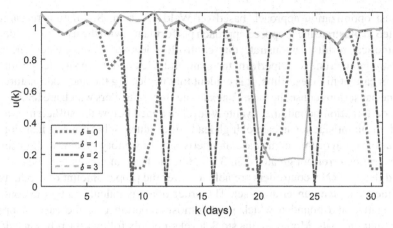

Fig. 7. Orders for controllers optimized using different perturbation bounds δ.

4.5 Test Size Influence on Testing Error

The next section is focused on the analysis of investigating the test size influence on the testing error. Figure 8 illustrates the obtained testing errors during the testing phase of RNN controllers.

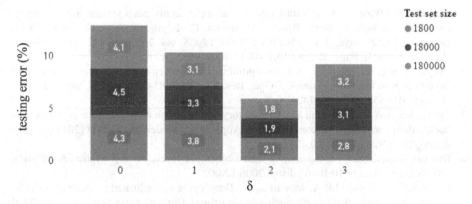

Fig. 8. Learning error for different lead-time perturbation bounds and test set sizes.

In Fig. 8 can be observed that testing error, for the highest perturbation bound for all test set sizes, is the smallest among the considered cases. On the contrary, the highest learning errors occur for the controller that controls the inventory system without considering any uncertainty. It can be noted that the testing error for all considered cases is in the range from 2.3% to 4.5%.

5 Conclusions

In brief, we developed a robust neural network controller to manage the perishable items in case of uncertain lead times. In order to optimize the developed model, we adopted the robust optimization approach based on Wald criterion. Simulation research was conducted to illustrate the proposed approach performance with the use of a real demand. Our numerical results demonstrate that controllers learned using greater uncertainty bounds are more prone to outperform the controllers learned using smaller perturbation bounds in case of high lead-time. It is evident that neglecting the uncertain nature of the lead-time has serious consequences. For example, for controllers which are learned using smaller perturbation bound, the inventory level dropped below the sufficient minimum of full demand satisfaction in case of high lead-time values. What is more, learning using an evolutionary algorithm in the case of a perishable inventory system with uncertainty provides testing error not greater than 3.8%. On the basis of conducted research it can be noted that the RNN controllers are able to order the proper amount of products in an exact time for a given uncertainty set. The order quantity calculated by the controllers is nonnegative and bounded which is of utmost importance in the case of practical implementation goals. Moreover, the stock level smoothly follows the reference demand value and do not cause any unnecessary overstocks. All these advantages are achieved in the environment of uncertain lead-time. Looking also at the limitations, our proposed approach can be extended in considering demand uncertainty and lead-time uncertainty at the same time. It is one of the main topics for our further researches.

References

1. Lau, H.C.W., Wong, C.W.Y.: Infrastructure of an agile supply chain system: a virtual agent modeling approach. In: Sloot, P.M.A., Abramson, D., Bogdanov, A.V., Gorbachev, Y.E., Dongarra, J.J., Zomaya, A.Y. (eds.) ICCS 2003. LNCS, vol. 2659, pp. 432–441. Springer, Heidelberg (2003). https://doi.org/10.1007/3-540-44863-2_43
2. Chu, J., Huang, K., Thiele, A.: A robust optimization approach to model supply and demand uncertainties in inventory systems. J. Oper. Res. Soc. **70**, 1885–1899 (2019). https://doi.org/10.1080/01605682.2018.1507424
3. Taleizadeh, A.A., Nematollahi, M.: An inventory control problem for deteriorating items with back-ordering and financial considerations. Appl. Math. Model. **38**, 93–109 (2014). https://doi.org/10.1016/j.apm.2013.05.065
4. Bienstock, D., Özbay, N.: Computing robust basestock levels. Discret. Optim. **5**, 389–414 (2008). https://doi.org/10.1016/j.disopt.2006.12.002
5. Sajadieh, M.S., Jokar, M.R.A., Modarres, M.: Developing a coordinated vendor-buyer model in two-stage supply chains with stochastic lead-times. Comput. Oper. Res. **36**, 2484–2489 (2009). https://doi.org/10.1016/j.cor.2008.10.001
6. Maiti, A.K., Maiti, M.K., Maiti, M.: Inventory model with stochastic lead-time and price dependent demand incorporating advance payment. Appl. Math. Model. **33**, 2433–2443 (2009). https://doi.org/10.1016/j.apm.2008.07.024
7. Gholizadeh, H., Fazlollahtabar, H.: Robust optimization and modified genetic algorithm for a closed loop green supply chain under uncertainty: case study in melting industry. Comput. Ind. Eng. **147**, 106653 (2020). https://doi.org/10.1016/j.cie.2020.106653
8. de Ruiter, F.J., Ben-Tal, A., Brekelmans, R.C., den Hertog, D.: Robust optimization of uncertain multistage inventory systems with inexact data in decision rules. Comput. Manag. Sci. **14**, 45–66 (2017)
9. Ben-Tal, A., Golany, B., Nemirovski, A., Vial, J.P.: Retailer-supplier flexible commitments contracts: a robust optimization approach. Manuf. Serv. Oper. Manag. **7**, 248–271 (2005). https://doi.org/10.1287/msom.1050.0081

10. Bertsimas, D., Thiele, A.: A robust optimization approach to inventory theory. Oper. Res. **54**, 150–168 (2006). https://doi.org/10.1287/opre.1050.0238
11. Kim, J., Do, C.B., Kang, Y., Jeong, B.: Robust optimization model for closed-loop supply chain planning under reverse logistics flow and demand uncertainty. J. Clean. Prod. **196**, 1314–1328 (2018). https://doi.org/10.1016/j.jclepro.2018.06.157
12. Rahdar, M., Wang, L., Hu, G.: A tri-level optimization model for inventory control with uncertain demand and lead time. Int. J. Prod. Econ. **195**, 96–105 (2018). https://doi.org/10.1016/j.ijpe.2017.10.011
13. Thorsen, A., Yao, T.: Robust inventory control under demand and lead time uncertainty. Ann. Oper. Res. **257**(1–2), 207–236 (2015). https://doi.org/10.1007/s10479-015-2084-1
14. Chołodowicz, E., Orłowski, P.: Development of new hybrid discrete-time perishable inventory model based on Weibull distribution with time-varying demand using system dynamics approach. Comput. Ind. Eng. **154**, 107151 (2021). https://doi.org/10.1016/j.cie.2021.107151
15. Thakkar, H., Ravalji, J.: Demand management of perishable products using supply chain management concept. Int. Conf. Ind. Eng. (2013)

Batch QR Factorization on GPUs: Design, Optimization, and Tuning

Ahmad Abdelfattah[1]($\boxtimes$), Stan Tomov[1], and Jack Dongarra[1,2,3]

[1] University of Tennessee, Knoxville, USA
{ahmad,tomov,dongarra}@icl.utk.edu
[2] Oak Ridge National Laboratory, Oak Ridge, USA
[3] University of Manchester, Manchester, UK

Abstract. QR factorization of dense matrices is a ubiquitous tool in high performance computing (HPC). From solving linear systems and least squares problems to eigenvalue problems, and singular value decompositions, the impact of a high performance QR factorization is fundamental to computer simulations and many applications. More importantly, the QR factorization on a batch of relatively small matrices has acquired a lot of attention in sparse direct solvers and low-rank approximations for Hierarchical matrices. To address this interest and demand, we developed and present a high performance batch QR factorization for Graphics Processing Units (GPUs). We present a multi-level blocking strategy that adjusts various algorithmic designs to the size of the input matrices. We also show that following the LAPACK QR design convention, while still useful, is significantly outperformed by unconventional code structures that increase data reuse. The performance results show multi-fold speedups against the state of the art libraries on the latest GPU architectures from both NVIDIA and AMD.

Keywords: Batch linear algebra · QR factorization · GPU computing

1 Introduction and Related Work

In the context of dense linear algebra, a batch routine performs a standard linear algebra algorithm on a batch of relatively small matrices. This kind of workload is quite different from operating on one large matrix. Many software packages, from both the industry and the research community, have been serving the latter form of workloads for many years. Examples include LAPACK [1], PLASMA [13], MAGMA [12], BLIS [19], and Intel's MKL [14]. Batch workloads, however, are relatively recent, and gained a lot of attention in many scientific communities. Applications include quantum chemistry [8], sparse direct solvers [21], astrophysics [16], and signal processing [6]. Vendor software libraries such as Intel's MKL [14], NVIDIA's cuBLAS [17], and AMD's hipBLAS [5] now provide many batch routines for several BLAS and LAPACK operations.

Batch routines often require a different mindset for performance optimization, especially on GPUs. Since we are dealing with small matrices, it is crucial

© The Author(s), under exclusive license to Springer Nature Switzerland AG 2022
D. Groen et al. (Eds.): ICCS 2022, LNCS 13350, pp. 60–74, 2022.
https://doi.org/10.1007/978-3-031-08751-6_5

to save as much memory traffic as possible. As an example, for very small matrices that fit in the register file of the GPU, fully unrolled and unblocked kernels can achieve a performance that is superior to any other approach [3]. For relatively larger matrices, however, different assumptions must be made in order to maintain a high performance across the size spectrum.

In this paper, we take the batch QR factorization as a case study for optimization on GPUs. We show that there is not a single design strategy that can serve all sizes efficiently. Each design strategy assumes a number of building blocks (e.g., GPU kernels) of the factorization, which might differ from the conventional LAPACK structure. This work is considered an improvement over the work by Haidar et al. [10], which is available in the MAGMA library.

2   Algorithmic Background

The QR factorization decomposes a dense matrix $A_{m \times n}$ into the product $Q_{m \times m} \times R_{m \times n}$, where Q is an orthogonal matrix, and R is upper triangular. Throughout the paper, we assume $m \geq n$. The standard LAPACK implementation does not compute Q explicitly. Upon completion, the matrix A is overwritten by the two matrices V and R, as shown in Fig. 1a. The matrix V is lower triangular with unit diagonals (not stored), such that each column v_i represents an elementary Householder reflector $H_i = I - \tau_i v_i v_i^T$, where τ is a scalar (stored separately). The Q factor is computed as $Q = \prod_i^n H_i$.

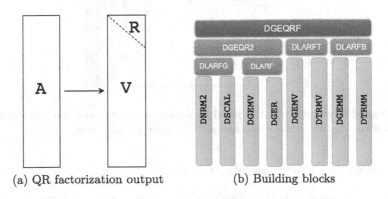

(a) QR factorization output   (b) Building blocks

Fig. 1. The LAPACK convention of the QR factorization

Assuming double precision, the standard LAPACK implementation is available in the `dgeqrf` routine, which has the building blocks shown in Fig. 1b. Both the `dgeqr2` and `dgeqrf` routines perform the QR factorization. However, `dgeqr2` is an *unblocked* design, meaning that it proceeds one column at a time, building the corresponding elementary reflector (`dlarfg`), and applying it to the rest of the matrix (`dlarf`). Therefore, `dgeqr2` is limited by the memory bandwidth of the hardware, since it relies on vector or matrix-vector operations only (BLAS level 1 and 2). On the other hand, `dgeqrf` is a *blocked* design. It uses `dgeqr2` to factorize a rectangular *panel*. The corresponding block of reflectors are applied to the trailing matrix using matrix-matrix (L3 BLAS) operations. The use of L3 BLAS enables dgeqrf to be compute-bound. The application of the block reflectors contains a preparatory stage (dlarft), during which a triangular factor T is computed from the V matrix and the scalars $\tau_i, i \in \{1, 2, \cdots, n\}$, such that $Q = I - V \times T \times V^T$. The last equation takes advantage of matrix multiplication (GEMM) when implicitly applying Q to the trailing matrix.

2.1 Nested Blocking

A standard QR factorization directly calls the unblocked panel factorization (dgeqr2). For a batch of relatively small matrices, the panel is thin, typically 4–8 in most cases. Thin panels lead to rank-k updates (batch GEMM) that are memory-bound. On the other hand, passing relatively wide panels directly to the memory-bound dgeqr2 also hinders the performance. The solution to this tradeoff is to use *nested blocking*, which is a well-known approach in LAPACK's blocked algorithms, despite not being used in the standard QR implementation. Figure 2 shows the general idea of nested blocking, where a wide panel is internally split during its factorization. Nested blocking increases the reliance on L3 BLAS operations (batch GEMM).

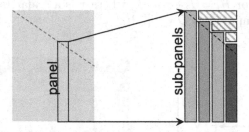

Fig. 2. Nested blocking in the QR panel factorization. The horizontal rectangles refer to parts of the matrix that are touched solely by the update step.

2.2 Computing the Triangular Factor T

The original implementation of the dlarft routine relies on two memory-bound operations, the matrix-vector product (dgemv), and the triangular matrix-vector product (dtrmv). For a block-reflector V of width nb, the factor $T_{nb \times nb}$ can be computed recursively as in Algorithm 1, where lines 2 and 3 update the same column of T using dgemv and dtrmv operations, respectively. However, previous work [11] has shown that all the calls to dgemv can be aggregated into one dgemm call, while the dtrmv calls remain roughly unchanged. While Algorithm 2 clearly shows a performance advantage over Algorithm 1, it needs preprocessing stages that may be costly for small matrices. In order to call dgemm, the matrix V must be separated from the R factor (refer to Fig. 1a). This requires (1) copying the R matrix into a workspace (dlacpy), (2) setting the upper triangular part of

V to zeros, with units on the diagonal (dlaset), and (3) another copy to bring
R back on top of V (dlacpy). In addition to the overhead of these calls, there
is also workspace management overhead. Note that there are other methods for
computing the block Householder transformations [18,20], which are beyond the
scope of this paper.

Algorithm 1: Classical dlarft	Algorithm 2: Improved dlarft
1 for $j=1$ to nb do 2 $T_{1:j-1,j} = -\tau_j V_{j:n,1:j-1}^T \times V_{j:n,j}$ 3 $T_{1:j-1,j} = T_{1:j-1,1:j-1} \times T_{1:j-1,j}$ 4 $T_{j,j} = \tau_j$ 5 end	1 $T_{1:nb,1:nb} = V_{1:n,1:n}^T \times V_{1:n,1:n}$ 2 for $j=1$ to nb do 3 $T_{1:j-1,j} =$ $-\tau_j T_{1:j-1,1:j-1} \times T_{1:j-1,j}$ 4 $T_{j,j} = \tau_j$ 5 end

3 Experimental Setup

Throughout the paper, we show the incremental performance improvements on
a system equipped with an NVIDIA Tesla A100-SXM4 GPU, which is clocked at
1.41 GHz and has 80 GB of memory. The GPU is hosted by an AMD EPYC 7742
64-Core Processor, clocked at 2.25 GHz. The CUDA version is 11.2. The final
performance results are collected on this system as well as on another system
equipped with an AMD Instinct MI100 GPU, which has 32 GB of memory, and
clocked at 1.5 GHz. The ROCM version is 4.5.0. The host CPU is an AMD
EPYC 7662 64-Core Processor, running at 3.25 Ghz. All the developments are
lined up to be released in the MAGMA library. Our solution will be referenced as
"MAGMA" in all the performance results. For NVIDIA GPUs, the performance
results are compared against the batch QR factorization in the cuBLAS library,
as well as against the open source KBLAS library [9]. For AMD GPUs, the
performance is compared against the hipBLAS library.

4 LAPACK-Style Design

Our goal is to maximize the batch QR factorization performance on any matrix
size and shape. A straightforward approach is to extend the primitive building
blocks in Fig. 1b to support a batch of matrices. There are two advantages to
this approach. First, it uses some of the existing batch BLAS routines, like batch
GEMM, which are often highly optimized by the vendor libraries, or by open
source libraries [2]. For the batch QR factorization in particular, the reliance on
an optimized batch GEMM routine guarantees performance portability across
different GPU architectures. Second, since the building blocks are assumed to
be LAPACK-compliant, the final implementation would support any matrix size
and shape. This is unlike some previous efforts that target application-specific
range of sizes [7,15].

Our first implementation of batching the building blocks is based on the
efforts by Haidar et al. [10]. It is improved by taking into account some of
the new features in the vendor libraries, especially the more optimized batch

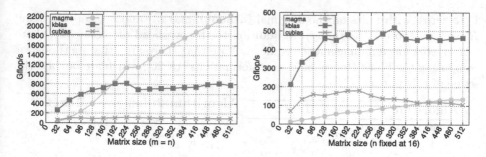

Fig. 3. Batch QR performance in double precision. The MAGMA design is based on batching the building blocks of Fig. 1b. Results are shown for square matrices (left) and tall-skinny matrices (right), using a Tesla A100 GPU.

GEMM kernels. Figure 3 shows the performance results on square and tall-skinny matrices. For square sizes, KBLAS outperforms the original MAGMA design for sizes less than 192. Otherwise, MAGMA has the best performance. Our profiling results show that only MAGMA calls batch GEMM underneath, which explains why its performance scales well, while cuBLAS and KBLAS stagnate.

Table 1. Time breakdown (%) for the original MAGMA design. Results are shown for double precision on the A100 GPU, with 1000 matrices per batch.

Category	Matrix size (m, n)			
	(128,16)	(64,64)	(128,128)	(256,256)
dgeqr2 kernels	**64.4**	**55.09**	**45.86**	30.28
dgemm	12.47	24.68	35.4	**56.23**
Auxiliary (dlacpy, dlaset, etc.)	21.28	15.93	14.48	10.48
trmv (for dlarft)	1.85	4.29	4.26	3.01

For tall-skinny matrices, both cuBLAS and KBLAS have a clear advantage over MAGMA. Our conclusion is that the MAGMA design favors wide matrices, where the trailing matrix update is rich in batch GEMM calls. For tall-skinny matrices, such an advantage is absent. In addition, the current panel implementation in MAGMA lacks optimizations for tall-skinny matrices. To emphasize this point, Table 1 shows the percentage of time spent in different parts of the MAGMA design for four selected sizes. It shows that the panel kernels contribute significantly to the total execution time. Therefore, we cannot rely on batch DGEMM alone in order to achieve high performance. The QR panel must undergo an extensive optimization. Since the dgeqr2 kernels are memory-bound, it is imperative to save memory traffic as much as possible. This can be achieved by merging multiple building blocks into a single execution context, which is often called *kernel fusion*. We use a *multi-level* kernel fusion, in which different parts of the algorithm are fused based on the matrix size.

5 Panel Optimization

We begin by designing a new GPU kernel for the panel factorization. The kernel caches a panel of size $m \times nb$ in the register file of the GPU and implements the unblocked factorization (dgeqr2). In a thread block, each thread possesses one row of the panel, so at least m threads are required for each thread block. There are two occasions where we perform a reduction operation across the columns of the panel. The first is during the generation of the Householder reflector, where the norm of the current column is computed. The second is when applying the reflector to the trailing matrix, i.e., $A = (I - \tau v v^T) \times A$. The product $v^T A$ requires a reduction operation across the columns of A. Since these reductions contradict with the thread-per-row assignment, a shared memory workspace is allocated to perform tree reductions. Note that the $v^T A$ product involves a multi-column reduction, for which we re-organize the threads into independent groups, and each group collaboratively reduces the assigned column. A key design aspect of this kernel is the use of compile-time constants. For example, the width of the panel nb must be known at compile time in order to avoid register spilling. It also helps the compiler unroll most of the loops inside the kernel. The kernel is instantiated for $1 \le nb \le nb_{max}$, where nb_{max} depends on the GPU resources as well as the compute precision.

The performance of this kernel is dependent on the height of the panel, since it requires one thread per row. For example, a panel of size $512 \times nb$ needs 512 threads. Assuming that all other resources are not a bottleneck, we can schedule four thread blocks at maximum per SM, due to a hardware limitation. Panels taller than 512 would cause at least 25% drop in the thread occupancy per SM. Depending on the width of the panel, other resources could be underutilized as well. The remedy to such a behavior is to relax the constraint on the number of threads. We propose a second kernel that stores the panel in shared memory instead. This enables us use any number of threads to factorize the panel. The proposed kernel assumes $nb \le$ #threads$\le m$. The tree reductions mentioned above are redesigned to work with any number of threads in that range. To prove our point, Fig. 4 shows the performance of the two kernels for a panel of width 4. For the shared memory kernel, we use 32, 64, and 128 threads. The figure shows that there is no clear winner, and that two decisions should be made before the panel factorization: (1) which kernel should be used (register vs. shared memory), and (2) if the shared memory kernel is used, how many threads should be used? All performance graphs in this figure have a staircase-like behavior. As the panel becomes taller, more resources are required, leading to drops in occupancy. In order to select the best performing kernel, we collect offline tuning data based on panel width, precision, and GPU architecture. These data are used to select such a kernel at run time.

Figure 5 shows the updated performance after incorporating the fused dgeqr2 kernels. For square matrices, the performance of MAGMA is improved by 12.9%–57.1%, while the speedups for the tall-skinny case are in the range 11.9%–41.4%. We generally observe that the smaller the matrix, the larger the speedup. This is expected, since the savings in memory traffic should be more critical for smaller

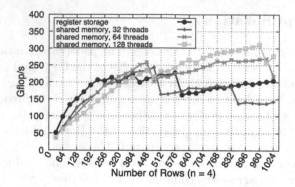

Fig. 4. Comparing different kernels for the fused `dgeqr2` step. Results are shown for double precision using a Tesla A100 GPU, with 1000 per batch.

problems. However, the general behavior against cuBLAS and KBLAS remains the same, except for the slightly earlier intersection points with the MAGMA performance graphs.

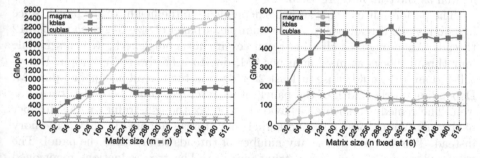

Fig. 5. Batch QR performance in double precision, with the fused panel kernels incorporated into the MAGMA design. Results are shown for square matrices (left) and tall-skinny matrices (right), using a Tesla A100 GPU.

In order to decide where the next optimization should be, we repeat the time breakdown experiment after using the fused panel kernels, which is shown in Table 2. The table shows that the panel kernels are no longer dominant in any of the four sizes. Another positive sign is the increased percentage of the `dgemm` kernel. However, the auxiliary kernels now contribute a noticeable amount of time, and even dominate the execution time for size 128×16. Recall that these auxiliary kernels are mostly called in a setup phase for computing the T factor (Sect. 2.2). In addition, if the size of T is small, calling batch GEMM multiple times on small matrices might be inefficient. We need to avoid these auxiliary kernels as much as possible, especially for tall-skinny sizes.

Table 2. Time breakdown (%) for the MAGMA design with fused geqr2. Results are shown for double precision on the A100 GPU, with 1000 matrices per batch.

Category	Matrix size (m, n)			
	(128,16)	(64,64)	(128,128)	(256,256)
dgeqr2 kernels	24.37	11.37	10.26	7.81
dgemm	26.39	**48.66**	**58.78**	**74.41**
Auxiliary (dlacpy, dlaset, etc.)	**45.29**	31.53	23.89	13.81
trmv (for dlarft)	3.95	8.44	7.07	3.96

6   Optimizing the Trailing Matrix Update

We acknowledge that we cannot use the batch dgemm kernel to compute the T factor for relatively thin matrices. At the same time, using the memory bound dgemv and dtrmv kernels is not expected to be a faster solution. Since this improvement is critical mostly for tall-skinny sizes, a candidate solution is to merge the dlarft and the dlarfb operations into one GPU kernel. However, since the fused kernels operate on the fastest memory levels of the GPU, the implementation can be simplified into applying the elementary reflectors directly to the trailing matrix (without forming the T factor). This strategy is partially similar to cuBLAS and KBLAS in the sense that they don't use the batch GEMM for the trailing matrix update. However, we limit its use for a certain width, as we discuss later in the paper. Algorithm 3 shows a pseudo code of the proposed kernel. It reads the output of dgeqr2 into shared memory, setting its upper triangular part to zeros, and its diagonal to ones. The factorized panel remains cached for the lifetime of the kernel. Assuming that the trailing matrix has a width $\bar{n}$, we loop over this width in a small step ib, so that the sub-trailing panel (tA[]) is cacheable in either the shared memory or the register file. We use a device routine implementation of the dlarf routine to apply each reflector in the panel. The dlarf routine has an optimized multi-column tree reduction and a parallel rank-1 update, which are the two main components required for the update. Finally, the tA[] buffer is written into the main memory, and a new sub-trailing panel is loaded. Similar to the fused panel kernel, there are two implementations of the update kernel, one that uses the register file for storing tA[] and uses a restricted number of threads, while the other uses shared memory only, and has a tunable number of threads.

An important point is that the fused panel and update kernels can be used to factorize the entire matrix, without utilizing the batch dgemm kernel. This decision is dependent on so many parameters, like the dimensions (m, n) of the matrix, the compute precision, and the GPU architecture. To achieve the best performance, we conducted a set of offline tuning sweeps that discover the *cut-off width*, below which we should use the fused panel/update kernels. The results from the tuning sweeps are stored in lookup tables. While we originally tuned the performance for the A100 GPU, it is straightforward to add lookup tables to

---

Algorithm 3: Pseudo code for the fused trailing panel update

```
1 pA[] ← read factorized panel in shared memory
2 pA[] ← dlaset(pA[], 'upper', 0, 'diag', 1) //device-routine
3 for =1 to n̄ step ib do
4    tA[] ← read the next block of columns from the trailing panel
5    for j=1 to ib do
6       │  tA ← dlarf(pA(:,j), tA[]) //device-routine
7    end
8    write tA[] back into memory
9 end
```

---

other GPUs. During the run time, the correct lookup table is used for deciding the best code path to execute. Figure 6 shows the performance of the MAGMA design after incorporating the new update kernel, where MAGMA is now able to outperform both cuBLAS and KBLAS across almost all sizes.

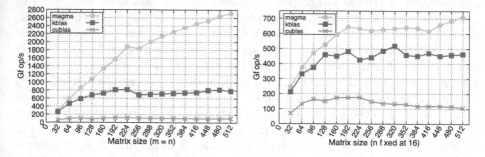

Fig. 6. Batch QR performance in double precision, with both the fused panel and fused update kernels incorporated into the MAGMA design. Results are shown for square matrices (left) and tall-skinny matrices (right), using a Tesla A100 GPU. Batch size = 1k.

7   Optimizations for Sub-warp Dimensions

A final optimization is possible when the entire matrix can be cached in the register file or in the shared memory. At this point, a fully fused and unrolled dgeqr2 is used. We have addressed this case in a previous work [4], but we discuss it here to complete the scope of the paper. The kernel has some similarities with the one described in Sect. 5, but it has some unique features. First, it uses a serial reduction for computing the norm of a column, and for the $v^T \times A$ product. For sub-warp dimensions, we found out that a serial reduction is often faster than a parallel reduction with repetitive synchronizations. Second, one warp can be involved in factorizing more than one matrix simultaneously. For example, a single warp can factorize four 8 × 8 matrices at the same time. Third, the code template is instantiated for every possible size. Without the loss of generality, we discuss square sizes up to 32 only. The obvious drawbacks to this approach is its applicability to a restricted range of sizes. It should also be instantiated for every possible (m, n) combination. However, its advantage is clear as shown in Fig. 7. Despite all the optimizations mentioned in the previous sections and in other libraries, the Figure shows significant speedups, up to 3.22× against the best competition.

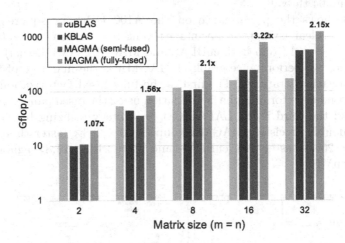

Fig. 7. Batch QR performance in double precision for tiny square matrices. Results are shown for a Tesla A100 GPU. Batch size = 10k. The speedup labels of the fully-fused design are calculated with respect to the best performance of the other three approaches.

8   The Big Picture

The optimized GPU kernels described in Sects. 4 through 7 are now put together into one solution. The factorization begins by a check for tiny matrices, for which the fully unrolled kernel (Sect. 7) can be used. Otherwise, it moves to a decision-making layer that determines whether to use a fused panel/update kernels for the entire factorization (Sects. 5 and 6).

- **If true**, another decision-maker determines which version of the panel/update to be used (in registers or in shared memory). The decision-maker also determines the number of threads in case the shared memory version is preferred.
- **If false**, the factorization proceeds with a LAPACK-style factorization utilizing batch GEMM.

The LAPACK-like implementation has a panel factorization step, during which it checks agains for the feasibility fused panel/update kernel. If they cannot be used (e.g. panel is too large), we fall back to a generic non-fused panel implementation. In either case, the factorization proceeds with computing the T factor and then calling batch GEMM to apply the block reflector to the trailing matrix. All the decision-making layers use a comprehensive set of offline performance benchmark results. The offline data resulting from these benchmarks are tabulated per GPU and per compute precision.

9   Final Performance Results

This section shows the final performance results on both NVIDIA and AMD GPUs. All results are shown for single and double precisions. We show the performance of the host CPU using OpenBLAS, which is called inside an OpenMP for loop using 64 threads.

Figure 8 shows the performance on the A100 GPU for square matrices. MAGMA has a clear asymptotic advantage thanks to the careful utilization of the batch GEMM kernel (from both cuBLAS and MAGMA's own kernel). As mentioned before, the performance graph of MAGMA is the marriage of three different factorization strategies. The first is the fully fused factorization for sizes ≤ 32, the second is performing the factorization using the fused panel/update kernel only, and the third is the LAPACK style strategy utilizing batch GEMM. For single/double precision, MAGMA is up to $2.3\times/3.3\times$ faster than KBLAS, up to $16.2\times/25.4\times$ faster than cuBLAS, and up to $21.9\times/14.8\times$ against Open-BLAS+OpenMP.

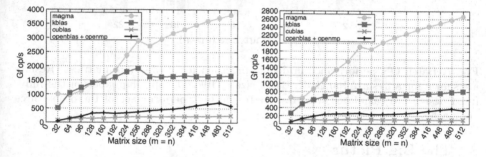

Fig. 8. Final performance of the batch QR factorization in single/double precision (left/right). Results are shown for square matrices using a Tesla A100 GPU. Batch size = 1k.

Figure 9 shows the final performance for tall-skinny matrices with exactly 16 columns. This test case represents problems that require the solution of least square systems. This test case corresponds to one factorization strategy in MAGMA, which is the fused panel and update kernels. But recall that MAGMA has two different kernels for each of the panel and update steps, and invokes the faster of the two depending on the matrix size. Similar to square sizes, both cuBLAS and the OpenBLAS with OpenMP are underperforming. Both MAGMA and KBLAS have the staircase-like behavior, which means that they both try to take advantage of the fast memory levels on the GPU, but face gradual degradation due to increased occupancy. However, MAGMA has an asymptotic advantage for single precision, and an overall advantage for double precision. This means that our solution has a better use of the available resources on the GPU. For single/double precision, MAGMA is up to $1.6\times/1.7\times$ faster than KBLAS, up to $5.8\times/7.4\times$ faster than cuBLAS, and up to $36.3\times/65.9\times$ against OpenBLAS+OpenMP.

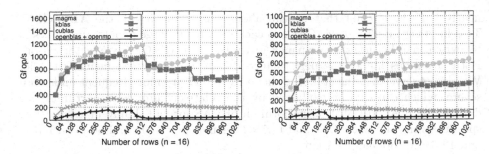

Fig. 9. Final performance of the batch QR factorization in single/double precision (left/right). Results are shown for tall-skinny matrices ($n = 16$) using a Tesla A100 GPU. Batch size = 1k.

Figures 10 and 11 show the corresponding results on the AMD MI100 GPU, where we compare the MAGMA performance against hipBLAS as well as Open-BLAS + OpenMP. To the best of our knowledge, KBLAS does not support AMD GPUs. We observe that the performance is lower than the A100 performance numbers. This is due to multiple reasons. **First**, the batch GEMM kernel on the A100 is better tuned for the use cases we need than on the MI100 GPU. **Second**, we notice that the fused kernels for performing the panel and the updates are also slower than on the A100. Our experience with porting our solution to AMD GPUs indicates that performing computations in the Local Data Share (LDS) memory is slower than the shared memory on NVIDIA GPUs. This is crucial to both the panel and the update kernels, since we perform many tree reduction in shared memory. MAGMA still outperforms hipBLAS for square sizes. The speedups range between 2.6× and 11.5× for single precision, and between 3.8× and 10.1× for double precision.

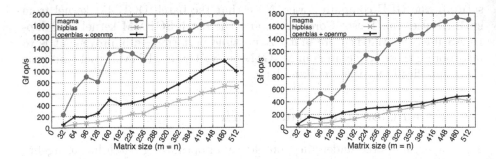

Fig. 10. Final performance of the batch QR factorization in single/double precision (left/right). Results are shown for square matrices using an AMD MI100 GPU. Batch size = 1k.

Another bottleneck on the MI100 GPU is the amount of the LDS memory available for one thread-block, which has a maximum of 64KB. This is nearly half the amount that we can allocate dynamically on the A100 GPU. This limits the ability of MAGMA to cache relatively large panels, and forces it to switch to either use thinner panels or to use the LAPACK-style factorization. Both situations hinder the performance due to the increased memory traffic. We also observe that the staircase shape in Fig. 11 are more frequent and more severe, which can also be explained by the relatively limited opportunities of data reuse. MAGMA still outperforms hipBLAS for tall-skinny matrices. The speedups range between 3.0× and 14.5× for single precision, and between 1.13× and 12.6× for double precision.

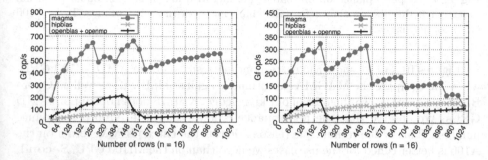

Fig. 11. Final performance of the batch QR factorization in single/double precision (left/right). Results are shown for tall-skinny matrices (n = 16) using an AMD MI100 GPU. Batch size = 1k.

In general, the asymptotic performance is not close to the GPU theoretical peak performances. This is mainly due to the focus on relatively small sizes, which limits the batch GEMM performance on both the A100 and the MI100 GPUs. The rank-updates use relatively small widths that are not enough to saturate the GPU compute power. Note that the batch DGEMM kernel from cuBLAS uses the Tensor Cores units, and the batch SGEMM kernel from hipBLAS uses the Matrix Core units. A possible performance improvement is to incorporate these accelerators in MAGMA's own batch GEMM kernel, and tune them specifically for these rank updates.

10   Conclusion and Future Work

This paper shows the underlying complexity of optimizing batch linear algebra operations on GPUs, taking the dense batch QR factorization as an example. We show that, depending on the problem size, there could be different strategies of performing the factorization. Since memory traffic is often critical to batch routines, fused kernels are used to efficiently utilize the memory bandwidth. However, kernel fusion increases the complexity of the overall solution, since it introduces new non-standard routines not found in BLAS or LAPACK. Our final solution for the batch QR factorization has three different strategies for execution, and within each strategy, there are multiple run-time decisions to select the best performing kernel. Future directions include investigating the performance regression on AMD GPUs, extension to variable-size batches, and considering more efficient algorithms for very tall and skinny matrices.

References

1. LAPACK - Linear Algebra PACKage. http://www.netlib.org/lapack/
2. Abdelfattah, A., Haidar, A., Tomov, S., Dongarra, J.: Performance, design, and autotuning of batched GEMM for GPUs. In: ISC High Performance 2016, Frankfurt, Germany, 19–23 June 2016, Proceedings, pp. 21–38 (2016)
3. Abdelfattah, A., Haidar, A., Tomov, S., Dongarra, J.: Factorization and inversion of a million matrices using GPUs: challenges and countermeasures. Procedia Comput. Sci. **108**, 606–615 (2017). ICCS 2017, Zurich, Switzerland
4. Abdelfattah, A., Haidar, A., Tomov, S., Dongarra, J.J.: Batched one-sided factorizations of tiny matrices using GPUs: challenges and countermeasures. J. Comput. Sci. **26**, 226–236 (2018)
5. hipBLAS. https://github.com/ROCmSoftwarePlatform/hipBLAS
6. Anderson, M., Sheffield, D., Keutzer, K.: A predictive model for solving small linear algebra problems in GPU registers. In: IEEE 26th International Parallel Distributed Processing Symposium (IPDPS) (2012)
7. Anzt, H., Dongarra, J., Flegar, G., Quintana-Ortí, E.S.: Batched Gauss-Jordan elimination for block-Jacobi preconditioner generation on GPUs. PMAM 2017, pp. 1–10. ACM, New York (2017)
8. Auer, A.A., et al.: Automatic code generation for many-body electronic structure methods: the tensor contraction engine. Mol. Phys. **104**(2), 211–228 (2006)
9. Boukaram, W.H., Turkiyyah, G., Ltaief, H., Keyes, D.E.: Batched QR and SVD algorithms on GPUs with applications in hierarchical matrix compression. Parallel Comput. (2017). https://doi.org/10.1016/j.parco.2017.09.001
10. Haidar, A., Dong, T., Luszczek, P., Tomov, S., Dongarra, J.: Batched matrix computations on hardware accelerators based on GPUs. IJHPCA **29**(2), 193–208 (2015)
11. Haidar, A., Tomov, S., Luszczek, P., Dongarra, J.: Magma embedded: towards a dense linear algebra library for energy efficient extreme computing. In: 2015 IEEE High Performance Extreme Computing Conference (HPEC), pp. 1–6, September 2015
12. MAGMA. http://icl.cs.utk.edu/magma/
13. PLASMA, October 2017. https://bitbucket.org/icl/plasma
14. Intel Math Kernel Library. http://software.intel.com/intel-mkl/
15. Kurzak, J., Anzt, H., Gates, M., Dongarra, J.: Implementation and tuning of batched Cholesky factorization and solve for NVIDIA GPUs. IEEE Trans. Parallel Distrib. Syst. **27**, 2036–2048 (2015)
16. Messer, O., Harris, J., Parete-Koon, S., Chertkow, M.: Multicore and accelerator development for a leadership-class stellar astrophysics code. In: Proceedings of "PARA 2012: State-of-the-Art in Scientific and Parallel Computing" (2012)
17. NVIDIA CUBLAS. https://developer.nvidia.com/cublas
18. Tomás Dominguez, A.E., Quintana Orti, E.S.: Fast blocking of householder reflectors on graphics processors. In: 2018 26th Euromicro International Conference on Parallel, Distributed and Network-Based Processing (PDP), pp. 385–393 (2018). https://doi.org/10.1109/PDP2018.2018.00068

19. Van Zee, F.G., van de Geijn, R.A.: BLIS: a framework for rapidly instantiating BLAS functionality. ACM TOMS **41**(3), 33 (2015). https://dl.acm.org/doi/10. 1145/2764454
20. Walker, Homer F.: Implementation of the GMRES method using householder transformations. SIAM J. Sci. Stat. Comput. **9**(1), 152–163 (1988). https://doi. org/10.1137/0909010
21. Yeralan, S.N., Davis, T.A., Sid-Lakhdar, W.M., Ranka, S.: Algorithm 980: sparse QR factorization on the GPU. ACM TOMS **44**(2), 17:1–17:29 (2017)

ChemTab: A Physics Guided Chemistry Modeling Framework

Amol Salunkhe$^{(\boxtimes)}$, Dwyer Deighan, Paul E. DesJardin, and Varun Chandola

University at Buffalo, Buffalo, NY 14260, USA
{aas22,dwyerdei,ped3,chandola}@buffalo.edu

Abstract. Modeling of turbulent combustion system requires modeling the underlying chemistry and the turbulent flow. Solving both systems simultaneously is computationally prohibitive. Instead, given the difference in scales at which the two sub–systems evolve, the two sub–systems are typically (re)solved separately. Popular approaches such as the *Flamelet Generated Manifolds* (FGM) use a two–step strategy where the governing reaction kinetics are pre–computed and mapped to a low–dimensional manifold, characterized by a few reaction progress variables (model reduction) and the manifold is then "looked–up" during the run–time to estimate the high–dimensional system state by the flow system. While existing works have focused on these two steps independently, we show that joint learning of the progress variables and the look–up model, can yield more accurate results. We propose *ChemTab* an architecture that learns jointly and demonstrate its superiority.

Keywords: Physics guided neural networks · DNN

1 Introduction

Modeling of turbulent flow combustion is central in the development of new combustion technologies in aviation, automotive and power generation [6]. Turbulent flow combustion combines two nonlinear and multi–scale phenomena: *turbulent flow* and *chemical reactions*. This coupling of the kinetic chemical reaction equations with the set of Navier-Stokes flow equations results in a problem that is too complex to be solved, at full resolution, by the current computational means. Even for a simple fuel such as methane, the combustion chemistry mechanism involves 53 species and 325 chemical reactions [19], and the numbers increase with increasing fuel complexity. Solving the details of such mechanisms during the flow simulation can consume up to 75% of the solution time [4].

In most cases, the large scale separation between the combustion chemistry/flame (typically sub millimeter/microsecond scale) and the characteristic turbulent flow (typically centimeter or meter/minute or hour scale) allows simplifying assumptions to be made that enable increased computational efficiency by (re)solving chemistry and flow separately [16]. In this paper, we focus on approximate methods that deal with handling the chemistry, and in particular,

D. Groen et al. (Eds.): ICCS 2022, LNCS 13350, pp. 75–88, 2022.
https://doi.org/10.1007/978-3-031-08751-6_6

the methods based on laminar flames [15]. Here the 1–D or single–species flame reactions are solved *a priori* and stored. During the flow simulation, these reactions are looked–up to estimate the high–dimensional thermochemical state of the system, as shown in Fig. 1.

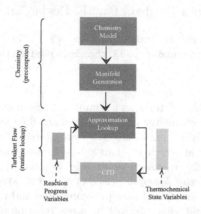

Fig. 1. (Re)solving systems separately

Most models developed for increased computational efficiency rely on the existence of a theoretical low–dimensional thermochemical state–space manifold to which the combustion chemistry can be mapped [11]. The central question then is, *how to efficiently model low–dimensional thermochemical manifolds that capture the relevant physics of the problem; and parametrize and approximate these manifolds which can then be accessed during turbulent flow simulations?*

While existing approaches (collectively referred to as *state–space parametrization* [16,17]) have been successful, they have primarily solved the two sub–problems – *progress variable generation* to characterize the manifold, and *manifold approximation* to perform the lookup during run–time, independently. This can result in sub–optimal solutions because the progress variables, learnt using methods such as *Principal Component Analysis* (PCA) [2,20], are not necessarily optimized to perform the run–time lookup. Similarly, while the traditional lookup approaches that use tabulation, or the recently proposed neural network based data–driven alternatives [1], facilitate efficient look–ups, the construction of the underlying data–structure or machine learning based model is not informed by the learning of the progress variables.

Our main hypothesis is that by simultaneously learning the progress variables and the manifold approximation (lookup model), we can achieve higher accuracy in terms of the estimation of the thermochemical state at run–time. But how does one combine the progress variable learning, an inherently linear mapping task, with a highly non–linear lookup model, while ensuring that the components influence each other during the learning phase? To that end, we propose a framework called *ChemTab*, in which the learning of these two components is formulated as a joint optimization task. An implementation of *ChemTab*, using a novel deep learning architecture, is proposed. The joint optimization includes a set of mathematical constraints that ensure that the progress variable learning is approximately similar to a PCA–type linear reduction, and, at the same time, can also predict the thermochemical state using a non–linear predictive component.

The deep learning implementation of ChemTab is shown to reduce the error by 73%, when compared to an existing tabulation based framework, in predicting one of the key thermochemical term, *source energy*, when applied to flames data for a Methane–Air fuel–oxidizer combination generated using the GRI–Mech 3.0 simulator. Moreover, the proposed architecture of ChemTab is shown to outperform a recently proposed state–of–art decoupled PCA+neural network based solution by 24%.

2 Related Work

In this section we provide a brief overview of existing in low–dimensional thermochemical manifold modeling, focusing more on data–driven methods. We note that there have been works that use physics–driven machine learning models for solving other physics problems [10,23], however, these methods generally focus on simpler physics and are not necessarily applicable in the domain of turbulent combustion.

Common approaches to low–dimensional thermochemical manifold modeling are combustion chemistry mechanism reduction and thermochemical state–space parametrization [18,20]. Chemistry mechanism reduction approach cannot be generalized and in the recent past state–space parametrization approach has been the most dominant method comprising of two phases *progress variable generation* and *manifold approximation*. For progress variable generation, existing methods have either used domain models or numerical methods.

Domain models like steady Laminar Flamelet Method (SLFM) [15], Flamelet–Generated Manifold (FGM) [21,22], Flamelet Progress Variable approach (FPVA) [7,17] and Flamelet–Prolongation of ILDM model (FPI) [5] theorize that a multi–dimensional flame can be considered as an ensemble of multiple one–dimensional locally laminar flames (flamelets). These flamelets are patametrized by a combination of conserved and reactive scalars [3,17,21,22]. A lot of research in this area builds on the principles laid out in [9] for progress variables regularization however the fundamental problem of generating adequate number of progress variables that capture the underlying physics is still open.

Numerical methods, like PCA, have shown significant promise for parametrization of the thermochemical state. PCA provides a method of generating reaction progress variables using the flamelet solutions, the state–space variables are still nonlinear functions of the reaction progress variables, and a nonlinear regression is learned to approximate the state–space manifold [2,12,13,20]. This purely numerical parametrization lack interpretability and may also not be generalizable enough due to variation capture maximization that may overlearn the numerical errors in the data. Linear Autoencoders have also been suggested [14] however this definition lacks incorporation of a principled approach to progress variable generation and thus may not be generalizable.

While domain based model have traditionally relied on tabular lookup, these are not scalable. Tabulated data occupies a larger portion of the available memory on every node where the flow simulation is computing. Also the searching and retrieval of this pre–tabulated data becomes increasingly expensive in a higher–dimensional space. For example, assuming a standard 3 progress variable discretization (200, 100, 50) with say 15 tabulated thermochemical state variables, we obtain a pre–computed combustion table of 120 Mb. The addition of a variable such as *enthalpy* with a very coarse discretization of 20 points, brings the size of the table to 2.4 Gb. To address the tabulation problem researchers like [1,24] build on the work of [8] to investigate the use of a neural networks for manifold approximation which replaces the Tabulation. The mapping between the progress variables (reduced dimensionality) and thermochemical state variables obtained using the flamelets solutions is learnt using a neural network. However, due to the highly non–linear, knotted and discontinuous nature of the lower dimensional manifolds formed by the progress variables generated *a priori* the accuracy gained by a neural network is not satisfactory.

3   ChemTab: Joint Learning Progress Variables and Manifold Approximation

To reduce the computational effort in coupled simulations, state–space parametrization approaches follow a two–phase strategy. First, parametrize and tabulate *a priori* the scalar evolution of a reactive turbulent environment by few progress variables that govern the scalar evolution in a laminar flame. Second, use a tabular lookup at run–time to determine the high–dimensional chemical state required by the CFD solver. For instance, the FGM approach replaces all species and temperature by a *mixture–fraction* and a single *reaction progress variable* or reaction progress parameter. In this study, we focus on state–space parametrization using *Unsteady Flamelet Generated Manifolds* or Unsteady FGMs [3]. We modify this approach in three ways: the progress variable generation is different, the manifold is not tabulated and lastly, the progress variables and manifold approximation are done jointly.

3.1   Background: Unsteady FGM

FGM is a widely used tabulated chemistry method and can deal with a range of complicated conditions. FGM model shares the same theoretical basis with flamelet approaches [15], in which a multi–dimensional flame can be considered as an ensemble of multiple one–dimensional flames. Generally FGM model used for combustion modeling follows three steps as shown below:

1. Calculation of the representative 1–D flamelets.

2. Transformation of 1–D flamelets solutions to progress variables space.
3. Retrieval of thermo–chemical variables from the FGM tables according to FGM control variables from CFD simulations.

Table 1. Definitions for terms used in Sect. 3.1

	Description		Description
Z_{mix}	Mixture fraction	T	Temperature of the mixture
C_{pv}	Progress variable	$\widehat{S}$	Reactive scalars source terms
Y	Species mass fraction	ψ	Non–linear function of $\widehat{Y}$ and Z_{mix}
$\dot{S}$	Species source terms	k	No. of species used to generate progress variables
ρ	Density of the mixture	p	Number of progress variables
$\widehat{Y}$	Reactive scalars	n	Number of data points
Le	Lewis number	$\dot{S}_i$	Source term of the ith species
μ	Viscosities	$h^0_{f,i}$	Heat of formation of the ith species
$\mathcal{D}_i$	Diffusivity of ith species	h	Total enthalpy
κ	Thermal conductivity	s	Total no. of species in mechanism
Pr	Prandtl number	ϕ	Non–linear function of Y
Sc	Schmidt number	ζ	Non–linear function of $\widehat{Y}$

Governing Equations. Conservation equations for mass, species, momentum and energy for the 1–D, fully compressible, and viscous flames, are given by:

$$\frac{\partial \rho}{\partial t} + \frac{\partial (\rho u_x)}{\partial x} = 0 \tag{1}$$

$$\frac{\partial (\rho Y_i)}{\partial t} + \frac{\partial \rho u_x Y_i}{\partial x} = \frac{\partial}{\partial x}\left(\rho \mathcal{D}_i \frac{\partial Y_i}{\partial x}\right) + \dot{S}_i \tag{2}$$

$$\frac{\partial (\rho u_x)}{\partial t} + \frac{\partial (\rho u_x^2)}{\partial x} = -\frac{\partial p}{\partial x} + \frac{\partial}{\partial x}\left(\mu \frac{\partial u_x}{\partial x}\right) \tag{3}$$

$$\frac{\partial (\rho e_t)}{\partial t} + \frac{\partial}{\partial x}(\rho u_x H_t) = \frac{\partial}{\partial x}\left(u_x \mu \frac{\partial u_x}{\partial x}\right) + \mu \frac{c_p}{Pr}\left(1 - \frac{1}{Le}\right)\frac{dT}{dx} \tag{4}$$

$$+ \frac{1}{Sc}\frac{dh}{dx} - \sum \dot{S}_i h^o_{f,i}$$

where the different terms are defined in Table 1.

We simplify the above equations making some well known assumptions. In 1D cartesian coordinates, the steady state solution to (1)–(4) is obtained only when the total mass flux is zero, i.e., velocity field is zero ($u_x = 0$) and so the four equations reduce to:

$$\frac{\partial}{\partial x}\left(\rho \mathcal{D}_i \frac{\partial Y_i}{\partial x}\right) + \dot{S}_i = 0 \tag{5}$$

$$\frac{\partial}{\partial x}\left(\kappa \frac{\partial T}{\partial x} + \sum \rho \mathcal{D}_i \frac{\partial Y_i}{\partial x} h_i\right) - \sum \dot{S}_i h^o_{f,i} = 0. \tag{6}$$

In (6), the final term in the energy equation is represented by the total sum of the product of all the source species and their respective heat of formation and is collectively called the source energy. Source energy is one of the crucial parameters in the combustion simulation and accurate chemistry description is required to define it. Prediction error of this term is used as the basis of comparison of our method against the other state of the art methods.

Flamelet Solutions. The data is generated by solving 1–D Steady State Flamelets differential equations in (6) using a finite volume PDE solver. The species Y and thermochemical state variables $\dot{S}$ are generated using the solver.

$$Y = \begin{bmatrix} Y_{11} & .. & .. & Y_{1s} \\ .. & .. & .. & .. \\ Y_{n1} & .. & .. & Y_{ns} \end{bmatrix}, \quad \dot{S} = \begin{bmatrix} S_{11} & .. & .. & S_{1s} \\ .. & .. & .. & .. \\ S_{n1} & .. & .. & S_{ns} \end{bmatrix}, \quad Z_{mix} = \begin{bmatrix} Z_{mix_1} \\ .. \\ Z_{mix_n} \end{bmatrix} \tag{7}$$

3.2   ChemTab

In ChemTab, the unsteady FGM approach is replaced with the following three steps:

1. Calculation of the representative 1D flamelets (data generation)
2. Using the data generated jointly generate Progress Variables (encoder) and Manifold Approximation (regressor) using ChemTab
3. Retrieval of thermo–chemical variables from the ChemTab–regressor according to progress variables from CFD simulations.

Formulation. The generated data described in (7) is then used by ChemTab. Conceptually the following equations summarize the relationships:

$$\dot{S} = \phi(Y) \tag{8}$$

$$S_{energy} = -\sum_i^s h^0_{f,i} * \dot{S}_i \tag{9}$$

The two sub–problems of state–space parametrization are formulated as a joint optimization problems as follows:

$$min(\sum_{i=1}^{k}\sum_{j=1}^{n} ||\dot{S}_{ij} - \zeta_i(\widehat{Y_j})||_t + \sum_{j=1}^{n} ||S_{energy} - \psi(\widehat{Y}, Z_{mix})||_t) \tag{10}$$

$$\text{s.t.} \quad t \in R \tag{11}$$

$$\underset{n \times p}{\widehat{Y}} = \underset{n \times s}{Y} \times \underset{s \times p}{W} \tag{12}$$

$$p << s \tag{13}$$

$$||W|| = 1 \tag{14}$$

$$W^T \times W = I \tag{15}$$

$$(\widehat{Y} \oplus Zmix)^T \times (\widehat{Y} \oplus Zmix) = I \qquad (16)$$

$$\dot{S} \approx \widehat{\dot{S}} = \zeta(\widehat{Y}, Z_{mix}) \qquad (17)$$

$$S_{energy} \approx \widehat{S_{energy}} = \psi(\widehat{Y}, Z_{mix}) \qquad (18)$$

The formulation described in Eq. 10 learns the optimal reactive scalars $C_{pv}s$ (described by the embedding $Y \times W$) that along with Z_{mix} form the progress variables. This is a linear dimensionality reduction problem such that the new basis retains the inherent physics in higher dimensions described by the non–linear relation between Y and $\dot{S}$. To facilitate the development of transport equations using the progress variables it is necessary that the embedding of the variables in the low–dimensional space be linear. The constraints on the linear embedding are inspired by the work of [9] and the key ideas from PCA.

Implementation. The joint optimization problem is solved using a Deep Neural Architecture. ChemTab jointly optimizes two neural networks for the tasks of reaction progress variable generation (*encoder*) and manifold approximation (*regressor*). The encoder network focuses on linear dimensionality reduction and creates a linear embedding for the input. The regressor network focuses on learning the manifold approximation: a regression function whose input is the linear embedding and the output are the desired thermo–chemical state variables

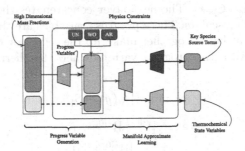

Fig. 2. ChemTab architecture

Table 2. Symbols used in Sect. 3.2

	Description		Description
f_θ	Prediction function	in	$s \times n$
y	Input/output matrix	s	Total no. of species in mechanism
W	Weight matrix	out	No. of thermo–chemical variables
b	Bias matrix	d_{in}	Input dimensions s
S	Themochemical state variables	L	No. of layers
σ	scalar/activation function	d_{out}	Output dimensions $s + 1$
o	Entry–wise operation	m	Number of neurons
n	Number of data points		

(Fig. 2 and Table 2).

$$f_\theta(y) = W^{[L-1]}\sigma \text{ o } (W^{[L-2]}\sigma \text{ o } (\ldots (W^{[1]}\sigma \text{ o }$$
$$(W^{[0]}y + b^{[0]}) + b^{[1]})\ldots) + b$$
$$\text{where, } W^{[l]} \in R^{m_{l+1} \times m_l} \tag{19}$$
$$b^{[l]} = R^{m_{l+1}} \quad ; \quad m_0 = d_{in} = d \quad ; \quad m_L = d_{out}$$

As described by (19) a Deep Neural Network can be conceptualized as a series of operations. The input of the network is the data for each of the species for each flame at each axial coordinate.

$$f_\theta^{[0]}(y) = y$$
$$f_\theta^{[1]}(y) = (W^{[0]}f_\theta^{[0]}(y))$$
$$f_\theta^{[2]}(y) = (f_\theta^{[2]}(y) \oplus Z_{mix}) \tag{20}$$
$$f_\theta^{[l]}(y) = \sigma \text{ o } (W^{[l-1]}f_\theta^{[l-1]}(y) + b^{[l-1]}) \ \forall \ l \ \text{ s.t. } \ 3 \le l \le L-1$$
$$f_\theta(y) = f_\theta^{[L]}(y) = \sigma \text{ o } (W^{[L-1]}f_\theta^{[L-1]}(y) + b^{[L-1]})$$

As described by (20) the network is a layer–wise composition. The input of the network is reduced at the first layer linearly: this creates the linear embedding/reacting scalars ($C_{pv}s$). The next layer concatenates the conserved scalar Z_{mix} with the reacting scalars. These progress variables are then fed to the next layer. The subsequent layers together make up the regressor that learns a non–linear function between the progress variables and the thermo–chemical state variables.

$$\arg\min_\theta \ |f_\theta(y) - \mathcal{S}|$$
$$s.t. \ \ W^{[0]T}W^{[0]} = I$$
$$\|W^{[0]}\| = 1 \tag{21}$$
$$f_\theta^{[2]}(y)^T f_\theta^{[2]}(y) = I$$

As described by (21), ChemTab minimizes the Mean Absolute Error in predicting the thermo–chemical state variables (Source Energy in the current work) while ensuring that the linear embedding conforms to the following constraints:

1. Embedding Weights w learnt are unit norm (UN)
2. Embedding Weights w learned for the species mass fractions Y_is are uncorrelated/orthogonal (WO)
3. The reaction progress variables are uncorrelated/orthogonal (AR)

The constraints in (21) will be also added to the objective in addition to the predictions of key source terms, corresponding to a few important species, which serve as the physics constraints.

Extensions. The current framework and implementation can be very easily extended to include the prediction of additional thermochemical state variables and the projection of the embedding to get back the high dimensional mass fractions. These can be implemented as two other neural networks and their respective prediction errors can be added to the objective function.

4   Experimentation and Results

In this section we explain the specifics of the data set used, the training strategy, impact of the number of C_{pv}, comparison with the existing framework and relevant machine learning methods and the performance of the best model in the context of the multiple objectives.

4.1   Dataset

The training data was generated by solving 1–D Steady State Flamelets differential equations using a finite volume PDE solver. GRI–Mech 3.0 is one of the widely used Methane mechanism to model the reaction kinetics. This mechanism consists of 53 chemical species and 325 reactions.

The Flamelet solver discretizes the domain into 200 grid points (200 observations on the axial coordinate) in between the fuel and the air boundary and 100 flame are solved to steady–state. To train the model 20,000 data points (100 flames and 200 grid points) for a single pressure setting are used. Some of the generated data that represent extinguished flames were discarded, which led to exclusion of approximately 3,500 data points.

We experiment the model training and evaluation using two strategies:

1. *50% Flamelets* – Train using data from 50% of flamelets selected randomly and test using data from the remaining 50% of the flamelets, and,
2. *50% Data points* – Train using 50% data points selected randomly, and test on remaining 50% data.

4.2   Evaluation

We use the Mean Absolute Error of the Source Energy across the entire dataset as the metric to compare the performance as described in Eq. (21).

4.3   Implementation and Settings

We implemented ChemTab using Tensorflow 2.3.0, Keras and Adam optimizer. Models were trained on a server with Nvidia Quadro RTX 5000 GPU and cuDNN 8.0 and CUDA 11.0. We performed a coarse grid search on the hyperparameters (dropouts, learning rate, early stopping, batch size) & standard model architecture (number of layers, number of nodes in the layers, activation functions). After the initial model architecture and hyper–parameter search, all subsequent models in the subsequent studies were trained for 500 epochs. Results are reported as average over 10 runs (Table 3).

Table 3. Model parameters

Parameter	Value	Parameter	Value
Learning rate	0.001	Number of layers	11
Momentum	0.5	Layer shapes	32|64|128|256|512
Dropout	5%	Activation Functions	ReLU
Early stopping	Yes	Number of epochs	500 (short run) | 20000 (long run)
Batch size	32	Network weight initialization	Uniform distribution

Table 4. ChemTab architectural variants

Abbreviation	Description
UN	Unit norm constraint on weights of the linear embedding
WO	Orthogonality constraint on weights of the linear embedding
AR	Orthogonality constraint on linear embedding concatenated with Z_{mix}
UN + WO	Unit norm constraint and orthogonality constraint on weights of the linear embedding
UN + AR	Unit norm constraint on the weights and orthogonality constraint on linear embedding concatenated with Z_{mix}
WO + AR	Orthogonality constraint on the weights and linear embedding concatenated with Z_{mix}
All	Unit norm and orthogonality constraint on the weights and linear embedding concatenated with Z_{mix}

4.4 Compared Methods

We compare the 7 variants of ChemTab with the relevant constraints on the Linear Embedding and the Progress Variables with a series of state–of–the–art baselines for Source Energy prediction Sect. 2.

4.5 Results

Current Framework Comparison. The current framework uses FGM based progress variables and Conformal Mapping based Tabulation and Lagrange Polynomial Interpolation based lookup. The tabulation was generated by using the entire data–set. The best MAE that the framework generated on the data–set was 2.243 E+09. The best ChemTab model trained on 50% of the data showed a 73% reduction in error. This reduction although high comes from the limitation of the current framework to include more than 2 progress variables and the realization of that through conformal mapping. We present a more principled comparison with the state–of–the–art methods in the next section.

Other Baseline Comparisons. We include *DNN–PVG(NL)–DNN* as reference although it cannot be used due to non–linear embedding. Similarly we

Table 5. Current state of the art methods and ChemTab variants

Method abbreviation	Progress variable generation
FGM–CPVG–DNN	FGM Constrained
PCA–PVG–DNN	PCA
DNN–PVG(NL)–DNN	Non–linear encoder
DNN–PVG(UL)–DNN	Unconstrained linear encoder
CT–PVG(ALL)–DNN	Physics constrained linear encoder Table 4
CT–PVG(UN)–DNN	Physics constrained (UN) linear encoder Table 4
CT–PVG(WO)–DNN	Physics constrained (WO) linear encoder Table 4
CT–PVG(AR)–DNN	Physics constrained (AR) linear encoder Table 4
CT–PVG(UN+WO)–DNN	Physics constrained (UN+WO) linear encoder Table 4
CT–PVG(UN+AR)–DNN	Physics constrained (UN+AR) linear encoder Table 4
CT–PVG(WO+AR)–DNN	Physics constrained (WO+AR) linear encoder Table 4

did not consider Gaussian Processes as there are several challenges with operationalization of Gaussian Process in our context and so we focus more on bench–marking against the relevant DNN based approaches (Table 5).

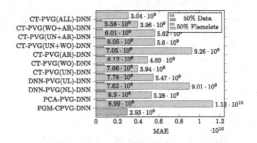

Fig. 3. MAE for source energy: data set split strategy

Figure 3 shows the results of an ablation study for both types of sampling strategies. When the trained using the sampled points, all models consistently do better than when trained using sampled flamelets. Essentially the flame is considered as an ensemble of multiple one–dimensional flamelets, each of which captures some of the highly nonlinear state–space and hence almost all models struggle in this training regime. ChemTab models still perform better and our assertions are that our constraints help in the generalization process. Our dataset is limited and so we limit ourselves to use only 50% of the data for training.

As we increase the number of C_{pv} the computational time of the flow simulation goes up, so we want to use the least number of C_{pv} while still capturing the essential physics. Figure 4 shows the MAE decreases with increase in the number of C_{pv} and then starts to increase again. As we add more C_{pv} the embedding has too many degrees of freedom and hence may start diverging.

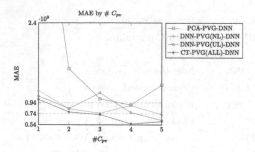

Fig. 4. MAE for source energy: Cpv ablation

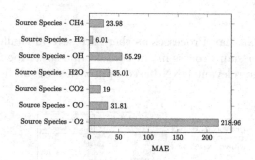

Fig. 5. MAE for key source terms – best model

Table 6. Constraints – best model

UN

w_1	w_2	w_3	w_4
1.004	1.005	1.001	0.998

WO

	w_1	w_2	w_3	w_4
w_1	1.004	-0.003	-0.002	0.005
w_2	-0.003	1.004	-0.003	0.002
w_3	-0.002	-0.003	1.001	0.001
w_4	-0.005	0.002	0.001	0.99

AR

	Z_{mix}	Cpv_1	Cpv_2	Cpv_3	Cpv_4
Z_{mix}	0.004	0.00	0.00	0.00	0.00
Cpv_1	0.00	0.008	-0.001	0.001	0.00
Cpv_2	0.00	-0.001	0.008	0.00	0.00
Cpv_3	0.00	0.001	0.00	0.007	-0.001
Cpv_4	0.00	0.00	0.00	-0.001	0.067

Best Model Performance. Table 6 shows the conformity of the constraints of Eq. 10. The first tabulation shows the (14) constraint conformity. The second tabulation shows the (15) constraint conformity and the third (16) constraint conformity. The (16) is also adequately satisfied as the constraint conformity is measured through covariance (Fig. 5).

Best Model Long Run Performance. We trained best model architecture on a 50% Data Points strategy for a long run of 20000 epochs and generated a MAE of 1.80E+08.

5 Conclusion

We propose ChemTab, a novel framework for jointly learning the progress variables and the manifold approximation. ChemTab follows the principle of physics guided neural networks [10], however no solutions exist that can directly benefit the combustion community. ChemTab outperforms the state–of–the–art state–space parametrization in combustion. Crucially, ChemTab generated reaction progress variables can be interpreted by examining the weight matrix, W, and thus, allow for physical insights into the systems being modeled. Incorporation of ChemTab into a flow simulation will be explored as part of future work.

Acknowledgments. Funded by the United States Department of Energy's (DoE) National Nuclear Security Administration (NNSA) under the Predictive Science Academic Alliance Program III (PSAAP III) at the University at Buffalo, under contract number DE–NA0003961.

References

1. Bhalla, S., Yao, M., Hickey, J.-P., Crowley, M.: Compact representation of a multidimensional combustion manifold using deep neural networks. In: Brefeld, U., Fromont, E., Hotho, A., Knobbe, A., Maathuis, M., Robardet, C. (eds.) ECML PKDD 2019. LNCS (LNAI), vol. 11908, pp. 602–617. Springer, Cham (2020). https://doi.org/10.1007/978-3-030-46133-1_36
2. Biglari, A., Sutherland, J.C.: An a-posteriori evaluation of principal component analysis-based models for turbulent combustion simulations. Combust. Flame **162**(10), 4025–4035 (2015)
3. Bojko, B.T., DesJardin, P.E.: Formulation and assessment of flamelet-generated manifolds for reacting interfaces. Combust. Flame **173**, 296–306 (2016)
4. El-Asrag, H.A.: A comparison between two different flamelet reduced order manifolds for non-premixed turbulent flames. In: Laminar & Turbulent Flames (2013)
5. Fiorina, B., Gicquel, O., Carpentier, S., Darabiha, N.: Validation of the FPI chemistry reduction method for diluted nonadiabatic premixed flames. Combust. Sci. Technol. **176**(5–6), 785–797 (2004)
6. Giusti, A., Mastorakos, E.: Turbulent combustion modelling and experiments: recent trends and developments. Flow Turbul. Combust. **103**(4), 847–869 (2019)

7. Ihme, M., Cha, C., Pitsch, H.: Prediction of local extinction and re-ignition effects in non-premixed turbulent combustion using a flamelet/progress variable approach. Proc. Combust. Inst. **30**, 793–800 (2005)
8. Ihme, M., Schmitt, C., Pitsch, H.: Optimal artificial neural networks and tabulation methods for chemistry representation in les of a bluff-body swirl-stabilized flame. Proc. Combust. Inst. **32**, 1527–1535 (2009)
9. Ihme, M., Shunn, L., Zhang, J.: Regularization of reaction progress variable for application to flamelet-based combustion models. J. Comput. Phys. **231**(23), 7715–7721 (2012)
10. Karpatne, A., et al.: Theory-guided data science: a new paradigm for scientific discovery from data. IEEE Trans. Knowl. Data Eng. **29**(10), 2318–2331 (2017)
11. Maas, U., Pope, S.B.: Implementation of simplified chemical kinetics based on intrinsic low-dimensional manifolds. In: Symposium (International) on Combustion, vol. 24, pp. 103–112. Elsevier (1992)
12. Malik, M.R., Isaac, B.J., Coussement, A., Smith, P.J., Parente, A.: Principal component analysis coupled with nonlinear regression for chemistry reduction. Combust. Flame **187**, 30–41 (2018)
13. Malik, M.R., Obando Vega, P., Coussement, A., Parente, A.: Combustion modeling using principal component analysis: a posteriori validation on Sandia flames D, E and F. Proc. Combust. Inst. **38**, 2635–2643 (2020)
14. Perry, B.A., de Frahan, M.T.H., Yellapantula, S.: Evaluation of co-optimized machine-learned manifolds for modeling premixed combustion (2021). https://ui.adsabs.harvard.edu/abs/2021APS..DFDF09003P/abstract
15. Peters, N.: Laminar diffusion flamelet models in non-premixed turbulent combustion. Prog. Energy Combust. Sci. **10**(3), 319–339 (1984)
16. Peters, N.: Turbulent combustion. IOP Publishing (2001)
17. Pierce, C.D., Moin, P.: Progress-variable approach for large-eddy simulation of non-premixed turbulent combustion. J. Fluid Mech. **504**, 73–97 (2004)
18. Rastigejev, Y., Brenner, M.P., Jacob, D.J.: Spatial reduction algorithm for atmospheric chemical transport models. Proc. Natl. Acad. Sci. **104**(35), 13875–13880 (2007)
19. Smith, G.P., et al.: GRI-mech 3.0 is an optimized mechanism designed to model natural gas combustion, including no formation and reburn chemistry
20. Sutherland, J.C., Parente, A.: Combustion modeling using principal component analysis. Proc. Combust. Inst. **32**(1), 1563–1570 (2009)
21. Van Oijen, J., De Goey, L.: Modelling of premixed laminar flames using flamelet-generated manifolds. Combust. Sci. Technol. **161**(1), 113–137 (2000)
22. van Oijen, J., Lammers, F., de Goey, L.: Modeling of complex premixed burner systems by using flamelet-generated manifolds. Combust. Flame **127**(3), 2124–2134 (2001)
23. Willard, J., Jia, X., Xu, S., Steinbach, M., Kumar, V.: Integrating scientific knowledge with machine learning for engineering and environmental systems (2021)
24. Zhang, Y., Xu, S., Zhong, S., Bai, X.S., Wang, H., Yao, M.: Large eddy simulation of spray combustion using flamelet generated manifolds combined with artificial neural networks. Energy AI **2**, 100021 (2020)

Establishing Metrics to Quantify Underlying Structure in Vascular Red Blood Cell Distributions

Sayan Roychowdhury[1]([✉]), Erik W. Draeger[2], and Amanda Randles[1]

[1] Department of Biomedical Engineering, Duke University, Durham, NC 27705, USA
{sayan.roychowdhury,amanda.randles}@duke.edu
[2] Center for Applied Scientific Computing, Lawrence Livermore National Laboratory, Livermore, CA, USA
draeger1@llnl.gov

Abstract. Simulations of the microvasculature can elucidate the effects of various blood flow parameters on micro-scale cellular and fluid phenomena. At this scale, the non-Newtonian behavior of blood requires the use of explicit cell models, which are necessary for capturing the full dynamics of cell motion and interactions. Over the last few decades, fluid-structure interaction models have emerged as a method to accurately capture the behavior of deformable cells in the blood. However, as computational power increases and systems with millions of red blood cells can be simulated, it is important to note that varying spatial distributions of cells may affect simulation outcomes. Since a single simulation may not represent the ensemble behavior, many different configurations may need to be sampled to adequately assess the entire collection of potential cell arrangements. In order to determine both the number of distributions needed and which ones to run, we must first establish methods to identify well-generated, randomly-placed cell distributions and to quantify distinct cell configurations. In this work, we utilize metrics to assess 1) the presence of any underlying structure to the initial cell distribution and 2) similarity between cell configurations. We propose the use of the radial distribution function to identify long-range structure in a cell configuration and apply it to a randomly-distributed and structured set of red blood cells. To quantify spatial similarity between two configurations, we make use of the Jaccard index, and characterize sets of red blood cell and sphere initializations.

Keywords: Red blood cells · Microvascular simulation · Cell packing

1 Introduction

Computational blood flow models are a powerful tool for answering biomedical questions. For microvessel simulations, where individual cell diameters are on the same order of magnitude as vessel size, the presence of cells plays a significant role in the non-Newtonian behavior of blood. In this regime, velocity

D. Groen et al. (Eds.): ICCS 2022, LNCS 13350, pp. 89–102, 2022.
https://doi.org/10.1007/978-3-031-08751-6_7

profile blunting has been observed due to the motion of cells towards the vessel centerline [1] and blood viscosity has been shown to be dependent on vessel diameter and hematocrit (volume percentage of cells in the blood) [2]. Additionally, cell-to-cell [3] and cell-to-vessel interactions [4] have been shown to affect the underlying blood flow profile. Therefore in small vessel simulations, blood must be modeled as a suspension of cells rather than a continuum fluid. Fluid-structure interaction (FSI) models, such as the immersed boundary method [5] or dissipative particle dynamics [6], which fully couple deformable particles with a background fluid, have been shown to accurately model cells in microfluidic [7] and microcirculatory systems [8]. Blood flow simulations using FSI models provide a wealth of information, as both microscopic and macroscopic quantities, such as individual cell position and deformation, and fluid pressure and velocity profiles, can be precisely tracked and studied over time [9,10]. More importantly, these models allow for the isolation and controlled variation of specific parameters such as cell size or stiffness, enabling researchers to probe the effects of individual parameters on the quantity of interest. Much of the *in silico* work in microvessels with cell FSI models has been focused on red blood cells (RBCs), including studies on the effects of cell deformability and shape [7,11,12], partitioning at junctions in the vasculature [13,14], aggregation mechanics [15], and development of a cell-depleted layer [16,17]. Simulation has also been used to study the motion of other particles in the presence of RBCs such as platelets [18,19], leukocytes [20,21], and circulating tumor cells [10,22–25].

While FSI models of cells in complex geometries are not new, advances in computational efficiency and capability [26–29] have only recently made this approach practical for comprehensive studies of realistic systems. The inclusion of explicit particles in particular introduces several new obstacles. The main challenge is simply one of statistics: the motion of particles diffusing through a vessel is an inherently stochastic process, thus trajectories must be sampled a sufficient number of times to capture average behavior. For example, when tracking cancer cells *in silico*, the distance to a vessel wall directly influences the cell's likelihood of adhesion [30] and subsequent escape into nearby tissue. We previously demonstrated the effects of varying cell positions while studying combinations of hemodynamic parameters and the motion of a tumor cell [22]. Even when all bulk fluid parameters were held constant, the trajectory of the tumor cell was found to vary significantly based on the relative configurations of neighboring cells.

In addition to increasing the overall computational cost, the need for a representative ensemble of starting configurations introduces new potential sources of error that must be managed. This challenge is particularly acute for systems with higher hematocrit values, where random coordinate generation must be done carefully to avoid artificial structure that would bias the observed dynamics. Similar to the well-known equilibration problem in molecular dynamics [31], flow simulations through tortuous vascular geometries have the added complication that one can not easily gather equilibrated statistics simply by running a closed system longer in time. Instead, one must generate a number of distinct sets of equilibrated starting points to be run independently [32]. To this end, we propose a method to generate many cell configurations and the use of the radial distribution function to characterize the structure in a particular configuration. The final challenge is to define quantitative metrics to rigorously compare individual cell configurations and to characterize the complete set as a whole. For this purpose, we propose the use of the Jaccard index to quantify spatial similarity between individual configurations as an appropriate metric for describing and comparing sets of cell configurations.

2 Methods and Metrics

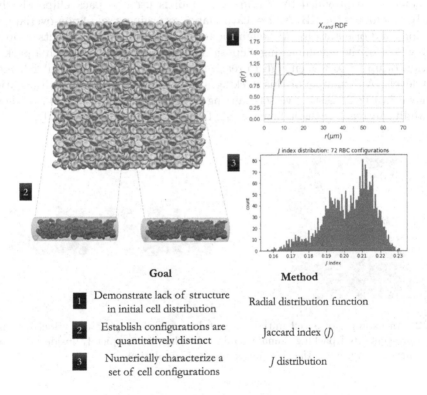

Fig. 1. Workflow for determining the best set of cell configurations to describe the ensemble. (1) The radial distribution function $g(r)$ is used to assess randomness in a distribution of cells. (2) The Jaccard index J is used to quantify spatial similarity between two configurations. (3) A set of pairwise J values are used to numerically describe a large set of cell configurations and presented as a distribution.

Our proposed workflow and associated metrics are shown in Fig. 1. In this section, we will describe both the methods we use to generate cell configurations as well as the associated metrics used to characterize them.

2.1 Generating Initial Configurations of Red Blood Cells in a Microvessel

Dense packing of non-overlapping shapes is a long-standing research problem of active interest [33–35]. Here we describe a procedure for generating and characterizing packed configurations of RBCs in arbitrary vessels at a set hematocrit. Rather than generating individual configurations on demand, we instead start with a large system of packed cells from which we can fill vessels of arbitrary size and shape. This technique has the advantage of letting us generate a packed domain in the simplest possible geometry prior to simulation while avoiding the code complexity an on-the-fly implementation would require. The source domain is created to be several times larger than the vessel of interest. The standalone implementation provided by Birgin *et al.* [36] is used to pack ellipsoids that tightly encompass the RBC's biconcave shape, returning a set of non-overlapping positions and orientations. Although the fully enclosed RBC represents approximately 70% of the encompassing ellipsoid volume, a distribution with a packing fraction of up to 60% is enough to reach the high end of microvascular hematocrit levels. An example of this packing and a corresponding cell initialization is shown in Fig. 2. Testing vessels ranging in diameter from 20 to 50 μm shows the ability to reach realistic hematocrits from 20% to 35% consistently.

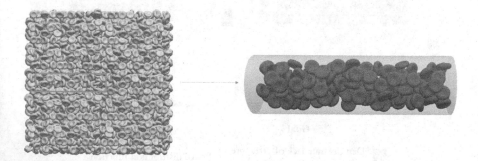

Fig. 2. An example of a cell initialization taken by submerging the vessel within a large, pre-generated packing domain. Only cells that fit completely inside the vessel are returned and used as the starting point for a simulation.

This approach of separately generating a packed source domain has the advantage of easily allowing for rigorous *a priori* analysis before performing expensive high performance computing (HPC) simulations. To avoid initializing FSI runs with non-physical starting configurations, the source bulk system must not have any long-range order consistent with crystalline packing. The radial distribution function $g(r)$ is a well-established metric in the simulation of fluids [37] used for confirming liquid structure, defined as:

$$g(r) = \frac{dn_c}{4\pi r^2 dr \rho} \tag{1}$$

where dn_c returns the number of cells within a shell of thickness dr and ρ is the bulk density. Long-range structure is reflected in the form of multiple peaks well beyond the average particle spacing. RBCs are assumed to start from a fully disordered liquid state, reflected by a $g(r)$ that quickly converges to a constant value of unity.

Once the bulk source geometry has been generated, individual configurations can be created by submerging the target vessel in the source domain at different locations and selecting all cells contained within. This process remains the same for both simple and complex geometries, establishing a straightforward method for generating many different configurations prior to running HPC simulations.

2.2   Quantifying Spatial Similarity Between Cell Configurations

After generating a set of multiple cell configurations, the next step is to verify that each of these packings are distinct by quantifying their spatial similarity to each other; specifically, the fraction of volume shared by two configurations. However, due to the irregular biconcave disk shape of RBCs, a simple analytical algorithm for overlap check given cell positions and orientation angles does not exist. Therefore in this calculation, a numerical method is utilized, where each configuration of RBCs is mapped to a 3-D grid, and overlap is calculated by the number of grid points shared. The Jaccard index, or intersection over union, is used to measure the similarity between two discrete sample sets, defined as:

$$J(C_i, C_j) = \frac{|C_i \cap C_j|}{|C_i \cup C_j|} \tag{2}$$

where C_i and C_j are independent samples of the same space. We propose the use of the Jaccard index to quantify the volume overlap between sets of RBCS by comparing the interior grid points. This is similar in approach to the algorithms used by the image segmentation community [38], such as the Dice similarity index.

Since $J(C_i, C_j)$ represents the percentage of overlapping cell volume between C_i and C_j, $J(C_i, C_j) = 100\%$ if two arrangements are identical and zero if there is no shared cell volume in space. A threshold value ϕ is chosen to label whether or not two configurations are correlated; if $J(C_i, C_j) > \phi$, the pair is marked similar. For example, two test configurations that contained the same group of cells shifted by a few tenths of a microns led to J over 90%, and would be marked as a similar pair. Because the likelihood of two configurations of cells both occupying a certain space increases with hematocrit, ϕ is not a static value, and is chosen on a per hematocrit basis.

A two-dimensional example for calculating J using RBCs is provided in Fig. 3. Each initialization contains a single cell marked red for configuration 1 and blue for configuration 2. The corresponding lattice points are marked with the color of containing cell. Once these two lattices are overlaid, the shared points are marked in yellow.

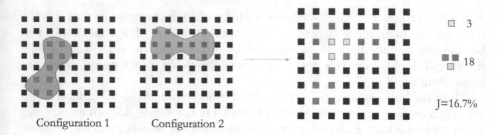

Configuration 1   Configuration 2

Fig. 3. A 2-D example for the calculation of Jaccard index J between two configurations of RBCs. After the RBCs are mapped to their corresponding lattices, there are 18 total points which contain a cell, of which 3 are shared in both configurations. $J = 3/18 = 16.7\%$ in this example. (Color figure online)

Another two-dimensional visual example is shown in Fig. 4, displaying distinct configurations of circles with significantly different J values. Compared to the base configuration, there is a clear difference in overlap, which can be identified visually and captured quantitatively through an analytical computation of J.

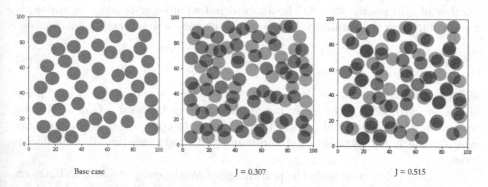

Base case $J = 0.307$ $J = 0.515$

Fig. 4. A 2-D example of differing Jaccard indices compared to a base configuration (left) using circular particles. Two other distinct configurations are generated and overlaid on the base configuration (middle and right). The middle configuration is less similar to the base case than the right configuration and can be confirmed visually by identifying by the overlapping violet regions and computed using J. Although the overlap is easier to visualize in 2-D, it is much easier to numerically identify similarity in 3-D space by using J.

Although these examples are displayed in two dimensions for clarity, our RBC configurations and future simulations are performed in three-dimensional space. The Jaccard index provides a numerical method to identify spatial similarity rather than a qualitative comparison.

Since the Jaccard index is applied between two particular configurations, J needs to be calculated on a pairwise basis before it can be used to quantify the entire distribution of configurations. For a set of configurations $S = \{C_1, ..., C_n\}$, we define J_S, the set of Jaccard similarity scores, as:

$$J_S = \{J(C_i, C_j) | i, j = 1...n, i \neq j\}. \tag{3}$$

To quantify the similarity of a particular configuration C_i with respect to all the others, the mean Jaccard index $\bar{J}(C_i)$ is calculated as:

$$\bar{J}(C_i) = \frac{1}{n-1} \sum_{j=1}^{n} J(C_i, C_j), j \neq i \tag{4}$$

for a set of n configurations. Given two similar configurations C and C' such that $J(C, C') > \phi$, and mean Jaccard indices such that $\bar{J}(C) < \bar{J}(C')$, configuration C' would be considered first for removal from the set.

3   Results and Discussion

3.1   Applying the Radial Distribution Function to Quantify Structure in a Single Packing

To test for the presence of long-range structure, the radial distribution function $g(r)$ is applied to two large, dense packings of RBCs. Both source arrangements are generated in a cubic domain of side length 200 μm with over 35,000 cells. A random distribution X_{rand} is created by packing the cube and then applying an external force to perturb the initial arrangement of cells, while a structured set of cells X_{struct} is produced by tessellating a small set of RBCs across the space. The radial distribution function is then applied to each set of cell centers splitting dr into 0.25 μm buckets, and the cell configurations and corresponding $g(r)$ functions are shown in Fig. 5.

Within X_{rand}, $g(r)$ contains a single peak near the lengthwise diameter of the RBC that quickly trails off to unity, indicative of a liquid-like, random distribution of particles. In the case of X_{struct}, multiple discrete peaks are visible, signifying the presence of long-range structure in the distribution of cells. A qualitative comparison between the two source domains can be performed visually, but the use of the radial distribution functions provide a quantitative confirmation for the presence of ordered structure.

Since the procedure to generate many cell configurations in a microvessel utilizes a subset of the cells in the large domain, it is important to confirm the randomness of the initial cell arrangements. The packing found in X_{struct} is non-physiological, and would generate many structured cell initializations as inputs to HPC simulations. Moving forward, we sample configurations from X_{rand} after confirming the lack of long-range structure in its distribution of cells.

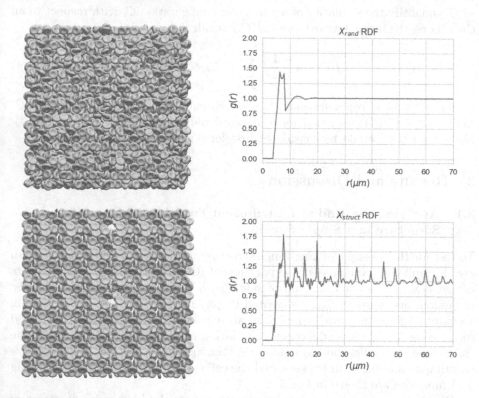

Fig. 5. A random distribution X_{rand} (top) and ordered distribution X_{struct} (bottom) of cells packed within a cube of side length 200 μm. The corresponding radial distribution functions are shown to the right. X_{rand}'s $g(r)$ shows a single peak and trails off to 1 quickly, analogous to a random liquid-like state, while X_{struct}'s $g(r)$ displays several peaks, indicating that the distribution contains a repetitive structure. Sampling cells from the random distribution provides a better initial set of the positions and orientations of red blood cells for running HPC simulations.

3.2   Utilizing Jaccard Index to Quantify Distributions of Cell Configurations

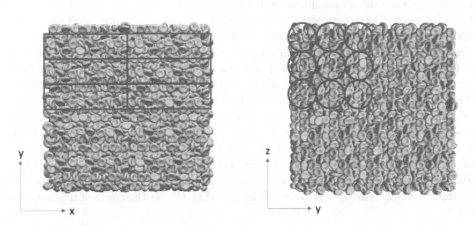

Fig. 6. There are 72 different non-overlapping initializations for a cylinder of diameter 30 μm and length 100 μm pulled from a packing domain of 200^3 μm³. Cells within this region are used to generate RBC configurations of initial positions and orientations at a target hematocrit of 25%. A subset of the cutouts are shown in each xy and yz planes for this set of initializations.

Non-overlapping cutouts representing an ideal microvessel geometry with diameter 30 μm and length 100 μm are created from X_{rand}. The dimensions of the source geometry allow for 72 independent configurations to be generated: 6 from the y- and z-planes, and 2 in the x-plane, as shown in Fig. 6. All configurations have a hematocrit of 25% with $N = 160$ RBCs on average. The Jaccard index is calculated using a grid spacing of 0.25 μm. A histogram of all pairwise Jaccard similarity index values is presented in Fig. 7a. For comparison, random configurations of 160 spheres were numerically generated at a 25% packing density in a cylinder with the same aspect ratio (see Fig. 7b–d). We note that increasing the number of configurations for better statistics gave smoother distributions but did not fundamentally change the shape.

As expected, the overlap index of configurations of randomly-placed spheres follows a normal distribution. The distribution of RBCs, on the other hand, is clearly skewed away from normal. This may be an artifact of the packing algorithm used to populate the source distribution X_{rand} or may be a funda-mental difference in how biconcave disks pack into a confined geometry; more work will be needed to elucidate the underlying cause. The magnitude of the average overlap differs significantly between the shapes as well despite all sys-tems having the same volume packing fraction. It should be noted that the RBC geometry likely has a systematic underestimation of the overlap due to discretiza-tion error, though this is not expected to be large. The spherical overlap was computed analytically as a function of distance between sphere centers. Figure 8 shows $\bar{J}$, the average pair overlap of a configuration with all other configurations. This provides a method to compare individual configurations' spatial coverage against the full set. We expect that both J_S and $\bar{J}$ distributions will change based on vessel geometry and hematocrit. However, this study establishes that a pairwise Jaccard index distribution can be used as a quantitative metric to describe a set of cell configurations, generated with the same packing fraction. We posit that selecting configurations with low $\bar{J}$ could be used to sample the configurational phase space more efficiently; this will be the topic of a follow-up study.

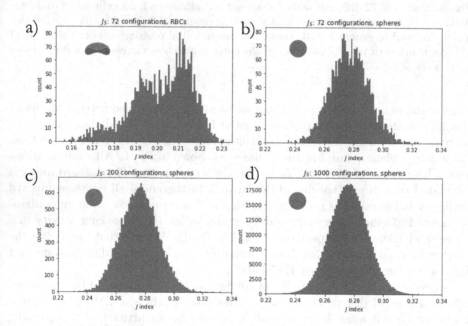

Fig. 7. The distribution J_S of pairwise J values for (a) RBC and (b-d) sphere config-urations. $N = 160$ objects.

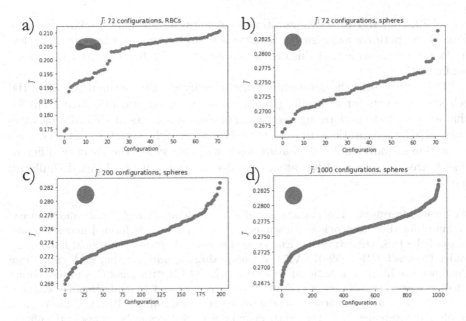

Fig. 8. Distribution of $\bar{J}$ on a per configuration basis for (a) RBC and (b-d) sphere configurations. $N = 160$ objects.

4 Conclusion

In the microvessel regime, FSI models are used to perform simulations with explicit RBCs to account for non-Newtonian effects. Now that recent computational advances have enabled the rise of large-scale FSI simulation studies, it is important to sufficiently sample the ensemble of potential cell arrangements to capture a macroscopic behavior. In order to select the minimum set of configurations that spans the parameter space, certain quantitative metrics must be established which (1) indicate that a particular configuration is a good starting point and (2) show that two separate arrangements are distinct and spatially uncorrelated. These parameters can then be used to define the space of possible configurations and determine which set of arrangements best span the space.

In this study, we apply the radial distribution function to particular configurations of RBCs to qualify whether a structured arrangement of cells exists in the distribution. We choose two large distributions of cells, one randomly placed and one structured, analogous to atoms in liquid- and solid-like materials, and show that this function is able to quantify the presence of long-range structure in cell positions. We also use the Jaccard index J to capture a quantitative representation of shared cell volume between two configurations. Taking the irregular shape of RBCs into account, we devise a numerical method that maps cells on to a 3-D lattice which is used to compute J. We then produce a set of 72 RBC configurations in a 30 μm diameter and 100 μm length microvessel from the randomly-distributed group of RBCs. For comparison, we also generate sets of 72, 200, and 1000 configurations of spheres at the same packing fraction. Finally, we perform pairwise J calculations and plot the distribution of Jaccard index values, showing that this metric can be used to define the space of particle configurations.

This study sets the groundwork for identifying the optimal set of initial cell arrangements for a specific group of simulation parameters. Next steps for this work include performing simulation studies with sets of spatially uncorrelated RBC configurations to determine how these affect certain outputs, such as motion of individual cells. Future work will also study the effect of different vessel sizes, shapes, and hematocrit on distributions of the Jaccard similarity index.

Acknowledgments. The authors would like to thank Daniel Puleri and Samreen Mahmud for their feedback and discussion. This work was performed under the auspices of the U.S. Department of Energy by Lawrence Livermore National Laboratory under Contract DE-AC52-07NA27344. Computing support for this work came from the Lawrence Livermore National Laboratory (LLNL) Institutional Computing Grand Challenge program. The work of Sayan Roychowdhury and Amanda Randles was supported by the National Science Foundation under award number 1943036. The content is solely the responsibility of the authors and does not necessarily represent the official views of the NSF.

References

1. Tangelder, G.J., Slaaf, D.W., Muijtjens, A., Arts, T., Oude Egbrink, M., Reneman, R.S.: Velocity profiles of blood platelets and red blood cells flowing in arterioles of the rabbit mesentery. Circ. Res. **59**(5), 505–514 (1986)
2. Pries, A.R., Neuhaus, D., Gaehtgens, P.: Blood viscosity in tube flow: dependence on diameter and hematocrit. Am. J. Physiol.-Heart Circ. Physiol. **263**(6), H1770–H1778 (1992)
3. Bishop, J.J., Nance, P.R., Popel, A.S., Intaglietta, M., Johnson, P.C.: Effect of erythrocyte aggregation on velocity profiles in venules. Am. J. Physiol.-Heart Circ. Physiol. **280**(1), H222–H236 (2001)
4. Panés, J., Perry, M., Granger, D.N.: Leukocyte-endothelial cell adhesion: avenues for therapeutic intervention. Br. J. Pharmacol. **126**(3), 537 (1999)
5. Peskin, C.S.: The immersed boundary method. Acta numerica **11**, 479–517 (2002)
6. Pivkin, I.V., Karniadakis, G.E.: Accurate coarse-grained modeling of red blood cells. Phys. Rev. Lett. **101**(11), 118105 (2008)
7. Krüger, T., Holmes, D., Coveney, P.V.: Deformability-based red blood cell separation in deterministic lateral displacement devices-a simulation study. Biomicrofluidics **8**(5), 054114 (2014)
8. Fedosov, D.A., Noguchi, H., Gompper, G.: Multiscale modeling of blood flow: from single cells to blood rheology. Biomech. Model. Mechanobiol. **13**(2), 239–258 (2013). https://doi.org/10.1007/s10237-013-0497-9
9. Bagchi, P.: Mesoscale simulation of blood flow in small vessels. Biophys. J . **92**(6), 1858–1877 (2007)
10. Pepona, M., et al.: Investigating the interaction between circulating tumor cells and local hydrodynamics via experiment and simulations. Cell. Mol. Bioeng. **13**(5), 527–540 (2020)

11. Lei, H., Karniadakis, G.E.: Quantifying the rheological and hemodynamic characteristics of sickle cell anemia. Biophys. J . **102**(2), 185–194 (2012)
12. Czaja, B., Gutierrez, M., Závodszky, G., de Kanter, D., Hoekstra, A., Eniola-Adefeso, O.: The influence of red blood cell deformability on hematocrit profiles and platelet margination. PLoS Comput. Biol. **16**(3), e1007716 (2020)
13. Balogh, P., Bagchi, P.: Analysis of red blood cell partitioning at bifurcations in simulated microvascular networks. Phys. Fluids **30**(5), 051902 (2018)
14. Yang, J., Yoo, S.S., Lee, T.R.: Effect of fractional blood flow on plasma skimming in the microvasculature. Phys. Rev. E **95**(4), 040401 (2017)
15. Zhang, J., Johnson, P.C., Popel, A.S.: Red blood cell aggregation and dissociation in shear flows simulated by lattice Boltzmann method. J. Biomech. **41**(1), 47–55 (2008)
16. Fedosov, D.A., Caswell, B., Popel, A.S., Karniadakis, G.E.: Blood flow and cell-free layer in microvessels. Microcirculation **17**(8), 615–628 (2010)
17. Katanov, D., Gompper, G., Fedosov, D.A.: Microvascular blood flow resistance: role of red blood cell migration and dispersion. Microvasc. Res. **99**, 57–66 (2015)
18. Vahidkhah, K., Diamond, S.L., Bagchi, P.: Platelet dynamics in three-dimensional simulation of whole blood. Biophys. J . **106**(11), 2529–2540 (2014)
19. Fitzgibbon, S., Spann, A.P., Qi, Q.M., Shaqfeh, E.S.: In vitro measurement of particle margination in the microchannel flow: effect of varying hematocrit. Biophys. J . **108**(10), 2601–2608 (2015)
20. Jain, A., Munn, L.L.: Determinants of leukocyte margination in rectangular microchannels. PLoS ONE **4**(9), e7104 (2009)
21. Freund, J.B.: Leukocyte margination in a model microvessel. Phys. Fluids **19**(2), 023301 (2007)
22. Roychowdhury, S., Gounley, J., Randles, A.: Evaluating the influence of hemorheological parameters on circulating tumor cell trajectory and simulation time. In: Proceedings of the Platform for Advanced Scientific Computing Conference, pp. 1–10 (2020)
23. Xiao, L.L., Lin, C.S., Chen, S., Liu, Y., Fu, B.M., Yan, W.W.: Effects of red blood cell aggregation on the blood flow in a symmetrical stenosed microvessel. Biomech. Model. Mechanobiol. **19**(1), 159–171 (2019). https://doi.org/10.1007/s10237-019-01202-9
24. Gounley, J., Draeger, E.W., Randles, A.: Numerical simulation of a compound capsule in a constricted microchannel. Procedia Comput. Sci. **108**, 175–184 (2017)
25. Balogh, P., Gounley, J., Roychowdhury, S., Randles, A.: A data-driven approach to modeling cancer cell mechanics during microcirculatory transport. Sci. Rep. **11**(1), 1–18 (2021)
26. Grinberg, L., et al.: A new computational paradigm in multiscale simulations: application to brain blood flow. In: Proceedings of 2011 International Conference for High Performance Computing, Networking, Storage and Analysis, pp. 1–5 (2011)
27. Ames, J., Puleri, D.F., Balogh, P., Gounley, J., Draeger, E.W., Randles, A.: Multi-GPU immersed boundary method hemodynamics simulations. J. Comput. Sci. **44**, 101–153 (2020)
28. Lu, L., Morse, M.J., Rahimian, A., Stadler, G., Zorin, D.: Scalable simulation of realistic volume fraction red blood cell flows through vascular networks. In: Proceedings of the International Conference for High Performance Computing, Networking, Storage and Analysis, pp. 1–30 (2019)

29. Gounley, J., Draeger, E.W., Randles, A.: Immersed boundary method halo exchange in a hemodynamics application. In: Rodrigues, J.M.F., et al. (eds.) ICCS 2019. LNCS, vol. 11536, pp. 441–455. Springer, Cham (2019). https://doi.org/10.1007/978-3-030-22734-0_32

30. Xiao, L., Liu, Y., Chen, S., Fu, B.: Effects of flowing RBCs on adhesion of a circulating tumor cell in microvessels. Biomech. Model. Mechanobiol. **16**(2), 597–610 (2017)

31. Stella, L., Melchionna, S.: Equilibration and sampling in molecular dynamics simulations of biomolecules. J. Chem. Phys. **109**(23), 10115–10117 (1998)

32. Gordiz, K., Singh, D.J., Henry, A.: Ensemble averaging vs. time averaging in molecular dynamics simulations of thermal conductivity. J. Appl. Phys. **117**(4), 045104 (2015)

33. Donev, A., et al.: Improving the density of jammed disordered packings using ellipsoids. Science **303**(5660), 990–993 (2004)

34. Malmir, H., Sahimi, M., Tabar, M.: Microstructural characterization of random packings of cubic particles. Sci. Rep. **6**(1), 1–9 (2016)

35. Miśkiewicz, K., Banasiak, R., Niedostatkiewicz, M., Grudzień, K., Babout, L.: An algorithm to generate high dense packing of particles with various shapes. In: MATEC Web of Conferences, vol. 219, p. 05004. EDP Sciences (2018)

36. Birgin, E.G., Lobato, R.D.: A matheuristic approach with nonlinear subproblems for large-scale packing of ellipsoids. Eur. J. Oper. Res. **272**(2), 447–464 (2019)

37. Allen, M.P., Tildesley, D.J.: Computer Simulation of Liquids. Oxford University Press, Oxford (2017)

38. Yeghiazaryan, V., Voiculescu, I.D.: Family of boundary overlap metrics for the evaluation of medical image segmentation. J. Med. Imaging **5**(1), 015006 (2018)

Coevolutionary Approach to Sequential Stackelberg Security Games

Adam Żychowski$^{(\boxtimes)}$ and Jacek Mańdziuk

Faculty of Mathematics and Information Science, Warsaw University of Technology,
Koszykowa 75, 00-662 Warsaw, Poland
{a.zychowski,j.mandziuk}@mini.pw.edu.pl

Abstract. The paper introduces a novel coevolutionary approach (CoEvoSG) for solving Sequential Stackelberg Security Games. CoEvoSG maintains two competing populations of players' strategies. In the process inspired by biological evolution both populations are developed simultaneously in order to approximate Stackelberg Equilibrium. The comprehensive experimental study based on over 500 test instances of two game types proved CoEvoSG's ability to repetitively find optimal or close to optimal solutions. The main strength of the proposed method is its time scalability which is highly competitive to the state-of-the-art algorithms and allows to calculate bigger and more complicated games than ever before. Due to the generic and knowledge-free design of CoEvoSG, the method can be applied to diverse real-life security scenarios.

Keywords: Coevolution · Security games · Cybersecurity

1 Introduction

New technologies bring new challenges. One of them is cybersecurity. In recent years, this topic has gained more and more importance [14] since more and more critical systems are connected to the Internet and increasingly many aspects of people's lives depend on reliable computer infrastructure. We are facing a constant arms race between defenders and attackers. One of the approaches to the issue of cybersecurity attacks is to model them as a non-cooperative game. This approach was applied, for instance, in intrusion detection problem in mobile ad-hoc networks [7], security-aware distributed job scheduling in cloud computing [5], detecting vulnerabilities in interbank network [6], planning deep packet inspections [29], and other. In particular, the Stackelberg Security Games (SSGs) recently gained lots of popularity due to a bunch of successful practical applications [19].

SSGs were successfully deployed not only in cybersecurity domain [20,26] but also in a wide range of real-world scenarios, e.g. scheduling Los Angeles International Airport canine patrols [8], protecting US Coast Guard's resources in Boston harbor [18], or preventing poaching in the Queen Elizabeth National Park in Uganda [4].

D. Groen et al. (Eds.): ICCS 2022, LNCS 13350, pp. 103–117, 2022.
https://doi.org/10.1007/978-3-031-08751-6_8

In SSG there are two asymmetrical players: the Defender and the Attacker. The Defender commits to their strategy first. Then, the Attacker, knowing the Defender's commitment, decides on their own strategy. The above order of strategy-related decisions favors the Attacker and mimics real-world scenarios in which the Attacker can observe the opponent's strategy (e.g. patrol schedules) and plan their attack accordingly.

The strategy chosen by the Defender is a *mixed one*, i.e. a probability distribution over all possible *pure* (i.e. simple deterministic) strategies [3]. The Attacker is aware of this distribution but has no knowledge about its specific materialization (the sequence of actions that will actually be played). The goal of SSG is to find *Stackelberg Equilibrium* (SE), i.e. the pair of players' strategies that fulfills the following assumption: changing strategy by any player will lead to his/her result deterioration.

In this paper, we consider sequential SSGs which means that each player's strategy consists of a sequence of actions to be executed (played) in consecutive time steps. In such SSGs, finding SE is an NP-hard problem [3]. For this reason, exact methods have limited applicability and are rarely implemented in real-world scenarios. Instead, a number of heuristics approaches were proposed in the literature, including the use of Evolutionary Algorithms (EAs) [12,28,29]. EAs are inspired by the process of biological evolution and consists in maintaining a population of potential solutions, which is iteratively modified by applying evolutionary operators: *mutation, crossover* and *selection*.

In this paper, we extend the previous EA approaches and propose the coevolutionary algorithm for solving SSGs (CoEvoSG). The method not only maintains a population of Defender's strategies (as EA-based approaches) but also a population of Attacker's strategies. Both populations compete with each other in the process of coevolution. In effect, the convergence to the near-optimal solution is much faster than in the state-of-the-art methods, which allows to solve larger and more complex games than ever before.

Contribution. The contribution of this paper is three-fold: (i) a novel coevolutionary algorithm (CoEvoSG) for Sequential Stackelberg Security Games, capable of finding optimal or near-optimal solutions is proposed, (ii) a comprehensive experimental study proves the efficacy of CoEvoSG and its ability to solve games of sizes and complexity that are beyond the capability of state-of-the-art methods, (iii) to our knowledge, application of coevolutionary algorithms to solving sequential SSGs, has never been considered before in the related literature.

2   Problem Definition

A sequential SSG is played by two players: the Defender (D) and the Attacker (A), and is composed of m time steps (moves). In each time step both players simultaneously choose their action to be performed. A *pure strategy* σ_P of player P ($P \in \{D, A\}$) is a list of his/her actions in consecutive time steps: $\sigma_P = (a_1, a_2, \ldots, a_m)$. If by Σ_P we denote a set of all possible pure strategies of P, then a probability distribution $\pi_P \in \Pi_p$ over Σ_P is the *mixed strategy* of P, where

Π_p is set of all his/her mixed strategies. For any pair of strategies (π_D, π_A) the expected payoffs for the players are defined and denoted by $U_D(\pi_D, \pi_A)$ and $U_A(\pi_D, \pi_A)$. Stackelberg Equilibrium is a pair of strategies (π_D, π_A) satisfying the following conditions:

$$\pi_D = \arg\max_{\bar{\pi}_D \in \Pi_D} U_A(\bar{\pi}_D, BR(\bar{\pi}_D)), \qquad BR(\pi_D) = \arg\max_{\pi_A \in \Pi_A} U_A(\pi_D, \pi_A).$$

The first equation chooses the best Defender's strategy π_D under the assumption that the Attacker always selects the best response strategy $(BR(\pi_D))$ to the Defender's committed strategy.

Furthermore, if there exists more than one optimal Attacker's response (with the same highest Attacker's payoff), the Attacker selects the one with the highest corresponding Defender's payoff, i.e. breaks ties in favor of the Defender. While this assumption may seem counterintuitive, the opposite way of breaking ties may lead to situations when equilibrium doesn't exist [24]. The above SE extension is known as Strong Stackelberg Equilibrium [1] and is adopted in this paper (as well as in the vast majority of SSG publications).

Both players choose their strategy at the beginning of the game (first the Defender and then the Attacker) and they cannot change it during the gameplay, i.e. in consecutive steps they follow actions encoded in the selected strategy irrespective of the opponent's moves (they are not aware of opponent's current and past actions). Conitzer et al. [3] proved that for each Defender's mixed strategy there exists at least one Attacker's pure strategy which maximizes their payoff. This property is commonly utilized by solutions proposed in the literature since it narrows the Attacker's response search space to only pure strategies.

3 Related Work

The methods of solving SSGs can be divided into two main groups: exact and approximate. Exact approaches are based on Mixed-Integer Linear Programming (MILP), which formulates SSG as an optimization problem with a specific target function and a set of linear integer constraints that must be fulfilled. MILP programs are usually computed by specially optimized software engines - *solvers*.

C2016. One of the most popular exact method is C2016 [23], which also bases on MILP but instead of directly computing SE, utilizes the Stackelberg Extensive-Form Correlated Equilibrium (SEFCE). In SEFCE, the Defender can send signals to the Attacker who has to follow them in their choice of strategy. C2016 uses a linear program for computing SEFCE and then modifies it by iteratively restricting the signals the Defender can send to the Attacker and converging to SE. In this article C2016 was used to calculate the reference optimal solutions.

O2UCT. Thee above-mentioned MILP approaches returns exact (optimal) solutions but suffer from exponential computation time and poor memory scalability, which makes them inefficient for large games. Thus, some approximate approaches have been recently proposed, e.g. O2UCT [10,11] which utilizes an Upper Confidence Bounds applied to Trees (UCT) algorithm [13] (a variant of Monte Carlo Tree search [21]). O2UCT is based on guided sampling of the Attacker's strategy space and optimizes the Defender's strategy under the assumption that the sampled Attacker's strategy is the optimal response. O2UCT scales visibly better than exact MILP-based solutions and returns close-to-optimal solutions for various types of games.

EASG. Another heuristic method, which is the most related to the approach presented in the paper, is Evolutionary Algorithm for Stackelberg Games (EASG) [27,28], which optimizes the Defender's payoff by evolving a population of candidate strategies. EASG starts off with a population that contains randomly selected pure Defender's strategies. Then, until the stop condition is not fulfilled, the population evolves in consecutive generations. In each generation, the following four operations are applied: crossover, mutation, evaluation, and selection.

Crossover combines two individuals randomly selected from a population by merging their pure strategies and halving their probabilities. Afterwards, the resultant chromosome is shortened (simplified) by deleting some of its pure strategies with a chance inversely proportional to their probabilities. The mutation operator changes one of the pure strategies encoded in the chromosome starting from a randomly selected time step. New actions are drawn from all feasible actions in a corresponding game state. The role of mutation is to boost exploration of the strategy space.

Next, each individual is assigned a fitness value which is the expected Defender's payoff. This step requires finding the optimal Attacker's response to the mixed Defender's strategy encoded in the chromosome. To this end, EASG iterates over all possible Attacker's pure strategies and selects the one with the highest Attacker's payoff. Due to the potentially large space of Attacker's pure strategies, the evaluation phase is the most time-consuming step of EASG.

Finally, in the selection phase, individuals with higher Defender's payoff are more likely to be selected to the next generation. The above evolutionary approach was successfully applied to various types of SSGs including games with moving targets [12] or games assuming Attacker's bounded rationality [29].

4   Coevolutionary Approach

Motivation. As we mentioned in the previous section, EASG evaluation process requires iterating over all possible Attacker's pure strategies in order to find the best one and calculate the expected Defender's payoff. This evaluation procedure is performed thousands of times (for each individual in each generation) which is infeasible (too time-consuming), except for small games.

Furthermore, in many SSG instances there exists a relatively small subset of Attacker's strategies that need actually to be considered when looking for the optimal response. Many of the Attacker's strategies can either be trivially qualified as weak (e.g. an attack at a well-protected target with low reward or a sequence of actions which does not lead to a target), or there are subsets of similar strategies and only one representative from each of them needs to be examined in order to find the best Attacker's response. However, for more complex games it is not possible to determine - within a reasonable time - which of the Attacker's strategies could be omitted, nor to define a representative subset of these strategies, due to the high dependence of this selection on the game topology (structure) and payoff distribution.

In order to address the issue of time-consuming evaluation process in EASG, we propose a novel coevolutionary approach which maintains two populations: one composed of the Defender's mixed strategies (as in EASG) and the other consisting of the Attacker's pure strategies. Strategies from the Attacker's population are used to evaluate the Defender's strategies. Instead of calculating the Defender's payoff against all possible Attacker's pure strategies, it is now calculated only versus a subset of the Attacker's strategies represented in the population. Both populations compete with each other, i.e. the Attacker's population attempts to find the best possible response to the strategies from the Defender's population and vice versa - the Defender's population tries to evolve the most effective strategies with respect to the response strategies encoded in the Attacker's population.

System Overview. A general overview of the CoEvoSG algorithm is presented in Fig. 1. Both populations are initialized with random pure strategies and then developed alternately. First, the Attacker's population is modified by evolutionary operators (crossover, mutation, and selection) through g_p generations. Then, the Defender's population is evolved through the same number of g_p generations. The above loop is repeated until the stop condition is not satisfied.

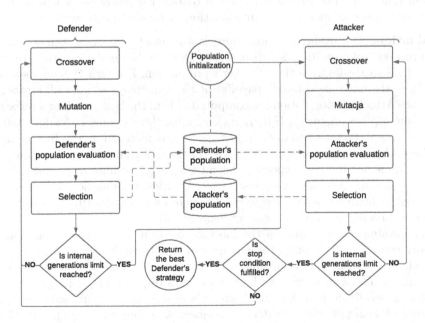

Fig. 1. A high-level overview of the CoEvoSG algorithm.

All evolutionary operators applied to the Defender's population are implemented in the same way as in EASG and briefly described in Sect. 3. Additional details can be found in [28]. The novel operators applied to the Attacker's population are described below.

Initialization. The Attacker's population contains N_A individuals. Each individual k represents a randomly selected pure Attacker's strategy, encoded as a list of actions in consecutive time steps: $\sigma_A^k = (a_1^k, a_2^k, \ldots, a_m^k)$. In each time step $t \in \{1, \ldots, m\}$ a_t^k is drawn uniformly from all feasible actions in a given state.

Crossover. Each individual from the Attacker's population is selected for a crossover with probability p_c. Selected individuals are paired randomly and for each pair, one-point crossover is performed, i.e. for strategies $\sigma_A^r = (a_1^r, a_2^r, \ldots, a_m^r)$ and $\sigma_A^s = (a_1^s, a_2^s, \ldots, a_m^s)$ the following two child individuals are created: $\sigma_A^{\prime r} = (a_1^r, \ldots a_i^r, a_{i+1}^s, \ldots, a_m^s)$ and $\sigma_A^{\prime s} = (a_1^s, \ldots a_i^s, a_{i+1}^r, \ldots, a_m^r)$, where $a_i^r = a_i^s$ is the first common action (in the same time step) in the parent strategies. If such an action does not exist, the crossover has no effect. For example, if an action is to choose a vertex in a game graph the player moves to, then $a_i^r = a_i^s$ would be the first common vertex on the paths defined by the parent strategies.

Mutation. Each individual is mutated with probability p_m. Mutation operator, starting from a randomly selected step, modifies all subsequent actions encoded in the chromosome. Each subsequent action is chosen randomly from all available actions in the current state. The result of mutation of strategy $\sigma_A^r = (a_1^r, a_2^r, \ldots a_m^r)$ is $\sigma_A^{\prime r} = (a_1^r, a_2^r, \ldots, a_{i-1}^r, a_i^{r'}, a_{i+1}^{r'}, \ldots, a_m^{r'})$, where i is the chosen time step. The role of mutation is to boost exploration of new areas in the search space by means of an introduction of random perturbations.

Evaluation. The evaluation procedure is the most important component of the proposed solution. Individuals from the Defender's population are evaluated against all strategies from the Attacker's population. For each Defender's strategy (π_D) the outcome (players' payoffs) of the gameplays against all strategies from the Attacker's population are computed. Then, the best Attacker's response is chosen: $\sigma_A^{best} = \arg\max_{\sigma_A} U_A(\pi_D, \sigma_A)$. Finally, the expected Defender's payoff against this Attacker's response ($U_D(\pi_D, \sigma_A^{best})$ is assigned as the fitness value of the evaluated Defender's strategy π_D. There is a chance that the above fitness value is not the true expected Defender's payoff because of the lack of the (overall) optimal Attacker's response in the Attacker's population. However, the expected algorithm's behavior is to evolve such a strategy (optimal response) in the coevolution process in subsequent generations.

The evaluation procedure of the individuals from the Attacker's population is more complicated. Usually, there is no single optimal Attacker's response for all Defender's strategies. Depending on the particular Defender's commitment (Defender's mixed strategy), the best Attacker's response may change.

It is generally desired that the Attacker's population is composed of optimal responses for all possible Defender's strategies. Assigning the average Attacker's payoff against all strategies from the Defender's population (or part of it) as fitness value may be a weak approach because a given Attacker's strategy is usually strong (optimal) only against specific Defender's strategies. Such Attacker's strategy needs to be preserved but averaging the payoffs will decrease the fitness of such a strategy posing a risk of omitting it in the selection process.

Hence, a better idea is to use the maximum metric. However, in Defender's population (in order to preserve its diversity) there exist also some weaker strategies. For those strategies most of the Attacker's strategies will lead to high Attacker's payoff and such an approach wouldn't allow distinguishing good Attacker's strategies from the bad ones (because all of them will get high fitness as a maximum payoff against one of the weak Defender's strategies). This observation discredits calculating maximum payoff against all Defender's strategies. On the other extreme, the Attacker's fitness value could be computed only against the best strategy from the Defender's population, but this would lead to degeneration (premature convergence) of the Attacker's population. All Attacker's strategies would tend to be an optimal response for a particular Defender's strategy, becoming vulnerable to other strategies from the Defender's population.

Consequently, an intermediate option was implemented, i.e. the Attacker's strategy fitness is the maximum of Attacker's payoffs against the N_{top} highest fitted individuals from the Defender's population (N_{top} is CoEvoSG parameter).

Selection. The selection process decides which individuals from the current population will be promoted to the next generation. At the beginning, e individuals with the highest fitness value are unconditionally transferred to the next generation. They are called *elite* and preserve the best-fitted solutions. Then, a *binary tournament* is repeatedly executed until the next generation population is filled with N_A individuals. For each tournament, two individuals are sampled (with replacement) from the current population (including those affected by crossover and/or mutation). The higher fitted chromosome wins (and is promoted to the next generation) with probability p_s (so-called selection pressure parameter). Otherwise, the lower-fitted one is promoted.

Stop Condition. The algorithm ends when at least one of the following conditions is satisfied: (a) CoEvoSG attained the maximum number of l_g generations, (b) no improvement of the best-found solution (Defender's payoff) was observed in consecutive l_c generations. Only generations referring to the Defender's population are considered when verifying the above conditions.

5 Experimental Setup

5.1 Benchmark Games

CoEvoSG was tested on two popular SSG benchmarks: FlipIt and Warehouse Games, previously used for testing state-of-the-art methods, e.g. in [11,17,23].

FlipIt Games. FlipIt Games (FIGs) [22] reflect certain cybersecurity scenarios. The Attacker attempts to gain control over some elements of network infrastructure (e.g. computers, routers, mobile devices) and the Defender can take actions to regain control of the infected units.

FIGs are played on directed graphs with n vertices, for a fixed number of m time steps. In each time step, players simultaneously select one vertex which they want to take control of (to *flip* the node). At the beginning, only some subset of vertices (entry nodes) is available for the Attacker. This mimics the scenario in which some part of the network infrastructure is publicly accessible from the outside (e.g. Internet). The Attacker starts penetrating the network from one of those entry nodes. Taking control over the vertex (flip action) is successful only if the two following conditions are fulfilled: (1) the player controls at least one of predecessor vertices (unless it is an entry node), (2) the current owner of this vertex does not take the flip action on it in the same time step.

At the beginning of the game, all vertices are controlled by the Defender. Each node has assigned two values: a reward (> 0) for controlling it, and a cost (< 0) of taking a flip attempt. The final player's payoff is calculated by summing the rewards in all nodes controlled by that player after each time step and the costs of all flip attempts (either successful or not). Figure 2 presents a sample FIG scenario.

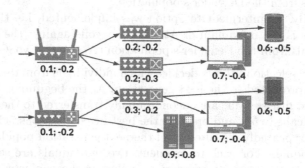

Fig. 2. Example FIG scenario with two entry nodes (routers) on the left. Numbers below each component denote a reward for controlling the node (left) and the cost of a flip attempt (right).

In the experiments, 280 FIG instances were generated randomly with the following parameters: $m \in \{3, 4, 5, 6, 8, 10, 15, 20\}$, $n \in \{5, 10, 15, 20, 25, 30, 40\}$. For each pair (m, n) 5 games were created with random payoffs (rewards drawn from $(0, 1)$, costs from $(-1, 0)$) and random graph structures (generated according to Watts-Strogatz model [25] with an average vertex degree $d_{avg} = 3$).

The experiments were performed in *No-Info* variant [2] which means that the players were not aware of whether their flip action succeeded or failed.

Warehouse Games. Warehouse Games (WHGs) [9] are inspired by real estate (warehouses or residential buildings) protection scenarios. The games are played on undirected graphs with n vertices, for m time steps. A subset of special vertices are called targets (T). Graph edges represent corridors and vertices symbolize rooms. At the beginning, the Defender and the Attacker are placed in the predetermined starting vertices. In each time step, each player's action consists in moving to one of the neighbor vertices (connected with an edge) or staying in the current vertex.

The game ends in one of the following circumstances: (a) both players are located in the same vertex v in the same time step - then, the Attacker is "caught" and the players are given payoffs associated with that vertex: $U_{D+}^v > 0$ (Defender) and $U_{A-}^v < 0$ (Attacker); (b) the Attacker reaches one of the targets $t \in T$ and is not caught (there is no Defender in this target) - in this case, the attack is successful and the players receive payoffs $U_{D-}^t < 0$ (Defender) and $U_{A+}^t > 0$ (Attacker); (c) none of above conditions is satisfied - both players receive a payoff of 0. Figure 3 presents a sample WHG scenario.

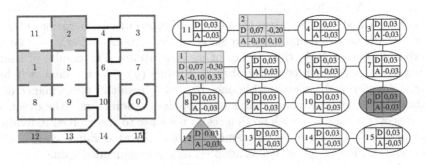

Fig. 3. An example WHG scenario: warehouse layout (left) and the corresponding graph (right) with payoffs of the players in the respective game outcomes. Green rectangular vertices are targets, a red triangle vertex and a blue circle vertex are the Attacker's and the Defender's starting points, respectively (Color figure online)

For CoEvoSG evaluation 240 WHG instances were generated with $m \subseteq \{3, 4, 5, 6, 8, 10, 15, 20\}$ and $n \in \{15, 20, 25, 30, 40, 50\}$ (5 games per each (m, n) pair). Players' payoffs were drawn from $[-1; 1]$. The number of targets depended on a graph size: $|T| = \lceil \frac{n}{5} \rceil$. Graphs were generated according to Watts-Strogatz random graphs model [25] with an average vertex degree $d_{avg} = 3$.

5.2 Parameterization

All common EASG and CoEvoSG parameters were set according to the recommendations proposed for EASG [28]. Namely, the Defender's population size $N_D = 200$, crossover probability $p_c = 0.8$, mutation probability $p_m = 0.5$, selection pressure $p_s = 0.9$, elite size $e = 2$, maximal number of generations $l_g = 1000$, maximal number of generations with no improvement $l_c = 20$. The parameters of evolutionary operators (mutation, crossover, selection) for the Attacker's population were assigned the same values as for the Defender's population, however, CoEvoSG requires several new parameters which need to be tuned. In order to find their recommended values, a set of parameter tuning experiments with 50 random games, different from the test WHG instances, were performed. FIGs have similar game structure and do not require separate parametrization.

The first tested parameter was the Attacker's population size (N_A). The following values were considered: $\{10, 20, 100, 200, 500, 1000, 2000, 5000\}$. The results (average Defender's payoff and computation time) are presented in Fig. 4a. Clearly, the bigger the Attacker's population size the better the results as the Defender's payoff is calculated more accurately. If the Attacker's population contained all possible Attacker's pure strategies, then the Defender's individuals evaluation would be an exact value (not an approximation) since the optimal Attacker's response would always be present in the Attacker's population. However, as stated previously one of the motivations for introducing coevolution is to speed up the Defender's strategies evaluation by checking them only against a representative subset of all Attacker's strategies. Thus, based on the presented results, $N_A = N_D = 200$ was set.

Another tested parameter was the number of consecutive generations for each player - g_p. Please recall that in CoEvoSG Defender's and Attacker's populations are evolved alternately in the batches of g_p generations. The results of tuning this parameter are presented in Fig. 4b. Small values ($g_p \leq 5$ - frequent switching between populations), as well as big ones ($g_p \geq 50$) result in performance deterioration. Infrequent switching makes one population dominant - the other one stagnates over a long time with no chances to response to the evolved individuals from the other population. At the same time, for all tested values computation time is similar. Hence, $g_p = 20$ was adopted as a recommended value.

The last tuned parameter was N_{top}, i.e. the number of the best individuals from the Defender's population involved in the Attacker's strategies evaluation. The result for $N_{top} \in \{1, 3, 5, 10, 20, 50, 100, 200\}$ presented in Fig. 4c confirm our previous conjecture formulated in in Sect. 4 about the harmfulness of using the whole Defender's population ($N_{top} = 200$). Also, small values of this parameter ($N_{top} < 5$) lead to weaker results due to the presence of some oscillations within the population. In the extreme case of $N_{top} = 1$ (evaluation of a given Attacker's strategy is based on the best Defender's strategy only), we observed that the Attacker's population quickly losses diversity/degenerates. Individuals in the population become similar to one another because they are optimized with respect to only one Defender's strategy. As a result the Attacker's population returns a good response only to this one specific Defender's strategy, and in the next coevolution phase the Defender's population is able to find with ease another strategy for which there is no good response in the Attacker's population. Afterwards, the whole Attacker's population again adapts to the new best Defender's strategy and "forgets" the previous ones. $N_{top} = 10$ appeared to be the best compromise between these two extremes (Fig. 4c).

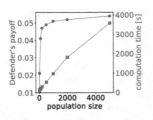 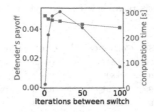 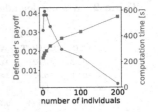

(a) Attacker's population size (N_A).

(b) Consecutive generations per population (g_p).

(c) Individuals to evaluate Attacker's strategy (N_{top}).

Fig. 4. Comparison of the average Defender's payoffs and computation times for CoEvoSG specific parameters: N_A, g_p, N_{top}.

Table 1. Average defender's payoffs with respect to the number of graph nodes.

	FIG					WHG			
n	C2016	O2UCT	EASG	CoEvoSG	n	C2016	O2UCT	EASG	CoEvoSG
5	0.890	0.887	0.886	0.886	15	0.052	0.051	0.051	0.050
10	0.854	0.851	0.847	0.845	20	0.054	0.053	0.052	0.050
15	0.811	0.807	0.802	0.798	25	0.048	0.046	0.045	0.043
20	-	0.784	0.780	0.772	30	-	0.044	0.042	0.039
25	-	-	0.754	0.746	40	-	-	0.040	0.036
30	-	-	-	0.730	50	-	-	-	0.029
40	-	-	-	0.722					

6   Results and Discussion

Payoffs. Tables 1 and 2 present average Defender's payoffs with respect to the number of graph nodes and time steps, resp., for the methods described in Sect. 3. Dashes mean that a particular algorithm was not able to compute some of the test instances within the limit of 100h per game. The results are averaged over 20 independent runs per game.

Presented outcomes show only minor differences between the evolutionary approach (EASG) and proposed coevolutionary algorithm (CoEvoSG). The average differences equal 0.0032 and 0.0020 for FIG and WHG instances, resp. Please note that EASG is a natural reference point for CoEvoSG since CoEvoSG approximates the Defender's payoff (in the evaluation procedure) while EASG computes it thoroughly. A relatively small difference in Defender's payoffs between the methods is a consequence of frequent existence (in over 84% of the cases) of the optimal Attacker's response in their population. The fitness function in such cases returns the same evaluation in both methods.

O2UCT slightly outperforms EASG and CoEvoSG but the differences are not statistically significant - p-values are 0.34 and 0.12, respectively (according to one-tailed t-test). For 23% of games, CoEvoSG returned better result than O2UCT whereas O2UCT was superior in 39% cases (for the remaining 38% of games the outcomes of both methods were the same).

Table 2. Average defender's payoffs with respect to the number of time steps.

FIG					WHG			
m	C2016	O2UCT	EASG	CoEvoSG	m	C2016	O2UCT	EASG	CoEvoSG
3	0.823	0.821	0.820	0.817	3	0.043	0.043	0.043	0.043
4	0.817	0.812	0.808	0.805	4	0.052	0.050	0.050	0.049
5	0.810	0.801	0.798	0.791	5	0.055	0.054	0.053	0.052
6	-	0.794	0.792	0.791	6	0.058	0.056	0.054	0.051
8	-	0.789	0.784	0.781	8	-	0.053	0.051	0.048
10	-	-	0.780	0.778	10	-	-	0.048	0.044
15	-	-	-	0.774	15	-	-	-	0.040
20	-	-	-	0.761	20	-	-	-	0.038

The exact MILP method (C2016) was able to solve 45 FIG and 60 WHG test instances within the allotted time. For these games, CoEvoSG returned the optimal strategy (a difference in Defender's payoff less than $\varepsilon = 0.0001$) in 29/45 (64%) and 38/60 (68%) cases, resp. The average differences between optimal results and CoEvoSG outcomes equaled 0.0137 (FIGs) and 0.0023 (WHGs).

Overall, CoEvoSG was able to solve much bigger games than any of the competitive methods, while returning only slightly weaker Defender's payoffs (whenever comparable).

Computation Scalability. Figure 5 illustrates computation time of tested methods with respect to the number of graph nodes and time steps. In all cases, the advantage of CoEvoSG is clear. The method preserves near-constant computation time irrespective of game size, while other methods scale approximately linearly (O2UCT and EASG) or exponentially (C2016). Computational complexity of CoEvoSG is approximately constant with respect to the graph size or the number of steps because the algorithm maintains the Defender's and the Attacker's populations of fixed size, independently of other game parameters.

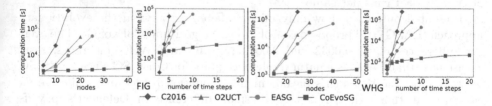

Fig. 5. Comparison of computation time (logarithmic scale) with respect to number of graph nodes and time steps for FIG (left) and WHG (right) games.

In summary, presented results demonstrate that despite slightly worse average Defender's payoffs the proposed coevolutionary approach, thanks to excellent time scalability, offers a viable alternative to both exact and approximate state-of-the-art methods, especially in the case of larger games which are beyond the capacity of the existing algorithms.

7   Conclusions

The paper proposes a novel coevolutionary algorithm for solving sequential Stackelberg Security Games. The method develops two competing populations of players' strategies by specially designed evolutionary operators.

Experimental evaluation performed on two well-established game types with more than 500 test instances have proven the efficacy of the proposed method - in the majority of test cases optimal solutions were found. The results are is on par with other approximate methods - O2UCT and EASG. However, the true strength of CoEvoSG lies in its time efficiency. It scales visibly better than all state-of-the-art methods and stands out with near-constant computation time irrespective of the game size. Thanks to this property CoEvoSG can be employed to solve arbitrarily large games which are beyond the capacity of the methods proposed hitherto. Moreover, the method is generic and can be easily adapted to other genres of Security Games. What's more, CoEvoSG is an *anytime* algorithm, i.e. is capable of returning a valid solution at any time of the execution process.

CoEvoSG can be directly applied to various real-life cybersecurity problems modelled by FlipIt Games, such as password reset policies, cloud auditing, or supervisory control and data acquisition in industrial internet of things [15].

Our future plans concentrate on extending CoEvoSG to games with multiple Defenders and/or Attackers [16] with the corresponding increase of the number of populations.

Acknowledgement. The project was funded by POB Research Centre Cybersecurity and Data Science of Warsaw University of Technology within the Excellence Initiative Program - Research University (ID-UB).

References

1. Breton, M., Alj, A., Haurie, A.: Sequential stackelberg equilibria in two-person games. J. Optim. Theor. Appl. **59**(1), 71–97 (1988)
2. Černý, J., Bošanský, B., Kiekintveld, C.: Incremental strategy generation for stackelberg equilibria in extensive-form games. In: Proceedings of the 19th ACM Conference on Economics and Computation, pp. 151–168 (2018)
3. Conitzer, V., Sandholm, T.: Computing the optimal strategy to commit to. In: Proceedings of the 7th ACM Conference on Electronic Commerce, pp. 82–90 (2006)
4. Fang, F., et al.: Deploying paws: field optimization of the protection assistant for wildlife security. In: Proceedings of the 28th Innovative Applications of Artificial Intelligence Conference (2016)
5. Gąsior, J., Seredyński, F.: Security-aware distributed job scheduling in cloud computing systems: a game-theoretic cellular automata-based approach. In: International Conference on Computational Science, pp. 449–462 (2019)
6. Guleva, V.Y.: Estimation of tipping points for critical and transitional regimes in the evolution of complex interbank network. In: International Conference on Computational Science, pp. 432–444 (2020)
7. Guo, Y., Zhang, H., Zhang, L., Fang, L., Li, F.: Incentive mechanism for cooperative intrusion detection: an evolutionary game approach. In: International Conference on Computational Science, pp. 83–97 (2018)

8. Jain, M., et al.: Software assistants for randomized patrol planning for the lax airport police and the federal air marshal service. Interfaces **40**(4), 267–290 (2010)
9. Karwowski, J., Mańdziuk, J.: A monte carlo tree search approach to finding efficient patrolling schemes on graphs. Eur. J. Oper. Res. **277**, 255–268 (2019)
10. Karwowski, J., Mańdziuk, J.: Stackelberg equilibrium approximation in general-sum extensive-form games with double-oracle sampling method. In: Proceedings of the 18th AAMAS Conference, pp. 2045–2047 (2019)
11. Karwowski, J., Mańdziuk, J.: Double-oracle sampling method for stackelberg equilibrium approximation in general-sum extensive-form games. In: Proceedings of the 34th AAAI Conference, vol. 34, pp. 2054–2061 (2020)
12. Karwowski, J., Mańdziuk, J., Żychowski, A., Grajek, F., An, B.: A memetic approach for sequential security games on a plane with moving targets. In: Proceedings of the 33rd AAAI Conference, vol. 33, pp. 970–977 (2019)
13. Kocsis, L., Szepesvári, C.: Bandit based monte-carlo planning. In: Fürnkranz, J., Scheffer, T., Spiliopoulou, M. (eds.) ECML 2006. LNCS (LNAI), vol. 4212, pp. 282–293. Springer, Heidelberg (2006). https://doi.org/10.1007/11871842_29
14. Lezzi, M., Lazoi, M., Corallo, A.: Cybersecurity for industry 4.0 in the current literature: a reference framework. Computers in Industry **103**, 97–110 (2018)
15. Liu, Z., Wang, L.: FlipIt Game model-based defense strategy against cyberattacks on SCADA systems considering insider assistance. IEEE Trans. Inf. Forensics Secur. **16**, 2791–2804 (2021)
16. Lou, J., Smith, A.M., Vorobeychik, Y.: Multidefender security games. IEEE Intell. Syst. **32**, 50–60 (2017)
17. Oakley, L., Oprea, A.: QFlip: an adaptive reinforcement learning strategy for the FlipIt security game. In: Alpcan, T., Vorobeychik, Y., Baras, J.S., Dán, G. (eds.) GameSec 2019. LNCS, vol. 11836, pp. 364–384. Springer, Cham (2019). https://doi.org/10.1007/978-3-030-32430-8_22
18. Shieh, E., et al.: Protect: a deployed game theoretic system to protect the ports of the united states. In: Proceedings of the 11th AAMAS Conference, pp. 13–20 (2012)
19. Sinha, A., Fang, F., An, B., Kiekintveld, C., Tambe, M.: Stackelberg security games: looking beyond a decade of success. In: Proceedings of the 27th IJCAI Conference, pp. 5494–5501 (2018)
20. Sinha, A., Nguyen, T.H., Kar, D., Brown, M., Tambe, M., Jiang, A.X.: From physical security to cybersecurity. J. Cybersecurity **1**(1), 19–35 (2015)
21. Świechowski, M., Godlewski, K., Sawicki, B., Mańdziuk, J.: Monte carlo tree search: a review of recent modifications and applications (2021). https://arxiv.org/abs/2103.04931
22. Van Dijk, M., Juels, A., Oprea, A., Rivest, R.L.: Flipit: the game of "stealthy takeover". J. Cryptology **26**(4), 655–713 (2013)
23. Čermák, J., Bošanský, B., Durkota, K., Lisý, V., Kiekintveld, C.: Using correlated strategies for computing stackelberg equilibria in extensive-form games. In: Proceedings of the 30th AAAI Conference, pp. 439–445 (2016)
24. Von Stengel, B., Zamir, S.: Leadership with commitment to mixed strategies. Technical Report, CDAM Research Report (2004)
25. Watts, D.J., Strogatz, S.H.: Collective dynamics of 'small-world' networks. Nature **393**(6684), 440–442 (1998)
26. Zhang, Y., Malacaria, P.: Bayesian stackelberg games for cyber-security decision support. Decis. Support Syst. **148**, 113599 (2021)
27. Żychowski, A., Mańdziuk, J.: A generic metaheuristic approach to sequential security games. In: Proceedings of the 19th AAMAS, pp. 2089–2091 (2020)

28. Żychowski, A., Mańdziuk, J.: Evolution of strategies in sequential security games. In: Proceedings of the 20th AAMAS Conference, pp. 1434–1442 (2021)
29. Żychowski, A., Mańdziuk, J.: Learning attacker's bounded rationality model in security games. In: Proceedings of the 28th ICONIP, vol. CCIS 1516, pp. 530–539 (2021)

FINCH: Domain Specific Language and Code Generation for Finite Element and Finite Volume in Julia

Eric Heisler$^{(\boxtimes)}$, Aadesh Deshmukh, and Hari Sundar

School of Computing, University of Utah, Salt Lake City, UT, USA
eric.heisler@utah.edu, u1369232@utah.edu, hari.sundar@utah.edu

Abstract. We introduce FINCH, a Julia-based domain specific language (DSL) for solving partial differential equations in a discretization agnostic way, currently including finite element and finite volume methods. A key focus is code generation for various internal or external software targets. Internal targets use a modular set of tools in Julia providing a direct solution within the framework. In contrast, external code generation produces a set of code files to be compiled and run with external libraries or frameworks. Examples include a `matlab` target, for smaller problems or prototyping, or `C++/MPI` based targets for larger problems needing scalability. This allows us to take advantage of their capabilities without needlessly duplicating them, and provides options tailored to the needs of the domain scientist. The modular design of FINCH allows ongoing development of these target modules resulting in a more extensible framework and a broader set of applications. The support for multiple discretizations, including finite element and finite volume methods, also contributes to this goal. Another focus of this project is complex systems containing a large set of coupled PDEs that could be challenging to efficiently code and optimize by hand, but that are relatively simple to specify using the DSL. In this paper we present the key features of FINCH that set it apart from many other DSL options, and demonstrate the basic usage and current capabilities through examples.

Keywords: Domain specific language · Code generation · Finite element method · Finite volume method · Parallel computing · Julia

1 Introduction

Solving partial differential equations (PDEs) numerically on a large scale involves a compromise between highly optimized code exploiting details of the problem or hardware, and extensible code that can be easily adapted to variations. Rapidly evolving technology and a shift to heterogeneous systems places a higher value on the latter, prompting a move away from hand-written code made by experts in high performance computing, to generated code produced through a high-level domain specific language (DSL). Another motivating factor is the realm of

D. Groen et al. (Eds.): ICCS 2022, LNCS 13350, pp. 118–132, 2022.
https://doi.org/10.1007/978-3-031-08751-6_9

medium-scale problems where good performance is needed, but the cost of developing optimal code may not be justified. At this scale it is up to domain scientists to develop their own software or piece it together from more general-purpose libraries. Finally, the choice of discretization method, like finite element(FE) or finite volume(FV), is significant in multiphysics systems where different aspects of the system are better handled by different methods.

In response, numerous DSLs for solving PDEs have been developed. On one end of the spectrum are high-level options such as Matlab Toolboxes and Comsol. They are general-purpose and don't require a high level of programming skill. As a trade-off, they lack customizability. The low-level code is often, by design, hidden from the user and difficult to modify.

At the opposite end are lower-level libraries such as Nektar++ [4] and deal.II [3] providing customizable components optimized for a specific purpose. They require more programming input and skill from the user. This also makes it harder to modify the code for variations, resulting in many of the limitations of hand-written code.

This work aims for a middle-ground, where most of the programming input is handled within the scope of a moderately high-level DSL while allowing low-level customization and in some cases direct code modification. Some options in this realm include Fenics [2] and Firedrake [25] for finite element methods, Open-FOAM [10] for finite volume methods, Devito [20] for finite difference methods, and many others focused on a specific type of problem or technique. There are also tools in Julia including DifferentialEquations.jl [24] which provides a broad environment of ordinary differential equation solvers with a Julia interface.

This work introduces FINCH, a DSL for solving PDEs. The framework aims to be discretization agnostic, and currently supports finite element and finite volume methods. The goal is to enable a domain scientist to create efficient code for problems ranging from small scale simulations on a laptop computer, to larger systems requiring scalability on modern supercomputers. Two key ideas to achieving this goal are a modular software design and generation for external software frameworks.

Rather than depending on a single, general-purpose code, a set of modules are used to grant the flexibility to adapt to problem requirements or resources. Some examples include various discretization methods such as FE, both CG and DG variants, and FV, as well as numerical tools such as PETSc's linear solvers, GPU based options, or matrix-free methods. The development of new modules opens up possibilities for optimization and new types of problems

Another strategy is the generation of code for various external software targets. This allows it to leverage the capabilities of existing software frameworks that are well suited to a type of problem. For example, the DENDRO library [8,9,26] provides an adaptive octree framework that is suitable for very large scale problems using distributed memory parallel techniques. Manually writing code for this framework requires high programming proficiency and familiarity with the software. FINCH provides a simpler interface to this resource while presenting the generated code to the user for modification or inspection. Another target is C++ using the AMAT [27] library which handles the mesh and data structure creation in Julia then utilizes a library of efficient parallel sparse matrix operations to compute the solution in an independent C++ program. The diversity of code generation targets allows constructing a set of tools suiting a user's needs.

FINCH is written completely in Julia, which is easy to use and has speed comparable to low-level languages such as C [16]. Julia is growing in popularity as a serious scientific computing language. It allows a simplified, intuitive interface without resorting to external C/C++/Fortran libraries as is common with Python-based DSLs. The metaprogramming features and wide selection of libraries also make Julia a convenient choice for FINCH.

2   Related Work

DSLs can be found in some form for countless mathematical and computational tasks. Some examples with a similar purpose and interface include the Unified Form Language(UFL) [1] and FreeFEM [12] used to write variational forms of PDEs. Components corresponding to test functions, trial functions, and other values are combined in expressions representing volume or facet integrals of elements. Since FINCH was originally developed for FE, a similar design was chosen. The internal representation involves categorizing terms of the expression depending on type of integral and linear vs. bilinear forms. The Julia-based FE DSL MetaFEM [29] also involves writing a variational form expression, though with a very different grammar.

In contrast, FINCH is designed to accommodate more general types of expressions and does not assume a variational form. It also allows custom operator definitions that act on the symbolic tensor arrays of entities in the expression. For example, when using a FV method, specialized flux operators can be defined and included in the PDE expression.

A relevant FV DSL is used by OpenFOAM [21], which again involves components such as variables and coefficients in an expression resembling the mathematical notation. This works with a predefined set of operations and is designed specifically for types of problems that commonly use FV methods. There is no notion of variational forms.

It is worth noting some modules of Dune [6], such as Dune-fem are designed for both FE and FV methods, but these are low-level interfaces that are difficult to compare to the higher-level DSLs described here.

The other aspect is code generation where the internal representation becomes numerical code. There are many code generation techniques for FE. Some exploit tensor product construction for high order FE [15,22,28]. Others use the independent nature of Discontinuous Galerkin methods to utilize GPUs [5] or vectorization [17]. The FE software FEniCS utilizes the set of tools FFC [18] and Dolfin [19]. There are also options for FV [23] and FD [21], though perhaps less common than for FE.

The code generation modules used by FINCH are specific to their target, and employ a variety of techniques accordingly. The modular design allows selection of ideal techniques either by the user or automatically depending on the target software, hardware, or problem details.

3   Domain Specific Language

The goal of a mathematical DSL is to provide an interface that closely resembles the notation used by domain scientists while reducing extraneous details and syntax needed by the underlying programming language. Many DSLs accomplish this in an object-oriented way by creating classes representing mathematical objects with a set of intuitive operations. We have adopted a similar strategy in which the basic components of the equations, such as unknown variables, coefficients and test functions, are very basic objects that include an array of symbolic components. For example, a 3-dimensional vector quantity u would correspond to the array $[u_1, u_2, u_3]$. Common arithmetic and differential operations are defined for these objects, and users can define their own custom operators that act on these symbolic arrays. It is also possible to use these basic operations to build packages of specialized operators for a class of problems.

As an example, the following code creates a vector-valued unknown variable u, a known scalar coefficient k defined by a function of coordinates (x, y, z, t), and a vector test function v which belongs to the same function space as u.

```
u = variable("u", type=VECTOR)
coefficient("k", "sin(pi*x)*y*z")
testSymbol("v", type=VECTOR)
```

The differential equations are written in terms of these objects. When using FE, this is done by writing the weak form of the equation in residual form. Note that integration over the volume is implied, and surface integrals can be specified by wrapping those terms in `surf(...)`. Here is an example of specifying a Poisson equation.

$$\text{Original PDE} \left| \nabla \cdot (a\nabla u) - f = 0 \right.$$
$$\text{Weak form} \left| -(a\nabla u, \nabla v) - (f, v) = 0 \right.$$
$$\text{FINCH input} \left| \texttt{-a*dot(grad(u),grad(v)) - f*v} \right.$$

When using FV, it is assumed that the equations are in a conservation form. The source and flux terms are given as input, and the time derivative of the variables is implied as shown in the following advection-reaction equation

$$\text{PDE} \left| \int_V \frac{du}{dt} dx = \int_V g(u, x) dx - \int_{\partial V} \mathbf{f}(u, x) \cdot \mathbf{n} ds \right.$$
$$\text{Source} \left| g(u, x) = ku \right.$$
$$\text{Flux} \left| \mathbf{f}(u, x) = u\mathbf{b} \right.$$
$$\text{FINCH input} \left| \texttt{source(k*u)} \right.$$
$$\left| \texttt{flux(u*b)} \right.$$

In addition to the standard operators such as *, −, dot and grad used above, a user can define new operators to put in these expressions. For example, the flux shown will result in a central flux approximation. To use a custom flux, one could define the operator `myFlux(u,b)`, and substitute that for `u*b` in the expression above. Note that this definition could be either a symbolic manipulation of the array for $u * b$ or a numerical callback function when needed.

3.1 User Input

Typically a Julia script will be written for a particular problem, but it is also possible to work interactively. A set of functions or macros are used to a) Set up the configuration, b) Specify the mesh, entities and equations, and c) Process data for output. A variety of example scripts are in the repository [13].

As an example, the following commands will configure a 2D unstructured grid using a fourth-order polynomial function space based on Lobatto-Gauss nodes, and generate code for a target specified in the `external_target_module.jl` file.

```
generateFor("external_target_module.jl")
domain(2, grid=UNSTRUCTURED)
functionSpace(space=LEGENDRE, order=4)
nodeType(LOBATTO)
```

In contrast to this, a user who is content with the defaults could provide as little as `domain(2)`.

Problem specification should start with a mesh. There are some simple mesh generation options built in. For example, to construct a uniform 50×20 grid of quadrilateral elements in a unit square domain with a separate boundary ID for each face, use the command: `mesh(QUADMESH,elsperdim=[50,20],bids=4)`

For more practical problems, external mesh generating software can be used to create a mesh file that is then imported into FINCH. Currently the GMSH(.msh) and MEDIT(.mesh) formats are supported.

Separating boundary regions for additional boundary conditions is done with the command `addBoundaryID(BID, onBdry)` where `BID` is a number to be assigned to that region, and `onBdry` is a function or expression of (x, y, z) that is true within the desired region.

For distributed memory parallelism it is necessary to partition the mesh. This is done internally using METIS via the Julia library METIS_jll. This will be done automatically according to the number of processes available through MPI, but can be configured as desired.

After setting up the scenario, entities such as variables and coefficients are defined and expressions for the equations are input as described above. Using the command `solve(u)` will then either generate the code files for external targets or run the internal solver to produce a solution. Considering the internal route, the solution will now be found in the `u.values` array, and is available for post-processing, visualizing, or output in a number of formats such as binary data or VTK files.

Indexed Entities - Some problems involve a set of several quantities that share the same type of equation with different parameters. Similarly, one may try to solve an equation over a range of parameters. In these cases indexed variables and coefficients greatly simplify the way the problem is specified and present an opportunity to reorganize the code in a more optimal way. As an example, consider a set of unknown quantities $u_{i,j}$ and a corresponding set of coefficients k_i belonging to the same type of diffusion equation. For brevity the dependence on j

is omitted, but could correspond, for example, to different boundary conditions.

$$\frac{d}{dt}u_{i,j} = k_i \Delta u_{i,j} \quad i = 1...20 \, , \, j = 1...40$$

It is cumbersome to individually write out the equations if there are many values of i and j. Rather, we can write one equation using indexed entities.

```
I  = index("I", range=[1,20])
J  = index("J", range=[1,40])
u  = variable("u", type=VAR_ARRAY, location=NODAL, index = [I,J])
k  = coefficient("k",k_array,type=VAR_ARRAY,location=NODAL,index=I)
weakForm(u, "Dt(u[I,J]*v) + k[I]*dot(grad(u[I,J]),grad(v))")
assemblyLoops(u, [I, J, "elements"])
```

The last line describing assembly loops instructs the code generator to nest the assembly loops in this order. In some cases it may be more efficient to parallelize an outer index loop before the elemental loop. The user can arrange this as desired.

3.2 Symbolic Representation

After entering the expressions for the equations, they are transformed into an intermediate symbolic representation. The entity symbols are replaced with arrays of corresponding tensor components, as discussed above, and the operators are applied to ultimately create a set of symbolic expressions. These expressions go through processing stages to separate known and unknown terms, simplify them, and identify time dependent terms. The resulting symbolic terms are in the form of computational graphs, based on Julia Expr trees, containing symbolic entity objects. These graphs are what is eventually passed to the code generation utilities.

This simple chart illustrates the process using the weak form input for a 2D Poisson equation. The input expression starts at the top, symbols are substituted, operators are applied, the terms are partitioned into groups and computational graphs are built with symbolic entities.

```
-a*dot(grad(u), grad(v)) - f*v
                  ↓
-[_a_1]*dot_op(grad_op([_u_1]), grad_op([_v_1]))-[_f_1]*[_v_1]
                  ↓
[-(_a_1*D_1__u_1*D_1__v_1 + _a_1*D_2__u_1*D_2__v_1)]+[-_f_1*_v_1]
                  ↓
    bilinear: [-_a_1*D_1__u_1*D_1__v_1 - _a_1*D_2__u_1*D_2__v_1]
    linear:   [-_f_1 * _v_1]
```

entities: $D\_1\_\_u\_1 = \frac{d}{dx}u_1$, $D\_2\_\_u\_1 = \frac{d}{dy}u_1$, etc.

The entity is essentially a symbol, like _u_, along with it's component index on the right, 1, and a collection of flags on the left, D_1_. The flags can have any value and will be interpreted by the relevant code generation module. For example, the flags CELL1_ and CELL2_ would be interpreted as values on respective sides of a face in a finite volume context.

4   Code Generation

The code generation step is where the process diverges. The details are specific to the generation target, but they essentially all perform the same two tasks. They must interpret the computational graph containing symbolic entities described above, generating their mathematical equivalent, and they must collect these calculations in a functional piece of code that performs the overall computation.

When designing a new target module, there are only three functions that must be provided. The first one, `get_external_language_elements`, provides basic language-specific info such as comment characters to aid with formatting. The second is `generate_external_code_layer`, which interprets the computational graph of the symbolic representation and generates code to perform the elemental calculations. The third function, `generate_external_files`, is responsible for creating all of the code files. It takes the elemental calculation from the second function and wraps it in the rest of the code to create a complete program including build files and instructions.

4.1   Elemental Computation

The elemental computation varies significantly with different targets, but to illustrate the process the 2D Poisson example from above will used as input. This type of problem will essentially need code for assembling elemental matrices and vectors embedded in an elemental loop. The elemental matrix will correspond to the bilinear terms,

```
-_a_1 * D_1__u_1 * D_1__v_1 - _a_1 * D_2__u_1 * D_2__v_1
```

Note that this symbolizes

$$\int_K \left(-a\frac{\partial u}{\partial x}\frac{\partial v}{\partial x} - a\frac{\partial u}{\partial y}\frac{\partial v}{\partial y} \right) dK$$

When discretized into polynomial basis functions at Gaussian integration points, this becomes

$$A_{jk}u_j = \sum_j u_j \sum_i w_i J_i \left(-a_i * \phi_{ij,x}\phi_{ik,x} - a_i * \phi_{ij,y}\phi_{ik,y} \right)$$

Where w_i are quadrature weights, J_i are geometric factors, $\phi_{ij,x}$ are x-derivatives of the jth basis functions at the ith quadrature points. The inner i sum can be arranged as a matrix expression.

$$Q_x^T W Q_x + Q_y^T W Q_y$$

With W being a diagonal matrix combining weights, geometric factors, and a_i for each quadrature point. Q_x combines geometric factors with precomputed matrices Q_R that essentially contain the basis function derivatives, $\frac{\partial \phi}{\partial R}$, at the quadrature points in a reference element, but in practice it will be more sophisticated as it will include a transformation from a nodal basis into a modal one to benefit from better properties. For details on this, please refer to [14].

When the code generator encounters a term like
`_a_1 * D_1__u_1 * D_1__v_1` it will recognize the three factors as coefficient, unknown, and test function respectively and make the associations
$$D\_1\_\_v\_1 \rightarrow Q_x^T$$
$$D\_1\_\_u\_1 \rightarrow Q_x$$
$$\_a\_1 \rightarrow a_i$$
and create code to perform those matrix operations. The way this calculation is implemented is up to the code generator, which provides opportunities for optimization. For example, when using uniform elements the Jacobian matrix only needs to be computed for one element and Q_x can be fully precomputed. Taking it one step further, if the coefficients in this term are also constant, the entire $Q_x^T W Q_x$ matrix can be precomputed.

Another opportunity for optimization depends on element type. For example, the DENDRO target exclusively uses hexahedral elements and exploits their symmetry by using the tensor product of one-dimensional operators. This saves on both arithmetic and memory costs.

When designing a new target or when taking advantage of some new hardware, these elemental calculations can be optimized in a modular way that makes the transition easy.

4.2   Global Computation

After handling the elemental computation, the next task is to combine these results into a global system. This is mainly where parallel strategies come into play. Since this typically involves looping over elements to assemble and solve a global linear system, the process can be parallelized using multithreading, distributed memory multiprocessing, and GPU techniques. Again, the details of this task may look completely different depending on the target and in many cases it is handled by the external software framework.

Note that when using FV the mathematics will be substantially different, but the overall structure of the computation is similar.

4.3   Modifying Generated Code

Advanced users may wish to inspect the generated code and make modifications by hand. In many cases there may be features of the problem that can be exploited for better performance that are not automatically included. For this purpose the elemental assembly code can be exported to a code file, modified as desired, and imported again to either run the calculation or generate the full code package for external targets. The commands for this are `exportCode` and `importCode`. Naturally, exporting should happen after the equations have been entered, and importing is done on a later run before solving.

Note that this code typically only contains the elemental assembly function. For even more control it is possible to also export and import the full assembly loop code for internal targets, but a good familiarity with the FINCH data structures is needed to take advantage of this. Similarly, since external targets are fully accessible as code files they can be modified as desired depending on the user's knowledge of the target software.

5   Performance Opportunities

Since one of the goals of FINCH is to take advantage of the capabilities of specialized external software, there are various strategies for parallelization, adaptivity, and efficient data structures available to achieve high performance. One example of this is the DENDRO target which offers distributed memory parallelism through MPI, adaptive mesh refinement, and proven large-scale scalability. It is ideal for problems that can benefit from very fine grained adaptive meshing, but is limiting in the possible domain geometry.

Another target is AMAT which is essentially a specialized linear algebra library providing very efficient algorithms for sparse linear systems. It also supports an assortment of parallelization strategies based on MPI, OpenMP, and GPU options.

The performance of both of these targets is explored in the Demonstrations section below.

5.1   Performance Within FINCH

The performance focus is not limited to external tools. The internal Julia targets can also make use of distributed and shared memory parallelism as well as efficient data organization options. When solving linear systems, the user can select a variety of tools beyond the defaults provided by Julia's LinearAlgebra package.

The simplest way to take advantage of these tools is with multithreading. FINCH automatically detects how many threads are available to the Julia instance and uses the native Julia package Threads to take advantage of this throughout the computation. To enable this feature a user simply needs to specify the number of threads when launching Julia. This is done with the argument -t n or --threads n to use n threads, or substitute auto in place of n to use the number of local CPU threads.

Distributed memory parallelism is provided by the Julia package MPI.jl which makes use of the system's available MPI implementation. Again, this is specified at launch using the system's MPI execution command. FINCH will detect how many processes are available and arrange the computation accordingly.

Partitioning is needed when using a distributed parallel strategy, and the most straightforward method is to partition the mesh evenly among the processes. This is accomplished using Metis through the METIS_jll package.

When using FV with higher order flux reconstructions several neighboring elements may be needed. This could potentially complicate the partitioning process because an irregular number of ghost elements would need to be maintained. To address this issue, partitioning is done on the initial course mesh with only nearest neighbor ghosts. Then the elements are refined in a consistent way depending on the flux order desired. The resulting finer mesh will include the needed number of ghosts in an efficient and reliable way. The refinement should be considered when planning a mesh utilizing this feature.

There are several choices when it comes to solving large, sparse linear systems. The default in FINCH is provided by the LinearAlgebra package which utilizes BLAS and LAPACK. Another option is PETSc, interfaced through PETSC.jl, which provides better performance in distributed parallel environments as demonstrated below. There is also a matrix-free option for certain targets that is particularly useful for large-scale problems where the cost of assembling a global matrix is prohibitive.

5.2 Cache Optimization

In addition to parallel techniques, the organization of data structures and the elemental loop ordering can improve performance through more efficient cache use. A number of data organization options are available in FINCHFor example, a mesh from the built-in mesh generation utility provides elements that are ordered lexicographically. In order to improve spatial locality, the elements can be rearranged either into a space-filling curve, such as a Hilbert or Morton curve, or into tiles.

To aid with this development and potentially provide a means for automated tuning, FINCH employs a cache simulator. Pycachesim [11] was chosen for this because it is light-weight and although it was developed for use in Python, the backend is written in C. The C library can be utilized directly by FINCH to roughly characterize the cache performance of a particular problem setup on a specified cache hierarchy.

The cache simulator is essentially another target for code generation. Rather than performing the mathematical computation, the approximate sequence of memory accesses is fed into the simulator. At the end of the computation the cache statistics are recorded and analyzed. This presents FINCH with a tool for tuning and measuring the effectiveness of changes in configuration.

6 Demonstration

The following example applications demonstrate some of the capabilities of FINCH and illustrate the performance aspects of the various tools and code generation targets. Since external targets rely on the performance capabilities of the target framework, please refer to their respective documentation for a more rigorous analysis. For example, the DENDRO framework has shown competitive scalability for large scale simulations [7,8].

6.1 Steady-State Advection-Diffusion-Reaction Equation

The following Eq. (1) is used to demonstrate a FE problem. We use several different sets of tools and compare them in terms of performance.

$$\nabla \cdot (D\nabla u) + \mathbf{s} \cdot \nabla u - cu = f \tag{1}$$

$$u(x \in \partial\Omega) = 0$$

$$\Omega = [0,1]^3, D = 1.1, c = 0.1, \mathbf{s} = (0.1, 0.1, 0.1)$$

Here all of the coefficients are given constant value, but we have intentionally generated them as functions of (x, y, z) to increase computational complexity. The motivation for this is to demonstrate the performance for a more practical problem while simplifying analysis with an exact solution. The function f was constructed such that u satisfies (2).

$$u(x, y, z) = \sin(3\pi x)\sin(2\pi y)\sin(\pi z) \tag{2}$$

The weak form expression provided to FINCH is

```
weakForm(u,
    "-D*dot(grad(u), grad(v)) + dot(s, grad(u))*v - c*u*v - f*v")
```

The discretization is continuous Galerkin with quadratic hexahedral elements.

Internal Target. With appropriate choice of mesh this is suitable for running on a typical computer, but for these tests we are using the Frontera supercomputer with dual socket Intel Xeon Platinum 8280 nodes having 56 cores. The execution time of different code generation targets and linear solvers were compared for a range of processor counts as shown in Fig. 1. For smaller problems running on only a few cores, the default Julia tools are an easy and viable option, though PETSc may be more efficient. The default method does not scale well in a distributed memory parallel context. For larger problems and many processors, PETSc and matrix-free are both good options. The figure shows that the matrix-free method is better when many processors are available. On the other hand, PETSc performed better for small process counts.

AMAT Target. The same problem was solved using the AMAT target. Code files, partitioned mesh data, reference element, and geometric factors were set up and exported from Julia. The code was compiled and run with the AMAT library using the precomputed data. AMAT provides options for assembling and solving the system including a direct PETSc solve or a hybrid matrix technique. MPI, OpenMP, and GPU tools are available. Figure 1(bottom left) compares the PETSc and hybrid versions based on MPI with the same hardware as above.

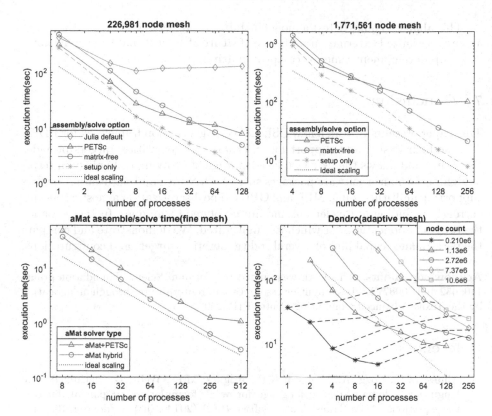

Fig. 1. *Top row:* internal target execution time on a coarse(*top left*) and fine mesh(*top right*) with different options. "Setup only" excludes assembly/solve time. *Bottom left:* AMAT's time on the fine mesh using PETSc and hybrid methods. Only assembly and solve time is included as setup is done separately within FINCH. *Bottom right:* Dendro's execution time for several mesh sizes. Black dashed lines show interpolated weak scaling contours. The blue dotted line is an ideal scaling based on the blue curve. (Color figure online)

Note that this execution time does not include the mesh creation and other setup. A comprehensive total would also include the compilation and file management time when using an external target, but they are omitted here.

DENDRO Target. Finally, the same problem was solved using the DENDRO target. Code files were generated in Julia then compiled using the DENDRO library. The resulting adaptively refined mesh produced by DENDRO contained between 0.21 million and 10.6 million nodes depending on input parameters. It was tested on the Notchpeak cluster at the University of Utah using two-socket Intel XeonSP Skylake nodes with 32 cores each. Figure 1(bottom right) demonstrates that the computation scales well for this problem on a fine mesh up to 256 processes.

This ability to quickly test a model in Julia before seamlessly transitioning to a more specialized external target is a key feature of FINCH that can significantly speed up development time for complex multiphysics problems.

7   Conclusion

This paper presents FINCH, a new DSL and code generation framework for PDEs. The modular design and support for external code generation targets provides versatility and allows the user to take advantage of evolving technology in terms of high performance software packages such as DENDRO, and hardware resources supporting multithreading, MPI, and GPUs. The discretization agnostic concept, currently including finite element and finite volume techniques, further expands the range of applications for which it is well suited. We demonstrate and compare the performance capability of several code generation targets and configurations.

Acknowledgements. This work was funded by National Science Foundation grants 1808652 and 2008772. The computing resources on Frontera were through an allocation by the Texas Advanced Computing Center PHY20033.

References

1. Alnæs, M.S., Logg, A., Ølgaard, K.B., Rognes, M.E., Wells, G.N.: Unified form language: a domain-specific language for weak formulations of partial differential equations. ACM Trans. Math. Softw. **40**(2) (2014). https://doi.org/10.1145/2566630
2. Alnaes, M.S., et al.: The fenics project version 1.5. Arch. Numer. Softw. **3**(100), 9–23 (2015). https://doi.org/10.11588/ans.2015.100.20553
3. Arndt, D., et al.: The deal.II library, version 9.2. J. Numer. Math. **28**(3), 131–146 (2020). https://doi.org/10.1515/jnma-2020-0043, https://dealii.org/deal92-preprint.pdf
4. Cantwell, C., et al.: Nektar++: an open-source spectral/HP element framework. Comput. Phys. Commun. **192**, 205–219 (2015). https://doi.org/10.1016/j.cpc.2015.02.008
5. Dorozhinskii, R., Bader, M.: Seissol on distributed multi-GPU systems: Cuda code generation for the modal discontinuous galerkin method. In: The International Conference on High Performance Computing in Asia-Pacific Region, pp. 69–82. HPC Asia 2021, ACM Press, New York (2021). https://doi.org/10.1145/3432261.3436753
6. Dune: Dune (2022). https://www.dune-project.org
7. Fernando, M., Neilsen, D., Lim, H., Hirschmann, E., Sundar, H.: Massively parallel simulations of binary black hole intermediate-mass-ratio inspirals. SIAM J. Sci. Comput. **42**(2), 97–138 (2019). https://doi.org/10.1137/18M1196972
8. Fernando, M., Neilsen, D., Sundar, H.: A scalable framework for adaptive computational general relativity on heterogeneous clusters. In: Proceedings of the ACM International Conference on Supercomputing, pp. 1–12. ICS 2019, ACM Press, New York (2019). https://doi.org/10.1145/3330345.3330346
9. Fernando, M., Sundar, H.: Dendro home page (2020). https://octree.org

10. Foundation, T.O.: Openfoam (2022). https://openfoam.org
11. Hammer, J.: Pycachesim: python cache hierarchy simulator (2001). https://github.com/RRZE-HPC/pycachesim
12. Hecht, F.: New development in freefem++. J. Numer. Math. **20**(3-4), 251–265 (2012), https://freefem.org/
13. Heisler, E., Deshmukh, A., Sundar, H.: Finch code repository (2022). https://github.com/paralab/Finch
14. Hesthaven, J.S., Warburton, T.: Nodal Discontinuous Galerkin Methods: Algorithms, Analysis, and Applications. Springer, New York (2008). https://doi.org/10.1007/978-0-387-72067-8
15. Homolya, M., Kirby, R.C., Ham, D.A.: Exposing and exploiting structure: optimal code generation for high-order finite element methods (2017). https://arxiv.org/abs/1711.02473
16. JuliaLang.org: Julia benchmarks (2021). https://julialang.org/benchmarks
17. Kempf, D., Heß, R., Müthing, S., Bastian, P.: Automatic code generation for high-performance discontinuous galerkin methods on modern architectures. ACM Trans. Math. Software **47**(1), 1–31 (2020). https://doi.org/10.1145/3424144
18. Kirby, R.C., Logg, A.: A compiler for variational forms. ACM Trans. Math. Softw. **32**(3), 417–444 (2006). https://doi.org/10.1145/1163641.1163644
19. Logg, A., Wells, G.N.: Dolfin: Automated finite element computing. ACM Trans. Math. Softw. **37**(2) (2010). https://doi.org/10.1145/1731022.1731030
20. Louboutin, M., et al.: Devito (v3.1.0): an embedded domain-specific language for finite differences and geophysical exploration. Geoscientific Model Dev. **12**(3), 1165–1187 (2019). https://doi.org/10.5194/gmd-12-1165-2019
21. Macià, S., Martínez-Ferrer, P.J., Mateo, S., Beltran, V., Ayguadé, E.: Assembling a high-productivity dsl for computational fluid dynamics. In: Proceedings of the Platform for Advanced Scientific Computing Conference, pp. 1–11. PASC 2019, ACM Press, New York (2019). https://doi.org/10.1145/3324989.3325721
22. McRae, A.T.T., Bercea, G.T., Mitchell, L., Ham, D.A., Cotter, C.J.: Automated generation and symbolic manipulation of tensor product finite elements. SIAM J. Sci. Comput. **38**(5), 25–47 (2016). https://doi.org/10.1137/15M1021167
23. Pietro, D.A.D., Gratien, J.M., Häberlein, F., Michel, A., Prud'homme, C.: Basic concepts to design a dsl for parallel finite volume applications: extended abstract. In: Proceedings of the 8th workshop on Parallel/High-Performance Object-Oriented Scientific Computing, pp. 1–12. POOSC 2009, ACM Press, New York (2009). https://doi.org/10.1145/1595655.1595658
24. Rackauckas, C., Nie, Q.: Differentialequations.jl-a performant and feature-rich ecosystem for solving differential equations in julia. J. Open Res. Softw. **5**(1), 15 (2017)
25. Rathgeber, F., et al.: Firedrake: automating the finite element method by composing abstractions. ACM Trans. Math. Softw. **43**(3), 1–27 (2016). https://doi.org/10.1145/2998441
26. Sundar, H., Sampath, R., Biros, G.: Bottom-up construction and 2:1 balance refinement of linear octrees in parallel. SIAM J. Sci. Comput. **30**(5), 2675–2708 (2008)
27. Tran, H., Sundar, H.: A scalable adaptive-matrix spmv for heterogeneous architectures. In: Proceedings of the IEEE International Parallel and Distributed Processing Symposium. IPDPS 2022, accepted for publication (2022)

28. Uphoff, C., Bader, M.: Yet another tensor toolbox for discontinuous galerkin methods and other applications. ACM Trans. Math. Software **46**(4), 1–40 (2020). https://doi.org/10.1145/3406835
29. Xie, J., Ehmann, K., Cao, J.: Metafem: a generic fem solver by meta-expressions (2021)

Classifying Anomalous Members in a Collection of Multivariate Time Series Data Using Large Deviations Principle: An Application to COVID-19 Data

Sreelekha Guggilam[1,2(✉)], Varun Chandola[1,2], and Abani K. Patra[3]

[1] Computational Data Science and Engineering, University at Buffalo (SUNY), Buffalo, USA
{sreelekh,chandola}@buffalo.edu
[2] Computer Science and Engineering, University at Buffalo (SUNY), Buffalo, USA
[3] Data Intensive Studies Center, Tufts University, Medford, USA
abani.patra@tufts.edu

Abstract. Anomaly detection for time series data is often aimed at identifying extreme behaviors within an individual time series. However, identifying extreme trends relative to a collection of other time series is of significant interest, like in the fields of public health policy, social justice and pandemic propagation. We propose an algorithm that can scale to large collections of time series data using the concepts from the theory of *large deviations*. Exploiting the ability of the algorithm to scale to high-dimensional data, we propose an online anomaly detection method to identify anomalies in a collection of multivariate time series. We demonstrate the applicability of the proposed *Large Deviations Anomaly Detection* (LAD) algorithm in identifying counties in the United States with anomalous trends in terms of COVID-19 related cases and deaths. Several of the identified anomalous counties correlate with counties with documented poor response to the COVID pandemic.

Keywords: Large deviations · Anomaly detection · High-dimensional data · Multivariate time series · Time series database

1 Introduction

Anomaly detection has been extensively studied over many decades across many domains [5] but remains difficult for comparisons across time series. This problem is critical to study policy responses in pandemic propagation, economics, social justice, climate change adaptation to name a few e.g. studying anomalous COVID-19 infection data trends across various countries, states or counties could identify successful public policies. Usual approaches to monitoring individual time series [16] and identifying sudden outbreaks or significant causal events

© The Author(s), under exclusive license to Springer Nature Switzerland AG 2022
D. Groen et al. (Eds.): ICCS 2022, LNCS 13350, pp. 133–149, 2022.
https://doi.org/10.1007/978-3-031-08751-6_10

cannot be used to detect gradual divergence or drift. In this paper, we propose a new anomaly detection algorithm *Large deviations Anomaly Detection* (LAD), for large/high-dimensional data and multivariate time series data. LAD uses the rate function from *large deviations theory* (LDT) [24] to deduce anomaly scores for identifying anomalies. Core ideas for the algorithm are inspired from an LDT projection theorem that allows better handling of high dimensional data. Unlike most high dimensional anomaly detection models, LAD does not use feature selection or dimensionality reduction, which makes it ideal to study multiple time series in an online mode. LAD model naturally segregates the anomalies at each time step while enabling comparison of multiple multivariate time series. Key advances of the novel LAD algorithm reported here are:[1]

1. *Large deviations Anomaly Detection* (LAD) algorithm is a scalable LDP based method, for scoring based anomaly detection.
2. LAD model can analyze large and high dimensional datasets without additional dimensionality reduction increasing accuracy and reducing cost.
3. Online extension of LAD can detect anomalies across many multivariate time series using an evolving anomaly score for each tracking developing behavior.
4. An empirical study of publicly available anomaly detection benchmark datasets to analyze robustness and performance on high dimensional and large datasets.
5. A detailed analysis of COVID-19 trends for US counties where we identify counties with anomalous behavior (See Fig. 1 for an illustration).

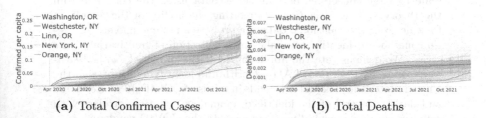

(a) Total Confirmed Cases   (b) Total Deaths

Fig. 1. Top 5 anomalous counties in USA identified by the LAD algorithm based on time-series data consisting of cumulative COVID-19 per-capita infections and deaths. The time-series for the non-anomalous counties are plotted (light-gray) in the background for reference. For the counties in New York, significant rise during early 2021 in confirmed cases (*left*) and high death rates, is detected. Washington and Linn County in Oregon are anomalous primarily due to steady low rates of infection.

2   Related Work

A large body of research exists on studying anomalies in high dimensional data [3]. Many anomaly detection algorithms use dimensionality reduction techniques

---

[1] An introductory pre-print version available as Guggilam et al. [11].

as a pre-processing step to anomaly detection. However, many high dimensional anomalies can only be detected in high dimensional problem settings and dimensionality reduction in such settings can lead to false negatives. Many methods exist that identify anomalies on high-dimensional data without dimensional reduction or feature selection, e.g. by using distance metrics. *Elliptic Envelope* (EE) [21] fits an ellipse around data centers by fitting a robust covariance estimates. *Isolation Forest* (I-Forest) [15] uses recursive partitioning by random feature selection and isolating outlier observations. *k nearest neighbor outlier detection* (kNN) [18] uses distance from nearest neighbor to get anomaly scores. *local outlier factor* (LOF) [4] uses deviation in local densities with respect to its neighbors to detect anomalies. *k-means--* [7] method uses distance from nearest cluster centers to jointly perform clustering and anomaly detection. *Concentration Free Outlier Factor* (CFOF) [2] uses a "reverse nearest neighbor-based score" which measures the number of nearest neighbors required for a point to have a set proportion of data within its envelope. In particular, methods like I-Forest and CFOF are targeted towards anomaly detection in high dimensional datasets. However, they are not tailored for evolving data.

Many score based anomaly detection algorithms have been designed to classify anomalies within individual time series. For instance, Twitter Ad Vec [14] are unsupervised study deviations from the data. Numenta [1] uses prediction errors to classify anomalies. Relative Entropy [25] compares entropy to identify anomalous observations. However, these algorithms are limited to studying only individual time series and not easily extended to an entire database of time series. Recently, large deviations theory has been widely applied in the fields of climate models [8], statistical mechanics [23], among others. Specially for analysis of time series, the theory of large deviations has proven to be of great interest over recent decades [17]. However, these methods are data specific, and often study individual time series. In most settings, real time detection of anomalies is needed to dispatch necessary preventive measures for damage control. Such problem formulation requires collectively monitoring a high dimensional time series database to identify anomalies in real time. While, the task of detecting anomalous time series in a collection of time series has been studied in the past [13], most of these works have focused on univariate time series and have not shown to scale to long time series data or provide limited explanation on why the identified trends are anomalous. Our proposed method addresses this.

3 Large Deviation Principle

Large deviations theory provides techniques to derive the probability of rare events[2] that have an asymptotically exact exponential approximation [9, 24]. The key concept of this theory is the Large Deviations Principle (LDP). The principle describes the exponential decay of the probabilities for the mean of random variables. To implement LDP on data with known distributions, it is important

---

[2] In our context, these rare events include outlier/anomalous behaviors.

to decipher the rate function $\mathcal{I}$. Cramer's Theorem provides the relation between $\mathcal{I}$ and the logarithmic moment generating function Λ[3].

Theorem 1 (Cramer's Theorem). *Let $X_1, X_2, \ldots X_n$ be a sequence of iid real random variables with finite logarithmic moment generating function, e.g. $\Lambda(t) < \infty$ for all $t \in \mathbb{R}$. Then the law for the empirical average satisfies the large deviations principle with rate $\epsilon = 1/n$ and rate function given by $\mathcal{I}(x) := \sup_{t \in \mathbb{R}} (tx - \Lambda(t)) \quad \forall t \in \mathbb{R}$.*

Thus, we get, $\lim_{n \to \infty} \frac{1}{n} \log \left(P \left(\sum_{i=1}^{n} X_i \geq nx \right) \right) = -\mathcal{I}(x), \quad \forall x > E[X_1]$. For more complex distributions, identifying the rate function using logarithmic moment generating function can be challenging. Many methods like contraction principle and exponential tilting exist that extend rate functions from one topological space that satisfies LDP to the topological spaces of interest [9]. For our work, we are interested in the Dawson-Gärtner Projective LDP, that generates the rate function using nested family of projections.

Theorem 2. *Dawson-Gärtner Projective LDP: Let $\{\pi^N\}_{N \in \mathbb{N}}$ be a nested family of projections acting on $\mathcal{X}$ s.t. $\cup_{N \in \mathbb{N}} \pi^N$ is the identity. Let $\mathcal{X}^N = \pi^N \mathcal{X}$ and $\mu_\epsilon^N = \mu_0 \circ (\pi^N)^{-1}, N \in \mathbb{N}$. If $\forall N \in \mathcal{N}$, the family $\{\mu_\epsilon^N\}_{\epsilon > 0}$ satisfies the LDP on $\mathcal{X}^N$ with rate function $\mathcal{I}^N$, then $\{\mu_\epsilon\}_{\epsilon > 0}$ satisfies the LDP with rate function I given by, $\mathcal{I}(x) = \sup_{N \in \mathbb{N}} \mathcal{I}^N(\pi^N x) \quad x \in \mathcal{X}$. Since $\mathcal{I}^N(y) = \inf_{\{x \in \mathcal{X} | \pi^N(x) = y\}} \mathcal{I}(x), y \in \mathcal{Y}$, the supremum defining $\mathcal{I}$ is monotone in N because projections are nested.*

The theorem allows extending the rate function from a lower to higher projection space. The implementation of this theorem in LAD model is seen in Sect. 4.

4   Methodology

Consider the case of multivariate time series data. Let $\{\mathbf{t_n}\}_{n=1}^{N}$ be a set of multivariate time series datasets where $\mathbf{t_n} = (\mathbf{t_{n,1}}, \ldots, \mathbf{t_{n,T}})$ is a time series of length T and each $\mathbf{t_{n,t}}$ has d attributes. The motivation is to identify anomalous $\mathbf{t_n}$ that diverge significantly from the non-anomalous counter parts at a given or multiple time steps. The main challenge is to design a score for individual time series that evolves in a temporal setting as well as enables tracking the initial time of deviation as well as the scale of deviation from the normal trend. As shown in following sections, our model addresses the problem through the use of rate functions derived from large deviations principle. We use the Dawson-Gärtner Projective LDP (See Sect. 4.2) for projecting the rate function to a low dimensional setting while preserving anomalous instances. The extension to temporal data (See Sect. 4.3) is done by collectively studying each time series data as one observation.

---

[3] The logarithmic moment generating function of a random variable X is defined as $\Lambda(t) = \log E[\exp(tX)]$.

4.1 Large Deviations for Anomaly Detection

Our approach uses a direct implementation of LDP to derive the rate function values for each observation. As the theory focuses on extremely rare events, the raw probabilities associated with them are usually very small [9,24]. However, the LDP provides a rate function that is used as a scoring metric for LAD.

Consider a dataset X of size n. Let $\mathbf{a} = \{\mathbf{a_1}, \ldots, \mathbf{a_n}\}$ and $\mathbf{I} = \{\mathbf{I_1}, \ldots, \mathbf{I_n}\}$ be anomaly score and anomaly label vectors for the observations respectively such that $a_i \in [0, 1]$ and $I_i \in \{0, 1\}$ $\forall i \in \{1, 2, \ldots, n\}$. By large deviations principle, we know that for a given dataset X of size n, $P(\bar{X} = p) \approx e^{-n\mathcal{I}(p)}$. Assuming that the underlying data is standard Gaussian distribution with mean 0 and variance 1, we can use the rate function for Gaussian data where $\mathcal{I}(p) = \frac{p^2}{2}$. Then the resulting probability that the sample mean is p is given by $P(\bar{X} = p) \approx e^{-n\frac{p^2}{2}}$. Now, in presence of an anomalous observation x_a, the sample mean is shifted by approximately x_a/n for large n. Thus, the probability of the shifted mean being the true mean is given by $P(\bar{X} = x_a/n) \approx e^{-\frac{x_a^2}{2n}}$. However, for large n and $|x_a| << 1$, the above probabilities decay exponentially which significantly reduces their effectiveness for anomaly detection. Thus, we use $\frac{x_a^2}{2n}$ as anomaly score for our model. Thus generalizing this, the anomaly score for each individual observation is given by $a_i = n\mathcal{I}(x_i)$ $\forall i \in \{1, 2, \ldots, n\}$.

4.2 LDP for High Dimensional Data

High dimensional data pose significant challenges to anomaly detection. Presence of redundant or irrelevant features act as noise making anomaly detection difficult. However, dimensionality reduction can impact anomalies that arise from less significant features of the datasets. To address this, we use the Dawson-Gärtner Projective theorem in LAD model to compute the rate function for high dimensional data. The theorem records the maximum value across all projections which preserves the anomaly score making it optimal to detect anomalies in high dimensional data. The model algorithm is presented in Algorithm 1.

Algorithm 1: Algorithm 1: LAD Model

Input: Dataset X of size (n, d), number of iterations N_{iter}, threshold th.

Output: Anomaly score $\mathbf{a}$

Initialization: Set initial anomaly score and labels $\mathbf{a}$ and $\mathbf{I}$ to zero vectors and, entropy matrix $E = 0_{(n,d)}$ where $0_{(n,d)}$ is a zero matrix of size (n, d).

for *each* $s \to 1$ **to** N_{iter} **do**
1. Subset $X_{sub} = X[I_i == 0]$
2. $X_{normalized}[:, d_i] = \frac{X[:,d_i] - X_{sub}[:,d_i]}{cov(X_{sub}[:,d_i])}, \forall d_i \in \{1, \ldots, d\}$
3. $E[i, :] = -X_{normalized}[i]^2 / 2n, \forall i$
4. $a_i = -max(E[i, :])$
5. $\mathbf{a} = \frac{\mathbf{a} - min(\mathbf{a})}{max(\mathbf{a}) - min(\mathbf{a})}$
6. $th = min(th, quantile(\mathbf{a}, \mathbf{0.95})$
7. $I_i = 1$ if $a_i > th, \forall i$

4.3 LAD for Time Series Data

Broadly, time series anomalies can be categorized to two groups [6]: (1) **Divergent trends/Process anomalies**: Time series with divergent trends that last for significant time periods fall into this group. Here, one can argue that generative process of such time series could be different from the rest of the non-anomalous counterparts, and (2) **Subsequence anomalies**: Such time series have temporally sudden fluctuations or deviations from expected behavior which can be deemed as anomalous. These anomalies occur as a subsequence of sudden spikes or fatigues in a time series of relatively non-anomalous trend. The online extension of the LAD model is designed to capture anomalous behavior at each time step. Based on the mode of analysis of the temporal anomaly scores, one can identify both divergent trends and subsequence anomalies. In this paper, we focus on the divergent trends (or process anomalies). In particular, we try to look at the anomalous trends in COVID-19 cases and deaths in US counties. Studies to collectively identify divergent trends and subsequence anomalies is being considered as a prospective future work.

In this section, we present an extension of the LAD model to multivariate time series data where we preserve the dependency temporal and across different features of the time series. Thus, as shown in Algorithm 2, a horizontal stacking of the data is performed. This allows collective study of temporal and non-temporal features. To preserve temporal dependency, the anomaly scores and labels are carried on to next time step where the labels are then re-evaluated.

As long term anomalies are of interest, time series with temporally longer anomalous behaviors are ranked more anomalous. The overall time series anomaly score A_n for each time series $\mathbf{t_n}$ can be computed as $A_n = \frac{\sum_{t=1}^{T} I[n,t]}{T}$

Algorithm 2: Algorithm 2: LAD for Time series anomaly detection

Input: Time series dataset $\{\mathbf{t_n}\}_{n=1}^{\mathbf{N}}$ of size (N, T, d), number of iterations N_{iter}, threshold th, window w.
Output: An array of temporal anomaly scores $\mathbf{a}$, an array of temporal anomaly labels I
Initialization: Set initial anomaly score and labels $\mathbf{a}$ and $\mathbf{I}$ to zero matrices of size (N, T) and, entropy matrix E to a zero matrix of size (N, T, d).

for each $t \to 1$ **to** T **do**
$\quad X = hstack(t_{n,t}^-)$ where $t_{n,t}^- = \{t_{n,t-w}, \ldots t_{n,t}\}$
$\quad I[i,t] = I[i, t-1]$
$\quad \mathbf{a}[:, \mathbf{t}] = \mathbf{a}[:, \mathbf{t} - \mathbf{1}]$
$\quad$ **for** each $s \to 1$ **to** N_{iter} **do**

$\quad\quad$ 1. Subset non-anomalous time series
$\quad\quad\quad X_{sub} = \{X[i,:] | I[i,t] == 0, \forall i\}$
$\quad\quad$ 2. $X_{normalized}[:, d_i] = \frac{X[:,d_i] - X_{sub}[:,d_i]}{cov(X_{sub}[:,d_i])}, \forall d_i \in$
$\quad\quad\quad \{1, 2, \ldots, d * w\}$
$\quad\quad$ 3. $E[i, :] = -X_{normalized}[i]^2 / 2n, \forall i$
$\quad\quad$ 4. $\mathbf{a}[i, \mathbf{t}] = -\mathbf{max}(\mathbf{E}[\mathbf{i}, :])$
$\quad\quad$ 5. $\mathbf{a}[:, \mathbf{t}] = \frac{\mathbf{a}[:,\mathbf{t}] - min(\mathbf{a}[:,\mathbf{t}])}{max(\mathbf{a}[:,\mathbf{t}]) - min(\mathbf{a}[:,\mathbf{t}])}$
$\quad\quad$ 6. $th = min(th, quantile(\mathbf{a}[:, \mathbf{t}], \mathbf{0.95})$
$\quad\quad$ 7. $I[i,t] = 1$ if $\mathbf{a}[\mathbf{i}, \mathbf{t}] > \mathbf{th}, \forall i$

$\forall n$. For a database of time series with varying lengths, the time series anomaly score is computed by normalizing with respective lengths. Similarly, the method can be extended to studying anomalies within an individual time series by breaking the series into a database of sub-sequences of a time series extracted via a sliding window. It must be noted that this approach allows for a retrospective classification of anomalies.

5 Experiments

In this section, we evaluate the performance of the LAD algorithm on multi-aspect datasets. The following experiments have been conducted to study the model: 1) Anomaly Detection Performance: LAD's ability to detect real-world anomalies as compared to state-of-the-art anomaly detection models is evaluated using the ground truth labels. 2) Handling Large Data: Scalability of the LAD model on large datasets (high observation count or high dimensionality) are studied. 3) COVID-19 Time Series Data.

5.1 Datasets

We consider a variety of publicly available benchmark data sets from Outlier Detection DataSets ODDS [19] (See Table 1) for the experimental evaluation. For anomaly detection within individual time series, we study univariate time series data from Numenta Benchmark Datasets[4]. Additionally, for the time series data, we use COVID-19 deaths and confirmed cases for US counties from John Hopkins COIVD-19 Data Repository [10]. The country level global data for COVID-19 trends was taken from the Our World in Data Repository [20].

5.2 Baseline Methods and Parameter Initialization

As described in Sect. 4, LAD falls under unsupervised learning regime targeted for high dimensional data, we do not compare with supervised algorithms. For this we consider *Elliptic Envelope* (EE) [21], *Isolation Forest* (I-Forest) [15][5], *local outlier factor* (LOF) [4], and *Concentration Free Outlier Factor* CFOF [2]. The CFOF and LOF models assign an anomaly score for each observation, while remaining methods provide an anomaly label. As above mentioned methods are parametric, we investigated a range of values for each parameter, and report the best results. For Isolation Forest, Elliptic Envelope and CFOF, the contamination value is set to the true proportion of anomalies in the dataset. To study anomaly detection in time series, the LAD model is compared with other score based time series anomaly detection algorithms like Twitter AD Vec (TAV) [14],

---

[4] http://numenta.com/press/numenta-anomaly-benchmark-nab-evaluates-anomaly-detection-techniques.htm.

[5] I-Forest model returns both anomaly scores and anomaly labels though we only present classification model since they outperforms score based schemes.

Table 1. Description of the benchmark data used for evaluation of the anomaly detection for high dimensional/large sample datasets and time series. N - number of instances, d - number of attributes and a - fraction of known anomalies in the data set.

Name	N	d	a
HTTP	567498	3	0.39%
MNIST	7603	100	9.207%
Arrhythmia	452	274	14.602%
Shuttle	49097	9	7.151%
Letter	1600	32	6.25%
Musk	3062	166	3.168%
Optdigits	5216	64	2.876%
Satellite Img.	6435	36	31.639%
Speech	3686	400	1.655%
SMTP	95156	3	0.032%
Satellite Img.-2	5803	36	1.224%
Forest Cover	286048	10	0.96%
KDD99	620098	29	29 0.17%

(a) High Dimensional and Large Sample Datasets

Dataset	N	a
EC2 CPU UTILIZATION 825CC2	4032	0.09%
EC2 NETWORK IN 257A54	4032	0.1%
EC2 CPU UTILIZATION 5F5533	4032	0.1%
EC2 CPU UTILIZATION AC20CD	4032	0.1%
EC2 CPU UTILIZATION 24AE8D	4032	0.1%
SPEED 7578	1127	0.1%
SPEED 6005	2500	0.1%
OCCUPANCY 6005	2380	0.1%
SPEED T4013	2495	0.1%
ART LOAD BALANCER SPIKES	4032	0.1%
EXCHANGE-3 CPM RESULTS	1538	0.1%
EXCHANGE-4 CPM RESULTS	1643	0.1%
TWITTER VOLUME KO	15851	0.1%
TWITTER VOLUME CVS	15853	0.1%
TWITTER VOLUME CRM	15902	0.1%
MACHINE TEMP. SYS. FAILURE	22695	0.1%
EC2 REQ. LATENCY SYS. FAILURE	4032	0.09%
CPU UTIL. ASG MISCONFIG.	18050	0.08%

(b) Benchmark Time Series

Skyline [22], Earthgecko Skyline (E.Skyline)[6], Numenta [1], Relative Entropy (RE) [25], Random Cut Forest (RCF) [12], Windowed Gaussian (WG). The LAD model relies on a threshold value to classify observations with scores the value as strictly anomalous. Though this value is iteratively updated, an initial value is required by the algorithm. In this paper, the initial threshold value for the experiment is set to 0.95 for all datasets. All the methods for anomaly detection benchmark datasets are implemented in Python and all experiments were conducted on a 2.7 GHz Quad-Core Intel Core i7 processor with a 16 GB.

5.3 Evaluation Metrics

As LAD is a score based algorithm, we study the ROC curves by comparing the True Positive Rate (TPR) and False Positive Rate (FPR), across various thresholds. The final ROC-AUC (Area under the ROC curve) is reported for evaluation. For anomaly detection within individual time series, we use the F-measure as the evaluation metric to study the overall performance of the model. Since all the models return anomaly scores, thresholds were used to classify observations as anomalous vs non-anomalous. Threshold was set to be the maximum score in the truly non-anomalous data for each model and the observations with scores higher than set threshold were labelled anomalous. This is to ensure that the model is able to distinguish anomalies from the rest of the data. For time series database anomaly detection, we present the final outliers and study their deviations from normal baselines under different model settings.

---

[6] https://github.com/earthgecko/skylin.

5.4   Anomaly Detection Performance

Table 2 shows the performance of LOF, I-Forest, EE, CFOF and LAD on anomaly detection benchmark datasets. Due to relatively large run-time[7], CFOF results are shown for datasets with samples less than 10k. For all the listed algorithms, results for best parameter settings are reported. The proposed LAD model outperforms other methods on most data sets. For larger and high dimensional datasets, it can be seen from Table 2 that the LAD model outperforms all the models in most settings.[8] It was interesting to note that the LAD model, despite being non-parametric (for a non-temporal setting), had a comparable if not better performance as compared to the LOF, EE, I-Forest and CFOF where multiple parameter setting were tested to derive the best fitting model. To study the LAD model's computational effectiveness, we study the computation time and scaling of LAD model on large and high dimensional datasets. Figure 2a shows the scalability of LAD with respect to the number of records against the time needed to run on the first k records of the KDD-99 dataset. Each record has 29 dimensions. Figure 2b shows the scalability of LAD with respect to the number of dimensions (linear-scale). We plot the time needed to run on the first 1, 2, ..., 29 dimensions of the KDD-99 dataset. The results confirm the linear scalability of LAD with number of records as well as number of dimensions.

Table 2. Comparing LAD with existing anomaly detection algorithms for large/high dimensional datasets using ROC-AUC as the evaluation metric.

(a) LAD scales linearly with the number of records for KDD-99 data

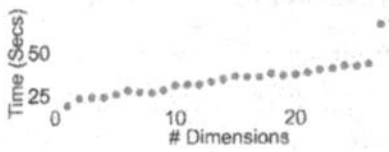

Data	LOF	I-Forest	EE	CFOF	LAD
SHUTTLE	0.52	0.98	0.96	–	**0.99**
SATIMAGE-2	0.57	0.95	0.96	0.70	**0.99**
SATIMAGE	0.51	0.64	**0.65**	0.55	0.6
KDD99	0.51	0.85	0.54	–	**1.0**
ARRHYTHMIA	0.61	0.67	0.7	0.56	**0.71**
OPTDIGITS	0.51	**0.52**	–	0.49	0.48
LETTER	0.54	0.54	0.6	**0.90**	0.6
MUSK	0.5	**0.96**	**0.96**	0.49	**0.96**
HTTP	0.47	0.95	0.95	–	**1.0**
MNIST	0.5	0.61	0.65	0.75	**0.87**
COVER	0.51	0.63	0.52	–	**0.96**
SMTP	**0.84**	0.83	0.83	–	0.82
SPEECH	0.5	**0.53**	0.51	0.47	0.47

(b) LAD scales linearly with the number of dimensions in KDD-99 data.

Fig. 2. LAD model scaling on large and high dimensional data

---

[7] The CFOF model is computationally expensive and its use is primarily for high-dimensional data. We restrict results to datasets with <10K observations.

[8] Lowest AUC values for the LAD model are observed for Speech and Optdigits data where multiple true clusters are noted.

5.5   Anomaly Detection in Individual Time Series

In Table 3, we compare the performance of the LAD model as compared to other score-based algorithms. In particular, it can be seen that LAD model with window length of 100 has the best anomaly detection performance as compared other methods in most datasets.

Table 3. Comparing LAD with existing anomaly detection algorithms for time series datasets using F-measure as the evaluation metric.

Data	WL = 10	WL = 50	WL = 100	TAV	Skyline	E.Skyline	Numenta	RE	RCF	WG
EC2 CPU UTIL. 825CC2	0.0	0.1	0.37	0.16	**0.45**	0.16	0.03	0.05	0.13	0.19
EC2 NETWORK IN 257A54	0.14	0.25	**0.33**	0.03	0.04	0.18	0.02	0.01	0.03	0.02
EC2 CPU UTIL. 5F5533	0.14	0.36	**0.57**	0.18	0.03	0.18	0.01	0.03	0.04	0.0
EC2 CPU UTIL. AC20CD	0.0	0.31	**0.33**	0.03	0.02	0.01	0.01	0.03	0.0	0.11
EC2 CPU UTIL. 24AE8D	0.09	0.12	**0.59**	0.01	0.01	0.0	0.0	0.0	0.0	0.01
SPEED 7578	0.26	0.29	**0.54**	0.19	0.08	0.05	0.05	0.08	0.02	0.17
SPEED 6005	0.15	**0.59**	**0.59**	0.04	0.11	0.11	0.03	0.04	0.04	0.01
OCCUPANCY 6005	0.08	0.29	**0.5**	0.01	0.01	0.01	0.01	0.01	0.01	0.0
SPEED T4013	0.27	**0.88**	0.45	0.15	0.16	0.02	0.04	0.03	0.13	0.14
ART LOAD BALANCER SPIKES	0.08	**0.16**	0.15	0.02	0.01	0.0	0.0	0.01	0.0	0.08
EXCHANGE-3 CPM RESULTS	0.0	0.4	**0.77**	0.01	0.01	0.01	0.01	0.03	0.01	0.01
EXCHANGE-4 CPM RESULTS	**0.21**	0.21	0.17	0.02	0.04	0.04	0.05	0.19	0.05	0.05
TWITTER VOL. KO	0.01	0.06	**0.11**	0.01	0.01	0.0	0.01	0.0	0.0	0.03
TWITTER VOL. CVS	0.04	0.06	**0.12**	0.03	0.01	0.01	0.01	0.01	0.01	0.03
TWITTER VOL. CRM	0.01	0.06	**0.11**	0.03	0.01	0.0	0.0	0.01	0.01	0.01
MACHINE TEMP SYS. FAIL.	0.02	0.04	0.08	**0.18**	0.03	0.01	0.0	0.02	0.03	0.0
EC2 REQ LATENCY SYS. FAIL.	0.2	**0.62**	0.35	0.15	0.04	0.15	0.02	0.15	0.03	0.02
CPU UTIL ASG MISCONFIG.	0.03	0.24	**0.83**	0.04	0.0	0.0	0.0	0.02	0.0	0.0

5.6   Anomaly Detection in Time Series Data

This section presents the results of LAD model on COVID-19 time series data at the US county level. Multiple settings were used to understand the data: 1. Deaths and confirmed case trends were considered for analysis. 2. Daily New vs Total Counts: Both total cases as well daily new cases were analyzed. 3. Complete history vs One Time Step: Two versions of the model were studied where data from previous time steps were and were not considered. By this, we tried to distinguish the impact of the history of the time series on identifying anomalous trends. 4. Univariate vs Multivariate Time Series data: To further understand the LAD model, the deaths and case trends were studied individually as a univariate time series as well as collectively in a multivariate time series data setting. 5. Time Series of Uniform vs Varying Lengths: Finally, all the above analyses were conducted on time series data with varying lengths. Here, for each county level time series, the time of first event was considered as initial time step to objectively study the relative temporal changes in trends. To bring all the counts to a baseline, the total counts in each time series were scaled to the respective county population. Missing information was replaced with zeros and counties with population less than 50k were eliminated from the study.

5.7   Discoveries: US COVID-19 Trends

In this section, we look at the daily new case and deaths in US counties trends in start of 2021. To rank the counties, anomaly scores between Jan 1–Mar 1 2021 were considered.

Complete history vs One Time Step. The full history setting considers the complete history of the time series and is aimed to capture most deviant trends over time. The one time step (or any smaller window) setting is more suitable to study deviations within the specific window. As we target long term deviating trends, the one time step setting returns trends that have stayed most deviant

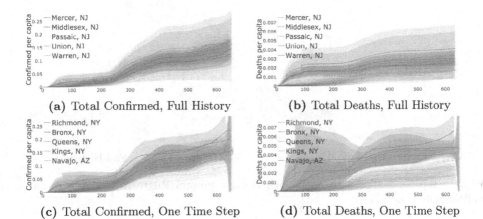

(a) Total Confirmed, Full History   **(b)** Total Deaths, Full History

(c) Total Confirmed, One Time Step   **(d)** Total Deaths, One Time Step

Fig. 3. Top 5 counties with anomalous trends: varying lengths, total counts, multivariate time series

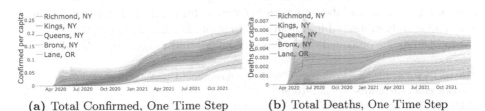

(a) Total Confirmed, One Time Step   **(b)** Total Deaths, One Time Step

Fig. 4. Top 5 counties with anomalous trends: uniform lengths, total counts, multivariate time series

throughout the entire time range. This can be seen in Figs. 3 and 4 where the one time step setting returns trends that have stayed deviant almost throughout the duration while the full history setting is able to capture significantly higher overall deviations from normal trends and therefore higher anomaly score. For instance, counties like Mercer (NJ), Union (NJ), that had extensive testing conducted[9] were captured in the one time step model as seen in Fig. 3c and 3d. Similarly, counties in NY observed a peak in early 2021[10], which was not captured as anomalous in the one time step model as seen in Figs. 1a and 1b.

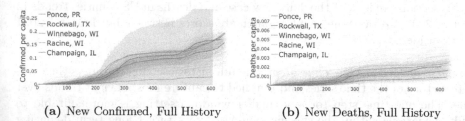

(a) New Confirmed, Full History   (b) New Deaths, Full History

Fig. 5. Top 5 counties with anomalous trends: varying lengths, daily new counts, multivariate time series

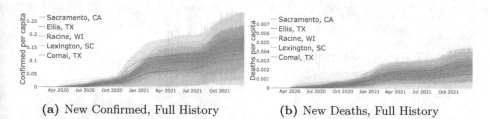

(a) New Confirmed, Full History   (b) New Deaths, Full History

Fig. 6. Top 5 counties with anomalous trends: uniform lengths, daily new counts, multivariate time series

Univariate vs Multivariate Time Series. In Figs. 3, 4, 5 and 6 we see the anomalous trends in multivariate time series, where total confirmed cases and deaths were collectively evaluated for anomaly detection. For instance, despite the near-normal trends in confirmed cases, Kings, Queens and Bronx (NY)[11] in Figs. 3c-

---

[9] https://www.nj.com/coronavirus/2021/12/more-covid-testing-sites-opening-as-cases-climb-here-are-9-places-to-go.html.

[10] https://www.newsday.com/news/health/coronavirus/coronavirus-long-island-deaths-vaccinations-1.50200404.

[11] https://www.nbcnewyork.com/news/coronavirus/nyc-mask-mandate-indoors-an-option-if-needed-mayor-says-as-23-nations-report-omicron/3428102/.

3d, were identified anomalous due to their the deviant death trends which significantly contributed to the anomaly scores. Figures. 7 and 8 show the use of univariate time series for detection of deviation in one feature.

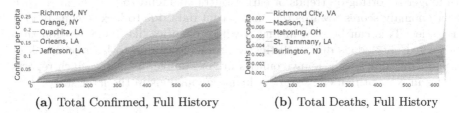

(a) Total Confirmed, Full History   (b) Total Deaths, Full History

Fig. 7. Top 5 counties with anomalous trends: varying lengths, total counts for univariate time series

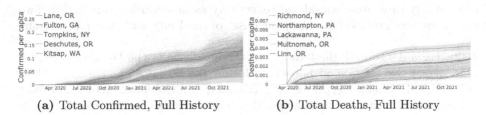

(a) Total Confirmed, Full History   (b) Total Deaths, Full History

Fig. 8. Top 5 counties with anomalous trends: uniform lengths, total counts for univariate time series

Daily New vs Total Counts. Figures 4 and 6, show anomalous trends in multivariate time series for total and daily new counts respectively. It can be seen that the anomaly score is relatively more erratic for new case trends as the data for new case and death counts is more erratic leading to fluctuating normal average and non-smooth anomaly scores. Similar behavior can be seen across Figs. 3 and 5. The LAD model on the daily new counts data was able to capture the escalation in Racine, Wisconsin in Fig. 6a and 6b during late 2020 when multiple meatpacking were tied to COVID-19 cases[12].

---

[12] https://www.jsonline.com/story/news/2020/11/25/meatpacking-plants-tied-more-covid-19-cases-than-known-new-bussiness-outbreak-data-shows/6376197002/.

Uniform Length vs Varying Length Time Series. The US county cases and deaths data consists of time series of uniform lengths. However, not all counties have events recorded in the early stages. Thus, studying the non-synchronized database creates a bias against counties with early reported cases. Also, counties with longer reporting on trends or earlier outbreaks tend to be associated with higher anomaly scores towards the most recent data due to lack of equally long time series. This can be seen in Figs. 4 where counties like Lane, Oregon that was flagged anomalous due to distinctively low cases due to later outbreak of the pandemic much after many counties in NY, unlike in Figs. 3 which reports counties in NY with an early start as highly anomalous in the later stages[13].

5.8   Global Trends and Emergence of Other COVID-19 Variants

Coronavirus Pandemic (COVID-19) Data from Our World in Data [20] for countries with population more than 5 million was used for the analysis. Trends in the daily new deaths per million and confirmed cases per million (7 day rolling average, right-aligned), biweekly growth rates in deaths and confirmed cases and case fatality rates were considered collectively as multivariate time series. Two end dates were studied to analyze the onset of the Delta and Omicron variants.

Delta Variant. To rank the trends post the incidence of the Delta variant (See Figs. 9a–9c), we considered behaviors during the 90 day period May 1 2021–July 29 2021. China, Egypt, Mexico, Tanzania and Columbia were found most anomalous. In particular, China and Mexico had low per capita weekly average deaths and confirmed cases. However, the case fatality rates were consistently high[14] indicating need for additional investigation to understand the root cause which may be under-reporting or reporting issues or presence of a new variant.

Omicron Variant. To study the Omicron variant, we looked at the 90 day period data September 23 2021 - December 21 2021 (See Figs. 9d–9f). UK, China have the most anomalous trends. Egypt, UK and Russia also have high anomaly scores[15]. However, in Egypt and Russia, the surge in cases was not due to the Omicron variant but due to earlier COVID wave that coincides with the it[16].

---

[13] https://time.com/5812569/covid-19-new-york-morgues/.

[14] https://www.marketwatch.com/story/new-daily-covid-19-cases-and-deaths-spike-to-6-week-highs-as-delta-variant-spreads-rapidly-11625673956.

[15] https://www.cnn.com/2021/12/13/uk/uk-omicron-infections-tidal-wave-gbr-intl/index.html.

[16] https://www.egyptindependent.com/egypt-has-not-passed-the-peak-of-the-covid-19-fourth-wave/, https://tass.com/society/1370957.

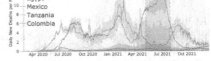

(a) Delta Variant: Daily new cases per million people (rolling 7-day average)

(b) Delta Variant: Daily new deaths per million people (rolling 7-day average)

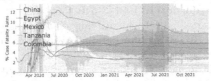

(c) Delta Variant: Case Fatality Rates

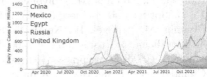

(d) Omicron Variant: Daily new cases per million people (rolling 7-day average)

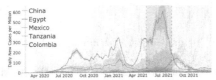

(e) Omicron Variant: Daily new deaths per million people (rolling 7-day average)

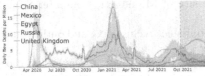

(f) Omicron Variant: Case Fatality Rates

Fig. 9. Top 5 anomalous country-level trends: delta and omicron variants

6   Conclusion

In this paper, we propose LAD, a novel scoring algorithm for anomaly detection in large/high-dimensional data. The algorithm successfully handles high dimensions by implementing large deviation theory. Our contributions include reestablishing the advantages of large deviations theory to large and high dimensional datasets. We present an online extension of the model aimed to identify anomalous time series in a multivariate time series data. The model shows vast potential in scalability and performance against baseline methods. The online LAD returns a temporally evolving score for each time series that allows us to study the deviations in trends relative to the complete time series database.

A potential extension to the model could include anomalous event detection for each individual time series. Another possible future work could be extending the model to enable anomaly detection in multi-modal datasets. Additionally, the online LAD model could be enhanced to use temporally weighted scores prioritizing recent events.

Acknowledgements. The authors would like to acknowledge University at Buffalo Center for Computational Research for computing resources and financial support of the National Science Foundation Grant numbers NSF/OAC 1339765 and NSF/DMS 1621853.

References

1. Ahmad, S., Lavin, A., Purdy, S., Agha, Z.: Unsupervised real-time anomaly detection for streaming data. Neurocomputing **2017**, 134–147 (2017)
2. Angiulli, F.: CFOF: a concentration free measure for anomaly detection. ACM TKDD **14**(1), 1–53 (2020)
3. Angiulli, F., Pizzuti, C.: Fast outlier detection in high dimensional spaces. In: Elomaa, T., Mannila, H., Toivonen, H. (eds.) PKDD 2002. LNCS, vol. 2431, pp. 15 27. Springer, Heidelberg (2002). https://doi.org/10.1007/3-540-45681-3_2
4. Breunig, M., Kriegel, H., Ng, R.T., Sander, J.: LOF: identifying density-based local outliers. In: Proceedings of 2000 ACM SIGMOD International Conference on Management of Data, pp. 93–104 (2000)
5. Chandola, V., Banerjee, A., Kumar, V.: Anomaly detection: a survey. Comput. Surv. **41**, 3 (2009)
6. Chandola, V., Cheboli, D., Kumar, V.: Detecting anomalies in a timeseries database. Technical Report 09–004. University of Minnesota, Computer Science Dept (2009)
7. Chawla, S., Gionis, A.: k-means: a unified approach to clustering and outlier detection. In: SDM (2013)
8. Dematteis, G., Grafke, T., Vanden-Eijnden, E.: Rogue waves and large deviations in deep sea. PNAS **115**(5), 855–860 (2018)
9. Den Hollander, F.: Large deviations, vol. 14. AMS (2008)
10. Dong, E., Du, H., Gardner, L.: An interactive web-based dashboard to track COVID-19 in real time. Lancet. Infect. Dis. **20**(5), 533–534 (2020)
11. Guggilam, S., Chandola, V., Patra, A.: Anomaly detection for high-dimensional data using large deviations principle. arXiv preprint arXiv:2109.13698 (2021)
12. Guha, S., Mishra, N., Roy, G., Schrijvers, O.: Robust random cut forest based anomaly detection on streams. In: ICML, pp. 2712–2721. PMLR (2016)
13. Homayouni, H., Ray, I., Ghosh, S., Gondalia, S., Kahn, M.G.: Anomaly detection in COVID-19 time-series data. SN Comput. Sci. **2**(4), 1–17 (2021)
14. Kejariwal, A.: Introducing practical and robust anomaly detection in a time series. Twitter Eng. Blog. Web **15** (2015)
15. Liu, F.T., Ting, K.M., Zhou, Z.-H.: Isolation-based anomaly detection. ACM TKDD **6**(1), 1–39 (2012)
16. Maleki, M., Mahmoudi, M., Wraith, D., Pho, K.: Time series modelling to forecast the confirmed and recovered cases of COVID-19. Travel Med. Infect. Dis. **37**(2020), 101742 (2020)
17. Mikosch, T., Wintenberger, O.: A large deviations approach to limit theory for heavy-tailed time series. Prob. Theory Related Fields **166**(1), 233–269 (2016)
18. Ramaswamy, S., Rastogi, R., Shim, K.: Efficient algorithms for mining outliers from large data sets. In: Proceedings of the 2000 ACM SIGMOD International Conference on Management of Data, pp. 427–438. ACM Press (2000)
19. Rayana, S.: ODDS Library (2016). http://odds.cs.stonybrook.edu
20. Ritchie, H., et al.: Coronavirus Pandemic (COVID-19). Our World in Data (2020)
21. Rousseeuw, P., Driessen, K.: A fast algorithm for the minimum covariance determinant estimator. Technometrics **41**(3), 212–223 (1999)
22. Stanway, A.: Etsy skyline. Online Code Repos (2013). https://github.com/etsy/skyline
23. Touchette, H.: The large deviation approach to statistical mechanics. Phys. Rep. **478**(1–3), 1–69 (2009)

24. Srinivasa Varadhan, S.R.: Large deviations and applications. SIAM (1984)
25. Wang, C., Viswanathan, K., Choudur, L., Talwar, V., Satterfield, W., Schwan, K.: Statistical techniques for online anomaly detection in data centers. In: 12th IFIP/IEEE International Symposium on Integrated Network Management (IM 2011) and Workshops, pp. 385–392. IEEE (2011)

Adaptive Regularization of B-Spline Models for Scientific Data

David Lenz[1]([✉]) [iD], Raine Yeh[2] [iD], Vijay Mahadevan[1] [iD], Iulian Grindeanu[1] [iD], and Tom Peterka[1] [iD]

[1] Argonne National Laboratory, Lemont, IL, USA
{dlenz,mahadevan,iulian,tpeterka}@anl.gov
[2] Google Inc., New York, NY, USA
raineyeh@google.com

Abstract. B-spline models are a powerful way to represent scientific data sets with a functional approximation. However, these models can suffer from spurious oscillations when the data to be approximated are not uniformly distributed. Model regularization (i.e., smoothing) has traditionally been used to minimize these oscillations; unfortunately, it is sometimes impossible to sufficiently remove unwanted artifacts without smoothing away key features of the data set. In this article, we present a method of model regularization that preserves significant features of a data set while minimizing artificial oscillations. Our method varies the strength of a smoothing parameter throughout the domain automatically, removing artifacts in poorly-constrained regions while leaving other regions unchanged. The behavior of our method is validated on a collection of two- and three-dimensional data sets produced by scientific simulations.

Keywords: B-Spline · Regularization · Functional approximation

1 Introduction

Data sets assembled from scientific simulations or experimental readings are often defined as a list of position-value pairs, where each data point consists of a measurement and the corresponding location of that measurement. These point locations can form structured grids, unstructured meshes, or unconnected point clouds, depending on the application. Methods for analyzing these data often apply only to particular layouts; usually, numerical analysis techniques become more complex as the geometry of the point locations becomes more general (e.g. point clouds). Even seemingly straightforward tasks such as interpolation can be computationally burdensome on unstructured point clouds and numerically

This work is supported by the U.S. Department of Energy, Office of Science, Advanced Scientific Computing Research under Contract DE-AC02-06CH11357, Program Manager Margaret Lentz.

inaccurate on highly nonuniform meshes. One way to avoid these challenges is by representing a data set with a mathematical function and then analyzing the function instead of the original data. This can substantially streamline the process of interpolation and differentiation away from data points, simplify visualization tasks, and make resampling the data almost trivial.

The focus of this article is the approximation of scientific data sets by (tensor-product) B-splines. B-splines are a family of smooth functions used ubiquitously throughout geometric modeling [14] and form the underpinnings of isogeometric analysis (IGA) [11]. Recent study has shown that large, complex data sets produced by scientific simulations at extreme scale can be effectively modeled by B-splines [16]. Similar results have also been obtained for nonuniform rational B-splines, or NURBS, which are a generalization of B-splines [15].

B-splines have a number of properties that make them useful as a functional representation of data. B-splines are high-order approximants, and evaluating, differentiating, and integrating a B-spline model is fast and numerically stable [3]. Crucially, differentiation and integration can be computed in closed-form and incur no additional loss of accuracy, unlike finite differences or Riemann sums. Thus, once a sufficiently accurate spline has been computed to represent the data, it is often more productive to analyze the functional model than the original data.

However, computing a best-fit B-spline requires solving a linear system which may be ill-conditioned or rank-deficient. A common cause of this ill-conditioning is an input data set that contains both sparse and dense patches of points in proximity to each other. Without additional effort, solving this system can produce a function that oscillates strongly between data points or even diverges in regions where input data are very sparse. This problem can be hard to detect automatically, since error metrics are usually defined in terms of the pointwise error between the original data and the model. Spurious oscillations occurring away from the input data will not be captured by these metrics.

To address these challenges, we developed a new method for fitting B-splines to unstructured data that reduces or eliminates oscillations while leaving critical features of the data set unchanged. Our method regularizes the solution to the B-spline fitting problem by adding a variable-strength smoothing parameter that automatically adapts based on characteristics of the input data set. This additional term smooths out spike artifacts in regions where the data set is very sparse but does not do any smoothing where data points are densely packed, thereby preserving accuracy in these regions. In addition, our method creates well-defined spline models even for data sets with irregular boundaries. No knowledge of the boundary is required; the method automatically handles areas outside the boundary that contain no data points.

The remainder of this paper is organized as follows. A review of related ideas and methods is given in Sect. 2. In Sect. 3, we provide a primer on the mathematical details used to describe B-splines throughout the paper. Our main result, a method for adaptive regularization of B-spline models, is described in Sect. 4. We then exhibit the performance of this method in Sect. 5 with a series of numerical examples. We summarize directions for further research in Sect. 6 and present conclusions in Sect. 7.

2   Related Work

Creating B-spline models to represent unstructured data sets is a particular example of scattered data approximation (SDA), a broad area of study concerned with defining continuous functions that interpolate or approximate spatially scattered inputs. SDA is often applied to image reconstruction problems, where an experimental or physical constraint prohibits the collection of uniformly-spaced samples, such as medical [1], seismic [5], or astronomical [18] imaging. An introductory comparison of SDA methods was compiled by Francis et al. [8].

Ill-conditioned numerical methods are a persistent challenge throughout the SDA literature, and a number of techniques have been proposed to increase numerical stability. Our approach is most similar to the variational methods of SDA, in which the magnitude of the approximating function's derivative (or "roughness") is minimized. The early work of Duchon [4] is a canonical example. Historically, roughness minimization has been achieved through the use of smoothing splines; a thorough exposition of smoothing splines can be found in the book by Gu [10]. The application of smoothing splines requires a trade-off between accuracy and roughness minimization, since aggressively penalizing roughness tends to degrade accuracy. Therefore, much work has been devoted to parametrizing this trade-off appropriately. Craven and Wahba [2] developed the influential "cross-validation" approach, which is expanded upon by Gu [9].

The functional approximation used in this article is based on global tensor-product B-splines, but a number of other spline-based regression methods have been proposed. Truncated thin-plate splines were used by Wood [19] to improve the efficiency of thin-plate regression splines while maintaining their characteristic stability. Lee et al. [13] utilized hierarchies of B-splines to fit unstructured data points, but the instability arising from sparse point distributions was not treated explicitly. Francis et al. [8] consider a two-step process for resampling unstructured point clouds with variable point density onto unstructured grids. While this method does not construct a functional approximation, it does show good performance as a resampling methodology.

Our novel adaptive regularization procedure was first explored in the dissertation of the second author [20]. The method is also directly inspired by the work of El-Rushaidat et al. [6], in which a two-level regularization process was introduced in the context of resampling unstructured data onto structured meshes. However, their method requires an ad-hoc selection of the criteria to switch between high and low regularization strengths, as well as an application-dependent overall level of smoothing. A notable contribution in our work is a continuously varying regularization strength (not two-level) that is adapted automatically.

3   Background on B-Splines

In this section, we provide a brief overview of the basic definitions and constructions necessary to describe B-spline models for scientific data. A thorough presentation on the fundamental theory of B-splines can be found in the books by de Boor [3] and by Piegl and Tiller [17].

3.1 B-Spline Curves

A one-dimensional B-spline curve of degree p in $\mathbb{R}^D$ is a parameterized curve

$$C(u) = \sum_{j=0}^{n-1} N_{j,p}(u) P_j, \tag{1}$$

where each $N_{j,p}$ is a piecewise-polynomial function of degree p, and each $P_j \in \mathbb{R}^D$ is a "control point" in D-dimensional space.

The B-spline basis functions, denoted $N_{j,p}$, are defined on the parameter space $[0,1] \subset \mathbb{R}$, which is divided by a nondecreasing sequence of "knots" $t_0 \le t_1 \le \ldots \le t_{n+p} \in [0,1]$. Each basis function $N_{j,p}$ is a bump function in $[t_j, t_{j+p+1}]$ and zero elsewhere.[1] In this paper, we will assume that the degree of the B-spline is fixed and drop the p subscript, instead denoting the jth function as N_j.

In order to simplify notation when describing high-dimensional tensor product splines, we use multi-indices to index quantities in multiple dimensions simultaneously. A multi-index $\alpha = (\alpha^1, \ldots, \alpha^d)$ is a d-tuple of nonnegative indices, where the sum of components of α is denoted $|\alpha| = \sum_k \alpha_k$.

We will often consider index sets for our multi-indices in the form of

$$A = \{\text{all } \alpha \in \mathbb{N}^d \text{ such that } 0 \le \alpha^k < n_k \text{ for } 1 \le k \le d\}, \tag{2}$$

where n_k are previously defined positive numbers. We impose a lexicographic ordering on these sets, and denote by $[\alpha]_A$ the index of α in the lexicographic ordering of A. In the following sections, we consider matrices in which each column corresponds to a multi-index. In this scenario, we list multi-indices in lexicographic order; thus, the multi-index α corresponding to the jth column satisfies $[\alpha]_A = j$.

3.2 Tensor Product B-Splines

Tensor product B-splines are a natural extension of B-spline curves to higher-dimensional manifolds, such as surfaces, volumes, and hypervolumes. Here, we denote by d the dimension of the tensor product volume and D the dimension of the ambient space (for instance, a 2D surface in 3D space would correspond to $d = 2, D = 3$). The parameter space for a d-dimensional tensor product B-spline is $[0,1]^d$, which is divided by d different knot vectors $\mathbf{t}_k = \{t_k^j\}_{j=0}^{n_k+p}$, $k = 1, \ldots, d$.

Given a tuple $u = (u^1, \ldots, u^d) \in [0,1]^d$, the tensor product basis functions are defined as

$$N_\alpha(u) = \prod_{k=1}^{d} N_{\alpha^k}^k(u^k). \tag{3}$$

where α is a multi-index as described above and $N_{\alpha^k}^k$ is the $(\alpha^k)^{th}$ basis function with respect to the knot vector $\mathbf{t}_k$. With $n_k + p + 1$ total knots in each dimension,

---

[1] We consider only "clamped" knot sequences in this paper; thus, the first $p+1$ knots are always 0 and the last $p + 1$ knots are always 1.

there are n_k basis functions in each dimension.[2] Therefore, the total number of tensor product basis functions is $n_{tot} = \prod_{k=1}^{d} n_k$, which is also the total number of control points for the tensor product spline.

A d-dimensional tensor product B-spline in $\mathbb{R}^D$ is a function of the form

$$C(u) = \sum_{\alpha \in A} N_\alpha(u) P_\alpha, \tag{4}$$

where A is the set of all basis functions and $P_\alpha \in \mathbb{R}^D$ for each α.

3.3 Optimal Control Points

Given a collection of knot vectors and polynomial degree, the best-fit B-spline to a given data set is determined by a linear least-squares minimization problem. Let $\{Q_i\}_{i=0}^{m-1}$ be the list of points in $\mathbb{R}^D$ to be approximated with a d-dimensional tensor product spline. For each $0 \leq i < m$, let $v_i \in [0,1]^d$ be the parameter tuple corresponding to the point Q_i. The optimal control points are determined by the least-squares minimization problem:

$$\{\hat{P}_j\} = \operatorname*{argmin}_{P_j} \sum_{i=0}^{m-1} \|Q_i - C(v_i)\|^2. \tag{5}$$

This minimization problem can be rewritten in normal form by differentiating the objective function in Eq. (5) with respect to each of the control points. The normal system reduces to the matrix equation $\mathbf{N}^T \mathbf{N} \mathbf{P} = \mathbf{N}^T \mathbf{Q}$, where

$$\mathbf{N}_{ij} = N_\alpha(v_i) \quad \text{where } [\alpha]_A = j, \qquad \mathbf{P}_{ij} = P_\alpha^j, \quad \text{where } [\alpha]_A = i, \qquad \mathbf{Q}_{ij} = Q_i^j. \tag{6}$$

The superscripts in the above equations index the components of the vectors P_α and Q_i. $\mathbf{N}$ is a $m \times n_{tot}$ matrix, often called the "B-spline collocation matrix," $\mathbf{P}$ is an $n_{tot} \times D$ matrix with each row containing a control point, and $\mathbf{Q}$ is a $m \times D$ matrix with each row containing an input point.

Typically, the matrix $\mathbf{N}$ is very sparse, and this system may be solved with an iterative method or sparse direct solver. However, as we show in the following section, this system is ill-conditioned when the sample density of the input points P_i varies from region to region.

4   Adaptive Regularization

A significant challenge when modeling unstructured data with tensor product B-splines is the (ill-)conditioning of the fitting procedure. Generally speaking, tensor product B-spline models can oscillate strongly due to overfitting in regions where input data is sparse (see Fig. 1). Our method of adaptive regularization

---

[2] Here we assume for simplicity that the degree of the B-spline is the same in each dimension, but the degree can vary in practice if desired.

produces a unique solution to systems which would otherwise be rank-deficient and improves the overall conditioning of the system. In practice, the adaptively regularized models possess fewer oscillatory artifacts and do not exhibit any divergent behavior in our testing. In contrast to standard regularization techniques, our method does not smooth out the model indiscriminately – instead, it regularizes only those regions that require smoothing.

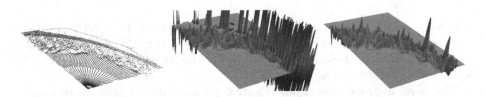

Fig. 1. Left: A data set with nonuniform point density. Center: Best-fit B-spline model without regularization. Right: B-spline model with adaptive regularization. The center image is cropped; spike artifacts in this model extend well outside the frame.

This technique employs a spatially-varying regularization strength that is computed automatically as a function of the relative positioning of input data points to the B-spline knots. In general, the regularization strength increases in regions with little to no input data and decreases (potentially to zero) in regions "saturated" with input points. When the regularization strength is zero throughout a region of the domain, no smoothing is performed in that region; therefore, any sharp features present in densely sampled regions of the domain will be preserved by the adaptive regularization procedure.

Standard roughness minimization can be formulated as a penalized least-squares minimization problem, similar to Eq. (5). The penalty term weights the size of the second derivative at each point with a new parameter, which we denote as $\lambda > 0$. The control points of the regularized spline are defined by:

$$\{\hat{P}_j\} = \operatorname*{argmin}_{P_j} \left(\sum_{i=0}^{m-1} \|Q_i - C(v_i)\|^2 + \lambda^2 \, S(C) \right), \tag{7}$$

where $S(C)$ approximates the size of the second derivatives of C.

Let $w_\alpha \in [0,1]^d$ be the parameter that maximizes the value of N_α. Let ∂^δ denote the partial derivative where the order of derivative in each dimension is given by the components of multi-index δ.[3] We define $S(C)$ to be

$$S(C) = \sum_{\alpha \in A} \sum_{|\delta|=2} \left\| \partial^\delta C(w_\alpha) \right\|^2. \tag{8}$$

Note that the summation above is a sum over all derivatives of order 2, including mixed partial derivatives.

---

[3] For example, $\partial^{(2,0)} f = \partial^2 f/\partial x_1^2$, and $\partial^{(0,2)} f = \partial^2 f/\partial x_2^2$, while $\partial^{(1,1)} f = \partial^2 f/(\partial x_1 \partial x_2)$.

Equation (7) can be converted into a system of equations in the same way as Eq. (5). The only additional step is computing the derivative of $S(C)$ with respect to the control points. In matrix form, the system is

$$\left(\mathbf{N}^T \, \lambda \mathbf{M}^T\right) \begin{pmatrix} \mathbf{N} \\ \lambda \mathbf{M} \end{pmatrix} \mathbf{P} = \mathbf{N}^T \mathbf{Q}, \qquad \text{where } \mathbf{M} = \begin{pmatrix} \mathbf{M}_{\delta_1} \\ \vdots \\ \mathbf{M}_{\delta_n} \end{pmatrix} \tag{9}$$

and $(\mathbf{M}_\delta)_{i,j} = \partial^\delta N_\beta(w_\alpha)$, $[\alpha]_A = i$, $[\beta]_A = j$. Intuitively, each column of $\mathbf{M}_\delta$ describes the ∂^δ partial derivative of an individual B-spline basis function. The matrix $\mathbf{M}$ is the concatenation of all the individual $\mathbf{M}_\delta$ matrices, where $|\delta| = 2$. $\mathbf{N}$, $\mathbf{P}$, and $\mathbf{Q}$ are defined as in Sect. 3.

The novel improvement of our adaptive regularization scheme is to modify the above system of equations by varying the size of λ for each column of $\mathbf{M}$. Since each column of this matrix corresponds to a B-spline basis function and control point, variation in the size of λ provides a mechanism to modify the smoothing conditions on each control point of the spline individually. Due to the local support property of B-splines, setting $\lambda_j = 0$ for control points in a given region "disables" the regularization in that region, while still allowing for smoothing to be applied elsewhere. Algebraically, we replace the scalar parameter λ by a diagonal matrix $\mathbf{\Lambda} = \text{diag}(\lambda_1, \ldots, \lambda_{n_{tot}})$, where each $\lambda_j \geq 0$, and consider the new linear system

$$\left(\mathbf{N}^T \, (\mathbf{M}\mathbf{\Lambda})^T\right) \begin{pmatrix} \mathbf{N} \\ \mathbf{M}\mathbf{\Lambda} \end{pmatrix} \mathbf{P} = \mathbf{N}^T \mathbf{Q}. \tag{10}$$

The value of each λ_i is computed automatically as a function of the relative positioning between input data points and B-spline knots.

To better control this function, we introduce a user-specified parameter called the "regularization threshold," denoted s^*. Changing the regularization threshold adjusts the criterion by which some regions of the domain are smoothed and others are not. As s^* increases, smoothing constraints will be applied to larger and larger regions in the domain.

Let s_j denote the jth column sum of $\mathbf{N}$ and $\tilde{s}_j$ the jth column sum of $\mathbf{M}$. Given $s^* \geq 0$, we define

$$\lambda_j = \frac{\max(s^* - s_j, 0)}{\tilde{s}_j}. \tag{11}$$

Thus, $\mathbf{\Lambda}$ is defined such that every column sum of $\begin{pmatrix} \mathbf{N} \\ \mathbf{M}\mathbf{\Lambda} \end{pmatrix}$ is no less than s^*.

Adapting the regularization strengths λ_j this way has a number of important results. If $s^* = 0$, then $\mathbf{\Lambda} = 0$ and the minimization becomes the usual least-squares problem. When s^* is small, λ_j will be zero unless the jth column sum of $\mathbf{N}$ is small, which is indicative of an ill-conditioned system. Here, adaptive regularization smooths out only those control points which are poorly constrained.

This formulation also explains why the adaptive regularization method preserves sharp, densely sampled features while smoothing out oscillatory artifacts.

In regions of the domain that are densely sampled, control points will be constrained by many data points and thus the corresponding column sum in $\mathbf{N}$ will be relatively large. By choosing s^* to be sufficiently small, all control points in this region will have a regularization strength of zero; i.e. $\lambda_j = 0$. Therefore, the best-fit spline in this region will not be artificially smoothed.

Finally, we remark that the adaptive regularization framework, while defined above in terms of second derivatives, can easily be extended to other derivatives. We find that constraining first and second derivatives simultaneously is particularly helpful when modeling data sets with no points at all in certain regions. This typically happens when the data represent an object with an interior hole or irregular boundary. To consider both first and second derivatives, the only change is to matrix $\mathbf{M}$ in Eq. (10). Originally, $\mathbf{M}$ is the concatenation of all matrices $\mathbf{M}_\delta$, where δ describes a second derivative. To minimize first derivatives as well, we change this to the concatenation of all $\mathbf{M}_\delta$ such that δ describes a first or second derivative. Both versions of the method are described in Sect. 5.

5   Results

We demonstrate the effectiveness of our method with a series of numerical experiments. First, we compare adaptive regularization to uniform regularization where the smoothing parameter has been chosen manually. Next, we study the reconstruction of an analytical signal from sparse samples with varying levels of sparsity. For each sparsity level, we report the error and condition number for and unregularized and adaptively regularized model. We then test the performance of adaptive regularization on data sets with no data in certain regions. In these problems, we construct a B-spline model that extrapolates into regions with no pointwise constraints, and check that the adaptive regularization method produces a reasonable result.

5.1   Data Sets

The performance of the adaptive regularization method was studied on a collection of two- and three-dimensional point clouds with different characteristics. Some data sets were sampled from analytical functions so that we could compute pointwise errors relative to a ground truth, while other data sets were generated by scientific experiments and simulations.

2D Polysinc. The polysinc data set is a two-dimensional point cloud sampling the function $f(x,y) = \text{sinc}\left(x^2 + y^2\right) \text{sinc}\left(2(x-2)^2 + (y+2)^2\right)$.[4] 360,000 point locations are uniformly sampled from the box domain $[-4\pi, 4\pi] \times [-4\pi, 4\pi]$, except at four disk-shaped regions where the sample rate is $50\times$ lower.

XGC Fusion. The XGC fusion data set represents a normalized derivative of electrostatic potential in a single poloidal plane of a Tokamak fusion simulation. The data set contains 56,980 points with an irregular boundary and was produced by the XGC code [12] in a simulation of the gyrokinetic equations.

---

[4]   We consider the unnormalized sinc function: $\text{sinc}(x) = \sin(x)/x$, with $\text{sinc}(0) = 1$.

CMIP6 Climate. The CMIP6 climate data set represents ocean surface temperature in a projected box region around Antarctica. The data set contains 585,765 points with a large hole (representing Antarctica) in the center and was produced by a Coupled Model Intercomparison Project (CMIP6) [7] simulation.

sahex Nuclear. The sahex nuclear data set is derived from a simulation of a single nuclear reactor component, produced with the SHARP toolkit [21]. The three-dimensional data are bounded by a hexagonal prism and point density is coarser in the z dimension than x and y. The data set contains 63,048 points.

5.2   Comparison of Adaptive Versus Uniform Regularization

Applying a uniform regularization strength to an entire model can produce unsatisfactory results, because sufficiently smoothing oscillatory artifacts can also smooth out sharp features. Figure 2 compares our adaptive regularization scheme (with $s^* = 6$) against three strengths of uniform regularization. The data in Fig. 2 is the XGC fusion data set, which contains sharp peaks in a ring but is flat inside the ring. Data are sparse or nonexistent outside the ring. The two images at right show a model with uniform regularization that is too weak, causing artifacts (top), or too strong, dampening the features (bottom). The best uniform regularization strength we could find is given at bottom-left, but even in this example the characteristic peaks in the data are smoothed down.

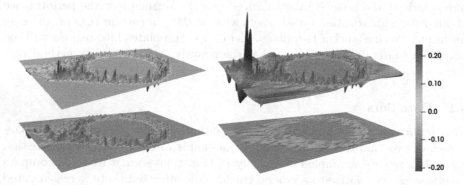

Fig. 2. Comparison of uniform vs adaptive regularization. Clockwise from top-left: Adaptive regularization, uniform regularization with $\lambda = 10^{-6}$, uniform regularization with $\lambda = 10^{-4}$, uniform regularization with $\lambda = 10^{-5}$.

5.3   Accuracy on Analytical Signals

To quantify the accuracy of B-spline models with adaptive regularization, we consider the oscillatory polysinc function with a highly nonuniform input data set (Fig. 3). We illustrate two B-spline models, one fit without regularization and one with our adaptive regularization ($s^* = 1$). Both models are degree four with a 300 × 300 grid of control points. Without regularization, the model diverges in the regions of low sample density; with adaptive regularization, the model produces an accurate representation even where sample density is low. A top-down view of the error profiles is given in the second row of Fig. 3. A close comparison of the ground-truth (top left) and adaptively-regularized spline (top right) shows that the spline model is not artificially smoothed in dense regions, even though the sparse regions are smoothed. In particular, our regularization procedure preserves the distinctive oscillations and peaks in the signal.

Fig. 3. Top row: Synthetic polysinc signal (left), model with no regularization (center), model with adaptive regularization (right). Bottom row: Top down view of input distribution (left), error profile with no regularization (center), error profile with adaptive regularization (right). Area of interest for error calculation is in red at bottom-left.

The degree of sparsity in the input data strongly influences the accuracy of a B-spline model. Table 1 lists the errors in each model for varying levels of sparsity in the voids. The errors are measured in a box around two voids (see Fig. 3) in order to pinpoint the behavior of the models in this region. When the point density is equal inside and outside of the voids (sparsity = 1.0), error for both models is low. As the voids become more sparse, the error in the unregularized model increases by four orders of magnitude while error in the adaptively regularized model stays essentially flat.

Data sparsity also affects the condition number of the least-squares minimization. Table 1 gives the condition number of the matrices $\mathbf{N}$ ('noreg') and $\binom{\mathbf{N}}{\mathbf{M}\mathbf{\Lambda}}$ ('reg') for each sparsity level. As sparsity is increased, the condition number of $\binom{\mathbf{N}}{\mathbf{M}\mathbf{\Lambda}}$ remains lower and steady but the condition number of $\mathbf{N}$ starts higher and eventually becomes infinite. Condition numbers were computed with the Matlab routine svds.

Table 1. Model errors as a function of sparseness. Maximum and L^2 (average) errors are computed for both adaptively regularized and unregularized models in the vicinity of two voids (see Fig. 3). Condition numbers for both minimization problems are reported at bottom.

Sparsity	0.02	0.08	0.16	0.32	0.64	1.00
Max Error (reg)	3.25e−2	2.89e−2	2.71e−2	2.39e−2	1.20e−2	1.16e−2
Max Error (no reg)	1.29e2	6.27e2	2.36e−1	3.65e−2	1.20e−2	1.16e−2
L^2 Error (reg)	1.93e−3	1.53e−3	1.13e−3	7.04e−4	5.56e−4	5.44e−4
L^2 Error (no reg)	2.17e0	4.68e0	3.42e−3	7.35e−4	5.56e−4	5.44e−4
Condition # (reg)	177	980	289	198	121	189
Condition # (noreg)	inf	inf	6.97e4	2.58e3	1.57e3	3.74e3

5.4   Extrapolation into Unconstrained Regions

When a large region of the domain does not contain any data points to constrain the best-fit B-spline problem, the least-squares minimization will be ill-posed and the resulting model can exhibit extreme oscillations. However, data sets with empty regions or "holes" are very common in scientific and industrial applications. For example, some climate models measure ocean temperatures or land temperatures, but not both simultaneously. Industrial simulations often model objects with irregular boundaries, and data from physics simulations are shaped by the locations of detectors.

Although empty regions are usually omitted in subsequent analysis, it is still important to understand and control the behavior of a B-spline model in empty regions. Extreme oscillations near the boundary of a hole can distort the derivative of the model away from this boundary. In addition, attempting to compute simple statistics about the model (minimum, maximum, mean) can be biased if the model exhibits unpredictable behavior in empty regions.

Figure 4 shows a scenario from a simulation of ocean temperatures. The data set is centered around the continent of Antarctica, for which no temperature values are given. Exhibited in the figure are two models, both of degree 2 with a 400×400 grid of control points. The unregularized model at center oscillates between $\pm 10^8$ along the coast (while the input data range from -2 to 10). In contrast, the regularized model (with $s^* = 5$) smoothly transitions at the coast to a near-constant value over the landmass. Regularization was performed with constraints on first and second derivatives, as described at the end of Sect. 4.

We next consider a three-dimensional data set representing power produced in a component of a nuclear reactor (Fig. 5). The data are contained inside a hexagonal prism, but the B-spline model is defined on the bounding box of this prism. Hence the corners of this box are devoid of any data. Without regularization, the least-squares minimization does not converge, so we exhibit only our regularized model (with $s^* = 10$) in Fig. 5. The adaptive regularization method (Fig. 5, center and right) produces an accurate model of the six interior "pins" and is well defined in the corner regions. Some artifacts are observed

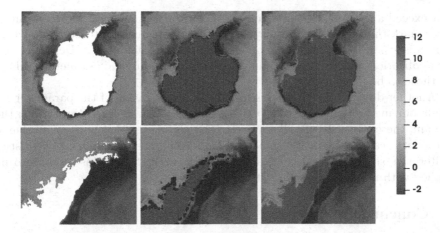

Fig. 4. Ocean temperature simulation around Antarctica. Top row: Input data on a hexagonal mesh (left), B-spline model without regularization (center), B-spline model with adaptive regularization (right). Bottom row: Detail view from the top row.

in the corners, but they are not significant enough to affect the interior of the model. Without regularization, the condition number of the system is infinite; with adaptive regularization, the condition number is 2.18×10^5. The model was produced by constraining first and second derivatives simultaneously.

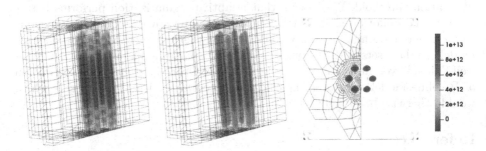

Fig. 5. Power production in a nuclear reactor simulation. Input data (left), spline model (center), spline model top view (right).

6   Future Work

In the construction of our method, we imposed constraints on the first and second derivatives of the B-spline in regions where the density of input data was low or vanishing. However, these additional constraints could have been based on different orders of derivative or been unrelated to derivatives altogether. For instance, a different type of constraint would be one that penalizes deviation from a known baseline value. Another option would be to penalize B-spline values that exceed a given range (for example, the original bounds of the input data). We remark that our procedure for adaptive regularization of tensor product B-splines is separate from the type of artificial constraint imposed. Depending on the application, different constraints may be more useful, and our method allows for those to be used instead.

Another direction for future research is an investigation of the parameter s^*. While our method automatically varies regularization strength throughout the domain, these strengths are all relative to the parameter s^*. Choosing a value of s^* too large can lead to overly-smoothed models. Further research into heuristics or iterative schemes to select s^* automatically would allow this method to be applied with no user interaction at all.

7 Conclusions

Modeling unstructured data sets with tensor product B-splines can be difficult due to the ill-conditioning of the fitting problem. In general, data sets with large variations in point density or regions without data exacerbate this problem to the point that artificial smoothing is necessary. However, smoothing an entire model can wash out sharp features in the data.

We introduced a regularization procedure for B-spline models that preserves features by adapting the regularization strength throughout the domain. Our method automatically varies the smoothing intensity as a function of input point density and relies on a single user-specified parameter, which we call the regularization threshold. We observe that adaptive regularization performs better than typical uniform regularization schemes that may over-smooth some regions while under-smoothing others. We also showed that our method can fit B-spline models to data sets with regions of extremely sparse point density and remain well-defined even in areas without data points. Overall, adaptive regularization of B-spline models produces smooth and accurate models for data sets which would otherwise be difficult to fit.

References

1. Arigovindan, M., Suhling, M., Jansen, C., Hunziker, P., Unser, M.: Full motion and flow field recovery from echo doppler data. IEEE Trans. Med. Imaging **26**(1), 31–45 (2006). https://doi.org/10.1109/TMI.2006.884201
2. Craven, P., Wahba, G.: Smoothing noisy data with spline functions. Numer. Math. **31**(4), 377–403 (1978). https://doi.org/10.1007/BF01404567
3. De Boor, C.: A Practical Guide to Splines. Applied Mathematical Sciences, vol. 27, Springer-Verlag, New York (2001)
4. Duchon, J.: Splines minimizing rotation-invariant semi-norms in Sobolev spaces. In: Schempp, W., Zeller, K. (eds.) Constructive Theory of Functions of Several Variables. LNM, vol. 571, pp. 85–100. Springer, Heidelberg (1977). https://doi.org/10.1007/BFb0086566
5. Duijndam, A., Schonewille, M., Hindriks, C.: Reconstruction of band-limited signals, irregularly sampled along one spatial direction. Geophysics **64**(2), 524–538 (1999). https://doi.org/10.1190/1.1444559

6. El-Rushaidat, D., Yeh, R., Tricoche, X.M.: Accurate parallel reconstruction of unstructured datasets on rectilinear grids. J. Visualization **24**(4), 787–806 (2021). https://doi.org/10.1007/s12650-020-00740-0

7. Eyring, V., et al.: Overview of the coupled model intercomparison project phase 6 (CMIP6) experimental design and organization. Geosci. Model Dev. **9**(5), 1937–1958 (2016). https://doi.org/10.5194/gmd-9-1937-2016

8. Francis, B., Viswanath, S., Arigovindan, M.: Scattered data approximation by regular grid weighted smoothing. Sādhanā **43**(1), 1–16 (2018). https://doi.org/10.1007/s12046-017-0765-y

9. Gu, C.: Cross-validating non-Gaussian data. J. Comput. Graph. Stat. **1**(2), 169–179 (1992). https://doi.org/10.1080/10618600.1992.10477012

10. Gu, C.: Smoothing Spline ANOVA Models. SSS, vol. 297. Springer, New York (2013). https://doi.org/10.1007/978-1-4614-5369-7

11. Hughes, T., Cottrell, J., Bazilevs, Y.: Isogeometric analysis: CAD, finite elements, NURBS, exact geometry and mesh refinement. Comput. Methods Appl. Mech. Eng. **194**(39), 4135–4195 (2005). https://doi.org/10.1016/j.cma.2004.10.008

12. Ku, S., Hager, R., Chang, C.S., Kwon, J., Parker, S.E.: A new hybrid-Lagrangian numerical scheme for gyrokinetic simulation of tokamak edge plasma. J. Comput. Phys. **315**, 467–475 (2016). https://doi.org/10.1016/j.jcp.2016.03.062

13. Lee, S., Wolberg, G., Shin, S.: Scattered data interpolation with multilevel b-splines. IEEE Trans. Visual Comput. Graphics **3**(3), 228–244 (1997). https://doi.org/10.1109/2945.620490

14. Lin, H., Maekawa, T., Deng, C.: Survey on geometric iterative methods and their applications. Comput. Aided Des. **95**, 40–51 (2018). https://doi.org/10.1016/j.cad.2017.10.002

15. Nashed, Y.S.G., Peterka, T., Mahadevan, V., Grindeanu, I.: Rational approximation of scientific data. In: Rodrigues, J.M.F., et al. (eds.) ICCS 2019. LNCS, vol. 11536, pp. 18–31. Springer, Cham (2019). https://doi.org/10.1007/978-3-030-22734-0_2

16. Peterka, T., Nashed, Y., Grindeanu, I., Mahadevan, V., Yeh, R., Trixoche, X.: Foundations of multivariate functional approximation for scientific data. In: Proceedings of 2018 IEEE Symposium on Large Data Analysis and Visualization (2018). https://doi.org/10.1109/LDAV.2018.8739195

17. Piegl, L., Tiller, W.: The NURBS Book, 2 edn. VISUALCOMM. Springer, Heidelberg (1997). https://doi.org/10.1007/978-3-642-59223-2

18. Vio, R., Strohmer, T., Wamsteker, W.: On the reconstruction of irregularly sampled time series. Publ. Astron. Soc. Pac. **112**(767), 74 (2000). https://doi.org/10.1086/316495

19. Wood, S.N.: Thin plate regression splines. J. R. Stat. Soc. Ser. B (Stat. Methodol.) **65**(1), 95–114 (2003). https://doi.org/10.1111/1467-9868.00374

20. Yeh, R.: Efficient knot optimization for accurate B-spline-based data approximation. Ph.D. thesis, Purdue University Graduate School (2020)

21. Yu, Y., Shemon, E., Mahadevan, V.S., Rahaman, R.O.: Sharp multiphysics tutorials. Technical report ANL/NE-16/1, Argonne National Lab. (ANL), Lemont, IL (United States) (2016). https://doi.org/10.2172/1250465

Content-Aware Generative Model for Multi-item Outfit Recommendation

Valery Volokha$^{(\boxtimes)}$ and Klavdiya Bochenina

ITMO University, Kronverksky Pr. 49 bldg. A, 197101 St. Petersburg, Russia
valierii.volokha@gmail.com

Abstract. Recently, deep learning-based recommender systems have received increasing attention of researchers and demonstrate excellent results at solving various tasks in various areas. One of the last growing trends is learning the compatibility of items in a set and predicting the next item or several ones by input ones. Fashion compatibility modeling is one of the areas in which this task is being actively researched. Classical solutions are training on existing sets and are learning to recommend items that have been combined with each other before. This severely limits the number of possible combinations. GAN models proved to be the most effective for decreasing the impact of this problem and generating unseen combinations of items, but they also have several limitations. They use a fixed number of input and output items. However, real outfits contain a variable number of items. Also, they use unimodal or multimodal data to generate only visual features. However, this approach is not guaranteed to save content attributes of items during generation. We propose a multimodal transformer-based GAN with cross-modal attention to simultaneously explore visual features and textual attributes. We also propose to represent a set of items as a sequence of items to allow the model to decide how many items should be in the set. Experimenting on FOTOS dataset at the fill-in-the-blank task is showed that our method outperforms such strong baselines as Bi-LSTM-VSE, MGCM, HFGN, and others. Our model has reached 0.878 accuracy versus 0.724 of Bi-LSTM-VSE, 0.822 of MGCM, 0.826 of HFGN.

Keywords: Outfit recommendations · Set recommendations · Multimodal recommendations · Generative Adversarial Networks (GAN) · Transformers · Recommender systems

1 Introduction

In recent years e-commerce has spread widely, especially under the influence of COVID-19 and related restrictions. A lot of people turned to online shopping over personal visits to shops, which led to an unbound variety of items to compare and combine during the shopping process. The fashion industry and online fashion marketplaces are one of the areas largely affected by this. Fashion compatibility modeling is of increasing interest for researchers and becomes a popular but challenging and contentious topic. There are a lot of downstream applications such as the outfit recommendation [1–11], the personalized

© The Author(s), under exclusive license to Springer Nature Switzerland AG 2022
D. Groen et al. (Eds.): ICCS 2022, LNCS 13350, pp. 164–177, 2022.
https://doi.org/10.1007/978-3-031-08751-6_12

fashion design [12–16], personal wardrobe creation [17, 18], fashion-oriented dialogue systems [19, 20], try-on [19, 20], and others. In this paper, we combined the task of personalized fashion design generation with an outfit recommendations task. We used a generative model to generate new personalized item representations but used them to find real equivalents to build outfit recommendations.

The rapid development of technologies and the increased computational power in the last decade allowed recommender systems to integrate into various areas of our life, including the fashion industry and e-commerce. Modern shopping apps assist and influence customer decisions. Therefore, this is becoming increasingly important to develop personalized and efficient recommender systems for choosing a set of clothes. The main aim of these systems is to automatically assess the compatibility of items and predict missing items of outfits. This area is actively researched, and there is impressive progress, but there are still has some unsolved problems, which limits the efficiency and the flexible usage of these systems.

Classical recommender systems learn to recommend items that have been already combined with input ones before, but it severely limits the compatibility of items and the variety of outfits. Several approaches tried to decrease the impact of this problem. Some of them applied noise to vector representations of input items. Others used variational autoencoders as a base of a model. They showed the effectiveness of recommendations but did not inspect the ability of the model to recommend items that had not been previously combined with the input ones. They only reduced the discontinuity of the latent space of the model, but it is not a complete solution, and the model is still fitted to recommend existing outfits but not to generate the most compatible items.

Generative adversarial network (GAN) based models are used to overcome this problem [12–16] but they also have some limitations. In particular, they use a fixed number of input and output items (primarily, one or two input items and a single output item) [12–16]. The main reason is that they frequently use noise as a placeholder for blanked items. A large amount of input noise makes the model unpredictable and reduces the influence of the input items to newly generated ones. Moreover, to the best of our knowledge, the presented GAN-based solutions aim to synthesize images of new items [12–16], but in the case of e-commerce, online shops, and recommending existing items that users can buy, there is no need to generate images directly.

Consequently, in this paper, we have focused on the compatibility modeling sets with a variable number of items by data of multiple modalities. Despite the fact that the data from several modalities are used in many approaches, to the best of our knowledge, most of these explore modalities separately in the field of recommending sets of items, including in the field of fashion. On the contrary, in our scheme, we have focused on capturing compatibility features between modalities simultaneously.

In this paper, we propose the following contributions to solve described problems:

- Different from the existing GAN-based methods which have a fixed number of input and output items, we propose to use a transformer-based GAN and represent a set of items as a sequence of items with start and end tags, similar to a sentence of words. This allows the model not only to generate a complete set of items but also to decide how many items should be in the set.

- To simultaneously explore multimodal features, we propose to use a cross-modal attention module in our transformer. Transformer architecture with self-attention and cross-modal attention allows GAN to jointly generate vectors of visual and textual features depending on multimodal input ones.

The rest of the paper is organized as follows. First, we described and discussed several related recommender-based and GAN-based studies. Second, we formulated a problem and described our proposed model and its parts. Then, we described conducted experiments, used dataset, compared with our model baselines, and comparison results. Finally, we summarized the contributions of this paper and obtained results.

2 Related Works

2.1 Recommender-Based Solutions

Plummer et al. (2018) proposed the method that embeds compatible items and outfits close to each other for searching similar items to the input ones and replacing items if necessary [21]. Tangsend et al. (2018) also embedded items and used a binary classifier to predict the compatibility of items within a set and to rank sets by compatibility [4]. Lu et al. (2019) used CNN to extract visual features from images of all types of items, and type-specific embeddings (each item type is associated with its own embedding) to project feature vectors depending on the type of item [7]. The authors proposed a fashion hashing network (FHN) which uses HashNet and BPR to assess the compatibility of computed feature vectors of a set. They personalize recommendations by adding a vector of user features. The problem with such approaches is that they only use visual features of items and ignore content attributes and descriptions. However, content information is very important and is near-always available in real conditions. For example, in the outfit recommendation task, black jeans and black leggings are close to each other in embedding space, but they are very different, and such content attribute as a material can help to separate them.

To overcome this problem, Xintong et al. (2017) proposed multimodal Bi-LSTM to sequentially predict the next item conditioned on previous ones to learn their compatibility relationships [1]. They also proposed a method to explore visual and textual together by projecting visual features to the space of textual attributes.

Cui et al. (2019) proposed multimodal graph-based neural network that optimizes Fashion Graph and uses the attention layer to compute the compatibility score [8]. The authors used deep convolutional neural network to extract visual features from images and one-hot-encoding to represent titles as boolean vector. They described a strategy to train the multimodal node-wise graph neural network (NGNN) and showed that their approach outperforms such baselines as unimodal GGNN, Bi-LSTM, and others.

Li et al. (2020) also proposed multi-model graph-based neural network [6] and unified two tasks: fashion compatibility modeling and personalized outfit recommendation. They proposed Hierarchical Fashion Graph Network (HFGN) to model relationships among users, items and outfits simultaneously. They also proposed an R-view attention map, which can capture the potential compatibility knowledge better. The authors demonstrated that their model outperforms such state-of-the-art methods as NGNN and FHN.

Cardoso et al. (2018) proposed multimodal embedding that takes item images, type, description and some content characteristics and projects them to interpreted categorical embedding [9]. The authors also introduced a hybrid architecture to compare and rank items that combines content-based and collaborative inputs as well as an embedding to project them. Elaine M.B. et al. (2019) proposed a method to assess the compatibility of items that use this embedding [22]. They showed how to recommend a highly scored set of items to a user by several input items. Their approach uses embeddings and applies dot product and softmax operations to calculate the compatibility score.

Sagar et al. (2020) proposed a multi-model method to personalized outfit recommendations with attribute-wise interpretability [10]. Their method is based on BPR and ranks triplets of items. The obtained vectors of features are integrated with the embedded user preferences vector and used by BPR to compute the compatibility score. The authors showed that their method outperforms such baselines as Bi-LSTM, VTBPR, GP-BPR, and others.

Yuan et al. (2018) proposed simple convolutional generative network for next item recommendation based on dilated CNN architecture [11]. The authors tried to implement the idea of learning short- and long-range dependencies between items using CNN. They stacked holed convolutional layers and used residual block structure. Results showed that the proposed generative model attains state-of-the-art accuracy.

2.2 GAN-Based Solutions

The main problem with the approaches proposed above is that they are trained to recommend items that were encountered in existed sets along with the items received as input. This severely limits the possible variety of combinations of items. Some authors tried to solve this problem by applying noise to the input items or using a variational autoencoder as the base of their model, but it does not solve the problem completely, but makes the latent space of the model less sparse. Compatible items that are not presented in the existing sets will still not be recommended together. To overcome this problem, it is proposed to use generative adversarial neural networks (GAN) that explore the items and the compatibility of items and learn to generate new items from noise. In order to offer real items to the user, the generated items are compared with the real ones from the target dataset.

Kang et al. (2017) proposed a method based on Siamese CNNs approach and showed how to use the proposed GAN model for outfit recommendations [13]. The model takes a text query and a history of user outfits and generates personalized fashion recommendations. The authors compared their method with some baselines and showed that it outperforms such base methods as WARP, FM, BPR (and some variants), and others.

Sudhir and Mithun (2019) combined the encoder-decoder architecture with GAN approach and proposed the model which takes a vector of features of input item image and random noise to generate a new item [12]. The noise is used to diversify the generated items.

Yu et al. (2019) also used encoder-decoder-based GAN to explore compatibility of items and generate compatible items for outfits [16]. An encoder-decoder-based generator was used to generate an item by an input one. Two same discriminators were used to compute the compatibility score of the built outfit and evaluate whether a real item was generated or not. The authors proposed to use the BPR-based method to compute compatibility by ranking a positive, a generated, a negative and a random compatible outfit. They showed that their method is more accurate than directly assessing by classifying or scoring the generated outfit.

Liu et al. (2019) proposed the Attribute-GAN model for clothing matching [14]. They added the second discriminator to assess attributes of synthetic and negative images. To extract attributes from the synthetic image, the model projects it to a vector of visual features, splits the vector into parts, and uses several dense layers to project them to attributes. Extracted one-hot encoded attributes and attributes of the negative item are used to calculate the loss. The real-fake discriminator is used to calculate the second part of the loss function.

Liu et al. (2020) proposed the multimodal method to generate an item by an input one [15]. It combines the encoder-decoder-based GAN and TextCNN to generate a new item by an input item. The encoder projects an input image to a vector of visual features as TextCNN projects an input description to a vector of context features. Vectors of features are concatenated and used by the decoder to generate a new item. The authors used a loss function that combines four parts: BPR loss, pixel difference between generated image and corresponding ground truth image from a dataset, compatibility of input and ground truth images, and compatibility of their descriptions.

The first problem with all of the GAN-based methods is that they use a fixed number of input and output items in the set. The models can only operate with the number of items specified during the training, and changing this value will require to retrain the model. The second problem is that the existing solutions either use only the visual features and extract content attributes from generated images, or process text separately from generation. However, such approaches do not guarantee keeping the attributes of the input items and the reliability of the evaluation of the attributes of the received items. We are trying to solve both problems mentioned above in this study.

3 Transformer-Based GAN for Outfit Recommendations

In this section, we describe the proposed method that tries to improve the compatibility of set items, to solve the problem of a fixed number of input and output items and to decrease the content features vanishing during generation. First, we define the problem and introduce the method of representing a set of items as a sequence of items. Then we describe our multimodal transformer block with cross-modal attention for simultaneous exploration of the compatibility of items inside a modality and between modalities. Finally, we present our multimodal transformer-based GAN with cross-modal attention.

3.1 Problem Formulation

Suppose we have some item domains $D = \{D_1, \ldots, D_N\}$ and a domain S of sets of items with a variable number of items. Each item $I_{i,j} \in D_i, i \in 1 \ldots N, j \in 1 \ldots |D_i|$

is associated with a visual image $Vis_{I_{i,j}}$ and a textual description $C_{I_{i,j}}$ (C as content). Since sets of items have a variable number of items, we extend them to a pre-defined maximum number of items M with a random normal noise $N = (0, 1)$ as placeholders, then an extended set of items $S_k = \left\{ \left(Vis_{I_1^k}, C_{I_1^k} \right), \left(Vis_{I_2^k}, C_{I_2^k} \right), \ldots, \left(Vis_{I_n^k}, C_{I_n^k} \right) \right\}$, $|S_k| = n$ can be described as $\hat{S}_k = \left\{ S_k, \left(Vis_{N_{n+1}^k}, C_{N_{n+1}^k} \right), \ldots, \left(Vis_{N_M^k}, C_{N_M^k} \right) \right\}$, where Vis_N^k, C_N^k are randomly sampled vectors of visual and textual features. We focused on devising an end-to-end multimodal generative compatibility modeling scheme Sch that is able to learn the compatibility c between a set of items and project an input noise to synthetic items $\tilde{I}$, by introducing the network G as follows:

$$
\begin{aligned}
G\left(\hat{S}_k | \Theta_G\right) &\to \tilde{S}_k; \\
\tilde{S}_k &= \left\{ S_k, \left(Vis_{\tilde{I}_{n+1}^k}, C_{\tilde{I}_{n+1}^k} \right), \ldots, \left(V_{\tilde{I}_M^k}, C_{\tilde{I}_M^k} \right) \right\}; \\
c_k &= Sch\left(\tilde{S}_k | \Theta^{Sch}\right),
\end{aligned} \tag{1}
$$

where Θ^G and Θ_{Sch} are a set of parameters to be learned of generator G and scheme Sch, c_k – is a compatibility score of generated set $\tilde{S}_k$.

3.2 Multimodal Transformer-Based GAN with Cross-Modal Attention

Multimodal Transformer Block. As the exploring of compatibility of items and the generation of new compatible items are the main tasks of our scheme, we can use the self-attention module, which is already presented in traditional transformer architecture [23]. It allows exploring the compatibility of items inside a single modality, for example, a visual or textual modality. However, each item is associated with data of multiple modalities. The usage of multi-way scheme, which explores each modality separately and then fuzes them, is not an optimal solution because the final vector contains features of multiple modalities but features of each modality are explored without others and are not connected with them.

To explore compatibility of items between the modalities, we propose to use a cross-modal attention (CMA) module firstly described in Click or tap here to enter text. Similar to the classical self-attention module, each sequence of items is represented as query (Q), key (K), value (V), however, K and V are swapped between modalities: $K_{Vis} \to K_C, K_C \to K_{Vis}, V_{Vis} \to V_C, V_C \to V_{Vis}$, where K_{Vis} and K_C are keys of a visual and textual modalities correspondingly, V_{Vis} and V_C are values of a visual and textual modalities correspondingly.

To exploit the advantages of self-attention and cross-modal attention modules simultaneously, we propose to stack these modules as shown in Fig. 1. First, we propose to forward vectors to the cross-modal attention module and obtain vectors of features that contain compatibility information between items of the same modality and between modalities.

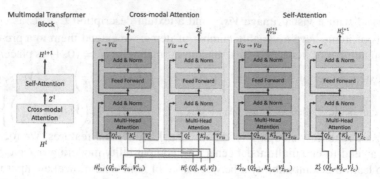

Fig. 1. Multimodal transformer block architecture (left) with cross-modal attention (center) and self-attention (right) modules.

Figure 1 shows the multimodal transformer block, which explores two modalities: a visual and a textual, but it can be easy to extend the scheme with additional modalities.

Multimodal Transformer-Based GAN. The problem of fitting a model to recommend items combined with input ones before is important in our opinion. It severely limits a variety of compatible items and, as a result, a variety and a number of sets. The obvious and efficient solution to smooth this problem is to use a GAN [12–16]. We propose to stack several multimodal transformer blocks described above and use them as a body of a generator of our GAN-based model, as shown in Fig. 2.

The generator G aims to translate the given sequence of items $\hat{S}_t$ aligned between modalities, which contains a noise as a placeholder to the missed items, to a compatible with non-noise items sequence of items $\hat{S}'_t$ with compatibility score c_t. The generator also aims to predict a variable number of items in a target sequence.

The standard method to learn the GAN-based generator is to use a real-fake discriminator. But, in fact, the traditional real-fake discriminator can only enforce the generator to produce realistic vectors. These vectors can be incompatible with input ones. In our context, we need not only to synthesize realistic vectors but also learn the compatibility of input vectors and synthesize new compatible vectors with them. To achieve this, we propose to use (in addition to a real-fake discriminator $Discr_{rf}$) a compatibility discriminator $Discr_{comp}$ as the guidance for compatibility modeling.

The body of discriminators is similar to the generator body. The real-fake discriminator takes a target sequence of items and projects it to a sequence of latent representations. Then, dense layer is used for each item of sequence to compute pre-item real-fake scores. Finally, real-fake scores are summed to compute a complete real-fake score and corresponding real-fake loss.

To overcome the problems of traditional GANs, [16, 25] proposed to use a combination of "relativistic discriminator" [25] and LSGAN [26] and described the following in Eq. 2 losses for the real-fake discriminator and generator.

$$L_{rf}^{Discr} = \frac{1}{2}\mathbb{E}_{o^r}\left[\left(s_{rf}\left(o^r\right) - \mathbb{E}_{o^f}s_{rf}\left(o^f\right) - 1\right)^2\right]$$

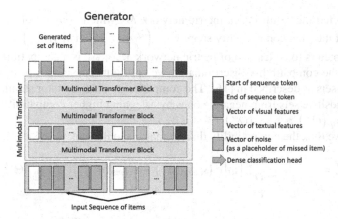

Fig. 2. Scheme of multimodal transformer. It is a base of generator and discriminators.

$$+ \frac{1}{2}\mathbb{E}_{o^f}\left[\left(s_{rf}\left(o^f\right) - \mathbb{E}_{o^r}s_{rf}\left(o^r\right) + 1\right)^2\right]\Big|\Theta_{rf}^{Discr}, \tag{2}$$

where s_{rf} is a real-fake score, $o^r \sim O^r$, $o^f \sim O^*$, O^r is a domain of real sets, O^* is a domain of synthesized sets of items, Θ_{rf}^{Discr} is a set of parameters to be learned of the real-fake discriminator. It can be seen that the discriminator keeps a margin between real and fake data. The generator should eliminate this gap by minimizing:

$$L_{rf}^G = \frac{1}{2}\mathbb{E}_{o^r}\left[\left(s_{rf}\left(o^r\right) - \mathbb{E}_{o^f}s_{rf}\left(o^f\right)\right)^2\right]$$
$$+ \frac{1}{2}\mathbb{E}_{o^f}\left[\left(s_{rf}\left(o^f\right) - \mathbb{E}_{o^r}s_{rf}\left(o^r\right)\right)^2\right]\Big|\Theta^G, \tag{3}$$

where Θ^G is a set of parameters to be learned of the generator. As [16] uses a single item as input, we re-defined the real-fake score function as a function to compute the real-fake score of a set. It calculates a score of a set as a summation of a per-item real-fake score of each item inside each modality as follows:

$$s_{rf}\left(\tilde{S}_k\right) = \sum_{i=0}^{|\tilde{S}_k|} s_{rf}(I_i^k) = \sum_{j=0}^{|\tilde{S}_k|}\left(s_{rf}(Vis_i^k) + s_{rf}(C_i^k)\right) \tag{4}$$

The compatibility discriminator similarly takes a target sequence and projects it into latent space. To obtain the compatibility score of a set of items by the latent representation, we have modified the scheme of computing the compatibility score by ranking, which is proposed in Click or tap here to enter text. We first take the element-wise product of each pair of items inside each modality and sum them to obtain a latent space representation z of the set:

$$z\left(\tilde{S}_k\right) = \sum_{i=0}^{|\tilde{S}_k|}\sum_{j=0}^{|\tilde{S}_k|} I_i^k \odot I_j^k, i \neq j; I_i^k \odot I_j^k = \left(Vis_i^k \odot Vis_j^k + C_i^k \odot C_j^k\right) \tag{5}$$

Then we fed the result into a metric network M, which consists of several dense layers, to get the final compatibility score $s_{comp}\left(\tilde{S}_k\right) = M\left(z\left(\tilde{S}_k\right)|\Theta_M\right)$, where Θ_M is a set of parameters to be learned of metric network M, s_{comp} is a compatibility score.

To train the compatibility discriminator, we split our dataset into positive O^+ and negative O^- sets as described in [16]. The compatibility discriminator should be able to distinguish positive sets from negative ones by assigning higher compatibility scores to positives $s_{comp}\left(O^+\right) > s_{comp}\left(O^-\right)$.

To achieve this, the compatibility discriminator should seek to reduce the loss:

$$L_{comp}^{Discr} = -\mathbb{E}_{\substack{o^+ \sim O^+ \\ o^- \sim O^-}} \left[\ln\left[\sigma\left(s_{comp}\left(o^+\right) - s_{comp}\left(o^-\right)\right)\right]\right] + \lambda_{\Theta_M}\left|\Theta_{comp}^{Discr}\right., \qquad (6)$$

where σ is the sigmoid function, λ is a regularization term, Θ_{rf}^{Discr} is a set of parameters to be learned of the compatibility discriminator, $\Theta_M \in \Theta_{rf}^{Discr}, o^+ \sim O^+, o^- \sim O^-$. To achieve this by generator, it should synthesize a set $\tilde{S}_k \sim O^*$ with a similar compatibility score as its positive set S_k^+. As a result, it should seek to reduce the loss:

$$L_{comp}^G = \frac{1}{2}\mathbb{E}_{o^+}\left[s_{comp}\left(o^+\right) - \mathbb{E}_{o^*}\left(s_{comp}\left(o^*\right)\right)^2\right]$$
$$+ \frac{1}{2}\mathbb{E}_{o^*}\left[s_{comp}\left(o^*\right) - \mathbb{E}_{o^+}\left(s_{comp}\left(o^+\right)\right)^2\right]\left|\Theta^G, \qquad (7)\right.$$

where $o^* \sim O^*$.

The overall architectures of the real-fake and the compatibility discriminators are shown in Fig. 3.

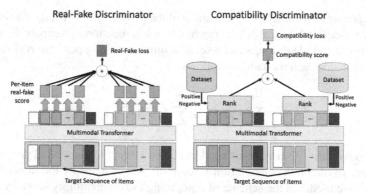

Fig. 3. Scheme of real-fake discriminator (left) and compatibility discriminator (right).

It should be noted that the multimodal transformers of discriminators have shared parameters $\Theta_{shared}^{Discr}, \Theta_{shared}^{Discr} \in \Theta_{rf}^{Discr}, \Theta_{shared}^{Discr} \in \Theta_{comp}^{Discr}$.

The final objective of our generator is to minimize loss function as follows in Eq. 8, and the final objective of our complete scheme is to minimize loss function as follows in Eq. 9:

$$L^G = \lambda_1 L_{rf}^G + \lambda_2 L_{comp}^G;$$ (8)

$$L = L^G(\lambda_1, \lambda_2) + \lambda_3 L_{rf}^{Discr} + \lambda_4 L_{comp}^{Discr},$$ (9)

where $\lambda_1, \lambda_2, \lambda_3, \lambda_4$ are model tradeoff parameters.

The training process can be described as follows. First, a step is made on the discriminators, after which they are updated to estimate the generator. A result of forwarding a batch to the generator is used to calculate losses of discriminators, but the losses are backwarded only to corresponding discriminators. Then the step is made on the generator with estimates from the discriminators. Similar to discriminators, the loss is backwarded only to the generator. This process continues until the generator converges.

4 Experiments

4.1 Dataset

Most of the previous works for outfit recommendation have used either Amazon data containing co-purchase information or Polyvore data containing outfits created by users. Co-purchase does not always mean that items are compatible as items are typically not bought with the intention of being worn together, but it is more likely to reflect a user's style preference [9, 22]. As a result, the estimate on this dataset is not reliable. Outfits in datasets obtained from Polyvore are built by users which gives a stronger signal of compatibility. They contain a variable number of items per outfit, visual and textual modalities, and some other data. They are fully suitable for evaluation on them in terms of the content, but they are outdated. Items from them are out of fashion and mostly not available. We decided to use the public dataset FOTOS which contains outfits and corresponding items [19, 20]. Each outfit and each item are associated with an image and metadata. Outfits contain a variable number of items, and they are created by users, similar to Polyvore. It consists of 10,988 compatible outfits and 20,318 items.

4.2 Baselines

To verify the effectiveness of proposed method, we compared it with the following baseline methods. **FHN** uses only visual features. It encodes them with category encoders and then learns one-hot encodings for item embeddings. The outfit score is the mean of pairwise compatibility scores of outfit items. **Bi-LSTM-VSE** is interpreting an outfit as a sequence of items and exploits the outfit compatibility by a bi-directional LSTM and visual-semantic consistency. **NGNN** is a node-wise graph-based neural network that optimizes a multimodal fashion graph to uncover the complex relationships among items and assess the compatibility score. **HFGN** is a hierarchical graph neural network to model relationships among users, items and outfits simultaneously. It uses message propagation across items and attention to better capture compatibility between items. The model contains two levels for exploring interactions between users and outfits, and between outfits and items correspondingly. For our experiment, we have used only the second one. **MGCM** is an autoencoder-based GAN model that uses deep CNN to extract visual features from images and TextCNN to extract textual features from descriptions. It explores the visual and textual features simultaneously.

It is worth noting some implementations details. MGCM generates images by default, but this is not necessary for this task. The model has been adapted to generating feature vectors instead of images. It also takes one item as input and generates one item as output. To generate one item by multiple ones, the encoded vectors of items have been averaged and the obtained vector have been used as input for the generator. MGCM, and our transformer-based GAN model are generating synthetic vectors of features. To assess the accuracy of models, we are comparing generated vector with vectors of candidates and choosing the closest one.

4.3   Evaluation and Results

The proposed method and baselines were compared on the fill-in-the-blank task. For each outfit in the dataset, a random item was selected as the blank. Similarly, three negative candidates were randomly selected for each outfit. The aim is to select the correct answer from four candidates to fill in the blank in the outfit. The accuracy of assessing the performance was proposed in Table 1. The best result is in bold, the second score is underlined, and the third score is in italic.

Table 1. Performance comparison on fill-in-the-blank task.

Method	FLTB (2 items)	FLTB (3 items)	FLTB (4 items)
FHN	0.697	0.669	0.669
Bi-LSTM-VSE	0.724	0.776	0.753
NGNN	0.791	0.765	0.765
HFGN	<u>0.826</u>	*0.801*	*0.800*
<u>MGCM</u>	*0.822*	<u>0.817</u>	<u>0.829</u>
Our	**0.878**	**0.863**	**0.911**

The results show that our proposed model outperforms the baselines. They also illustrate that GAN-based methods outperform the classical ones, despite the fact that GAN learn to generate new items rather than choose from existing ones.

Figures 4. and 5. show some examples where our model predicts blanked items better than others compared. Fashion and style are very complex concepts, and preferences can differ from person to person. In our opinion, the figures show that HFGN and MGCM models emphasized various aspects of the input items, while our model evaluated the overall style of the received items and their attributes and generated the most appropriate item for the set.

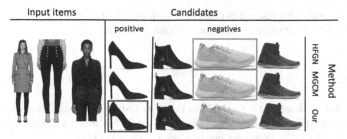

Fig. 4. Example of fill-in-the-blank predictions of blanked "shoes" item by input "outer", "bottom" and "top" items.

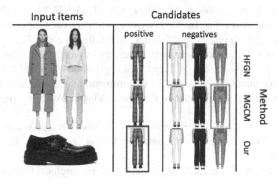

Fig. 5. Example of fill-in-the-blank predictions of blanked "bottom" item by input "outer", "top" and "shoes" items

5   Conclusions

In this paper, we propose transform-based GAN methods for multimodal generation of set of items by input ones. We have tried to solve some problems of existing GAN-based solutions such as the problem of a fixed number of input and output items and the problem of losing content attribute features during the generation process.

We have proposed to interpret a set of items as a sequence of items with starting and ending tag. It allows a model to control a number of items which a set of items contains.

The multimodal transformer with cross-attention module have been proposed as a body of our GAN-based model to overcome the problem of losing a content attribute features during generation. It allows the model to explore and generate visual and textual features simultaneously.

We have compared the proposed model with some strong baseline models and shown that our proposed method outperforms baselines at the fill-in-the-blank task on the FOTOS dataset. Our model has reached 0.878 accuracy in filling one blanked item by one of four candidates (one is positive and three are negatives) by one input item. Versus 0.724 of Bi-LSTM-VSE, 0.822 of MGCM, 0.826 of HFGN. Similarly, our model outperforms other baseline models with a different number of input items.

Acknowledgements. This research is financially supported by The Russian Science Foundation, Agreement №17-71-30029 with co-financing of Bank Saint Petersburg.

References

1. Han, X., Wu, Z., Jiang, Y.G., Davis, L.S.: Learning fashion compatibility with bidirectional LSTMs. In: MM 2017 - Proceedings of the 2017 ACM Multimedia Conference, July 2017, pp. 1078–1086 (2017). https://doi.org/10.1145/3123266.3123394
2. Song, X., Feng, F., Liu, J., Li, Z., Nie, L., Ma, J.: NeuroStylist: neural compatibility modeling for clothing matching. In: Proceedings of the 25th ACM International Conference on Multimedia, pp. 753–761 (2017). https://doi.org/10.1145/3123266.3123314
3. Yang, X., Ma, Y., Liao, L., Wang, M., Chua, T.S.: TransNFCM: translation-based neural fashion compatibility modeling. In: 33rd AAAI Conference on Artificial Intelligence, AAAI 2019, 31st Innovative Applications of Artificial Intelligence Conference, IAAI 2019 and the 9th AAAI Symposium on Educational Advances in Artificial Intelligence, EAAI 2019, December 2018, pp. 403–410 (2018). https://doi.org/10.1609/aaai.v33i01.3301403
4. Tangseng, P., Yamaguchi, K., Okatani, T.: Recommending outfits from personal closet. In: 2017 IEEE International Conference on Computer Vision Workshop (ICCVW), pp. 2275–2279 (2017)
5. Lu, Z., Hu, Y., Chen, Y., Zeng, B.: Personalized outfit recommendation with learnable anchors. In: 2021 IEEE/CVF Conference on Computer Vision and Pattern Recognition (CVPR), pp. 12722–12731 (2021)
6. Li, X., Wang, X., He, X., Chen, L., Xiao, J., Chua, T.S.: Hierarchical fashion graph network for personalized outfit recommendation. In: SIGIR 2020 - Proceedings of the 43rd International ACM SIGIR Conference on Research and Development in Information Retrieval, May 2020, pp. 159–168 (2020). https://doi.org/10.1145/3397271.3401080
7. Lu, Z., Hu, Y., Jiang, Y., Chen, Y., Zeng, B.: Learning binary code for personalized fashion recommendation. In: Proceedings of the IEEE Computer Society Conference on Computer Vision and Pattern Recognition, June 2019, pp. 10554–10562 (2019). https://doi.org/10.1109/CVPR.2019.01081
8. Cui, Z., Li, Z., Wu, S., Zhang, X., Wang, L.: Dressing as a whole: outfit compatibility learning based on node-wise graph neural networks. In: The Web Conference 2019 - Proceedings of the World Wide Web Conference, WWW 2019, February 2019, pp. 307–317 (2019). https://doi.org/10.1145/3308558.3313444
9. Cardoso, A., Daolio, F., Vargas, S.: Product characterisation towards personalisation: learning attributes from unstructured data to recommend fashion products. In: Proceedings of the ACM SIGKDD International Conference on Knowledge Discovery and Data Mining, March 2018, pp. 80–89 (2018). https://doi.org/10.1145/3219819.3219888
10. Sagar, D., Garg, J., Kansal, P., Bhalla, S., Shah, R.R., Yu, Y.: PAI-BPR: personalized outfit recommendation scheme with attribute-wise interpretability. In: Proceedings - 2020 IEEE 6th International Conference on Multimedia Big Data, BigMM 2020, August 2020, pp. 221–230 (2020). https://doi.org/10.1109/BigMM50055.2020.00039
11. Yuan, F., Karatzoglou, A., Arapakis, I., Jose, J.M., He, X.: A simple convolutional generative network for next item recommendation. In: WSDM 2019 - Proceedings of the 12th ACM International Conference on Web Search and Data Mining, vol. 19, pp. 582–590 (2018). https://doi.org/10.1145/3289600.3290975
12. Kumar, S., das Gupta, M.: c$^+$GAN: complementary fashion item recommendation. In: KDD 2019: Workshop on AI for Fashion, June 2019. https://arxiv.org/abs/1906.05596v1. Accessed 21 Jan 2022

13. Kang, W.C., Fang, C., Wang, Z., McAuley, J.: Visually-aware fashion recommendation and design with generative image models. In: Proceedings - IEEE International Conference on Data Mining, ICDM, November 2017, pp. 207–216 (2017). https://doi.org/10.1109/ICDM. 2017.30

14. Liu, L., Zhang, H., Ji, Y., Jonathan Wu, Q.M.: Toward AI fashion design: an attribute-GAN model for clothing match. Neurocomputing **341**, 156–167 (2019). https://doi.org/10.1016/J. NEUCOM.2019.03.011

15. Liu, J., Song, X., Chen, Z., Ma, J.: MGCM: multi-modal generative compatibility modeling for clothing matching. Neurocomputing **414**, 215–224 (2020). https://doi.org/10.1016/J.NEU COM.2020.06.033

16. Yu, C., Hu, Y., Chen, Y., Zeng, B.: Personalized fashion design. In: 2019 IEEE/CVF International Conference on Computer Vision (ICCV). October 2019, pp. 9045–9054 (2019). https://doi.org/10.1109/ICCV.2019.00914

17. Hsiao, W.L., Grauman, K.: Creating capsule wardrobes from fashion images. In: Proceedings of the IEEE Computer Society Conference on Computer Vision and Pattern Recognition, December 2017, pp. 7161–7170 (2017). https://doi.org/10.1109/CVPR.2018.00748

18. Dong, X., Jing, P., Song, X., Xu, X.S., Feng, F., Nie, L.: Personalized capsule wardrobe creation with garment and user modeling. In: MM 2019 - Proceedings of the 27th ACM International Conference on Multimedia, October 2019, pp. 302–310 (2019). https://doi.org/ 10.1145/3343031.3350905

19. Zheng, N., Song, X., Niu, Q., Dong, X., Zhan, Y., Nie, L.: Collocation and try-on network: whether an outfit is compatible. In: MM 2021 - Proceedings of the 29th ACM International Conference on Multimedia, October 2021, pp. 309–317 (2021). https://doi.org/10.1145/347 4085.3475691

20. Dong, X., Wu, J., Song, X., Dai, H., Nie, L.: Fashion compatibility modeling through a multi-modal try-on-guided scheme. In: Proceedings of the 43rd International ACM SIGIR Conference on Research and Development in Information Retrieval, July 2020, pp. 771–780 (2020). https://doi.org/10.1145/3397271.3401047

21. Vasileva, M.I., Plummer, B.A., Dusad, K., Rajpal, S., Kumar, R., Forsyth, D.: Learning type-aware embeddings for fashion compatibility. In: Ferrari, V., Hebert, M., Sminchisescu, C., Weiss, Y. (eds.) ECCV 2018. LNCS, vol. 11220, pp. 405–421. Springer, Cham (2018). https://doi.org/10.1007/978-3-030-01270-0_24

22. Bettaney, E.M., Hardwick, S.R., Zisimopoulos, O., Chamberlain, B.P.: Fashion outfit generation for e-commerce. In: Dong, Y., Ifrim, G., Mladenić, D., Saunders, C., VanHoecke, S. (eds.) ECML PKDD 2020. LNCS (LNAI), vol. 12461, pp. 339–354. Springer, Cham (2021). https://doi.org/10.1007/978-3-030-67670-4_21

23. Vaswani, A., Shazeer, N., Parmar, N., et al.: Attention is all you need. Adv. Neural Inf. Process. Syst. 5999–6009 (2017). https://arxiv.org/abs/1706.03762v5. Accessed 21 Jan 2022

24. Cheng, Y., Wang, R., Pan, Z., Feng, R., Zhang, Y.: Look, listen, and attend: co-attention network for self-supervised audio-visual representation learning. In: MM 2020 - Proceedings of the 28th ACM International Conference on Multimedia, vol. 20, pp. 3884–3892 (2020). https://doi.org/10.1145/3394171.3413869

25. Jolicoeur-Martineau, A.: The relativistic discriminator: a key element missing from standard GAN. In: 7th International Conference on Learning Representations. ICLR 2019, July 2018. https://arxiv.org/abs/1807.00734v3. Accessed 21 Jan 2022

26. Mao, X., Li, Q., Xie, H., Lau, R.Y.K., Wang, Z., Smolley, S.P.: Least squares generative adversarial networks. In: Proceedings of the IEEE International Conference on Computer Vision, October 2016, pp. 2813–2821 (2016). https://doi.org/10.1109/ICCV.2017.304

Retrofitting Structural Graph Embeddings with Node Attribute Information

Piotr Bielak[(✉)][iD], Daria Puchalska, and Tomasz Kajdanowicz[iD]

Department of Artificial Intelligence, Wrocław University of Science and Technology,
Wrocław, Poland
piotr.bielak@pwr.edu.pl

Abstract. Representation learning for graphs has attracted increasing
attention in recent years. In this paper, we define and study a new problem of learning attributed graph embeddings. Our setting considers how
to update existing node representations from structural graph embedding
methods when some additional node attributes are given. To this end, we
propose Graph Embedding RetroFitting (GERF), a method that deliv-
ers a compound node embedding that follows both the graph structure
and attribute space similarity. Unlike other attributed graph embedding
methods, GERF is a novel representation learning method that does
not require recalculation of the embedding from scratch but rather uses
existing ones and retrofits the embedding according to neighborhoods
defined by the graph structure and the node attributes space. Moreover,
our approach keeps the same embedding space all the time and allows
comparing the positions of embedding vectors and quantifying the impact
of attributes on the representation update. Our GERF method updates
embedding vectors by optimizing the invariance loss, graph neighbor
loss, and attribute the neighbor loss to obtain high-quality embeddings.
Experiments on WikiCS, Amazon-CS, Amazon-Photo, and Coauthor-
CS datasets demonstrate that our proposed algorithm receives similar
results compared to other state-of-the-art attributed graph embedding
models despite working in retrofitting manner.

Keywords: Graph embedding · Attributed graphs · Graph embedding
retrofitting

1 Introduction

Machine learning methods have been studied in a variety of applications and
data types, including images and video (computer vision), text (natural language processing), audio or time-series data, among many others. Since most
downstream ML models expect a vector from a continuous space as input, representation learning methods have been developed to create those representation
vectors (embeddings) automatically. While there are many embedding methods

© The Author(s), under exclusive license to Springer Nature Switzerland AG 2022
D. Groen et al. (Eds.): ICCS 2022, LNCS 13350, pp. 178–191, 2022.
https://doi.org/10.1007/978-3-031-08751-6_13

traditional data types, such as word2vec [11] and FastText [4] for text, or ResNet [7] and EfficientNet [14] for images, this task is much more difficult for graph-structured data. A simple concatenation of unimodal representations (graph structure and node attributes) is often not sufficient, as it does not consider the mutual relationships between modalities. Therefore, the main challenge for such methods is discovering the interrelationship between multiple modalities to create a coherent representation that will integrate the multimodal information.

Problem Statement. Consider a situation in which data changing over time is analyzed on an ongoing basis. In the first case, the structure of the network remains unchanged, but the attributes of the nodes are constantly changing – an example may be a network of connected weather sensors. Conversely, the values of the node attributes can be constant, but the structure of the graph changes, e.g., in a telephone network, where the edge denotes the currently ongoing call. In both situations, graph embedding models that consider both the network structure and node attributes can be used, however, if one of these modalities does not change over time, this may not be the best solution. Especially, in the first of the above-mentioned situations, it may be more advantageous to generate the structural graph embeddings once, and then use a method that would modify them depending on the current values of the attributes, somehow incorporating information from the attribute space into the structural embedding space. The simplest solution would be to simply concatenate both vectors, but the resulting representation would be neither consistent nor low-dimensional.

Goal. The aim of this work is to develop an algorithm that will enhance (retrofit) existing structural node embeddings by incorporating information from the attribute space. That is, based on the node attributes, it will appropriately modify the embedding vectors derived from the structural graph representation learning methods. The new embedding vectors returned by such method should provide better performance in downstream tasks than by using naive approaches (like concatenation of structural embeddings with node attribute vectors).

Contributions. We summarize our contributions as follows: (1) we introduce a new problem in the area of graph representation learning, in which a structural network embedding is updated (retrofitted) according to node attributes, (2) we propose a novel method (GERF) for unsupervised representation learning on graphs which combines information from the space of structural embeddings and node attributes while maintaining low dimensionality of the representation vectors, (3) we perform experiments demonstrating competitive quality of the proposed GERF method compared to other approaches, (4) we make our code and experimental pipeline publicly available to ensure reproducibility: https://github.com/pbielak/gerf/.

2   Related Work

The problem of graph representation learning (GRL) has received a lot of attention in recent years in the machine learning community. The main goal is to learn low-dimensional continuous vector representations (embeddings), which can be later used for specific downstream prediction tasks such as node classification or link prediction. We can distinguish two groups of embeddings in GRL methods: (i) structural embeddings that take into account information extracted from the network structure only, such as the neighborhood of nodes proximities, and (ii) attributed embeddings which, apart from the relationships in the network structure, also reflect the similarity in the node feature space.

2.1   Structural Representation Learning Methods

Early GRL methods built low-dimensional node embeddings that reflect the structure of the network. Among them, the most frequently referenced and used are: DeepWalk [12], Node2vec [6], LINE [15] and SDNE [16]. **DeepWalk** [12] samples node sequences using random walks and passes them into the Skip-gram model [11] (a word embedding method). **Node2vec** [6] extends DeepWalk by developing a biased random walk procedure to explore diverse neighborhoods by interpolating between a breadth-first (BFS) and depth-first (DFS) graph search algorithms. **LINE** [15] is a scalable method that learns node representations by preserving the first-order (similar embeddings of neighbor nodes) and second-order graph proximities (similar embeddings of nodes sharing the same neighborhood). **SDNE** [16] also focuses on preserving the first-order and second-order proximities. However, it uses an autoencoder approach to map the highly non-linear underlying network structure to latent space.

2.2   Attributed Graph Embedding Methods

The structure of the network is given by connections between objects. However, there are many other possible sources of information. Additional node attributes can be given in the form of a vector representation of their content, which in the case of classic methods such as *bag-of-words* model or *tf-idf* is an additional challenge because these vectors are usually sparse. Methods designed to learn representations in attributed networks include TADW [17], FSCNMF [3], DANE [5] and ANRL [18]. **TADW** [17] (**T**ext-**A**ssociated **D**eep**W**alk) shows that DeepWalk is equivalent to matrix factorization and proposes its text-associated version. **FSCNMF** [3] is based on non-negative matrix factorization and produces node embeddings that are consistent with the graph structure and nodes' attributes. The structure-based embedding matrix serves as a regularizer when optimizing the attribute-based embedding matrix and vice-versa.

3   Notations and Problem Definition

Graph. A graph G is a pair $G = (V, E)$, where $V = \{v_1, \ldots, v_{|V|}\}$ is a set of nodes and $E \subseteq V \times V$ is the set of edges that connect node pairs, i.e., each edge e_{ij} is a pair (v_i, v_j). The graph connectivity can be represented as an adjacency matrix $\mathbf{A} \in \{0, 1\}^{|V| \times |V|}$ with element A_{ij} indicating the existence of an edge (v_i, v_j).

Attributed Graph. Apart from the structure of connections between objects, this kind of graph has additional information about each of the nodes, i.e., each node has an assigned feature vector (also called the attribute vector). An attributed graph is a 3-tuple $G = (V, E, \mathbf{X})$, where V and E follow the previous definition. $\mathbf{X} \in \mathbb{R}^{|V| \times d_X}$ is a matrix that encodes all node attributes information, and $\mathbf{x}_i$ describes the attributes associated with node v_i.

Attribute Proximity. We can analyze the similarity between nodes not only based on the network structure but also in the attribute space. Given a network $G = \{V, E, \mathbf{X}\}$, the attribute proximity of two nodes v_i and v_j is determined by the similarity of $\mathbf{x}_i$ and $\mathbf{x}_j$. Note that these are two separate spaces to analyze. The similarity of two nodes in the graph structure does not imply their similarity in the attribute space and vice versa. Thus, the representation learning methods for attributed graphs should take into account dependencies in both spaces and coherently combine them.

Node Representation Learning. Given a network $G = (V, E)$ (or $G = (V, E, \mathbf{X})$ in the case of an attributed network), the goal is to represent every graph node $v_i \in V$ as a low-dimensional vector $\mathbf{z}_i$ (called node embedding) by learning a mapping function $f : v_i \rightarrow \mathbf{z}_i \in \mathbb{R}^{d_Z}$, where $d_Z << |V|$, such that important network properties are preserved in the embedding space (e.g. structural and semantic graph information). Overall, the node embeddings are stored as a node embedding matrix $\mathbf{Z} \in \mathbb{R}^{V \times d_Z}$. If two nodes are similar in the graph structure (they are connected or share neighbors), or have similar attribute values, their learned embeddings should also be similar.

Attribute-Based Neighborhood. We can easily define the neighborhood of a node in the network as the set of other nodes that are connected to it by an edge. However, there are no clear relationships between objects within the attribute space itself. To combine information from the attribute space (which objects are closer to each other in this space and which are further) with structural relationships, it is necessary to first define the so-called *attribute-based neighborhood*. Based on the attribute matrix $\mathbf{X}$, for each node in the network G, its k nearest neighbors in this space were found based on the Euclidean distance metric, with k being equal to the number of neighbors of this node in the network G. The neighborhood of the node v_i in the attribute space, defined in this way, will be denoted by $\mathcal{N}_\mathbf{X}(v_i)$. Therefore, $\forall_i \ |\mathcal{N}(v_i)| = |\mathcal{N}_\mathbf{X}(v_i)|$.

4   Graph Embedding RetroFitting (GERF)

In this section, we describe our proposed Graph Embedding RetroFitting model that allows to update existing structural node embeddings $\mathbf{Z}$ with the node attribute information $\mathbf{X}$, resulting in the retrofitted node embeddings $\mathbf{Z}^*$. Figure 1 shows the overall processing pipeline of our method.

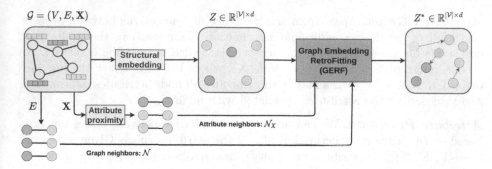

Fig. 1. Our proposed graph embedding retrofitting (GERF) method uses information about the structure of the graph (graph neighbors) and the node attribute space (attribute neighbors) to retrofit a structural embedding Z into one that incorporates node attribute information Z^*.

4.1 Objective Function

Our model is based on the optimization of the $\mathbf{Z}^* \in \mathbb{R}^{V \times d}$ matrix. The objective function of our proposed GERF model takes the following form:

$$\mathcal{L}(\mathbf{Z}^*) = (1 - \lambda_G - \lambda_X) \sum_{i=1}^{n} \|\mathbf{z}_i^* - \mathbf{z}_i\|^2$$
$$+ \lambda_G \sum_{i=1}^{n} \sum_{j:\, v_j \in \mathcal{N}(v_i)} \frac{\|\mathbf{z}_i^* - \mathbf{z}_j^*\|^2}{|\mathcal{N}(v_i)|} + \lambda_X \sum_{i=1}^{n} \sum_{j:\, v_j \in \mathcal{N}_{\mathbf{X}}(v_i)} \frac{\|\mathbf{z}_i^* - \mathbf{z}_j^*\|^2}{|\mathcal{N}_{\mathbf{X}}(v_i)|}, \quad (1)$$

where $\mathbf{Z} = (\mathbf{z}_1, \ldots, \mathbf{z}_n)$ are the pre-trained structural embeddings for each node, $\mathbf{Z}^* = (\mathbf{z}_1^*, \ldots, \mathbf{z}_n^*)$ are the new embeddings combining multimodal information (from both spaces), and $\lambda_G > 0$ and $\lambda_X > 0$ are non-negative method hyperparameters that control the importance of the structural and attribute similarity, respectively.

With the purpose of the work in mind, one can easily explain the intuition behind each component in the objective function and why it should be included there. Since the method is intended to enhance the space of structural embeddings by incorporating information from the attribute space, it is necessary to include an a component in the objective function that will "keep the embeddings in place". That is, make sure that the new embeddings do not deviate significantly from their original values because this would lead to a complete loss of information from this space. Hence, what is needed is a component which is later referred to as **invariance loss**:

$$\mathcal{L}_I(\mathbf{Z}^*) = \sum_{i=1}^{n} \|\mathbf{z}_i^* - \mathbf{z}_i\|^2. \quad (2)$$

Further, in order to combine information from the network structure and node attributes, for each node its neighborhood in both of these spaces is considered, as defined earlier. In order for the representation of each node to be similar to the representation of objects close to it in both spaces, it is necessary to define the loss related to the distance between the embeddings in the network, called **graph neighbor loss**:

$$\mathcal{L}_G(\mathbf{Z}^*) = \sum_{i=1}^{n} \sum_{j:\, v_j \in \mathcal{N}(v_i)} \frac{1}{|\mathcal{N}(v_i)|} \|\mathbf{z}_i^* - \mathbf{z}_j^*\|^2, \tag{3}$$

as well as an analogous component that controls the distances between node embeddings based on their attributes, called **attribute neighbor loss**:

$$\mathcal{L}_X(\mathbf{Z}^*) = \sum_{i=1}^{n} \sum_{j:\, v_j \in \mathcal{N}_{\mathbf{X}}(v_i)} \frac{1}{|\mathcal{N}(v_i)|} \|\mathbf{z}_i^* - \mathbf{z}_j^*\|^2. \tag{4}$$

By combining Eqs. 2–4, it is possible to write the formula in Eq. 1 in a different, simpler form:

$$\mathcal{L}(\mathbf{Z}^*) = (1 - \lambda_G - \lambda_X)\mathcal{L}_I(\mathbf{Z}^*) + \lambda_G \mathcal{L}_G(\mathbf{Z}^*) + \lambda_X \mathcal{L}_X(\mathbf{Z}^*). \tag{5}$$

4.2 Optimization

The Adam optimizer [8] was used to minimize the objective function from Eq. 1. One can easily derive the formula for the first derivative of the function $\mathcal{L}$ with respect to one vector $\mathbf{z}_i^*$ as follows:

$$\frac{\partial \mathcal{L}}{\partial \mathbf{z}_i^*} = 2 \left(1 - \lambda_G - \lambda_X\right)(\mathbf{z}_i^* - \mathbf{z}_i)$$

$$+ 2\lambda_G \sum_{j:\, v_j \in \mathcal{N}(v_i)} \frac{\mathbf{z}_i^* - \mathbf{z}_j^*}{|\mathcal{N}(v_i)|} - 2\lambda_G \sum_{j:\, v_i \in \mathcal{N}(v_j)} \frac{\mathbf{z}_j^* - \mathbf{z}_i^*}{|\mathcal{N}(v_j)|}$$

$$+ 2\lambda_X \sum_{j:\, v_j \in \mathcal{N}_{\mathbf{X}}(v_i)} \frac{\mathbf{z}_i^* - \mathbf{z}_j^*}{|\mathcal{N}_{\mathbf{X}}(v_i)|} - 2\lambda_X \sum_{j:\, v_i \in \mathcal{N}_{\mathbf{X}}(v_j)} \frac{\mathbf{z}_j^* - \mathbf{z}_i^*}{|\mathcal{N}_{\mathbf{X}}(v_j)|}.$$

The matrix $\mathbf{Z}^*$ is initialized with the values of $\mathbf{Z}$. Currently, the values of λ_G and λ_X hyperparameters are determined based on grid search and simultaneous analysis of the model results in downstream tasks. However, more advanced techniques could be proposed for this purpose, e.g. based on the properties of the network structure and attribute space, which is planned for future work.

4.3 Summary

The proposed method allows the creation of a coherent representation for nodes in the network based on their attribute values and pre-trained structural embeddings. It assumes that there are dependencies between objects in the attribute space. Objects are considered adjacent if the distance between their attribute vectors is sufficiently small compared to others. The main advantages of this method are intuitive assumptions and simplicity of operation, as it allows for easy integration of multimodal information from the network structure and node attributes in the form of a low-dimensional representation vector.

5 Experimental Setup

We perform an analysis of selected graph representation learning methods in the node classification downstream task. We compare attributed graph embedding models (TADW, FSCNMF, DGI), structural embeddings (node2vec, LINE, SDNE), and the ones modified by the proposed GERF method and a few other baselines with each other. Additionally, a simple approach is tested that completely ignores the network structure and uses only node attributes for the prediction. Four real-world benchmark datasets are employed.

5.1 Datasets

We employ four real-wold benchmark datasets from the *PyTorch-Geometric* [2] library. The statistics are provided in Table 1.

- **WikiCS** [10] is a network of Computer Science-related Wikipedia articles with edges denoting references between those articles. Each article belongs to one of 10 subfields (classes) and has features computed as averaged GloVe embeddings of the article content. We use the first provided train/val/test data splits without any modifications (we recompute the embeddings 10 times).
- **Amazon Computers** (Amazon-CS), **Amazon Photos** [9] are two networks extracted from Amazon's co-purchase data. Nodes are products and edges denote that these products were often bought together. Based on the reviews, each product is described using a Bag-of-Words representation (node features). There are 10 and 8 product categories (node classes), respectively. There are no data splits available for those datasets, so we generate a random train/val/test split (10%/10%/80%) for each one.
- **Coauthor-CS** is a network extracted from the Microsoft Academic Graph [13]. Nodes are authors, and edges denote a collaboration of two authors. Each author is described by the keywords used in their articles (Bag-of-Words representation; node features). There are 15 author research fields (node classes). Similar to the Amazon datasets, there is no data split provided, so we generate a random train/val/test split (10%/10%/80%).

Table 1. Datasets statistics.

Name	Nodes	Edges	Features	Classes
WikiCS	11,701	216,123	300	10
Amazon computers	13,752	245,861	767	10
Amazon photos	7,650	119,081	745	8
Coauthor-CS	18,333	81,894	6,805	15

5.2   Embedding Methods

To be able to make a qualitative comparison of embedding methods and their ability to compress highly non-linear dependencies in a low-dimensional space, it was concluded that the same size of embedding would be assumed for each of the methods. Thus, each of the algorithms produces 128-dimensional representation vectors. The exact settings for the structural embedding methods are listed below:

- **node2vec** – the same default parameters were adopted for each of the datasets because the quality of the representation vectors obtained in this way was satisfactory (note that the priority of the work is not to achieve the highest possible results of the compared algorithms, but to show that the method proposed in Sect. 4 is able to improve them, therefore little importance was given to the search for the best set of hyperparameters). We use the following settings: batch size – 128, learning rate – 0.01, walk length – 20, number of walks per node – 10, context size – 10, and number of negative samples – 1. Besides, the algorithm parameters p and q have been set to 1, which means that it is effectively DeepWalk.
- **LINE** – the number of epochs was set to 10 for all datasets, and the mini-batch size was set to 128 (WikiCS, Coauthor-CS), 4096 (Amazon-CS), 256 (Amazon-Photos). Such settings were required as otherwise, the training resulted in collapsed embeddings or NaN values in the embedding vectors. The number of samples in the negative sampling procedure was set to 1 for all datasets.
- **SDNE** – these embeddings showed the worst quality in the downstream task, therefore a hyperparameter grid search was performed, searching for the optimal values of the α, β, ν parameters, as well as the number of epochs and the size of the autoencoder hidden layer. Based on preliminary experiments, the number of epochs was set to 50 and the hidden layer size to 256 for all datasets. The values of method parameters α was chosen to be 10^{-4}, β was left at default 5 and ν_1, ν_2 were set to 10^{-5}, 10^{-4}, respectively.

As the proposed GERF method combines information from the network structure and the attribute space, it can be compared with attributed representation learning methods for graphs. We choose the following:

- **TADW** – the learning rate was set to 0.01 and the number of epochs to 20, the other parameters were taken as defaults from the *Karate Club* implementation [1].
- **FSCNMF** – the number of epochs was set to 500, other settings are also default from the *Karate Club* implementation.
- **DGI** – we use a single layer Graph Convolutional (GCN) encoder network with PReLU activation and train it using the Adam optimizer with a learning rate of 0.001.

In addition to the above-mentioned representation learning methods, an approach in which the graph structure is completely ignored and only node attributes are used for prediction is additionally tested. Such a representation of nodes is high-dimensional and extremely sparse, and in experiments, it will be referred to as **features**.

5.3 Baselines

To check the quality of the proposed method, which allows for modifying the existing structural embeddings based on node attributes, it is compared with several baseline methods:

- **Concat** – concatenation of the structural embedding and the attribute vector for each node (note the large dimension size of such a representation),
- **ConcatPCA** – in this method, the dimensionality of the concatenated structural embedding and the feature vector is reduced to the size of the embedding only, using *Principal Component Analysis*,
- **MLP** – a simple autoencoder architecture, with an encoder consisting of three linear layers: the first one with the size of $d_X + d_Z$ neurons with the ReLU activation function, another with the size of $(d_X + d_Z)//2$ also with the ReLU activation function, and the last one with the size of d_Z with the Tanh activation. The decoder is a single linear layer that takes a vector from a hidden space with dimension d_Z and returns a vector of size $d_X + d_Z$, which should be the best reconstruction of the input vector to the encoder. The autoencoder was trained for 20 epochs with the mini-batch size of 128, and Adam with learning rate of 0.001 was used as the optimizer.

5.4 GERF Hyperparameters

The hyperparameter values (λ_G, λ_X) of the proposed GERF method were determined by performing a grid search (see Table 2). For each of the hyperparameters, we checked the following values: $0, 0.1, \ldots, 1.0$, while preserving the overall hyperparameter constraints $(\lambda_G + \lambda_X \leq 1)$. Moreover, the learning rate was set to 10^{-1}, while the number of epochs was set to 100.

Table 2. Best found GERF hyperparameter values (λ_G, λ_X) for all datasets.

Dataset	node2vec	LINE	SDNE
	(λ_G, λ_X)	(λ_G, λ_X)	(λ_G, λ_X)
WikiCS	(0.1, 0.4)	(0.4, 0.4)	(0.3, 0.4)
Amazon-CS	(0.2, 0.3)	(0.3, 0.2)	(0.8, 0.0)
Amazon-Photo	(0.2, 0.4)	(0.3, 0.3)	(0.5, 0.2)
Coauthor-CS	(0.2, 0.5)	(0.6, 0.3)	(0.9, 0.0)

While performing the grid search, we collected the node classification performance for each hyperparameter setting (not only the best one). We present the influence of each hyperparameter on the overall node classification performance in Fig. 2 (for the WikiCS dataset). We notice that the results for both hyperparameters form a convex function with a single value that maximizes the downstream task performance. We also note that the proposed GERF method is robust in terms of hyperparameters. However, it is worth optimizing them as we observed up to 5 pp dispersion in the hyperparameter combinations impact on AUC (depending on the structural embeddings – LINE, node2vec, SDNE).

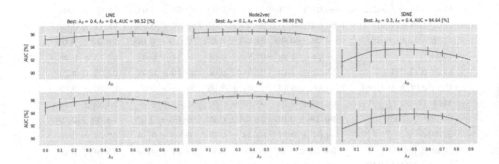

Fig. 2. Evaluation of different hyperparameter (λ_G, λ_X) values of the proposed GERF method in the node classification task on the WikiCS dataset. We present the mean and standard deviation of AUC (validation split) for each of the hyperparameter values. Note that while keeping one parameter fixed, we compute the AUC statistics over all possible values of the other hyperparameter.

6   Node Classification

Setup. The embeddings returned by the attributed representation learning methods, structural embeddings (themselves and enhanced by the proposed GERF method and baselines), as well as node attribute vectors, were compared in the node classification task. We compute the embeddings of both datasets 10 times to mitigate the random nature of the methods and their optimization procedure for both the structural and attributed graph representation learning methods. Each of those 10 embeddings is processed by the baselines and the proposed GERF model. We use a L_2 regularized logistic regression (from the *scikit-learn* package) trained on the embedding vectors (input) and the class information (output). The maximum number of iterations was set to 250, other parameters were left with their default values.

The classification results in terms of the AUC metric are shown in Table 3 For each of the three structural node embedding methods – node2vec, LINE and SDNE – as well as the embeddings updated by the baselines and our proposed GERF method – the best result is marked in bold. We report both the mean and standard deviation over 10 embedding recalculations.

Table 3. Node classification results in terms of the mean and standard deviation of the AUC metric over 10 recomputations of embeddings. For each structural embedding method (node2vec, LINE and SDNE) and their updated versions (by baselines and our proposed GERF method) we mark the best result in bold.

Method	WikiCS	Amazon-CS	Amazon-Photo	Coauthor-CS
Features	94.79 ± 0.00	90.10 ± 0.00	94.58 ± 0.00	98.16 ± 0.00
node2vec	93.97 ± 0.15	98.22 ± 0.04	98.64 ± 0.04	98.33 ± 0.04
Concat (node2vec)	96.25 ± 0.10	98.23 ± 0.04	98.65 ± 0.04	98.38 ± 0.04
ConcatPCA (node2vec)	96.01 ± 0.11	98.22 ± 0.04	98.64 ± 0.04	98.34 ± 0.04
MLP (node2vec)	96.03 ± 0.08	98.29 ± 0.04	98.66 ± 0.04	98.41 ± 0.04
GERF (node2vec)	**96.28 ± 0.09**	**98.65 ± 0.04**	**99.18 ± 0.03**	**99.23 ± 0.02**
LINE	91.74 ± 0.20	97.63 ± 0.06	98.44 ± 0.08	93.13 ± 0.28
Concat (LINE)	95.02 ± 0.15	97.65 ± 0.06	98.45 ± 0.08	93.69 ± 0.26
ConcatPCA (LINE)	94.68 ± 0.14	97.64 ± 0.06	98.44 ± 0.08	93.16 ± 0.28
MLP (LINE)	94.90 ± 0.12	97.55 ± 0.07	98.45 ± 0.06	93.39 ± 0.24
GERF (LINE)	**96.18 ± 0.05**	**98.28 ± 0.05**	**99.06 ± 0.03**	**98.39 ± 0.05**
SDNE	74.94 ± 0.71	88.24 ± 0.41	90.89 ± 0.34	67.05 ± 1.05
Concat (SDNE)	**94.14 ± 0.27**	88.81 ± 0.41	91.34 ± 0.33	**93.86 ± 0.68**
ConcatPCA (SDNE)	93.63 ± 0.31	88.46 ± 0.43	91.16 ± 0.33	92.44 ± 0.73
MLP (SDNE)	93.75 ± 0.27	87.84 ± 0.42	90.55 ± 0.30	68.13 ± 0.86
GERF (SDNE)	92.97 ± 0.74	**97.49 ± 0.07**	**98.43 ± 0.08**	87.37 ± 2.86
TADW	90.65 ± 0.00	58.71 ± 0.00	55.91 ± 0.00	81.33 ± 0.00
FSCNMF	84.24 ± 0.00	49.93 ± 0.00	49.56 ± 0.00	50.14 ± 0.00
DGI	93.54 ± 0.17	78.20 ± 0.55	90.02 ± 0.54	98.48 ± 0.06

Discussion. The first thing to note is the high score of the model that only uses node attributes to predict the class label and completely ignores information from the network structure despite the extremely large dimensionality of such a representation. In all cases, using only node attributes (features) for their classification gave better or similar results than the attributed representation learning methods. However, it should be noted that due to the long time of operation of these algorithms, the hyperparameter grid search was not performed, and their default values were adopted, which could have an impact on the results obtained.

Structural embedding vectors for WikiCS performed even worse than the attributed ones, particularly those learned using SDNE. However, the quality of the predictions increased significantly (in the case of SDNE, one could even say drastically) as they were processed by the baselines or the proposed GERF method, which can be seen in the increase of the AUC measure even by almost 18% points (comparing SDNE and GERF (SDNE) embeddings for WikiCS)!

In general, in each case, the proposed GERF method allowed the incorporation of information from the attribute space into the structural embedding space, improving the quality of prediction in this downstream task. Surprisingly good results were achieved with the use of this method on the node2vec and LINE embeddings, where it turned out to be not only better than the dedicated attributed methods, but also than all the proposed baselines, and in a significant way.

In the case of the SDNE embeddings group and the WikiCS and Coauthor-CS datasets, the baselines (*Concat*, *ConcatPCA* and *MLP*) methods turned out to be better than the proposed GERF method, but it is worth noting that both have some disadvantages. The concatenation of the attribute vector and structural embedding, which has proven to be best is highly dimensional and inconsistent with the structural embedding part. On the other hand, the *MLP* refiner, based on a simple autoencoder architecture, can exhibit problems when reconstructing sparse attribute vectors. All things considered, the results obtained by the proposed GERF method are satisfactory. While maintaining a low-dimensional representation, which allows saving memory, it achieves results similar or better to other methods, which depends on the quality of the underlying structural embeddings.

7 Conclusions

In this paper, we introduced a new graph representation learning problem setting, where given already precomputed structural node embeddings, we want to update them accordingly to node attributes, in such a way that the resulting embedding will preserve the information from both the structure and attributes. We proposed a novel graph embedding retrofitting model (GERF), which solves this problem by optimizing a compound loss function, which includes an invariance loss (keeping the new embedding close to the structural one), a graph neighborhood loss (which pushes embedding of neighboring nodes closer together) and a attribute neighborhood loss (which decreases the distance of embeddings of nodes with similar attributes). We evaluate this method on four real-world benchmark datasets (WikiCS, Amazon-CS, Amazon-Photo and Coauthor-CS), comparing it to attributed graph representation learning methods and other baselines and find that our method allows to enhance structural embeddings and results in better downstream node classification performance. In all cases, our method achieves the best results compared to other attribute aware embedding methods as well as for all datasets. In future, we want to find a way to automatically determine the hyperparameters (λ_G and λ_X) of our method based on the available graph data.

References

1. Karate Club: An API Oriented Open-source Python Framework for Unsupervised Learning on Graphs. https://github.com/benedekrozemberczki/karateclub
2. PyTorch geometric main page. https://pytorch-geometric.readthedocs.io/en/latest/index.html

3. Bandyopadhyay, S., Kara, H., Kannan, A., Murty, M.: FSCNMF: Fusing Structure and Content via Non-negative Matrix Factorization for Embedding Information Networks, April 2018
4. Bojanowski, P., Grave, E., Joulin, A., Mikolov, T.: Enriching word vectors with subword information. Trans. Assoc. Comput. Linguist. 5, 135–146 (2017)
5. Gao, H., Huang, H.: Deep attributed network embedding. In: Proceedings of the Twenty-Seventh International Joint Conference on Artificial Intelligence, IJCAI 2018, pp. 3364–3370 (2018). https://doi.org/10.24963/ijcai.2018/467
6. Grover, A., Leskovec, J.: Node2vec: scalable feature learning for networks. In: Proceedings of the 22nd ACM SIGKDD International Conference on Knowledge Discovery and Data Mining, KDD 2016, pp. 855–864. Association for Computing Machinery, New York (2016). https://doi.org/10.1145/2939672.2939754
7. He, K., Zhang, X., Ren, S., Sun, J.: Deep residual learning for image recognition. In: 2016 IEEE Conference on Computer Vision and Pattern Recognition (CVPR), pp. 770–778 (2016)
8. Kingma, D.P., Ba, J.: Adam: a method for stochastic optimization. In: Bengio, Y., LeCun, Y. (eds.) 3rd International Conference on Learning Representations, ICLR 2015, San Diego, CA, USA, 7–9 May 2015, Conference Track Proceedings (2015)
9. McAuley, J., Targett, C., Shi, Q., van den Hengel, A.: Image-based recommendations on styles and substitutes. In: Proceedings of the 38th International ACM SIGIR Conference on Research and Development in Information Retrieval, pp. 43–52. Association for Computing Machinery (2015). https://doi.org/10.1145/2766462.2767755
10. Mernyei, P., Cangea, C.: Wiki-CS: a wikipedia-based benchmark for graph neural networks. arXiv preprint arXiv:2007.02901 (2020)
11. Mikolov, T., Chen, K., Corrado, G., Dean, J.: Efficient estimation of word representations in vector space. In: Bengio, Y., LeCun, Y. (eds.) 1st International Conference on Learning Representations, ICLR 2013, Scottsdale, Arizona, USA, 2–4 May 2013, Workshop Track Proceedings (2013)
12. Perozzi, B., Al-Rfou, R., Skiena, S.: Deepwalk: online learning of social representations. In: Proceedings of the 20th ACM SIGKDD International Conference on Knowledge Discovery and Data Mining, KDD 2014, pp. 701–710. ACM (2014)
13. Sinha, A., et al.: An overview of microsoft academic service (MAS) and applications. In: Proceedings of the 24th International Conference on World Wide Web, WWW 2015 Companion, pp. 243–246. Association for Computing Machinery, New York (2015). https://doi.org/10.1145/2740908.2742839
14. Tan, M., Le, Q.: EfficientNet: rethinking model scaling for convolutional neural networks. In: Chaudhuri, K., Salakhutdinov, R. (eds.) Proceedings of the 36th International Conference on Machine Learning. Proceedings of Machine Learning Research, vol. 97, pp. 6105–6114. PMLR, 09–15 June 2019
15. Tang, J., Qu, M., Wang, M., Zhang, M., Yan, J., Mei, Q.: LINE: large-scale information network embedding. In: Proceedings of the 24th International Conference on World Wide Web, May 2015. https://doi.org/10.1145/2736277.2741093. http://dx.doi.org/10.1145/2736277.2741093
16. Wang, D., Cui, P., Zhu, W.: Structural deep network embedding. In: Proceedings of the 22nd ACM SIGKDD International Conference on Knowledge Discovery and Data Mining, KDD 2016, pp. 1225–1234. Association for Computing Machinery, New York (2016). https://doi.org/10.1145/2939672.2939753

17. Yang, C., Liu, Z., Zhao, D., Sun, M., Chang, E.Y.: Network representation learning with rich text information. In: Proceedings of the 24th International Conference on Artificial Intelligence, IJCAI 2015, pp. 2111–2117. AAAI Press (2015)
18. Zhang, Z., et al.: ANRL: attributed network representation learning via deep neural networks. In: Proceedings of the Twenty-Seventh International Joint Conference on Artificial Intelligence, IJCAI 2018, pp. 3155–3161. International Joint Conferences on Artificial Intelligence Organization, July 2018. https://doi.org/10.24963/ijcai.2018/438

Hierarchical Ensemble Based Imbalance Classification

Jie Xie[1,2] , Mingying Zhu[3,4(✉)] , and Kai Hu[2]

[1] School of Computer and Electronic Information and the School of Artificial Intelligence, Nanjing Normal University, Nanjing, China
[2] Key Laboratory of Advanced Process Control for Light Industry (Ministry of Education), School of Internet of Things Engineering, Jiangnan University, Wuxi 214122, China
[3] School of Economics, Nanjing University, 22 Hankou Road, Nanjing 210093, Jiangsu, China
zhumy@nju.edu.cn
[4] Hopkins-Nanjing Center for Chinese and American Studies, Johns Hopkins University, 162 Shanghai Road Nanjing University, Nanjing 210093, Jiangsu, China

Abstract. In this paper, we propose a hierarchical ensemble method for improved imbalance classification. Specifically, we perform the first-level ensemble based on bootstrap sampling with replacement to create an ensemble. Then, the second-level ensemble is generated based on two different weighting strategies, where the strategy having better performance is selected for the subsequent analysis. Next, the third-level ensemble is obtained via the combination of two methods for obtaining mean and covariance of multivariate Gaussian distribution, where the oversampling is then realized via the fitted multivariate Gaussian distribution. Here, different subsets are created by (1) the cluster that the current instance belongs to, and (2) the current instance and its k nearest minority neighbors. Furthermore, Euclidean distance-based sample optimization is developed for improved imbalance classification. Finally, late fusion based on majority voting is utilized to obtain final predictions. Experiment results on 15 KEEL datasets demonstrate the great effectiveness of our proposed method.

Keywords: Imbalance learning · Bootstrap sampling · Gaussian-based oversampling · Hierarchical ensembel · Fraud detection

1 Introduction

A class imbalance has been a practical problem of data mining and machine learning. The three cases that are specifically listed here are animal sound classification [1], fraud detection [2], and software defect prediction [3]. For the standard algorithm, the disadvantage of imbalanced problems is that the positive class (minority class) is easily misclassified. However, those positive samples are often more important. Take fraud detection for instance, it is prevalent across various domains such as banking, government, and medical and public sectors. However, the number of positive samples are difficult to be obtained. Meanwhile, in a driving maneuver recognition system, lane change is often regarded as one of the most common causes of accidents [4]. However, for a standard driving dataset, there are fewer lane change events than other maneuvers. Therefore, it is necessary to investigate imbalance learning for solving real-world problems.

Numerous methods have been developed to treat imbalanced datasets, which can be divided into three categories: (1) data resampling [5,6], (2) algorithm modification [7–9], and (3) ensemble methods [10–12]. Among those categories, ensemble methods are the important area that prove to improve the classification performance. Given a specific learning task, one single learner often cannot guarantee the result of the imbalance classification. Instead, a fusion including multiple classifiers can generate different outputs. Then, selected models are fused aiming to improve the final classification performance and avoid the choice of those worst classifiers. To create the ensemble, previous techniques such as data resampling and algorithm modification are used. Furthermore, the ensemble of data resampling achieves the state-of-the-art performance [3,19–21].

In this study, we propose a hierarchical ensemble oversampling for improved imbalance classification. In particular, our contribution to this work can be summarized as follows: (1) A hierarchical ensemble is proposed for the imbalance classification, which includes bootstrap sampling for creating the first-level fusion, the second-level instance weighting-based fusion. (2) Bootstrap sampling with replacement is used to create the ensemble and avoid the worst classifiers for improving the performance. (3) Two different weighting strategies are compared and discussed. (4) k-means clustering is used to generate a select minority class subset for oversampling based on multivariate Gaussian distribution.

2   Related Work

Various methods have been proposed in recent years. [22] proposed a novel evolutionary cluster-based oversampling ensemble. This framework combined a novel cluster-based synthetic data generation method with an evolutionary algorithm to create an ensemble. [3] used different oversampling techniques to build an ensemble classifier that can reduce the effect of low minority samples in the defective data. [23] proposed a three-way decision model by considering the differences in the cost of selecting key samples. Here, the ensemble was created by applying Constructive Covering Algorithm to divide the minority samples. [24] predicted default events by analyzing different ensemble classification methods that empowered the effects of the synthetic minority oversampling technique (SMOTE) used in the preprocessing of the imbalanced microcredit dataset. [25] investigated heterogeneous ensembles for imbalance learning. [21] applied an entropy-based method to minority samples to create an ensemble. Then, minority samples in various views were combined with the majority sample, where SMOTE is further used. [26] integrated ensemble learning with the union of a margin-based undersampling and diversity-enhancing oversampling. Here the ensemble was created by applying bootstrap sampling to all samples. [27] proposed an imbalanced classification ensemble method, where a weighted bootstrap method was introduced to generated sub-datasets containing diverse local information.

3   The Proposed Method

Our proposed hierarchical ensemble-based imbalance classification system consists of four main steps: (1) bootstrap sampling with replacement, (2) minority instance selection and weighting, (3) oversampling based on multivariate Gaussian distribution, and (4) Euclidean distance-based sample selection (Fig. 1).

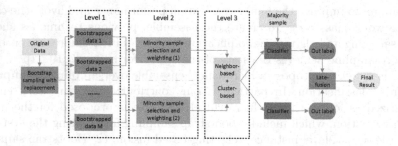

Fig. 1. Flowchart of our proposed work using hierarchical-ensemble based imbalance classification

3.1   Bootstrap Sampling

Bootstrap sampling is a sampling method that uses random sampling with replacement. Previous studies have demonstrated that bootstrap sampling with data resampling is an excellent strategy in dealing with class imbalance issues [20,26]. Specifically, bootstrap sampling with replacement ensures the independent training of each base classifier. Furthermore, bootstrap sampling can change the specific imbalance characters, which can thus reduce the variance of the loss functions of the base classifier. In this study, bootstrap sampling with replacement is used to create the ensemble (first-level fusion).

3.2   Instance Selection and Weighting Strategy

For the oversampling, the selection of minority instances is important for the classification results. Previous studies used various features to select instance for oversampling including classification error [28] and k-disagree neighbors (kDN) [29]. The kDN denotes the number of disagree class within k neighbors of a specific instance. For classification error, each instance of the train set is classified by Decision Tree (DT). The DT is run five times using 10-fold cross-validation to obtain the final classification error. Naturally, the sample carrying *more information* will be oversampled more times. However, the definition of *more information* is confusing based on the analysis of previous studies. [5] oversampled those borderline minority samples. In contrast, [29] selected those safe samples for oversampling. Both methods prove to be effective in oversampling. Here, we believe that *datasets with different characters work well using different sample selection and weighting strategies.*

After obtaining kDN and *error* of sample X_i, they are first normalized to [0, 1]. Then two different weighting strategies are used to calculate sample hardness (Eq. 1 and Eq. 2). The first weighting strategy (Eq. 1) assumes that those difficult minority samples carry more information, while the second weighting strategy (Eq. 2) assumes that those safe minority samples carry more information. The selection of the weighting strategy is based on the average performance of each ensemble. The value of calculated hardness is between 0 and 1.

$$I_1(X_i) = (\frac{kDN(X_i)}{max(kDN(X_i))} + \frac{error(X_i)}{max(error(X_i))}) * 0.5 \tag{1}$$

$$I_2(X_i) = 1 - I_1(X_i) \tag{2}$$

Finally, the number of instances, which will be over-sampled for each minority instance is calculated as follows:

$$E(X_i) = \frac{e^{I(X_i)}}{\sum_{i=1}^{|N_i^{min}|} e^{I(X_i)}} (|N_i^{maj}| - |N_i^{min}|) \tag{3}$$

where $|N_i^{maj}|$ and $|N_i^{min}|$ denote the number of minority and majority class instances in K neighbor instances, respectively, $I(X_i)$ is $I_1(X_i)$ or $I_2(X_i)$.

When the number of majority instances of K neighbors for a minority instance is K, the minority instance will be regarded as noise. However, if all minority instances are regarded as noise, the number of instances, which will be over-sampled for each minority instance, is calculated as follows:

$$E(X_i) = \frac{1}{|N_i^{min}|} (|N_i^{maj}| - |N_i^{min}|) \tag{4}$$

In addition to minority class instance selection and weighting, we use the kDN value of the majority class instance for noisy instance removal. Here, those majority class instances with a kDN value of K will be removed for the subsequent analysis. The value of K is set to 5. To control the quality of generated instances, we first generate $\lambda * E(X_i)$ new instances, then select $E(X_i)$ instances whose Euclidean distances are the smallest. Here, the value of λ is set to 50.

3.3   Multivariate Gaussian Distribution Based Oversampling

After obtaining the hardness of minority samples, we use multivariate Gaussian distribution [30] for new instance generation since most real data follows

Gaussian distribution. In this study, multivariate Gaussian distribution fits the minority class subset for instance oversampling.

Specifically, multivariate Gaussian distribution $N(\mu_i, \Sigma_i)$ first fits the selected subset of the current instance using two strategies, which are described as follows: In the first strategy, we combine both current and its closest instance for calculating μ_i and Σ_i. In the second strategy, we first apply k-means clustering to minority clusters. Then, μ_i and Σ_i are obtained from the cluster, which the current instance belongs to. After obtaining μ_i and Σ_i, the synthetic instances are generated using the following equation:

$$f_X(x_1, ..., x_k) = \frac{exp(-\frac{1}{2}(X - \mu)^T \Sigma^{-1}(X - \mu))}{\sqrt{(2\pi)^k |\Sigma|}} \tag{5}$$

For both strategies, the number of synthetic samples of each minority class instance is split as follows:

$$E(X_i) = \theta * E_1(X_i) + (1 - \theta) * E_2(X_i) \tag{6}$$

where θ is used to balance the number of synthetic samples by the combination of current and closest samples $(1 - \theta) * E_1(X_i)$, and the clusters $\theta * E_2(X_i)$. E_1 and E_2 denote the number of synthetic samples for neighbor- and cluster-based Gaussian oversampling, respectively. When θ is set to 1, it represents cluster-based Gaussian oversampling. In contrast, it stands for neighbor-based Gaussian oversampling when θ is set to 0. The setting of θ will be investigated in Experiment section.

For the Gaussian oversampling, we use the subsets obtained by GMM to calculate μ_i and Σ_i. After generating $\lambda * E(X_i)$ new instances, we further select $E(X_i)$ instances whose Euclidean distances are the smallest.

3.4   Hierarchical Ensemble

In the first level, we create an ensemble $(D_{m=1}^M)$ by applying bootstrap sampling with replacement to all samples. Then, we compare the averaged performance of two different weighting strategies, and the one with better performance in the second level is selected as follows.

$$\bar{D} = \begin{cases} avg(D_{m=1}^{M/2}) > avg(D_{m=M/2+1}^M) \\ avg(D_{m=1}^{M/2}) <= avg(D_{m=M/2+1}^M) \end{cases} \tag{7}$$

where $avg(\cdot)$ denotes the averaged performance of selected ensembles, and $\bar{D}$ is the selected ensembles for the subsequent analysis. Next, multivariate Gaussian distribution based oversampling is used to obtain synthetic instances, where the subsets are obtained by Gaussian Mixture Models. Finally, performance-based majority voting is used to make a final prediction for each test sample.

4 Datasets and Experiment Setting

In total, we selected 15 datasets from the KEEL data repository [32] for evaluating our proposed method. Here, those 15 datasets exhibited a considerable variety in sample number, feature number, and IR. Table 1 shows the detailed properties of the 15 selected KEEL datasets including the number of samples for both majority and minority classes, feature dimension, and imbalance ratio (IR). For all datasets, each method is evaluated with a 5-fold cross-validation procedure. For fairness and uniformity, we use an ensemble size of 40 and a Decision Tree (DT) as the base learning model in all experiments. All experiments are implemented using the Python library scikit-learn [33] with default parameters unless stated otherwise.

Table 1. Summary of datasets for evaluating and comparing the proposed method. Here, the data is sorted by the name, IR denotes imbalance ratio.

No.	Dataset	Samples	Majority samples	Minority samples	Features	IR
1	car-good	1728	1659	69	6	24.04
2	dermatology-6	358	338	20	34	16.90
3	ecoli-0-3-4_vs_5	200	180	20	7	9.00
4	glass0	214	144	70	9	2.06
5	haberman	306	225	81	3	2.78
6	iris0	150	100	50	4	2.00
7	kddcup-buffer_overflow_vs_back	2233	2203	30	41	73.43
8	kr-vs-k-zero-one_vs_draw	2901	2796	105	6	26.63
9	page-blocks-1-3_vs_4	472	444	28	10	15.86
10	pima	768	500	268	8	1.87
11	segment0	2308	1979	329	19	6.02
12	shuttle-2_vs_5	3316	3267	49	9	66.67
13	vehicle3	846	634	212	18	2.99
14	vowel0	988	898	90	13	9.98
15	yeast-0-2-5-7-9_vs_3-6-8	1004	905	99	8	9.14

To demonstrate the effectiveness of our proposed method, several ensemble-based techniques are included for the comparison: (1) undersampling-based ensemble: EasyEnsemble, RUSBoost, BalancedBagging, SelfPacedEnsemble. (2) oversampling-based ensemble: OverBoost, SMOTEBoost, OverBagging, and SMOTEBagging.

4.1 Evaluation Metrics

For imbalance classification, accuracy often does not well reflect the overall performance of a proposed classification method. Usually, other metrics are adopted for better comparing different classification methods. In our study, G-mean and Area Under the Curve (AUC) are selected as the performance measures, since they are trade-off metrics between the correctly classified positive and negative instances. The definition of G-mean is

$$G - mean = \sqrt{Sens \cdot Spec} \tag{8}$$

where $Sens = \frac{TP}{TP+FN}$ and $Spec = \frac{TN}{TN+FP}$. TP (true positive) is the number of correctly classified positive instances. FN (false negative) is the number of the positive instances that were misclassified as negative. FP (false positive) is the number of negative instances that were misclassified as positive. TN (true negative) is the number of correctly classified negative instances.

For AUC, it can be calculated as follows:

$$AUC = \frac{Sens + Spec}{2} \tag{9}$$

5   Experiments

In this section, we first investigate the effect of hyperparameters of our proposed method. Then, we will compare our proposed method with two types of ensemble-based methods: undersampling-based ensembles and oversampling-based ensembles.

5.1   The Effect of Hyperparameters

Tables 2 show the averaged AUC and G-mean over 15 datasets. Comparing classification results in terms of cluster size and θ, although the difference between different combinations of k (k-means clustering) and θ is not significant, the performance using a cluster size of 2 and θ of 0.8 is the best which will be selected for the subsequent analysis. Furthermore, we plot the performance for all datasets (Fig. 2). We can observe that similar performance is obtained for all datasets over all combinations of k and θ.

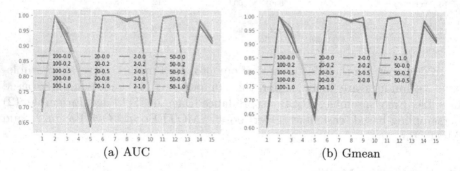

(a) AUC   (b) Gmean

Fig. 2. Classification AUC and G-mean of different cluster sizes and θ for 15 datasets. Here, x-axis denotes dataset index.

5.2   Our Method vs. Undersampling-Based Ensembles

Tables 3 and 4 show the classification AUC and G-mean between our method and undersampling-based ensembles. The result highlighted in bold indicates better AUC—G-mean of our method against all other undersampling-based ensemble

Table 2. Averaged AUC and Gmean of different cluster sizes and θ.

	Cluster size	θ				
		0	0.2	0.5	0.8	1
AUC	2	0.8933	0.8972	0.8945	**0.8992**	0.8964
	20	0.8933	0.8968	0.8950	0.8955	0.8978
	50	0.8933	0.8965	0.8951	0.8958	0.8985
	100	0.8933	0.8965	0.8951	0.8958	0.8984
Gmean	2	0.8850	0.8907	0.8871	**0.8922**	0.8895
	20	0.8850	0.8896	0.8882	0.8884	0.8910
	50	0.8850	0.8893	0.8883	0.8888	0.8917
	100	0.8850	0.8893	0.8883	0.8888	0.8915

methods in 9—9 datasets out of 15, followed by SelfPacedEnsemble with only 6—6 wins.

For BalancedBagging, the ensemble is created based on the undersampling technique using a negative binomial distribution. EasyEnsemble is to create an ensemble set by iteratively applying random under-sampling. Although the complexity of those methods is low, the performance of those methods is worse due to the information loss in the undersampling step. Furthermore, oversampling has been demonstrated to be better than undersampling in terms of the ROC curve [34]. Therefore, our proposed multivariate Gaussian distribution-based ensemble oversampling can obtain better performance.

Table 3. Comparison of AUC between our method vs. undersampling-based ensembles.

No.	BalancedBagging	EasyEnsemble	RUSBoost	SelfPacedEnsemble	Ours
1	0.8219	**0.8264**	0.7142	0.6974	0.6937
2	**0.9985**	0.9735	0.9706	**0.9985**	**0.9985**
3	0.9306	0.9083	0.8639	0.9000	**0.9611**
4	**0.8273**	0.8203	0.7692	0.8128	0.8094
5	0.6078	0.6128	0.5271	0.5867	**0.6629**
6	1.0000	1.0000	1.0000	1.0000	1.0000
7	0.9964	0.9995	0.9982	1.0000	1.0000
8	**0.9859**	0.9843	0.9846	0.9735	0.9823
9	0.9876	0.9899	0.9721	**0.9978**	0.9955
10	0.7237	0.7213	0.6654	**0.7239**	0.7130
11	0.9901	0.9896	0.9911	0.9939	**0.9942**
12	1.0000	0.9989	1.0000	1.0000	1.0000
13	0.7511	**0.7865**	0.6358	0.7558	0.7612
14	0.9605	0.9666	0.9805	0.9683	**0.9883**
15	0.9154	0.9052	0.8785	0.8987	**0.9276**
Rank	5	3	2	6	9

Table 4. Comparison of G-mean between our method vs. undersampling-based ensembles.

No.	BalancedBagging	EasyEnsemble	RUSBoost	SelfPacedEnsemble	Ours
1	0.8041	**0.8080**	0.6247	0.4791	0.6022
2	**0.9985**	0.9717	0.9690	**0.9985**	**0.9985**
3	0.9282	0.9050	0.8374	0.8907	**0.9595**
4	**0.8260**	0.8194	0.7539	0.8115	0.8060
5	0.5941	0.6061	0.4495	0.5732	**0.6587**
6	**1.0000**	**1.0000**	**1.0000**	**1.0000**	**1.0000**
7	0.9963	0.9995	0.9982	**1.0000**	**1.0000**
8	**0.9858**	0.9841	0.9845	0.9731	0.9822
9	0.9875	0.9898	0.9712	**0.9977**	0.9955
10	0.7200	0.7178	0.6406	**0.7224**	0.7120
11	0.9901	0.9896	0.9911	0.9939	**0.9942**
12	**1.0000**	0.9989	**1.0000**	**1.0000**	**1.0000**
13	0.7478	**0.7837**	0.5681	0.7527	0.7596
14	0.9592	0.9659	0.9802	0.9661	**0.9882**
15	0.9135	0.9035	0.8702	0.8948	**0.9266**
Rank	5	3	2	6	9

5.3 Our Method vs. Oversampling-based Ensembles

Tables 5 and 6 show the classification AUC and G-mean between our method and oversampling-based ensembles. The results highlighted in bold indicate better AUC—G-mean of our method against all other regular ensemble methods in 11—11 datasets out of 15. For SMOTEBagging, it has 7—8 wins in terms of both AUC and G-mean. For OverBagging, it has 6—5 wins in terms of both AUC and G-mean.

Figure 3 plots the comparison between our method VS. oversampling-based ensembles in terms of average performance and winning time. Our proposed method achieves the best performance. In terms of those datasets having higher IR than 20, our method hits 3 of 4 highest AUC and G-mean values respectively. For 9 datasets whose IR are less than 10, our method hits the 7 highest AUC and G-mean values. This result indicates that our proposed method can work well for both high and low IR.

Table 5. Comparison of AUC between our method vs. oversampling-based ensembles.

No.	OverBoost	OverBagging	SMOTEBoost	SMOTEBagging	Ours
1	0.6355	**0.6064**	0.5760	0.6203	**0.6937**
2	**0.9985**	**0.9985**	**0.9985**	**0.9985**	**0.9985**
3	0.9139	0.8417	0.9056	0.8639	**0.9611**
4	**0.7842**	**0.8195**	0.7701	0.8195	0.8094
5	0.5820	0.5151	0.5593	0.5414	**0.6629**
6	**1.0000**	**1.0000**	**1.0000**	**1.0000**	**1.0000**
7	**1.0000**	**1.0000**	**1.0000**	**1.0000**	**1.0000**
8	**0.9647**	0.9691	0.9474	**0.9827**	0.9823
9	0.9811	**0.9978**	0.9811	**0.9978**	0.9955
10	0.6243	0.7024	0.6673	**0.7341**	0.7130
11	0.9889	0.9896	0.9924	0.9899	**0.9942**
12	**1.0000**	**1.0000**	**1.0000**	**1.0000**	**1.0000**
13	0.6695	**0.6415**	0.6761	0.7122	**0.7612**
14	0.9289	0.9405	0.9567	0.9511	**0.9883**
15	0.8685	0.8782	0.8768	0.8965	**0.9276**
Rank	4	6	4	7	11

Table 6. Comparison of Gmean between our method vs. oversampling-based ensembles.

No.	OverBoost	OverBagging	SMOTEBoost	SMOTEBagging	Ours
1	0.4008	**0.3643**	0.3148	0.4272	**0.6022**
2	**0.9985**	**0.9985**	**0.9985**	**0.9985**	**0.9985**
3	0.9042	0.8216	0.8960	0.8365	**0.9595**
4	**0.7777**	**0.8147**	0.7586	**0.8165**	0.8060
5	0.5357	0.4199	0.5270	0.4760	**0.6587**
6	**1.0000**	**1.0000**	**1.0000**	**1.0000**	**1.0000**
7	**1.0000**	**1.0000**	**1.0000**	**1.0000**	**1.0000**
8	**0.9635**	0.9680	0.9458	**0.9825**	0.9822
9	0.9803	**0.9977**	0.9803	**0.9977**	0.9955
10	0.6080	0.6826	0.6619	**0.7257**	0.7120
11	0.9888	0.9896	0.9924	0.9898	**0.9942**
12	**1.0000**	**1.0000**	**1.0000**	**1.0000**	**1.0000**
13	0.6427	**0.5812**	0.6596	0.6927	**0.7596**
14	0.9257	0.9377	0.9544	0.9487	**0.9882**
15	0.8600	0.8706	0.8705	0.8908	**0.9266**
Rank	4	5	4	8	11

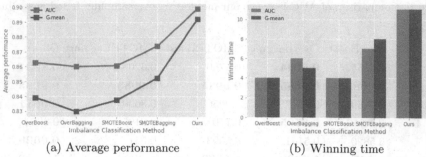

(a) Average performance   (b) Winning time

Fig. 3. Classification AUC and G-mean of different ensemble methods in terms of averaged result and winning time.

6   Conclusion and Future Work

In this study, we propose an ensemble oversampling framework, which shows significant improvement over state-of-the-art algorithms using a total of 15 imbalance datasets. The framework consists of a novel hierarchical ensemble utilizing bootstrap sampling, instance selection and weighting, and multivariate Gaussian distribution-based oversampling. For multivariate Gaussian distribution-based oversampling, the parameters, μ, and Σ are calculated using (1) clusters, (2) current, and its k nearest minority neighbors. The final predict label is obtained by combining selected ensemble members to deliver relatively superior results.

In future work, we are considering including an evolutionary algorithm for adaptively selecting ensemble members. In addition, we will extend the proposed approach to cover multi-class imbalanced datasets.

Acknowledgment. This work is supported by National Natural Science Foundation of China (Grant No: 61902154 and 72004092). This work is also partially supported by Natural Science Foundation of Jiangsu Province (Grant No: BK2019043526), Jiangsu Province Post Doctoral Fund (Grant No: 2020Z430), and China Postdoctoral Science special Foundation (Grant No. 2021T140281).

References

1. Kahl, S., et al.: Large-scale bird sound classification using convolutional neural networks. In: CLEF (Working Notes), vol. 1866 (2017)
2. Zhu, H., Liu, G., Zhou, M., Xie, Y., Abusorrah, A., Kang, Q.: Optimizing weighted extreme learning machines for imbalanced classification and application to credit card fraud detection. Neurocomputing **407**, 50–62 (2020)
3. Huda, S., et al.: An ensemble oversampling model for class imbalance problem in software defect prediction. IEEE Access **6**, 24184–24195 (2018)
4. Shawky, M.: Factors affecting lane change crashes. IATSS Res. **44**, 155–161 (2020)

5. Han, H., Wang, W.-Y., Mao, B.-H.: Borderline-SMOTE: a new over-sampling method in imbalanced data sets learning. In: Huang, D.-S., Zhang, X.-P., Huang, G.-B. (eds.) ICIC 2005. LNCS, vol. 3644, pp. 878–887. Springer, Heidelberg (2005). https://doi.org/10.1007/11538059_91

6. Kang, Q., Chen, X., Li, S., Zhou, M.: A noise-filtered under-sampling scheme for imbalanced classification. IEEE Trans. Cybernet. **47**, 4263–4274 (2016)

7. López, V., Fernández, A., Moreno-Torres, J.G., Herrera, F.: Analysis of preprocessing vs. cost-sensitive learning for imbalanced classification. Open problems on intrinsic data characteristics. Expert Syst. Appl. **39**, 6585–6608 (2012)

8. Liu, Y., Lu, H., Yan, K., Xia, H., An, C.: Applying cost-sensitive extreme learning machine and dissimilarity integration to gene expression data classification. Comput. Intell. Neurosci. **2016** (2016)

9. Khan, S.H., Hayat, M., Bennamoun, M., Sohel, F.A., Togneri, R.: Cost-sensitive learning of deep feature representations from imbalanced data. IEEE Trans. Neural Netw. Learn. Syst. **29**, 3573–3587 (2017)

10. Li, J., Fong, S., Wong, R.K., Chu, V.W.: Adaptive multi-objective swarm fusion for imbalanced data classification. Inf. Fus. **39**, 1–24 (2018)

11. Chen, R., Guo, S.-K., Wang, X.-Z., Zhang, T.-L.: Fusion of multi-RSMOTE with fuzzy integral to classify bug reports with an imbalanced distribution. IEEE Trans. Fuzzy Syst. **27**, 2406–2420 (2019)

12. Yang, J., Xie, G., Yang, Y.: An improved ensemble fusion autoencoder model for fault diagnosis from imbalanced and incomplete data. Control Eng. Pract. **98**, 104358 (2020)

13. Liu, X.-Y., Wu, J., Zhou, Z.-H.: Exploratory undersampling for class-imbalance learning. IEEE Trans. Syst. Man Cybernet. Part B (Cybernetics) **39**, 539–550 (2008)

14. Seiffert, C., Khoshgoftaar, T.M., Van Hulse, J., Napolitano, A.: RUSBoost: a hybrid approach to alleviating class imbalance. IEEE Trans. Syst. Man Cybernet. A Syst. Hum. **40**, 185–197 (2009)

15. Wang, S., Yao, X.: Diversity analysis on imbalanced data sets by using ensemble models. In: 2009 IEEE Symposium on Computational Intelligence and Data Mining, pp. 324–331. IEEE (2009)

16. Chen, C., Liaw, A., Breiman, L.: Using random forest to learn imbalanced data, Technical report (2004)

17. Chawla, N.V., Lazarevic, A., Hall, L.O., Bowyer, K.W.: SMOTEBoost: improving prediction of the minority class in boosting. In: Lavrač, N., Gamberger, D., Todorovski, L., Blockeel, H. (eds.) PKDD 2003. LNCS (LNAI), vol. 2838, pp. 107–119. Springer, Heidelberg (2003). https://doi.org/10.1007/978-3-540-39804-2_12

18. Maclin, R., Opitz, D.: An empirical evaluation of bagging and boosting. In: AAAI/IAAI 1997, pp. 546–551 (1997)

19. Khoshgoftaar, T.M., Van Hulse, J., Napolitano, A.: Comparing boosting and bagging techniques with noisy and imbalanced data. IEEE Trans. Syst. Man Cybernet. A Syst. Hum. **41**, 552–568 (2010)

20. Galar, M., Fernandez, A., Barrenechea, E., Bustince, H., Herrera, F.: A review on ensembles for the class imbalance problem: bagging-, boosting-, and hybrid-based approaches. IEEE Trans. Syst. Man Cybernet. C (Appl. Rev.) **42**, 463–484 (2011)

21. Dongdong, L., Ziqiu, C., Bolu, W., Zhe, W., Hai, Y., Wenli, D.: Entropy-based hybrid sampling ensemble learning for imbalanced data. Int. J. Intell. Syst. **36**, 3039–3067 (2021)

22. Lim, P., Goh, C.K., Tan, K.C.: Evolutionary cluster-based synthetic oversampling ensemble (eco-ensemble) for imbalance learning. IEEE Trans. Cybernet. **47**, 2850–2861 (2016)
23. Yan, Y.T., Wu, Z.B., Du, X.Q., Chen, J., Zhao, S., Zhang, Y.P.: A three-way decision ensemble method for imbalanced data oversampling. Int. J. Approx. Reason. **107**, 1–16 (2019)
24. Gicić, A., Subasi, A.: Credit scoring for a microcredit data set using the synthetic minority oversampling technique and ensemble classifiers. Expert Syst. **36**, e12363 (2019)
25. Zefrehi, H.G., Altınçay, H.: Imbalance learning using heterogeneous ensembles. Expert Syst. Appl. **142**, 113005 (2020)
26. Chen, Z., Duan, J., Kang, L., Qiu, G.: A hybrid data-level ensemble to enable learning from highly imbalanced dataset. Inf. Sci. **554**, 157–176 (2021)
27. Yuan, B.-W., Zhang, Z.-L., Luo, X.-G., Yu, Y., Zou, X.-H., Zou, X.-D.: OIS-RF: a novel overlap and imbalance sensitive random forest. Eng. Appl. Artif. Intell. **104**, 104355 (2021)
28. Chongomweru, H., Kasem, A.: A novel ensemble method for classification in imbalanced datasets using split balancing technique based on instance hardness (sBAL_IH). Neural Comput. Appl. **33**, 1–22 (2021)
29. Xie, Y., Qiu, M., Zhang, H., Peng, L., Chen, Z.: Gaussian distribution based oversampling for imbalanced data classification. IEEE Trans. Knowl. Data Eng. (2020)
30. Biernacki, C., Celeux, G., Govaert, G.: Choosing starting values for the EM algorithm for getting the highest likelihood in multivariate gaussian mixture models. Comput. Stat. Data Anal. **41**, 561–575 (2003)
31. Crump, M.J., Navarro, D., Suzuki, J.: Answering questions with data: introductory statistics for psychology students (2019)
32. Alcalá-Fdez, J., et al.: Keel data-mining software tool: data set repository, integration of algorithms and experimental analysis framework. J. Multiple-Valued Logic Soft Comput. **17** (2011)
33. Pedregosa, F., et al.: Scikit-learn: machine learning in Python. J. Mach. Learn. Res. **12**, 2825–2830 (2011)
34. Chawla, N.V., Bowyer, K.W., Hall, L.O., Kegelmeyer, W.P.: Smote: synthetic minority over-sampling technique. J. Artif. Intell. Res. **16**, 321–357 (2002)

Numerical Approximation
of the One-Way Helmholtz Equation
Using the Differential Evolution Method

Mikhail S. Lytaev$^{(\boxtimes)}$ ⓘ

St. Petersburg Federal Research Center of the Russian Academy of Sciences,
14-th Linia, V.I., No. 39, Saint Petersburg 199178, Russia
mikelytaev@gmail.com

Abstract. This paper is devoted to increasing the computational efficiency of the finite-difference methods for solving the one-way Helmholtz equation in unbounded domains. The higher-order rational approximation of the propagation operator was taken as a basis. Computation of appropriate approximation coefficients and grid sizes is formulated as the problem of minimizing the discrete dispersion relation error. Keeping in mind the complexity of the developed optimization problem, the differential evolution method was used to tackle it. The proposed method does not require manual selection of the artificial parameters of the numerical scheme. The stability of the scheme is provided by an additional constraint of the optimization problem. A comparison with the Padé approximation method and rational interpolation is carried out. The effectiveness of the proposed approach is shown.

Keywords: Wave propagation · Helmholtz equation · Differential evolution · Optimization · Rational approximation

1 Introduction

Despite the constant increase in the computing power, the numerical solution of many mathematical physics equations remains a very resource-intensive operation. Fast and efficient numerical schemes usually use complex approximations and require quite sophisticated software implementation. Even more nontrivial is the question of the stability of complex numerical schemes and the determination of the limits of their applicability. Despite the existence of general-purpose numerical schemes, such as the finite element method, the construction and detailed study of a new numerical scheme is required in each specific case.

This study is aimed at improving the efficiency of the computer simulation methods for wave propagation in large integration domains without boundaries. Such kind of problems arise in computational hydroacoustics [5,18], tropospheric radio wave propagation [12,16,22], geophysics [21], optics and quantum mechanics [6]. Despite the different nature of the physical phenomena occurring in these scientific domains, the underlying mathematical models are to a certain extent universal [20].

The existing numerical methods for solving this class of problems have two significant disadvantages, which are fully manifested when trying to implement them as part of complex software systems. Firstly, they depend on several artificial computational parameters, which are usually selected manually by experts. The expert does not have reliable mechanisms for verifying the adequacy of the selected parameters, which can lead to errors. Secondly, they usually do not take into account the specific parameters of the propagation environment in an optimal way, which leads to significant overspending of computing resources and a decrease in the relevance of the results obtained.

To answer these questions, in this paper it is proposed to use stochastic methods [8] to optimize numerical schemes. The use of stochastic methods for constructing numerical schemes has been an actively developing scientific direction over the past few years. In particular, the physics-informed neural networks (PINN) method is actively developing [10,13]. In the PINN method, the solution of the equation is sought in the form of a deep neural network. The method proposed in this paper assumes the preservation of the numerical scheme structure, only its coefficients and parameters are upgraded.

This article is a continuation of the work on improving the numerical schemes for solving the Helmholtz equation. Previously, a method for finding optimal computation parameters for the Padé approximation was proposed [14]. Works [15,16] discovered the possibility of increasing the performance of existing schemes by using more suitable rational approximations.

The paper is organized as follows. The next section briefly describes the problem statement and numerical scheme based on rational approximation. Section 3 provides a discrete dispersion analysis of the numerical scheme under consideration. Section 4 is devoted to the optimization of coefficients and parameters of the scheme using the differential evolution method. Section 5 shows the application of the proposed method to the wedge diffraction problem and comparison with other rational approximation methods.

2   Rational Approximation of the One-Way Helmholtz Equation

We are seeking the solution to the two-dimensional scalar Helmholtz equation

$$\frac{\partial^2 \psi}{\partial x^2} + \frac{\partial^2 \psi}{\partial z^2} + k^2 n^2 (x, z) \psi = 0, \tag{1}$$

where $\psi(x, z)$ is the wave field, $k = 2\pi/\lambda$ is the wavenumber, λ is the wave length, $n(x, z)$ is the refractive index. It is assumed that the length in x of the propagation medium is much larger than the height in z.

The wave field is generated by an initial condition of the form

$$\psi(0, z) = \psi_0(z),$$

where $\psi_0(z)$ is a known function.

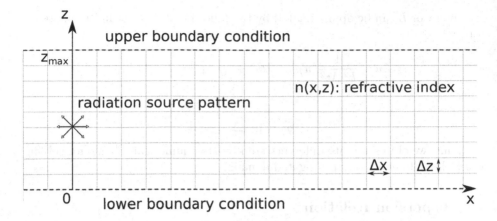

Fig. 1. Schematic description of the considered problem [16].

The considered problem is schematically shown in Fig. 1.

Following [9,16] and neglecting to backscatter, we obtain the so-called one-way Helmholtz equation, written as a step-by-step solution along x-axis

$$u^{n+1} = \exp\left(ik\Delta x\left(\sqrt{1+L}-1\right)\right)u^n \tag{2}$$

$$u(x,z) = e^{-ikx}\psi(x,z),$$

$$u^n(z) = u(n\Delta x, z).$$

As suggested in [4], we apply a rational approximation of order $[n/m]$ the propagation operator (2)

$$\exp\left(ik\Delta x\left(\sqrt{1+L}-1\right)\right) \approx \frac{\prod_{l=1}^n 1+a_l L}{\prod_{l=1}^m 1+b_l L} = \prod_{l=1}^p \frac{1+a_l L}{1+b_l L}, \tag{3}$$

$$Lu = \frac{1}{k^2}\frac{\partial^2 u}{\partial z^2} + \left(n^2(x,z)-1\right)u,$$

meaning of coefficients $a_1\ldots a_p, b_1\ldots b_p$ will be clarified in the following sections.

Rational approximation (3) makes it possible to represent the action of the propagation operator (2) as a sequence of one-dimensional differential equations

$$\begin{cases} (1+b_1 L)v_1^n = (1+a_1 L)u^{n-1} \\ (1+b_l L)v_l^n = (1+a_l L)v_{l-1}^n \quad l=2,\ldots,p-1, \\ \ldots \\ (1+b_p L)u^n = (1+a_p L)v_{p-1}^n. \end{cases}$$

where $v_1\ldots v_{p-1}$ are some auxiliary functions.

Operator L can be approximated by the following 2nd order finite-difference schema

$$Lu \approx \frac{1}{k^2 \Delta z^2} \left[u_{j-1} - 2u_j + u_{j+1} \right] + \left(n_j^2 - 1 \right) u_j,$$

where

$$u_j = u(j\Delta z).$$

Thus, we obtain a finite-difference step-by-step numerical scheme for solving the Helmholtz equation in elongated domain.

3   Dispersion Relation

The accuracy and stability of the numerical scheme under consideration will be analyzed using discrete dispersion relations [3]. To do this, it is enough to consider how a plane wave passes through the numerical scheme. Substitute a two-dimensional plane wave of the form

$$E\left(x, z\right) = \exp\left(i\tilde{k}_x x + ik_z z \right),$$

where $k_z = k\sin\theta$ is the vertical wavenumber, θ is the angle between the direction of the wave and x-axis. For simplicity, we further consider the case of a homogeneous medium $(n(x, z) \equiv 1)$. Then, the discrete horizontal wavenumber $\tilde{k}_x$ takes the form [14]

$$\tilde{k}_x \left(k_z, \Delta x, \Delta z, a_1 \ldots a_p, b_1 \ldots b_p \right) = k + \frac{\ln \prod_{l=1}^p t_l}{i\Delta x}, \tag{4}$$

$$t_l = \frac{1 - \frac{4a_l}{(k\Delta z)^2} \sin^2 \left(\frac{k_z \Delta z}{2} \right)}{1 - \frac{4b_l}{(k\Delta z)^2} \sin^2 \left(\frac{k_z \Delta z}{2} \right)}.$$

Latter expression is also known as the discrete dispersion relation.

Horizontal wavenumber for the original Helmholtz Eq. (1) is expressed as follows

$$k_x \left(k_z \right) = \begin{cases} \sqrt{k^2 - k_z^2}, & |k_z| \leq k, \\ i\sqrt{k_z^2 - k^2}, & |k_z| > k. \end{cases} \tag{5}$$

4   Optimization of the Numerical Scheme

It is common to use the Padé approximation method [2] to obtain coefficients $a_1 \ldots a_p, b_1 \ldots b_p$ of rational approximation (3). The Padé approximation is local one and works well in the vicinity of zero propagation angle. As the propagation angle increases, its accuracy drops rapidly. To tackle this issue, it was previously proposed to use a class of rational approximation methods on an interval [16].

These methods make it possible to achieve uniform accuracy on the desired interval of propagation angles.

In this paper, the coefficients of the numerical scheme are defined by minimizing the difference between the real and discrete dispersion relation. The resulting optimization problems have a very complex structure and are hardly solvable by known deterministic optimization methods. In this connection, it is reasonable to apply evolutionary algorithms to solve them. Due to their stochastic nature, they are able to solve very complex optimization problems.

Our perspective of optimization of a numerical scheme is an increasing the speed of calculations without increasing the computational resources or reducing the applicability limits of the scheme.

4.1   Differential Evolution Method

In this work, we will use the differential evolution method [19] as one of the most well-known representatives of evolutionary algorithms. Let's briefly describe its main features.

Consider the following minimization problem

$$f\left(x\right) \to \min, \, x \in D \subset \mathbb{R}^n.$$

Fitness function f is a black box for the differential evolution method, no restrictions are imposed on it. The only thing required is to be able to compute its value in any point on its domain.

At each iteration, the algorithm generates a new generation of vectors by randomly combining vectors from the previous generation. For each vector x_i three different vectors v_1, v_2 and v_3 are randomly selected among the vectors from the old generation and a new mutant vector is produced

$$v = v_1 + F \cdot \left(v_2 - v_3\right), \tag{6}$$

where $F \in [0, 2]$ is a parameter called mutation. The mutation can be set by a constant or selected randomly at each iteration. Note that the formula for calculating the mutant vector (6) may differ depending on the strategy. Number of vectors in population (population size) is also a parameter of the method.

Then the crossover operation is applied: some coordinates are randomly replaced by the corresponding coordinates from the mutant vector v. If the new vector turns out to be better, then it passes to the next generation, otherwise the old one remains.

The unconditional optimization problem was considered above, but the differential evolution method can be generalized to take into account arbitrary nonlinear constraints [11] of the form

$$a \le g\left(x\right) \le b.$$

We further use the implementation of the differential evolution method from the SciPy library [1].

4.2 Unconditional Optimization

Let's start with the simplest optimization problem. Consider that values of the computational grid sizes Δx and Δz are fixed. Assume also that we know maximum propagation angle θ_{max}. The numerical scheme should minimize the difference between the real (5) and discrete (4) dispersion relation for all propagation angles from interval $[0, \theta_{max}]$. Formally, this can be written as follows

$$\text{argmin}_{a_1 \ldots a_p, b_1 \ldots b_p} \tag{7}$$
$$\left[\max_{k_z \in [0, k_z^{max}]} \frac{1}{k} |\tilde{k}_x (k_z, \Delta x, \Delta z, a_1 \ldots a_p, b_1 \ldots b_p) - k_x (k_z) | \right],$$

where $k_z^{max} = k \sin \theta_{max}$.

Consider an example of solving optimization problem (7) with the following parameters: $\Delta x = 50\lambda$, $\Delta z = 0.25\lambda$, rational approximation order is equal to $[6/7]$, $\theta_{max} = 22°$. Figure 2 demonstrates the dependence of the discrete dispersion relation error on the propagation angle for various rational approximations. It is clearly observable that the proposed method gives a much more accurate solution than the Padé approximation and the rational interpolation method [16]. Note that since the computational grid and the order of approximation are the same in all three cases, the complexity of propagation computations is equivalent.

Table 1 compares various strategies of the differential evolution method for the given example. The *randtobest1bin* strategy proved to be the best in this example. A high crossover probability leads to faster convergence. Reducing the range of mutation selection increases the rate of convergence, but sometimes it leads to a less optimal solution.

Table 1. Comparison of various optimization strategies. $\Delta x = 50\lambda$, $\Delta z = 0.25\lambda$, approximation order is equal to $[6/7]$, $\theta_{max} = 22°$. Population size is equal to 20.

Strategy	Mutation	Crossover probability	Error	Number of iterations
currenttobest1exp	[0, 2]	1.0	1e−5.77	8089
currenttobest1exp	[0.5, 1]	1.0	1e−5.77	4541
currenttobest1exp	[0.5, 1]	0.7	1e−2.86	>10000
best1bin	[0.5, 1]	1.0	1e−4.68	>10000
best2exp	[0.5, 1]	1.0	1e−1.90	>10000
rand2exp	[0.5, 1]	1.0	1e−1.83	>10000
best1exp	[0.5, 1]	1.0	1e−5.77	5856
rand1exp	[0.5, 1]	1.0	1e−1.9	>10000
randtobest1bin	[0.5, 1]	1.0	1e−5.77	3808
currenttobest1bin	[0.5, 1]	1.0	1e−5.77	6074

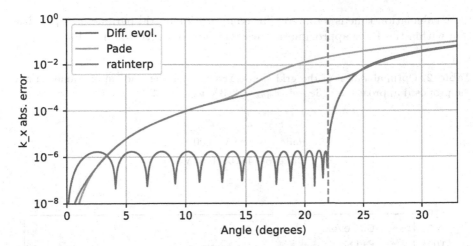

Fig. 2. Dependence of the discrete dispersion relation error on the propagation angle for the Padé approximation, rational interpolation and the proposed method. In all cases $\Delta x = 50\lambda$, $\Delta z = 0.25\lambda$, approximation order is equal to $[6/7]$, $\theta_{max} = 22°$.

4.3 Optimization for a Given Accuracy

In the previous example, we optimized a numerical scheme based on the known parameters of the computational grid. Usually, these parameters are unknown in advance and need to be determined. We will assume that we know acceptable error ε at distance x_{max} from the start propagation point. In this case, we need to find such parameters of the numerical scheme that would maximize the values of the grid steps (and accordingly minimize the computational time) while providing the specified accuracy. Bearing in mind that the error accumulates at each iteration of the step-by-step method, we come to the following conditional optimization problem

$$\mathrm{argmax}_{\Delta x, \Delta z, a_1 \ldots a_p, b_1 \ldots b_p} \left[\Delta x \Delta z \right],$$

on condition

$$\max_{k_z \in [0, k_z^{max}]} \frac{1}{k\Delta x} |\tilde{k}_x \left(k_z, \Delta x, \Delta z, a_1 \ldots a_p, b_1 \ldots b_p \right) - k \left(k_z \right)| < \frac{\varepsilon}{x_{max}}.$$

Table 2 demonstrates the optimal values of the grid steps for the proposed method and the Padé approximation method with the required accuracy $\varepsilon = 3e - 4$ at distance $x_{max} = 3e3\lambda$. Figure 3 depicts the dependence of the discrete dispersion relation error on the propagation angle at distance $x_{max} = 3e3\lambda$. It can be seen that in order to achieve the same accuracy, Padé approximation method requires a much thicker computational grid, and, accordingly, more calculation time is required. The computational grid parameters optimization within the Padé approximation was previously suggested in [14].

Table 2. Optimal values of the grid steps Δx и Δz for the Padé approximation and the proposed approach. $\varepsilon = 3e - 4$, $x_{max} = 3e3\lambda$, $\theta_{max} = 22°$.

	Δx	Δz
Padé	10.8λ	0.005λ
Diff.evol	46.9λ	0.67λ

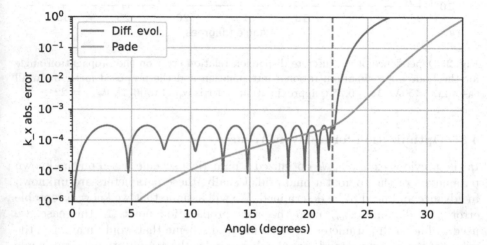

Fig. 3. Dependence of the discrete dispersion relation error on the propagation angle at distance $x_{max} = 3e3$ for the Padé approximation ($\Delta x = 10.8\lambda$, $\Delta z = 0.005\lambda$) and the proposed approach ($\Delta x = 46.9\lambda$, $\Delta z = 0.67\lambda$). In all cases approximation order is equal to [6/7], $\theta_{max} = 22°$.

4.4 Stability Condition

One of the most sophisticated issues in the development of numerical schemes is the determination of their stability conditions. Note that the discrete dispersion relation analysis is equivalent to the Von Neumann stability analysis. Thus, the stability condition is written as follows

$$\forall k_z \; \text{Im}\left(\tilde{k}_x \left(k_z \right) \right) > 0.$$

It is equivalent to the following condition

$$\left| \prod_{l=1}^{p} \frac{(k\Delta z)^2 - 4a_l x}{(k\Delta z)^2 - 4b_l x} \right| < 1, \; x \in [0; 1]. \tag{8}$$

Note that in this case, it is no longer sufficient to fulfill the condition only for the propagation angles of interest. Failure to comply with this condition contributes to the exponential growth of evanescent waves arising during diffraction [12].

This condition can also be taken into account within the differential evolution method. Figure 4 and 5 demonstrate the dependence of the horizontal wave number $\tilde{k}_x$ on vertical wavenumber k_z without stability condition (8) and with its accounting. Parameters from the previous subsection were used. The effect of the stability condition on the resulting solution will be demonstrated in the next section.

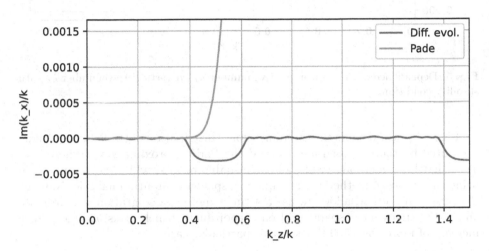

Fig. 4. Dependence of the horizontal wave number $\tilde{k}_x$ on vertical wavenumber k_z without stability condition.

5   Numerical Results

We will demonstrate the application of the proposed method to the classical wedge diffraction problem. The harmonic wave source is located at an altitude of 200 m and emits a signal at a frequency of 1 GHz. A wedge with a height of 200 m is located at a distance of 1500 m from the source. A transparent boundary condition is imposed on the upper boundary of the computational domain [7,17]. The wedge is approximated by a staircase function [12].

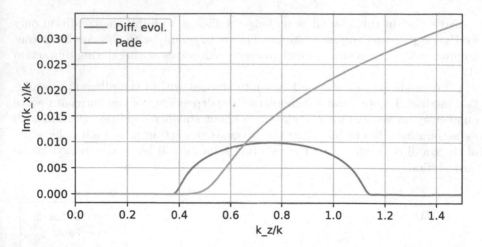

Fig. 5. Dependence of the horizontal wave number $\tilde{k}_x$ on vertical wavenumber k_z with stability condition.

Figure 6 and 7 depicts the two-dimensional distribution of the field amplitude, computed by the proposed method and the Padé approximation method. It is clear that both methods yield indistinguishable results, while the computation using the proposed method is faster due to a sparser computational grid. Namely, the proposed method allows to use a 4 times more sparse grid on x coordinate and a 130 times more sparse grid on z coordinate which gives a performance increase of more than 500 times in this particular case.

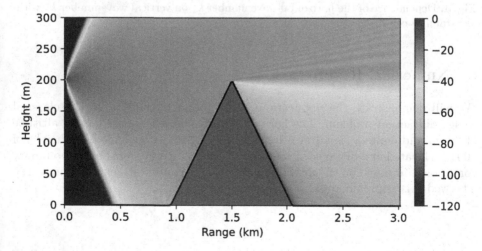

Fig. 6. Wedge diffraction. Spatial distribution of the filed amplitude ($20 \log |\psi|$), obtained by the proposed method. $\varDelta x = 46.9\lambda$, $\varDelta z = 0.67\lambda$, approximation order is equal to [6/7].

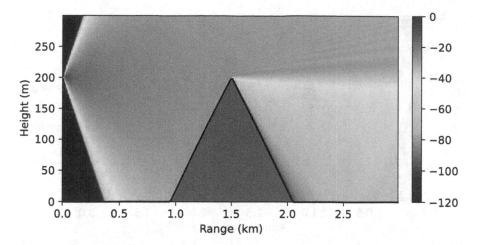

Fig. 7. Wedge diffraction. Spatial distribution of the filed amplitude $(20\log|\psi|)$, obtained by the Padé method. $\Delta x = 10.8\lambda$, $\Delta z = 0.005\lambda$, approximation order is equal to $[6/7]$.

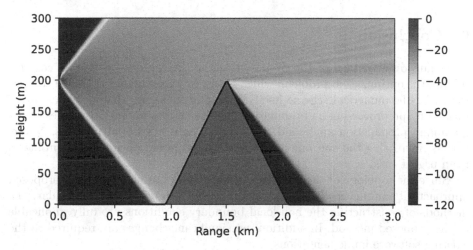

Fig. 8. Wedge diffraction. Spatial distribution of the filed amplitude $(20\log|\psi|)$, obtained by the Padé method. $\Delta x = 46.9\lambda$, $\Delta z = 0.67\lambda$, approximation order is equal to $[6/7]$.

Figure 8 shows the field distribution obtained by the Padé method on a sparse grid. It is clearly seen that in this case, the Padé method gives an incorrect solution in the diffraction zone behind the obstacle. Finally, Fig. 9 shows the field distribution calculated by the proposed method without taking into account the stability condition (8). One can see that the solution actually diverged due to the exponentially growing field components with high propagation angles.

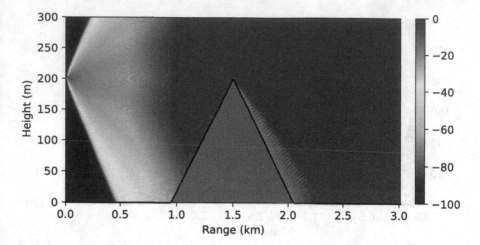

Fig. 9. Wedge diffraction. Spatial distribution of the filed amplitude ($20 \log |\psi|$), obtained by the proposed method without stability condition. $\Delta x = 46.9\lambda$, $\Delta z = 0.67\lambda$, approximation order is equal to [6/7].

6   Conclusion

The main disadvantage of the proposed method is the computational cost for solving the optimization problem, which is several times higher than the computation by the numerical scheme itself. Generally speaking, this problem can be tackled by preprocessing and tabulating the coefficients for various values of the maximum propagation angle and the required accuracy of calculations. Nevertheless, increasing the convergence rate of the proposed optimization problems is an urgent task.

Since the topology of the numerical scheme does not change in the proposed approach, it automatically obtains many useful properties. In particular, the methods of constructing the non-local boundary conditions are fully applicable to the proposed method. In addition, no significant changes are required to the existing software implementations.

Note that the proposed approach goes far beyond solving the Helmholtz equation. Similarly, it is possible to optimize almost any higher-order numerical scheme with a number of coefficients and computational parameters. With the classical approach, a numerical scheme is first developed, and then its accuracy and stability are analyzed. In the proposed approach, the required properties of the numerical scheme can be specified a priori.

Other configurations of the numerical scheme should be investigated, as well as other optimization methods should be applied in future studies.

Acknowledgements. This study was supported by the Russian Science Foundation grant No. 21-71-00039.

References

1. Scipy (2022). https://scipy.org/
2. Baker, G.A., Graves-Morris, P.: Pade Approximants, vol. 59. Cambridge University Press, Cambridge (1996)
3. Bekker, E.V., Sewell, P., Benson, T.M., Vukovic, A.: Wide-angle alternating-direction implicit finite-difference beam propagation method. J. Light. Technol. **27**(14), 2595–2604 (2009)
4. Collins, M.D.: A split-step Pade solution for the parabolic equation method. J. Acoust. Soc. Am. **93**(4), 1736–1742 (1993)
5. Collins, M.D., Siegmann, W.L.: Parabolic Wave Equations with Applications. Springer, New York (2019). https://doi.org/10.1007/978-1-4939-9934-7
6. van Dijk, W.: Efficient explicit numerical solutions of the time-dependent Schrödinger equation. Phys. Rev. E **105**(2), 025303 (2022)
7. Ehrhardt, M., Zisowsky, A.: Discrete non-local boundary conditions for split-step Padé approximations of the one-way Helmholtz equation. J. Comput. Appl. Math. **200**(2), 471–490 (2007)
8. Eiben, A.E., Smith, J.E., et al.: Introduction to Evolutionary Computing, vol. 53. Springer, Heidelberg (2003). https://doi.org/10.1007/978-3-662-05094-1
9. Fishman, L., McCoy, J.J.: Derivation and application of extended parabolic wave theories. I. The factorized Helmholtz equation. J. Math. Phys. **25**(2), 285–296 (1984)
10. Karniadakis, G.E., Kevrekidis, I.G., Lu, L., Perdikaris, P., Wang, S., Yang, L.: Physics-informed machine learning. Nat. Rev. Phys. **3**(6), 422–440 (2021)
11. Lampinen, J.: A constraint handling approach for the differential evolution algorithm. In: Proceedings of the 2002 Congress on Evolutionary Computation. CEC 2002 (Cat. No. 02TH8600), vol. 2, pp. 1468–1473. IEEE (2002)
12. Levy, M.F.: Parabolic Equation Methods for Electromagnetic Wave Propagation. The Institution of Electrical Engineers, UK (2000)
13. Lu, L., Meng, X., Mao, Z., Karniadakis, G.E.: DeepXDE: a deep learning library for solving differential equations. SIAM Rev. **63**(1), 208–228 (2021)
14. Lytaev, M.S.: Automated selection of the computational parameters for the higher-order parabolic equation numerical methods. In: Gervasi, O., et al. (eds.) ICCSA 2020. LNCS, vol. 12249, pp. 296–311. Springer, Cham (2020). https://doi.org/10.1007/978-3-030-58799-4_22
15. Lytaev, M.S.: Chebyshev-type rational approximations of the one-way Helmholtz equation for solving a class of wave propagation problems. In: Paszynski, M., Kranzlmüller, D., Krzhizhanovskaya, V.V., Dongarra, J.J., Sloot, P.M.A. (eds.) ICCS 2021. LNCS, vol. 12742, pp. 422–435. Springer, Cham (2021). https://doi.org/10.1007/978-3-030-77961-0_35
16. Lytaev, M.S.: Rational interpolation of the one-way Helmholtz propagator. J. Comput. Sci. **58**, 101536 (2022)
17. Lytaev, M.S.: Nonlocal boundary conditions for split-step Padé approximations of the Helmholtz equation with modified refractive index. IEEE Antennas Wirel. Propag. Lett. **17**(8), 1561–1565 (2018)
18. Petrov, P.S., Ehrhardt, M., Trofimov, M.: On decomposition of the fundamental solution of the Helmholtz equation over solutions of iterative parabolic equations. Asymptot. Anal. **126**(3–4), 215–228 (2022)
19. Price, K.V.: Differential evolution. In: Zelinka, I., Snášel, V., Abraham, A. (eds.) Handbook of Optimization. Intelligent Systems Reference Library, vol. 38. Springer, Heidelberg (2013). https://doi.org/10.1007/978-3-642-30504-7_8

20. Samarskii, A.A., Mikhailov, A.P.: Principles of Mathematical Modelling: Ideas, Methods, Examples. Taylor and Francis, London (2002)

21. Schreiber, M., Loft, R.: A parallel time integrator for solving the linearized shallow water equations on the rotating sphere. Numerical Linear Algebra Appl. **26**(2), e2220 (2019)

22. Zhou, H., Chabory, A., Douvenot, R.: A fast wavelet-to-wavelet propagation method for the simulation of long-range propagation in low troposphere. IEEE Trans. Antennas Propag. **70**, 2137–2148 (2021)

Iterative Solution for the Narrow Passage Problem in Motion Planning

Jakub Szkandera[✉] and Ivana Kolingerová

Department of Computer Science and Engineering, Faculty of Applied Sciences,
University of West Bohemia, Univerzitni 8, 30614 Plzen, Czech Republic
{szkander,kolinger}@kiv.zcu.cz

Abstract. Finding a path in a narrow passage is a bottleneck for randomised sampling-based motion planning methods. This paper introduces a technique that solves this problem. The main inspiration was the method of exit areas for cavities in protein models, but the proposed solution can also be used in another context. For data with narrow passages, the proposed method finds passageways for which sampling-based methods are not sufficient, or provides information that a collision-free path does not exist. With such information, it is possible to quit the motion planning computation if no solution exists and its further search would be a loss of time. Otherwise, the method continues to sample the space with sampling-based method (a RRT algorithm) until a solution is found or the maximum number of iterations is reached. The method was tested on real biomolecular data - *dcp* protein - and on artificial data (to show the superiority of the proposed solution on better-imagined data) with positive results.

Keywords: Motion planning · Sample-based algorithms · Rapidly exploring random tree · Narrow passage · Bottleneck · Binary search

1 Introduction

Motion planning (finding a collision-free path for a moving object between at least two locations in an obstacle-filled environment) has always been one of the essential research areas. Now it is coming to the forefront even more due to the rapid development of applications, such as the control of autonomous vehicles and sophisticated robots. Fast and reliable solutions in real or near-real time are necessary to cope with such applications. Geometric methods such as Voronoi diagrams for computing centerlines can be used for moving objects (generally referred to as agents) of very simple shapes. However, these methods are inappropriate if the object is complex or even flexible or if the environment is complicated or even dynamically changing. In this case, more sophisticated and general methods should be used.

Motion planning is generally interpreted using the concept of configuration space, which is the set of all existing configurations, where each configuration represents a unique position and rotation of the navigated object. The combination of

D. Groen et al. (Eds.): ICCS 2022, LNCS 13350, pp. 219–232, 2022.
https://doi.org/10.1007/978-3-031-08751-6_16

position and rotation is called the degrees of freedom. The dimension of the problem and its associated complexity increases with the number of degrees of freedom. For example, the configuration of a navigated object may be up to a 6-dimensional vector describing its position and rotation (both properties have 3 vector components) in 3D space. The total number of configurations in the configuration space is huge. It is impossible to process them all in a reasonable time, so randomised sampling-based methods are used to select and process specific configurations.

Methods based on random sampling [14, 17] randomly generate or select configurations to be tested whether they are collision-free. If the generated configuration is in collision with the environment, it is rejected, and the method generates a new random configuration. The non-colliding configuration is added to the path-finding structure - the so-called roadmap that approximates the configuration space's free regions. Graph-based path planning methods [10, 15] can then be applied to the roadmap. Using roadmaps is an efficient way to find a collision-free passage through the environment, usually in a reasonable (often near-real) time. As already mentioned, the biggest pitfall of the random sampling-based algorithms is the narrow passage problem, as, by random sampling, it is difficult to hit such a place properly.

The proposed solution to the narrow passage problem is an improvement of our previous solution [22], based on a combination of Voronoi diagrams, binary search, randomisation, and the idea of so-called exit areas [19], an instrument for motion planning in protein molecular models. Our previous solution still had the pitfall caused by the randomisation, so it was impossible to decide whether a collision-free path existed or not. The solution proposed in this paper removes this pitfall by a divide and conquer approach. The decision has almost 100% certainty, with the only limitation being the accuracy of the computer representation of real numbers.

The structure of the article is as follows. A description of ideas behind the existing motion planning methods that are widely used for agent navigation in configuration space, and their various modifications to address specific problems, is given in Sect. 2. A detailed description of the idea behind the proposed solution and the solution itself for motion planning, including a description of the algorithm that unambiguously determines whether or not a solution exists in the narrow passage, is in Sect. 3. Section 4 contains experiments and results on artificially generated and real bio-molecular data. Section 5 concludes the paper.

2   Related Work

The motion planning algorithms based on the randomized sampling of the environment can be divided into two groups. The former group, Rapidly Exploring Random Tree (RRT) [17], produces a tree structure representing the collision-free part of the environment. The latter, Probabilistic Roadmaps (PRM) [14], creates a graph structure representing the collision-free part of the environment. The idea of PRM algorithms [9] is to generate a set of random samples, which are simultaneously tested for collisions (sample in a collision is removed, a collision-free sample is preserved). Next, if the edge does not pass through an environmental obstacle, the closest collision-free samples are connected by an edge. Finally, we use one of the graph path planning algorithms such as A$\sim$ [10] or the more complex D$\sim$ Lite [15] to find the passage through the environment. The set of obstacles can be sufficient input for a PRM-based algorithm since the algorithm itself does not require knowledge of the initial and final configurations. Still, their knowledge can be used in some heuristics to refine the sampling.

A simplified version of the Probabilistic Roadmap algorithm called sPRM [14] is used mainly for the analysis of follow-up algorithms. Its additional advantage is that it finds the asymptotically optimal path. The sPRM is also the basis of the PRM* [13] algorithm, which uses a heuristic function to minimise the length of roadmaps, where potential samples to connect are selected from neighbourhoods with radii $r > 0$, which can be defined, e.g., as a function of sample dispersion [17]. Other PRM-based methods attempt to relax the collision constraints [2,11], leading to a simplification of the overall complexity of the problem, by reducing the agent volume, e.g., by making the agent thinner [11] or by scaling down the size of agent [2]. The resulting found roadmap is only an approximate solution, and, therefore, this solution is iteratively corrected [2] to achieve the original solution. The next PRM method [11] goes one step further and also thins the obstacles themselves, where the level of thinning is determined by a binary search at each step of the algorithm.

The second group of random sampling algorithms was designed for use in models with a number of complex physical constraints. In contrast to PRM, it generates a tree structure instead of a graph, after which this group of algorithms is called the rapidly exploring random tree (RRT) [17]. The input to RRT is a set of constraints (similar to PRM) plus an initial configuration that is the root of the initialised tree structure t_{main}. Creating a tree structure instead of a graph has several undeniable advantages. First, the generated tree incrementally expands towards unexplored regions of the configuration space. Second, it simplifies the path planning part of the process since backtracking is sufficient to find the resulting path. The idea of the whole RRT algorithm contains three basic steps that are cyclically repeated. Randomly generating a new sample in the configuration space is the first step. The next step is to steer the new sample near the nearest list tree t_{main}. The last step is to check if the sample is collision-free. If it is, the sample is added to the t_{main} tree; otherwise, it is rejected.

There are many modifications for the RRT family of algorithms, primarily targeted at special problems. RRT* [13] is an optimised RRT that finds the optimal solution using a heuristic function. It can provide the shortest possible path to the goal in the best case. The problem is that the shortest possible path is guaranteed when the number of samples approaches infinity, which is unrealistic in practice, but the found path is optimal. The dynamic environment can be solved using RRT^X [21] which is the first proposed asymptotically optimal sampling-based replanning algorithm.

Both approaches share the well-known problem of finding a collision-free passage through narrow passages. For PRM algorithms, many different modifications have been introduced to address the narrow passage problem [7]. For example, for low-dimensional configuration spaces, it is possible to exploit this by generating a large number of random samples near obstacles or around the medial axis of the environment [16]. On the other hand, the RRT family of algorithms can work with a medial axis, towards which MARRT (Medial Axis RRT) [6] tries to shift the newly generated samples. The resulting tree structure then does not cover the entire collision-free configuration space, but follows the central axis. Geometry-based methods [23] that sample along a precomputed path can also be used. Another option to solve the narrow passage problem is more detailed sampling around obstacles, such as NP-RRT* (Narrow passage RRT*) [3], or combination of different abstraction levels [20].

Most motion planning methods are primarily developed for the navigation of mechanical objects (robots, autonomous vehicles, etc.), but these methods have applications in other important research areas. For example, the problem of ligand navigation in protein is motion planning in molecular simulations, which inspired our proposed solution. It is possible to use Probabilistic Roadmaps, which help to speed up molecular dynamics simulations, to sample the configuration space of protein [1]. The atomic boundaries of a ligand often lead to sampling in a high-dimensional configuration space, which is an inappropriate problem for the group of PRM algorithms.

A more suitable choice of a planner is an RRT-based algorithm that can even handle motion planning for a flexible ligand [5]. The performance of RRT is greatly affected by the ability to generate new configurations for the high-dimensional space problem, which can be very time-consuming, but ML-RRT (Manhattan-like RRT) copes with this problem [8]. This method has also been modified for flexible ligands [5]. A solution to this problem can also be achieved by projecting the high-dimensional space of the roadmap back into 3D space [4]. Another challenge is to find a passage through a dynamic protein [24] in which individual paths cannot only change their shape but also arise and disappear.

The Voronoi diagram is used not only for better motion planning [23] but also for calculating cavities and their exit areas in protein models [19]. The Voronoi vertex and edge graph capture all possible trajectories of spherical probes avoiding collisions between spherical obstacles (the atoms of the protein model). A probe located in the cavity of a protein model cannot reach the outer space without collision unless the probe radius is reduced to allow the probe to escape from the cavity. By analysing a graph of Voronoi vertices and edges, the exact location where the probe will be located (the primary exit location), and the exact value of the probe radius to which the probe must shrink to escape the cavity can be calculated. The edge of the Voronoi diagram along which the probe could escape from the cavity is then removed, and the shrinking of the probe continues until all remaining exits from the cavity are discovered. The groups constructed from the intersecting probes at the cavity exit locations are referred to as exit areas.

3   Proposed Solution

The proposed solution combines three different approaches - exit areas [18], RRT [17] and the idea of binary search - into one sophisticated motion planning method. Our previous proposed solution [22] combined only exit areas and the RRT method, which was very helpful for detecting narrow passages so that they can be processed in more detail. The biggest problem of the previous solution is that it was not possible to unambiguously decide whether a collision-free agent location exists. Detecting the collision-free location of an agent in a narrow passage was still guided by random pattern generation, so it could happen that an existing collision-free solution was not found. The newly proposed solution removes this pitfall and essentially determines whether an agent's collision-free placement in a narrow passage is possible.

A few modifications to the RRT algorithm are required in the proposed solution. Before the main RRT cycle starts, we need to calculate the exit areas that tell us the exact location of the most problematic places (narrow passage) in the data. By incorporating and modifying the binary search, we can now unambiguously determine if a collision-free path exists through the narrow passage. If this is not possible, we can temporarily close the narrow passage, e.g., by using temporary barriers to avoid the RRT sampling the narrow passage unnecessarily. Otherwise, we have essentially found a collision-free path through one of the worst possible places in the data. Once the narrow passages are located and processed, the main RRT cycle can be run to find a collision-free path through the data.

Let us illustrate in 2D the main idea of the proposed solution to unambiguously determine whether there is a collision-free position in the narrow passage or not. First, we define the principal direction - the orientation of the agent (Fig. 1a). The principal direction can be defined in the same way, which must be uniform for each flexible agent configuration. In our case, the agent A is a ligand consisting of several spheres, which are stored in a list of spheres. Its principal direction $\vec{v_a}$ can be defined, for example, as the principal direction vector between the center of the central sphere s_c taken as the origin of the local coordinate system of the agent and the center of the first defined sphere s_1 inside the structure representing the agent, if the first sphere is not also the central sphere. In case $s_1 = s_c$, another sphere from the structure needs to be used (e.g., the second or the last one). In the next step, we place the agent in the center of the narrow passage whose exact location is known by computing the exit areas $v_{exit}^i, i = 1, ..., n$, whose center is the center of the narrow passage. The agent is in its initial position (without rotation), and now we need to find the correct collision-free rotation of the agent. We divide the imaginary space of rotations into m sectors, e.g., $m = 4$, where the boundary between the sectors corresponds to the initial angular rotation of the agent. By rotating the agent we get 4 rotated agents a_0, a_{90}, a_{180} and a_{270} whose principal direction is rotated by $0°$, $90°$, $180°$ and $270°$ in the narrow passage (Fig. 1b). For each of these rotated agents, we compute a collision function w, which can be arbitrarily defined (e.g. number of collision spheres, size of volume in collision, sum

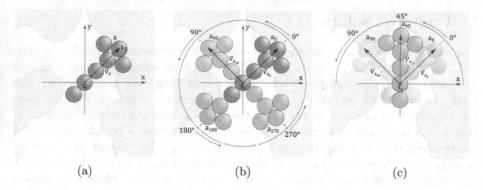

(a)   (b)   (c)

Fig. 1. The main idea of proposed solution where transparent grey represents protein and white collision free tunnel. (a) Example of agent illustrated as set of orange spheres in the narrow passage of protein, (b) Initial agent rotation to calculate collision-free rotation where each initial agent rotation has a different color for better predictability, (c) First iteration in 2D where the initial agent rotations are illustrated with transparent colors. (Color figure online)

of collision distances, etc.), and then we sum every two adjacent collision functions $(w(a_0) + w(a_{90}), w(a_{90}) + w(a_{180})$, etc.). We rank the summed collision functions from the smallest to the largest and process the sector with the smallest summed collision function value, e.g., agents a_0 and a_{90}. We calculate the new rotation of the agent $w(a_{45})$ that is in the middle of the processed sector, i.e. 45°, (Fig. 1c) and calculate the collision function $w(a_{45})$. We select the half of the sector that has the smaller value of the sum of the collision functions $(w(a_0) + w(a_{45})$ or $w(a_{45}) + w(a_{90}))$. We repeat this halving until we obtain a collision-free result or reach the chosen maximum number of iterations. If a collision-free result is not found, we process the second initial sector in the same way. In case the result is not found in the second sector, continue with the third, etc.

The proposed solution can also be generalised for 3D. The whole idea behind the algorithm is exactly the same. The only difference is that in 2D only one angular variable is needed to rotate the agent and create sectors, which is insufficient information in 3D. To achieve similar sectors in 3D, it is convenient to use parametric equations of a sphere to achieve sectors which look like plates, see Fig. 2a. The corners of the plate are the principal direction of the initial agent rotation for the sector (Fig. 2b). Dividing the plate gives 4 sub-plates (Fig. 2c) for which 5 new agent rotations $(\vec{v}_{p_{12}}, \vec{v}_{p_{13}}, \vec{v}_{p_c}, \vec{v}_{p_{24}}$ and $\vec{v}_{p_{34}})$ need to be computed in order to compute the summed collision functions $(w(p_1) + w(p_{12}) + w(p_{13}) + w(p_c))$, $w(p_{12}) + w(p_2) + w(p_c) + w(p_{24})$, etc.). The sub-plate with the smallest value of the summed collision functions is further divided into smaller parts. For each selected sub-plate, it is then necessary to compute next 5 new angular rotations and agent collision functions, which can be computationally intensive especially

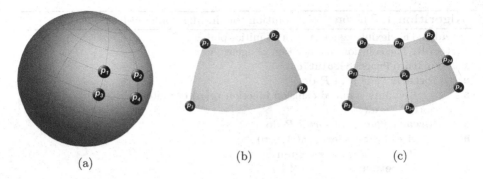

Fig. 2. (a) Space of rotation in 3D, (b) Initial spherical plate, (c) First iteration in 3D

when using poor data structures. Experimentally, we have found that it is sufficient to compute only the agent rotations to the center of the p_c plate, since the sum of the two collision functions (i.e., $w(p_1) + w(p_c)$, $w(p_2) + w(p_c)$, etc.) is sufficient to determine the appropriate sub-plate for further processing and each weighting function is then just the sum of the collision function at one corner of the plate and at its center.

We incorporated this solution into our previously proposed solution [22] to find and sample narrow passages in more detail. The structure of the whole algorithm and the proposed solution for unambiguous narrow passages resolution is then as follows. First, we find all exit areas in the data representing narrow passages. We try to place an agent in each found exit area so that it does not collide with the environment (Algorithm 1). To do this, we divide the imaginary sphere of possible agent rotations into plates. We compute the collision function $w(p)$ for each plate p (Algorithm 1, lines 2–3) and sort the plates according to the value of the collision function $w(p)$ from smallest to largest (Algorithm 1, line 4), and then process all plates. Cycling through all the plates (Algorithm 1, lines 5–8) is only used if, in some artificial case, the same value of the collision function existed in almost all angular rotations.

If the method does not find a collision-free rotation of the agent, then there is no collision-free path for the agent through this narrow passage and we can fill the narrow passage with a temporary obstacle. If it is possible to place the agent in the narrow passage without collision with the environment, we run a tiny RRT algorithm to map just the neighborhood of the narrow passage. We try to fix each collision pattern with our proposed solution. We do this for each exit area found, yielding a set of tree structures whose root is at the center of the narrow passage. The main RRT algorithm is then run, which searches for a path through the protein from the protein's active site (initial position). In case the main tree is close to some other tree that has been formed around the narrow passage, we merge these two tree structures into one, which solves the narrow passage problem very easily and quickly. The algorithm terminates when we have found a way out of the protein or after a specified maximum number of iterations.

Algorithm 1: The proposed solution for finding collision-free position

Data: The flexible agent A, List of initial plates P
Result: The collision-free rotation of agent A

1 **Algorithm** ProposedSolution
2   **foreach** *Plate p in P* **do**
3   Compute summed collision function $w(p)$ of plate p
4   Sort P by $w(p)$
5   **foreach** *Plate p in sorted P* **do**
6         $A \leftarrow$ ProcessSection($A, 1, p$)
7   **if** A *is collision-free* **then**
8   **return** A
9   **return** $\emptyset$
10 **Procedure** ProcessPlate(*The flexible agent A, Current iteration i, Plate p*)
11   **if** $i = max_{iteration}$ **then**
12   **return** A
13     $A_c \leftarrow$ Rotate A so its orientation is pointing to the middle of p
14   **if** A_c *is collision-free* **then**
15   **return** A
16   Sum up $w(A_c)$ with each corner value separately
17   Select the sub-plate s of p with minimum summed collision function $w(s)$
18     $A \leftarrow$ ProcessSection($A, i + 1, s$)
19   **return** A

---

4 Experiments and Results

Both real data and artificial data were used to test the proposed method. Within the real data, we used protein data, which are also freely available in the data bank, and flexible ligands as agents whose meaningfulness was consulted by biomolecular experts. To present the results, we used the *dcp* protein to represent the real data. The artificial data were used to illustrate the strengths and weaknesses of the method. All experiments were performed on a computer with the CPU Intel® Core™ i7-7700K (4.2 GHz) and 64 GB 2400 MHz RAM, the proposed method was implemented in C#. Also the images in this section were created using CAVER Analyst 2.0 [12] unless otherwise noted.

The results of the proposed method for real data are difficult to compare with existing methods since they specialize in different problems, so we compare them with our previous solution (pRRT) to show a significant improvement for finding a collision-free path through the narrow passage.

The experiment is on real protein data, specifically the protein *dcp* (each atom has different color - hydrogen light grey, carbon dark grey, oxygen red, etc.), which can be seen in Fig. 3a. Figure 3b shows the cut through protein *dcp* with few highlighted narrow passages (green points) which are visible in this particular cut. The example of one configuration of the flexible agent is shown in Fig. 3c. The result of the experiment is a comparison of the collision-free path through narrow passages found using the previous solution and the current solution

Fig. 3. (a) Tested data of protein *dcp*, (b) *dcp* protein cut with highlighted narrow passages, (c) The example of one configuration of flexible agent *A*. (Color figure online)

(Fig. 4 which is created by our own computation system). In both images, the *dcp* protein is invisible so that all narrow passages and the starting point of the main cycle of the RRT algorithm can be seen. The results do not show the finding of the entire tree structure and the resulting passes through the protein, but only the completed processing of the narrow passages and the processing of their closest surroundings. The starting point of the motion planning is the vertex colored in green. The narrow passages are the vertices colored red and blue. If the vertex is blue, the algorithm has found a collision-free agent location in that narrow passage. If no collision-free location was found, the vertex is red. White vertices represent the found collision-free samples by the RRT algorithm in the surroundings of the narrow passage. The individual vertices are then connected to each other by an orange edge that uniquely identifies where the agent can move from its current location. Figure 4a shows the processed narrow passages of *dcp* protein with the previous solution (pRRT). The result of the same problem by the currently proposed method (cRRT) is shown in Fig. 4b. As can be seen, the cRRT method succeeded in the collision-free placement of the flexible agent even in the narrow passages where the pRRT method failed. One narrow passage remained without finding a collision-free location, which, unlike the results of the pRRT method, does not mean that we cannot climb the path but that there is no collision-free passage for the agent under test through this narrow passage. It is the most significant advantage of the proposed solution.

Table 1 contains a comparison of three algorithms - standard RRT, pRRT, and cRRT. Testing was performed 1000 times on the real biomolecular data mentioned above; each algorithm was timed to two minutes. Each iteration had a different random generator seed to guarantee a different result. At the same time, the same seed was used for all algorithms (the first run with seed x for all algorithms, the second run with seed y, etc.). Table 1 shows how many times a given algorithm was better than the other two algorithms in the evaluated property, where the properties being assessed were a higher number of paths found and shorter pathfinding time. The current proposed cRRT solution always found more paths, namely five, than the other two algorithms. The reason is that

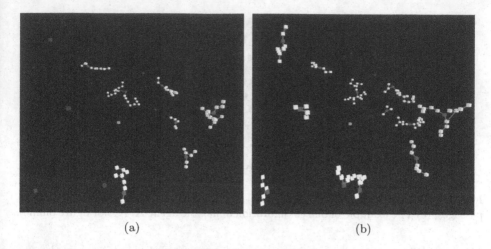

(a)   (b)

Fig. 4. (a) Result narrow passage trees of previous RRT solution, (c) Result narrow passage trees of current RRT solution. (Color figure online)

Table 1. Comparison of the RRT algorithm, our previous (pRRT) and current (cRRT) modified versions - tested on the real bio-molecular data

Algorithm	Number of paths	Time of the found path				
		First	Second	Third	Fourth	Fifth
RRT	0	127	78	1	0	0
pRRT	0	243	217	273	48	0
cRRT	1000	630	705	726	952	1000

RRT finds only the first two paths that are wide enough for this algorithm, and pRRT finds a different third path and, in most cases, a fourth path but fails on the fifth path. It is also interesting to note that the RRT algorithm was once able to find the third path while being faster than the other algorithms, a purely random result. The time results of pathfinding between the pRRT and cRRT methods usually differed by units of milliseconds.

The following experiment was to determine the narrow passage threshold for each method. That is, to find how much wider the narrow passage must be for the method to have no problem finding a path through the narrow passage. For this experiment, we made artificial data (Fig. 5a) in the shape of a cube; each wall has a passage with different radius (e.g. first wall has passage with $r = 1$, second passage with $r = 1.4$, etc.). The radius of the smallest possible passage corresponds to the radius of one sphere of the agent (Fig. 5b). The initial position is in the middle of the cube (Fig. 5c); the ideal number of paths found is the same as the number of walls of the cube, i.e., six. The result is in Table 2, which indicates how many times one of the RRT algorithms was able to find a path through the narrow passages in the wall. For each algorithm,

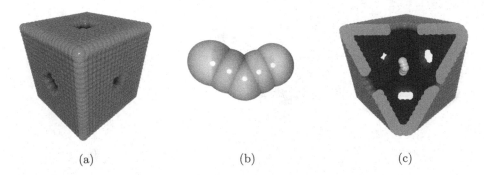

(a)   (b)   (c)

Fig. 5. (a) The artificial testing data with different radii of narrow passages, (b) The example of flexible agent A, (c) Data cut with starting position of the agent A

Table 2. Collision-free paths through narrow passages with different radii sizes

Algorithm	Narrow passage number (radius)					
	First (1)	Second (1.3)	Third (1.6)	Fourth (1.9)	Fifth (2.2)	Sixth (2.5)
RRT	0	0	0	0	476	1000
pRRT	0	0	671	1000	1000	1000
cRRT	1000	1000	1000	1000	1000	1000

we repeated the measurements 1000 times with a agent whose sphere radius is $r = 1$. In all cases, the standard RRT algorithm finds only the largest narrow passage ($r = 2.5$), which is the expected result since the method does not use any auxiliary information or structures. With a sufficiently long sampling (up to 2 min) of the configuration space, it also finds a path through the second-largest narrow passage ($r = 2.2$) in almost half of the cases. The previous solution pRRT performs much better, always finding a path through the narrow passage in half of the cases. The best solution is the proposed solution, which finds a collision-free path through every narrow passage, even the smallest possible.

The previous results have shown that the proposed method finds a passage through even the tightest possible narrow passage. Therefore, now we focus on the method itself to see what operations and how many iterations are needed to find one collision-free position. For this experiment, we have created artificial data in the form of a tunnel (Fig. 6a) into which we are trying to place a rod-shaped agent (Fig. 6b). Before each algorithm run, the data and the agent itself were randomly rotated (each with a different rotation). The experiment was repeated 500 times; the method searched for collision-free rotations of the agent (Fig. 6c). The results of this experiment are shown in Table 3. We tried this experiment for different numbers of initial spherical plates, which must be sorted further by the collision function and processed. As we can see, as the number of plates increases, the computation time and finding the correct solution also increase. It makes sense because more initial agent rotations and collision functions must be calculated.

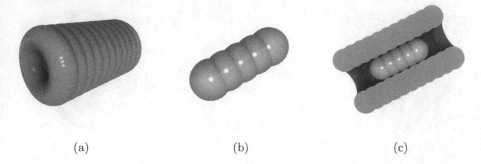

(a)   (b)   (c)

Fig. 6. (a) The artificial testing data of tunnel, (b) The example of agent A, (c) Data cut with starting position of agent A.

Table 3. Result of different initial conditions of the proposed solution cRRT

Number of plates	Plate order			Plate iteration			Time [ms]
	First	Second	Higher	First	Second	Higher	
2	500	0	0	251	68	181	27.871
6	413	56	31	370	43	87	34.158
15	387	93	20	401	67	32	65.966
28	341	107	52	384	71	45	139.466
45	326	108	66	398	58	44	204.932

In most cases, the solution was found in the first spherical plate and even in the first possible iteration. However, it is surprising that the method works very well and very fast also for a minimal number of plates, namely two. All solutions were found in the first plate, which makes sense since the data is symmetric. However, the method needs to be adjusted appropriately, as the larger the initial plate, the greater the chance that the method iterates into the wrong part of the plate. In this case, we need to add a stack or priority queue and process the rejected part of the plate as well.

We conclude the section with three essential observations from testing the proposed method. The first is that the method is severely limited by the accuracy of the computer representation of decimal numbers. It is vital when calculating the rotation of an agent from one position to another, so one needs to have a perfect understanding of working with decimals on the computer or use a suitable library (e.g., Unity can handle this). The next observation is that the most appropriate collision function for biomolecular data composed of spheres turned out to be the total volume of the agent that is in collision with its surroundings. The last is that the unambiguous result of whether or not a collision-free path exists is closely related to the maximum number of iterations. Experiments have shown that for protein data it is longest in the fourth iteration.

5   Conclusion

A modified version of the RRT algorithm, cRRT, based on our previous method, whose improvement for collision-free pathfinding consists mainly in finding collision-free paths through narrow passages, is presented in this paper. In addition to finding the path itself, the method also provides the exact location of the narrow passages, which it can investigate in detail. This information is crucial, and our modified RRT algorithm can decide whether the narrow passage path exists. Since the proposed solution must additionally compute the position of the exit areas and sample their neighborhoods, it is not suitable for data without narrow passages - it will find the correct path, but probably more slowly than the original RRT. Once the data contains narrow passages, the proposed solution excels and provides better results than our original solution and many times better results than the original RRT algorithm. Moreover, the method can find collision-free narrow passages even for data that are only touching (agent with protein).

Acknowledgement. This work was supported by the Ministry of Education, Youth and Sports of the Czech Republic, the project SGS-2022-015 New Methods for Medical, Spatial and Communication Data.

References

1. Amato, N.M., Dill, K.A., Song, G.: Using motion planning to map protein folding landscapes and analyze folding kinetics of known native structures. J. Comput. Biol. **10**(3–4), 239–255 (2003)
2. Bayazit, O.B., Xie, D., Amato, N.M.: Iterative relaxation of constraints: a framework for improving automated motion planning. In: 2005 IEEE/RSJ International Conference on Intelligent Robots and Systems, pp. 3433–3440. IEEE (2005)
3. Belaid, A., Mendil, B., Djenadi, A.: Narrow passage RRT*: a new variant of RRT. Int. J. Comput. Vis. Robot. **12**(1), 85–100 (2022)
4. Cortés, J., Barbe, S., Erard, M., Siméon, T.: Encoding molecular motions in voxel maps. IEEE/ACM Trans. Comput. Biol. Bioinform. (TCBB) **8**(2), 557–563 (2011)
5. Cortés, J., Le, D.T., Iehl, R., Siméon, T.: Simulating ligand-induced conformational changes in proteins using a mechanical disassembly method. Phys. Chem. Chem. Phys. **12**(29), 8268–8276 (2010)
6. Denny, J., Greco, E., Thomas, S., Amato, N.M.: MARRT: medial axis biased rapidly-exploring random trees. In: 2014 IEEE International Conference on Robotics and Automation (ICRA), pp. 90–97. IEEE (2014)
7. Elbanhawi, M., Simic, M.: Sampling-based robot motion planning: a review. IEEE Access **2**, 56–77 (2014)
8. Ferré, E., Laumond, J.-P.: An iterative diffusion algorithm for part disassembly. In: Proceedings of the 2004 IEEE International Conference on Robotics and Automation, vol. 3, pp. 3149–3154. IEEE (2004)
9. Geraerts, R., Overmars, M.H.: A comparative study of probabilistic roadmap planners. In: Boissonnat, J.-D., Burdick, J., Goldberg, K., Hutchinson, S. (eds.) Algorithmic Foundations of Robotics V. STAR, vol. 7, pp. 43–57. Springer, Heidelberg (2004). https://doi.org/10.1007/978-3-540-45058-0_4

10. Hart, P.E., Nilsson, N.J., Raphael, B.: A formal basis for the heuristic determination of minimum cost paths. IEEE Trans. Syst. Sci. Cybern. **4**(2), 100–107 (1968)
11. Hsu, D., Sánchez-Ante, G., Cheng, H.-L., Latombe, J.-C.: Multi-level free-space dilation for sampling narrow passages in PRM planning. In: Proceedings 2006 IEEE International Conference on Robotics and Automation 2006. ICRA 2006, pp. 1255–1260. IEEE (2006)
12. Jurcik, A., et al.: CAVER analyst 2.0: analysis and visualization of channels and tunnels in protein structures and molecular dynamics trajectories. Bioinformatics **34**(20), 3586–3588 (2018)
13. Karaman, S., Frazzoli, E.: Sampling-based algorithms for optimal motion planning. Int. J. Robot. Res. **30**(7), 846–894 (2011)
14. Kavraki, L.E., Kolountzakis, M.N., Latombe, J.-C.: Analysis of probabilistic roadmaps for path planning. IEEE Trans. Robot. Autom. **14**(1), 166–171 (1998)
15. Koenig, S., Likhachev, M.: D* Lite. In: AAAI/IAAI, pp. 476–483 (2002)
16. Kurniawati, H., Hsu, D.: Workspace-based connectivity oracle: an adaptive sampling strategy for PRM planning. In: Akella, S., Amato, N.M., Huang, W.H., Mishra, B. (eds.) Algorithmic Foundation of Robotics VII. STAR, vol. 47, pp. 35–51. Springer, Heidelberg (2008). https://doi.org/10.1007/978-3-540-68405-3_3
17. LaValle, S.M.: Planning Algorithms. Cambridge University Press, Cambridge (2006)
18. Manak, M.: Voronoi-based detection of pockets in proteins defined by large and small probes. J. Comput. Chem. **40**(19), 1758–1771 (2019)
19. Manak, M., Anikeenko, A., Kolingerova, I.: Exit regions of cavities in proteins. In: 2019 IEEE 19th International Conference on Bioinformatics and Bioengineering, BIBE, pp. 1–6. IEEE Computer Society (2019)
20. Orthey, A., Toussaint, M.: Section patterns: efficiently solving narrow passage problems in multilevel motion planning. IEEE Trans. Rob. **37**(6), 1891–1905 (2021)
21. Otte, M., Frazzoli, E.: RRTX: Asymptotically optimal single-query sampling-based motion planning with quick replanning. Int. J. Robot. Res. **35**(7), 797–822 (2016)
22. Szkandera, J., Kolingerová, I., Maňák, M.: Narrow passage problem solution for motion planning. In: Krzhizhanovskaya, V.V., et al. (eds.) ICCS 2020. LNCS, vol. 12137, pp. 459–470. Springer, Cham (2020). https://doi.org/10.1007/978-3-030-50371-0_34
23. Vonásek, V., Faigl, J., Krajník, T., Přeučil, L.: A sampling schema for rapidly exploring random trees using a guiding path. In: Proceedings of the 5th European Conference on Mobile Robots, vol. 1, pp. 201–206 (2011)
24. Vonásek, V., Jurčík, A., Furmanová, K., Kozlíková, B.: Sampling-based motion planning for tracking evolution of dynamic tunnels in molecular dynamics simulations. J. Intell. Robot. Syst. **93**(3), 763–785 (2019)

A Productive and Scalable Actor-Based Programming System for PGAS Applications

Sri Raj Paul[1][✉], Akihiro Hayashi[2], Kun Chen[2], and Vivek Sarkar[2]

[1] Intel Corporation, Austin, USA
sriraj.paul@intel.com
[2] Georgia Institute of Technology, Atlanta, USA
{ahayashi,kunfz,vsarkar}@gatech.edu

Abstract. The Partitioned Global Address Space (PGAS) model is well suited for executing irregular applications on cluster-based systems, due to its efficient support for short, one-sided messages. Separately, the actor model has been gaining popularity as a productive asynchronous message-passing approach for distributed objects in enterprise and cloud computing platforms, typically implemented in languages such as Erlang, Scala or Rust. To the best of our knowledge, there has been no past work on using the actor model to deliver both productivity and scalability to PGAS applications on clusters.

In this paper, we introduce a new programming system for PGAS applications, in which point-to-point remote operations can be expressed as fine-grained asynchronous actor messages. In this approach, the programmer does not need to worry about programming complexities related to message aggregation and termination detection. Our approach can also be viewed as extending the classical Bulk Synchronous Parallelism model with fine-grained asynchronous communications within a phase or superstep. We believe that our approach offers a desirable point in the productivity-performance space for PGAS applications, with more scalable performance and higher productivity relative to past approaches. Specifically, for seven irregular mini-applications from the Bale benchmark suite executed using 2048 cores in the NERSC Cori system, our approach shows geometric mean performance improvements of $\geq 20\times$ relative to standard PGAS versions (UPC and OpenSHMEM) while maintaining comparable productivity to those versions.

Keywords: Actors · Communication aggregation · Conveyors · OpenSHMEM · PGAS · Selectors

1 Introduction

In today's world, performance is improved mainly by increasing parallelism, thereby motivating the critical need for programming systems[1] that can deliver

---

[1] Following standard practice, we use the term, "programming system", to refer to both compiler and runtime support for a programming model.

© The Author(s), under exclusive license to Springer Nature Switzerland AG 2022
D. Groen et al. (Eds.): ICCS 2022, LNCS 13350, pp. 233–247, 2022.
https://doi.org/10.1007/978-3-031-08751-6_17

both productivity and scalability for parallel applications. The *Actor Model* [4] is the primary concurrency mechanism in languages such as Erlang and Scala, and is also gaining popularity in modern system programming languages such as Rust. Large-scale cloud applications [3] from companies such as Facebook and Twitter that serve millions of users are based on the actor model. Actors express communication using "mailboxes" [16]; the term, "selector" [11], has been used to denote an actor with multiple mailboxes. The actor runtime maintains a separate logical mailbox for each actor. Any actor or non-actor, can send messages to an actor's mailbox. An important property of communication in Actors/Selectors is their inherent asynchrony, i.e., there are no global constraints on the order in which messages are processed in mailboxes.

The Partitioned Global Address Space (PGAS) model [20] is well suited to such irregular applications due to its efficient support for short, non-blocking one-sided messages. However, a key challenge for PGAS applications is the need for careful aggregation and coordination of short messages to achieve low overhead, high network utilization, and correct termination logic. Communication aggregation libraries such as Conveyors [17] can help address this problem by locally buffering fine-grain communication calls and aggregating them into medium/coarse-grain messages. However, the use of such aggregation libraries places a significant burden on programmer productivity and assumes a high expertise level.

In this paper, we introduce a new programming system for PGAS applications, in which point-to-point remote operations can be expressed as fine-grained asynchronous actor messages. In this approach, the programmer does not need to worry about programming complexities related to message aggregation and termination detection. Further, the actor model also supports the desirable goal of migrating computation to where the data is located, which is beneficial for many irregular applications [15].

Our approach can also be viewed as extending the classical Bulk Synchronous Parallelism (BSP) model with fine-grained asynchronous communications within a phase or superstep. Many current HPC execution models have been influenced by the simplicity and scalability of the BSP model, which consists of "supersteps" separated by barriers executing on homogeneous processors. However, the increasing degree of heterogeneity and performance variability in exascale machines has motivated the need for including asynchronous computations within a superstep so as to reduce the number of barriers performed and the total time spent waiting at barriers.

Specifically, this paper makes the following contributions:

1. An extension of the BSP model to a Fine-grained-Asynchronous Bulk-Synchronous Parallelism (FA-BSP) model.
2. A new PGAS programming system which extends the actor/selector model to enable asynchronous communication with automatic message aggregation for scalable performance.
3. Development of a source-to-source translator that translates our lambda-based API for actors to a more efficient class-based API.

4. Our results show a geometric mean performance improvement of 25.59× relative to the UPC versions and 19.83× relative to the OpenSHMEM versions, while using 2048 cores in the NERSC Cori system on seven irregular mini-applications from the Bale suite [17,18]

2   Background: Communication in PGAS Applications

In this section, we summarize two fundamental messaging patterns in PGAS applications, namely *read* and *update*, as well as the Conveyors library that can be used to aggregate messages. Since the focus of our work is on scalable parallelism, we assume a Single Program Multiple Data (SPMD) model in which each processing element (PE) starts by executing the same code with a distinct rank, as illustrated in the following code examples.

Listing 1.1. An OpenSHMEM program that reads data from a distributed array.

```
1 for(i = 0; i < n; i++){
2   int col = index[i] / shmem_n_pes();
3   int pe = index[i] % shmem_n_pes();
4   gather[i] = shmem_g(data+col, pe);
5 }
```

Listing 1.2. An OpenSHMEM program that creates a histogram by updating a distributed array.

```
1 for(i = 0; i < n; i++) {
2   int spot = index[i] / shmem_n_pes();
3   int PE = index[i] % shmem_n_pes();
4   shmem_atomic_inc(histo+spot, PE);
5 }
```

2.1   Read Pattern

In this pattern, each PE sends a request for data from a dynamically identified remote location and then processes the data received in response to the request. An OpenSHMEM version of a program using this pattern is shown in Listing 1.1. This program reads values from a distributed array named **data** and stores the retrieved values in a local array named **gather** based on global indices stored in a local array named **index**. The corresponding operation can also be performed in MPI using MPI_Get.

2.2   Update Pattern

In this pattern, each PE updates a remote location at an address that is computed dynamically. An OpenSHMEM program that updates remote locations is shown in Listing 1.2. This program updates a distributed array named **histo** based on global indices stored in each PE's local **index** array using atomic increment, thereby creating a histogram. The corresponding operation can also be performed in MPI using MPI_Accumulate or MPI_Get_Accumulate or MPI_Fetch_and_op.

2.3   Conveyors

Conveyors [17] is a C-based message aggregation library built on top of conventional communication libraries such as SHMEM, MPI, and UPC. It provides the following three basic operations:

1. convey_push: locally enqueue a message for delivery to a specified PE.
2. convey_pull: attempts to fetch a received message from the local buffer.
3. convey_advance: enables forward progress by transferring buffers.

It is worth noting that both push and pull operations can fail (return false) due to resource constraints. push can fail due to a lack of available buffer space, and pull can fail due to a lack of an available item. Due to these failures, push and pull operations must always be placed in a loop that ensures that the operations are retried. Further, advance needs to be called to ensure progress and to also help with termination detection. These complexities place a significant burden on programmer productivity and assumes a high expertise level. Table 1 demonstrates that user-directed message aggregation with Conveyors can achieve much higher performance compared to non-blocking operations in state of the art communication libraries/systems, some of which includes automatic message aggregation [6]. Analysis using Rice University's HPCToolkit [2] showed that the conveyors version reduced stall cycles by an order of magnitude compared to the OpenSHMEM version. As a result, we decided to use Conveyors as a lower-level library for automatic message aggregation in our programming system.

Table 1. Absolute performance numbers in seconds using best performing variants for Read and Update benchmarks on 2048 PEs (64 nodes with 32 PEs per node) in the Cori supercomputer which performs 2^{23} ($\approx$8 million) reads and updates.

Communication system	Non blocking	Read (sec)	Update (sec)
OpenSHMEM (cray-shmem 7.7.10)	N	35.5	NA
OpenSHMEM NBI (cray-shmem 7.7.10)	Y	4.2	4.3
UPC (Berkley-UPC 2020.4.0)	N	22.6	23.9
UPC NBI (Berkley-UPC 2020.4.0)	Y	19.7	NA
MPI3-RMA (OpenMPI 4.0.2)	Y	25.8	88.9
MPI3-RMA (cray-mpich 7.7.10)	Y	8.3	>300
Charm++ (6.10.1, gni-crayxc w/TRAM)	Y	21.3	9.7
Conveyors (2.1 on cray-shmem 7.7.10)	Y	2.3	0.5

3   Our Approach

3.1   Fine-grained-Asynchronous Bulk-Synchronous Parallelism (FA-BSP) Model

The classical Bulk-Synchronous Parallelism (BSP) [19] model consists of "supersteps" separated by barriers executing on homogeneous processors. Each processor only performs local computations and asynchronous communications in a superstep, and the role of the barrier is to ensure that all communications in a superstep have been completed before moving to the next superstep. However, the increasing degree of heterogeneity and performance variability in modern cluster machines has motivated the need for including asynchronous computations within a superstep so as to reduce the number of barriers performed and the total time spent waiting at barriers. To that end, we propose extending BSP to a Fine-grained-Asynchronous Bulk-Synchronous Parallelism (FA-BSP) model, as follows.

Our proposal is to realize the FA-BSP model by building on three ideas from past work in an integrated approach. The first idea is the actor model, which enables distributed asynchronous computations via fine-grained active messages while ensuring that all messages are processed atomically within a single-mailbox actor. For FA-BSP, we extend classical actors with multiple symmetric mailboxes for scalability, and with automatic termination detection of messages initiated in a superstep. The second idea is message aggregation, which we believe should be performed automatically to ensure that the FA-BSP model can be supported with performance portability across different systems with different preferences for message sizes at the hardware level due to the overheads involved. The third idea is to build on an asynchronous tasking runtime within each node, and to extend it with message aggregation and message handling capabilities.

3.2   High-Level Design of Programming System

Our primary approach to delivering both productivity and scalability for PGAS applications is by building a programming system based on the actor model that also supports automatic message aggregation and termination detection. Relative to the Conveyors approach, we would like to remove the burden of the user having to worry about about 1) the lack of available buffer space (`convey_push`), 2) the lack of an available item (`convey_pull`), and 3) the progress and termination of communications (`convey_advance`). We believe that the use of the actor/selector model is well suited for this problem since its programming model productively enables the specification of fine-grained asynchronous messages. Some key elements of the high-level design are summarized below

Abstracting *Buffers* as *Mailboxes*. We observe that buffer operations can be elevated to actor/selector mailbox operations with much higher productivity. For example, the `convey_push` operation on a buffer can be elevated to an actor/selector **send** operation, and a `convey_pull` operation can be made implicit in an actor/selector's message processing routine, while leaving it to our programming system to handle buffer/item failures and progressing/terminating communications among actors/selectors. More details on how our runtime handles failure scenarios are given in Sect. 4.3.

An important design decision for scalability is to treat a collection of mailboxes as a distributed object so that the mailboxes can be partitioned across PEs, analogous to how memory is partitioned in the PGAS programming model. This partitioned global actor design allows users to access a target actor's mailbox conveniently, instead of having to search for the corresponding actor object across multiple nodes as is done in many actor runtime systems. Thus we differ from classical actors through the usage of partitioned global mailbox.

Among the two patterns discussed in Sect. 2, the *read* patterns differs from the *update* pattern in that it involves communication in two directions, namely request and response. In this case we can use 'Selector' [11], which is an actor with multiple mailboxes.

Progress and Termination. In general, the Actors/Selectors model provides an `exit` [13] operation to terminate actors/selectors. While it may seem somewhat natural to expose this operation to users, one problem with this termination semantics is that it requires users to ensure that all messages in the incoming mailbox are processed (or received in some cases) before invoking `exit`, which adds additional complexities even for the simplest mini-applications such as Histogram Listing 1.2. To mitigate this burden, we added a relaxed version of `exit`, which we call `done`, to enable the runtime do more of the heavy lifting. The semantics of `done` is that users tell the runtime that the PE on which a specific actor/selector object resides will not send any more messages in the future to a particular mailbox, so the runtime can still keep the corresponding actor/selector alive so it can continue to receive messages and process them.

3.3 User-facing API

Based on the discussions in Sect. 3.2, we provide a C/C++ based actor/selector programming framework as shown in Listing 1.3.

Listing 1.3. Actor/Selector Interface with partitioned global mailboxes.

```
1  //L: lambda type
2
3  class Actor<L> {
4    void send(int PE, L msg);
5    void done();
6  };
7
8  class Selector<N, L> { // N mailboxes
9    void send(int mailbox_id,
10             int PE, L msg);
11   void done(int mailbox_id);
12 };
```

Listing 1.4. Actor version of the Update benchmark (Histogram) using lambda.

```
1  Actor h_actor;
2  for(int i=0; i < n; i++) {
3    int spot = index[i] / shmem_n_pes();
4    int remote_PE = index[i] % shmem_n_pes();
5    h_actor.send(remote_PE,
6               [=](){histo[spot]+=1;});
7  }
8  h_actor.done();
```

Update:[2] Listing 1.4 shows our version of the histogram benchmark. We use C++ lambdas to succinctly describe both the message and its processing routine. The main program creates an Actor object as a collective operation in Line: 1, which is used for communication. Then to create the histogram, it finds the target PE in Line:4 and local index within the target in Line: 3 from the global index. Then it sends a message lambda to the target PE's mailbox using the send API. Once the target PE's mailbox gets the message, the actor invokes it,

---

[2] Showing only update pattern due to page limitation. Read pattern is in the artifact.

which updates the `histo` array. Note that the lambda automatically captures the value of `spot` inside it. Also, the code for the lambda does not need to be communicated since it is compiled ahead of time and available on nodes.

3.4 Class-based API

While lambdas help with productivity by automatically capturing variables from the environment and enabling the developer to write routines with in-line message-handling logic instead of separate functions, lambda-based operations can incur additional overhead relative to direct method calls. To avoid this overhead, we also created a class-based version of our APIs (Listing 1.5) that gives the user more control regarding what data needs to be communicated and also allows for automatic translation from the lambda API to the class-based API. In the class based version, user need to express message handling using the `process` API and explicitly construct the message used in `send` API.

Listing 1.5. Actor/Selector class-based interface with partitioned global mailboxes.

```
 1  class Actor<T> {
 2    void process(T msg, int PE);
 3    void send(T msg, int PE);
 4    void done();
 5
 6    Actor() {
 7      mailbox[0].process = this->process;
 8    }
 9  };
10  class Selector<N,T> { // N mailboxes
11    void process_0(T msg, int PE);
12    ...
13    void process_N_1(T msg, int PE);
14    void send(int mailbox_id, T msg, int PE);
15    void done(int mailbox_id);
16
17    Selector() {
18      mailbox[0].process = this->process_0;
19      ...
20      mailbox[N-1].process = this->process_N_1;
21    }
22  };
```

4 Implementation

In this section, we discuss the implementation of the selector runtime prototype created by extending HClib [9], a C/C++ Asynchronous Many-Task (AMT) Runtime library. We first discuss our execution model in Sect. 4.2 and then describe our extensions to the HClib runtime to support our selector runtime in Sect. 4.3.

4.1 HClib Asynchronous Many-Task Runtime

Habanero C/C++ library (HClib) [9] is a lightweight asynchronous many-task (AMT) programming model-based runtime. It uses a lightweight work-stealing scheduler to schedule the tasks. HClib uses a persistent thread pool called workers, on which tasks are scheduled and load balanced using lock-free concurrent deques. HClib exposes several programming constructs to the user, which in turn helps them to express parallelism easily and efficiently.

A brief summary of the relevant APIs is as follows:

1. `async`: Used to create asynchronous tasks dynamically.

2. `finish`: Used for bulk task synchronization. It waits on all tasks spawned (including nested tasks) within the scope of the finish.
3. `promise` and `future`: Used for point-to-point inter-task synchronization in C++11 [7]. A promise is a single-assignment thread-safe container, that is used to write some value and a future is a read-only handle for its value. Waiting on a future causes a task to suspend until the corresponding promise is *satisfied* by putting some value to the promise.

4.2 Execution Model

Figure 1 shows the high level structure of the execution model for our approach from the perspective of PE j, shown as process[j], with memory[j] representing that PE's locally accessible memory. This local memory includes partitions of global distributed data, in accordance with the PGAS model. Users can create as many tasks as required by the application, which are shown as *Computation Tasks*. For the communication part, each mailbox corresponds to a *Communication Task*. All tasks get scheduled for execution on to underlying worker threads. For example, if an application uses a selector with two mailboxes and an actor/selector with one mailbox, it corresponds to three communication tasks—two for the selector and one for the actor. All computation and communication tasks are created using the HClib [9] Asynchronous Many-Task (AMT) runtime library.

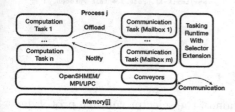

Fig. 1. The execution model showing internal structure of tasks and mailboxes within a single PE.

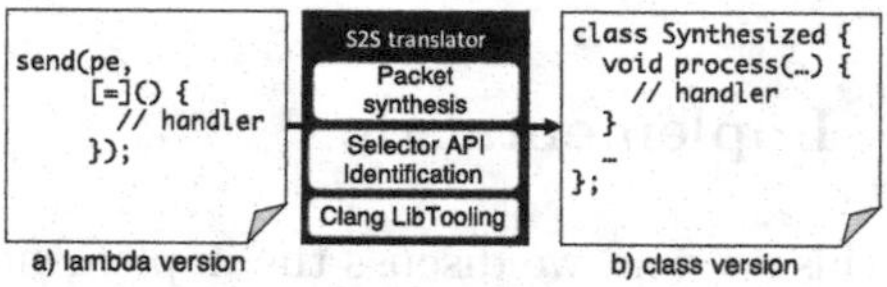

Fig. 2. Source-to-source translator from lambda version to class based version.

To enable asynchronous communication, the computation tasks offload all remote accesses on to the communication tasks. When the computation task sends a message, it is first pushed to the communication task associated with the mailbox using a local buffer. Eventually, the communication task uses the conveyors library to perform message aggregation and actual communication. Currently we use a single worker thread that multiplexes all the tasks. When a mailbox receives a message, the mailbox's process routine is invoked.

It is worth noting that users are also allowed to directly invoke other communication calls outside the purview of our Selector runtime. For example, the user application can directly invoke the OpenSHMEM barrier or other collectives.

4.3 Selector Runtime

The implementation details presented are based on the class-based interface introduced in Sect. 3.4, since our results were obtained by converting the lambda API to the class-based API using the translator described in Sect. 4.4. As mentioned earlier, we hide the low-level details of Conveyors operations from the programmer and incorporate them into our Selector runtime instead. To reiterate, such details include maintaining the progress and the termination of communication as well as handling 1) the lack of available buffer space, and 2) the lack of an available item. This enables users to only stick with the send(), done(), and process() APIs. The implementation details of these APIs are as follows:

Selector.send(): We map each mailbox to a conveyor object. Each send in a mailbox gets eventually mapped to a conveyor_push. Note that the send does not directly invoke the conveyor_push because we want to relieve the computation task on which the application is running from dealing with the failure handling of conveyor_push. Instead, this API adds a packet with the message and receiver PE's rank to a small local buffer[3] that is based on the Boost Circular Buffer library [8]. The packet is later picked up by the communication task associated with the mailbox and is passed into a conveyor_push operation. Whenever the mailbox's local circular buffer gets filled, the runtime automati-

---

Algorithm 1. Worker loop associated with each mailbox

```
 1: while buff.isempty() do
 2:     yield()                                    ▷ yield until message is pushed to buffer
 3: end while
 4: pkt ← buffer[0]
 5: while convey_advance(conv_obj, is_done(pkt)) do
 6:     for i ← 0 to buffer.size-1 do
 7:         pkt ← buffer[i]
 8:         if is_done(pkt) then
 9:             break
10:         end if
11:         if convey_push(conv_obj, pkt.data,pkt.rank) then
12:             break
13:         end if
14:     end for
15:     buffer.erase(0 to i)
16:     while convey_pull(conv_obj, &data, &from) do
17:         create computation_task(
18:             process(data, from)
19:         )
20:     end while
21:     yield()
22: end while
23: end_promise.put(1)                             ▷ To signal completion of mailbox
```

---

[3] This local buffer is different from the Conveyor's internal buffer.

cally passes control to the communication task, which drains the buffer, thereby allowing us to keep its size fixed.

Selector.done(): Analogous to `send`, when `done` is invoked, we enqueue a special packet to the mailbox that denotes the end of sending messages from the current PE to that mailbox.

Selector.process(): When the communication task receives a data packet through `conveyor_pull`, the mailbox's process routine is invoked.

Worker Loop: The selector runtime creates a conveyor object for each mailbox and processes them separately within its own worker loop, as shown in Algorithm 1. When a mailbox is started, it creates a corresponding conveyor object (`conv_obj`) and a communication task that executes the algorithm shown in Algorithm 1. Initially, the communication task waits for data packets in the mailbox's local buffer, which gets added when the user performs a `send` from the mailbox partition. During this polling for packets from the buffer, the communication task yields control to other tasks, as shown in Line 2. Once the data is added to the buffer, it breaks out of the polling loop and starts to drain elements from the buffer in Line 6. It then pushes each element in the buffer to the target PE in Line 11 until push fails. Then it removes all the pushed items from the buffer and starts the pull cycle. It pulls the received data in Line 16 and creates a computation task, which in turn invokes the mailbox's process method, as shown in Line 18. As mentioned before, in case there is only one worker that is shared by all the tasks, we invoke the process method directly without the creation of any computation task. Once we come out of the processing of the received data, the task yields so that other communication tasks can share the communication worker.

Once the user invokes `done`, a special packet is enqueued to the buffer. When this special packet is processed, the `is_done` API in Line 5 returns true, thereby informing the conveyor object to start its termination phase. Once the communication of all remaining items is finished, the `convey_advance` API returns false, thereby exiting the work loop. Finally the communication task terminates and signals the completion of the mailbox using a variable of type `promise` named as `end_promise`, as shown in Line 23. The signaling of the promise schedules a dependent cleanup task which informs all dependent mailboxes about the termination of the current mailbox. This task also manages a counter to find out when all the mailboxes in the selector have performed cleanup, to signal the completion of the selector itself using a `future` variable associated with the selector. Since the selector runtime is integrated with the HClib runtime, the standard synchronization constructs in AMT runtimes such as `finish` scope and `future` can be used by the user to coordinate with the completion of the selector. Other dependent tasks can use the `future` associated with the selector to wait for its completion. Users can also wait for completion by using a `finish` scope.

4.4 Source-to-Source Translation from Lambda-Based to Class-Based Messaging

While the use of C++ lambda expressions further simplifies writing remote message handlers (Sect. 3.3), during experiments the performance of the lambda-based version was found to be 2× slower than that of the class-based version (Sect. 3.4). This motivates us to perform automatic source-to-source translation from the lambda version to the class version to improve productivity without this performance loss. This kind of translation could be beneficial to other lambda-based libraries as well.

Figure 2 illustrates the end-to-end flow for the translation. The translator is a standalone tool built on top of Clang LibTooling. First, it identifies the use of the send API with a lambda expression by utilizing Clang LibTooling's AST traversal APIs. For each lambda, it analyzes captured variables to synthesize a packet structure that is used for the class-based version. Then, it synthesizes a class declaration with a message handler and a packet struct type for actor messages.

5   Evaluation

This section presents the results of an empirical evaluation of our selector runtime system on a multi-node platform to demonstrate its performance and scalability. The goal of our evaluation is twofold:

1. to demonstrate that our selector-based programming system based on the FA-BSP model can be used to express a range of irregular mini-applications, and
2. to compare the performance of our approach with that of UPC, OpenSHMEM and Conveyors versions of these mini-applications.

Platform: We ran the experiments on the Cori supercomputer located at NERSC. In Cori, each node has two sockets, with each socket containing a 16-core Intel Xeon E5-2698 v3 CPU @ 2.30 GHz (Haswell). For inter-node connectivity, Cori uses the Cray Aries interconnect with Dragonfly topology that has a global peak bisection bandwidth of 45.0 TB/s. We use one worker thread per PE rank for the experiments; since the mini-applications have sufficient parallelism across PE ranks, there was no motivation to use multiple worker threads within a single PE rank. The Conveyors library was compiled using cray-shmem for our experiments since cray-shmem provided the best performance based on our evaluation in Table 1.

Mini-applications: We used all seven mini-applications in Bale [17,18] that have Conveyors versions for our study. Bale can be seen as a proxy for key components in an irregular application that involve a large number of irregular point-to-point communication operations.

Experimental Variants:

1. **UPC:** This version is written using UPC.
2. **OpenSHMEM:** This version is written using OpenSHMEM.
3. **Conveyor:** This version directly invokes the Conveyors APIs, which includes explicit handling of failure cases and communication progress.
4. **Selector:** This version uses the class-based version of the Selector API introduced in this paper, obtained by automatic translation from the lambda version as described in Sect. 4.4.

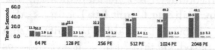

(a) Histogram mini-application with 10,000,000 updates per PE on a distributed table with 1,000 elements/PE.

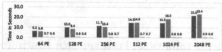

(b) Index-gather mini-application with 10,000,000 reads per PE on a distributed table with 100,000 elements/PE.

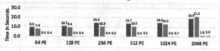

(c) Permute-matrix mini-application with 100,000 rows of the matrix/PE with an average of 10 nonzeros per row.

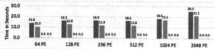

(d) Random-permutation mini-application with 1,000,000 elements per PE.

(e) Topological-sort mini-application with 100,000 rows of the matrix/PE with an average of 10 nonzeros per row.

(f) Transpose-matrix mini-application with 100,000 rows of the matrix per PE with an average of 10 nonzeros per row.

(g) Triangle-counting mini-application with 10,000 rows of the matrix per PE with an average of 35 nonzeros per row.

Fig. 3. Comparison of execution time of the UPC, OpenSHMEM, conveyor and selector variants (lower is better for the Y-axis).

In Figs. 3(a) to 3(g), the Y-axis shows the weak scaling average execution time of five runs in seconds, so smaller is better. From the figures, we can see that the Conveyor versions perform much better than their UPC and OpenSHMEM counterparts. For the 2048 PE/core case, the Conveyor versions show a geometric mean performance improvement of 27.77× relative to the UPC and 21.52× relative to the OpenSHMEM versions, across all seven mini-applications.

This justifies our decision to use the Conveyors library for message aggregation in our Selector-based approach. Overall, we see that the Selector version also performs much better than the UPC/OpenSHMEM versions and close to the Conveyor version. For the 2048 PE/core case, the Selector versions show a geometric mean performance improvement of 25.59× relative to the UPC and 19.83× relative to the OpenSHMEM versions, and a geometric mean slowdown of only 1.09× relative to the Conveyor versions. These results confirm the performance advantages of our approach, while the productivity advantages can be seen in the simpler programming interface for the Selector versions relative to the Conveyor versions.

Table 2 shows the source lines of code (SLOC) for different versions of the kernel of each mini-application, as measured by the CLOC utility [1]. The table convincingly shows that moving to the Actor/Selector model results in lower SLOC values relative to the Conveyor model, which in turn demonstrates higher productivity for the Actor/Selector model.

Table 2. Kernel size of each mini-application in terms of source lines of code.

	Histogram	Index-gather	Permute-matrix	Random-permutation	Topological-sort	Transpose	Triangle-counting
UPC	18	16	37	41	72	43	43
OpenSHMEM	19	17	51	43	92	50	49
Conveyor	30	40	108	111	148	83	61
Actor/Selector	21	25	78	99	130	69	53

6   Related Work

The actor is the primary concurrency mechanism in Scala, however it is not scalable for HPC workloads [5]. The Chare abstraction in Charm++ [14] has taken inspiration from the Actor model, and is also designed for scalability. As indicated earlier, the performance of Charm++ is below that of Conveyors (and hence that of our approach) for the workloads studied in this paper.

In the past, there has been much work on optimizing the communication of PGAS programs through communication aggregation. Jenkins *et al.*. [12] created the Chapel Aggregation Library (CAL) which aggregates user-defined data using an Aggregator object. UPC [6] performs automatic message aggregation to improve the performance of fine-grained communication but is unable to achieve performance compared to user-directed message aggregation.

7   Conclusions and Future Work

This paper proposes a scalable programming system for PGAS runtimes to accelerate irregular distributed applications. Our approach is based on the actor/selector model, and introduces the concept of a *Partitioned Global Mailbox*. Our programming system also abstracts away low-level details of message aggregation (e.g., manipulating local buffers and managing progress and termination) so that the programmer can work with a high-level selector interface. Our Actor runtime is more than a message-aggregation system since it also supports user-defined active messages, which can support the migration of computation closer to data that is beneficial for irregular applications. For the 2048 PE case, our approach show a geometric mean performance improvement of 25.59× relative to the UPC versions, 19.83× relative to the OpenSHMEM versions, and a geometric mean slowdown of only 1.09× relative to the Conveyors versions. These results suggest that the FA-BSP model offers a desirable point in the productivity-performance spectrum, with higher performance relative to PGAS models such as UPC and OpenSHMEM and higher productivity relative to the use of low-level hand-coded approaches for communication management and message aggregation.

In future, it would be interesting to explore compiler extensions to automatically translate from the natural version to our selector version, thereby directly improving the performance of natural PGAS programs. We would also like to improve our performance result reporting based on the paper [10].

Artifact. https://github.com/srirajpaul/hclib/tree/bale_actor/modules/bale_actor

References

1. cloc. http://manpages.ubuntu.com/manpages/man1/cloc.1.html
2. Adhianto, L., et al.: HPCTOOLKIT: tools for performance analysis of optimized parallel programs. Concurr. Comput. Pract. Exp. **22**(6), 685–701 (2010). https://doi.org/10.1002/cpe.1553
3. Agha, G.: Actors programming for the mobile cloud. In: 2014 IEEE 13th International Symposium on Parallel and Distributed Computing, pp. 3–9 (2014). https://doi.org/10.1109/ISPDC.2014.31
4. Agha, G.A.: ACTORS - A Model of Concurrent Computation in Distributed Systems. MIT Press Series in Artificial Intelligence, MIT Press (1990)
5. Charousset, D., et al.: Revisiting actor programming in c++. Comput. Lang. Syst. Struct. **45**(C), 105–131 (2016). https://doi.org/10.1016/j.cl.2016.01.002
6. Chen, W.Y.: Building a Source-to-Source UPC-to-C translator. EECS Department, University of California, Berkeley (Tech. rep. (2004)
7. cplusplus.com: Future (2020). http://www.cplusplus.com/reference/future/
8. Gaspar, J.: Boost. Circular Buffer. https://www.boost.org/doc/libs/1_72_0/doc/html/circular_buffer.html. Accessed 20 Apr 2020
9. Grossman, M., et al.: A pluggable framework for composable HPC scheduling libraries. In: 2017 IEEE International Parallel and Distributed Processing Symposium Workshops (IPDPSW), pp. 723–732. IEEE Computer Society (2017). https://doi.org/10.1109/IPDPSW.2017.13
10. Hoefler, T., Belli, R.: Scientific benchmarking of parallel computing systems: Twelve ways to tell the masses when reporting performance results. In: Proceedings of the International Conference for High Performance Computing, Networking, Storage and Analysis, SC 2015, pp. 1–12 (2015). https://doi.org/10.1145/2807591.2807644
11. Imam, S.M., Sarkar, V.: Selectors: actors with multiple guarded mailboxes. In: Proceedings of the 4th International Workshop on Programming based on Actors Agents & Decentralized Control, pp. 1–14. ACM (2014). https://doi.org/10.1145/2687357.2687360
12. Jenkins, L., et al.: Chapel aggregation library (CAL). In: 2018 IEEE/ACM Parallel Applications Workshop, Alternatives To MPI, PAW-ATM SC 2018, pp. 34–43. IEEE (2018). https://doi.org/10.1109/PAW-ATM.2018.00009
13. Joyner, M.: Introduction to the actor model (2020). https://wiki.rice.edu/confluence/download/attachments/4435861/comp322-s20-lec21-slides-wide.pdf
14. Kalé, L.V., et al.: CHARM++: a portable concurrent object oriented system based on C++. In: Object-Oriented Programming Systems, Languages, and Applications (OOPSLA). ACM (1993). https://doi.org/10.1145/165854.165874
15. Kogge, P.M., Kuntz, S.K.: A case for migrating execution for irregular applications. In: Workshop on Irregular Applications: Architectures and Algorithms, IA3@SC 2017, pp. 6:1–6:8. ACM (2017). https://doi.org/10.1145/3149704.3149770
16. Koster, J.D., Cutsem, T.V., Meuter, W.D.: 43 years of actors: a taxonomy of actor models and their key properties. In: Proceedings of the 6th International Workshop on Programming Based on Actors, Agents, and Decentralized Control, pp. 31–40. ACM (2016). https://doi.org/10.1145/3001886.3001890
17. Maley, F.M., DeVinney, J.G.: Conveyors for streaming many-to-many communication. In: Workshop on Irregular Applications: Architectures and Algorithms, IA3 SC 2019, pp. 1–8. IEEE (2019). https://doi.org/10.1109/IA349570.2019.00007

18. Maley, F.M., DeVinney, J.G.: A collection of buffered communication libraries and some mini-applications. https://github.com/jdevinney/bale (2020)
19. Valiant, L.G.: A bridging model for parallel computation. Commun. ACM **33**(8), 103–111 (1990). https://doi.org/10.1145/79173.79181
20. Yelick, K.A., et al.: Productivity and performance using partitioned global address space languages. In: Parallel Symbolic Computation, PASCO 2007, International Workshop, pp. 24–32, 27–28 July 2007. https://doi.org/10.1145/1278177.1278183

Is Context All You Need?
Non-contextual vs Contextual Multiword
Expressions Detection

Maciej Piasecki[⊠][iD] and Kamil Kanclerz[iD]

Department of Artificial Intelligence, Wrocław University of Science and Technology,
Wrocław, Poland
{maciej.piasecki,kamil.kanclerz}@pwr.edu.pl

Abstract. Effective methods of the detection of multiword expressions
are important for many technologies related to Natural Language Processing. Most contemporary methods are based on the sequence labeling
scheme, while traditional methods use statistical measures. In our approach, we want to integrate the concepts of those two approaches. In
this paper, we present a novel weakly supervised multiword expressions
extraction method which focuses on their behaviour in various contexts.
Our method uses a lexicon of Polish multiword units as the reference
knowledge base and leverages neural language modelling with deep learning architectures. In our approach, we do not need a corpus annotated
specifically for the task. The only required components are: a lexicon of
multiword units, a large corpus, and a general contextual embeddings
model. Compared to the method based on non-contextual embeddings,
we obtain gains of 15% points of the macro F1-score for both classes and
30% points of the F1-score for the incorrect multiword expressions. The
proposed method can be quite easily applied to other languages.

Keywords: Natural Language Processing · Multiword expressions ·
Detection of multiword expressions · Contextual embeddings

1 Introduction

Multiword expressions (henceforth MWEs) are defined in different ways in literature, e.g. see the overview in [29]. In this work, we consider MWEs from the
lexicographic point of view as lexical units that "has to be listed in a lexicon" [12]
and we focus on methods of automated extraction of MWEs from text corpora to
be included in a large semantic lexicon as *multi-word lexical units*. Summarising
a bit the definition of [29], MWEs are "lexical items decomposable into multiple
lexemes", "present idiomatic behaviour at some level of linguistic analysis" and
"must be treated as a unit" and, thus, should be described in a semantic lexicon,
e.g. *skrzynia biegów* ('gearbox', lit. box of gears), *pogoda ducha* (≈'optimism',
'good attitude', lit. 'weather of spirit'). A similar definition was adopted in the
PARSEME Shared Task resource [30,31]. As we target the construction of a

D. Groen et al. (Eds.): ICCS 2022, LNCS 13350, pp. 248–261, 2022.
https://doi.org/10.1007/978-3-031-08751-6_18

general lexicon expressing good coverage for lexical units occurring frequently enough in a very large corpus – and we will test our approach against such a resource, see Sect. 3 – we need also to take into account *multiword terms*, i.e. [29] "specialised lexical units composed of two or more lexemes, and whose properties cannot be directly inferred by a non-expert from its parts because they depend on the specialised domain". Several MWE properties are postulated that can guide the extraction process, like arbitrariness, institutionalisation, limited semantic variability (especially non-compositionality and non-substitutability), domain-specificity and limited syntactic variability [29]. As we are interested in the lexicon elements, the frequency of potential MWEs should be taken into account and MWEs are indeed in some way specific with respect to the frequency of co-occurrence of its components. Extraction of MWEs and their description in a semantic lexicon (at least as a reference resource) is important for many NLP applications like semantic indexing, knowledge graph extraction, vector models, topic modelling etc. Due to the specific properties of MWEs as whole units, their automated description by the distributional semantics method, e.g. embeddings, is not guaranteed, especially in the case of MWEs of lower frequency.

Traditionally, MWEs extraction is preceded by finding collocations (frequent word combinations) by statistical or heuristic *association measures* and filtering them by syntactic patterns. Recent methods follow *sequence labelling* scheme and try to explore the specific behaviour of MWEs as language expressions in text. Due to our objective, we aim at combining the best of the two worlds. We propose a new weakly supervised method for MWE extraction from large text corpora that explores their peculiar properties as elements of language structures across various contexts. The proposed method combines neural language modelling with deep learning and a lexicon of MWEs as a knowledge base, i.e. the sole source of supervision. In contrast to many methods from literature, we neither need a corpus laboriously annotated with MWE occurrences, nor language models specially trained for this task. We investigated and combined non-contextual representation of MWEs as lexical units and their contextual representation as elements of the sentence structures. For the latter purpose, we leverage deep neural contextual embeddings to describe the peculiarities of the semantic but also syntactic behaviour of MWEs in contrast to the behaviour of their components. What is more, the evidence for the whole MWEs and their components can be collected from different sentences in the corpus, not only those including whole MWEs. Our method can be quite easily adapted to any language, the only required elements are: a large corpus, an initial lexicon of MWEs, and a general contextual embeddings model. The proposed method, after training, may be applied to a list of collocations extracted from a corpus by association measures to distinguish MWEs from mere collocations.

2   Related Work

Initially, MWE recognition methods were based on statistical association measures based on co-occurrence statistics in text corpora [12] for weighting collocations as potential MWEs. Many association measures were examined and combined into complex ones, e.g. by a neural network [27]. Syntactic information from parsing was used in counting statistics or post-filtering collocations [36]. Morpho-syntactic tagging and lexico-syntactic constraints were also used instead of parsing [7]. For Polish, association measures combined by a genetic algorithm and expanded with lexico-syntactic filtering were used to extract potential MWEs [28]. Several systems for MWE extraction were proposed, combining different techniques, e.g. *mwetoolkit* by Ramisch [29] combines statistical extraction and morpho-syntactic filtering, but also describes collocations with feature vectors to train Machine Learning (ML) classifiers. Lexico-syntactic patterns, measures, length and frequency can be a feature source in ML-based MWE extraction [37]. Linguistic patterns were used to extract potential MWEs and post-filter out incorrect ones after association measures [2]. MWEs were also detected by tree substitution grammars [14] or finite state transducers [16].

Recently, attention was shifted to supervised ML and MWE extraction as a sequence labelling problem, e.g. [9], where corpora are annotated on the level of words, typically, BIO annotation format [32]: B – a word begins an MWE, I is inside, O – outside. Sequence labelling approaches can also be combined with heuristic rules [35] or supersenses of nouns or verbs [18]. Such heuristics are applied to extract linguistic features from texts for training a Bayesian network model [8]. Convolutional graph networks and self-attention mechanisms can be used to extract additional features [33]. There are many challenges related to the nature of the MWEs, e.g.: discontinuity – another token occurs between the MWE components or overlapping – another MWE occurs between the components of the given MWEs. To counteract this, a model based on LSTM, the long short-term memory networks and CRF is proposed [4]. The model from [38] combines two learning tasks: MWE recognition and dependency parsing in parallel. The approach in [21] leverages feature-independent models with standard BERT embeddings. mBERT was also tested, but with lower results. An LSTM-CRF architecture combined with a rich set of features: word embedding, its POS tag, dependency relation, and its head word is proposed in [39].

MWEs can be also represented as subgraphs enriched with morphological features [6]. Graphs can be next combined with the *word2vec* [24] embeddings to represent word relations in the vector space and then used to predict MWEs on the basis of linguistic functions [3]. Morphological and syntactic information can be also delivered to a recurrent neural network [19]. Saied et al. [34] compared two approaches to MWE recognition within a transition system: one based on a multilayer perceptron and the second on a linear SVM. Both utilise only lemmas and morphosyntactic annotations from the corpus and were trained and tested on PARSEME Shared Task 1.1 data [30].

However, such sequence labeling approaches focus on word positions and orders in sentences, and seem to pay less attention to the semantic incompatibility of MWEs or semantic relations between their components. Furthermore, sequence labeling methods do not emphasize the semantic diversity of MWE occurrence contexts. Thus, they overlook one of the most characteristic MWE factors: components of a potential MWE co-occur together regardless of the context. It allows us to distinguish a lexicalised MWE from a mere collocation or even a term strictly related to one domain. To the best of our knowledge, the concept of using deep neural contextual embeddings to describe the semantics of the MWEs components and the semantic relations between them in a detection task has not been sufficiently studied, yet. Moreover, due to the sparsity of the MWEs occurrences in the corpus, the corpus annotation process is very time consuming and can lead to many errors and low inter-annotator agreement. For this reason, we propose a lexicon-based corpus annotation method. On the basis of the assumption that the vast majority of MWEs are monosemous, e.g. the set of more than 50k MWEs in plWordNet [11], we performed an automated extraction of sentences containing the MWE occurrences and treated all sentences including a given MWE as representing the same multiword lexical unit.

3   Dataset

For evaluation, we used MWEs from plWordNet [11] marked as *multi-word lexical units* [23]. In addition, we utilised as negative data multiword lemmas removed from plWordNet as non-lexicalised over the years by the linguists. There is no information about all collocations considered for adding to plWordNet, but those that were once erroneously included must be more tricky ones. plWordNet contains 53,978 two-word MWEs and 6,369 longer than 2 words for Polish. English WordNet includes 59,079 two-word MWEs and 10,649 longer than 2 words. In the Polish part of the PARSEME corpus, there are also 3,427 two-word MWEs and 568 MWEs longer than 2 words (in the English part respectively, 457 two-word MWEs and 85 ones longer). Due to this high numerical prevalence of two-word MWEs, we concentrate on them in this paper. Two sample representations were compared: *non-contextual* and *contextual*. In the first case – a baseline – the representation is derived from word embeddings vectors. In the latter case of the contextual representation, we used the KGR10 Polish corpus [20], one of the largest Polish corpora (4,015,569,051 tokens, 18,084,712 unique ones) with a rich variety of text types.

3.1   plWordNet-based Non-Contextual Dataset

Context-free representation was built for both correct MWEs and incorrect 'MWEs' using the *fastText* skipgram model [5] (trained on the KGR10 corpus). It concatenates embeddings of the MWE components with vectors of differences between them. Figure 1 and Eq. 1 show the generation of non-contextual MWE representation emb_{NC} from the vectors w_1 and w_2 of the component words. Such representation, including the difference of vectors, has been inspired by the sample representation used in the NLI domain and also in semantic relations extraction [13]. Moreover, the concatenation of the difference vector along with the word embeddings was also used to represent word relations in [22].

$$emb_{NC}(w_1, w_2) = \overrightarrow{w_1} \oplus \overrightarrow{w_2} \oplus (\overrightarrow{w_1} - \overrightarrow{w_2}) \tag{1}$$

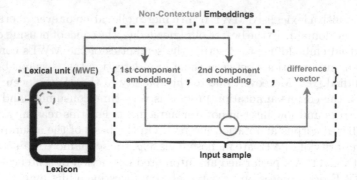

Fig. 1. Non-Contextual MWE representation generation

3.2   KGR10-based Contextual Dataset

For the contextual MWE representation, 687,900 sentences were extracted from the KGR10 corpus. Components of the correct MWEs were detected in 648,481 sentences and the incorrect in 39,419. We started by detecting the MWE component among the lemmas occurring in sentences. If lemmas of multiple MWEs were detected in a sentence, then it was associated with each of them as separate *training samples*, see Algorithm 1. In order to test the performance of our method in detecting sentences containing MWE components, we prepared 4 randomly selected samples of 100 found sentences each. They were verified by linguists who found that 99% of all sentences contained correct MWE components.

Algorithm 1. Procedure of obtaining sentences (s) from the corpus (C), if they include MWEs or their components by comparing sentence word lemmas ($l_i \in [l_0, l_1, \ldots, l_n]$) to the list ($M$) of lemmatised MWEs ($m_j \in [m_0, m_1, \ldots, m_k]$)

1: *sentence_list* ← [ ]
2: **for** $s \in C$ **do**
3:     **for** $l_i \in s$ **do**
4:         **for** $m_j \in M$ **do**
5:             **if** $l_i \in m_j$ **then**
6:                 *sentence_list*.insert(s)
7:             **end if**
8:         **end for**
9:     **end for**
10: **end for**
11: **return** *sentence_list*

Equation 2 describes the generation of contextual embeddings for MWEs as sample representations: an MWE embedding ($S_{m_{sent}}$) in the sentence context ($\overrightarrow{m_{sent}}$) is an average of the WordPiece subtoken vectors ($\overrightarrow{\nu_s}$) related to the MWE components. Next, we subsequently replaced the MWE occurrences in sentences with each of their components and obtained their contextual embeddings $(\overrightarrow{c_{sent}})$ by averaging the corresponding subtoken vectors representations $(\overrightarrow{\nu_s})$ related to the substituted components $(S_{c_{sent}})$, see Eq. 3. The final contextual embedding (emb_C) of a training sample related to a sentence $(sent)$ containing MWE (m) and one of its components (c) is described in Eq. 4. For each MWE occurrence, we generated the contextual embeddings corresponding to each of its components separately.

$$\overrightarrow{m_{sent}} = \frac{\sum_{s \in S_{m_{sent}}} \overrightarrow{\nu_s}}{|S_{m_{sent}}|} \tag{2}$$

$$\overrightarrow{c_{sent}} = \frac{\sum_{s \in S_{c_{sent}}} \overrightarrow{\nu_s}}{|S_{c_{sent}}|} \tag{3}$$

$$emb_C(c, m, sent) = \overrightarrow{c_{sent}} \oplus \overrightarrow{m_{sent}} \oplus (\overrightarrow{m_{sent}} - \overrightarrow{c_{sent}}) \tag{4}$$

We aim at observing the difference between the contextual embedding of a whole MWE and each of its components across sentences. Thus, we calculated the difference vector between the representation of the complete expression and its component in the context of a sentence as is illustrated in Fig. 2.

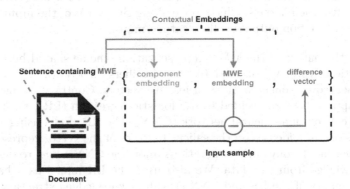

Fig. 2. MWE contextual representation generation.

4   Methods for Multiword Expression Detection

We assume that the context plays a significant role in the MWE detection. The first dataset from Sect. 3.1 contains training samples of non-contextual MWE representations (MWE vector, component vectors, and the difference vector). In this task, classifiers should focus on the semantic differences between the vector representations of the MWE components and the entire MWE. This is focused on non-compositional character of genuine MWEs. An incorrect 'MWE' example of a *materiał opatrunkowy* (en. 'bandage cloth') is in fact compositional in contrast to a correct, genuine MWE: *głos serca* (en. lit. 'heart's voice'), whose semantics cannot be inferred from its component meanings.

In contrast to the first (baseline) non-contextual representation, the dataset from Sect. 3.2 includes samples of contextual MWE representations (contextual vectors of MWE components and the entire MWE, plus the difference vector). In this case, the task of classifiers is to decide on the correctness of an expression on the basis of knowledge extracted from the contexts of the expression occurrences and the interaction between the contexts and the semantic representation of the whole MWEs and their components. An example of an incorrect 'MWE' is *barwnik naturalny* (en. 'a natural pigment'), which is compositional in any context and an example of correct MWE is *ojciec chrzestny* (en. 'a godfather'), which is non-compositional and when occurs in different contexts, its components should receive significantly different contextual vectors from the MWE vector.

We prepared three different model architectures to measure the influence of context knowledge on the classification of collocations as MWEs:

- **Logistic Regression (LR)** – a statistical model, which utilizes the logistic function to model the probability of a discrete binary dependent variable,
- **Random Forest (RF)** – an ensemble learning method, aggregating multiple decision trees by calculating the mode of their predictions,
- **Convolutional Neural Network (CNN)** – a deep learning architecture, using convolution kernels, which move along the vector of the input data and provide translation outputs called feature maps.

Due to the nature of the MWE representation scheme shared between both representation types, we decided to use classifiers that work well with samples represented by concatenations of feature vectors. Contrary to the sequence labelling approaches, we decided to use logistic regression (LR), random forest (RF), and convolutional neural network (CNN). The RF model using an ensemble of decision trees focuses on the salient features of the vector representations. On the other hand, convolution operations allow the CNN model to derive additional knowledge from the data. We also used the LR model as a baseline to verify the quality of the RF and CNN classifiers knowledge extraction.

In the contextual representation, Sect. 3.2, a single collocation or its component may occur in multiple sentences, so the same collocation may occur in several samples. As the vast majority of MWEs are monosemous, e.g. plWordNet [11], we leveraged this fact by preparing several voting strategies that aggregate the decisions of a selected model related to the same collocation:

- **Occurrence Classification (OC)** – each collocation occurrence is classified on its own, i.e. a separate decision, independent of the other occurrences, is made solely on the text of the given context,
- **Majority Voting (MV)** – predictions for all occurrences of a given collocation are collected and the final decision is made by majority voting,
- **Weighted Voting (WV)** – as the previous one, but the overall decision is made by weighted voting with confidence levels of a classifier as weights.

5 Experiments

The task selected for all conducted experiments is a single-task binary classification, where each classifier had to predict the correct label out of the 2 available for the given expression as a potential MWE. We used the HerBERT model [25] to generate contextual embeddings as it is considered as one of the best transformer models trained and evaluated on texts in Polish. Implementations of the LR and RF classifiers come from the scikit-learn library [26], and CNN from the TensorFlow library [1]. The CNN architecture consists of three convolutional layers each followed by the pooling layer and the dropout layer, and is shown in Fig 3. To counter the impact of class imbalance in both datasets of samples (53,978 to 5,598 for the non-contextual one and 648,481 to 39,419 for the contextual one), we used the F1-macro measure to estimate the performance quality of classifiers. Moreover, we used the weighted loss function, depending on the number of instances of a given class in the training set. In addition, we applied 4 different variants of the SMOTE method (SMOTE [10], SVM-SMOTE [10], Borderline SMOTE [15], and ADASYN [17]) to generate additional synthetic training samples on the basis of the real sample embeddings. To avoid data leakage, we utilized the lexical split to counteract the risk of the same MWE appearing in both the training and test sets. We applied the 10-fold cross-validation in every experiment and used statistical tests to measure the significance of the differences between the models. We used the independent samples t-test with the Bonferroni correction if its assumptions were fulfilled. Otherwise the non-parametric Mann-Whitney U test was applied.

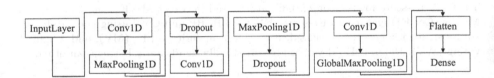

Fig. 3. Convolutional neural network classifier structure.

6 Results

Table 1 shows results averaged over ten folds for methods based on non-contextual and contextual representations. The expanded contextual knowledge resulted in significant improvements in the prediction quality of each classifier. The increase in the macro F1-score measure caused by the use of the contextual embeddings in comparison to non-contextual ones is presented in Fig. 4. The highest gain of 15% can be observed for the CNN model, as it was able to extract the most knowledge from the HerBERT embeddings due to its highest complexity.

Figure 5 shows the performance improvement in the case of incorrect MWEs. In the case of the RF classifier, the use of contextual embeddings resulted in more than sixfold and, in the case of CNN – over thirtyfold improvement in detection of incorrect expressions.

Table 1. F1-score values for incorrect MWEs (Inc F1), correct MWEs (Cor F1) and macro F1-score (F1) for non-contextual (N-C) and contextual (C) embeddings; models: LR, RF and CNN; values in **bold** are statistically significantly better in a given pair.

Model	Embedding	Inc F1	Cor F1	F1
LR	N-C	0.31	0.82	0.56
	C	**0.32**	**0.95**	**0.64**
RF	N-C	0.05	0.92	0.49
	C	**0.30**	**0.94**	**0.62**
CNN	N-C	0.01	0.96	0.48
	C	**0.31**	0.96	**0.63**

The performance of all voting strategies combined with each classifier is shown in Table 2. The use of weighted voting improved the value of the macro F1-score for the RF and CNN models by 2 and 4% points, respectively, in relation to the results for occurrence classification. Moreover, the F1-score measure for incorrect MWEs increased by 6% points for the CNN classifier. The improvement in evaluation performance may reflect the effect of using weighted voting to counteract the overfitting of more complex models, as this strategy benefits the most from the assumed monosemous nature of MWEs.

Table 2. F1-score for the contextual dataset and incorrect MWEs (Inc F1), correct MWEs (Cor F1) and macro F1-score (F1) for LR, RF and CNN using three different voting strategies: occurrence classification (OC), majority voting (MV), and weighted voting (WV). Dataset. **Bold** values are statistically significantly better than others.

Model	Voting	Inc F1	Cor F1	F1
LR	OC	0.32	0.95	0.64
	MV	**0.33**	0.95	0.64
	WV	**0.33**	0.95	0.64
RF	OC	0.30	0.94	0.62
	MV	**0.33**	**0.95**	**0.64**
	WV	**0.33**	**0.95**	**0.64**
CNN	OC	0.31	0.96	0.63
	MV	0.36	0.96	0.66
	WV	**0.37**	**0.97**	**0.67**

The evaluation results of different SMOTE methods used to counteract the class imbalance in the contextual representation samples, Sect. 3.2, are in Table 3. The use of SVM-SMOTE and Borderline SMOTE methods improved F1-score of the CNN model for incorrect MWEs by 14%. It also improved the overall F1-score by 5%. The CNN model was able to extract the most knowledge from the synthetic samples generated by the SMOTE methods due to the fact that it has the most complex architecture among all used classifiers.

Table 3. F1-score values for incorrect MWEs (Inc F1), correct MWEs (Cor F1), and macro F1-score (F1) LR, RF and CNN models trained on contextual embeddings based on the KGR10-based dataset with the use of four different SMOTE techniques: SMOTE, SVM-SMOTE, Borderline SMOTE, and ADASYN and no SMOTE (None). Values in **bold** are statistically significantly better than others.

SMOTE method	LR			RF			CNN		
	Inc F1	Cor F1	F1	Inc F1	Cor F1	F1	Inc F1	Cor F1	F1
None	0.30	0.94	0.62	0.30	0.94	0.62	0.17	**0.98**	0.58
SMOTE	0.31	**0.95**	0.63	0.30	0.94	0.62	0.27	0.93	0.60
SVM-SMOTE	**0.32**	**0.95**	**0.64**	0.30	**0.95**	0.62	**0.31**	0.95	**0.63**
Borderline SMOTE	0.31	**0.95**	0.63	0.30	0.94	0.62	**0.31**	0.96	**0.63**
ADASYN	0.31	**0.95**	0.63	0.30	0.94	0.62	0.30	0.94	0.62

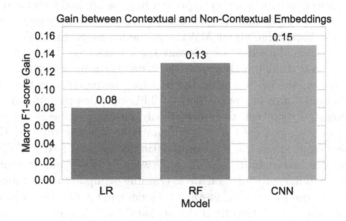

Fig. 4. F1-score improvement for the contextual vs non-contextual representations.

7 Discussion

One of the most important advantages of our method based on contextual representations is its ability to transform any text collection into a dataset, even if it has no annotations. We can leverage a MWE annotated corpus, but also any text collection, e.g. from web scraping. A seed MWE lexicon for a given language is enough. Time-consuming and expensive corpus annotation is avoided. Moreover, it is easier to maintain high quality in a collection such as a lexicon, which can be annotated by several linguists, and metrics such as inter-annotator agreement

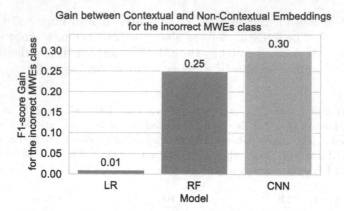

Fig. 5. F1-score increase for the incorrect MWEs class between the evaluation results for models trained contextual MWEs embeddings and the non-contextual ones.

can be easily calculated. Such a transformation of lexicon-based knowledge into a dataset enables the use of deep neural network models requiring a large number of training samples. Several linguistic resources can be also merged – both annotated texts, as well as lexicons. Our approach may be applied to texts in different languages, both to obtain multilingual collections and to apply transfer learning to facilitate the knowledge about MWEs in one language to MWE recognition in another language. This may be relevant for low-resource languages. Another advantage of contextual representation is faster training and prediction compared to sequence labeling methods. In our case, the model gets the full sample representation only once before prediction. This shortens the inference time.

Non-contextual representations based only on word embeddings result in a smaller dataset with less noise and significantly reduce the training time. This approach also emphasizes the non-compositional nature of the MWEs, as the model focuses on the semantic differences between an MWE and its components.

Providing a full representation of a training sample to the model in one step enabled the use of SMOTE methods. Generating synthetic samples carries the risk of too much deviation from the actual data. This phenomenon has the greatest impact on sequence labeling methods that are vulnerable to outliers.

Our CNN method, pre-trained on contextual embeddings with weighted voting, applied to MWE recognition in the Polish part of the PARSEME corpus (mostly verbal) achieved significantly better results than the best results reported during Edition 1.2 of the PARSEME Shared Task [31] – our RF classifier using weighted voting scored 0.5244 on the macro F1-score measure, while the best result for PARSEME Edition 1.2 is 0.4344 macro F1-score, which indicates a promising potential of our method. It is worth to emphasise that there is no overlap between the training set of our method and the PARSEME set of MWEs.

8   Conclusions and Future Work

Context plays a crucial role in MWE detection. Our three classifiers achieved significantly better results, with the CNN one on the top, with contextual embeddings than with the non-contextual ones. The context provided additional information on the MWE semantics, which improved the quality of the predictions. This is related to the non-compositional nature of the MWEs, the meaning of which cannot be inferred from the meanings of their components. The noncontextual representation forced the models to focus only on the nonstructural aspect meanings of the component meanings, but significantly reduced the training time. It may be more applicable in practice, when the training time and inference time are more important than the quality of prediction. This method is also faster to prepare as it requires no corpus data. On the other hand, the method based on contextual embeddings allows transforming any set of texts with the use of dictionary knowledge into an annotated corpus containing occurrences of the MWEs and their components. The model, by examining the semantic differences between the component and the entire expression, takes into account the variability of the context, which should allow for the extraction of the MWE meaning following the assumption of its monosemous character.

The use of SMOTE methods was possible, because, in our setup, the model receives full data about the training sample in one step. The use of sequential methods with synthetic data generated by the SMOTE methods would carry too high a risk of overfitting the model due to the noise caused by synthetic fragments of the training sequences, potentially very different from the original data. In our approach, the generated synthetic data significantly improved the effectiveness of recognition of incorrect cases. In future work, we want to apply our methods in the multilingual MWEs detection, and to explore the transfer learning mechanism in a language-independent MWE detection.

Acknowledgements. This work was financed by the National Science Centre, Poland, project no. 2019/33/B/HS2/02814.

References

1. Abadi, M., Agarwal, A., Barham, P., et al.: TensorFlow: large-scale machine learning on heterogeneous systems (2015). https://www.tensorflow.org/
2. Agrawal, S., Sanyal, R., Sanyal, S.: Hybrid method for automatic extraction of multiword expressions. Int. J. Eng. Technol. **7**, 33 (2018)
3. Anke, L.E., Schockaert, S., Wanner, L.: Collocation classification with unsupervised relation vectors. In: Proceedings of the 57th Annual Meeting of the ACL (2019)
4. Berk, G., Erden, B., Güngör, T.: Deep-BGT at PARSEME shared task 2018: bidirectional LSTM-CRF model for verbal multiword expression identification. In: Proceedings of the Joint Workshop on Linguistic Annotation, Multiword Expressions and Constructions (LAW-MWE-CxG-2018), pp. 248–253. ACL (2018)
5. Bojanowski, P., Grave, E., Joulin, A., et al.: Enriching word vectors with subword information. Trans. ACL **5**, 135–146 (2017)

6. Boros, T., Burtica, R.: GBD-NER at PARSEME shared task 2018: Multi-word expression detection using bidirectional long-short-term memory networks and graph-based decoding. In: Proceedings of the Joint Work. on Linguistic Annotation, Multiword Expressions and Constructions (LAW-MWE-CxG-2018), pp. 254–260. ACL (2018)
7. Broda, B., Derwojedowa, M., Piasecki, M.: Recognition of structured collocations in an inflective language. Syst. Sci. **34**(4), 27–36 (2008)
8. Buljan, M., Šnajder, J.: Combining linguistic features for the detection of Croatian multiword expressions. In: Proc. of the 13th Workshop on Multiword Expressions (MWE 2017), pp. 194–199. ACL (2017)
9. Chakraborty, S., Cougias, D., Piliero, S.: Identification of multiword expressions using transformers (2020)
10. Chawla, N.V., Bowyer, K.W., Hall, L.O., et al.: Smote: synthetic minority over-sampling technique. J. Artif. Intell. Res. **16**, 321–357 (2002)
11. Dziob, A., Piasecki, M., Rudnicka, E.K.: plwordnet 4.1 - a linguistically motivated, corpus-based bilingual resource. In: Proceedings of the Tenth Global Wordnet Conference: 23–27 July 2019, Wrocław (Poland), pp. 353–362 (2019)
12. Evert, S.: The Statistics of Word Cooccurrences: Word Pairs and Collocations. Ph.D. thesis, Institut für maschinelle Sprachverarbeitung, Univ. of Stuttgart (2004)
13. Fu, R., Guo, J., et al.: Learning semantic hierarchies via word embeddings. In: Proceedings of the 52nd Annual Meeting of the Association for Computational Linguistics, pp. 1199–1209. Baltimore, Maryland (2014)
14. Green, S., de Marneffe, M.C., Manning, C.D.: Parsing models for identifying multiword expressions. Comput. Linguist. **39**(1), 195–227 (2013)
15. Han, H., Wang, W.Y., Mao, B.H.: Borderline-smote: a new over-sampling method in imbalanced data sets learning. In: International Conference on Intelligent Computing, pp. 878–887 (2005)
16. Handler, A., Denny, M., Wallach, H., et al.: Bag of what? simple noun phrase extraction for text analysis. In: NLP+CSS@EMNLP (2016)
17. He, H., Bai, Y., Garcia, E.A., Li, S.: Adasyn: adaptive synthetic sampling approach for imbalanced learning. In: 2008 IEEE International Joint Conference on Neural Networks, pp. 1322–1328. IEEE (2008)
18. Hosseini, M.J., Smith, N.A., Lee, S.I.: UW-CSE at SemEval-2016 task 10: detecting multiword expressions and supersenses using double-chained conditional random fields. In: Proceedings of the 10th International Workshop on Semantic Evaluation (SemEval-2016), pp. 931–936. ACL (2016)
19. Klyueva, N., Doucet, A., Straka, M.: Neural networks for multi-word expression detection. In: Proceedings of the 13th Workshop on Multiword Expressions (MWE 2017), pp. 60–65. ACL (2017)
20. Kocoń, J., Gawor, M.: Evaluating kgr10 polish word embeddings in the recognition of temporal expressions using bilstm-crf. Schedae Informaticae **27** (2018)
21. Kurfalı, M.: TRAVIS at PARSEME shared task 2020: how good is (m)BERT at seeing the unseen? In: Proceedings of the Joint Workshop on Multiword Expressions and Electronic Lexicons, pp. 136–141. ACL (2020)
22. Levy, O., Remus, S., et al.: Do supervised distributional methods really learn lexical inference relations? In: Proceedings of the 2015 Conference of the North American Chapter of ACL: Human Language Technologies, pp. 970–976 (2015)
23. Maziarz, M., Szpakowicz, S., Piasecki, M.: A procedural definition of multi-word lexical units. In: Mitkov, R., Angelova, G., Boncheva, K. (eds.) Proceedings of the International Conference Recent Advances in Natural Language Processing - RANLP'2015, pp. 427–435. INCOMA Ltd. (2015)

24. Mikolov, T., Chen, K., Corrado, G., et al.: Efficient estimation of word representations in vector space. In: Bengio, Y., LeCun, Y. (eds.) 1st International Conference on Learning Representations, ICLR 2013, pp. 2–4. Arizona, USA, May, Scottsdale (2013)
25. Mroczkowski, R., Rybak, P., Wróblewska, A., et al.: HerBERT: efficiently pretrained transformer-based language model for Polish. In: Proceedings of the 8th Workshop on Balto-Slavic Natural Language Processing, pp. 1–10. ACL (2021)
26. Pedregosa, F., Varoquaux, G., Gramfort, A., et al.: Scikit-learn: machine learning in python. J. Mach. Learn. Res. **12**, 2825–2830 (2011)
27. Pečina, P.: Lexical association measures and collocation extraction. Lang. Res. Eval. **44**, 137–158 (2010)
28. Piasecki, M., Wendelberger, M., Maziarz, M.: Extraction of the multi-word lexical units in the perspective of the wordnet expansion. In: Proceedings of the International Conference Recent Advances in Natural Language Processing (2015)
29. Ramisch, C.: Multiword Expressions Acquisition. TANLP, Springer, Cham (2015). https://doi.org/10.1007/978-3-319-09207-2
30. Ramisch, C., et al.: Edition 1.1 of the PARSEME shared task on automatic identification of verbal multiword expressions. In: Proceedings of the Joint Workshop on Linguistic Annotation, Multiword Expressions and Constructions (2018)
31. Ramisch, C., Savary, A., Guillaume, B., et al.: Edition 1.2 of the PARSEME shared task on semi-supervised identification of verbal multiword expressions. In: Proceedings of the Joint Workshop on Multiword Expressions and Electronic Lexicons (2020)
32. Ramshaw, L., Marcus, M.: Text chunking using transformation-based learning. In: Third Workshop on Very Large Corpora (1995)
33. Rohanian, O., Taslimipoor, S., Kouchaki, S., et al.: Bridging the gap: attending to discontinuity in identification of multiword expressions. In: Proceedings of the 2019 Conference of the North American Chapter of the ACL: Human Language Technologies, Volume 1 (Long and Short Papers), pp. 2692–2698. ACL (2019)
34. Saied, H.A., Candito, M., Constant, M.: Comparing linear and neural models for competitive MWE identification. In: Proceedings of the 22nd Nordic Conference on Computational Linguistics, pp. 86–96. Linköping University Electronic Press (2019)
35. Scholivet, M., Ramisch, C.: Identification of ambiguous multiword expressions using sequence models and lexical resources. In: Proceedings of the 13th Workshop on Multiword Expressions (MWE 2017), pp. 167–175 (2017)
36. Seretan, V.: Syntax-Based Collocation Extraction, Text, Speech and Language Technology, vol. 44. Springer, Netherlands (2011). https://doi.org/10.1007/978-94-007-0134-2
37. Spasić, I., Owen, D., Knight, D., et al.: Unsupervised multi-word term recognition in Welsh. In: Proceedings of the Celtic Language Technology Workshop, pp. 1–6 (2019)
38. Taslimipoor, S., Bahaadini, S., Kochmar, E.: MTLB-STRUCT @Parseme 2020: capturing unseen multiword expressions using multi-task learning and pre-trained masked language models. In: Proceedings of the Joint Workshop on Multiword Expressions and Electronic Lexicons, pp. 142–148. ACL (2020)
39. Yirmibeşoğlu, Z., Güngör, T.: ERMI at PARSEME shared task 2020: embedding-rich multiword expression identification. In: Proceedings of the Joint Workshop on Multiword Expressions and Electronic Lexicons, pp. 130–135. ACL (2020)

Out-of-Distribution Detection in High-Dimensional Data Using Mahalanobis Distance - Critical Analysis

Henryk Maciejewski[✉] [ID], Tomasz Walkowiak [ID], and Kamil Szyc [ID]

Wroclaw University of Science and Technology, Wrocław, Poland
{henryk.maciejewski,tomasz.walkowiak,kamil.szyc}@pwr.edu.pl

Abstract. Convolutional neural networks used in real-world recognition must be able to detect inputs that are Out-of-Distribution (OoD) with respect to the known or training data. A popular, simple method is to detect OoD inputs using confidence scores based on the Mahalanobis distance from known data. However, this procedure involves estimating the multivariate normal (MVN) density of high dimensional data using the insufficient number of observations (e.g., the dimensionality of features at the last two layers in the ResNet-101 model are 2048 and 1024, with ca. 1000–5000 examples per class for density estimation). In this work, we analyze the instability of parametric estimates of MVN density in high dimensionality and analyze the impact of this on the performance of Mahalanobis distance-based OoD detection. We show that this effect makes Mahalanobis distance-based methods ineffective for near OoD data. We show that the minimum distance from known data beyond which outliers are detectable depends on the dimensionality and number of training samples and decreases with the growing size of the training dataset. We also analyzed the performance of modifications of the Mahalanobis distance method used to minimize density fitting errors, such as using a common covariance matrix for all classes or diagonal covariance matrices. On OoD benchmarks (on CIFAR-10, CIFAR-100, SVHN, and Noise datasets), using representations from the DenseNet or ResNet models, we show that none of these methods should be considered universally superior.

Keywords: Out-of-distribution detection · Mahalanobis distance · Convolutional neural networks

1 Introduction

Machine learning systems used in real-world recognition tasks need to classify inputs far from the known or training data as unrecognized or Out-of-Distribution (OoD). This is important in image or text recognition, where it is infeasible to train models for all categories encountered in open-world recognition. Recognition of OoD samples is vital in safety-critical applications or

D. Groen et al. (Eds.): ICCS 2022, LNCS 13350, pp. 262–275, 2022.
https://doi.org/10.1007/978-3-031-08751-6_19

incremental-learning systems [4,5,13,19]. However, popular models for image or text classification, e.g., ResNet, DenseNet, EfficientNet, are still vulnerable to OoD or adversarial examples that are easily recognized by humans [3,9,14,20]. This is despite high classification accuracy realized on the benchmark datasets (e.g., top-1 accuracy on the ImageNet is ca. 90 [15]).

Many current approaches recognize OoD inputs using confidence scores obtained from class-conditional posterior distributions. A popular method, due to its simplicity, is to use multivariate Gaussian distributions as models of class-conditional distributions [1,11,16,18]. This approach leads to estimating the uncertainty of prediction using Mahalanobis distance.

However, these procedures rely on the estimation of probability density in high-dimensional data. The dimensionality of the representations generated by CNNs used for image classification is usually ca 10^3. E.g., the dimensionality of features at the last layer of the ResNet-101 [7] is 2048, and of the EfficientNet-B3 is 1536. Class conditional distributions are estimated from training data, typically based on an insufficient number of examples, e.g., using 5000 observations per class in the CIFAR-10 dataset. The purpose of this work is the analysis of the quality of such parametric density estimates in high-dimensional data and the impact of errors in density estimation on the performance of the Mahalanobis distance-based OoD detection.

Our contributions are the following.

- We analyzed the instability of estimated densities in high-dimensional data. We showed, using simulated data, that the generative MVN models fitted to the training data are far from the testing samples from the same distribution. Hence, OoD detection based on this model will tend to reject testing samples as outliers. We analyzed this effect as a function of dimensionality and training sample size.
- We analyzed the limitations of Mahalanobis distance-based OoD detection: we showed that due to the model estimation error, near OoD samples are not distinguishable from known data. The minimum distance from known data beyond which outliers are detectable depends on the dimensionality of the features and the training sample size and decreases for larger training samples.
- We analyzed simple modifications of the method used to reduce the impact of model fitting errors: Mahalanobis distance using one covariance matrix shared by all classes of known data, or using diagonal covariance matrices. We illustrate the performance of these methods on OoD benchmarks, with the CIFAR-10 as in-distribution and the CIFAR-100, the SVHN, and the Noise datasets as OoD, and with features generated by different CNN models. We showed that none of these Mahalanobis distance-based methods should be declared universally best, as the performance depends on the characteristics of benchmark datasets. On some benchmarks, Mahalanobis distance-based OoD detectors are outperformed by simple methods, which use the Euclidean or standardized Euclidean distance.

2   OoD Detection with Mahalanobis Distance - Simulation Study

2.1   Method - Using Mahalanobis Distance for OoD Detection

Using Mahalanobis distance as the score for OoD detection relies on the estimation of multivariate Gaussian (MVN) distribution as a model of class-conditional posterior distribution. Here we briefly summarize the method. Given the known (in-distribution) dataset $X_c \subset R^d$ for the class $c \in C = \{1, 2, \ldots, m\}$, with N_c examples, we estimate the model $\mathcal{N}(\mu_c, \Sigma_c)$ with the mean vector $\mu_c = \frac{1}{N_c} \sum_{x \in X_c} x$ and the covariance matrix $\Sigma_c = \frac{1}{N_c} \sum_{x \in X_c} (x - \mu_c)(x - \mu_c)^\top$.

Given a test sample u, Mahalanobis distance to the MVN model of class c is computed as

$$d_{Mah,c}(u) = \sqrt{(u - \mu_c)^\top \Sigma_c^{-1}(u - \mu_c)}. \tag{1}$$

The confidence score used to label the sample u as in-distribution or OoD is calculated as $s(u) = -\min_{c \in C} d_{Mah,c}(u)$.

To minimize errors due to unreliable estimation of Σ_c in high dimensional data, some works (e.g., [11,16]) assume that all m classes share the common covariance matrix, estimated from the larger sample of size $N = \sum_c N_c$ as $\Sigma = \frac{1}{N} \sum_{c \in C} \sum_{x \in X_c} (x - \mu_c)(x - \mu_c)^\top$.

The Mahalanobis distance of a test sample u to class c is then computed as

$$d_{MahUF,c}(u) = \sqrt{(u - \mu_c)^\top \Sigma^{-1}(u - \mu_c)}. \tag{2}$$

Other modifications/simplifications of this procedure assume that the covariance matrix $\Sigma_c = V_c$ is the diagonal matrix with diagonal components calculated as variances of features computed over samples in X_c. Then the distance of a test sample u to the MVN model of class c is calculated as the standardized Euclidean distance:

$$d_{SEuc,c}(u) = \sqrt{(u - \mu_c)^\top V_c^{-1}(u - \mu_c)}. \tag{3}$$

Finally, the distance of a sample u to the model of class c can be computed as the Euclidean distance $d_{Euc,c}(u) = |u - \mu_c|_2$, (which implies that all variances in V_c in Eq. 3 are equal 1).

2.2   Non-robust Estimation of MVN Model in High Dimensional Data

We performed a simulation study in which we analyzed the instability of the MVN model of in-distribution data as a function of dimensionality and sample size. We generated n training and n testing observations from the MVN distribution in d dimensions, with the mean at $[0]_d$ and with uncorrelated variables with variance 1. We estimated the MVN model from the training sample and compared the distances of the training and testing samples from the model.

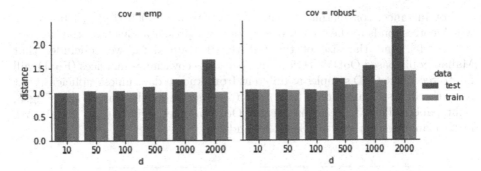

Fig. 1. Mean mahalanobis distances of train and test samples to the MVN model fitted to the train samples. Train and test samples drawn from the same MVN. Number of samples $n = 5000$. Note: distance shown here is the squared mahalanobis distance divided by d.

Results as a function of dimensionality d, for fixed sample size $n = 5000$ are shown in Fig. 1. We used the scikit learn library MLE estimator of covariance (referred to as *empirical*), and the Minimum Covariance Determinant estimator (MCD) [17], referred to as *robust* due to its resistance to outliers. We observe that when the dimensionality of data grows, the test data tend to lie significantly further off the model than the train data. Since this effect is more prominent with the robust estimator, we conclude that the robust estimator is not appropriate for high-dimensional data. Note that this observation holds even if $n > 5d$, a condition deemed to guarantee a low error of the MCD estimator.

In Fig. 2 and 3 we analyze the effect of the growing sample size. This analysis can be used to determine, for a given dimensionality of features d, the required size of training data to guarantee a robust model of known data.

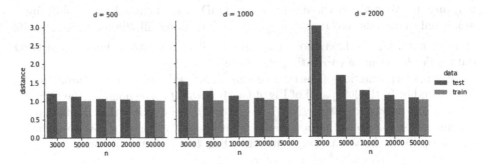

Fig. 2. Mean mahalanobis distances of train and test samples to the MVN model fitted to the train samples, as a function of the size of samples n and dimensionality of data d. Note: distance is the squared Mahalanobis distance divided by d.

For instance, considering the case $d = 2000$, $n = 5000$ (Fig. 3, left panel), which corresponds to the dimensionality of the representations from the ResNet-101 model, and the size of the CIFAR-10 train data, we conclude, that Mahalanobis-based OoD detection with per-class covariance matrices (Eq. 1) will fail to recognize OoD samples as different from known data unless sufficiently far from the in-distribution data ($d_{Mah,c} > 62$). Increasing the sample size (Fig. 3, right panel) allows to recognize nearer OoD samples (with $d_{Mah,c} > 50$). We further analyze this effect in Sects. 2.3 and 3.4.

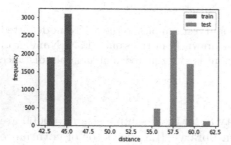 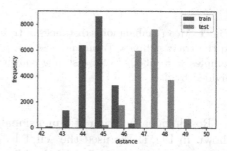

Fig. 3. Distribution of mahalanobis distance of train and test samples to the MVN model fitted to the train data, dimensionality $d = 2000$, sample size $n = 5000$ (left), $n = 20000$ (right)

2.3 Non-robust MVN Model Used for OoD Detection

In the second experiment, we compare the Mahalanobis distance of in-distribution test data and OoD data to the MVN model fitted to the train data. We model known data as in Sect. 2.2, and OoD data as MVN with mean μ shifted from $[0]_d$ by r, i.e. $|\mu - [0]_d| = r$, and with uncorrelated variables with variance 1. We realized three schemes of OoD data, denoted *ood1*: shift by r along only one axis; *ood3*: $\mu = [\frac{r}{\sqrt{d}}]_d$, i.e. shift along all the axes; *ood2*: shift along $\frac{d}{2}$ axes. (As we later show, the scheme effect is visible if known and OoD data differ in terms of correlation structure).

Results as a function of the sample size n and OoD shift r are summarized in Figs. 4 and 5. In the left panel of Fig. 4 (with n and d corresponding the CIFAR-10 training data and the ResNet-101 features), we observe that Mahalanobis distance is unable to distinguish in-distribution test and OoD data. To quantify the dissimilarity between groups shown in Fig. 4, we use the measure

$$\Delta(group1, group2) = \frac{\bar{X}_1 - \bar{X}_2}{\sqrt{s_1^2 + s_2^2}}, \tag{4}$$

where $\bar{X}_i$ and s_i are the mean and its standard error in group i. Note that this measure is used as the test statistic in Welch's t-test. We observe that with growing n, $\Delta(test, train)$ decreases, and $\Delta(ood, test)$ increases. Hence, with increasing sample size, the model stabilizes and leads to better separation between in- and OoD data.

In Fig. 5 we analyze the effect of shift r on the separability of in- and OoD data. We observe that for sufficiently far OoD data (e.g., $r = 32$), the inaccuracy in MVN model estimation no longer matters: OoD samples are significantly more distant from the model than the test in-distribution samples. We argue that this effect accounts for the success or failure of the Mahalanobis distance-based OoD detection in CNN benchmarks, as further analyzed in Sect. 3.4.

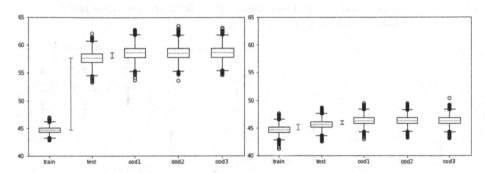

Fig. 4. Distribution of mahalanobis distance of train, test and OoD samples to the MVN model fitted to the train data, dimensionality $d = 2000$, OoD shift $r = 8$, sample size $n = 5000$ (left), $n = 50000$ (right). Dissimilarity between distances: $\Delta(test, train) =$ 708 (left), 204 (right); $\Delta(ood1, test) = 39$ (left), 154 (right). Large training samples lead to more robust models of in-distribution data (difference between train and test data decreases), and better separability of in- and OoD data (difference between test and ood increases).

Finally, in Fig. 6, we signal the effect of feature correlation on OoD performance. We assume correlated in-distribution and uncorrelated OoD data. We observe that if the correlation schemes of in-distribution and OoD data differ, the Mahalanobis distance-based separability of OoD and known data improves, and distances to the model of $ood1, 2, 3$ schemes become significantly different, hence in this case, the distance from the model depends on the direction of OoD shift.

2.4   Mahalanobis Distance-based vs. Nonparametric Outlierness Factor-based OoD in High Dimensional Data

Since the estimation of density in high-dimensional data is generally considered unattractive [6], we want to empirically show that the confidence scores obtained from density estimates lead to the limited performance of OoD detection. On the

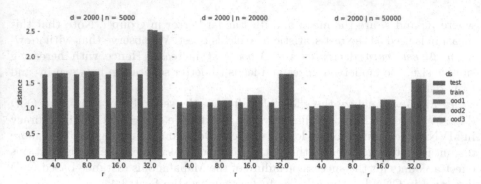

Fig. 5. Comparing mahalanobis distances of train, test and OoD samples, as a function of sample size n, OoD shift r, for dimensionality $d = 2000$. Note: distance is the squared Mahalanobis distance divided by d.

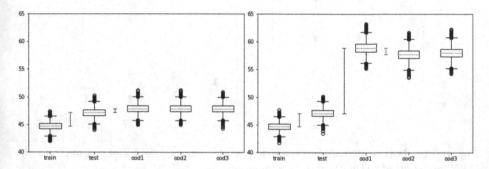

Fig. 6. Effect of correlated features: distribution of mahalanobis distance of train, test and OoD samples to the MVN model, for dimensionality $d = 2000$, OoD shift $r = 8$, sample size $n = 20000$, uncorrelated features (left); 1000 features correlated with coefficient 0.5. (right). Dissimilarity between distances: $\Delta(test, train) = 330$ (left and right); $\Delta(ood1, test) = 95$ (left), 1278 (right). If in- and OoD data differ in correlation structure, separability of OoD and in-distribution data improves. Note that with correlated data, distance of different outlier groups $ood1, 2, 3$ to the model is significantly different, e.g., $\Delta(ood1, ood2) = 122$ (right), whereas in previous examples (left panel and Fig. 5) differences between ood schemes were not significant.

other hand, outlier or out-of-distribution detection in high-dimensional data can be performed reasonably well using scores obtained from the Local Outlierness Factor (LOF) method [2] (see Sect. 3.3 for technical details of LOF).

We performed a simple simulation study in which we compared the performance of OoD detection (see Sect. 2.5 for technical details of the used metrics) based on confidence scores obtained using the Mahalanobis distance vs confidence scores obtained with the LOF algorithm. As in-distribution (known) data, we generated two clusters from the MVN distribution in d dimensions, with the mean at $[0]_d$ and $[-1]_d$ and uncorrelated variables with variance 1. As OoD, we used a cluster with mean at $[\frac{r}{\sqrt{d}}]_d$, uncorrelated, with variance 1.

Confidence scores were calculated as the $MahUF$ distance (see Eq. 2) between a test sample and the closest class conditional Gaussian distribution, which can be interpreted as the log of the probability density of the test sample. In the alternative approach, confidence scores were obtained as local outlierness factors (LOF) calculated for test samples with respect to the closest cluster of known data.

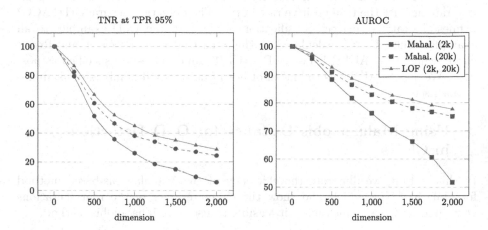

Fig. 7. Comparison of the averaged TNR at TPR 95 % and the AUROC for $MahUF$ and LOF on simulation data. The inlier set consists of two classes, simulated by MVN with variance 1 and distance 1. The outlier is also simulated by MVN moved from the closest inlier class by a given distance (r=8). The number of training (inlier) examples was set to 2k (solid) and 20k (dashed line), with 2k for test inliers and outliers. The results for 2k and 20k for LOF are undistinguished. By increasing the complexity of the problem, expanding the input data dimension, the LOF method is much more stable and achieves better results. The dimension above 2000 is common in the last layers of CNNs. Moreover, LOF is less prone to the changes of the training set size.

We observe that for $d = 2000$ with 1000 training points per class (there are two classes), the Mahalanobis procedure no longer detects outliers (AUROC $\approx$ 50%), while LOF is more reliable (AUROC $> 80\%$). Where for 10k training points per class, the Mahalanobis procedure gives results closer to the LOF ones. It shows how the number of examples is important for the Mahalanobis approach.

2.5 Evaluation Metrics

In the evaluation of OoD performance, we follow the approach used in [8] where the outlier detection is considered a binary classification. The outliers are defined as the positive class and the closed set examples (test set) as the negative class. The confidence score allows the binary classification. In the result presentation, we used the standard metrics: TNR at TPR 95%, AUROC, DTACC, and

AUPR - the higher the values of all metrics, the better the OoD detection is. The True Negative Rate at 95% True Positive Rate (TNR at TPR 95%) can be interpreted as the probability of correctly classifying the Out-of-Distribution examples when the In-Distribution (test) samples are classified as high as 95%. The Area Under Receiver Operating Characteristic curve (AUROC) defines the OoD method's ability to discriminate between cases (test examples) and non-cases (OoD examples). It can be calculated by the area under the false positive rate against the true positive rate curve. The detection accuracy (DTACC) defines the ratio of correct classification of the test and OoD examples to all examples. The AUPR is calculated by the Area Under the Precision and Recall curve, where test (AUPR In) or OoD (AUPR Out) images are specified as positive. We denote AUPR as the mean of both due to equal numbers of examples in both sets.

3   Using Mahalanobis Distance for OoD Detection in CNNs

In this section, we illustrate the efficiency of the Mahalanobis-based method for OoD in CNN models and show the characteristics of the representations generated by CNNs which make it feasible to use the Mahalanobis method.

3.1   Datasets and CNN Models

We used popular benchmark datasets successfully used in OoD detection in computer vision: the CIFAR-10, the CIFAR-100, the SVHN, and the Noise. The CIFAR-10[1] dataset contains $60,000$ 32×32 color images divided into 10 classes: airplane, automobile, bird, cat, deer, dog, frog, horse, ship, and truck. There are $5,000$ images per class in the training set and $1,000$ in the test set. CIFAR-100 is similar - there are 100 classes (disjoint from the CIFAR-10 classes) with 500 and 100 images per class, respectively, train and test subsets. SVHN contains real-world images of Street View House Numbers from Google Street View - they are easily distinguishable for humans compared to CIFARs ones. The Noise dataset consists of randomly generated images - a theoretically straightforward recognition task. To evaluate OoD methods, we used the testing partitions of the in-distribution datasets and the given Out-of-Distribution dataset, with a 1:1 proportion of known and unknown samples.

We trained two models, ResNet-101 [7] trained on CIFAR-10 (achieving 94.75% accuracy) and DenseNet-169 [10] trained on CIFAR-100 (achieving 74.04% accuracy). These model architectures were chosen due to their high popularity in commercial applications and OoD detection problems. We used the classic method of feature extraction from deep models, which uses vectors after applying the Global Average Pooling [12] on the last convolutional layer. Dimensionality of feature vectors is $2,048$ (for ResNet-101), and $1,664$ (for DenseNet-169). The procedure in our experiments is as follows: (1) train the model using

---

[1] https://www.cs.toronto.edu/~kriz/cifar.html.

training subset, (2) extract features from the model for images in the training subset, (3) fit the in-distribution model for OoD detection, (4) evaluate confidence scores (or distances) for in- and OoD test data.

3.2   Analysis of Mahalonobis Distances for the CIFAR-10

First, we analyze the characteristics of features generated by ResNet-101 for the CIFAR-10 train and test data and OoD data. We estimated the distribution of distances from the class centers (calculated on the train data sets) to different groups of data. Results for one of the CIFAR-10 classes are shown in Fig. 8. We can notice the large difference between train and test data in the case of Mah distance. It is the same phenomenon as discussed in Sect. 2.5, i.e., insufficient number of data. Moreover, one can see the relative shift of the noise set (black curves) position with respect to the SVHN and the CIFAR-100 (green and red) for distances with the full covariance matrix (Mah and $MahUF$) and with limited ($SEuc$) or non-existing (Euc) one. This suggests that the type of distance (still from the same family) may have a big influence on the OoD detection.

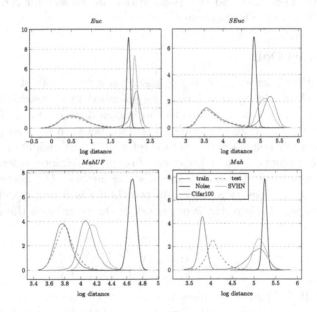

Fig. 8. Log distance distribution for different distance metrics (Euc, $SEuc$, $MahUF$, and Mah). Distances are from the center of one of the CIFAR-10 labels to train and test data for the same label, and also to outlier data sets (Noise, SVHN, and CIFAR-100). Plots represent the probability density function of log distances obtained by the gaussian kernel-density estimator with scott's bandwidths.

Table 1. Difference between train and test data distances to the model of one the CIFAR-10 class, and the difference averaged over all the CIFAR-10 classes. Differences between train and test distances are measured using Welch's t-test statistic - Eq. 4.

Distance	CIFAR-10 (ResNet-101)		CIFAR-100 (DenseNet-169)	
	Class 1	Avg over all classes	Class 1	Avg over all classes
Euc	5.11	8.86	6.02	5.66
$SEuc$	**4.95**	**8.72**	**5.77**	5.17
$MahUF$	6.63	8.81	6.41	**4.5**
Mah	39.36	40.12	14.41	15.03

Table 1 shows the Welch's t-test statistic comparing the distances of the train and test data for a selected CIFAR-10 class (the same class as used in Fig. 8), and averaged over all classes for each of the analyzed Mahalanobis distances. It can be seen that Mah gives the largest values of statistic t suggesting that the train and test population means are the most distant (the same conclusion as from Fig. 8). The train and test populations are closest for $SEucl$ but the differences to Euc and $SEuc$ are small.

3.3 LOF-based OoD

The Local Outlier Factor [2] (LOF) is based on an analysis of the local density of points. It works by calculating the so-called local reachability density $LRD_k(x, X)$ of input x with regard to the known dataset X. LRD is defined as an inverse of an average reachability distance between a given point, its k-neighbors, and their neighbors (for details refer to [2]). K-neighbors ($N_k(x, X)$) includes a set of points that lie in the circle of radius k-distance, where k-distance is the distance between the point, and it's the farthest k^{th} nearest neighbor ($||N_k(x, X)|| >= k$). The local outlier factor (LOF) is formally defined as the ratio of the average LRD of the k-neighbors of the point x to the LRD of the point.

$$d_{LOF}(u) = \frac{\sum_{x \in N_k(u,X)} LRD_k(x, X)}{||N_k(u, X)|| LRD_k(u, X)} \tag{5}$$

Intuitively, if the point is an inlier, the ratio of the average LRD of neighbors is similar to the LRD of the point. Therefore, the LOF is around 1. For outliers, it should be above 1 since the density of an outlier is smaller than its neighbor density.

3.4 OoD Experiments

In Table 2, we demonstrate the performance of the Mahalanobis based OoD detection using popular CNN architectures: ResNet-101 (trained on CIFAR-10) and DenseNet-169 (trained on the CIFAR-100). We used three outlier data sets:

Table 2. The comparison of analysed OoD methods for CIFAR-10 and CIFAR-100. Note that there is no best or worst method.

In-dist (Model)	OOD	Method	TNR at TPR 95%	AUROC	DTACC	AUPR
CIFAR-10 (ResNet-101)	Noise	Euc	97.12	98.81	96.21	98.56
		$SEuc$	45.92	94.57	94.71	91.24
		$MahUF$	100.00	100.00	100.00	100.00
		Mah	100.00	100.00	100.00	100.00
		LOF	100.00	99.90	99.30	99.89
	SVHN	Euc	55.03	93.00	86.75	92.52
		$SEuc$	13.55	86.59	82.57	83.85
		$MahUF$	53.84	91.13	83.26	90.96
		Mah	41.34	89.33	82.25	88.77
		LOF	57.80	93.00	86.43	92.73
	CIFAR-100	Euc	41.64	87.38	80.69	86.49
		$SEuc$	35.75	86.30	80.19	84.84
		$MahUF$	24.07	78.18	71.38	77.40
		Mah	37.59	85.91	78.55	85.16
		LOF	47.84	87.22	79.49	86.74
CIFAR-100 (DenseNet-169)	Noise	Euc	99.98	98.99	98.02	98.32
		$SEuc$	100.00	100.00	100.00	100.00
		$MahUF$	100.00	100.00	100.00	100.00
		Mah	100.00	100.00	100.00	100.00
		LOF	81.05	95.91	95.81	92.99
	SVHN	Euc	12.24	75.21	70.65	73.95
		$SEuc$	37.30	85.85	78.41	85.07
		$MahUF$	29.11	81.82	74.30	80.93
		Mah	19.37	81.31	75.73	79.74
		LOF	24.48	83.46	76.86	81.96
	CIFAR-10	Euc	15.48	73.29	68.39	71.19
		$SEuc$	9.30	69.77	66.40	68.07
		$MahUF$	8.05	65.65	62.22	63.97
		Mah	10.08	69.42	65.98	68.33
		LOF	13.51	73.36	68.28	71.71

Noise, SVHN, and CIFAR-100 (for ResNet-101) or CIFAR-10 (for DenseNet-169). We evalauetd of the four versions of Mahalonobis distances ($Euc, SEuc$, $MahUF$, Mah) presented in Sect. 2.1. LOF is shown as an alternative, non-parametric approach.

The Noise dataset is very well detected as OoD. However, there are some problems (TNR worse then 50%) in case of $SEuc$ for ResNet-101. The SVHN and CIFARs datasets are harder to be detected as OoD for DenseNet-169 than in the case of ResNet-101.

Comparison of OoD methods gives no straightforward conclusions. There is no best or worst method. We can find a failure scenario (when a method is much worse than the best one) for each of the analyzed methods and a situation when a given method significantly outperforms others. For example, Euc fails for DenseNet-169 and SVHN, and outperforms others for the CIFAR-10 and the same model, $SEuc$ fails for ResNet-101 and Noise, and outperforms for SVHN and DenseNet-169, $MahUF$ fails in CIFAR-10/DenseNet-169, and Mah in SVHN/DenseNet-169. Our results suggest that we should carefully state that the given method is the best since the results (OoD metric) strongly depend on data (CNN features), so not only on image data sets but also network architecture and the process of model training.

4   Conclusion

In this paper, we analyzed the performance of the Mahalanobis distance-based OoD detection method in high-dimensional data. This method is popular due to its simplicity, but it relies on parametric density estimates in high-dimensional data. We analyzed the instability of MVN estimates of density and showed that this issue leads to the intrinsic limitation of this method: near OoD samples are not distinguishable from known data. For fixed dimensionality of features and the size of training data, we can estimate the minimum distance from known data beyond which outliers are detectable. We showed that this distance decreases with the growing number of training samples.

We also analyzed common modifications of the method used to mitigate the density estimation errors: Mahalanobis distance with single covariance matrix shared by all classes in known data, or standardized Euclidean distance with diagonal covariance matrices. We compared the performance of these methods using OoD benchmarks with CIFAR-10 as in-distribution vs. CIFAR-100, SVHN, Noise as OoD, and CIFAR-100 vs. CIFAR-10, SVHN, and Noise datasets. We showed that none of these methods should be seen as universally superior, as the performance of OoD detectors depends on the benchmark dataset and the CNN model used to generate representations.

References

1. Bendale, A., Boult, T.: Towards Open World Recognition. In: Proceedings of the IEEE Conference on Computer Vision and Pattern Recognition, pp. 1893–1902 (2015)
2. Breunig, M.M., Kriegel, H.P., Ng, R.T., Sander, J.: Lof: identifying density-based local outliers. SIGMOD Rec. **29**(2), 93–104 (2000). https://doi.org/10.1145/335191.335388
3. Chakraborty, A., Alam, M., Dey, V., Chattopadhyay, A., Mukhopadhyay, D.: A survey on adversarial attacks and defences. CAAI Trans. Intell. Technol. **6**(1), 25–45 (2021). https://doi.org/10.1049/cit2.12028

4. Eykholt, K., et al.: Robust physical-world attacks on deep learning visual classification. In: 2018 IEEE Conference on Computer Vision and Pattern Recognition, CVPR 2018, Salt Lake City, UT, USA, 18–22 June 2018, pp. 1625–1634. IEEE Computer Society (2018). https://doi.org/10.1109/CVPR.2018.00175

5. Feng, D., Rosenbaum, L., Dietmayer, K.: Towards safe autonomous driving: capture uncertainty in the deep neural network for lidar 3d vehicle detection. In: 2018 21st International Conference on Intelligent Transportation Systems (ITSC), pp. 3266–3273. IEEE (2018)

6. Hastie, T., Tibshirani, R., Friedman, J.: The Elements of Statistical Learning: Data Mining, Inference, and Prediction. Springer, New York (2009). https://doi.org/10.1007/978-0-387-21606-5

7. He, K., Zhang, X., Ren, S., Sun, J.: Deep residual learning for image recognition. In: Proceedings of the IEEE Conference on Computer Vision and Pattern Recognition, pp. 770–778 (2016)

8. Hendrycks, D., Mazeika, M., Dietterich, T.: Deep anomaly detection with outlier exposure. Proceedings of the International Conference on Learning Representations (2019)

9. Hendrycks, D., Zhao, K., Basart, S., Steinhardt, J., Song, D.: Natural adversarial examples. arXiv preprint arXiv:1907.07174 (2019)

10. Huang, G., Liu, Z., Van Der Maaten, L., Weinberger, K.Q.: Densely connected convolutional networks. In: Proceedings of the IEEE Conference on Computer Vision and Pattern Recognition, pp. 4700–4708 (2017)

11. Lee, K., Lee, K., Lee, H., Shin, J.: A simple unified framework for detecting out-of-distribution samples and adversarial attacks. In: Proceedings of the 32nd International Conference on Neural Information Processing Systems, pp. 7167–7177. NIPS 2018, Curran Associates Inc., Red Hook, NY (2018)

12. Lin, M., Chen, Q., Yan, S.: Network in network. arXiv preprint arXiv:1312.4400 (2013)

13. McAllister, R., et al.: Concrete problems for autonomous vehicle safety: Advantages of bayesian deep learning. In: Proceedings of the 26th International Joint Conference on Artificial Intelligence, pp. 4745–4753. IJCAI 2017, AAAI Press (2017)

14. Nguyen, A., Yosinski, J., Clune, J.: Deep neural networks are easily fooled: high confidence predictions for unrecognizable images. In: Proceedings of the IEEE Conference on Computer Vision and Pattern Recognition, pp. 427–436 (2015)

15. Pham, H., Dai, Z., Xie, Q., Luong, M.T., Le, Q.V.: Meta pseudo labels. In: IEEE Conference on Computer Vision and Pattern Recognition (2021). https://arxiv.org/abs/2003.10580

16. Ren, J., Fort, S., Liu, J., Roy, A.G., Padhy, S., Lakshminarayanan, B.: A simple fix to mahalanobis distance for improving near-ood detection. arXiv preprint arXiv:2106.09022 (2021)

17. Rousseeuw, P.J.: Least median of squares regression. J. Am. Stat. Assoc. **79**(388), 871–880 (1984)

18. Sehwag, V., Chiang, M., Mittal, P.: Ssd: a unified framework for self-supervised outlier detection. In: International Conference on Learning Representations (2021). https://openreview.net/forum?id=v5gjXpmR8J

19. Sharif, M., Bhagavatula, S., Bauer, L., Reiter, M.K.: Accessorize to a crime: real and stealthy attacks on state-of-the-art face recognition. In: Proceedings of the 2016 ACM SIGSAC Conference on Computer and Communications Security, pp. 1528–1540 (2016)

20. Zhou, Z., Firestone, C.: Humans can decipher adversarial images. Nature Commun. **10**(1), 1–9 (2019)

Multi-contextual Recommender Using 3D Latent Factor Models and Online Tensor Decomposition

Basem Suleiman[1], Ali Anaissi[1(✉)], Muhammad Johan Alibasa[2], and Harrison Truong[1]

[1] School of Computer Science, University of Sydney, Sydney, Australia
{ali.anaissi,basem.suleiman,harrison.truong}@sydney.edu.au
[2] School of Computing, Telkom University, Bandung, Indonesia
alibasa@telkomuniversity.ac.id

Abstract. Traditional approaches to recommendation systems involve using collaborative filtering and content-based techniques which make use of the similarities between users and items respectively. Such approaches evolved to encapsulate model-based latent factor (LF) algorithms that use matrix decomposition to ingest a user-item matrix of ratings to generate recommendations. In this paper, we propose a novel approach based on 3D LF model and tensor decomposition method for devising personalized recommendations driven from additional contextual features. We also present our stacking method for a tensor generation prior to incorporating LF models. We validate our proposed personalized recommender using two real-world datasets. Our experimental results showed that additional contextual features can help to personalize recommendations while maintaining similar or better performance compared to conventional 2D LF methods. Furthermore, our results demonstrate the importance of the quality of the contextual features to be used in 3D LF models. In addition, our experiments show the effective performance of our new stacking method on computation time and accuracy of the proposed recommender.

1 Introduction

Data on interactions between users and products is a major resource that is increasingly collected and used by all online platforms today. The development of recommendation systems has been effective in assisting consumers in sifting through seemingly limitless options to more efficiently decide on products or services to consume [9]. Recommendation systems have proliferated throughout various industries, emerging in products and applications such as Netflix, Spotify, and Amazon with an abundance of literature released by both academia and industry [4]. As a result, there have been constant advancements to more accurately predict user preferences whilst optimizing for scalability and efficiency

[3, 22]. Since the Netflix Prize Competition in 2006 [13], there has been a substantial increase in the use of 2D Latent Factor models to generate recommendations based on each user's explicit ratings.

In this paper, we build upon the use of 2D latent factor models in recommendation systems. Specifically, we aim to create 3D LF models that are capable of generating more personalized recommendations by incorporating additional information in the form of contextual features whilst maintaining benchmarked performances in accuracy. Due to advancements in machine learning techniques, the availability of more data and processing power, higher rank-decomposition methods on higher-order matrices (i.e. tensors) are now possible, where n-rank tensors can be decomposed using methods such as CP decomposition and Tucker decomposition. In this paper, we introduce a 3D LF model and CP decomposition method to help produce personalized recommendations. Furthermore, we investigate the performance of our approach to determine whether incorporating a third dimension, referred to as context can result in similar or better accuracy compared to conventional 2D methods. As such, the contributions made by this paper are: (1) A novel approach based on 3D Latent Factor model and tensor decomposition method for devising personalized recommendations driven from additional contextual information. (2) A novel stacking method for tensor generation prior to incorporating LF models. (3) Empirical validation of the performance of our 3D LF models and its dependency on the contextual features chosen using two real-world datasets.

2 Related Work

In recommender systems, Content-Based Filtering (CB) method is based solely on how a user interacts with specific content and involves creating a user profile using past behaviour to classify content on whether a user would or would not consume the content [14]. Collaborative Filtering (CF), on the other hand, groups together users who interact similarly with content (e.g., provide the same ratings for the same movies) and generates recommendations based on content others in the same group have also interacted with [18]. A pure CB system does not require other user information compared to CF and thus can cater specifically to the niche interests and viewing behaviours of the user. However, it requires an extra amount of data on the content itself to compute similarities to make effective recommendations. CB recommenders also tend to overspecialize, resulting in recommendations that only cater for a user's historical preferences but do not facilitate the discovery of different content.

To overcome the above disadvantages, hybrid recommender approaches were proposed to provide a more optimal solution [6]. Hybrid systems can be effectively initiated by utilizing the user profile data gained from a CB system, thereby overcoming the cold start problem experienced by CF systems. With visual items, the continued collection of user usage information will also allow content to be filtered out (if users have watched a video already) and recommended (if they want to continue watching) [4]. Over-specialization of CB systems is also negated as the discovery of new content is facilitated by CF systems.

Other techniques for improving recommendation systems involve the use of 2D LF Latent Factor (LF) models. LF Models are a branch of collaborative filtering that ingest a user-item matrix of ratings to predict unknown ratings via dimensionality reduction. It was first proposed as a case study on a movie dataset and an e-commerce dataset [17]. The success of LF models was proven during the Netflix Prize competition [13]. Since then, this area of research has proliferated, with accuracy and performance continuously optimized by adjusting the methodologies that applied Singular Value Decomposition (SVD) to the user-item matrix. This included different SVD methods such as the development of Regularised SVD [13] and the use of neighboring computational techniques alongside SVD such as Boltzmann Machines with SVD [16].

Other studies investigated recommendation systems and tensor decomposition where it is suggested that adding context, or additional features, could improve the accuracy of model-based recommender systems [8,19]. For example, when watching movies, other contextual features such as the "time of day watched" or "whom you watch with i.e. with a group or solely with your partner" may contain valuable information that will affect the ratings given by a user. Further research builds on this idea of contextually-aware recommender systems and has proposed other techniques that ingest higher level matrices beyond the user-item matrix. This includes models that deal with explicit feedback using factorisation machines [15] and multi-layer tensor factorisation methods [20].

3 Methodology

Datasets: In this paper, two datasets will be used to address the research problem. The first dataset is the 100K MovieLens dataset, which describes ratings (ranging from 1 to 5), given by 943 users across 1682 movies. The data is a sparse matrix where not all users have rated all movies. The sparsity for this dataset is 6.3%. The second dataset is the Book Crossing dataset, which describes ratings from 1 to 10, given by 278,858 users across 271,379 books with 1,149,780 ratings that are both implicit and explicit. The rating data becomes a user-item matrix with a sparsity of 0.00152%. The dataset was cleaned by removing users from the dataset that had missing/incorrect age and year of publication values.

To overcome the scalability issues when implementing the latent factor algorithm for ratings predictions, a subset of the dataset is extracted. The Book Crossing dataset, with over 1 million ratings, is substantially larger than the 100,000 ratings provided by the MovieLens dataset. After basic cleansing of data, the following steps were taken to create a subset of the Book Crossing dataset. All zero ratings were removed (handling implicit feedback is beyond the scope of this paper). The top 1000 most actively rated users was identified. From these users, the 2000 most rated books was extracted. From this list, all users who have less than 10 ratings in the new subset of data were removed.

Application of 2D LF Models Using SVD in Recommendation Context: Before describing the challenge of sparsity in the dataset and the following experiments, an overview of 2D and 3D matrix and tensor decomposition will be given respectively. The 2D SVD (Singular Value Decomposition) in a recommendation context is initiated by the generation of a user-item matrix from a list of ratings provided by users [3,22]. Since not every user can consume or rate every available item, several missing values in the user-item matrix will need to be filled before prediction. These null values will ultimately correspond to the ratings that will be predicted by the recommendation algorithm.

To apply SVD, imputation methods are used to fill in the sparse user-item matrix. It is evident that imputation methods can impact the quality of prediction and will consequently improve upon the final predicted ratings of each user to each item [3]. Once imputed, SVD is applied to deconstruct the original matrix into smaller matrices describing a series of latent factors (consisting of eigenvalues and eigenvectors). A subset of these latent factors will then be selected to reconstruct a new matrix. The newly reconstructed matrix will house the predicted ratings of each user for each particular item.

It should be noted that SVD has been investigated in the related literature for different purposes [3,22]. This includes using (a) different types of SVD, (b) different rank values to select the best subset of latent factors for matrix reconstruction, and (c) different imputation methods to fill in missing values. This is important as seen from the Netflix Prize in attaining optimal results.

Application of 3D Latent Factor Models in a Recommendation Context: 3D tensor factorization in a recommendation context, unsurprisingly, shares a similar process compared to 2D tensor factorization. However, there are some notable differences. First, the requirement for datasets with additional data (besides ratings) to be used as the third dimension (contextual feature). This may include dimensions such as time of consumption, gender, age, or demographic information. Theoretically, any dimension that has the potential to discover latent relationships between users, items, and their ratings can be used as the third dimension. The second difference is the type of tensor decomposition technique used. The two common types of tensor decomposition techniques are CP Decomposition [10] and Tucker Decomposition [21]. CP Decomposition will be used in our method because it was identified that no one had used CP Decomposition in this type of application previously. The difference is the imputation method. Regarding the imputation step, which is found also in 2D SVD, imputation may potentially be required twice during 3D tensor factorization. Once, when generating the user-item matrix, and the second time when creating the user-item-context matrix. This represents an additional challenge and will be addressed in one of the experiments undertaken during this paper.

Imputation Methods: The datasets used in our study have a high level of sparsity as a result of not every user being able to consume every available item. Imputation methods are, therefore, required to fill these missing values and allow the application of decomposition techniques. The imputation methods adopted in our approach are: (1) *Filling with Zeros* where all missing values are filled with zeros. This method is typically the most basic and least effective imputation method. Due to its simplicity, it will be used as the simplest benchmark to compare results between 2D SVD and 3D CP Decomposition. (2) *Filling with*

Item Averages where missing values of each item will be filled in using the average of all ratings received for that item (movie or book). (3) *Filling with User Averages* where the missing values of each user will be filled in using the average of all ratings given by that user. These imputation methods are chosen to show how robust the performance of a 3D latent factor model is when predicting ratings and compared to a 2D latent factor model.

Contextual Features: The choice of contextual features is an independent variable that will affect the results of the experiments conducted in this paper. For the MovieLens dataset, the two contextual features used are Age and Gender as they will form the context component of the user-item-content matrix and will allow the 3D latent factor model to generate more personalized predictions. The contextual features used for the Book Crossing dataset are Age and Year of publication which can be important to personalize recommendations.

Nesterov CP Decomposition (NeCPD): Here, we present our tensor decomposition method which adopts NeCPD. Given an Nth-order tensor $\mathcal{X} \in \mathbb{R}^{I_1 \times \cdots \times I_N}$, NeCPD initially divides the CP problem into a convex N sub-problems since its loss function L is non-convex problem which may have many local minima. For simplicity, we present our method based on three-way tensor data. However, it can be naturally extended to handle a higher-order tensor.

In a three-way tensor $\mathcal{X} \in \Re^{I \times J \times K}$, $\mathcal{X}$ can be decomposed into three matrices $A \in \Re^{I \times R}$, $B \in \Re^{J \times R}$ and $C \in \Re^{K \times R}$, where R is the latent factors. Following the SGD, we need to calculate the partial derivative of the loss function L defined in Eq. 1 with respect to the three modes A, B and C alternatively.

The rational idea of the SGD algorithm depends only on the gradient information of $\frac{\partial L}{\partial w}$. In such non-convex setting, this partial derivative may encounter data points with $\frac{\partial L}{\partial w} = 0$ even though it is not at a global minimum. These data points are known as saddle points which may detente the optimization process to reach the desired local minimum if not escaped [1]. These saddle points can be identified by studying the second-order derivative (aka Hessian) $\frac{\partial L}{\partial w}^2$. Theoretically, when the $\frac{\partial L}{\partial w}^2(x; w) \succ 0$, x must be a local minimum; if $\frac{\partial L}{\partial w}^2(x; w) \prec 0$, then we are at a local maximum; if $\frac{\partial L}{\partial w}^2(x; w)$ has both positive and negative eigenvalues, the point is a saddle point. The second-order methods guarantee convergence, but the computing of Hessian matrix $H^{(t)}$ is high, which makes the method infeasible for high dimensional data and online learning. Ge *et al.* [1] show that saddle points are very unstable and can be escaped if we slightly perturb them with some noise. Based on this, we use the perturbation approach which adds Gaussian noise to the gradient. This reinforces the next update step to start moving away from that saddle point toward the correct direction. After a random perturbation, it is highly unlikely that the point remains in the same band and hence it can be efficiently escaped (i.e., no longer a saddle point). Since we are interested in fast optimization process due to online settings, we further incorporate Nesterov's method into the perturbed-SGD algorithm to accelerate the convergence rate.

$$L(\mathcal{X}, A, B, C) = \min_{A,B,C} \| \mathcal{X} - \sum_{r=1}^{R} A_{ir} \circ B_{jr} \circ C_{kr} \|_{f}^{2}, \tag{1}$$

$$A^{(t+1)} := A^{(t)} + \eta^{(t)} \nu^{(A,t)} + \epsilon - \beta \|A\|_{L_1} \tag{2}$$

where

$$\nu^{(A,t)} := \gamma \nu^{(A,t-1)} + (1 - \gamma)\frac{\partial L}{\partial A}(X^{(1,t)}, A^{(t)}) \tag{3}$$

Recently, Nesterov's Accelerated Gradient (NAG) [11] has received much attention for solving convex optimization problems [2,5,12]. It introduces a smart variation of momentum that works slightly better than standard momentum. This technique modifies the traditional SGD by introducing velocity ν and friction γ, which tries to control the velocity and prevents overshooting the valley while allowing faster descent.

Algorithm 1: NeCPD algorithm

NeCPD
Input: Tensor $X \in \Re^{I \times J \times K}$, number of components R
Output: Matrices $A \in \Re^{I \times R}$, $B \in \Re^{J \times R}$ and $C \in \Re^{K \times R}$

   1: Initialize A, B, C
   2: Repeat
     3: Compute the partial derivative of A, B and C
     3: Compute ν of A, B and C using Equation 3
     4: Update the matrices A, B and C using Equation 2
   6: until convergence

Our idea behind Nesterov's [11] is to calculate the gradient at a position that we know our momentum is about to take us instead of calculating the gradient at the current position. In practice, it performs a simple step of gradient descent to go from $w^{(t)}$ to $w^{(t+1)}$, and then it shifts slightly further than $w^{(t+1)}$ in the direction given by $\nu^{(t-1)}$. In this setting, we model the gradient update step with NAG as shown in Eq. 2 and 3 where ϵ is a Gaussian noise, $\eta^{(t)}$ is the step size, and $\|A\|_{L_1}$ is the regularization and penalization parameter into the L_1 norms to achieve smooth representations of the outcome and thus bypassing the perturbation surrounding the local minimum problem. The updates for $(B^{(t+1)}, \nu^{(B,t)})$ and $(C^{(t+1)}, \nu^{(C,t)})$ are similar to the aforementioned ones. With NAG, our method achieves a global convergence rate of $O(\frac{1}{T^2})$ comparing to $O(\frac{1}{T})$ for traditional gradient descent. Based on the above models, we present our NeCPD Algorithm 1.

Experiment 1 Setup: the aim of this is to evaluate whether our proposed 3D latent factor model using CP Decomposition can generate personalized predictions (via the incorporation of the contextual feature) whilst maintaining rating prediction accuracy compared to a 2D SVD approach. To predict ratings using CP Decomposition, a 3D tensor describing the user-item-context matrix was prepared from the standard user-item dataset. This involved adding the contextual feature along the third axis of the user-item matrix. For the Age contextual feature of the MovieLens dataset, different age buckets were tested to determine whether this would affect performance. These age buckets were 0–18, 19–50, and > 50 [7]. It was kept at 3, keeping the user-item-context matrix to a dimension of 3 along the context axis.

Once the 3D tensors were generated, the train and test sets were created using five fold cross-validation. The missing values of the train set were first imputed using an imputation methodology before CP Decomposition was applied. Choosing different rank values (denoted as K in the Results), the resulting two matrices were then reconstructed to obtain the predicted values for each user-item combination. The predicted values were then compared to their corresponding values in the test set and the mean absolute error was calculated. The experiment is repeated for the different imputation methods (zeros, item average, user average) and for different rank values to identify the combination of variables that result in the optimal performance for ratings prediction. The rank values indicate the degree of information retained during the decomposition process. The results obtained from Experiment 1 are then finally compared to the results observed by Ghazanfar et al. [3], who performed a modified version of 2D SVD on the same dataset to predict ratings.

Experiment 2 Setup: the aim of this is to compare the effectiveness of using different contextual features on rating prediction using CP Decomposition. In this experiment, both datasets were used to compare the impact of using Age, Gender and Rating Avidness (MovieLens) and Age and Year of Publication (Book Crossing). Before experimenting with the Book Crossing dataset, a subset of the original Book Crossing data was created to overcome the scalability issues encountered when using the full dataset. The process of sub-setting the dataset is described in the *Datasets*.

Using the new subset of data for Book Crossing (or complete data for Movie-Lens), a user-item-context matrix was created similar to Experiment 1 by splitting the contextual factor into three different buckets of values. These groupings are shown in the Results section. With regards to the imputation method and rank values, only the item-average imputation method and a rank value of 15 are used respectively. This is based on the results obtained from Experiment 1, which depicted these methods as being the most optimal in achieving best MAE when predicting ratings. After the user-item-context matrices are generated, they are further split into a train and test set for five-fold cross validation. We applied our CP Decomposition on the train set in order to generate predictions and the results are compared to the actual ratings stored in the test set. The results are then recorded for further discussion and evaluation. Experiment 2 is then repeated, with either different Age groups or different contextual features.

Experiment 3 Setup: upon the analysis of results obtained from prior experiments (Experiment 1 and 2), a different, our novel approach to applying tensor decomposition to the user-item-context tensor was tested. Specifically, the user-item-context was generated differently, with the aim of this experiment to identify whether this novel method was capable of delivering better MAE compared to the method used in the prior experiments. To generate the new user-item-context tensor, the following actions were performed: (1) the number of users belonging to each context was made equal and (2) each layer of users was then stacked behind each other.

To illustrate the difference, if there were 900 users and 100 items and 3 contextual values. Method 1 (used in Experiment 1 and 2) would generate a user-item-context tensor that has 900 rows, 100 items and 3 layers. The new method used in this experiment will generate a user-item-context tensor that has 300 rows, 100 items and 3 layers. During the generation of the matrix, the imputation method (item averages) and context (Age) were kept consistent for the two datasets. Experiment 3 was then repeated 3 times, with a randomised order for each user within each layer to overcome bias. Results were recorded at the end of each activity for further discussion and evaluation.

Table 1. Results from Experiment 1 - part 1

MovieLens dataset							
Imputation method	Context	K = 2	K = 5	K = 8	K = 10	K = 12	K = 15
Zeros	Age	2.712	2.525	2.513	2.406	2.393	2.382
Zeros	Gender	2.697	2.471	2.401	2.351	2.331	2.313
Item average	Age	0.927	0.793	0.786	0.782	0.781	0.779
Item average	Gender	0.803	0.791	0.787	0.78	0.776	0.763
User average	Age	0.94	0.795	0.912	0.788	0.786	0.784
User average	Gender	0.824	0.793	0.789	0.785	0.78	0.769

Table 2. Results from Experiment 1 - part 2

MovieLens dataset						Ghazanfar et al. results [3]		
Imputation method	Context	K=8	K=10	K=12	K=15	Imputation method	K	MAE
Zeros	Gender	2.401	2.351	2.331	2.313	Zeros	12	2.321
Item average	Gender	0.787	0.78	0.776	0.763	Item average	10	0.774
User average	Gender	0.789	0.785	0.78	0.769	User average	8	0.778

Evaluation Methodology: The evaluation methodology used in this paper consists of five-fold cross validation with a train test split ratio of 0.8 and 0.2

to each user's ratings. It must be highlighted that instead of the conventional removal of 20% of all users from the full dataset, 20% of a user's ratings are retained in a test set instead. This means that all users will be present in both train and test sets. This is required due to the nature of CP Decomposition, which requires the same dimension of users and items for prediction (you can not predict a user's ratings if they are not in the matrix at the point of decomposition).

Once the train set has been deconstructed using CP decomposition with the chosen rank value, the resulting matrices are then reconstructed to generate predictions for each user and each item. The Mean Absolute Error (MAE) is calculated by averaging all corresponding values in the test set. This is then repeated using each fold during cross validation to calculate the average MAE.

4   Results

Experiment 1 Results: from Table 1, it can be seen that Gender slightly outperformed Age in mean absolute error across almost all iterations. Similarly, Item Averages outperforms Zeros and User Average imputation methods. There is a substantial difference between Zeros and the item/user average results. This is expected, as the zeros imputation method is understood across the literature to be the most basic method and used for benchmarking purposes only.

From Table 2, it can be observed that the application of our CP Decomposition algorithm is capable of achieving slightly better results than the 2D results reported by Ghazanfar et al. [3]. The use of Gender outperformed the benchmarked results, delivering an improvement of 0.1 in MAE across all imputation methods. Specifically, this result is observed at a rank value of 15 whereas 2D SVD reported optimal results at 8, 10, and 12 for the different imputation methods (zeros, item average, and user average respectively). It can be noted that

Table 3. Results from Experiment 2 - MovieLens dataset

MovieLens dataset							
Context	Buckets	K = 2	K = 5	K = 8	K = 10	K = 12	K = 15
Age	0–18	0.927	0.793	0.786	0.782	0.781	0.779
	19–50						
	50						
Age	0–26	0.931	0.794	0.786	0.783	0.782	0.782
	27–38						
	38						
Gender	M / F	0.803	0.791	0.787	0.78	0.776	0.763
Avidness of ratings	1–40	0.804	0.812	0.789	0.782	0.78	0.779
	41–120						
	120						

using a higher rank value is more beneficial to 3D tensor factorization because there is more information to retain and take advantage of compared to 2D SVD approaches. It should be noted, however, that in Ghazanfar et al.'s research [3], it was reported that a large range of rank values was tested, and the reported rank values resulted in the best performance.

Experiment 2 Results: similar to Experiment 1 results, Table 3 shows that Gender continues to be the most effective contextual feature, followed by Age and Avidness. What is more interesting to observe across Table 3 and 4 is the performance between the two age buckets, with the context using Ages and a threshold of 18 and 50 outperforming the Age context that equally distributes the number of ratings across each bucket. The two extra contexts, Avidness and Year of Publication perform relatively poorly compared to the other features, which is an indication that more fine-tuning is required when choosing contextual features. These results show that the choice of contextual feature is highly important when using tensor decomposition to predict ratings.

Experiment 3 Results: the results from Tables 5 and 6 show that at a rank value of 2, an improved MAE is observed from both datasets and converge to the initial method as the rank value increases. Consequently, our proposed method shows better performance at a lower rank value which may prove beneficial during implementation as lower rank values require less computational time and effort to run. At a high level, it can be concluded that our proposed method of tensor generation can potentially be used to predict ratings, however, there are a number of limitations to consider as discussed in Sect. 5.

Table 4. Results from Experiment 2 - book crossing dataset

Book crossing dataset						
Context	Buckets	K = 2	K = 5	K = 8	K = 10	K = 15
Age	0–18	1.521	1.326	1.320	1.319	1.315
	19–50					
	50					
Age	0–29	2.891	1.328	1.323	1.322	1.315
	30–37					
	38					
Year of publication	1995	2.832	2.843	2.834	2.836	2.841
	1995–2000					
	2000					

5 Discussion

Experiment 1: Experiment 1 was performed by changing a number of different variables. These included two different contextual features (Age and Gender), three imputation methods (Zeros, Item Average and User Average), and a range of rank values (2, 5, 8, 10, 15). Experimentation using these iterations surfaced a number of interesting results.

Firstly, Gender slightly outperformed Age in mean absolute error across almost all iterations. By analyzing the distribution of ratings as a percentage of total ratings, a Male/Female split shows that Males provide a higher concentration of 3s and 4s compared to Females, who in turn tend to provide more extreme results in the form of 1s and 5s. Looking at the Rating Distribution by Age buckets (< 18, 19–50, > 50), we see the younger audience giving the more extreme values (1s and 5s), the middle-aged audience giving more 3s, and the older audience giving more 4s. The above variations in ratings may contribute to the reason why Gender outperforms Age, however, a few caveats should be noted. The trends highlighted above are very slight and are not apparent when observing the rating distribution. The Gender buckets and Age buckets have a different number of raters within each group and are imbalanced. For example, the rating distribution shows that there are more males than females. Similarly, the younger age group has 54, the middle group has 784 and the older group has 105. It is also unclear how dominant the effects are of having a heavily skewed

Table 5. Results from Experiment 3 - MovieLens dataset

Book crossing dataset						
Tensor generation method	Context (Age)	K = 2	K = 5	K = 8	K = 10	K = 15
Conventional method	0–29	0.927	0.793	0.786	0.782	0.779
	30–37					
	37					
Stacking method	0–29	0.901	0.802	0.792	0.788	0.786
	30–37					
	37					

Table 6. Results from Experiment 3 - book crossing dataset

Book crossing dataset						
Tensor generation method	Context (Age)	K = 2	K = 5	K = 8	K = 10	K = 15
Conventional method	0–29	3.521	1.323	1.321	1.319	1.31
	30–37					
	37					
Stacking method	0–29	1.323	1.32	1.321	1.312	1.311
	30–37					
	37					

middle group. This relationship is further explored in Experiment 2. Since only one model (CP Decomposition) is in scope for this paper, it is unclear whether the model itself has a bias towards groups that skew towards a certain rating. For example, it is possible that the model has a bias towards ratings with higher concentrations of 3s and 4s as opposed to more extreme values. As this relationship is better surfaced by the Gender context, is it possible that the model thus naturally tends towards this relationship when predicting unknown ratings? This is another possible avenue for future work.

An important conclusion to draw from this, however, is that the choice of contextual feature is important when exploring the use of 3D tensor factorization. This is further explored in Experiment 2. Lastly, the optimal imputation method out of the three explored was item averages. This was consistent with the findings of Ghazanfar et al. [3], who also found that item average performed the best when compared to user averages and zeros. Due to this observation, the item averages imputation method was adopted for all following experiments in this paper.

The second part of analysis for Experiment 1 results was the comparison of results to Ghazanfar et al.'s research [3], who looked into the optimal results of 2D SVD using different imputation methods. Analyzing the results from Table 2, it is clear that our 3D tensor factorization is capable of producing an overall improved result compared to 2D SVD. There are, however, practical implications of these results to be discussed. An improvement in MAE of 0.1 in the context of a 1–5 rating can be arguable insignificant when building a recommendation system. As a result, it is identified that further experimentation in the form of different contextual features may be required to realize larger improvements for 3D tensor factorization in its current form to be beneficial from a practical perspective. During experimentation, it was also observed that increasing the rank value also increased computational time. When implementing such a solution in the real world, a trade-off between computational time and performance must be considered and could be overcome with more powerful machines and processing power. The results from Experiment 1 as a whole confirm that our 3D tensor factorization has the potential to incorporate more information to deliver results that are arguably more personalized whilst maintaining the performance in accuracy compared to 2D methods. With this confirmation, Experiment 2 was undertaken to critically evaluate the effectiveness of different contextual features.

Experiment 2: The aim of Experiment 2 was to critically evaluate the effectiveness of using different contextual features to predict ratings using CP decomposition. Different variations of experiments were performed on two datasets, MovieLens and Book Crossing. Specific to the MovieLens results, the results show that Gender continues to be the best contextual feature out of those tested, with a consistently lower MAE across most rank values compared to Age and Rating Avidness. However, it is the comparison between the different Age buckets that is more interesting, with the 18 and 50 buckets consistently outperforming the "equal distribution" buckets across both datasets. This raises an interesting hypothesis as to whether it is more effective to split buckets into an equal number of ratings per bucket or rather to split buckets based on their inherent relationships (such as users who tend to rate lower, the rate at extremes, or the rate at averages). Based on theoretical discussions and the results recorded from Experiment 2, it is suggested that the latter is more effective in providing better rating prediction performance. The use (and poor MAE performance) of Avidness and Year of Publication can also be used to support the previous argument and to conclude the results of Experiment 2. Both contextual features were split based on an equal distribution of ratings and showed poor performance compared to the other features. Additionally, it is clear upon analysis of all results in Experiment 2 that the choice of the contextual feature remains a highly important decision that will impact the overall performance of the MAE.

Experiment 3: Experiment 3 was performed in order to investigate an alternative method of creating the user-item-context matrix without requiring a second imputation step during tensor generation (and subsequent) factorization. In order to reduce the time taken to perform this experiment, item averages were used as the imputation method, similar to Experiment 2. To maintain robustness, different rank values were tested across the two available datasets. As mentioned in the Results section, the results show that at a rank value of 2, an improved MAE is observed from both datasets and converges to the initial method as the rank value increases. There are, however, theoretical and practical inconsistencies with this method, regardless of the positive performance observed. As described in the Methodology Section, the method stacks users behind each other, which creates an artificial relationship between users along the third (contextual) dimension. For example, if user 4 is stacked behind user 1, then tensor factorization will create a component matrix for users that will attempt to capture that information. Secondly, there is a practical limitation to applying this method. In order to stack users behind each other, the number of users belonging to each contextual value or layer must be the same. In the real world, it is highly unlikely that the numbers of users belonging to each contextual feature will be the same. Based on these observations, it can be seen that there are a number of limitations to the alternate method of generating user-item-context tensors. If these limitations can be overcome, there is potential for this method to be further revised and improved based on the initial results seen from Experiment 3. These limitations are further described in the next section under limitations and future works.

Limitations and Future Works: The scalability of the CP Decomposition algorithm was found to be limited in its ability to process large datasets. The complete Book Crossing dataset was too large for the algorithm to handle and therefore, a subset of the dataset was used in this paper. Improvements to the algorithm or the use of more computational power may have influenced the results of the Book Crossing dataset and may have allowed for different observations.

Exploring the use of implicit feedback to improve rating prediction would be a very big area of future work for this paper. In the Book Crossing dataset, implicit feedback (whether a user consumed the item but did not rate it) was given. An interesting avenue for exploration would be to use implicit feedback as input for improved recommendations. Exploring this area as a future option would be extremely beneficial as it is vastly more likely in the real world to capture implicit feedback rather than explicit feedback. By developing an algorithm that can incorporate implicit feedback, the number of scenarios where this algorithm could be applied and the amount of training data would increase dramatically.

Finally, evaluation of the paper's experiments without using Mean Absolute Error should also be incorporated in future work. In this paper, due to the number of uncertainties with the new technique, the methodology, evaluation methods, and datasets, it was difficult to incorporate a different metric of evaluation. MAE is highly utilized in the academic literature and therefore well documented, but is not a practical method for evaluating recommendation systems that are implemented in real work. A very strong improvement to this paper would be to explore the qualitative performance of 3D tensor factorization compared to conventional 2D conventional methods.

References

1. Ge, R., Huang, F., Jin, C., Yuan, Y.: Escaping from saddle points-online stochastic gradient for tensor decomposition. In: Learning Theory, pp. 797–842 (2015)
2. Ghadimi, S., Lan, G.: Accelerated gradient methods for nonconvex nonlinear and stochastic programming. Math. Program. **156**(1–2), 59–99 (2016)
3. Ghazanfar, M.A., Prügel-Bennett, A.: The advantage of careful imputation sources in sparse data-environment of recommender systems: generating improved svd-based recommendations. Informatica (Slovenia) **37**, 61–92 (2013)
4. Gomez-Uribe, C.A., Hunt, N.: The Netflix recommender system: algorithms, business value, and innovation. ACM Trans. Manage. Inf. Syst. (TMIS) **6**(4), 1–19 (2015)
5. Guan, N., Tao, D., Luo, Z., Yuan, B.: Nenmf: an optimal gradient method for nonnegative matrix factorization. Trans. Signal Process. **60**(6), 2882–2898 (2012)
6. Isinkaye, F.O., Folajimi, Y.O., Ojokoh, B.A.: Recommendation systems: principles, methods and evaluation. Egyptian Inf. J. **16**(3), 261–273 (2015)
7. Karatzoglou, A., Amatriain, X., Baltrunas, L., Oliver, N.: Multiverse recommendation: n-dimensional tensor factorization for context-aware collaborative filtering. In: Proceedings of the 4th ACM Conference on Recommender Systems, pp. 79–86 (2010)
8. Kolda, T.G., Sun, J.: Scalable tensor decompositions for multi-aspect data mining. In: IEEE International Conference on data Mining, pp. 363–372. IEEE (2008)
9. Kumar, P., Thakur, R.S.: Recommendation system techniques and related issues: a survey. Int. J. Inf. Technol. **10**(4), 495–501 (2018). https://doi.org/10.1007/s41870-018-0138-8
10. Lebedev, V., Ganin, Y., Rakhuba, M., Oseledets, I., Lempitsky, V.: Speeding-up convolutional neural networks using fine-tuned CP-decomposition. arXiv:1412.6553 (2014)
11. Nesterov, Y.: Introductory Lectures on Convex Optimization: A Basic Course, vol. 87. Springer, New York (2013). https://doi.org/10.1007/978-1-4419-8853-9
12. Nitanda, A.: Stochastic proximal gradient descent with acceleration techniques. In: Advances in Neural Information Processing Systems, pp. 1574–1582 (2014)

13. Paterek, A.: Improving regularized singular value decomposition for collaborative filtering. In: Proceedings of KDD Cup and Workshop, vol. 2007, pp. 5–8 (2007)
14. Pazzani, M.J., Billsus, D.: Content-based recommendation systems. In: Brusilovsky, P., Kobsa, A., Nejdl, W. (eds.) The Adaptive Web. LNCS, vol. 4321, pp. 325–341. Springer, Heidelberg (2007). https://doi.org/10.1007/978-3-540-72079-9_10
15. Rendle, S.: Factorization machines. In: 2010 IEEE International Conference on Data Mining, pp. 995–1000. IEEE (2010)
16. Salakhutdinov, R., Mnih, A., Hinton, G.: Restricted boltzmann machines for collaborative filtering. In: Proceedings of International Conference on Machine Learning, pp. 791–798 (2007)
17. Sarwar, B., Karypis, G., Konstan, J., Riedl, J.: Item-based collaborative filtering recommendation algorithms. In: Proceedings of the 10th International Conference on World Wide Web, pp. 285–295 (2001)
18. Schafer, J.B., Frankowski, D., Herlocker, J., Sen, S.: Collaborative filtering recommender systems. In: Brusilovsky, P., Kobsa, A., Nejdl, W. (eds.) The Adaptive Web. LNCS, vol. 4321, pp. 291–324. Springer, Heidelberg (2007). https://doi.org/10.1007/978-3-540-72079-9_9
19. Sun, J., Tao, D., Papadimitriou, S., Yu, P.S., Faloutsos, C.: Incremental tensor analysis: Theory and applications. ACM Trans. Knowl. Discov. Data (TKDD) 2(3), 11 (2008)
20. Tang, X., Bi, X., Qu, A.: Individualized multilayer tensor learning with an application in imaging analysis. J. Am. Stat. Assoc. 115(530), 836–851 (2020)
21. Yang, L., Fang, J., Li, H., Zeng, B.: An iterative reweighted method for tucker decomposition of incomplete tensors. Signal Process. 64(18), 4817–4829 (2016)
22. Zhou, X., He, J., Huang, G., Zhang, Y.: SVD-based incremental approaches for recommender systems. J. Comput. Sys. Sci. 81(4), 717–733 (2015)

Efficient Computational Algorithm for Stress Analysis in Hydro-Sediment-Morphodynamic Models

Alia Al-Ghosoun$^{(\boxtimes)}$, Ashraf S. Osman, and Mohammed Seaid

Department of Engineering, University of Durham, South Road,
Durham DH1 3LE, UK
algsoon2006@yahoo.com

Abstract. Understanding of complex stress distributions in lake beds
and river embankments is crucial in many designs in civil and geotechni-
cal engineering. We propose an accurate and efficient computational algorithm for stress analysis in hydro-sediment-morphodynamic models. The
governing equations consists of the linear elasticity in the bed topography
coupled to the shallow water hydro-sediment-morphodynamic equations.
Transfer conditions at the bed interface between the water surface and
the bedload are also developed using frictional forces and hydrostatic
pressures. A hybrid finite volume/finite element method is implemented
for the numerical solution of the proposed model. Well-balanced discretization of the gradient fluxes and source terms is formulated for the
finite volume and the treatment of dry areas in the model is discussed in
the present study. The finite element method uses quadratic elements on
unstructured meshes and interfacial forces are samples on the common
nodes for finite volume and finite element grids. Numerical results are
presented for a dam-break problem in hydro-sediment-morphodynamic
models and the computed solutions demonstrated the ability of the proposed model in accurately capturing the stress distributions for erosional
and depositional deformations. In addition, the coupled model is accurate, very efficient, well-balanced and it can solve complex geometries.

Keywords: Stress analysis · Finite element method · Finite volume
method · Shallow water flows · Sediment transport · Morphodynamics

1 Introduction

Water movement over an erodible bed in either steady or unsteady conditions
can scour particles off the bed and transport them some distance [13]. These
particles can either travel as suspended sediment which is immersed in the flow
of water itself or as bedload, where sediment tumbles across of the bed [10].
Understanding the dynamics of sediment transport and erosion-deposition processes is important in different applications like road cuts, embankments and

D. Groen et al. (Eds.): ICCS 2022, LNCS 13350, pp. 291–304, 2022.
https://doi.org/10.1007/978-3-031-08751-6_21

dams designs [6]. Erosion and deposition of soil comprise one of the major concerns in studying soil properties. Spacial and temporal informations of soil erosion processes is required to reflect the pattern of sediment transport during different environmental conditions [14]. Indeed, previous knowledge of the factors affecting soil erosion is very useful in different morphodynamic applications like dam-break, dam removal and storms [1]. In recent years, the investigations of soil erosion and deposition through the development of different algorithms have been rapidly increased [16]. These algorithms depend on different equations, some of them depends on the fundamental energy transport equations [5], sediment flux equation [9] and the steady-state continuity equations for deposition [12]. However, there are still many models and techniques which suffer from a range of problems, such as over-estimation due to parameters in compliance with the initial conditions and the assumptions unsuitability to the present case alongside with the existence of uncertainty in the system parameters [15].

Sediment is continually subjected to physical stress in the environment. These stresses are of paramount importance to geomorphologists because they are a driver of geomorphic change. These stresses can describe forces applied to the soil, that result in sediment deformation or fracture [7]. Many excellent methods have been developed to quantify these stresses in soil. The most accurate of these are conducted within the laboratory using specialist equipment. However, cohesive sediment undergoes significant changes in sediment properties when it is cored, transported, stored and finally analysed in the laboratory, and these changes can significantly alter its shear strength [11]. Different techniques have been using to quantify the stresses in soil. In the current study, a coupled finite element/finite volume method for solving soil stresses over erodible beds is proposed. The governing equations consist of the one-dimensional non-linear shallow water equations for the water flow and a two-dimensional linear elasticity model for the bed deformation. Deformations in the topography can be caused as a result of the hydrostatic pressure distribution and the frictional force obtained from the shallow water movement. Coupling conditions at the interface are also investigated in this study and a well-balanced finite volume method using non-uniform grids is implemented without the requirements of the interpolation procedures at the interface between the finite element and finite volume nodes. On the other hand, a force is sampled from the hydrostatic pressure and applied on the bed surface to be used in the stress analysis. Numerical results for both the bed-load and stress distributions are presented in this study for a dam-break problem over erodible bed.

This paper is organized as follows. In Sect. 2 we present the governing equations used for the hydro-sediment-morphodynamic models. Section 3 is devoted to the development of an efficient computational algorithm for solving the coupled system. We formulate the hybrid finite element method/finite volume method. In this section we also discuss the coupling conditions at the interface. In Sect. 4, we examine the numerical performance of the proposed method using several examples of hydro-sediment-morphodynamic problems. Our new approach is demonstrated to enjoy the expected efficiency as well as the accuracy. Concluding remarks are summarized in Sect. 5.

2 Equations for Hydro-Sediment-Morphodynamic Models

In the present study, we assume a longitudinal one-dimensional shallow water-sediment flow over an erodible bed composed of uniform, non-cohesive sediment of particle diameter d_s. Therefore, the governing shallow water hydro-sediment-morphodynamic equations can be derived by directly applying the Reynolds transport theorem in fluid dynamics assuming a hydrostatic pressure and the flow is almost horizontal with the vertical component of the acceleration is vanishingly small. The model consists of mass and momentum conservation laws for the water-sediment mixture and separate mass conservation laws for sediment and bed material. The resulting system of equations can be expressed in the standard well-structured conservation form as

$$\frac{\partial \mathbf{q}}{\partial t} + \frac{\partial \mathbf{F(q)}}{\partial x} = \mathbf{Q(q)} + \mathbf{S(q)}, \tag{1}$$

where the vector of unknowns $\mathbf{q}$ and the flux vector $\mathbf{F(q)}$ are

$$\mathbf{q} = \begin{pmatrix} h \\ hv \\ hc \\ Z \end{pmatrix}, \qquad \mathbf{F(q)} = \begin{pmatrix} hv \\ hv^2 + \dfrac{1}{2}gh^2 \\ hvc \\ \dfrac{q_b}{1-p} \end{pmatrix},$$

and the source vector $\mathbf{S(q)}$ is

$$\mathbf{S(q)} = \begin{pmatrix} \dfrac{\mathcal{E}-\mathcal{D}}{1-p} \\ -\dfrac{(\rho_0-\rho)(\mathcal{E}-\mathcal{D})v}{\rho(1-p)} - gh\dfrac{n_b^2 v\,|v|}{h^{4/3}} \\ \mathcal{E}-\mathcal{D} \\ -\dfrac{\mathcal{E}-\mathcal{D}}{1-p} \end{pmatrix}.$$

The differential source term $\mathbf{Q((q)}$ is defined as

$$\mathbf{Q(q)} = \begin{pmatrix} 0 \\ -gh\dfrac{\partial Z}{\partial x} - \dfrac{(\rho_s-\rho_w)}{2\rho}gh^2\dfrac{\partial c}{\partial x} \\ 0 \\ 0 \end{pmatrix}.$$

Here, v is the depth-averaged water velocity, h the water depth, Z the bottom topography, g the gravitational acceleration, p the porosity, ρ_w the water density, ρ_s the sediment density, c is the depth-averaged concentration of the suspended sediment, n_b is the Manning roughness coefficient, $\mathcal{E}$ and $\mathcal{D}$ represent the entrainment and deposition terms in upward and downward directions, respectively. The density of the water-sediment mixture ρ and the density of the saturated bed ρ_0 are defined by

$$\rho = \rho_w(1-c) + \rho_s c, \qquad \rho_0 = \rho_w p + \rho_s(1-p). \tag{2}$$

It should be noted that although there are various bedload transport formulae which were empirically proposed based on laboratory or fieldwork datasets, none can be universally applied due to the range and varying distribution of grain sizes. In this study, we consider the Grass formula [8]

$$q_b = A_g v^3, \tag{3}$$

where $A_g \in [0,1]$ is a dimensionless constant usually obtained experimentally by accounting for the diameter of the particles and the kinematic viscosity. For values of A_g close to zero, the model shows a weak interaction between the sediment bottom and the fluid. However, for values of A_g close to one, the interaction between the sediment bottom and the fluid is strong.

To determine the entrainment and deposition terms in the above equations we assume a non-cohesive sediment and we use empirical relations reported in [4]. Thus,

$$D = w\alpha_c(1 - \alpha_c c)^m c, \tag{4}$$

where w is the settling velocity of a single particle in tranquil water

$$\omega = \sqrt{\left(\frac{13.95\nu}{d}\right)^2 + 1.09 sgd} - \frac{13.95\nu}{d}, \tag{5}$$

with ν is the kinematic viscosity of the water, d the averaged diameter of the sediment particle, m an exponent indicating the effects of hindered settling due to high sediment concentrations and it is computed using the Reynolds number of the particle as

$$m = 4.45 Re^{-0.1}, \qquad Re = \frac{wd}{\nu}.$$

To ensure that the near-bed concentration does not exceed $(1-p)$, the coefficient α_c is computed by [4]

$$\alpha_c = \min\left(2, \frac{1-p}{c}\right).$$

For the entrainment of a cohesive material, the following relation is used

$$E = \begin{cases} \varphi \dfrac{\theta - \theta_c}{h} vd^{-0.2}, & \text{if } \theta \geq \theta_c, \\ 0, & \text{otherwise}, \end{cases} \tag{6}$$

where φ is a coefficient to control the erosion forces determined by

$$\varphi = \varphi_c \frac{560\,(1-p)\,\nu^{0.8}}{3\,(sg)^{0.4}\,\theta_c},$$

with φ_c is a dimensionless value that depends on the phenomenon to recreate. Here, θ_c is a critical value of Shields parameter for the initiation of the sediment motion and θ is the Shields coefficient defined by

$$\theta = \frac{u_*^2}{sgd}, \tag{7}$$

with u_* is the friction velocity defined as

$$u_* = \sqrt{\frac{\tau}{\rho}},$$

where τ is the threshold stress of bottom computed using

$$\tau = \frac{g\rho n_b^2 v \, |v|}{h^{1/3}}. \tag{8}$$

In (7), s is the submerged specific gravity of sediment given by

$$s = \frac{\rho_s}{\rho_w} - 1.$$

In the present study, we are interested in developing a robust analysis of stresses on the beds generated by sediment transport. There are different models for describing the bed deformation, due to its complex nature only few theoretical models exist which use idealized and simplified assumptions. Most of the deformation models, which are used in practice, are static and usually only work in particular context with no sediment transport included in their formulations and for this reason there is not yet a universally accepted theory of stress analysis by sediment transport. However, we can in general describe the stress distribution in the erodible bed through two processes. The sediment can move in a layer close to the bottom topography which is known as bed load and is characterized by a rolling and sliding movement, or the flow can cause the sediment to separate completely from the bottom in which case it is referred to as suspended load and in this case the sediment is transported as a concentration of the water column and will later be deposited in the bottom.

Let us consider a two-dimensional bed domain Ω with smooth boundary $\partial\Omega$, the equilibrium governing equations of linear elasticity read [2]

$$\frac{\partial \sigma_x}{\partial x} + \frac{\partial \tau_{xz}}{\partial z} = f_x,$$

$$\frac{\partial \sigma_z}{\partial z} + \frac{\partial \tau_{xz}}{\partial x} = f_z, \tag{9}$$

where σ_x and σ_z are the normal stresses in the x- and z-direction, respectively. Here, τ_{xz} is the shear stress, f_x and f_z are the external forces in the x- and z-direction, respectively. The displacement vector is denoted by $\boldsymbol{u} = (u_x, u_z)^{\top}$ and the infinitesimal strain tensor is defined by

$$\epsilon = \frac{1}{2}\left(\nabla \boldsymbol{u} + (\nabla \boldsymbol{u})^{\top}\right). \tag{10}$$

The system has been solved subjected to the following boundary conditions

$$\sigma = \sigma_s, \quad \text{on} \quad \Gamma_i,$$

$$u = 0, \quad \text{on} \quad \Gamma,$$

(11)

where σ_s is the sediment stress on the interfecial boundary Γ_i. In the current study, the constitutive relation is defined as

$$\sigma = \mathbf{D}\, \epsilon,$$

(12)

where the stress vector σ and the constitutive matrix $\mathbf{D}$ are given as

$$\sigma = \begin{pmatrix} \sigma_x \\ \sigma_z \\ \tau_{xz} \end{pmatrix}, \quad \mathbf{D} = \frac{E}{(1+\nu)(1-2\nu)} \begin{pmatrix} 1-\nu & \nu & 0 \\ \nu & 1-\nu & 0 \\ 0 & 0 & \dfrac{1-2\nu}{2} \end{pmatrix},$$

with E is the Young modulus and ν is the Poisson ratio characterizing the bed material. Note that other constitutive relations in (12) can also be applied in the proposed stress analysis in hydro-sediment-morphodynamic without major modifications in our formulation. It should also be stressed that although the Eqs. (9) are static, the interface boundary Γ_i between the water and bed depends on time.

3   A Coupled Finite Element/Finite Volume Method

To solve the equations for the considered hydro-sediment-morphodynamic model we proposed a coupled finite element/finite volume method for which transfer conditions are transmitted at the interface Γ_i. A well-balanced one-dimensional finite volume method is used for the sediment transport equations whereas an unstructured two-dimensional finite element method is used for the elasticity equations. Coupling conditions at the interface are also discussed in this section.

3.1   Well-balanced Finite Volume Solution of Sediment Transport

For the time integration of the system (1) we divide the time interval into subintervals $[t_n, t_{n+1}]$ with variable size Δt_n such that $t_n = t_{n-1} + \Delta t_n$, $n = 1, 2, \ldots$ and $t_0 = 0$. We use the notation $\mathbf{q}^n(x)$ to denote the discrete solution $\mathbf{q}(t_n, x)$. In the current work, we use the splitting operator to deal with the differential source terms $\mathbf{Q}(\mathbf{q})$ and the non-differential source term $\mathbf{S}(\mathbf{q})$ in (1). The splitting procedure consists of the following two steps:

Step 1: Solve for $\tilde{\mathbf{q}}$

$$\frac{\tilde{\mathbf{q}} - \mathbf{q}^n}{\Delta t_n} + \frac{\partial \mathbf{F}(\mathbf{q}^n)}{\partial x} = \mathbf{Q}(\mathbf{q}^n).$$

(13)

Step 2: Solve for $\mathbf{q}^{n+1}$

$$\frac{\mathbf{q}^{n+1} - \widetilde{\mathbf{q}}}{\Delta t_n} = \mathbf{S}\left(\widetilde{\mathbf{q}}\right). \tag{14}$$

For the space discretization we discretize the one-dimensional space domain in non-uniform control volumes $\left[x_{i-\frac{1}{2}}, x_{i+\frac{1}{2}}\right]$ with length Δx_i and we use the notation $\mathbf{q}_i^n$ to denote the space-averaged of $\mathbf{q} = \mathbf{q}(t, x)$ in the cell $\left[x_{i-\frac{1}{2}}, x_{i+\frac{1}{2}}\right]$ at time t_n, and $\mathbf{q}_{i+\frac{1}{2}}^n$ are the intermediate solutions at $x_{i+\frac{1}{2}}$ at time t_n,

$$\mathbf{q}_i^n = \frac{1}{\Delta x_i} \int_{x_{i-\frac{1}{2}}}^{x_{i+\frac{1}{2}}} \mathbf{q}(t_n, x)\, dx, \qquad \mathbf{q}_{i+\frac{1}{2}}^n = \mathbf{q}\left(t_n, x_{i+\frac{1}{2}}\right).$$

Integrating the system (13) over the space-time control domain $\left[x_{i-\frac{1}{2}}, x_{i+\frac{1}{2}}\right] \times [t_n, t_{n+1}]$, one obtains the following fully discrete system

$$\mathbf{q}_i^{n+1} = \mathbf{q}_i - \frac{\Delta t_n}{\Delta x_i}\left(\mathbf{F}_{i+\frac{1}{2}}^n - \mathbf{F}_{i-\frac{1}{2}}^n\right) + \Delta t_n \mathbf{Q}_i^n, \tag{15}$$

where $\mathbf{F}_{i\pm\frac{1}{2}}^n = \mathbf{F}\left(\mathbf{q}_{i\pm\frac{1}{2}}^n\right)$ are the numerical fluxes at $x = x_{i\pm\frac{1}{2}}$ and time $t = t_n$, and $\mathbf{Q}_i^n$ is the space-averaged of the source term $\mathbf{Q}$ defined as

$$\mathbf{Q}_i^n = \frac{1}{\Delta x_i} \int_{x_{i-\frac{1}{2}}}^{x_{i+\frac{1}{2}}} \mathbf{Q}(\mathbf{q})\, dx. \tag{16}$$

The spatial discretization (15) is complete when the numerical fluxes $\mathbf{F}_{i\pm 1/2}^n$ and the source term $\mathbf{Q}_i^n$ are reconstructed. Generally, this step can be carried out using any finite volume method developed in the literature for solving hyperbolic systems of conservation laws, see for example [2]. In the present study, we consider the Roe reconstruction defined as

$$\mathbf{F}_{i+\frac{1}{2}}^n = \frac{1}{2}\left(\mathbf{F}(\widehat{\mathbf{q}}_{i+1}^n) + \mathbf{F}(\widehat{\mathbf{q}}_i^n)\right) + \frac{1}{2}\mathbf{A}\left(\widehat{\mathbf{q}}_{i+\frac{1}{2}}^n\right)\left(\widehat{\mathbf{q}}_i^n - \widehat{\mathbf{q}}_{i+1}^n\right), \tag{17}$$

where the averaged state $\widehat{\mathbf{q}}_{i+\frac{1}{2}}^n$ is calculated as

$$\widehat{\mathbf{q}}_{i+\frac{1}{2}}^n = \begin{pmatrix} \dfrac{h_i^n + h_{i+1}^n}{2} \\ \dfrac{\sqrt{h_i^n}\, u_i^n + \sqrt{h_{i+1}^n}\, u_{i+1}^n}{\sqrt{h_i^n} + \sqrt{h_{i+1}^n}} \\ \dfrac{\sqrt{h_i^n}\, c_i^n + \sqrt{h_{i+1}^n}\, c_{i+1}^n}{\sqrt{h_i^n} + \sqrt{h_{i+1}^n}} \\ \dfrac{Z_i^n + Z_{i+1}^n}{2} \end{pmatrix}, \tag{18}$$

and the Roe matrix in (17) is defined as $\mathbf{A} = \mathbf{R}\Lambda\mathbf{R}^{-1}$ with

$$
\mathbf{R} = \begin{pmatrix} 1 & 1 & 1 & 1 \\ \widehat{u} & \widehat{\lambda}_2 & \widehat{\lambda}_3 & \widehat{\lambda}_4 \\ \widehat{c} - \dfrac{2\widehat{\rho}}{(\rho_s - \rho_w)} & \widehat{c} & \widehat{c} & \widehat{c} \\ 0 & \dfrac{(\widehat{\lambda}_2 - \widehat{u})^2 - g\widehat{h}}{g\widehat{h}} & \dfrac{(\widehat{\lambda}_3 - \widehat{u})^2 - g\widehat{h}}{g\widehat{h}} & \dfrac{(\widehat{\lambda}_4 - \widehat{u})^2 - g\widehat{h}}{g\widehat{h}} \end{pmatrix}, \quad (19)
$$

$$
\Lambda = \begin{pmatrix} \widehat{\lambda}_1 & 0 & 0 & 0 \\ 0 & \widehat{\lambda}_2 & 0 & 0 \\ 0 & 0 & \widehat{\lambda}_3 & 0 \\ 0 & 0 & 0 & \widehat{\lambda}_3 \end{pmatrix},
$$

with the four eiegenvalues

$$
\lambda_1 = u, \qquad \lambda_2 = 2\sqrt{-Q}\cos\left(\frac{1}{3}\theta\right) + \frac{2}{3}u,
$$

$$
\lambda_3 = 2\sqrt{-Q}\cos\left(\frac{1}{3}(\theta + 2\pi)\right) + \frac{2}{3}u, \qquad (20)
$$

$$
\lambda_4 = 2\sqrt{-Q}\cos\left(\frac{1}{3}(\theta + 4\pi)\right) + \frac{2}{3}u,
$$

where $\theta = \arccos\left(\dfrac{R}{\sqrt{-Q^3}}\right)$, with

$$
Q = -\frac{1}{9}\left(u^2 + 3g(h + h\xi)\right), \qquad R = \frac{u}{54}\left(9g(2h - h\xi) - 2u^2\right).
$$

For the discretization of the source term $\mathbf{Q}_i^n$ we implement a well-balanced reconstruction investigated in [2]. Thus, the well-balanced discretization of $\mathbf{Q}_i^n$ is achieved by in splitting the integral in (16) over the two sub-cells $\left[x_{i-\frac{1}{2}}, x_i\right]$ and $\left[x_i, x_{i+\frac{1}{2}}\right]$ of the control volume $\left[x_{i-\frac{1}{2}}, x_{i+\frac{1}{2}}\right]$ as

$$
\mathbf{Q}_i^n = \frac{1}{\Delta x_i}\left(\frac{(x_i - x_{i-1})}{2}\mathbf{Q}_{i-\frac{1}{2}}^L + \frac{(x_{i+1} - x_i)}{2}\mathbf{Q}_{i+\frac{1}{2}}^R\right), \qquad (21)
$$

where $\mathbf{Q}_{i-\frac{1}{2}}^L$ and $\mathbf{Q}_{i+\frac{1}{2}}^R$ are the space-averaged of the source term $\mathbf{Q}$ in the sub-cells $\left[x_{i-\frac{1}{2}}, x_i\right]$ and $\left[x_i, x_{i+\frac{1}{2}}\right]$ defined as

$$
\mathbf{Q}_{i-\frac{1}{2}}^L = \begin{pmatrix} 0 \\ -g\dfrac{h_i + h_{i-1}}{2}(Z_i - Z_{i-1}) \\ 0 \\ 0 \end{pmatrix}, \quad \mathbf{Q}_{i-\frac{1}{2}}^R = \begin{pmatrix} 0 \\ -g\dfrac{h_{i+1} + h_i}{2}(Z_{i+1} - Z_i) \\ 0 \\ 0 \end{pmatrix}.
$$

It is evident that for small water depths, the bed friction term dominates the other terms in the momentum equation. This is mainly due to the presence of the term $h^{\frac{4}{3}}$ in the dominator of τ in (8). To overcome this drawback we use a semi-implicit time integration of the source term $\mathbf{S}$ in (14) as

$$\frac{h^{n+1} - \widetilde{h}}{\Delta t_n} = 0,$$

$$\frac{(hv)^{n+1} - \left(\widetilde{h}\widetilde{v}\right)}{\Delta t_n} = -gn_b^2 \frac{(hv)^{n+1} |\widetilde{v}|}{\left(\widetilde{h}\right)^{\frac{4}{3}}}, \tag{22}$$

where $\widetilde{h}$ and $\widetilde{v}$ are the water height and velocity obtained from the first step (13) of the splitting procedure. Solving the second equation in (22) for $(hv)^{n+1}$ yields

$$(hv)^{n+1} = \frac{\left(\widetilde{h}\widetilde{v}\right)}{1 + \Delta t_n gn_b^2 |\widetilde{v}| / \left(\widetilde{h}\right)^{\frac{4}{3}}}. \tag{23}$$

As in most explicit time integration schemes, the time step in our finite volume method is selected using a Courant-Friedrichs-Lewy (CFL) condition. In our simulations, the Courant number Cr is fixed and Δt_n is chosen at each time step according to the following CFL condition

$$\Delta t_n = Cr \frac{\min_i (\Delta x_i)}{\max_{k=1,2,3,4} \left(\left|\widehat{\lambda}_k^+\right|, \left|\widehat{\lambda}_k^-\right|\right)}, \tag{24}$$

where $\widehat{\lambda}_k^\pm$ are the eignevalues (20) computed using the space-averaged solutions in the control volume $\left[x_{i-\frac{1}{2}}, x_{i+\frac{1}{2}}\right]$ and its two neighbouring cells.

3.2   Unstructured Finite Element Solution of Elasticity

The starting point for the finite element method is the variational formulation of the strain energy in the domain Ω. Thus, multiplying the strong form of x-direction equation in (9) by an arbitrary weight function ϕ_x and integrate over the domain yields

$$\int_\Omega \frac{\partial \sigma_x}{\partial x} \phi_x \, dx + \int_\Omega \frac{\partial \tau_{xz}}{\partial z} \phi_x \, dx - \int_\Omega f_x \phi_x \, dx = 0.$$

Using the Green-Gauss theorem, the above equation becomes

$$\oint_{\partial\Omega} \sigma_x n_x \phi_x \, dx - \int_\Omega \frac{\partial \phi_x}{\partial x} \sigma_x \, dx + \oint_{\partial\Omega} \tau_{xz} n_z \phi_x \, dx - \int_\Omega \frac{\partial \phi_x}{\partial z} \tau_{xz} \, dx - \int_\Omega f_x \phi_x \, dx = 0,$$

where $\boldsymbol{x} = (x, z)^{\top}$ and $\mathbf{n} = (n_x, n_z)^{\top}$ is the outward unit normal on $\partial\Omega$ with $\partial\Omega = \Gamma \cup \Gamma_i$. Using the x-component of the traction $\mathcal{T}_x = \sigma_x n_x + \tau_{xz} n_z$, the above equation can be written as

$$\oint_{\partial\Omega} \mathcal{T}_x \phi_x \, d\boldsymbol{x} - \int_{\Omega} \left(\frac{\partial\phi_x}{\partial x} \sigma_x + \frac{\partial\phi_x}{\partial z} \tau_{xz} \right) d\boldsymbol{x} - \int_{\Omega} f_x \phi_x \, d\boldsymbol{x} = 0. \qquad (25)$$

Similar steps applied to the z-direction equation in (9) give

$$\oint_{\partial\Omega} \mathcal{T}_z \phi_z \, d\boldsymbol{x} - \int_{\Omega} \left(\frac{\partial\phi_z}{\partial x} \tau_{xz} + \frac{\partial\phi_z}{\partial z} \sigma_z \right) d\boldsymbol{x} - \int_{\Omega} f_z \phi_z \, d\boldsymbol{x} = 0, \qquad (26)$$

where $\mathcal{T}_z = \sigma_z n_z + \tau_{xz} n_x$. Adding the two Eqs. (25) and (26) yields

$$\oint_{\partial\Omega} (\mathcal{T}_x \phi_x + \mathcal{T}_z \phi_z) \, d\boldsymbol{x} - \int_{\Omega} (f_x \phi_x + f_z \phi_z) \, d\boldsymbol{x}$$
$$- \int_{\Omega} \left(\frac{\partial\phi_x}{\partial x} \sigma_x + \frac{\partial\phi_x}{\partial z} \tau_{xz} + \frac{\partial\phi_z}{\partial z} \sigma_z + \frac{\partial\phi_z}{\partial x} \tau_{xz} \right) d\boldsymbol{x} = 0,$$

which can be reformulated in a vector form as

$$\int_{\Omega} \widehat{\boldsymbol{\phi}} \cdot \boldsymbol{\sigma} \, d\boldsymbol{x} = \oint_{\partial\Omega} \boldsymbol{\phi}^{\top} \cdot \boldsymbol{\mathcal{T}} \, d\boldsymbol{x} + \int_{\Omega} \boldsymbol{\phi}^{\top} \cdot \mathbf{f} \, d\boldsymbol{x}, \qquad (27)$$

where $\boldsymbol{\phi} = (\phi_x, \phi_z)^{\top}$, $\boldsymbol{\mathcal{T}} = (\mathcal{T}_x, \mathcal{T}_z)^{\top}$ and $\widehat{\boldsymbol{\phi}} = \left(\frac{\partial\phi_x}{\partial x}, \frac{\partial\phi_z}{\partial z}, \frac{\partial\phi_x}{\partial z} + \frac{\partial\phi_z}{\partial x} \right)^{\top}$. To solve the weak form (27) with the finite element method, the domain Ω is discretized into a set of elements where the solution is approximated in terms of the nodal values U_j and the polynomial basis functions $N_j(x, z)$ as

$$\boldsymbol{u}(x, z) = \sum_{j=1}^{N_d} \mathbf{U}_j N_j(x, z), \qquad (28)$$

where N_d is the number of mesh nodes. In the present work, we consider quadratic triangular elements with six nodes for which the elementary matrices are assembled into a global system of equations

$$\mathbf{Ku} = \mathbf{b}, \qquad (29)$$

where $\mathbf{K}$ is the global stiffness matrix, $\boldsymbol{u}$ is the nodal displacement vector and $\boldsymbol{b}$ is the force vector. In our simulations, the matrix $\mathbf{K}$ is decomposed into an $LUL^{\top}$ factorization, then the solution is reduced to backward/forward substitutions after updating the right-hand side vector $\boldsymbol{b}$ at every time step.

3.3   Implementation of Coupling Conditions at the Interface

One of the advantages in using non-uniform grids in the finite volume solution is to avoid interpolations at the interface for interchange coupling conditions.

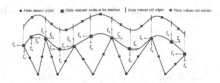

Fig. 1. An illustration of finite element and finite volume nodes at the interface.

Here, the selected control volumes in the finite volume methods coincide with the finite element nodes on the interface as shown in Fig. 1. At each time step coupling conditions are transferred on the interface for both models to update the solutions for the displacement u, water height h and water velocity v. In the present work, the deformed finite element nodes on the interface are used to reconstruct the bed $Z(x,t)$ for the shallow water Eqs. (1). Here, a triangular finite element with three nodes on the interface yields two non-uniform control volumes the edges of which are the three nodes and their centers are obtained by averaging the coordinates of these nodes, compare Fig. 1. We also assume that once the deformation occurs, the time variation in these coordinates is negligible and therefore no need for interpolation procedures to reconstruct the bed topography in the finite volume method. This bed profile is used in the finite volume solution of the flow system to obtain the water height h^{n+1} and the water velocity v^{n+1} at the next time level t_{n+1}. For coupling conditions from the water flow to the bed on the interface, the forces f_x and f_z in the elasticity Eqs. (9) are reconstructed at each time step. Here, the horizontal force f_x in the x-direction is updated using the friction term as

$$f_x = -gn_b^2 h^{n+1} \frac{v^{n+1}\left|v^{n+1}\right|}{(h^{n+1})^{\frac{4}{3}}}. \tag{30}$$

The vertical force f_z in the z-direction is reconstructed at each time step using the change in the hydrostatic pressure as

$$p^{n+1} = -\rho g \frac{h^{n+1} - h^n}{\Delta t}_n,$$

and at each node of the three finite element nodes located on the interface, the force f_z is distributed using the integral form as

$$f_z^{(1)} = \int_{-1}^{1} -\frac{1}{2}\xi(1-\xi)\, p^{n+1}\frac{\hbar}{2}\, d\xi \;=\; \frac{1}{6}p^{n+1}\hbar,$$

$$f_z^{(2)} = \int_{-1}^{1} (1-\xi^2)\, p^{n+1}\frac{\hbar}{2}\, d\xi \;=\; \frac{2}{3}p^{n+1}\hbar, \tag{31}$$

$$f_z^{(3)} = \int_{-1}^{1} \frac{1}{2}\xi(1+\xi)\, p^{n+1}\frac{\hbar}{2}\, d\xi \;=\; \frac{1}{6}p^{n+1}\hbar,$$

where $\hbar$ is the edge length of the considered element on the interface. It is easy to verify that $f_z^{(1)} + f_z^{(2)} + f_z^{(3)} = p^{n+1}\hbar$. The total force f_z in the z-direction

Table 1. Parameters used in simulations for the dam-break problem.

Quantity	Reference value	Quantity	Reference value
ρ_w	$1000\ kg/m^3$	ν	$1.2 \times 10^{-6}\ m^2/s$
ρ_s	$2650\ kg/m^3$	n_b	$0.03\ s/m^{1/3}$
g	$9.81\ m/s^2$	p	0.4
φ	$0.015\ m^{1.2}$	θ_c	0.045

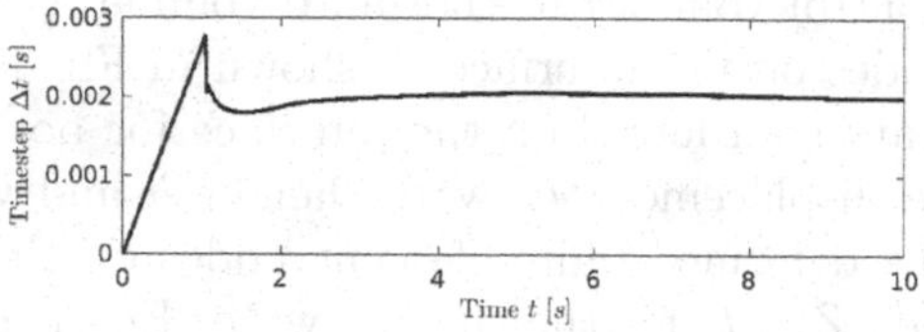

Fig. 2. Time evolution of the time step Δt using the CFL condition (24).

is obtained by accumulating the elemental forces on the overlapping nodes, see Fig. 1 for an illustration.

4   Numerical Results

We solve the problem of a dam-break over erodible bed studied in [3] using the parameters listed in Table 1. All the simulations are performed using a mesh with 100 gridpoints (unless stated) and numerical results are displayed at time $t = 10\ min$ using a time step adjusted according to the CFL condition (24) with $Cr = 0.75$. The obtained time evolution of the time step Δt is presented in Fig. 2 confirming that it does not overpass 3×10^{-3} for this problem. Figure 3 depicts the mesh used in our simulations before and after deformation. Based on a mesh convergence study not reported here for brevity, an unstructured triangular mesh with 1749 quadratic elements and 3763 nodes is used in our simulations as it offers a compromise between accuracy and efficiency in the numerical method.

   To validate our results to experiment data for this example, we present in Fig. 4 the results obtained at time $t = 10$. The agreement between the numerical simulations and experimental measurements in this figure is fairly good. The water free-surface and the erodible bed are well predicted by the proposed numerical approach. Obviously, the computed results for both water height and bed profile verify the stability and the shock capturing properties of the numerical method for this dam-break problem over a wet bed. Figure 5 depicts the distribution of the normal stress component σ_z and the shear stress τ_{xz}. It is clear that maximum values of stresses are located on the bed surface where the erosion has taken place. The deformed bed has also been accurately resolved using our finite element method. Under the considered conditions, stress distributions exhibit symmetrical features in both stresses. Furthermore, no mesh distortion has been detected in all results obtained for this dam-break problem.

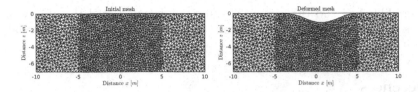

Fig. 3. Initial mesh (left) and deformed mesh at time $t = 10$ s (right).

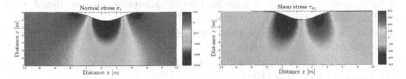

Fig. 4. Comparison between the numerical results and experiments at time $t = 10$ s.

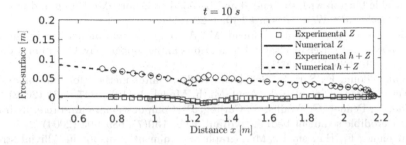

Fig. 5. Normal stress σ_z (left) and shear stress τ_{xz} at time $t = 10$ s (right).

It should be stressed that results from the proposed coupled model should be compared with observations of laboratory free-surface flows and bed deformations for this problem. However, as of now, no data is available to carry out this comparison study. Thus, at the moment we can only perform numerical simulations and verify that results are plausible and consistent.

5   Concluding Remarks

An accurate and efficient computational algorithm is presented in this study for stress analysis in hydro-sediment-morphodynamic models. The linear elasticity equations for the bed topography are coupled to the shallow water hydro-sediment-morphodynamic equations to form a consistent model for the considered problems. At the bed interface between the water surface and the bedload, transfer conditions are also developed using frictional forces and hydrostatic pressures. As a numerical solver we implement a hybrid finite volume/finite element method. The method is well-balanced for solutions of the shallow water equations and uses quadratic elements on unstructured meshes for the elasticity equations. Numerical results are presented for the problem of a dam-break over erodible bed and the computed solutions demonstrated the ability of the proposed model to accurately capture the stress distributions for erosional and depositional deformations.

References

1. Aguirre, J., Castro, M., Luna, T.: A robust model for rapidly varying flows over movable bottom with suspended and bedload transport: Modelling and numerical approach. Adv. Water Resour. **140**, 1–22 (2020)
2. Al-Ghosoun, A., Osman, A., Seaid, M.: A computational model for simulation of shallow water waves by elastic deformations in the topography. Commun. Comput. Phys. **29**, 1095–1124 (2021)
3. Benkhaldoun, F., Sari, S., Seaid, M.: A flux-limiter method for dam-break flows over erodible sediment beds. Appl. Math. Model. **36**(10), 4847–4861 (2012)
4. Cao, Z., Pender, G., Wallis, S., Carling, P.: Computational dam-break hydraulics over erodible sediment bed. J. Hydraul. Eng. **130**(7), 689–703 (2004)
5. Engelund, F., Hansen, E.: Monograph on sediment transpot in Allivial streams TUDelf. Northernlands **37**, 1–14 (1997)
6. Escalante, C., Luna, T., Castro, M.: Non-hydrostatic pressure shallow flows: GPU implementation using finite volume and finite difference scheme. Int. J. Comput.-Aided Eng. Softw. **338**, 631–659 (2018)
7. Grabowski, R.: Measuring the shear strength of cohesive sediment in the field. Geomorphol. Tech. **3**, 1–7 (2014)
8. Grass, A.: Sediment transport by waves and currents. Technical Report FL29, SERC. London Cent. Mar (1981)
9. Hairsine, B., Rose, W.: Modelling water erosion due to overland flow using physical principles. Water Resour. **28**, 237–243 (1992)
10. Hajigholizadeh, M., Melesse, A., Fuentes, H.: Erosion and sediment transport modelling in shallow waters: a review on approaches, models and applications. Int. J. Environ. Res. Public Health **15**, 1–24 (2018)
11. Hazelden, J., Boorman, A.: Soils and managed retreat in south east England. Soil use Manage. **30**, 180–189 (2001)
12. Merritt, S., Letcher, A., Jakeman, J.: A review of erosion and sediment transport models. Environ. Model Softw. **18**, 761–799 (2003)
13. Merritt, W., Letcher, R., Jakeman, A.: A review of erosion and sediment transport models. Environ. Model. Softw. **18**, 761–799 (2003)
14. Recanatesi, F., Ripa, A., Leoni, M.: Assessment of software runoff management practices and PMPs under soil sealing: a study case in a peri-urban watershed of the metroplitan area of Rome. J. Environ. Manag. **201**, 6–18 (2017)
15. Reisinger, A., Gibeaut, J., Tissot, P.: Estuarine suspended sediment dynamics: observations derived from over a decade of Satellite data. Front. Marine Sci. **4**, 1–10 (2017)
16. Takken, I., Govers, V., Nachtergaele, G., Steegen, J.: The effect of tillage-induced roughness on runoff and erosion patterns. Geomorphology **37**, 1–14 (2001)

Enhancing Computational Steel Solidification by a Nonlinear Transient Thermal Model

Fatima-Ezzahrae Moutahir[1]([✉]), Youssef Belhamadia[2], Mofdi El-Amrani[1],
and Mohammed Seaid[3]

[1] Mathematics and Applications Laboratory, FST, Abdelmalek Essaadi University,
Tangier, Morocco
fatimaezzahra94.im@gmail.com
[2] Department of Mathematics and Statistics, American University of Sharjah,
Sharjah, United Arab Emirates
[3] Department of Engineering, University of Durham,
South Road, Durham DH1 3LE, UK

Abstract. Designing efficient steel solidification methods could contribute to a sustainable future manufacturing. Current computational models, including physics-based and machine learning-based design, have not led to a robust solidification design. Predicting phase-change interface is the crucial step for steel solidification design. In the present work, we propose a simplified model for thermal radiation to be included in the phase-change equations. The proposed model forms a set of nonlinear partial differential equations and it accounts for both thermal radiation and phase change in the design. As numerical solver we implement a fully implicit time integration scheme and a Newton-type algorithm is used to deal with the nonlinear terms. Computational results are presented for two test examples of steel solidification. The findings here could be used to understand effect of thermal radiation in steel solidification. Combining the present approach with physics-based computer modeling can provide a potent tool for steel solidification design.

Keywords: Steel solidification · Phase change · Thermal radiation · Computational design

1 Introduction

Melting and solidification processes are natural phenomena and occur in many industrial processes such as crystal growth, continuous casting and metal welding among others. During solidification the phase front travels at the interface between the liquid and solid materials. In all applications that involve high temperature, radiation is expected to greatly influence the thermal features and it cannot be neglected. Experimental predictions of the impact of radiation in materials during the solidification process can be very demanding and laborious. Although the new development of modern engineering technologies, accurate prediction of effects of radiative heat transfer in this type of phase-change

© The Author(s), under exclusive license to Springer Nature Switzerland AG 2022
D. Groen et al. (Eds.): ICCS 2022, LNCS 13350, pp. 305–317, 2022.
https://doi.org/10.1007/978-3-031-08751-6_22

materials still faces several complex issues and can be experimentally demanding and challenging. It is well-known that any experimental system intended for investigation always involves meticulous design and subsequent procurement of materials, fabrication or construction of the system, which necessitates heavy financial resources and involves practically more time, see for instance [1–3]. Hence, computational simulations are commonly preferred for designing, modelling and simulation of thermal systems. This numerical investigation helps to accumulate functional data, and identify operating conditions or environment at which the best performance of a workable system could be obtained. Computational simulations therefore can play a crucial role and provide accurate and effective thermal predictions in this class of applications. The present work aims to develop robust computational tools for highly accurate simulations of steel solidification processes.

Many developed methods considered mathematical models based on the enthalpy formulation for simulating the phase-change in the materials. These types of phase change models have been coupled with the natural convection and the mechanical deformation to account for the fluid flow dynamics and internal cracks respectively (see [3–7] among others). Existence of a weak solution of heterogeneous Stefan problem using enthalpy formulation is presented in [8]. Coupling radiation with phase-change models is very complicated and highly demanding. A full radiative heat transfer model consists of an integro-differential equations that are spatially, spectrally, and directionally dependent [9, 10]. These equations are therefore extremely difficult to solve, especially when coupled with the energy transport equation and phase-change closures. However, for optically thick materials with high scattering effects, the thermal radiation can be well approximated with the Rosseland model proposed in [11]. This simplification significantly reduces the computational costs compared to solving the full radiative heat transfer model, see for example [12]. In the present study, we are interested in coupling a class of phase-change models [13, 14] with the Rosseland diffusion approximation of radiative heat transfer. The phase change model employed is considered as intermediate formulation between the enthalpy and the so-called phase-field formulations, where a phase parameter that takes constant values in the solid and liquid phases is employed. The Rosseland approximation includes thermal radiation into the system through a nonlinear diffusion term with convective boundary conditions. The coupled system is expected to provide an accurate representation for radiation transport in both participating and non-participating optically thick media. For the numerical solution of the coupled Rosseland-phase-change model, we propose a consistent finite difference method using staggered grids. The Newton's method is employed to deal with the non-linearity in the mathematical model and two-dimensional numerical results are presented for two test examples to illustrate the effects of radiation on the steel solidification.

This paper is organized as follows. In Sect. 2 we introduce the mathematical equations used for modelling steel solidification. Formulation of the proposed fully implicit method is presented in Sect. 3. Section 4 is devoted to numerical results for two test examples for steel solidification problems. Concluding remarks are presented in Sect. 5.

2   Mathematical Models for Steel Solidification

In general applications, modelling steel solidification involves a computational domain $\Omega = \Omega_l(t) \cup \Omega_s(t) \cup \Gamma(t)$ with time-dependent liquid domain $\Omega_l(t)$, solid domain $\Omega_s(t)$ and the interface $\Gamma(t)$ between both domains. The material properties are expected to vary from one state to another according for the interface location. In this case, the set of governing equations is difficult to solve and one way to overcome the numerical difficulties is to use a new formation over the entire computational domain Ω. In the current work, we reformulate the system using the enthalpy formulation and the semi-phase-field technique described in [13,14] among others. Thus, given a bounded two-dimensional domain $\Omega \subset \mathbb{R}^2$ with Lipschitz continuous boundary $\partial\Omega$ and a time interval $[0, T]$, we focus on solving the time-dependent heat equation coupled with the phase change

$$\eta(\phi)\frac{\partial T}{\partial t} + \rho L \frac{\partial F_\epsilon(T)}{\partial t} - \nabla \cdot (\mathcal{K}_c(\phi)\nabla T) = 0, \qquad (\mathbf{x}, t) \in \Omega \times [0, T],$$

$$\mathcal{K}_c(\phi)\mathbf{n}(\hat{\mathbf{x}}) \cdot \nabla T + \hbar_c(\phi)(T - T_b) = 0, \qquad (\hat{\mathbf{x}}, t) \in \partial\Omega \times [0, T], \quad (1)$$

$$T(\mathbf{x}, 0) = T_0(\mathbf{x}), \qquad \mathbf{x} \in \Omega,$$

where $\mathbf{n}(\hat{\mathbf{x}})$ denotes the outward normal in $\hat{\mathbf{x}}$ with respect to $\partial\Omega$, $T(x, t)$ is the temperature field, T_b the boundary temperature, T_0 the initial temperature, ρ the density, L the latent heat of fusion and ϕ is the regularized phase-field function defined as

$$\phi = F_\epsilon(T) = \frac{1}{2} - \frac{1}{2}\tanh\left(\frac{T_f - T}{\epsilon}\right),$$

with T_f is the melting temperature and ϵ is a small parameter selected such that the resulting function F_ϵ is differentiable. In (1),

$$\eta(\phi) = \rho_s c_s + \phi\left(\rho_l c_l - \rho_s c_s\right), \qquad \mathcal{K}_c(\phi) = K_s + \phi\left(K_l - K_s\right),$$

uation

$$\hbar_c(\phi) = \hbar_s + \phi\left(\hbar_l - \hbar_s\right), \qquad \alpha(\phi) = \alpha_s + \phi\left(\alpha_l - \alpha_s\right),$$

where ρ_i, c_i, K_i, $\hbar_i$ and α_i are respectively, the density, specific heat, thermal conductivity, convective heat transfer coefficient and hemispheric emissivity of the phase i with subscripts s and l refer to the solid and liquid phases.

In the present study, to enhance steel solidification modelling we include thermal radiation in grey optically thick media. For a weakly semitransparent medium with large scattering such as steel, an asymptotic expansion for the radiative transfer equation yields the equilibrium diffusion or the Rosseland approximation [11]. When coupled to the phase change equations (1), the Rosseland approximation yields

$$\eta(\phi)\frac{\partial T}{\partial t} + \rho L \frac{\partial F_\epsilon(T)}{\partial t} - \nabla \cdot (\mathcal{K}_r(\phi)\nabla T) = 0, \qquad (\mathbf{x}, t) \in \Omega \times [0, T],$$

$$\mathcal{K}_c(\phi)\mathbf{n}(\hat{\mathbf{x}}) \cdot \nabla T + \hbar_c(\phi)(T - T_b) = \alpha(\phi)\pi\left(B(T_b) - B(T)\right), \qquad (2)$$

$$T(\mathbf{x}, 0) = T_0(\mathbf{x}), \qquad \mathbf{x} \in \Omega,$$

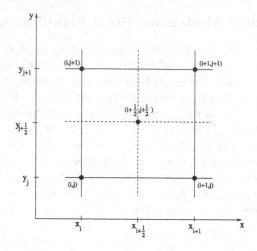

Fig. 1. The staggered grid used for the space discretization.

where $B(T)$ is the spectral intensity of the black-body radiation defined by the Planck function as

$$B(T) = \sigma_R T^4,$$

with $\sigma_R = 5.67 \times 10^{-8}$ is the Stefan-Boltzmann constant. Here, the conduction coefficient $\mathcal{K}_t(T)$ is defined as a function of the temperature by

$$\mathcal{K}_r(\phi) = \mathcal{K}_c(\phi) + \frac{4\pi}{3\kappa(\phi)} \frac{\partial B}{\partial T},$$

where κ is the absorption coefficient. It should be stressed that the Rosseland approach is widely accepted as an accurate model for radiation transport in both participating and non-participating optically thick media. These equations do not have analytical solutions for phase change and their numerical solutions lead to computationally demanding problems due to the nonlinear diffusion and the presence of internal and external thermal boundary layers.

3   Numerical Methods for the Nonlinear System

Different numerical methods can be used for solving systems (1) and (2) (see chapter 4 in [15]). The convergence analysis of iterative methods resulted from implicit time discretization of the enthpay formulation coupled with finite difference of the space variables is studied in [16]. Since it is easier to combine upwinding with finite volume discretization than other methods, we consider in this study a space discretization based on volume control and cell averaging. For the time integration we implement a fully implicit backward second-order scheme allowing for large stability in the simulations. Hence, we divide the time interval into subintervals $[t_n, t_{n+1}]$ of equal length Δt and $t_n = n\Delta t$. For simplicity, we discretize the spatial domain into cells with sizes $(\Delta x)_i$ and $(\Delta y)_j$

in the x and y directions, respectively. We also define the maximum cell size $h = \max_{ij}((\Delta x)_i, (\Delta y)_j)$ and the averaged gridpoints as shown in Fig. 1 by

$$(\Delta x)_{i+\frac{1}{2}} = x_{i+1} - x_i, \qquad (\Delta y)_{j+\frac{1}{2}} = y_{j+1} - y_j,$$

$$x_{i+\frac{1}{2}} = \frac{x_{i+1} + x_i}{2}, \qquad y_{j+\frac{1}{2}} = \frac{y_{j+1} + y_j}{2}.$$

Using the notation W_{ij}^n to denote the approximation value of the function W at time $t = t_n$ and the gridpoint (x_i, y_j), the semi-discrete form of the system (2) reads

$$\eta(\phi_{i+\frac{1}{2}j+\frac{1}{2}}^{n+1})\mathcal{D}_t^2 T_{i+\frac{1}{2}j+\frac{1}{2}}^{n+1} + \rho L \mathcal{D}_t^2 \phi_{i+\frac{1}{2}j+\frac{1}{2}}^{n+1} - \mathcal{D}_h^2 (\mathcal{K}_r T)_{ij}^{n+1} = 0, \qquad (3)$$

where the temporal difference $\mathcal{D}_t^2$ is defined as

$$\mathcal{D}_t^2 W^{n+1} = \frac{3W^{n+1} - 4W^n + W^{n-1}}{2\Delta t},$$

and the spatial difference operator $\mathcal{D}_h^2$ is given by $\mathcal{D}_h^2 = \mathcal{D}_x^2 + \mathcal{D}_y^2$ with

$$\mathcal{D}_x^2(\mathcal{K}W)_{ij} = \frac{\mathcal{K}_{i,j} + \mathcal{K}_{i+1j}}{2} \frac{W_{i+1j} - W_{ij}}{(\Delta x)_{i+\frac{1}{2}}^2} - \frac{\mathcal{K}_{t_{i-1j}} + \mathcal{K}_{ij}}{2} \frac{W_{ij} - W_{i-1j}}{(\Delta x)_{i+\frac{1}{2}}^2},$$

$$\mathcal{D}_y^2(\mathcal{K}W)_{ij} = \frac{\mathcal{K}_{i,j} + \mathcal{K}_{ij+1}}{2} \frac{W_{ij+1} - W_{ij}}{(\Delta y)_{j+\frac{1}{2}}^2} - \frac{\mathcal{K}_{ij-1} + \mathcal{K}_{ij}}{2} \frac{W_{ij} - W_{ij-1}}{(\Delta y)_{j+\frac{1}{2}}^2},$$

with the cell averages of a function W are given by

$$W_{i+1j} = \frac{1}{(\Delta x)_{i+\frac{1}{2}}} \int_{y_j}^{y_{j+1}} W(x_i, y)dy,$$

$$W_{ij+1} = \frac{1}{(\Delta y)_{j+\frac{1}{2}}} \int_{x_i}^{x_{i+1}} W(x, y_j)dx, \qquad (4)$$

$$W_{ij} = \frac{1}{(\Delta x)_{i+\frac{1}{2}}(\Delta y)_{j+\frac{1}{2}}} \int_{x_i}^{x_{i+1}} \int_{y_j}^{y_{j+1}} W(x, y)dxdy.$$

Here, the function value of $W_{i+\frac{1}{2}j+\frac{1}{2}}$ at the cell centre is simply approximated by bilinear interpolation as

$$W_{i+\frac{1}{2}j+\frac{1}{2}} = \frac{W_{ij} + W_{i+ij} + W_{ij+1} + W_{i+1j+1}}{4},$$

and the discrete phase-field function $\phi_{i+\frac{1}{2}j+\frac{1}{2}}^{n+1}$ in (3) is defined by

$$\phi_{i+\frac{1}{2}j+\frac{1}{2}}^{n+1} = F_\epsilon \left(T_{i+\frac{1}{2}j+\frac{1}{2}}^{n+1} \right).$$

Algorithm 1

1: Given $\mathcal{F}$, tolerance τ and initial guess $T^{(0)}$ chosen to be the solution at the previous
   time step, the Newton-GMRES algorithm for solving (2) uses the following steps:
   (we denote by GMRES($\mathcal{A}$,q,$\mathbf{z}^{(0)}$,τ) the result of GMRES algorithm applied to
   linear system $\mathcal{A}\mathbf{z} = \mathbf{q}$ with initial guess $\mathbf{z}^{(0)}$ and tolerance τ).
2: **for** $k = 0, 1, \ldots$ **do**
3:    Compute the residual

$$\mathcal{H}(T^{(k)}) = T^{(k)} - \mathcal{F}(T^{(k)}).$$

4:    Solve using GMRES

$$\mathbf{d}^{(k)} = \text{GMRES}\big(\mathcal{H}'(T^{(k)}), -\mathcal{H}(T^{(k)}), \mathbf{d}^{(0)}, \tau^{(k)}\big).$$

5:    Update the solution
$$T_L^{(k+1)} = T^{(k)} + \xi\mathbf{d}^{(k)}.$$

6:    Check the convergence

$$\text{if } \left(\|T^{(k+1)}\|_{L^2} \leq \tau\right) \qquad \text{stop.}$$

7: **end for**

Similarly, the gradient in the boundary condition in (2) is approximated by
upwinding without using ghost points. For example, on the left boundary of the
domain, the boundary discretization is

$$-\mathcal{K}_c\left(\phi_{\frac{1}{2}j+\frac{1}{2}}^{n+1}\right)\frac{T_{\frac{3}{2}j+\frac{1}{2}}^{n+1} - T_{\frac{1}{2}j+\frac{1}{2}}^{n+1}}{(\Delta x)_{\frac{1}{2}}} + \hbar_c\left(\phi_{\frac{1}{2}j+\frac{1}{2}}^{n+1}\right)\left(T_{\frac{1}{2}j+\frac{1}{2}}^{n+1} - T_b\right)$$
$$= \alpha\left(\phi_{\frac{1}{2}j+\frac{1}{2}}^{n+1}\right)\pi\left(B(T_b) - B\left(T_{\frac{1}{2}j+\frac{1}{2}}^{n+1}\right)\right), \qquad (5)$$

and similar work has to be done for the other boundaries. All together, the above
discretization leads to a nonlinear system reformulated as a fixed point problem
for the temperature T as

$$T = \mathcal{F}(T). \qquad (6)$$

The Newton's method applied to (6) results in the following iteration

$$T^{(k+1)} = T^{(k)} - \mathcal{H}'(T^{(k)})^{-1}\mathcal{H}(T^{(k)}), \qquad (7)$$

where $\mathcal{H}'_L$ is the system Jacobian approximated by a difference quotient of the
form

$$\mathcal{H}'(T^{(k)})\mathbf{w} \approx \frac{\mathcal{H}(T^{(k)} + \delta\mathbf{w}) - \mathcal{H}(T^{(k)})}{\delta}. \qquad (8)$$

If a GMRES method [17] is used to compute the Newton's direction then, at
each time step Algorithm 1 is carried out to update the solution T^{n+1}. Here
$\|\cdot\|_{L^2}$ denotes the discrete L^2-norm. The Newton step ξ, the tolerance $\tau^{(k)}$ to

Table 1. Convergence results for the accuracy example using different time steps Δt at time $t = 1$.

Δt	L^∞-error	Rate	L^1-error	Rate	L^2-error	Rate
0.2	1.1140E−01	−	1.0460E−01	−	7.0500E−02	−
0.1	3.1200E−02	1.8361	2.9300E−02	1.8359	1.9700E−02	1.8394
0.05	8.1000E−03	1.9456	7.6000E−03	1.9468	5.1000E−03	1.9496
0.025	2.0000E−03	2.0179	1.9000E−03	2.0000	1.3000E−03	1.9720

Table 2. Convergence results for the accuracy example using different space steps $h = \Delta x = \Delta y$ at time $t = 1$.

h	L^∞-error	Rate	L^1-error	Rate	L^2-error	Rate
0.2	6.2000E−03	−	6.3000E−03	−	6.4000E−03	−
0.1	8.9546E−04	2.7916	9.8414E−04	2.6553	1.2000E−03	2.4150
0.05	2.1259E−04	2.0746	2.3479E−04	2.0675	2.8667E−04	2.0656
0.025	5.3530E−05	1.9897	5.9890E−05	1.9710	1.8668E−05	2.0162

stop the inner iterations in GMRES, and the difference increment δ in (8) are selected according to backtracking linesearch, Eisenstat-Walker and Hardwired techniques. We refer to [18] for detailed discussions on these techniques. Three to five Newton's iterations were necessary to achieve convergence with a residual norm less than 10^{-6}. The choice of above mentioned time discretization is based on our previous experience [19], where this scheme provided better numerical solutions for other type of interface problems. It should be noted that when using uniform structured meshes, the above spatial discretization is equivalent to the well-established central finite difference which provides good numerical results.

4   Results and Discussions

In this section, we present numerical results for the proposed Rosseland-phase-change simplified model to examine the effect of radiation in materials under phase change. Two test problems are considered to demonstrate the performance of the proposed method. We first start with a manufactured solution to study the convergence of our algorithm. Then, the effect of radiation on the temperature distributions is examined in two solidification examples in the process of continuous casting of steel.

4.1   Accuracy Example

As a first test example, we consider a two-dimensional analytical solution to discuss the order of convergence in space and time of the proposed method. Since

Table 3. Thermophysical parameters used for the steel solidification.

$\mathcal{K}_c$	34 W/mK
c	691 J/KgK
ρ	7400 Kg/m^3
L	272000 J/Kg
T_f	1809 K
T_0	2000 K
T_b	300 K
h_c	1648.5 J/(Kg K)
κ	10 m^{-1}
α	0.0001

the analytical solution does not represent a specific physical meaning, we take only numerical values without considering any units. The problem is solved in a squared domain $\Omega = [0, 1] \times [0, 1]$ subject to boundary and the initial conditions explicitly calculated such that the analytical solution of (1) is given by

$$T_{ex}(t, x, y) = e^{-t^2 - x^2 - y^2}.$$

The nonlinear diffusion coefficient is given by

$$K_c(\phi) = T^2.$$

We consider the following relative $L^\infty-$, L^1- and L^2-error norms

$$\| e \|_{L^\infty} = \frac{\|T - T_{ex}\|_{L^\infty}}{\|T_{ex}\|_{L^\infty}}, \quad \| e \|_{L^1} = \frac{\|T - T_{ex}\|_{L^1}}{\|T_{ex}\|_{L^1}}, \quad \| e \|_{L^2} = \frac{\|T - T_{ex}\|_{L^2}}{\|T_{ex}\|_{L^2}},$$

where T is the numerical solution and T_{ex} is the analytical solution computed at the final time $t = 1$. First, to test the convergence of the time discretization scheme, we consider four time steps of different sizes using a fine uniform mesh with $\Delta x = \Delta y = 0.002$. The obtained results are listed in Table 1 along with their corresponding convergence rates. Similarly, to examine the convergence of the spatial discretization we consider four space steps of different sizes using a fine time step $\Delta t = 0.0001$. The obtained results are listed in Table 2. As it can be seen the proposed method preserves the second-order accuracy for both space and time for this test problem.

4.2   Steel Solidification: Case 1

In all applications that involve high temperature, radiation is expected to greatly influence the thermal features and it cannot be neglected. In this example, we will examine the effect of radiation on the temperature profile in a solidification example during the process of continuous casting of steel. The mathematical model

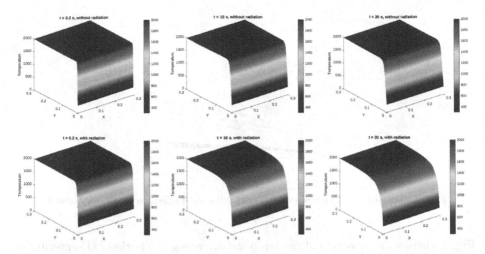

Fig. 2. Temperature without radiation (first row) and with radiation (second row) obtained for Case 1 at time $t = 0.2\,\mathrm{s}$ (first column), $t = 10\,\mathrm{s}$ (second column) and $t = 20\,\mathrm{s}$ (third column).

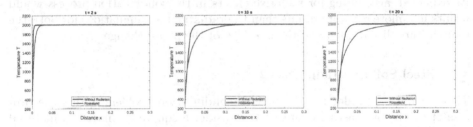

Fig. 3. Cross-sections of the temperature with and without radiation along the horizontal centerline for Case 1 at time $t = 0.2\,\mathrm{s}$, $t = 10\,\mathrm{s}$ and $t = 20\,\mathrm{s}$.

(1) is used to simulate the solidification examples without radiation while, the presence of radiation is simulated using the proposed simplified Rosseland-phase-change model (2). In this first case, we consider a square $0.3\,\mathrm{m} \times 0.3\,\mathrm{m}$ material and the thermophysical properties are assumed to be the same in both the solid and liquid phases. The initial temperature of molten steel is 2000 K, which is higher than the melting temperature of 1809 K. We consider Dirichlet boundary condition $(T = 300\,\mathrm{K})$ in the left side while homogeneous Neumann boundary condition is considered for all others sides of the computational domain. The thermophysical properties employed for the numerical simulations are summarized Table 3. A uniform mesh with 100×100 cells is used in our computations and the time step $\Delta t = 0.005\,\mathrm{s}$ is considered in this section.

Figure 2 presents the time evolution of temperature for both cases without and with radiation at three different instants $t = 0.2\,\mathrm{s}$, 10 s and 20 s. As it can be seen, the presence of radiation has affected the temperature profile as well as the position of the liquid-solid interface. This can be clearly seen in

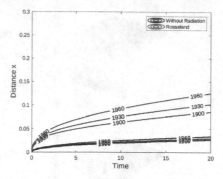

Fig. 4. Temperature contours with and without radiation along for Case 1.

Fig. 3 where a cross-section of the temperature along the horizontal centerline is displayed for the considered times. Furthermore, to clearly illustrate the effect of radiation on the temperature, Fig. 4 shows the isolines corresponding to $T = 1900\,\mathrm{K}$, $T = 1930\,\mathrm{K}$ and $T = 1960\,\mathrm{K}$ for both cases, with and without radiation. As expected, accounting for radiative effects in the solidification process would results in a more accurate results than the radiationless simulations for both the temperature distribution and the interface of the phase change.

4.3   Steel Solidification: Case 2

This example is similar to the previous solidification problem during the process of continuous casting of steel. We consider a square $0.2\,\mathrm{m} \times 0.2\,\mathrm{m}$ material

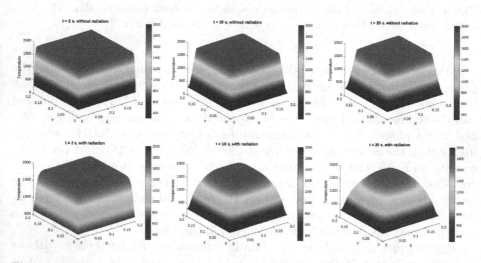

Fig. 5. Temperature without radiation (first row) and with radiation (second row) obtained for Case 2 at time $t = 0.2\,\mathrm{s}$ (first column), $t = 10\,\mathrm{s}$ (second column) and $t = 20\,\mathrm{s}$ (third column).

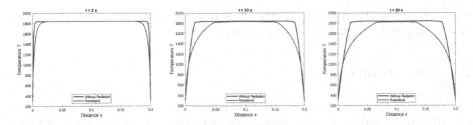

Fig. 6. Cross-sections of the temperature with and without radiation along the horizontal centerline for Case 1 at time $t = 0.2$ s, $t = 10$ s and $t = 20$ s.

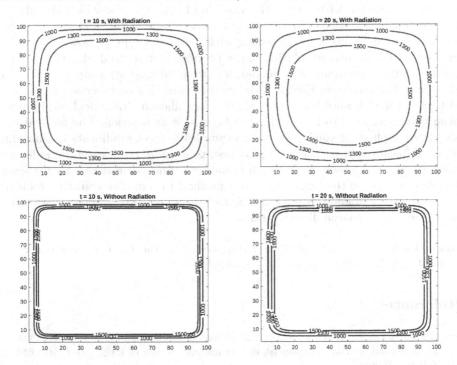

Fig. 7. Temperature contours with and without radiation along for Case 2.

using similar thermophysical properties. However, the boundary conditions in this case are different, where we consider Robbin boundary in all the computational domain boundaries as given by Eq. (2). The time evolution of temperature for both simulations, with and without radiation is presented in Fig. 5 at three different instants $t = 0.2$ s, 10 s and 20 s. As in the previous example, the presence of radiation has affected both, the temperature profile and the position of the liquid-solid interface. This is clearly depicted in Fig. 6 where a cross-section of the temperature along the horizontal center-line is plotted. We also present in Fig. 7 the isolines corresponding to temperature values of $T = 1000$ K, $T = 1300$ K and $T = 1500$ K. This figure clearly shows that the presence of radiation affects the temperature distribution and therefore it should not be neglected in phase-change applications that involve high temperature.

5   Conclusions

In this study, we have presented a class of computational techniques for enhancing steel solidification by a nonlinear transient thermal model. The governing equations consist of a nonlinear heat transfer equations with a phase-field function to account for phase change in the model. The thermal radiation effects are included in the model by using the Rosseland approach for which the phase-change properties appear in the optical parameters of the material. A fully implicit time integration scheme along with a Newton-type algorithm is implemented for the numerical solution of the proposed model to deal with the nonlinear terms. In our numerical simulations, we have used structured meshes for the space discretization. However, the method can also be extended to the use of unstructured meshes based on a similar formulation. Numerical results have been presented for a test example with known exact solution. The method has also been applied for solving two test examples in steel solidification using different diffusion values. The presented results support our expectations for an accurate and stable behaviour for all radiative regimes considered. Future work will concentrate on the extension of this method to radiative transfer problems in phase-change domains using full radiative model on unstructured meshes and using high-order spatial discretizations.

Acknowledgements. Financial support provided by the Royal Society under the contract IES-R2-202078 is gratefully acknowledged.

References

1. Choudhary, S.K., Mazumdar, D.: Mathematical modelling of fluid flow, heat transfer and solidification phenomena in continuous casting of steel. Steel Res. **66**(5), 199–205 (1995)
2. Klimeš, L., Štětina, J.: A rapid GPU-based heat transfer and solidification model for dynamic computer simulations of continuous steel casting. J. Mater. Process. Technol. **226**, 1–14 (2015)
3. Koric, S., Hibbeler, L.C., Thomas, B.G.: Explicit coupled thermo-mechanical finite element model of steel solidification. Int. J. Numer. Meth. Eng. **78**(1), 1–31 (2009)
4. El Haddad, M., Belhamadia, Y., Deteix, J., Yakoubi, D.: A projection scheme for phase change problems with convection. Comput. Math. Appl. **108**, 109–122 (2022)
5. Belhamadia, Y., Fortin, A., Briffard, T.: A two-dimensional adaptive remeshing method for solving melting and solidification problems with convection. Numer. Heat Transfer Part A Appl. **76**(4), 179–197 (2019)
6. Belhamadia, Y., Kane, A., Fortin, A.: An enhanced mathematical model for phase change problems with natural convection. Int. J. Numer. Anal. Model. **3**(2), 192–206 (2012)

7. Koric, S., Thomas, B.G.: Thermo-mechanical models of steel solidification based on two elastic visco-plastic constitutive laws. J. Mater. Process. Technol. **197**(1), 408–418 (2008)

8. Roubicek, T.: The Stefan problem in heterogeneous media. Annales de l'Institut Henri Poincaré (C) Non Linear Anal. **6**(6), 481–501 (1989)

9. Seaid, M., Klar, A., Pinnau, R.: Numerical solvers for radiation and conduction in high temperature gas flows. Flow Turbul. Combust. **75**(1), 173–190 (2005)

10. Seaid, M.: Multigrid Newton-Krylov method for radiation in diffusive semitransparent media. J. Comput. Appl. Math. **203**(2), 498–515 (2007)

11. Rosseland, S.: Theoretical Astrophysics. Atomic Theory and the Analysis of Stellar Atmospheres and Envelopes. Clarendon Press, Oxford (1936)

12. Larsen, E.W., Thömmes, G., Klar, A., Seaid, M., Götz, T.: Simplified PN approximations to the equations of radiative heat transfer and applications. J. Comput. Phys. **183**(2), 652–675 (2002)

13. Belhamadia, Y., Fortin, A., Chamberland, É.: Anisotropic mesh adaptation for the solution of the Stefan problem. J. Comput. Phys. **194**(1), 233–255 (2004)

14. Belhamadia, Y., Fortin, A., Chamberland, É.: Three-dimensional anisotropic mesh adaptation for phase change problems. J. Comput. Phys. **201**(2), 753–770 (2004)

15. Alexiades, V., Solomon, A.D.: Mathematical Modeling of Melting and Freezing Processes. CRC Press Taylor & Francis Group, London (1993)

16. White, R.E.: A numerical solution of the enthalpy formulation of the Stefan problem. SIAM J. Numer. Anal. **19**(6), 1158–1172 (1982)

17. Saad, Y., Schultz, M.H.: GMRES: a generalized minimal residual algorithm for solving nonsymmetric linear systems. SIAM J. Sci. Stat. Comput. **7**(3), 856–869 (1986)

18. Brown, P.N., Saad, Y.: Hybrid Krylov methods for nonlinear systems of equations. SIAM J. Sci. Stat. Comput. **11**(3), 450–481 (1990)

19. Belhamadia, Y.: A time-dependent adaptive remeshing for electrical waves of the heart. IEEE Trans. Biomed. Eng. **55**(2), 443–452 (2008)

Weakly-Supervised Cell Classification for Effective High Content Screening

Adriana Borowa[1,2](✉)(iD), Szczepan Kruczek[3], Jacek Tabor[1](iD),
and Bartosz Zieliński[1,2](iD)

[1] Faculty of Mathematics and Computer Science, Jagiellonian University,
Łojasiewicza 6, 30-348 Kraków, Poland
ada.borowa@student.uj.edu.pl
{jacek.tabor,bartosz.zielinski}@uj.edu.pl
[2] Ardigen SA, Podole 76, 30-394 Kraków, Poland
[3] Selvita SA, Bobrzyńskiego 14, 30-348 Krakow, Poland
szczepan.kruczek@selvita.com

Abstract. High Content Screening (HCS) allows for a complex cell analysis by combining fluorescent microscopy with the capability to automatically create a large number of images. Example of such cell analysis is examination of cell morphology under influence of a compound. Nevertheless, classical approaches bring the need for manual labeling of cell examples in order to train a machine learning model. Such methods are time- and resource-consuming. To accelerate the analysis of HCS data, we propose a new self-supervised model for cell classification: Self-Supervised Multiple Instance Learning (SSMIL). Our model merges Contrastive Learning with Multiple Instance Learning to analyze images with weak labels. We test SSMIL using our own dataset of microglia cells that present different morphology due to compound-induced inflammation. Representation provided by our model obtains results comparable to supervised methods proving feasibility of the method and opening the path for future experiments using both HCS and other types of medical images.

Keywords: High Content Screening · Weakly-supervised learning ·
Self-supervised learning · Multiple Instance Learning

1 Introduction

High Content Screening (HCS) is used to identify changes in a cell phenotype caused by e.g. small molecules or RNA [12]. The main application of HCS is the

The works are carried out under contract no. POIR.01.01.01-00-0878/19-00, as: "HiS-cAI - Development of cell-based phenotypic platform based on high content imaging system integrated with artificial intelligence data analysis for neuroinflammatory and fibrosis drug discovery", co-financed by the European Regional Development Fund under the Smart Growth Operational Programme, Submeasure 1.1.1.: Industrial research and development work implemented by enterprises.

D. Groen et al. (Eds.): ICCS 2022, LNCS 13350, pp. 318–330, 2022.
https://doi.org/10.1007/978-3-031-08751-6_23

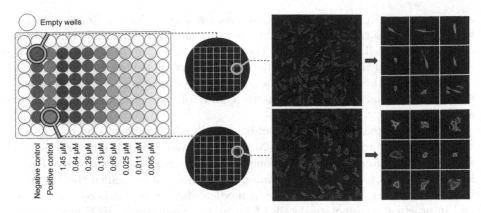

Fig. 1. High Content Screening data: Starting from left, the first part of the figure present an entire HCS plate containing 96 wells. Wells marked red contain negative control samples – without any compounds – cells are not inflamed and they are elongated with rather smooth borders and visible long processes. Wells marked green contain positive control samples with LPS induced inflammation cells, they get smaller, rounder, and develop short 'spikes'. Wells marked blue contain LPS and CLI-095 compound, the darker the color the higher the CLI-095 concentration and the bigger influence of the compound – cells are less inflamed and more similar to the positive control. Each well contains 49 microscopic images as presented in the middle of the figure. On the right, we present examples of final microscopic images and patches extracted from them. Top patches present cells from negative control and the bottom patches present cells from positive control. (Color figure online)

drug discovery process where it accelerates screening of potential therapeutic compounds. Images are created using fluorescent reagents which mark specific cell structures, proteins, or DNA in a cell to measure its characteristic, e.g. chromatin in a nuclei [16], or perform more complicated tasks, like mode of action recognition [2].

Our work focuses on morphological changes in a microglia cell as a marker of inflammation. Activated microglia cells are key mediators of the chronic neuroinflammatory process, which is associated with the pathogenesis of many neurodegenerative disorders. During activation, apart from changes in the expression of surface antigens and production of pro-inflammatory factors, microglia also change their morphology, which is characterized by ameboid shape with many shortened processes [30,33]. Analyzing morphological changes can be, therefore, relevant in the determination of microglia activation and, as a consequence, can be useful for development of bioassays for the drug discovery process.

In our experiment, cells are treated with bacterial lipopolysaccharides (LPS) to induce inflammation and then treated with CLI-095 [17] to decrease it. To measure the influence of compounds, a biologist performs a quantitative analysis of cells to find how many cells are inflamed or active. Such analysis is performed semi-automatically because it requires a human to provide examples of both active and inactive cells to a HCS analysis software[1]. Then, a linear machine learning model is trained to classify cells using morphological features, such as color intensity, cell shape and size. The main disadvantage of this procedure is an expensive and time-consuming step of manual cell labelling.

The goal of this work is to introduce a weakly-supervised method that can classify active and inactive cells, using a label of the entire image instead of single cells.

HCS images contain a variable number of cells which additionally react differently in response to the compound. Nevertheless, only one label, describing cell activity derived from a compound concentration, is assigned to an image. Problem with an input data point containing multiple instances, called the Multiple Instance Learning [10], often occurs in medical data, where it is too expensive and impractical to annotate details of an image. Similarly, in HCS images it is impossible to annotate every single cell for training purposes. In consequence, MIL is a perfect approach for such a problem, it is typically a weakly-supervised learning task with a goal of predicting a label for an entire image or a bag of data points. However, we want to broaden this idea by using MIL in a self-supervised setup to create an image representation that can be used to classify single instances. Our contributions are as follows:

- We propose a method for an instance level cell classification based on image level labels and apply it to HCS images.
- We introduce a novel position-aware Self-Supervised Multiple Instance Learning method (SSMIL) that combines Contrastive Learning and Multiple Instance Learning approaches.
- We demonstrate usability of our method with detailed tests conducted by both Machine Learning researchers and molecular biologists.

2 Related Works

Firstly, we present current research in the field of High Content Screening, both using Deep Learning and classic approaches. Later, we summarize works on weakly-supervised and self-supervised learning with focus on Contrastive Learning.

2.1 High Content Screening

There are two main streams of High Content Screening image analysis: using classic hand-crafted features and deep learning methods. The first one utilizes morphological features that describe single cell and population-wide characteristics. Almost all manufacturers of HCS imagers provide image analysis software that uses such features, but noticeably more and more often Deep Learning modules are also provided. One of the most popular software is CellProfiler [5] which is free and open source. CellProfiler features were used for breast cancer detection

---

[1] https://www.perkinelmer.com/product/harmony-4-8-office-hh17000001.

by creating hidden representations from an autoencoder model that reconstruct those hand-crafted features [21]. Also, they were used with regression models to predict phenotype [19,35] and with Random Forests [31] for compound functional predictions. In recent years researchers also started to use Deep Learning methods for HCS analysis. [34] used a convolutional autoencoder to find abnormal cells by comparison to control images using deep learning features and [24] used variatonal autoencoders to show variation in cell phenotypes. At the same time, [14] used GANs for cell modelling. Convolutional neural networks are very widely used, [1,3,22,23] and [29] even show that ImageNet-trained network can be successfully used to create features without fine-tuning. Additionally, some multi-scale CNN-based architectures were developed, [9,13,20]. Deep learning shows great results and the only disadvantage so far is lack of easy biological interpreatability of Deep Learning features.

2.2 Weakly-Supervised Learning

Weakly-supervised methods arise from the need to train models in case of lack of manual labels because, most often, only partial labels are available, e.g. in whole-slide images. CLAM [27] approaches this problem by performing instance-level clustering and then attention pooling in pathology classification. Similar problem to ours occurs in satellite images which are large and hard to annotate, it was tackled by use of stacked discriminative sparse autoencoder (SDSAE) [36]. Moreover, WELDON [11] automatically selects relevant regions by top-instance scoring.

2.3 Self-supervised Learning

Self-supervised methods can create meaningful image representation without labels using so-called Contrastive Learning. Its main goal is to create similar representation for similar images, e.g. created by different augmenting of the same image. This is a base for Contrastive Predictive Coding (CPC) [28] which introduces contrastive loss and SimCLR [6,7] which proposes NT-Xent loss. BYOL [15] transforms that idea by assigning online and target network, online network is trained continuously, while weights of the target network are the moving average of the first one. On the other hand, SwAV [4] utilizes swapped prediction mechanism and clustering, while SimSiam [8] simplifies contrastive idea by removing projection head from one of the paths and show that both representations (projected and not) are similar. So far, contrastive learning was used in MIL problems to create representation by training patch-level models [25,26]. Our model utilizes it to compare image-level representation but pooling information from the model on the instance level to train a classifier for patches.

3 Data

HCS images are acquired from a high throughput microplate using an imager. Samples are prepared with multiple fluorescent dies which results in images with various numbers of channels, even up to 6. In our experiment, cells were stained with two dyes: Hoechst 33342 and CellMask Deep Red Plasma Membrane. Microplate, or simply plate, has multiple wells organized into rows and columns. Each well consists of multiple images assigned to fields. In our case, the plate contains almost half a million cells with on average 150 cells per field and, in consequence, per image. Our plate is organized as follows (see Fig. 1): 2 columns contain controls (negative control does not contain any compound, positive control contains LPS which induces inflammation, called cell activity), 8 consecutive columns contain CLI-095 compound in decreasing concentration: from $1.45\mu M$ to $0.005\mu M$. CLI-095 is meant to decrease inflammation: the higher the concentration the smaller the inflammation (activity). In our training procedure, we treat negative control as having maximum concentration and positive control as having minimum concentration.

3.1 Data Acquisition

BV2 microglia cells (Elabscience) were maintained in Dulbecco's modified Eagle's medium (DMEM) with 4.5 g/l glucose containing 5% (v/v) fetal bovine serum (Gibco), 1 mM sodium pyruvate (Gibco), 2mM L-glutamine (Gibco), 100 U/ml penicillin and 100 μg/ml streptomycin (Gibco) at 37° C in a humidified atmosphere containing 5% CO2. For study of morphological changes, BV2 cells were plated in assay medium (DMEM with 4.5 g/l glucose containing 1% (v/v) fetal bovine serum, 1 mM sodium pyruvate, 2mM L-glutamine , 100 U/ml penicillin and 100 μg/ml streptomycin) in poly-D-lysine coated CellCarrier Ultra 96-well plate (PerkinElmer) at density of 2000 cells/well. 24 h later, cells were preincubated with different concentrations of CLI-095 in assay medium for 30 min, followed by stimulation with 0.1 μg/ml ultrapure LPS (Invivogen) and appropriate concentrations of CLI-095 in assay medium for further 24 h. Final concentration of DMSO (CLI-095 solvent) was normalized to 0.0725%. After stimulation, cells were fixed with 4% formaldehyde for 25 min and washed with PBS 3 times. Then, nuclei were stained with Hoechst 33342 (TOCRIS) at the concentration of 5 μg/ml in PBS for 5 min and washed with PBS 3 times for 5 min. Then, cellular membrane was stained with CellMask Deep Red Plasma membrane (Invitrogen) at the dilution of 1:3000 in PBS for 10 min and washed with PBS 3 times for 5 min. Plates were then sealed and images were acquired with Operetta CLS (PerkinElmer) high content imaging system at conditions provided in Table 1. Images were initially analyzed with use of Harmony software with Phenologic module (PerkinElmer).

3.2 Data Labelling

Harmony is a software used along HCS Imager. It helps with image acquisition and experiment orchestration. Harmony has a basic image analysis function and can find and segment cells as well as provide morphological characteristics. Based on those, it uses Linear Regression (provided in the Phenologic module) to divide cells into up to 6 classes. Human operator selects training examples for each class and software marks the rest of the data with those classes. We use Harmony

Table 1. Conditions of image acquisition.

General settings		
Autofocus	Two Peak	
Objective	20x with water immersion, NA 1.0	
Mode	Confocal	
Binning	2	
Channel settings		
	Hoechst 3334	CellMask Deep Red Plasma Membrane
Excitation	355–385 nm	615–645 nm
Emission	430–500 nm	655–760 nm
Time	60 ms	60 ms
Power	40%	20%
Height	$-6.0\,\mu$m	$-11.0\,\mu$m

labels as the ground truth. Cells might be in a transitional state which makes labels noisy and even a human operator is not able to distinguish between them. To assure correctness of the label, we take into account regression scores, which are also provided by Harmony, and use only cells with top and bottom 5% of values.

4 Methods

We introduce a novel position-aware method called SSMIL, which we compare to three baselines, two of which are supervised and one is self-supervised. Supervised methods include:

- Convolutional neural network (CNN) trained to classify cell activity based on image-level labels. We use ResNet-18 with single patches as an input and image labels.
- Attention-based Multiple Instance Learning Pooling (AbMILP) which aggregates all patches from an image to create one representation vector describing the entire image [18]. Additionally, we test its extension called Self-Attention Attention-based MIL Pooling (SA-AbMILP) [32].

A self-supervised baseline method is SimCLR [6] which is a contrastive method trained on patches containing single cells. The backbone of all methods is ResNet-18 to remove an influence of the CNN architecture.

4.1 Position-Aware Self-Supervised Multiple Instance Learning Method (SSMIL)

SSMIL, presented in Fig. 2, combines Contrastive Learning and Multiple Instance Learning approaches. Following paragraphs describe each step of the model trained in an end-to-end manner.

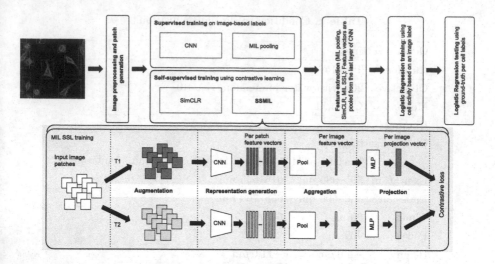

Fig. 2. Training pipeline: In the preprocessing step, we normalize images and remove those with quality issues, create patches and filter them due to size. In the training step, supervised models use compound's concentration as the output of a model and self-supervised models are trained in the contrastive manner. Then, we extract features from the ultimate (SSL) or penultimate (SL) layer of the CNN and we train Logistic regression using those features. LR is trained using cell activity based on the label of the entire image. Finally, we test our representation using ground-truth obtained with Harmony software. Additionally, we present the pipeline of the SSMIL model. Patches are augmented using two sets of transformations: $T1$ and $T2$, then passed through CNN to obtain patch representation which is aggregated using the pooling model to create representation of an entire image. Lastly this representation is passed through the projection module (MLP), and we calculate contrastive loss.

Augmentation. An image is split into patches generated for each previously detected cell. Then patches are transformed using two sets of augmentations, $T1$ and $T2$, to create two sets of patches.

Representation Generation. Each patch is passed through a convolutional neural network to obtain its representation. In our setup, we use ResNet-18 without the classification layer. A patch representation is enriched by adding patch position at the end of the feature vector pooled from CNN. This information takes advantage of the fact that cells in the sample may influence each other, e.g. close cells might squeeze each other.

Aggregation. In the aggregation step, all patch feature vectors of the same image are pooled to create one feature vector representing an entire image. In this work, we use attention-based MIL pooling [18]:

$$z = \sum_{k=1}^{K} a_k h_k, \tag{1}$$

where h_k is embedding of kth patch and

$$a_k = \frac{\exp(\mathbf{w}^T \tanh(\mathbf{V}\mathbf{h}_k^T))}{\sum_{j=1}^{K} \exp(\mathbf{w}^T \tanh(\mathbf{V}\mathbf{h}_j^T))} \tag{2}$$

with $\mathbf{w}$ and $\mathbf{V}$ as trainable parameters.

Projection. Image feature vector is projected as according to [6] it is beneficial to use projection instead of feature vector. Projection corresponds to a 3-layer MLP.

Contrastive Loss. Finally, we calculate contrastive loss using NT-Xent loss [6]:

$$L_{NT-Xent}(\mathbf{x}_i, \mathbf{x}_j) = -\log \frac{\exp(sim(\mathbf{x}_i, \mathbf{x}_j)/\tau)}{\sum_{k \neq i}^{2N} \mathbb{1}_{[k \neq i]} \exp(sim(\mathbf{x}_i, \mathbf{x}_k)/\tau} \tag{3}$$

where τ is temperature coefficient, sim is cosine similarity, and N is a batch size.

5 Experimental Setup

Training and testing procedure for all models is presented in Fig. 2 and is described in consecutive paragraphs.

Train-Test Split. We utilize the plate's structure to create a train-test split. The HCS plate is arranged in wells by rows and columns. Each well contains multiple fields (in our case 49). Our dataset has samples in 6 rows and we used 5 of them to select subsets. Rows are used in 5-fold split to create training (4 rows) and testing (1 row) datasets. Testing datasets are then filtered and we use only patches that have very high or low regression values, as given by Harmony software, to assure high confidence in labels used in testing. Our dataset is imbalanced because some cells die during the experiment due to induced inflammation.

Data Preprocessing and Patch Generation. Cells are detected using Harmony software which also provides a bounding box enclosing a cell. We create 224×224 patches with the cell in the middle. Patches that are too close to the image border (112 pixels or less) and smaller than $500px^2$ are discarded to remove debris and cells that might not be correctly segmented. Finally, patches are resized to 112×112 pixels and normalized.

Supervised and Self-supervised Training. Supervised models are trained to classify patches (CNN) or images (AbMILP, SA-AbMILP) using activity labels based on the compound concentration. We assign active to columns with highest concentration of negative control and inactive to the lowest concentration columns and positive control. Self-supervised models (SimCLR, SSMIL) are trained in contrastive matter without any knowledge about the label. All models are trained using Adam optimizer with learning rate $lr = 0001$. CNN is trained with batch size 128, SimCLR with batch size 1024. AbMILP, SA-AbMILP, and SSMIL are trained with batch size 8 which translates to on average 1200 patches. We use color jitter and rotation augmentations to minimize size changes of cells.

Feature Extraction. Next, we create a patch representation from AbMILP, SA-AbMILP, SimCLR, and SSMIL models. The patch representation is given by feature vectors pooled from the last layer of CNN.

Logistic Regression Training. In this step, we train Linear Regression using a patch representation to classify the cell activity. For training we use columns with the highest and lowest concentration to assure that the label is as correct as possible.

Logistic Regression Testing. The last step of the procedure is testing. Linear regression models, trained in the previous step, are tested against labels obtained from Harmony+Phenologic software. CNN model is tested by using its direct prediction.

6   Results

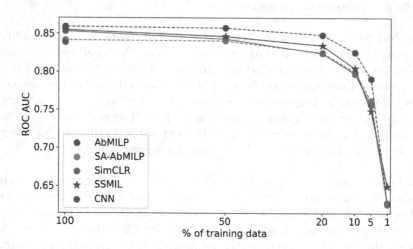

Fig. 3. Results: we present ROC AUC values for models trained with various fractions of training dataset. Our model SSMIL achieves on par results with supervised models and when trained on only 1% training data achieves 3.4% higher ROC AUC than other models, even supervised ones.

Results of our methods as well as baseline methods are presented in Fig. 3. We present ROC AUC (Receiver Operating Characteristic Area Under Curve) values for Linear Regression models trained with 100%, 50%, 20%, 10%, 5% and 1% of the training data. Using 100% of training data SSMIL achieves ROC AUC of 0.8547 ± 0.0129 while AbMILP and SA-AbMILP have ROC AUC of 0.8593 ± 0.0104 and 0.8415 ± 0.0147, respectively. This shows that all models achieve very similar results and that by using self-supervised methods, we can obtain as good representation as in supervised setup. What is more, in all setups we achieve ROC AUC greater than 0.8 when we decrease size of training dataset to 20% of data or even 10% in case of SSMIL and supervised AbMILP. Worth noting is the fact that our model, SSMIL, achieves on average 3.4% better results than other models when trained only on 1% of data with SSMIL having ROC AUC of 0.6498 ± 0.0102 while other methods have ROC AUC of 0.6283 ± 0.0076, 0.6265 ± 0.0125, and 0.6283 ± 0.0088 (AbMILP, SA-AbMILP, and SimCLR respectively).

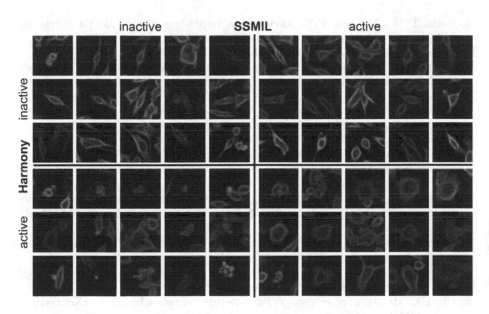

Fig. 4. Qualitative confusion matrix between Harmony labels and SSMIL predictions. SSMIL returns incorrect predictions for active cells mostly when those cells are destroyed or there are clumps of multiple cells. Similarly, the model incorrectly predicts inactive cells when they are too close to each other. In some cases we can see that cells labeled as inactive by Harmony are round and processes are not visible.

Figure 4 presents a qualitative confusion matrix which shows cells that were correctly and incorrectly predicted by the SSMIL model. We notice that SSMIL gives wrong predictions when faced with cells destroyed during the experiment. Additionally, the model can be confused by clumps of cells which can be mitigated by better segmentation which on its own can be a challenge when cells are touching.

7   Conclusions

We presented a weakly-supervised problem of cell classification using HCS data and compared 4 approaches to solve this problem in both supervised manner with CNN or MIL pooling, and using self-supervised methods, SimCLR and SSMIL proposed by us. We show that using either patch or image level Contrastive Learning provides representation as good as training a supervised model. Finally, we introduced a new contrastive learning method, SSMIL, that creates patch-level representations using aggregation and image-level labels and can be successfully trained even with an imbalanced data or a small number of training samples.

References

1. Ando, D.M., McLean, C.Y., Berndl, M.: Improving phenotypic measurements in high-content imaging screens. bioRxiv (2017). https://doi.org/10.1101/161422, https://www.biorxiv.org/content/early/2017/07/10/161422
2. Bickle, M.: The beautiful cell: high-content screening in drug discovery. Anal. Bioanal. Chem. **398**, 219–226 (2010). https://doi.org/10.1007/s00216-010-3788-3
3. Caicedo, J., McQuin, C., Goodman, A., Singh, S., Carpenter, A.: Weakly supervised learning of single-cell feature embeddings, vol. 2018, pp. 9309–9318 (2018). https://doi.org/10.1109/CVPR.2018.00970
4. Caron, M., Misra, I., Mairal, J., Goyal, P., Bojanowski, P., Joulin, A.: Unsupervised learning of visual features by contrasting cluster assignments. arXiv preprint arXiv:2006.09882 (2020)
5. Carpenter, A., et al.: CellProfiler: image analysis software for identifying and quantifying cell phenotypes. Genome Biol. **7**, R100 (2006). https://doi.org/10.1186/gb-2006-7-10-r100
6. Chen, T., Kornblith, S., Norouzi, M., Hinton, G.: A simple framework for contrastive learning of visual representations (2020)
7. Chen, T., Kornblith, S., Swersky, K., Norouzi, M., Hinton, G.: Big self-supervised models are strong semi-supervised learners (2020)
8. Chen, X., He, K.: Exploring simple Siamese representation learning. arXiv preprint arXiv:2011.10566 (2020)
9. Datta, K., et al.: Training multiscale-CNN for large microscopy image classification in one hour. High Perform. Comput. 463–477 (2019). https://doi.org/10.1007/978-3-030-34356-9_35
10. Dietterich, T.G., Lathrop, R.H., Lozano-Pérez, T.: Solving the multiple instance problem with axis-parallel rectangles. Artif. Intell. **89**(1), 31–71 (1997). https://doi.org/10.1016/S0004-3702(96)00034-3, https://www.sciencedirect.com/science/article/pii/S0004370296000343
11. Durand, T., Thome, N., Cord, M.: WELDON: weakly supervised learning of deep convolutional neural networks. In: Proceedings of the IEEE Conference on Computer Vision and Pattern Recognition (CVPR), June 2016
12. Giuliano, K., et al.: High-content screening: a new approach to easing key bottlenecks in the drug discovery process. J. Biomol. Screen. **2**, 249–259 (1997). https://doi.org/10.1177/108705719700200410

13. Godinez, W.J., Hossain, I., Lazic, S.E., Davies, J.W., Zhang, X.: A multi-scale convolutional neural network for phenotyping high-content cellular images. Bioinformatics **33**(13), 2010–2019 (2017). https://doi.org/10.1093/bioinformatics/btx069
14. Goldsborough, P., Pawlowski, N., Caicedo, J.C., Singh, S., Carpenter, A.E.: Cyto-GAN: generative modeling of cell images. bioRxiv (2017). https://doi.org/10.1101/227645, https://www.biorxiv.org/content/early/2017/12/02/227645
15. Grill, J.B., et al.: Bootstrap your own latent: a new approach to self-supervised learning. arXiv preprint arXiv:2006.07733 (2020)
16. Haney, S., Lapan, P., Pan, J., Zhang, J.: High-content screening moves to the front of the line. Drug Disc. Today **11**, 889–894 (2006). https://doi.org/10.1016/j.drudis.2006.08.015
17. Ii, M., et al.: A novel cyclohexene derivative, ethyl (6r)-6-[n-(2-chloro-4-fluorophenyl)sulfamoyl]cyclohex-1-ene-1-carboxylate (tak-242), selectively inhibits toll-like receptor 4-mediated cytokine production through suppression of intracellular signaling. Molec. Pharmacol. **69**(4), 1288–1295 (2006). https://doi.org/10.1124/mol.105.019695, https://molpharm.aspetjournals.org/content/69/4/1288
18. Ilse, M., Tomczak, J., Welling, M.: Attention-based deep multiple instance learning. In: Dy, J., Krause, A. (eds.) Proceedings of the 35th International Conference on Machine Learning. Proceedings of Machine Learning Research, vol. 80, pp. 2127–2136. PMLR, 10–15 July 2018. https://proceedings.mlr.press/v80/ilse18a.html
19. Janosch, A., Kaffka, C., Bickle, M.: Unbiased phenotype detection using negative controls. SLAS Disc. Adv. Sci. Drug Disc. **24**(3), 234–241 (2019). https://doi.org/10.1177/2472555218818053, pMID: 30616488
20. Janssens, R., Zhang, X., Kauffmann, A., Weck, A., Durand, E.: Fully unsupervised deep mode of action learning for phenotyping high-content cellular images, July 2020. https://doi.org/10.1101/2020.07.22.215459
21. Kandaswamy, C., Silva, L.M., Alexandre, L.A., Santos, J.M.: High-content analysis of breast cancer using single-cell deep transfer learning. J. Biomol. Screen. **21**(3), 252–259 (2016). https://doi.org/10.1177/1087057115623451, pMID: 26746583
22. Kensert, A., Harrison, P., Spjuth, O.: Transfer learning with deep convolutional neural networks for classifying cellular morphological changes. SLAS Disc. Adv. Life Sci. R&D **24**, 247255521881875 (2019). https://doi.org/10.1177/2472555218818756
23. Kraus, O., et al.: Automated analysis of high-content microscopy data with deep learning. Molec. Syst. Biol. **13**, 924 (2017). https://doi.org/10.15252/msb.20177551
24. Lafarge, M.W., Caicedo, J.C., Carpenter, A.E., Pluim, J.P., Singh, S., Veta, M.: Capturing single-cell phenotypic variation via unsupervised representation learning. In: Cardoso, M.J., et al (eds.) Proceedings of The 2nd International Conference on Medical Imaging with Deep Learning. Proceedings of Machine Learning Research, vol. 102, pp. 315–325. PMLR, 08–10 July 2019. https://proceedings.mlr.press/v102/lafarge19a.html
25. Li, B., Li, Y., Eliceiri, K.W.: Dual-stream multiple instance learning network for whole slide image classification with self-supervised contrastive learning. In: Proceedings of the IEEE/CVF Conference on Computer Vision and Pattern Recognition (CVPR). pp. 14318–14328, June 2021
26. Lu, M.Y., Chen, R.J., Wang, J., Dillon, D., Mahmood, F.: Semi-supervised histology classification using deep multiple instance learning and contrastive predictive coding (2019)

27. Lu, M.Y., Williamson, D.F.K., Chen, T.Y., Chen, R.J., Barbieri, M., Mahmood, F.: Data efficient and weakly supervised computational pathology on whole slide images (2020)
28. van den Oord, A., Li, Y., Vinyals, O.: Representation learning with contrastive predictive coding (2019)
29. Pawlowski, N., Caicedo, J.C., Singh, S., Carpenter, A.E., Storkey, A.: Automating morphological profiling with generic deep convolutional networks. bioRxiv (2016). https://doi.org/10.1101/085118, https://www.biorxiv.org/content/early/2016/11/02/085118
30. Perry, V.H., Nicoll, J.A.R., Holmes, C.: Microglia in neurodegenerative disease. Nat. Rev. Neurol. 6(4), 193–201 (2010). https://doi.org/10.1038/nrneurol.2010.17
31. Rose, F., Basu, S., Rexhepaj, E., Chauchereau, A., Nery, E., Genovesio, A.: Compound functional prediction using multiple unrelated morphological profiling assays. SLAS Technol. Transl. Life Sci. Innov. 23, 247263031774083 (2017). https://doi.org/10.1177/2472630317740831
32. Rymarczyk, D., Borowa, A., Tabor, J., Zieliński, B.: Kernel self-attention in deep multiple instance learning (2021)
33. Sarkar, S., et al.: Characterization and comparative analysis of a new mouse microglial cell model for studying neuroinflammatory mechanisms during neurotoxic insults. NeuroToxicology 67, 129–140 (2018). https://doi.org/10.1016/j.neuro.2018.05.002
34. Sommer, C., Hoefler, R., Samwer, M., Gerlich, D.: A deep learning and novelty detection framework for rapid phenotyping in high-content screening. bioRxiv (2017). https://doi.org/10.1101/134627
35. Way, G., et al.: Predicting cell health phenotypes using image-based morphology profiling, July 2020. https://doi.org/10.1101/2020.07.08.193938
36. Yao, X., Han, J., Cheng, G., Qian, X., Guo, L.: Semantic annotation of high-resolution satellite images via weakly supervised learning. IEEE Trans. Geosci. Remote Sens. 54(6), 3660–3671 (2016). https://doi.org/10.1109/TGRS.2016.2523563

Which Visual Features Impact the Performance of Target Task in Self-supervised Learning?

Witold Oleszkiewicz[1]([✉])(ID), Dominika Basaj[2](ID), Tomasz Trzciński[1,2,3](ID), and Bartosz Zieliński[3,4](ID)

[1] Warsaw University of Technology, plac Politechniki 1, Warszawa, Poland
witold.oleszkiewicz@pw.edu.pl
[2] Tooploox, Teczowa 7, Wrocław, Poland
[3] Faculty of Mathematics and Computer Science, Jagiellonian University,
Łojasiewicza 6, Kraków, Poland
[4] Ardigen, Podole 76, Kraków, Poland

Abstract. Self-supervised methods gain popularity by achieving results on par with supervised methods using fewer labels. However, their explaining techniques ignore the general semantic concepts present in the picture, limiting to local features at a pixel level. An exception is the visual probing framework that analyzes the vision concepts of an image using probing tasks. However, it does not explain if analyzed concepts are critical for target task performance. This work fills this gap by introducing amnesic visual probing that removes information about particular visual concepts from image representations and measures how it affects the target task accuracy. Moreover, it applies Marr's computational theory of vision to examine the biases in visual representations. As a result of experiments and user studies conducted for multiple self-supervised methods, we conclude, among others, that removing information about 3D forms from the representation decrease classification accuracy much more significantly than removing textures.

Keywords: Explainability · Self-supervision · Probing tasks

1 Introduction

Visual representations are critical in many computer vision and machine learning applications. The spectrum of these applications is broad, starting with visual search [21] to image classification [16] and visual question answering [3]. However, supervised representation learning requires a large amount of labeled data, usually time-consuming and expensive. Hence, self-supervised methods gain popularity, achieving results on par with supervised methods using fewer labels [6,8,13].

Along with the increasing proliferation of self-supervised methods for representation learning, there is a growing interest in developing methods that allow

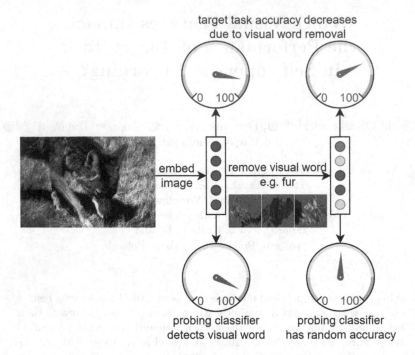

Fig. 1. Amnesic visual probing removes a specific visual concept (here corresponding to fur) from the self-supervised representation of an image (here corresponding to a wolf). As a result, the probing classifier cannot detect the presence of fur in the representation, and the target task accuracy decreases. The level of decrease represents the importance of the considered concept.

the interpretation of the resulting representation space and draw conclusions regarding the information it conveys. However, most of them focus on supervised approaches and study local features at a pixel level [2,20]. At the same time, the general semantic concepts present in the image are often overlooked, and their influence on model decisions is unknown. From this perspective, an exception is visual probing [4] that analyzes the vision concepts of an image using probing tasks. The probing tasks provide information about the presence of visual concepts in the representations but do not explain if they are critical for target task performance.

In this work, we overcome this limitation, providing a method that investigates the importance of visual features in the context of target task performance, referring to the amnesic probing [10] used in natural language processing (NLP). We remove information about particular visual concepts from image representations using the Iterative Nullspace Projection [19] and measure how it affects the target task accuracy. In addition, we conduct user studies to describe the visual concepts using Marr's computational theory of vision [17]. As a consequence, we can examine the biases in image representations.

Our contributions can be summarized as follows:

- We propose amnesic visual probing, a method for analyzing which visual features impact the performance of a target task.
- We apply Marr's computational theory of vision to examine the biases in visual representations.
- We conduct a complete user study and assign automatically generated visual concepts to one of six visual features from Marr's computational theory of vision.

2 Related Works

Our work corresponds to two research areas: self-supervised learning and probing tasks. We briefly cover the latest achievements in these two topics in the following paragraphs.

Self-Supervised Image Representations. Image representations obtained in a self-supervised manner are increasingly popular due to the competitive performance compared to supervised approaches. It is because they leverage the power of datasets without label annotations. One of the methods, called MoCo v1 [14], is based on a dictionary treated as a queue of data samples. It contains two encoders for query and keys, which are matched by contrastive loss. This queue enables to use of a large dictionary of examples previously limited to the batch size. SimCLR v2 [8] is another powerful method, which builds upon its predecessor, SimCLR [7] that maximizes the agreement between two views of the same sample by contrastive loss. In [8], the authors use a deeper and thinner backbone (ResNet-152 3x), deepen the projection head, which is not removed after contrastive training, and adapt memory mechanism from MoCo to increase the pool of negative examples. SwAV [6] takes advantage of contrastive methods. However, it compares clusters of data instead of single examples. The consistency between clusters, which can be seen as views of the same data sample, is enforced by learning to predict one view from another. In contrast to the above methods, BYOL [13] does not use the explicitly defined contrastive loss function, so it does not need negative samples. Instead, it uses two neural networks, referred to as online and target networks, that interact and learn the representation of the same image from each other.

Probing Tasks. The probing tasks originally come from Natural Language Processing (NLP). Their objective is to discover the characteristics interpretable by humans, which are encoded in the representation obtained by neural networks [5]. Probing is usually a simple classifier applied to trained representations like word embeddings. The probing classifier predicts whether the linguistic phenomenon that we want to verify exists or not. The probing classifiers in the NLP research community are popular tools for inspecting the internals of representations. However, some recent work extends the usability of probing tasks by introducing the concept of amnesic probing [10] to measure the influence of the phenomenons on the target task performance.

Although probing tasks are popular in NLP, they only recently have been adapted to the Computer Vision (CV) domain in [4] based on the mapping defined between NLP and CV domains. These visual probing tasks allow one to gain intuition about the knowledge conveyed in the representation by the various self-supervised methods. However, there is no clear consensus on their impact on the target task performance.

3   Methods

This section introduces amnesic visual probing (AVP), a tool for explaining visual representations. It analyzes how important are particular visual concepts for a target task. Therefore, to define AVP, we first provide visual concepts (here called Visual Words, VW) and then obtain their meaning. Finally, we remove information about VW from the representation and analyze how it influences a target task.

Generating Visual Words. To generate visual words, we use the established ACE algorithm [12]. It starts by dividing the image into superpixels using the SLIC algorithm [1]. Because different superpixel sizes are preferred, we run the algorithm three times with different parameters and obtain three sets with 15, 50, and 80 superpixels for each image. Then, we pass all the superpixels through the network trained on ImageNet to obtain their representations. These representations are clustered separately for each class using the k-means algorithm with $k = 25$ (infrequent and unpopular clusters are removed as described in [12]). Clusters obtained this way could be directly used as visual words. However, so many visual words would be impractical due to the similarity between concepts of ImageNet classes. Therefore, to obtain a credible dictionary with visual words shared between different classes, we filter out concepts with the smallest TCAV score [15] and cluster the remaining 6,000 ones using the k-means algorithm into $N = 50$ new clusters. These N clusters are visual words that form our visual language (see Fig. 2).

Cognitive Vision Systematic. To obtain the meaning of the generated visual words, we use cognitive visual systematic [18] based on Marr's computational theory of vision [17]. According to Marr's theory, three levels of visual representations play an essential role in perception and discovering essential features of visible objects. These are the primal sketch, the 2.5D sketch, and the 3D model representation. The primal sketch is a two-dimensional image representation that uses light intensity changes, edges, colors, and textures. The 2.5D sketch represents mostly two-dimensional shapes, and the 3D model representation allows an observer to imagine the spatial object features based on its two-dimensional image. We will analyze six visual features from Marr's theory: brightness, color, texture, and lines (all primal sketch), shape (2.5D sketch), and form (3D model representation). We conduct user studies to establish the relationship between these features and individual visual words (see Fig. 3).

Fig. 2. Sample visual words, each represented by a row of 5 superpixels.

Amnesic Visual Probing. We want to remove the information about a visual word from the representations and analyze how they differ from the original ones. For this purpose, we divide an image into superpixels, pass them through the network to obtain their representations, and assign them to the closest visual word. Then, we *define Word Content labels* $z_i \in \{0,1\}^N$ for representations $x_i \in \mathbb{R}^d$, where $z_i[j] = 1$ means that at least one superpixel of i-th image is assigned to j-th visual word.

Then, we *remove information about j-th visual word from a representation* x_i. For this purpose, we adapt an algorithm called Iterative Nullspace Projection (INLP) [19]. The probing classifier for $z_i[j]$ is parameterized by the matrix W_0. We first construct a projection matrix P_0 such that $W_0(P_0 x_i) = 0$ for all representations x_i (using method from [19]). Then, we iteratively train additional classifiers W_1 and perform the same procedure until no linear information

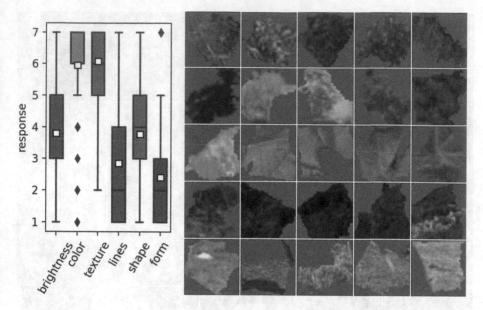

Fig. 3. Sample visual word (corresponding to grass) and its distributions of Likert scores obtained from user studies. One can observe that users mostly decided to assign this word to color and texture from the Marr's computational theory of vision.

regarding $z_i[j]$ remains in x_i, i.e., until the chance of predicting the presence of a j-th visual word by the linear model is random. As a result we obtain a matrix $P_n \cdot P_{n-1} \cdot \ldots \cdot P_0$ which, when applied to representation, removes information about visual word j.

Finally, one can analyze changes in target task performance after removing information about a particular visual word. In this case, a target task is defined as multi-class classification with labels $y_i \in \{1, \ldots, k\}$, where $k = 1000$ is the number of ImageNet's classes. It is trained and tested for two types of representations, original and with removed visual word information.

4   User Studies

To understand the meaning of visual words, we conduct user studies with 97 volunteers (64 males, 32 females, and 2 others aged 25 ± 7 years), including 71.1% students or graduates of computer science and related fields. Users completed an online survey with the number of questions corresponding to the number of visual words. We presented 12 typical (randomly chosen) superpixels for each visual word, and we asked to what extent a particular visual feature was essential for its creation. In reference to Marr's computational theory of vision [17] (see Sect. 3), six features were taken into consideration: brightness, color, texture, lines (all primal sketch), shape (2.5D sketch) and form (3D model representation). We

---

Algorithm 1. Amnesic visual probing (AVP)

Require: X – set of image representations, Y – set of target labels, Z – set of visual
words labels, C – codebook of visual words,
getNullSpaceProj(X, Z) – returns projection matrix that removes information about
a visual word from representations,
trainValProb(X, Z) – trains model on probing task and returns validation accuracy,
trainValTarget(X, Y) – trains model on target task and returns validation accuracy
for each: $c \in C$
$\quad X_{proj} \leftarrow X$
$\quad$**repeat**
$\quad\quad P \leftarrow$ getNullSpaceProj(X_{proj}, Z)
$\quad\quad X_{proj} \leftarrow P X_{proj}$
$\quad\quad acc_{prob} \leftarrow$ trainValProb(X_{proj}, Z)
$\quad$**until** $acc_{prob} \geq \frac{1}{2}$
$\quad acc_{target} \leftarrow$ trainValTarget(X, Y)
$\quad acc_{target}^{\neg c} \leftarrow$ trainValTarget(X_{proj}, Y)
$\quad influence^c = acc_{target}^{\neg c} - acc_{target}$

---

use the Likert scale with seven numerical responses from 1 to 7, corresponding
to insignificant and key features, respectively.

Before completing the survey, users got familiarized with the examples of
visual words with particular features selected by a trained cognitivist. They also
completed two training trials to become familiar with the main task. Moreover,
completing the task was not limited in time. Finally, due to the high number
of visual words, assessing all 50 visual words would be tedious for the users.
Therefore, we have prepared four questionnaire versions (one with twenty visual
words and three with ten visual words) and assigned them to users randomly.

Based on the user studies results, we ranked the most representative visual
words for each of the six features: brightness, color, texture, lines, shape, and
form. We used those rankings to obtain detailed results of the amnesic visual
probing.

5 Experimental Setup

Models. We examine four self-supervised methods (MoCo v1 [14], SimCLR v2 [8],
BYOL [13], and SwAV [6]), with a publicly available implementation based on
the ResNet-50 (1x) architecture, trained on the entire ImageNet dataset[1]. We
use the penultimate layer of ResNet-50 to generate representations with a length
of 2048.

---

[1] We use the following implementations of the self-supervised methods: https://github.
com/{google-research/simclr, yaox12/BYOL-PyTorch, facebookresearch/swav, face-
bookresearch/moco}.

Data and Target Task. We consider ImageNet [9] classification as the target task, but our approach could also be applied to other tasks. In order to get the classification model, we freeze the self-supervised trained model and fine-tune an ultimate fully-connected layer for 100 epochs. We conduct our experiments with a standard train/validation split.

Removing Visual Words. Interventions that remove visual words are parametrized by 2048 × 2048 matrices applied to self-supervised representations. We obtain these matrices with our adaptation of the INLP algorithm, where we iterate until the probing classifier (detecting a visual word) achieves random accuracy.

Metric. We consider the difference in top-5 classification accuracy before and after the intervention. For each self-supervised method, we carry out a series of interventions, removing the information about successive visual words from the ranking obtained based on the user studies (see Sect. 4). For each of the six features, we start with visual words considered as crucial for a given feature.

6   Results

As shown in Table 1, *removing visual words from self-supervised representations reduces the top-5 accuracy* of the target task. It is expected because, as presented in [4], image representation contains semantic knowledge. However, depending on a self-supervised model and a type of visual word, the level of degradation significantly differs. In the case of SimCLR v2, visual words related to the shape and form have the most significant influence on the classifier decisions. For BYOL, brightness and form have the greatest influence. Results for SimCLR and BYOL are also similar because they are least sensitive to texture removal from the representations. In contrast, MoCo and SwAV are the least sensitive to removing shape. In the case of MoCo, we also observe the most significant decrease in classification accuracy when removing forms, while the performance of SwAV is the most sensitive to color removal.

In Fig. 4, we present the most important visual words (according to our user studies) for each of the six visual concepts from Marr's computational theory of vision. These are visual words that we first remove from the representation.

In general, except for MoCo v1, representations are the least sensitive to removing textures from representations, which is inconsistent with what is found in [11]. Also, the two-dimensional shape is the most influential feature only for the classifier using the SimCLR v2 model. On the other hand, on average, removing visual words corresponding to the three-dimensional form and color from the self-supervised representation causes the most significant drop in the classification accuracy.

In Fig. 5, we present the change of target task accuracy when removing the successive most important visual words of the considered Marr's visual features (obtained with user studies). In general, the classification accuracy decreases as we remove the successive visual words. There are only a few exceptions to this, most notable in the case of SimCLR v2. We notice that in some cases, after removing two or three visual words from a given category, deleting the next ones causes only a slight further decrease in accuracy. It happens, for example, when removing visual words related to shape from SwAV representations or texture from SimCLR v2 representation. We also notice that in the case of three models (except MoCo v1), initially, when removing a small number of visual words, the most significant loss of accuracy occurs when removing the simplest visual features such as brightness (BYOL and SwAV) and color (SwAV and SimCLR v2). However, as we remove more visual words, the impact of removing more complex visual words corresponding to three-dimensional forms increases. This result may be because three-dimensional forms are more diverse and heterogeneous than colors and brightness.

Table 1. Removing visual words from the self-supervised representations influences the top-5 accuracy. The results are presented for six visual concepts from Marr's computational theory of vision. For each visual feature we remove five visual words according to the ranking obtained based on the user studies. The colors denote higher (dark blue) or lower (light blue) accuracy drop (in percentage points). These results demonstrate the biases in the self-supervised representations.

	Top-5 acc.	Decrease in top-5 acc.					
	No interv.	Remove visual words					
		Bright.	Color	Texture	Lines	Shape	Form
MoCo v1	82.5	−3.09	−4.27	−3.84	−4.04	−2.98	−4.73
SimCLR v2	86.0	−2.00	−2.44	−1.60	−1.68	−2.51	−2.51
BYOL	86.5	−3.99	−3.37	−2.35	−2.75	−2.36	−3.49
SwAV	92.4	−2.20	−2.94	−1.56	−1.85	−1.00	−2.09

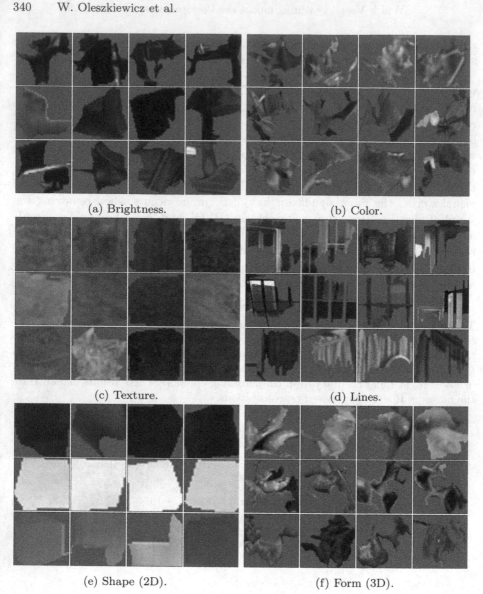

(a) Brightness.

(b) Color.

(c) Texture.

(d) Lines.

(e) Shape (2D).

(f) Form (3D).

Fig. 4. The most important visual words (according to our user studies) for each of the six visual concepts from Marr's computational theory of vision.

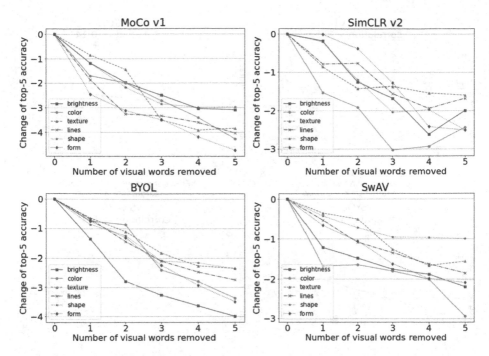

Fig. 5. Decrease in top-5 accuracy (in percentage points) when removing the information about successive visual words according to the ranking obtained based on the user studies, presented for six visual concepts from Marr's computational theory of vision.

Amnesic Visual Probing vs. Word Content Probing Task. The correlation between the results of amnesic visual probing and the Word Content (WC) probing task is relatively weak, as presented in Fig. 6. The Pearson correlation coefficient ranges from 0.14 for SimCLR v2 to 0.52 for MoCo v1. In Fig. 6 we can see that although the WC probing task shows that there is a similar level of information about the visual words corresponding to lines and forms in Sim-CLR's representation, removing forms from this representation causes a much more significant decrease of target task accuracy than removing lines. The same relationship regarding lines and forms is also valid for BYOL, in which case the correlation between target task accuracy and WC results is the largest among the examined methods, even though it is still weak.

In general, this weak correlation supports the thesis that the WC probing task focuses on what visual words are encoded in the representation, but it does not assess how this information is used. Therefore, we conclude that the *Word Content probing task cannot be directly used to evaluate target task accuracy, which justifies the introduction of amnesic visual probing.* Nevertheless, WC is still needed for amnesic visual probing to analyze the representation and should be considered as a complementary tool.

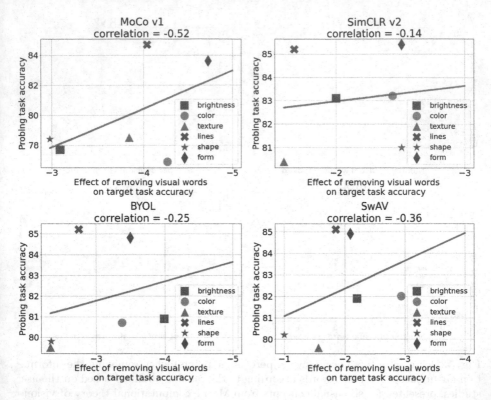

Fig. 6. There is a weak correlation between the results of amnesic visual probing (in percentage points) and the Word Content (WC) probing task (in percents). It means that WC cannot be directly used to evaluate target task accuracy. Hence introducing the amnesic visual probing is justified.

7 Conclusions

The visual probing framework provides interesting insight into the self-supervised representations. However, this insight does not correspond to the performance of the target task. Hence, we propose Amnesic Visual Probing (AVP) to analyze the visual concepts that influence the target task. Thanks to preserving the semantic taxonomy of visual words from the visual probing framework, we can use AVP to examine and compare the biases of individual self-supervised methods. Finally, the user studies allow us to describe those biases using six visual features from Marr's computational theory of vision.

Acknowledgments. This research was funded by Foundation for Polish Science (grant no POIR.04.04.00-00-14DE/18-00 carried out within the Team-Net program co-financed by the European Union under the European Regional Development Fund), National Science Centre, Poland (grant no 2020/39/B/ST6/01511). The authors have applied a CC BY license to any Author Accepted Manuscript (AAM) version arising from this submission, in accordance with the grants' open access conditions. Dominika

Basaj was financially supported by grant no 2018/31/N/ST6/02273 funded by National Science Centre, Poland.

References

1. Achanta, R., Shaji, A., Smith, K., Lucchi, A., Fua, P., Süsstrunk, S.: Slic superpixels compared to state-of-the-art superpixel methods. IEEE Trans. Pattern Anal. Mach. Intell. **34**(11), 2274–2282 (2012)
2. Adebayo, J., Gilmer, J., Muelly, M., Goodfellow, I., Hardt, M., Kim, B.: Sanity checks for saliency maps. arXiv preprint arXiv:1810.03292 (2018)
3. Antol, S., et al.: VQA: visual question answering. In: Proceedings of the IEEE International Conference on Computer Vision, pp. 2425–2433 (2015)
4. Basaj, D., et al.: Explaining self-supervised image representations with visual probing. In: IJCAI-21, pp. 592–598, August 2021. https://doi.org/10.24963/ijcai.2021/82
5. Belinkov, Y., Glass, J.: Analysis methods in neural language processing: a survey. Trans. Assoc. Computat. Linguist. **7**, 49–72 (2019)
6. Caron, M., Misra, I., Mairal, J., Goyal, P., Bojanowski, P., Joulin, A.: Unsupervised learning of visual features by contrasting cluster assignments. In: Proceedings of Advances in Neural Information Processing Systems (NeurIPS) (2020)
7. Chen, T., Kornblith, S., Norouzi, M., Hinton, G.: A simple framework for contrastive learning of visual representations. In: III, H.D., Singh, A. (eds.) Proceedings of the 37th International Conference on Machine Learning. Proceedings of Machine Learning Research, vol. 119, pp. 1597–1607. PMLR, 13–18 July 2020
8. Chen, T., Kornblith, S., Swersky, K., Norouzi, M., Hinton, G.E.: Big self-supervised models are strong semi-supervised learners. In: Larochelle, H., Ranzato, M., Hadsell, R., Balcan, M.F., Lin, H. (eds.) Advances in Neural Information Processing Systems, vol. 33, pp. 22243–22255. Curran Associates, Inc. (2020)
9. Deng, J., Dong, W., Socher, R., Li, L.J., Li, K., Fei-Fei, L.: ImageNet: a large-scale hierarchical image database. In: 2009 IEEE Conference on Computer Vision and Pattern Recognition, pp. 248–255. IEEE (2009)
10. Elazar, Y., Ravfogel, S., Jacovi, A., Goldberg, Y.: Amnesic probing: behavioral explanation with amnesic counterfactuals. Trans. Assoc. Comput. Linguist. **9**, 160–175 (03 2021)
11. Geirhos, R., Narayanappa, K., Mitzkus, B., Bethge, M., Wichmann, F.A., Brendel, W.: On the surprising similarities between supervised and self-supervised models. arXiv preprint arXiv:2010.08377 (2020)
12. Ghorbani, A., Wexler, J., Zou, J., Kim, B.: Towards automatic concept-based explanations. arXiv preprint arXiv:1902.03129 (2019)
13. Grill, J.B., et al.: Bootstrap your own latent - a new approach to self-supervised learning. In: Larochelle, H., Ranzato, M., Hadsell, R., Balcan, M.F., Lin, H. (eds.) Advances in Neural Information Processing Systems, vol. 33, pp. 21271–21284. Curran Associates, Inc. (2020)
14. He, K., Fan, H., Wu, Y., Xie, S., Girshick, R.: Momentum contrast for unsupervised visual representation learning. In: 2020 IEEE/CVF Conference on Computer Vision and Pattern Recognition (CVPR), pp. 9726–9735 (2020). https://doi.org/10.1109/CVPR42600.2020.00975
15. Kim, B., Wattenberg, M., Gilmer, J., Cai, C., Wexler, J., Viegas, F., et al.: Interpretability beyond feature attribution: quantitative testing with concept activation

vectors (TCAV). In: International Conference on Machine Learning, pp. 2668–2677. PMLR (2018)

16. Krizhevsky, A., Sutskever, I., Hinton, G.E.: ImageNet classification with deep convolutional neural networks. Adv. Neural. Inf. Process. Syst. **25**, 1097–1105 (2012)

17. Marr, D.: Vision: A Computational Investigation into the Human Representation and Processing of Visual Information. Henry Holt and Co., Inc, New York, NY, USA (1982)

18. Oleszkiewicz, W., et al.: Visual probing: cognitive framework for explaining self-supervised image representations. CoRR abs/2106.11054 (2021). https://arxiv.org/abs/2106.11054

19. Ravfogel, S., Elazar, Y., Gonen, H., Twiton, M., Goldberg, Y.: Null it out: guarding protected attributes by iterative nullspace projection. In: Proceedings of the 58th Annual Meeting of the Association for Computational Linguistics, pp. 7237–7256. Association for Computational Linguistics, Online, July 2020. https://doi.org/10.18653/v1/2020.acl-main.647

20. Simonyan, K., Vedaldi, A., Zisserman, A.: Deep inside convolutional networks: Visualising image classification models and saliency maps. arXiv preprint arXiv:1312.6034 (2013)

21. Sivic, J., Zisserman, A.: Video Google: efficient visual search of videos. In: Ponce, J., Hebert, M., Schmid, C., Zisserman, A. (eds.) Toward Category-Level Object Recognition. LNCS, vol. 4170, pp. 127–144. Springer, Heidelberg (2006). https://doi.org/10.1007/11957959_7

DITA-NCG: Detecting Information Theft Attack Based on Node Communication Graph

Zhenyu Cheng[1], Xiaochun Yun[4(✉)], Shuhao Li[1,2,3], Jinbu Geng[1], Rui Qin[1], and Li Fan[1]

[1] Institute of Information Engineering, Chinese Academy of Sciences, Beijing, China
{chengzhenyu,lishuhao,gengjinbu,qinrui,fanli}@iie.ac.cn
[2] School of Cyber Security, University of Chinese Academy of Sciences, Beijing, China
[3] Key Laboratory of Network Assessment Technology, Chinese Academy of Sciences, Beijing, China
[4] National Computer Network Emergency Response Technical Team/Coordination Center of China, Beijing, China
yunxiaochun@cert.org.cn

Abstract. The emergence of information theft poses a serious threat to mobile users. Short message service (SMS), as a mainstream communication medium, is usually used by attackers to implement propagation, command and control. The previous detection works are based on the local perspective of terminals, and it is difficult to find all the victims and covert attackers for a theft event. In order to address this problem, we propose DITA-NCG, a method that globally detects information theft attacks based on node communication graph (NCG). The communication behavior of a NCG's node is expressed by both call detail record (CDR) vectors and network flow vectors. Firstly, we use CDR vectors to implement social subgraph division and find suspicious subgraphs with SMS information entropy. Secondly, we use network flow vectors to distinguish information theft attack graphs from suspicious subgraphs, which help us to identify information theft attack. Finally, we evaluate DITA-NCG by using real world network flows and CDRs , and the result shows that DITA-NCG can effectively and globally detect information theft attack in mobile network.

Keywords: Information theft · CDR · Network flow · Node communication graph

1 Introduction

Smartphones, as the most popular mobile devices, provide a convenient communication way for users and always save amounts of users' information. With the explosive growth of mobile communications, users' privacy security issues are becoming more prominent. The rapid increase of app usage makes smartphones as prime targets for attackers to steal users' information.

As it is reported that most of information theft attacks use short message service (SMS) to spread and use mobile network to send back users' information. Swift Cleaner [23] reported by Trend Micro receives SMS commands to execute remote command, steal information, send short messages, etc. FakeSpy [15] is an information stealer delivered by SMS, which steals text messages, account information, contacts, and call records.

Due to the heavy use of SMS transmission in information theft events, call detail record (CDR) data may be an important factor to identify the attacks. Researchers have done some analyses with CDRs. For example, there is a survey [6] that focuses on the analysis of massive CDR datasets and mainly studies the structure of social networks and human mobility. Some CDR researches also focus on the users' privacy security, but most of them only care phone calls [18] or mobility patterns [11,25]. We consider that we can use CDRs to find out some relationships among these communication nodes, which can help us to identify information theft attack graphs.

In order to protect users from information theft, researchers have done a lot of works to analyse malwares. Static analysis methods like signature-based detection [12] and content-based detection [5] always need some priori knowledge before analyse. Dynamic analysis methods like behavior-based detection [13] usually root the system or make a sand box for detection. There are also some works that use mobile network traffic to analyse users' behavior patterns or apps' features. But few of them focus on detecting information theft. Besides, these works are based on the local perspective of terminals, and it is hard to find all the victims and covert attackers in an information theft event.

In this paper, we propose DITA-NCG, which is a model globally identifying mobile network information theft attacks based on node communication graph (NCG). We use real data to verify the accuracy and validity of our detection model. The result of experiment shows that the detection model can effectively identify information theft attack graphs. The main contributions of this paper are summarized as follows:

- We construct suspicious subgraphs of NCG based on information entropy. By using CDR vectors to achieve social subgraph division, we calculate SMS content length information entropy for each subgraph. The lower value of information entropy means a subgraph is more suspicious.
- We identify information theft attack graphs based on network flow vectors. We employ Convolutional Neural Networks (CNN) to detect network flows of nodes, which are selected from a suspicious subgraph, and then use Support Vector Machine (SVM) to determine whether a new graph is an information theft attack graph or not.
- We evaluate our model by using real world CDRs (200,522 short message CDRs) and network flows (37,384 information theft network flows and 61,635 benign network flows). The result shows that the detection model achieves a 90.55% accuracy, a 92.02% precision and a 94.25% recall.

The rest of this paper is organized as follows: Sect. 2 highlights the related works. The detailed content of our method is described in Sect. 3. We present the performance of detecting information theft attack graphs in Sect. 4. In Sect. 5, we draw conclusions of this paper and discuss several limitations of our work.

2   Related Work

Some researchers have investigated the malware identification and private information tracking using network traffic. Taylor et al. [21] implemented Appscanner, which could achieve automatic fingerprinting and real-time identification of Android apps from their encrypted network traffic. Conti et al. [10] designed a system that can identify the specific action when a user was performing on mobile apps, by analyzing the statistical features of Android encrypted traffic. Ren et al. [19] proposed Recon to reveal and control personal identifiable information leaks in mobile network traffic, in which the key/value pairs were used for detection. Wang et al. [22] proposed a malware detection method to identify malware by using the text semantics of network traffic. Alam et al. [4] proposed DroidDomTree to detect malware, which mined the dominance tree of API calls to find similar patterns in Android applications.

There are also some works about CDR data research. Zang et al. [25] proposed an approach to infer, from 30 billion call records, the "top N" locations for each user and correlate this information with publicly-available side information such as census data. Bogomolov et al. [7] presented an approach to predict crime in a geographic space from multiple data sources, particularly in mobile phone and demographic data. Sultan et al. [20] proposed a framework for classifying mobile traffic patterns, which was based on the spatiotemporal analysis of CDR data. Zhang et al. [26] presented a large-scale characterization of fake base station spam ecosystem, which investigated how fraudulent messages are constructed to trap users.

Through summarizing the existing research works as above, we know that although there are some studies focusing on detecting malwares, it is lack of using network flows to detect attack events. And most of CDR data researches focus on human mobility or data analysis. It is lack of using CDRs to identifying information theft attacks.

3   Modeling and Methodology

We propose DITA-NCG to globally detect information theft attacks. Figure 1 shows the process of detecting information theft attack graphs. And we will introduce our detection model in two phases: data pre-process and attack detection.

3.1   Data Pre-process

First of all, the prototype data need to be reconstructed and filtered. The CDR data generate CDR vectors, and the network traffic data generate network flow vectors.

CDR Vector Generation. A CDR is a data record produced by a telephone communications or other telecommunication transactions. CDR contains various attributes [14,16,17], such as: originating telephone number, terminating telephone number, call duration, International Mobile Subscriber Identity (IMSI) number, International Mobile station Equipment Identity (IMEI) number, disposition or the results of the call, call type, etc. In actual modern practice, CDRs are much more detailed.

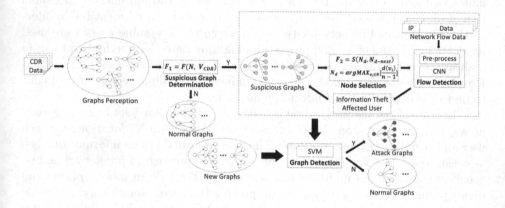

Fig. 1. The model of detecting information theft attack graphs.

In information theft events, attackers always use SMS to achieve command and control. And these messages always have the same content and format. Since the information theft apps spread and command through SMS, we take the data redundancy elimination for CDRs. Firstly, we reserve the originating and terminating phone numbers since the phone number is essential. Secondly, we get the start-time, end-time and duration from the time stamp. Thirdly, we remain the telecommunication type which can determine whether the communication is "send" or "received". At last, the SMS content' length is retained, which is an important characteristic for detecting malicious communication nodes, and we will explain it in Sect. 4. The other data segments like IMSI, IMEI, location data, etc., are discarded. In order to protect users' privacy, phone numbers and SMS contents are masked and desensitized.

We extract some features and statistics to generate the CDR vector, which contains: originating number, terminating number, send time, content length, average content length and time interval for the node, out-degree and incoming-degree of originating and terminating number.

Flow Vector Generation. Information theft attackers usually use apps to steal users' personal information through network traffic, so we want to analyse network flows to detect information theft. From our previous research works [8,9], we recognized that: (1) Different flows have different lengths (packet number) in a solitary complete session. (2) The flows' content sizes are much different from each other. (3) The ratio of incoming traffic and outgoing traffic is distinct for different flows. Therefore, we decide to use the network flows to detect information theft attack.

In this phase, we restructure the traffic data and divide them into a time series of packets. The network flow is regarded as a fundamental entity, and we define it as a sequence of TCP/IP packets ordered by time in a single session. Because really valued data are just parts of network flows, flow filtering is necessary to further improve the effectiveness of extracted features. We discard worthless information by taking domain filtering, packet filtering and so on.

Except for regular features, we also select some new features such as values of Dynamic Time Warping (DTW). DTW is an algorithm used to measure the similarity between two temporal sequences. It is recursively defined as:

$$min\_dtw = MIN(DTW(i-1,j), DTW(i-1,j-1), DTW(i,j-1)) \tag{1}$$

$$DTW(i,j) = local\_distance(i,j) + min\_dtw \tag{2}$$

We use $DTW(s,t)$ to compare the "shapes" of network flow s and t. To reduce the computation burden of calculating DTW, a leader flow is elected for each app, and this can be expressed as:

$$arg\ MIN_{f_l \in F}\left(\sum_{i=1}^{n} DTW(f_l, f_i)\right) \tag{3}$$

The flow f_l represents the leader flow of F, and f_i represents other flows.

As a result, a network flow vector contains: the values of DTW, duration, and number of packets, flow length, average packet length, average packet interval for different flow type (incoming, outgoing and complete).

3.2 Attack Detection

Information theft attack graph detection can help us globally find all the relative nodes in information theft events. Figure 2 illustrates a case of information theft attack. The attacker uses SMS to lure users to download malwares, which always contains a malicious URL. After receiving the malicious SMS, the user may not be controlled to resend the mal-SMS to all of his/her contacts. Because the usually contacted people have a higher degree of credibility, the infected user will be forced to send the mal-SMS to his/her frequent contacts. However, if a user does not read the message or download the suspicious app, he/she will be safe in this attack. When a user trusts the source of SMS and click the URL, he/she will access a server controlled by the attacker and automatically download the malwares. Then his/her smartphone will be infected and send the same mal-SMS to others.

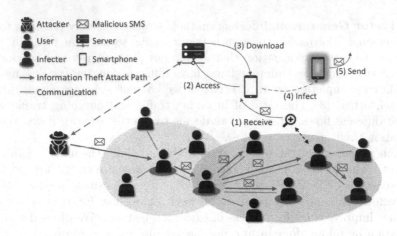

Fig. 2. The case of information theft attack.

In this phase, our model achieves the goal of detecting attack graphs of information theft. After the pre-process, we use the CDR vectors to generate NCG, which is based on the phone numbers of SMS. In graphs perception, the model gets a lot of communication subgraphs. Then the detecting model uses the similarity matching function to identify suspicious subgraphs and normal subgraphs. The function is defined as:

$$F_1 = F(N, V_{CDR}) \tag{4}$$

N represents the node set of a divided subgraph, and V_{CDR} represents the set of CDR vectors of N. Function $F()$ calculates the information entropy of the communication subgraph and use the result to detect suspicious subgraphs. The information entropy of the subgraph is defined as:

$$H(M) = -\sum_{m \in M} p(m) \log p(m) \tag{5}$$

M represents the node's SMS content length set, and m represents a separate SMS content length. $p(m)$ indicates the probability of m occurrence in M. We use 2 as the base of the logarithm. $H(M)$ is the information entropy of SMS content length of the subgraph, and its value gets large, the SMS content lengths are volitile, and the information gets more confusing and complex. If the information entropy value gets lower, the subgraph will be more suspicious.

After obtaining the suspicious subgraphs, DITA-NCG will choose some nodes of a suspicious subgraph to further detect whether the graph is normal or malicious. The node selection function is used to select communication nodes for detection, and it is defined as:

$$F_2 = S(N_d, N_{d-next}) \tag{6}$$

$$N_d = \arg MAX_{v_i \in N}[d(v_i)/(n-1)] \tag{7}$$

N_d represents the node that has the max degree centrality in the graph, and N_{d-next} is the next hop communicating nodes of N_d. $d()$ is the function to calculate the node's degree. For the selected nodes, DITA-NCG detects each node's network flows to identify whether it is truly under information theft attack or not.

In network flow detection, we use CNN for detecting. As one of the most popular deep learning algorithms, CNN overcomes the difficulties in feature extraction and is good at extracting local features. And in our previous work [9], we have already verified the effectiveness of CNN. In this paper, we use the network flow vectors as input. And then we set 4 filters with the kernel size of 3. At last, we use the "sigmoid" function to squash the single-unit output layer. After training, we take the number 0 as the label of normal flows, and the number 1 as the label of information theft flows. From the result of flow detection, we can judge whether a suspicious subgraph is an information theft attack graph or not.

Due to the time limit of detection, we train an SVM classifier to detect all of the information theft attack graphs by using the classified graph sets. In machine learning, SVM is a supervised learning model with associated learning algorithms. When the new graphs are added to the training samples, the model will retrain and update the classifier. The process of our model's detection is described in Algorithm 1.

Algorithm 1. The algorithm of detecting information theft attack graphs

---

Input:
　　Node vector data: $N = \{N_0, N_1, ..., N_n\}$.
Output:
　　Classified graphs: $G = \{G_{normal}, G_{attack}\}$.
1: Pre-processing and normalization N;
2: Establish the relationships of N to generate the communication graphs: G_{all};
3: Suspicious graph determination: $G_{suspicious}$;
4: **while** true **do**
5: 　**if** $G_{suspicious}$!= NULL **then**
6: 　　Select and remove one subgraph $graph_s$ from $G_{suspicious}$;
7: 　　Random select nodes form $graph_s$;
8: 　　Network flow detection with CNN;
9: 　　**if** label == 1 **then**
10: 　　　Classify the graph into G_{attack};
11: 　　**else**
12: 　　　Classify the graph into G_{normal};
13: 　　**end if**
14: 　**else**
15: 　　All the subgraphs of $G_{suspicious}$ is classified;
16: 　　break;
17: 　**end if**
18: **end while**
19: Use SVM to detect new graphs with G;

---

4   Experiment and Evaluation

A server computer (Intel Core i7-4790 3.60 GHz with 8 GB DDR3 RAM) is used
as a control center, which is running Windows 7 with two network cards. It is
emulated as a router to receive and save network traffic data. Some smartphones
are used to generate traffic data, like Galaxy Note 4 (SM-N9100) and Xiaomi 4
(MI 4LTE) both running the Android 6.0.1 operation system. And a computer
is used to run several virtual devices to simulate traffic generation. These virtual
devices run with different versions of system. A sever is used to save and analyse
data with a GPU GeForce GTX Titan X.

4.1   Experimental Data

We collect 113 information theft app samples provided by CNCERT/CC (National
Computer Network Emergency Response Technical Team/Coordination Center of
China). But we find that most of them can not run properly. Because these apps
appear a long time, their C&C servers can not be connected, and they can not steal
information by their program logic. And due to the update of operation system version, some of them can not be installed correctly. Therefore, we extract the information theft modules from these apps and recode them into ITM-capsule. Also we
design a simulated C&C server for remote control. ITM-capsule steals the users'
information such as phone identification information, contacts list, call history,
SMS, etc. As shown in Fig. 3, ITM-capsule runs in the Android system to communicate with the server and transfer users' information. The process of information theft is: (1) ITM-capsule collects phone basic information (i.e. phone number,
IMEI, MAC, etc.), and sends it to the C&C server. (2) The sever checks for presence of the phone's information. If there is no information about this phone, the
database will create a new ID. (3) The server returns the ID to the phone. (4) ITM-
capsule adds the ID to all the next transferred contents to ensure the information
unity and sends them to the sever. (5) The sever saves the received information
into the database. At last, when the whole transmission procedure ends, i.e., the
sever has not received new data for a long time, it will send a stop command to
stop ITM-capsule's work.

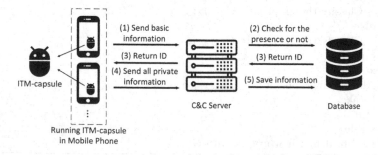

Fig. 3. The information theft process of ITM-capsule.

We also download lots of information theft apps from VirusShare [2], which is a repository of malware samples accessible for security researchers, and this set contains 293 samples. Furthermore, we collect 1,012 apps from YingYongBao, which is a popular third party app market of Tencent [1]. As we know that not all the collected apps are benign, we upload these apps to VirusTotal [3] to ensure the credibility of the training set. As a result, we get 997 benign apps. Using our own designed network traffic collection platform, the information theft apps and ITM-capsule generate 7.8 GB traffic data, and the benign apps generate 6.5 GB traffic data. Then we obtain 37,384 and 61,635 network flows respectively from information theft and benign traffic data. To ensure the independence of capture process, we tag all the flows for different devices and times. As a result, there are 11,752 clusters of network flows tagged.

We get a set of CDRs through an internet service provider. It should be noted that the CDR data is desensitized and masked. We promise that all the users' private information is protected and we do not use SMS contents in the experiment, so there is no ethical issues. The CDR dataset is 5.2 GB, and it contains 3 days' 986,752 records. After discarding records of phone call and base station, etc., we extract 200,522 SMS CDRs, which belong to 10,329 phone numbers. And there are 1,846 phone numbers tagged with suspicious label. The format of the processed data is shown in the Table 1. "OPN" means originating phone number, and "TPN" means terminating phone number.

Table 1. The format of CDR dataset

No.	Type	OPN	TPN	content_len	time_stamp
1	7	68d81229d5ec9c...	b04b1fa4a5b3c0...	2	1512954871
2	7	b04b1fa4a5b3c0...	68d81229d5ec9c...	12	1513214984
3	8	b58765baf88359...	f9195b5d19f7b7...	66	1513046068
4	8	903eabe1b70d2b...	f9195b5d19f7b7...	26	1513130349
5	7	c2c9376a350ee9...	b8f02594b747d3...	17	1513146196
...	...	...	...	...	...

We have analysed two cases of information theft apps, Cckun and xxShenQi [24]. Combining with CDR dataset, Fig. 4 shows the comparison of the SMS content length of different node type examples. In this figure, "ad_1" and "ad_2" represent two advertisement node types; "sus_1" and "sus_2" represent two suspicious node types. The figure illustrates that SMS content length is an important feature for classification. If the node's SMS content lengths are convergent, its information entropy value will be low, and this node will be more suspicious. From the examples of Fig. 4, we can realise that suspicious nodes' SMS content length is similar to malicious one's. How to distinguish advertisement nodes and malicious nodes is really a question. To solve this problem, we use network flows to determine whether a node is malicious or not.

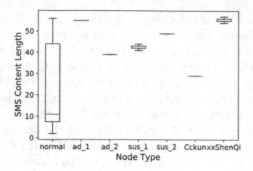

Fig. 4. The comparison of the SMS content length.

Based on the real proportion of these actual information theft events, we match the network flow vectors and CDR vectors by labels. From the collected data (11,752 clusters of tagged network flows and 10,329 phone numbers of CDRs), we select 10,000 as the number of nodes. To ensure the randomness and generality of our research, we randomly select 10,000 phone numbers' CDRs and devices' network flows to make a node vector set. The set contains 10,000 nodes and 79,637 communication paths, and 1,722 infected nodes and 18,915 information theft propagation paths. There are also 1,247 advertisement communication paths, which add some extraneous factors for simulating reality scenarios.

4.2   Experimental Result and Analysis

Because it is almost impossible to deal with the network flows of millions of nodes in the tolerable time. We need to find out the suspicious nodes firstly. We can establish NCG with the generated CDR vectors, and our model can lock on suspicious graphs with extracted features. The next step is to identify whether these nodes are really in information theft events or just are advertisement nodes or some other third service nodes. It is practicable to detect information theft flows and identify node type by using the network flow vectors. After that, we can check the other nodes of the communication paths by victims' CDRs. Figure 5 shows an example of detecting information theft attack graph. The big red spot is a attacker that sends malicious SMS (No. 1 and 5 indicate the different propagation paths), and the small red spots are victims, who are infected but not controlled to resend malicious SMS. The yellow spots are also infected, moreover, they send malicious SMS to their contacts (like No. 2, 3 and 4 indicating). The green spots are advertiser and their target users. With these confirmed spots, we can identify the information theft attack graphs.

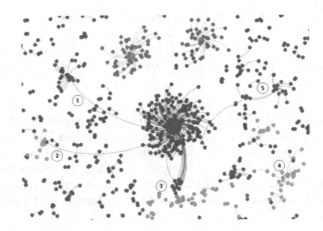

Fig. 5. An example of information theft attack graph detection.

In addition, we analyse the training set size requirement. Figure 6 illustrates different values with different sizes of the training set. The values of recall, F-Measure, precision and false positive rate (FPR) are indicated with different lines. We can find out that when the training set size reaches 2,000, the values of metrics get stable and effective. Therefore, we set 2,000 as the size of the training set and 8,000 for the testing set.

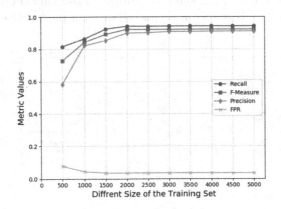

Fig. 6. The recall, F-Measure, precision and FPR values with different sizes of the training set.

Table 2 shows the results of accuracy, precision, recall, FPR and F-Measure. The detection rate for information theft attack graphs reaches 94.25% whereas the misjudgment rate for normal paths is only 3.53%. All the values demonstrate that our model is effective in identifying information theft attack graphs in mobile network.

Table 2. The performance of DITA-NCG

	Accuracy	Precision	Recall	FPR	F-Measure
Value	0.9055	0.9102	0.9425	0.0353	0.9260

The chosen of SVM in our model is based on the comparison with other representative machine learning algorithms, such as Random Forest, Naive Bayes and AdaBoost. Figure 7 illustrates that different algorithms have different performances. Particularly, SVM has a higher accuracy, precision, recall and F-Measure in this work, while its FPR is lower than others. Overall, the results show that SVM is better to solve the problems of detecting information theft attack graphs.

4.3 Comparison with Other Methods

Due to the differences of research perspective and experimental data, almost all the popular detection methods can not run with our experimental data. Therefore, we make a qualitative analysis to compare our model with others. Table 3 shows the comparison between our model and other methods, such as TaintDroid [13], AppScanner [21] and Recon [19]. D_1, D_2 and D_3 separately represents information theft detection, mobile network traffic detection and attack events detection. "No Root" and "No Sand-box" mean that there is no rooting system and no establishing a sand-box during detection. "Perspective" shows the detection method is based on the global or local perspective.

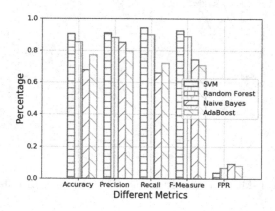

Fig. 7. The comparison between SVM and other machine learning algorithms.

Table 3. The comparison with other methods

	D_1	D_2	D_3	No root	No sand-box	Perspective
DITA-NCG	√	√	√	√	√	Global
TaintDroid	√	–	–	–	–	Local
AppScanner	–	√	–	√	√	Local
Recon	√	√	–	√	–	Local

5   Conclusion and Discussion

Information theft attacks pose a significant threat to the security of smartphone users. In order to globally find out the victims and attackers of information theft events, we propose DITA-NCG. We use SMS information entropy to identify suspicious subgraphs, and use CNN to detect the network flows of the selected nodes in suspicious subgraphs. At last, we use SVM to detect information theft attack graphs from new ones. We also develop a simulated information theft app, ITM-capsule, to solve the problem of C&C server invalidation and system version incompatibility. In the experiment, we evaluate the performance of DITA-NCG, which shows that our method achieves a high rate in accuracy, precision and recall. And it should be noted that all personally identifiable information (PII) in the dataset used in our experiment have been anonymized. The result indicates that DITA-NCG is effective to detect mobile network information theft attack globally, and it has a great potential to be applied in real scenarios for internet service providers.

However, there are some limitations about the proposed model. We use network flows and CDRs to generate vectors for identification, but most of the data are collected after information theft attacks. Although we can detect the information theft, how to prevent and block the attacks is remained to be solved in the future. And we will continue the research to establish a blocking mechanism to perfect the detecting model.

There are also some other ways to spread malwares, like uploading a fake app to third party markets, using phishing web sites, etc. These propagation ways may not spread rapidly or not have a bad influence, although they are really exist in the world. In the further research, we will comprehensively consider the information theft propagation.

The automation needs to be improved for DITA-NCG. And the model still needs to be ran with the data of real information theft events, although we have established a simulated communication node set based on some real samples.

Acknowledgement. This work is supported by the National Key Research and Development Program of China (Grant No. 2019YFB1005201). We would also like to thank the reviewers for the thorough comments and helpful suggestions.

References

1. App market of yingyongbao. https://android.myapp.com/ (2021)
2. Virusshare. https://virusshare.com/ (2021)
3. Virustotal. https://www.virustotal.com/ (2021)
4. Alam, S., Alharbi, S.A., Yildirim, S.: Mining nested flow of dominant APIs for detecting android malware. Comput. Netw. **167**, 107026 (2020)
5. Arzt, S., et al.: Flowdroid: precise context, flow, field, object-sensitive and lifecycle-aware taint analysis for android apps. ACM Sigplan Notices **49**(6), 259–269 (2014)
6. Blondel, V.D., Decuyper, A., Krings, G.: A survey of results on mobile phone datasets analysis. EPJ Data Sci. **4**(1), 1–55 (2015). https://doi.org/10.1140/epjds/s13688-015-0046-0
7. Bogomolov, A., Lepri, B., Staiano, J., Oliver, N., Pianesi, F., Pentland, A.: Once upon a crime: towards crime prediction from demographics and mobile data. In: Proceedings of the 16th International Conference on Multimodal Interaction, pp. 427–434. ACM (2014)
8. Cheng, Z., Chen, X., Zhang, Y., Li, S., Sang, Y.: Detecting information theft based on mobile network flows for android users. In: 2017 International Conference on Networking, Architecture, and Storage (NAS), pp. 1–10. IEEE (2017)
9. Cheng, Z., Chen, X., Zhang, Y., Li, S., Xu, J.: MUI-defender: CNN-Driven, network flow-based information theft detection for mobile users. In: Gao, H., Wang, X., Yin, Y., Iqbal, M. (eds.) CollaborateCom 2018. LNICST, vol. 268, pp. 329–345. Springer, Cham (2019). https://doi.org/10.1007/978-3-030-12981-1_23
10. Conti, M., Mancini, L.V., Spolaor, R., Verde, N.V.: Analyzing android encrypted network traffic to identify user actions. IEEE Trans. Inf. Forensics Secur. **11**(1), 114–125 (2016)
11. De Montjoye, Y.A., Hidalgo, C.A., Verleysen, M., Blondel, V.D.: Unique in the crowd: the privacy bounds of human mobility. Sci. Rep. **3**, 1376 (2013)
12. Desnos, A., et al.: Androguard: Reverse engineering, malware and goodware analysis of android applications. https://code.google.com/p/androguard/153 (2013)
13. Enck, W., et al.: TaintDroid: an information flow tracking system for real-time privacy monitoring on smartphones. Commun. ACM **57**(3), 99–106 (2014)
14. Horak, R.: Telecommunications and Data Communications Handbook. Wiley (2007). https://books.google.com/books?id=dO2wCCB7w9sC
15. N,B.: Fakespy - android information stealing malware attack to steal text messages, call records & contacts. https://gbhackers.com/fakespy/ (2019)
16. Petersen, J.: The Telecommunications Illustrated Dictionary. CRC Press advanced and emerging communications technologies series, CRC Press (2002). https://books.google.com/books?id=b2mMzS0hCkAC
17. Peterson, K.: Business Telecom Systems: A Guide to Choosing the Best Technologies and Services. Taylor & Francis (2000). https://books.google.com/books?id=W79R0niNU5wC
18. Ratti, C., Sobolevsky, S., Calabrese, F., Andris, C., Reades, J., Martino, M., Claxton, R., Strogatz, S.H.: Redrawing the map of great Britain from a network of human interactions. PLoS ONE **5**(12), e14248 (2010)
19. Ren, J., Rao, A., Lindorfer, M., Legout, A., Choffnes, D.: ReCon: revealing and controlling PII leaks in mobile network traffic. In: Proceedings of the 14th Annual International Conference on Mobile Systems, Applications, and Services, pp. 361–374. ACM (2016)

20. Sultan, K., Ali, H., Ahmad, A., Zhang, Z.: Call details record analysis: a spatiotemporal exploration toward mobile traffic classification and optimization. Information **10**(6), 192 (2019)
21. Taylor, V.F., Spolaor, R., Conti, M., Martinovic, I.: AppScanner: automatic fingerprinting of smartphone apps from encrypted network traffic. In: 2016 IEEE European Symposium on Security and Privacy (EuroS&P), pp. 439–454. IEEE (2016)
22. Wang, S., Yan, Q., Chen, Z., Yang, B., Zhao, C., Conti, M.: Detecting android malware leveraging text semantics of network flows. IEEE Trans. Inf. Forensics Secur. **13**(5), 1096–1109 (2017)
23. Wu, L.: First kotlin-developed malicious app signs users up for premium sms services. http://t.cn/EMSyiof (2019)
24. Yun, X., Li, S., Zhang, Y.: SMS worm propagation over contact social networks: modeling and validation. IEEE Trans. Inf. Forensics Secur. **10**(11), 2365–2380 (2015)
25. Zang, H., Bolot, J.: Anonymization of location data does not work: a large-scale measurement study. In: Proceedings of the 17th Annual International Conference on Mobile Computing and Networking, pp. 145–156. ACM (2011)
26. Zhang, Y., et al.: Lies in the air: characterizing fake-base-station spam ecosystem in china. In: Proceedings of the 2020 ACM SIGSAC Conference on Computer and Communications Security, pp. 521–534 (2020)

Boosted Ensemble Learning Based on Randomized NNs for Time Series Forecasting

Grzegorz Dudek[(✉)] [iD]

Electrical Engineering Faculty, Czestochowa University of Technology,
Czestochowa, Poland
grzegorz.dudek@pcz.pl

Abstract. Time series forecasting is a challenging problem particularly when a time series expresses multiple seasonality, nonlinear trend and varying variance. In this work, to forecast complex time series, we propose ensemble learning which is based on randomized neural networks, and boosted in three ways. These comprise ensemble learning based on residuals, corrected targets and opposed response. The latter two methods are employed to ensure similar forecasting tasks are solved by all ensemble members, which justifies the use of exactly the same base models at all stages of ensembling. Unification of the tasks for all members simplifies ensemble learning and leads to increased forecasting accuracy. This was confirmed in an experimental study involving forecasting time series with triple seasonality, in which we compare our three variants of ensemble boosting. The strong points of the proposed ensembles based on RandNNs are very rapid training and pattern-based time series representation, which extracts relevant information from time series.

Keywords: Boosted ensemble learning · Ensemble forecasting · Multiple seasonality · Randomized NNs · Short-term load forecasting

1 Introduction

Ensemble methods are considered to be a cornerstone of modern machine learning [1]. They are commonly used for regression and classification problems. Ensembling is also a very effective way of increasing the predictive power of forecasting models. Combining many base models improves the final forecasting accuracy as well as the stability of the response when compared to a single model approach. Success in ensemble learning depends on the proper flexibility of the ensemble members and the trade-off between their performance and diversity [2]. It is also determined by the way learners are generated at the successive stages of ensembling and the method employed to combine them.

Supported by grant 020/RID/2018/19 from the Polish Minister of Science and Higher Education titled "Regional Initiative of Excellence", 2019-22.

The effectiveness of ensembling in forecasting is evidenced by the fact that in the most renowned forecasting competition, M4 [3], of the 17 most accurate models, 12 used ensembling in some form [4]. The winning submission, which is a hybrid model combining exponential smoothing and long short-term memory, used three types of ensembling simultaneously [5]: combining results of the stochastic training process, bagging, and combining multiple runs.

To improve the performance of ensemble learning many approaches have been proposed such as stacking [6], bagging [7], boosting [8], negative correlation learning [9], snapshot ensembles [10], and horizontal and vertical ensembles developed for deep learning [11]. Boosting, which this work focuses on, is a general ensemble technique that involves sequentially adding base models to the ensemble where subsequent models correct the performance of prior models. This approach is very effective as evidenced by the high ranking positions of boosted models such as XGBoost [12], i.e. ensemble of decision trees with regularized gradient boosting, in competitions such as those organized by Kaggle. Boosting also has many applications in the forecasting field. Some examples are: [13], where an ensemble of boosted trees is used for bankruptcy prediction; [14], where XGBoost is combined with a Gaussian mixture model for monthly streamflow forecasting; [15], where AdaBoost is applied as a component of a hybrid model for multi-step wind speed forecasting; and [16], where a natural gradient boosting algorithm is applied for solar power probabilistic forecasting. This last work highlights a valuable advantage of ensembling. It can produce probabilistic forecasts, i.e. the distribution of the forecasted variable in the future.

In this study, we propose a boosted version of our ensemble of randomized neural networks (RandNNs) for complex time series forecasting [17]. In [17], we focused on strategies for controlling the diversity of ensemble members. The members were trained independently. Here, we construct an ensemble sequentially. When a new member is added to the ensemble, it learns by taking into account the results of the ensemble members so far. We consider three methods of boosting. The contribution of this study is threefold:

1. We propose new methods of ensemble boosting: ensemble learning based on corrected targets and ensemble learning based on opposed response. They are both employed to ensure similar tasks for all ensemble members. This unification of the tasks justifies the use of identical base models in terms of architecture and hyperparameters at all stages of ensembling.
2. We develop three ensemble learning approaches for forecasting complex time series with multiple seasonality. They are based on RandNNs, pattern-based time series representation and three methods of boosting: based on residuals, corrected targets and opposed response.
3. We empirically compare the performance of the proposed ensemble methods on challenging short-term load forecasting problems with triple seasonality, and conclude that the opposed response-based approach outperforms its competitors in terms of accuracy and sensitivity to hyperparameters.

The rest of the work is organized as follows. In Sect. 2, we present a base model, RandNN. Details of the proposed three methods of boosted ensemble learning are described in Sect. 3. The experimental framework used to evaluate and compare the proposed ensemble methods is described in Sect. 4. Finally, Sect. 5 concludes the work.

2   Base Forecasting Model - RandNN

As a base model, we use a single hidden layer feedforward NN with m logistic sigmoid hidden nodes [18]. In randomized learning, the weights of hidden nodes are selected randomly from a uniform distribution and symmetrical interval $U = [-u, u]$. The biases of these nodes are calculated, according to recent research [19], based on the weights as follows: $b_j = -\mathbf{a}_j^T \mathbf{x}_j^*$, where $\mathbf{a}_j$ is the vector of weights for the j-th hidden node, and $\mathbf{x}_j^*$ is one of the training patterns selected at random (see [19] and [20] for details, justification and other variants). This way of generating hidden nodes places the sigmoids into the input feature space limited to some hypercube, avoiding their saturated parts. Moreover, the sigmoids are distributed according to the data distribution. These improve the aproximation and generalization abilities of the model [19,20].

The hidden node sigmoids are combined linearly by the output nodes: $\varphi_k(\mathbf{x}) = \sum_{j=1}^{m} \beta_{j,k} h_j(\mathbf{x})$, where $h_j(\mathbf{x})$ is the output of the j-th hidden node, and $\beta_{j,k}$ is the weight between j-th hidden and k-th output nodes. The only learnable parameters are the output weights. They are calculated from $\boldsymbol{\beta} = \mathbf{H}^+\mathbf{Y}$, where $\mathbf{H}$ is a matrix of the hidden layer outputs, $\mathbf{H}^+$ denotes its Moore-Penrose generalized inversion, and $\mathbf{Y}$ is a matrix of target output patterns. $\mathbf{H}$ is a nonlinear feature mapping from n-dimensional input space to m-dimensional projection space (usually $m \gg n$). Note that this projection is random. Due to the fixed parameters of the hidden nodes, the optimization problem in randomized learning (selection of weights $\boldsymbol{\beta}$) becomes convex and can be easily solved by the standard least-squares method.

The forecasting model based on RandNN, which we adapt from [18], has two more components: encoder and decoder. Let us consider time series $\{E_k\}_{k=1}^{K}$ with multiple seasonality. The encoder transforms this series into input and output patterns expressing seasonal sequences of the shortest length. The input patterns, $\mathbf{x}_i = [x_{i,1}, \ldots, x_{i,n}]^T$, represent sequences $\mathbf{e}_i = [E_{i,1}, \ldots, E_{i,n}]^T$, while the output patterns, $\mathbf{y}_i = [y_{i,1}, \ldots, y_{i,n}]^T$, represent forecasted sequences $\mathbf{e}_{i+\tau} = [E_{i+\tau,1}, \ldots, E_{i+\tau,n}]^T$, where n is a period of the seasonal cycle (e.g. 24 h for daily seasonality), $i = 1, \ldots, K/n$ is the sequence number, and $\tau \geq 1$ is a forecast horizon. The patterns are defined as follows:

$$\mathbf{x}_i = \frac{\mathbf{e}_i - \bar{e}_i}{\widetilde{e}_i}, \qquad \mathbf{y}_i = \frac{\mathbf{e}_{i+\tau} - \bar{e}_i}{\widetilde{e}_i} \tag{1}$$

where $\bar{e}_i$ is the mean value of sequence $\mathbf{e}_i$, and $\widetilde{e}_i = \sqrt{\sum_{t=1}^{n}(E_{i,t} - \bar{e}_i)^2}$ is a measure of sequence $\mathbf{e}_i$ dispersion.

Input patterns $\mathbf{x}_i$ represent successive seasonal sequences which are centered and normalized. They have a zero mean, and the same variance and unity length. Thus they are unified and differ only in shape. In contrast, patterns $\mathbf{y}_i$ are not globally unified and can express additional seasonality, e.g. when time series include both daily and weekly seasonalities, the former is expressed in the y-pattern shape, while the latter is expressed in y-pattern level and dispersion (compare x- and y-patterns in Fig. 2 and see discussion in [18]). In the case of such seasonalities, we build forecasting models that learn from data representing the same days of the week, e.g. for test query pattern x representing Monday, training set Φ consists of x-patterns representing all historical Mondays from the data and y-pattern representing the Tuesdays following them (assuming $\tau = 1$).

The decoder based on the y-pattern forecasted by the network, $\widehat{\mathbf{y}}$, and coding variables describing the query sequence, $\widetilde{e}$ and $\overline{e}$, using transformed Eq. (1) for y, calculates the forecasted seasonal sequence:

$$\widehat{\mathbf{e}} = \widehat{\mathbf{y}}\widetilde{e} + \overline{e} \qquad (2)$$

Remarks:

1. RandNN was designed for forecasting time series with multiple seasonalities. In the case of one seasonality, y-patterns express only one seasonality, and there is no need to decompose the forecasting problem. In the case of no seasonality, the input pattern length should be selected experimentally, while the y-pattern length is equal to the forecast horizon.
2. The bounds of the interval for random weights, u, correspond to the maximum sigmoid slope. To increase interpretability, let us express the bounds u using the slope angles α_{max} [20]: $u = 4\tan\alpha_{max}$, and treat α_{max} as a hyperparameter. The second hyperparameter is the number of hidden nodes, m. Both hyperparameters decide about the bias-variance tradeoff of the model and should be tuned to the complexity of the target function.
3. Advantages of RandNN are very fast training and the simplification of the forecasting task due to pattern representation. In [18], it was shown that RandNN can compete with fully-trained NN in terms of forecasting accuracy, but is much faster to train.

3 Boosting of Ensemble Learning

3.1 Ensemble Learning Based on Residuals

An ensemble based on residuals, which is a simplified variant of a gradient boosting algorithm [21], is constructed sequentially. In the first step, a base model learns on the training set $\{(x_i, y_i)\}_{i=1}^{N}$ (we consider a scalar input and output for simplicity) and fits to original data. Let us denote this model $f_1(x)$ and the ensemble model including one member $F_1(x) = f_1(x)$. In the second step, the second base model is added, $f_2(x)$, such that the sum of this model with the previous one, $F_2(x) = F_1(x) + f_2(x)$, is the closest possible to the target y. In the successive steps, further base models are added with the same expectation. In the k-th step, the function fitted by the ensemble is $F_k(x) = F_{k-1}(x) + f_k(x)$. Thus, the error in this step, $MSE_k = \frac{1}{N}\sum_{i=1}^{N}(y_i - F_k(x_i))^2$, can be written as:

$$MSE_k = \frac{1}{N} \sum_{i=1}^{N} \left(f_k(x_i) - r_{k-1,i} \right)^2 \qquad (3)$$

where $r_{k-1,i} = y_i - F_{k-1}(x_i)$ is a residual between the target and the response of the ensemble of $k - 1$ members.

Equation (3) clearly shows that the base model added to the ensemble at the k-th stage fits to the residuals between the target and the ensemble built at stage $k - 1$. So, it attempts to correct the errors of its predecessors. The training set for the base model in the k-th stage is $\{(x_i, r_{k-1,i})\}_{i=1}^{N}$. The final ensemble response is the sum of all member responses: $F_K(x) = \sum_{k=1}^{K} f_k(x)$.

Algorithm 1. EnsR: Boosting based on residuals

Input: Base model f (RandNN), Training set $\Phi = \{(\mathbf{x}_i, \mathbf{y}_i)\}_{i=1}^{N}$, Ensemble size K
Output: RandNN ensemble F_K
Procedure:
for $k = 1$ **to** K **do**
    Learn f_k based on Φ
    Calculate ensemble response $F_k(\mathbf{x}_i) = \sum_{l=1}^{k} f_l(\mathbf{x}_i), i = 1, ..., N$
    Determine residuals $\mathbf{r}_{k,i} = \mathbf{y}_i - F_k(\mathbf{x}_i), i = 1, ..., N$
    Modify training set $\Phi = \{(\mathbf{x}_i, \mathbf{r}_{k,i})\}_{i=1}^{N}$
end for

Algorithm 1 and Fig. 1 demonstrate the process of building an ensemble based on residuals (EnsR) for pattern-based forecasting. Learner 1 (RandNN) learns the original target function on Φ. Each subsequent learner learns the residuals between the target patterns $\mathbf{y}$ and the aggregated outputs of its predecessors shown in the lower panel of Fig. 1. Note that in EnsR, each learner can have a different problem to solve, expressing different features. Learner 1 learns the specific patterns of seasonal cycles $\mathbf{y}$, while the next learners learn completely different tasks, i.e. the residuals, which do not have such distinct patterns as $\mathbf{y}$ and have a large stochastic component. This inconsistency between the problems solved by the learners can affect negatively the final result, especially in the common case of using identical base models (the same architecture and hyperparameters) at every stage of ensembling.

3.2　Ensemble Learning Based on Corrected Targets

To unify the problems solved by the learners at successive stages of ensembling, we modify the EnsR framework as follows. During the sequential process, at stage k, the base model $f_k(x)$ is added to the ensemble. Ensemble response at this stage is the average of k learners: $F_k(x) = \frac{1}{k} \sum_{l=1}^{k} f_l(x)$. The loss function can be expressed as:

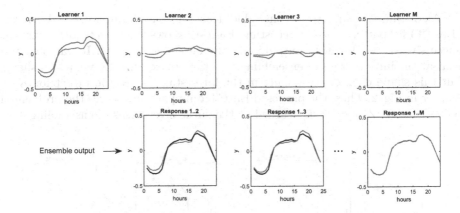

Fig. 1. EnsR: Responses produced by individual learners (upper panel) and the ensemble (lower panel). Target responses of individual learners in blue, target responses of the ensemble in black, real responses in red. (Color figure online)

$$MSE_k = \frac{1}{N} \sum_{i=1}^{N} \left(y - \left(\frac{1}{k} \sum_{l=1}^{k-1} f_l(x) + \frac{f_k(x)}{k} \right) \right)^2$$

$$= \frac{1}{Nk} \sum_{i=1}^{N} \left(f_k(x) - \left(ky - \sum_{l=1}^{k-1} f_l(x) \right) \right)^2 \tag{4}$$

As can be seen from this equation, the difference between ky and the sum of the previous learners is the new target to which the k-th learner is fitted. This target expresses pattern y corrected by aggregated residuals of the previous learners: $y + \sum_{l=1}^{k-1} (y - f_l(x))$. If the aggregated residuals are much smaller compared to the y-pattern (we expect this), the new targets at all stages of ensembling have a similar shape to the y-pattern. So, all learners have similar tasks to solve and it is justified for them to have the same architecture and hyperparameters. This unburdens us from the awkward and time-consuming task of selecting the optimal model at each stage of ensembling.

Algorithm 2. EnsCT: Boosting based on corrected targets

Input: Base model f (RandNN), Training set $\Phi = \{(\mathbf{x}_i, \mathbf{y}_i)\}_{i=1}^{N}$, Ensemble size K
Output: RandNN ensemble F_K
Procedure:
for $k = 1$ **to** K **do**
  Learn f_k based on Φ
  Calculate ensemble response $F_k(\mathbf{x}_i) = \frac{1}{k} \sum_{l=1}^{k} f_l(\mathbf{x}_i), i = 1, ..., N$
  Determine corrected targets $\mathbf{y}'_{k,i} = (k+1)\mathbf{y}_i - kF_k(\mathbf{x}_i), i = 1, ..., N$
  Modify training set $\Phi = \{(\mathbf{x}_i, \mathbf{y}'_{k,i})\}_{i=1}^{N}$
end for

Algorithm 2 summarizes ensemble learning based on corrected targets (EnsCT) for pattern-based forecasting. Figure 2 shows the targets for learners at successive stages, learners' outputs and the ensemble outputs. It was observed that at the initial stages of ensembling, the targets express the y-pattern shape, but this shape degrades gradually in the later stages - see the target for K-th learner in Fig. 2. Thus, the proposed EnsCT only partially solves the problem of inconsistency between tasks learned at the successive stages of ensembling.

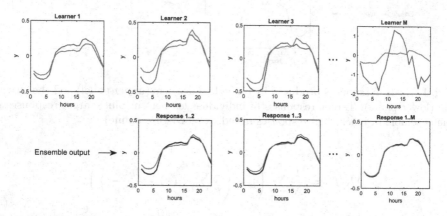

Fig. 2. EnsCT: Responses produced by individual learners (upper panel) and the ensemble (lower panel). Target responses of individual learners in blue, target responses of the ensemble in black, real responses in red. (Color figure online)

3.3   Ensemble Learning Based on Opposed Response

To prevent the degradation of the target shapes in the subsequent steps of ensembling, we propose ensemble learning based on opposed response (EnsOR). The first step of the ensemble building procedure is the same as in EnsR and EnsCT. At this stage, the base model $f_1(x)$ learns on original training set Φ. The ensemble response is $F_1(x) = f_1(x)$. The residual is calculated, $r_1 = y - F_1(x)$, and the "opposed" response pattern is determined as follows:

$$\hat{y}'_1 = y + r_1 = 2y - F_1(x) \tag{5}$$

The opposed response pattern expresses the target pattern augmented by the opposed error produced by the ensemble (see Fig. 3). The average of the response pattern $F_1(x)$ and the opposed response pattern $\hat{y}'_1$ gives the target pattern y. In the next step, base model $f_2(x)$ learns on training set $\{(x_i, \hat{y}'_{1,i})\}_{i=1}^{N}$. Thus it learns the opposed response to reduce the ensemble residual. The ensemble response is calculated as the average of learners: $F_2(x) = \frac{f_1(x)+f_2(x)}{2}$. These operations are repeated in the following steps (see Algorithm 3). Namely, in step k, the opposed response pattern determined at stage $k - 1$, $\hat{y}'_{k-1} = 2y -$

$F_{k-1}(x)$, becomes the target pattern for learner $f_k(x)$. The ensemble response is the average of k learners and the loss function at stage k takes the form:

$$MSE_k = \frac{1}{N} \sum_{i=1}^{N} \left(f_k(x_i) - \hat{y}'_{k-1,i} \right)^2 \qquad (6)$$

where $\hat{y}'_{k-1,i} = y_i + r_{k-1,i} = 2y_i - F_{k-1}(x_i)$ is the opposed response pattern at stage $k - 1$.

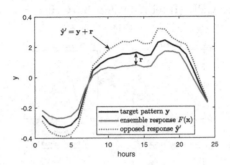

Fig. 3. Construction of the opposed response pattern.

Algorithm 3. EnsOR: Boosting based on opposed response

Input: Base model f (RandNN), Training set $\Phi = \{(\mathbf{x}_i, \mathbf{y}_i)\}_{i=1}^{N}$, Ensemble size K
Output: RandNN ensemble F_K
Procedure:
for $k = 1$ to K do
   Learn f_k based on Φ
   Calculate ensemble response $F_k(\mathbf{x}_i) = \frac{1}{k} \sum_{l=1}^{k} f_l(\mathbf{x}_i), i = 1, ..., N$
   Determine opposed response $\hat{\mathbf{y}}'_{k,i} = 2\mathbf{y}_i - F_k(\mathbf{x}_i), i = 1, ..., N$
   Modify training set $\Phi = \{(\mathbf{x}_i, \hat{\mathbf{y}}'_{k,i})\}_{i=1}^{N}$
end for

Figure 4 shows learners' and ensemble responses in the following steps of ensembling. Dashed lines express the opposed responses which become targets for the based models in the next steps (blue lines). Note that the shape of the initial target pattern $\mathbf{y}$ is maintained until the last stage. Thus, all learners learn on similar data, and using identical base models at each stage of ensembling means no objections can be raised, unlike in the case of EnsR.

4   Experimental Study

In this section, we verify our proposed methods of ensemble boosting on four time series forecasting problems. These comprise short-term electrical load forecasting problems for four European countries: Poland (PL), Great Britain (GB), France (FR) and Germany (DE) (data was collected from www.entsoe.eu). The hourly load time series express three seasonalities: yearly, weekly and daily. The data period is four years, from 2012 to 2015. Atypical days such as public holidays were excluded from the data (between 10 and 20 days a year). The forecasting problem is to predict the load profile (24 hourly values) for each day of 2015 based on historical data. The forecast horizon is one day, $\tau = 1$. The number of ensemble members was $K = 50$. As a performance metric we use mean absolute percentage error (MAPE). All algorithms were implemented in Matlab 2021b and run on a ten-core CPU (Intel i7-6950x, 3.0 GHz, 48 GB RAM).

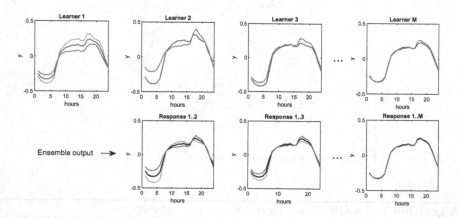

Fig. 4. EnsOR: Responses produced by individual learners (upper panel) and the ensemble (lower panel). Target responses of individual learners in blue, target responses of the ensemble in black, real responses in red, and opposed responses in dashed red. (Color figure online)

In the first experiment, we evaluate ensemble sensitivity to the base model hyperparameters: number of hidden nodes m and size of the interval for hidden weights α_{max}. Figure 5 shows the impact of hyperparameters on the test MAPE. Note that EnsR is the most sensitive to hyperparameters, while EnsOR is the least sensitive. To quantitatively compare the sensitivities of the ensemble variants, as a rough measure of sensitivity to a given hyperparameter, we define standard deviation of the test MAPE for the ensemble with optimal values of other hyperparameters:

$$S_m = Std(MAPE(\mathbf{y}, F(\mathbf{x}, \alpha_{max}^*, m))) \tag{7}$$

$$S_{\alpha_{max}} = Std(MAPE(\mathbf{y}, F(\mathbf{x}, \alpha_{max}, m^*))) \tag{8}$$

where the optimal hyperparameter values are marked with asterisks.

From Fig. 5, we can see that the accuracy of EnsR and EnsCT deteriorates quickly with the number of hidden nodes and interval U size, when these hyperparameters exceed their optimal values. This deterioration is related to the gradual loss of generalization for higher values of m and α_{max}. Deterioration for

EnsOR is much slower, which means that this approach is more resistant to overtraining.

Table 1 shows optimal hyperparameters, test and training errors and compares the sensitivities of ensembling variants. The lowest values are in bold. Table 2 clearly shows that EnsOR performs best. This method gave the lowest errors for each dataset and had the lowest sensitivity to both hyperparameters. By contrast, EnsR performed worst in terms of error and sensitivity. Note that the optimal hyperparameter values are smaller for EnsR and EnsCT than for EnsOR. This means that the base models in these boosting variants need to be less flexible (weaker) to prevent overfitting. Nevertheless, EnsR and EnsCT do not perform as well as EnsOR.

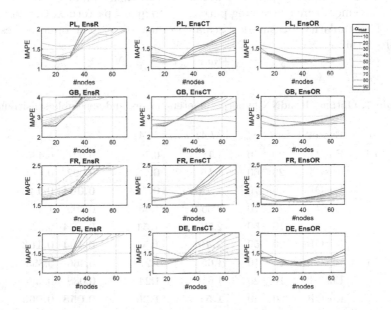

Fig. 5. Ensemble sensitivity to RandNN hyperparameters.

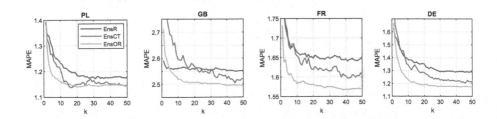

Fig. 6. Ensemble error at successive stages.

Figure 6 demonstrates test MAPE in successive steps of ensembling. We can observe from this figure that for EnsOR the error converges faster with k than for other ensemble variants. The convergence curve is also smoother for EnsOR than for its competitors. For them, in many cases, adding a new member to the ensemble causes a temporary increase in error. Moreover, it was observed for EnsR that if the hyperparameters are too high (greater than optimal), which means a flexible base model, the error starts to successively increase with subsequent ensembling stages. This phenomenon is to a lesser degree observed for EnsCT, but not observed for EnsOR. Figure 7 demonstrates this problem for PL data and $m = 100$, $\alpha_{max} = 70$.

To improve further ensemble learning based on opposed response, we introduce weights for training patterns which express their similarity to the query pattern. The more similar training pattern $\mathbf{x}_i$ to query pattern $\mathbf{x}$, the higher the weight. The similarity measure is a scalar product, $\mathbf{x}^T\mathbf{x}_i$. The opposed response takes the form:

$$\hat{\mathbf{y}}'_{k,i} = \mathbf{y}_i + w_i\mathbf{r}_{k,i}, i = 1, ..., N \tag{9}$$

Table 1. Optimal RandNN hyperparameters, errors and ensemble sensitivity.

Data	Ensemble	m^*	α^*_{max}	$MAPE_{tst}$	$MAPE_{trn}$	S_m	$S_{\alpha_{max}}$
PL	EnsR	20	40	1.18	0.76	0.641	0.436
	EnsCT	20	60	1.15	**0.72**	0.092	0.174
	EnsOR	40	70	**1.14**	**0.72**	**0.064**	**0.110**
GB	EnsR	20	10	2.55	1.55	0.509	1.531
	EnsCT	10	50	2.52	**1.51**	**0.104**	0.628
	EnsOR	20	50	**2.49**	**1.51**	**0.104**	**0.163**
FR	EnsR	10	10	1.65	1.27	0.466	0.853
	EnsCT	10	20	1.61	1.24	0.088	0.436
	EnsOR	20	40	**1.57**	**1.20**	**0.058**	**0.069**
DE	EnsR	10	40	1.29	1.37	0.774	0.572
	EnsCT	10	70	1.21	1.26	0.138	0.141
	EnsOR	40	80	**1.17**	**1.05**	**0.036**	**0.063**

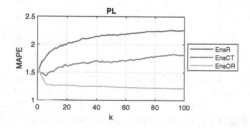

Fig. 7. Ensemble error at successive stages for too flexible base models.

where $\mathbf{r}_{k,i} = \mathbf{y}_i - F_k(\mathbf{x}_i)$ is a residual vector for the i-th training pattern and $w_i \in [0, 1]$ is the weight of this pattern.

In the base variant of EnsOR, the weights for all patterns are equal to one. In typical ensemble learning, the weights are zero (the base models at each stage of ensembling learn on the original training set Φ; such an approach we considered in [17]). By introducing weights, we try to balance these two approaches.

As a weighting function, we consider four variants. In the simplest one, EnsOR1, we assume that the weighting function g is just the scalar product:

$$g(\mathbf{x}, \mathbf{x}_i) = \mathbf{x}^T \mathbf{x}_i \tag{10}$$

To avoid negative values of (10) we can also use $g(\mathbf{x}, \mathbf{x}_i) = \frac{1}{2}(\mathbf{x}^T \mathbf{x}_i + 1)$ or replace negative values with zeros.

In the second variant, EnsOR2, we sort the training patterns according to similarity to the query pattern, from the most to the least similar. Let $r = 1, ..., N$ be the rank of the training patterns in the similarity ranking. The weighting function expresses the linear dependence of the weight on the rank:

$$g(\mathbf{x}, \mathbf{x}_i) = 1 + \frac{1 - r_i}{N} \tag{11}$$

In the third variant, EnsOR3, the weighting function expresses the non-linear dependence of the weight on the rank:

$$g(\mathbf{x}, \mathbf{x}_i) = \left(1 + \frac{1 - r_i}{N}\right)^d \tag{12}$$

where $d > 1$ (we assume $d = 4$).

In the fourth variant, EnsOR4, the most similar training patterns to the query pattern have unity weights, while the others have zero weights:

$$g(\mathbf{x}, \mathbf{x}_i) = \begin{cases} 1 & \text{if } \mathbf{x}_i \in \Xi_\kappa(\mathbf{x}) \\ 0 & \text{if } \mathbf{x}_i \notin \Xi_\kappa(\mathbf{x}) \end{cases} \tag{13}$$

where $\Xi_\kappa(\mathbf{x})$ denotes a set of κ nearest neighbors of query pattern $\mathbf{x}$ in Φ.

An example of weights assigned to the training patterns by the above weighting functions are shown in Fig. 8.

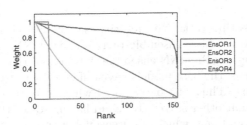

Fig. 8. Examples of weights assigned to training pattern in different weighing variants.

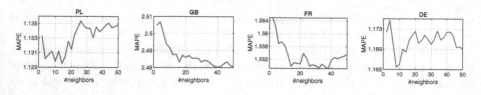

Fig. 9. Ensemble error for EnsOR4 depending on κ.

Figure 9 allows us to evaluate the impact of the number of nearest neighbours κ in EnsOR4 on the test MAPE. The optimal value of κ depends on the data set: for PL $\kappa = 14$, for GB $\kappa = 40$, for FR $\kappa = 38$, and for DE $\kappa = 8$.

Table 2. Results for EnsOR with weighted patterns.

Data	EsnOR1	EnsOR2	EnsOR3	EnsOR4	RandNN [17]	Ens1 [17]
PL	1.1419	1.1381	1.1329	**1.1295**	1.3206	1.1417
GB	2.4935	2.4804	2.4822	**2.4803**	2.6126	2.5148
FR	1.5668	1.5570	1.5524	**1.5492**	1.6711	1.5690
DE	1.1743	1.1711	1.1683	**1.1655**	1.3809	1.1811

Table 2 compares the proposed methods of pattern weighting. In this table results for individual RandNN and Ens1 are also shown. Ens1 is a classical (not boosted) ensemble of RandNN. Ens1 constructs the final forecast as an average of the responses of RandNN, which learn simultaneously on the same training set Φ [17]. As can be seen from this table, EnsOR4 outperformed its competitors as well as classical ensembling Ens1. The much higher error for individual RandNN justifies fully ensembling. It is worth mentioning that comparing Ens1 with other forecasting models, including statistical models (ARIMA, exponential smoothing, Prophet) and machine learning models (MLP, SVM, ANFIS, LSTM, GRNN, nonparametric models), reported in [17] clearly shows that Ens1 outperforms all its competitors in terms of accuracy.

5   Conclusion

Forecasting complex time series with multiple seasonalities is a challenging problem, but one we solve using ensemble of randomized NNs. We propose three methods of boosting the RandNN ensemble. We showed that in ensemble learning based on residuals, the base models have different tasks to solve at successive stages of ensembling. Thus, the base model which is optimal at a given stage may not be optimal at other stages. To avoid having to select an optimal model at each stage of ensembling, which is unreasonable and too time-consuming, we propose to unify the tasks solved at all stages. Doing so allows us to use the same base model (the same architecture and hyperparameters), RandNN in our case, which is optimal for all stages. To unify the tasks, we propose ensemble learning based on corrected targets and ensemble learning based on opposed response. The latter proved to be more resistant to task degradation at subsequent stages.

The experimental studies performed on four forecasting problems expressing triple seasonalities confirmed that the opposed response-based approach outperforms its competitors in terms of forecasting accuracy as well as sensitivity to both the base model hyperparameters and ensemble size. Further improvement of the winning solution was achieved by weighting the training patterns according to their similarity to the query pattern.

In further research, we plan to develop the opposed response-based approach for other types of learners, e.g. decision trees, and other types of problems (regression, classification).

References

1. Reeve, H.W.J., Brown, G.: Diversity and degrees of freedom in regression ensembles. Neurocomputing **298**, 55–68 (2018)
2. Brown, G., Wyatt, J.L., Tino, P.: Managing diversity in regression ensembles. J. Mach. Learn. Res. **6**, 1621–1650 (2005)
3. Makridakis, S., Spiliotis, E., Assimakopoulos, V.: The M4 competition: results, findings, conclusion and way forward. Int. J. Forecasting **34**(4), 802–808 (2018)
4. Atiya, A.F.: Why does forecast combination work so well? Int. J. Forecasting **36**(1), 197–200 (2020)
5. Smyl, S.: A hybrid method of exponential smoothing and recurrent neural networks for time series forecasting. Int. J. Forecasting **36**(1), 75–85 (2020)
6. Wolpert, D.H.: Stacked generalization. Neural Netw. **5**(2), 241–259 (1992)
7. Breiman, L.: Bagging predictors. Mach. Learn. **24**(2), 123–140 (1996)
8. Drucker, H.: Boosting using neural nets. In: Sharkey, A. (ed.) Combining Artificial Neural Nets: Ensemble and Modular Learning. Springer (1999)
9. Chen, H., Yao, X.: Regularized negative correlation learning for neural network ensembles. IEEE T Neur. Net. Lear. **20**(12), 1962–1979 (2009)
10. Huang, G., et al.: Snapshot ensembles: Train 1, get M for free. arXiv:1704.00109 (2017)
11. Xie, J., Xu, B., Zhang, C.: Horizontal and vertical ensemble with deep representation for classification. arXiv:1306.2759 (2013)
12. Chen, T., Guestrin, C.: XGBoost: a scalable tree boosting system. In: Proceedings of the 22nd ACM SIGKDD International Conference on Knowledge Discovery and Data Mining, pp. 785–794 (2016)
13. Zieba, M., Tomczak, S.K., Tomczak, J.M.: Ensemble boosted trees with synthetic features generation in application to bankruptcy prediction. Expert Syst. Appl. **58**(1), 93–101 (2016)
14. Ni, L., et al.: Streamflow forecasting using extreme gradient boosting model coupled with Gaussian mixture model. J. Hydrol. **586**, 124901 (2020)
15. Li, Y., et al.: Smart wind speed forecasting approach using various boosting algorithms, big multi-step forecasting strategy. Renew. Energ. **135**, 540–553 (2019)
16. Mitrentsis, G., Lens, H.: An interpretable probabilistic model for short-term solar power forecasting using natural gradient boosting. Appl. Energ. **309**, 118473 (2022)

17. Dudek, G., Pełka, P.: Ensembles of randomized neural networks for pattern-based time series forecasting. In: Mantoro, T., Lee, M., Ayu, M.A., Wong, K.W., Hidayanto, A.N. (eds.) ICONIP 2021. LNCS, vol. 13110, pp. 418–430. Springer, Cham (2021). https://doi.org/10.1007/978-3-030-92238-2_35
18. Dudek, G.: Randomized neural networks for forecasting time series with multiple seasonality. In: Rojas, I., Joya, G., Català, A. (eds.) IWANN 2021. LNCS, vol. 12862, pp. 196–207. Springer, Cham (2021). https://doi.org/10.1007/978-3-030-85099-9_16
19. Dudek, G.: Generating random weights and biases in feedforward neural networks with random hidden nodes. Inform. Sci. **481**, 33–56 (2019)
20. Dudek, G.: Generating random parameters in feedforward neural networks with random hidden nodes: drawbacks of the standard method and how to improve it. In: Yang, H., Pasupa, K., Leung, A.C.-S., Kwok, J.T., Chan, J.H., King, I. (eds.) ICONIP 2020. CCIS, vol. 1333, pp. 598–606. Springer, Cham (2020). https://doi.org/10.1007/978-3-030-63823-8_68
21. Mason, L., Baxter, J., Bartlett, P.L., Frean, M.: Boosting algorithms as gradient descent. In: Solla S.A., et al. (eds.) Advances in Neural Information Processing Systems, vol. 12, pp. 512–518. MIT Press (1999)

Exploring Ductal Carcinoma In-Situ to Invasive Ductal Carcinoma Transitions Using Energy Minimization Principles

Vivek M. Sheraton[1,2,3]($\boxtimes$) and Shijun Ma[4]

[1] HEALTHTECH NTU, Interdisciplinary Graduate School, Nanyang Technological University, Singapore, Singapore
v.s.muniraj@uva.nl
[2] Institute for Advanced Study, University of Amsterdam, Amsterdam, The Netherlands
[3] Center for Experimental and Molecular Medicine, UMC Location University of Amsterdam, Amsterdam, The Netherlands
[4] School of Chemical and Biomedical Engineering, Nanyang Technological University, Singapore, Singapore

Abstract. Ductal carcinoma in-situ (DCIS) presents a risk of transformation to malignant intraductal carcinoma (IDC) of the breast. Three tumor suppressor genes *RB*, *BRCA1* and *TP53* are critical in curtailing the progress of DCIS to IDC. The complex transition process from DCIS to IDC involves acquisition of intracellular genomic aberrations and consequent changes in phenotypic characteristics and protein expression level of the cells. The spatiotemporal dynamics associated with breech of epithelial basement membrane and subsequent invasion of stromal tissues during the transition is less understood. We explore the emergence of invasive behavior in benign tumors, emanating from altered expression levels of the three critical genes. A multiscale mechanistic model based on Glazier-Graner-Hogeweg method-based modelling (GGH) is used to unravel the phenotypical and biophysical dynamics promoting the invasive nature of DCIS. Ductal morphologies including comedo, hyperplasia and DCIS, evolve spontaneously from the interplay between the gene activity parameters in the simulations. The spatiotemporal model elucidates the cause-and-effect relationship between cell-level biological signaling and tissue-level biophysical response in the ductal microenvironment. The model predicts that *BRCA1* mutations will act as a facilitator for DCIS to IDC transitions while mutations in *RB* act as initiator of such transitions.

Keywords: Glazier-Graner-Hogeweg model · BRCA1 · Ductal morphologies

1 Introduction

Mutations in the intraductal epithelial cells of the mammary gland result in unrestrained proliferation of the cells resulting in ductal carcinoma in situ (DCIS) [1]. This condition is classified as a non-invasive lesion. These epithelial cells are confined within the lumen and therefore are restrained from spreading outside the duct to the surrounding tissues.

D. Groen et al. (Eds.): ICCS 2022, LNCS 13350, pp. 375–388, 2022.
https://doi.org/10.1007/978-3-031-08751-6_27

Further progression of DCIS and disruption of the basal membrane results in malignant condition known as invasive ductal carcinoma (IDC) [2]. After the onset of IDC, the cancer cells invade other parts of the breast tissue and turn metastatic. Transformation from DCIS to life threatening IDC has long been a subject of clinical research [3–5]. Studies have suggested that DCIS is a precursor for IDC and on average 40% of patients with DCIS subsequently develop IDC [6]. Luminal B1 tumors [7] have shown faster progression from DCIS to IDC than luminal A, triple negative and HER2 type tumors. HER2 positive tumors are found to transit from DCIS to IDC slower by staying in the DCIS state longer. Though, specific pathways and biomarkers for this transformation are yet to be discovered. Logullo et al. [8] analyzed epithelial to mesenchymal transition (EMT) markers for their association with DCIS to IDC transformation and found that c-met and TGFβ1 had positive association with the tumor transformation. However, most EMT biomarkers did not yield significant prognosis value. One important characteristic associated with IDC is the presence of intra-tumor morphological and genetic heterogeneity. "Evolutionary bottleneck" was suggested as an outcome of progression from DCIS to IDC by Cowell et al. [2], to explain the genetic heterogeneity observed in IDCs. This transformation is a complex process involving multiple mutations resulting in a highly heterogenous tumor microenvironment. p53 overexpression [9] has been observed in both DCIS and IDC, with the overexpression leading to lower mitotic index and apoptotic index in luminal cell phenotype and the opposite in stem cell phenotype. There is evidence [10, 11] suggesting that DCIS lesions and IDC tumors have deactivated the retinoblastoma gene, ie., loss of Rb function and that the Rb pathway is a likely regulator of the transformations. Rb has also been shown to be a deciding factor in the recurrence of DCIS [11] through overexpression of p16ink4a. Mutations in the *BRCA* genes (the tumor suppressor gene widely associated with breast cancer) and the gene encoding p53 (*TP53*) have been found to occur together in both DCIS and IDC tumors [12]. Kumar et al. [13] studied the combined effect of the defects of these three major tumor suppressors genes, *RB*, *BRCA1* and *TP53*. They observed that simultaneous deactivation of these pathways resulted in formation of highly metastatic invasive breast cancer tumors in their mouse models and concluded that the pathways have a combinatorial effect on the progress and evolution of the tumor growth. The resulting tumors were found to have heterogenous morphology suggesting a product of "Evolutionary bottleneck" similar to studies of Cowell et al. [2] Thus, alterations in these three pathways would result in formation of highly invasive tumors and could possibly act as precursors in DCIS to IDC transformations.

Histological staining of tissue sections is traditionally used to study DCIS-IDC transformation, yet it is rather difficult to understand the complexities associated with the disease and it is less feasible to design experiments to track the progress of tumor evolution. Computational modelling of biological cells has become a useful tool in predicting the outcome of tumor growth, angiogenesis, and tissue morphologies in-silico [14, 15]. Multiscale agent based models have been accepted widely due to their versatility in handling multiple cell phenotypes and therefore tissue heterogeneity [16, 17]. Qiao et al. [18] developed an agent based model to simulate the proliferation of multiple myeloma tumor cells and analyzed the effect of drugs on the population of osteoclasts and osteoblasts. They modelled cells as agents and drugs as apoptosis inducers. The cells were modelled to undergo apoptosis at variable rates based on the quantity of drugs they are in contact with. The effect of tyrosine inhibitor kinases (TKIs) on brain cancer was studied using a similar multiscale model by Sun et al. [19] They incorporated EGFR signaling in their model using partial differential equations to simulate various phenotypes observed in the tumor. The developmental stages of DCIS also have been simulated using agent-based modeling. The model developed by Macklin et al. [20] was able to provide insight into the formation of necrotic core and calcification regions. Though these models can quantify the tumor proliferation and apoptosis, they lack proper energy based realistic cell allocations from mitosis or cell motility. These mesoscale interactions determine the spatial distribution and localized effects associated with the position of tumor cells. A better way of modelling the cells in an energy optimistic way with cell motility while retaining the individual characteristics of cells is through Glazier-Graner-Hogeweg method-based modelling (GGH). Boghaert et al. [21] used GGH to simulate the progression of DCIS growth. Due to the inherent energy minimization principle of GGH, they were able to simulate four different forms of DCIS morphologies, namely micropapillary, cribriform, solid and comedo. These morphologies would not have evolved in other agent-based models due to the absence of localized energy interactions. However, the model was unable to predict beyond the transition from DCIS to IDC and explain the reasons for the observed phenotypic heterogeneity in the clinical IDC sections.

In this study, we use GGH to develop a model to elucidate on the transition from DCIS and IDC stage. We consider the critical three tumor suppressor pathways involving p53, Rb and BRCA to simulate the phenotypic variations associated with the underlying genotypic changes. We make valid assumptions based on available literature data to interpret the effects of genotypic changes into physical model parameters. By combining phenotypic changes, genotypic influences on cell proliferation and energy minimization principles we model the intraductal luminal epithelial cells and myoepithelial cells to elaborate the mechanism of DCIS to IDC transformation and the associated morphological heterogeneity.

2 Methods and Materials

2.1 Model Description

Two basic cell types, epithelial cells and myoepithelial cells were modeled. The former cell type forms the inner layer and the latter stays along the periphery as shown in Fig. 1. Individual cells were modelled as collection of lattice points on the simulation grid. The Glazier-Graner-Hogeweg (GGH) method-based model was implemented using an open-source software framework called CompuCell3D v3.7.5. The developed method is an extension of the model developed by Boghaert et al. [21] to simulate DCIS. The cells in the simulation are free to deform based on the energy constraints listed in Eq. 1 at each monte-carlo step (mcs). The cells deform at lattice point level through lattice point-copy attempts [22]. The probability of success for a lattice point copy attempt is defined by equation one, where, $\sigma\left(\overrightarrow{i}\right)$ denotes the lattice point occupied by a cell σ, $\overrightarrow{i'}$ represents the new lattice point where the lattice point copy attempt is supposed occur, ΔH is the change in energy of the system and T_m is the temperature or fluctuation amplitude of the system. The lattice point-level deformations, on a longer time scale, constitute the motility of the cells. Volume constraints were imposed on the cells to ensure that the cells stay within permissible levels of volume increase or decrease using Eq. 2. The term E_v denotes the volume energy of the cell, V_{cell} denotes the total number of lattice points occupied by the cell, V_T denotes the target volume of the cell and λ_v denotes the volume potential (similar to a spring constant). A surface area constraint was introduced in the model to prevent the cells from evolving into biologically unreasonable shapes. The terms E_s, S_{cell}, S_T and λ_s denote the surface energy, total number of lattice points on the surface (perimeter in 2D), target surface area and the surface potential respectively. In addition, to restrict the cells from disintegrating and assuming fractured morphologies, a penalty function E_p was imposed [22]. This parameter retains individual cells as a single entity. The focal point plasticity (FPP) of the cell and contact energies between the cells are dictated by the Eqs. 4 and 5. Here, lij is the distance between cells at positions i and j, Lij is the target distance between the cells, and λij is the FPP potential. We used the physically equivalent values as suggested by Boghaert et al. [21], to simulate the cells' attraction-repulsion potential and adhesion forces. The energy arising from a lattice point copy attempt 'H' is therefore given by Eq. 6. We assume the cells to be spherical with a diameter of 15 μm.

$$P\left(\sigma\left(\vec{i}\right) \to \sigma\left(\vec{i'}\right)\right) = \begin{cases} \left[exp\left(-\frac{\Delta H}{T_m}\right)\right], \Delta H > 0 \\ 1, \Delta H \le 0 \end{cases} \tag{1}$$

$$E_v = \sum_\sigma \lambda_v (V_{cell}(\sigma) - V_T(\sigma))^2 \tag{2}$$

$$E_s = \sum_\sigma \lambda_s (S_{cell}(\sigma) - S_T(\sigma))^2 \tag{3}$$

$$E_c = \sum_{i,j} J(\tau_{\sigma(i)}, \tau_{\sigma(j)})\left(1 - \delta_{\sigma(i),\sigma(j)}\right) \tag{4}$$

$$E_f = \sum_{i,j} \lambda_{i,j}\left(l_{i,j} - L_{i,j}\right)^2 \tag{5}$$

$$H = E_v + E_s + E_c + E_f \tag{6}$$

Each cell has three major biological parameters associated with it, the DNA damage level, oxidative stress, and proliferation potential. All these biological parameters were modelled as non-dimensional continuous variables from 0 to 1.0. Equation 6 describes the change in DNA damage level within the cells. Cells accumulate DNA damage 'd' in a stochastic manner with a probability f_{dd} [23] as shown in Eq. 7. C_B indicates the expression level of BRCA. The proliferation potential 'p' was assumed to decrease with increase in number of epithelial neighbors, since a crowded microenvironment would result in less nutrient available for cell growth. To calculate the number of neighbors, we used a neighbor order of 1.5 times the cell radius for epithelial (Ne) and myoepithelial cell neighbors (Nm). Also, the proliferation potential was modelled to increase with DNA damage accumulated in the cancer cells. The increased proliferation potential of a DNA damaged cell can be considered as an inherent property of a mutated cell with higher survivability capacity [24]. Proliferation potential was calculated using Eq. 8. Where, β, k_{os}, C_{Rb} and O represent specific growth rate, oxidative stress coefficient and oxidative stress respectively. The oxidative stress within the cells was assumed to increase from crowding of cell neighbors (decrease in nutrient availability) and the cell type of these neighbors [25]. Epithelial contribution to oxidative stress was considered to be negligible for a cell with fewer than 5 neighbors. The oxidative stress contribution from myoepithelial neighbors was also considered to reduce with increase in number of neighbors as dictated by Eq. 9. This assumption is based on the ready availability nutrient for epithelial cells that are located near myoepithelial cells, as observed by Norton et al. [26]. In Eq. 9, the constants ω_1 and ω_2 are the weighting fractions for oxidative stress contribution from the two types of neighbor cells, W is the maximum contribution to oxidative stress from the neighbors and 'a' stress generation rate coefficient.

The gene activity and protein expression levels control the cell cycle by indirectly controlling the three major cell survivability parameters. RB functions as a tumor suppressor gene by being a negative regulator of cell proliferation. Cells which have suffered DNA damage or accumulated oxidative stress are arrested in their G1 phase by the activity of RB [27]. In our model, this gene activity is modelled by considering Rb (Rb protein or pRb) as a regulator of cell proliferation potential. Thus, in the simulation, Rb reduces and eventually prevents the proliferation of cells that have accumulated DNA damage and oxidative stress as defined in Eq. 8. The extent of Rb effect is dependent on its activity potential in the cell 'C_{Rb}'.

In normal cells, the DNA damage must be limited to avoid run-off cell proliferations. Proliferation reduction in DNA damaged cells is handled by BRCA which is involved in DNA repair [28]. In cases of breast cancer, the inactivation of BRCA genes have been the major reason for DNA damage and accumulation of multiple mutations. We therefore modelled BRCA activity 'CB' as a modifier of the DNA damage levels in the cells as shown in Eq. 7.

If the DNA damage or oxidative stress in cells exceed a critical level, then apoptosis is initiated based on the lethal cell DNA damage level (d_l) and probability dictated by the effectivity of p53 (C_{p53}) [29] respectively. Cells with apoptosis index 'I_A' values larger than maximum cell apoptotic index value 'I_{Ac}' are marked for removal from the simulation domain. These cells are removed if they continue to express I_A values greater than I_{Ac} for more than 'N_A' simulation time steps. Similarly, in case of mitosis, cells divide in simulation if the mitosis index 'I_M' (Eq. 11) is higher than the maximum cell mitotic index value 'I_{Mc}'

$$d = f_{dd}[1 - C_B] \tag{7}$$

$$\frac{dp}{dt} = \beta(1 - k_{os}C_{Rb}O + C_{Rb}d), \text{ with } \max(p) = 1.0 \tag{8}$$

$$\frac{dO}{dt} = \begin{cases} W[\omega_1 e^{-aN_m} + \omega_2(1 - e^{-a(N_e - N_{max})})], & \text{if} N_e \geq N_{max} \\ W[\omega_1 e^{-aN_m}], & \text{else} \end{cases} , \text{ with } \max(O) = 1.0 \tag{9}$$

$$I_A = O * C_{p53} \tag{10}$$

$$I_M = p * C_{p53} \tag{11}$$

The cells were allowed to evolve based on the constraints dictated by Eq. 6 throughout the simulation. The calculations based on activity level Eqs. 7 to 11 were computed for every 100 mcs. This cycle ensures that the cells relax for sufficient time after their growth, thereby, assuring minimal energy state of the system. Each 100 mcs corresponds to 0.25 h real time. The simulations were carried out for a maximum of 100000 mcs (250 h) or until the maximum cell count in the simulations reached 1000, whichever event occurred earlier. Parametric values used in the simulations are summarized in Table 1.

Table 1. Key parametric values used in the simulations

Parameter	Notation	Value (units)
Time step	mcs	2.5×10^{-3} (h)
Volume potential	λ_v	5–10
Surface potential	λ_s	2
Temperature	T_m	10
DNA damage probability	f_{dd}	0.5
Oxidative stress contribution from myoepithelial neighbors	ω_1	0.75
Oxidative stress contribution from epithelial neighbors	ω_2	0.25
Maximum neighbors	N_{max}	5
Oxidative stress coefficient	k_{os}	0.67
Stress generation rate coefficient	a	0.1

3 Results and Discussions

The model simulations with maximum levels (1.0) of C_B, C_{p53} and C_{Rb} should result in a normal ductal structure. Maximum levels of activity should prevent apoptosis evasion or proliferation runoff of the cells. These structures should possess a single layer of myoepithelial cells enclosing one or two layers of epithelial cell with minimal distortions to ductal morphology. Ensuring that the model establishes and maintains the above-mentioned structure throughout the simulation duration is the first step in verifying if the model assumptions represent in-vivo dynamics. As hypothesized, the simulations carried out with maximum expression levels are found to produce benign normal ductal structures as shown in Fig. 1a. The structure is found to be in a dynamic equilibrium. Meaning, the aging cells are removed with simulation progress and are replaced by new daughter cells. This cycle repeats itself throughout the entire simulation duration. The variation in formation of new cells and live epithelial cell count during the course of simulation for this case are shown in Fig. 1b and 1c respectively. The intertwined effects of apoptosis and mitosis can be seen to regulate the ductal level cell population homeostasis. Thus,

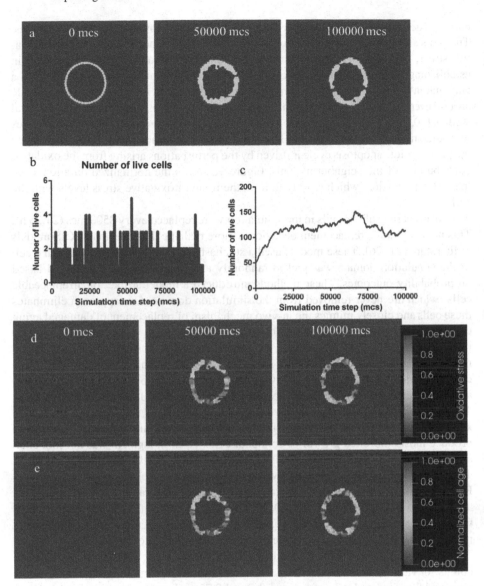

Fig. 1. Model simulation results for normal ductal structure formation at maximum (1.0) C_B, C_{p53} and C_{Rb} values. (a) Panel showing spatial changes in ductal structure (white color indicates epithelial cells and red color indicates myoepithelial cells) at 0, 50000 and 100000 mcs, (b) temporal variations in number of new cell formations, (c) temporal variations in total epithelial cell count, panel showing spatial variations of (d) oxidative stress and (e) normalized cell age at 0, 50000 and 100000 mcs. (Color figure online)

the simulations capture the in-vivo co-operative effect of BRCA, RB and P53 on the regulation of ductal development and functioning. A cell-field plot of oxidative stress levels is presented in Fig. 1d. Oxidative stress levels are found to vary throughout the entire structure suggesting the absence of localized overcrowding in these structures. The plots also capture the elimination of any cells that have accumulated critical oxidative stress. The major fixed parameters used in the study, W and β play a crucial role in establishing this mitosis and apoptosis balance. The value of W was chosen in such a way that in the absence of DNA damage and oxidative stress, on average the cells divide around every 17 h. This proliferation duration is similar to those reported in experimental studies [30, 31]. β-value is an analog of W-value for oxidative stress, both these values were quantitatively set to be equal. Hence, the dynamic equilibrium established between the proliferation-apoptosis cycle is driven by the perturbations arising from the oxidative contributions of the neighboring cells. Figure 1e shows the normalized duration since the last cell division, which closely follows the trend of oxidative stress levels with the cells.

On an average all the cells in the simulation were replaced every 8500 mcs (21.25 h). This novel dynamic replacement behavior is more realistic than previous DCIS models in literature [21, 26]. These models used a stochastic way of apoptosis, where in a cell in the simulation domain was picked randomly at fixed intervals and removed based on probability outcomes. These methods introduced a model artifact of 'irreplaceable cells' which are never removed from the simulation domain. Our approach eliminates these cells and closely mimics the in-vivo mechanism of replacement of damaged aging cells with new cells. The values of C_B, C_{p53} and C_{Rb} were varied from 0.25 to 1.0 in increments of 0.25 in a combined manner in the simulations. Figure 2a shows the number of cases with less than 125 surviving cells at the end of simulations. These cases, in general, produce a ductal configuration very similar to the normal state, that is, a single layer of myoepithelial cells binding fewer than three layers of epithelial cells. The x-axis in Fig. 2a denotes the cases where the values of C_B, C_{p53} and C_{Rb} were kept maximum (1.0). As observed, simulations with maximum values of C_{Rb} tend to produce most controlled growth of epithelial cells. Even though Rb was not modelled as a direct influencer of apoptosis, it is found to be a major driver of cell aging control. This simulation result also correlates with other experimental observations where lower Rb expression levels have been implicated with risk of ipsilateral breast event (IBE) [32] and DCIS to IDC transformations [10, 11]. It should be noted that there are other cases where even with maximum expression of Rb, the cells proliferated in an uncontrolled manner. This means that C_{Rb} is not the sole controlled of apoptosis-mitosis equilibrium.

To further unravel the intertwined effects of C_B and C_{p53}, 2D density plots of cell counts for various fixed values of C_{Rb} are used as shown in Fig. 2b,2c and 2d. It can be observed that for fixed values of $C_{Rb}= 0.5$ and 0.75, reduced cell population levels are seen for C_{p53} value of 0.25, irrespective of the C_B value. These values are lower than their counterparts with C_{p53} values 0.5 and 0.75. This outcome is unexpected since a reduced C_{p53} value would mean more chances of cell survival as dictated by Eq. 10. Although counterintuitive, this observation can be attributed to the decreased levels of both apoptotic and mitotic indices of these cells (I_A and I_M) at minimum C_{p53} value. This results in a pathway-race between proliferation and apoptosis. Hence, for fixed values of C_{Rb} at 0.5 and 0.75, effect of apoptosis is more pronounced in simulations with C_{Rb} value of 0.25 than simulations with C_{Rb} values greater than 0.25. The above observations do not hold true for simulations with fixed C_{Rb} and C_{p53} values of 1.0

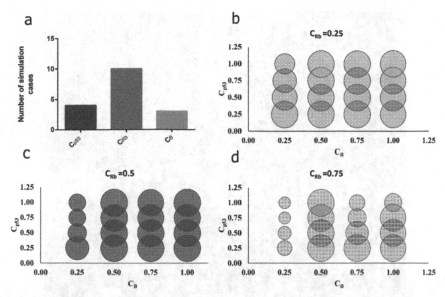

Fig. 2. Variation of cell count for different parametric values. (a) Number of cases with cell count less than 125 and maximum parameter values (1.0) Total cell count at the end of simulations for different values of C_{p53} and C_B and fixed value C_{Rb} of (b) 0.25, (c) 0.5 and (d) 0.75. The circle size indicates the number of cells, with maximum cell count (1000) corresponding to the maximum circle diameter.

and 0.5 respectively. In these cases, the total live cell population is found to be lower than simulations with fixed C_{Rb} of 1.0 and C_{p53} values of other than 0.5. Interestingly, simulations with C_{p53} value of 0.5 are the only simulations with more than 150 cells at maximum C_{Rb} expression levels. Thus, the system produces non-linear response to C_{p53} effectivity levels and the outcomes are dependent on the interplay of all three parameters (C_{p53}, C_{Rb} and A_B).

Heightened levels of cell proliferation and survival alone cannot be considered as indicators of invasive transformations. Elevated proliferation and cell survival are characteristics of both DCIS and IDC tumors. To be characterized invasive, the epithelial cell populations should proliferate enough to fill the space between the ducts and penetrate the myoepithelial layer to invade surrounding tissues. The simulation results were examined for morphological differences arising from variations in the activity parameters. Four unique ductal structures were obtained from the parametric simulation studies. The structures are shown in Fig. 3a–3d. They can be classified as (1) normal ductal configuration, (2) ductal hyperplasia, (3) solid or comedo and (4) invasive ductal configuration. The spatial development of these configurations is shown in Fig. 3a–3d respectively. In ductal hyperplasia structures (Fig. 3b), layers of epithelial cells are found to extend from the myoepithelial wall towards the ductal core. Such structures are generally categorized as benign in-vivo. However, they are considered as risk factors for breast cancer development. Further uncontrolled proliferation of cells in hyperplasia structures will result in formation of solid or comedo structures such as the one shown in Fig. 3c. These structures are categorized as high-grade DCIS. If left untreated these structures can progress

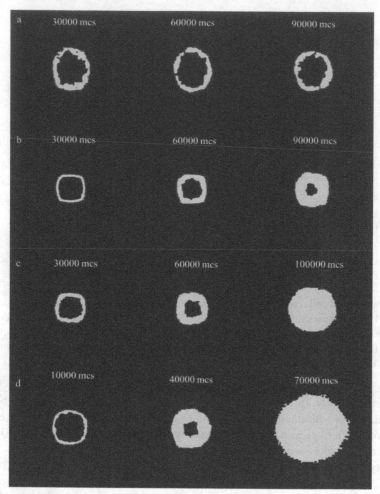

Fig. 3. Ductal structures formed during the simulations. (a) Formation of normal duct structure with parameter values of $C_B = 1$, $C_{p53} = 1$ and $C_{Rb} = 1$, (b) formation of ductal hyperplasia structure with parameter values of $C_B = 1$, $C_{p53} = 0.25$ and $C_{Rb} = 0.75$, (c) formation of solid structure with parameter values of $C_B = 1$, $C_{p53} = 0.25$ and $C_{Rb} = 0.5$ and (d) formation of invasive ductal carcinoma structure with parameter values of $C_B = 0.25$, $C_{p53} = 0.5$ and $C_{Rb} = 0.75$. The white spots indicate epithelial cells and red spots indicate myoepithelial cells. (Color figure online)

to invasive ductal carcinoma. Figure 3d shows the invasive ductal carcinoma formed in simulation. The major difference between solid and invasive structures is the presence of cells that have penetrated the myoepithelial layer. Invasive ductal carcinoma (IDC) structures are malignant in nature, after penetration, the IDC cells invade the local tissue and establish a population there. In most simulation cases, the penetration of cells is aided by the intraductal pressure or energy build up. This energy build-up makes the myoepithelial chain unstable resulting in local adhesion failures and epithelial cell break away.

Multiple structures can form for a single value of C_B, C_{p53} and C_{Rb} parameter in the parametric combinations. To better understand the contribution of each of these values towards formation of different ductal structures, Fig. 4 summarizes the number of instances of formation of various structures at different expression levels. Conclusively, all variational combinations are found to predominantly produce IDC structure. This means DCIS to IDC transformations are possible from all variations in C_{p53} and C_{Rb}. From simulation data, for the case of variations in C_B values, IDC transformations are possible only if coupled with variation in expression levels of other parameters. This observation suggests a promotional role for BRCA in IDC transformations rather than as an initiator. C_{Rb} is found to produce most of the normal ductal structures in the simulation at the maximum expression level. Therefore, Rb should act as a major restraint in DCIS to IDC transformations. In addition, minimum level $C_{Rb}(0.25)$ is found to solely produce IDC structures. This shows the crucial effect of C_{Rb} reduction on formation of IDC structures. The non-linear nature of the responses evoked from C_{p53} variations is further evident in Fig. 4c. p53 effectivity level of 0.5 is found to produce solely of IDC structures. The number of structures produced at this effectivity level is even higher than the number of structures produced at a lower effectivity level of 0.25. As previously discussed, this phenomenon can be attributed to the pathway-race condition between apoptosis and cell proliferation.

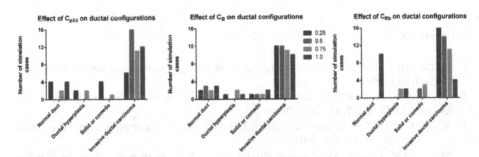

Fig. 4. Formation of various ductal structures for different values of C_B, C_{p53} and C_{Rb}. The color legend indicates various activity levels of the simulation parameters.

4 Conclusions

We developed a GGH method-based model for simulating the biological and biophysical interactions occurring between cells in the mammary ducts. The numerical model was used to explore the effects of mutation of three major tumor suppressor genes (RB, TP53 and BRCA1) on the ductal microenvironment. Through the model simulations, we captured the spatiotemporal changes occurring in the tumor microenvironment, which are instigated by the altered gene expression levels in the ductal cells. The developed model has captured the dynamics of development of various ductal structures. These structures include normal duct, ductal hyperplasia, comedo and invasive ductal carcinoma structure. Cells in the simulations were replaced periodically by new cells similar to in-vivo tissue cell population homeostasis. DCIS to IDC transformation is found to be initiated through variations in expression levels and effectivity of Rb and p53 respectively. The model simulations suggest a promoter role for BRCA in progression from DCIS to IDC. In a word, our model simulates the transformation of DCIS to IDC of breast cancer, and demonstrates the influence of impaired activities of BRCA, RB and P53 on such transitions.

Gene sequencing techniques and clinical sample processing methods have advanced dramatically in recent years. Technologies including microarray, next generation sequencing and ddPCR enable clinicians and scientists to detect mutations and expression levels of thousands of genes from liquid biopsy or tissues samples of patients. However, the lack of efficient analysis tools and modeling systems limits our understanding of the sequencing and detection data. Here we have developed a numerical model to simulate the combinational influence of BRCA, Rb and p53 activities on DCIS to IDC transformation. Our model could be used to analyze the mutation status and expression levels of BRCA, Rb and p53, to predict the disease progression and survival of patients. Drugs that directly target BRCA, Rb and p53 or inhibit the related pathways have been approved for breast cancer treatment, such as Palbociclib, Ribociclib and Olaparib. The simulation model can act as a tool for experimental hypotheses testing. Based on the mutations and expression data of the genes, further development would enable gene expression data-driven personalized selective inhibitor drug scheduling.

References

1. Gorringe, K.L., Fox, S.B.: Ductal carcinoma in situ biology, biomarkers, and diagnosis. Front Oncol **7**, 248 (2017)
2. Cowell, C.F., et al.: Progression from ductal carcinoma in situ to invasive breast cancer: revisited. Mol Oncol **7**, 859–869 (2013)
3. Ma, X.-J., et al.: Gene expression profiles of human breast cancer progression. Proc. Natl. Acad. Sci. **100**, 5974–5979 (2003)
4. Aubele, M.M., et al.: Accumulation of chromosomal imbalances from intraductal proliferative lesions to adjacent in situ and invasive ductal breast cancer. Diagn. Mol. Pathol. **9**, 14–19 (2000)
5. Volinia, S., et al.: Breast cancer signatures for invasiveness and prognosis defined by deep sequencing of microRNA. Proc. Natl. Acad. Sci. **109**, 3024–3029 (2012)
6. Boughey, J.C., Gonzalez, R.J., Bonner, E., Kuerer, H.M.: Current treatment and clinical trial developments for ductal carcinoma in situ of the breast. Oncologist **12**, 1276–1287 (2007)
7. Kurbel, S., Marjanović, K., Dmitrović, B.: A model of immunohistochemical differences between invasive breast cancers and DCIS lesions tested on a consecutive case series of 1248 patients. Theor. Biol. Med. Model. **11**, 29 (2014)
8. Brentani: Concomitant expression of epithelial-mesenchymal transition biomarkers in breast ductal carcinoma: association with progression. Oncology Reports 23 (2009)
9. Megha, T., et al.: p53 mutation in breast cancer. Correlation with cell kinetics and cell of origin. J. Clin. Pathol. **55**, 461–466 (2002)
10. Gauthier, M.L., et al.: Abrogated response to cellular stress identifies DCIS associated with subsequent tumor events and defines basal-like breast tumors. Cancer Cell **12**, 479–491 (2007)

11. Witkiewicz, A.K., et al.: Association of RB/p16-pathway perturbations with DCIS recurrence: dependence on tumor versus tissue microenvironment. Am. J. Pathol. **179**, 1171–1178 (2011)
12. Wang, X., et al.: p53 alteration in morphologically normal/benign breast luminal cells in BRCA carriers with or without history of breast cancer. Hum Pathol **68**, 22–25 (2017)
13. Kumar, P., et al.: Cooperativity of Rb, Brca1, and p53 in malignant breast cancer evolution. PLoS Genet. **8**, e1003027 (2012)
14. Palm, M.M., Merks, R.M.: Large-scale parameter studies of cell-based models of tissue morphogenesis using CompuCell3D or VirtualLeaf. Tissue Morphogenesis: Methods and Protocols 301–322 (2015)
15. Anderson, A.R., Chaplain, M.: Continuous and discrete mathematical models of tumor-induced angiogenesis. Bull. Math. Biol. **60**, 857–899 (1998)
16. Thorne, B.C., Bailey, A.M., Peirce, S.M.: Combining experiments with multi-cell agent-based modeling to study biological tissue patterning. Brief. Bioinform. **8**, 245–257 (2007)
17. Bailey, A.M., Thorne, B.C., Peirce, S.M.: Multi-cell agent-based simulation of the microvasculature to study the dynamics of circulating inflammatory cell trafficking. Ann. Biomed. Eng. **35**, 916–936 (2007)
18. Qiao, M., Wu, D., Carey, M., Zhou, X., Zhang, L.: Multi-scale agent-based multiple myeloma cancer modeling and the related study of the balance between osteoclasts and osteoblasts. PLoS ONE **10**, e0143206 (2015)
19. Sun, X., Zhang, L., Tan, H., Bao, J., Strouthos, C., Zhou, X.: Multi-scale agent-based brain cancer modeling and prediction of TKI treatment response: incorporating EGFR signaling pathway and angiogenesis. BMC Bioinformatics **13**, 218 (2012)
20. Macklin, P., Edgerton, M.E., Thompson, A.M., Cristini, V.: Patient-calibrated agent-based modelling of ductal carcinoma in situ (DCIS): from microscopic measurements to macroscopic predictions of clinical progression. J. Theor. Biol. **301**, 122–140 (2012)
21. Boghaert, E., Radisky, D.C., Nelson, C.M.: Lattice-based model of ductal carcinoma in situ suggests rules for breast cancer progression to an invasive state. PLoS Comput. Biol. **10**, e1003997 (2014)
22. Swat, M.H., Thomas, G.L., Belmonte, J.M., Shirinifard, A., Hmeljak, D., Glazier, J.A.: Multiscale modeling of tissues using CompuCell 3D. Methods Cell Biol. **110**, 325–366 (2012)
23. Rattan, S.I.: Theories of biological aging: genes, proteins, and free radicals. Free Radical Res. **40**, 1230–1238 (2006)
24. Davis, J.D., Lin, S.-Y.: DNA damage and breast cancer. World J. Clin. Oncol. **2**, 329–338 (2011)
25. Gaweł-Bęben, K., et al.: TMEFF2 shedding is regulated by oxidative stress and mediated by ADAMs and transmembrane serine proteases implicated in prostate cancer. Cell Biology International n/a-n/a
26. Norton, K.-A., Popel, A.S.: An agent-based model of cancer stem cell initiated avascular tumour growth and metastasis: the effect of seeding frequency and location. J. R. Soc. Interface **11**, 20140640 (2014)
27. Giacinti, C., Giordano, A.: RB and cell cycle progression. Oncogene **25**, 5220–5227 (2006)
28. Yoshida, K., Miki, Y.: Role of BRCA1 and BRCA2 as regulators of DNA repair, transcription, and cell cycle in response to DNA damage. Cancer Sci. **95**, 866–871 (2004)
29. Fridman, J.S., Lowe, S.W.: Control of apoptosis by p53. Oncogene **22**, 9030 (2003)
30. Corbin, E.A., Adeniba, O.O., Cangellaris, O.V., King, W.P., Bashir, R.: Evidence of differential mass change rates between human breast cancer cell lines in culture. Biomed. Microdevice **19**(1), 1–7 (2017). https://doi.org/10.1007/s10544-017-0151-x

31. Starcevic, S.L., Diotte, N.M., Zukowski, K.L., Cameron, M.J., Novak, R.F.: Oxidative DNA damage and repair in a cell lineage model of human proliferative breast disease (PBD). Toxicol. Sci. **75**, 74–81 (2003)
32. Knudsen, E.S., et al.: Retinoblastoma and phosphate and tensin homolog tumor suppressors: impact on ductal carcinoma in situ progression. JNCI: J. Natl. Can. Inst. 104, 1825–1836 (2012)

Scaling the PageRank Algorithm for Very Large Graphs on the Fugaku Supercomputer

Maxence Vandromme[1]([✉]), Jérôme Gurhem[2], Miwako Tsuji[3], Serge Petiton[1,2], and Mitsuhisa Sato[3]

[1] Univ. Lille, UMR 9189 - CRIStAL, CNRS, 59000 Lille, France
maxence_vandromme@msn.com, serge.petiton@univ-lille.fr
[2] USR 3441 - Maison de la Simulation, CNRS, Saclay, France
[3] RIKEN Center for Computational Science, Kobe, Japan

Abstract. The PageRank algorithm is a widely used linear algebra method with many applications. As graphs with billions or more of nodes become increasingly common, being able to scale this algorithm on modern HPC architectures is of prime importance. While most existing approaches have explored distributed computing to compute an approximation of the PageRank scores, we focus on the numerical computation using the power iteration method. We develop and implement a distributed parallel version of the PageRank. This application is deployed on the supercomputer Fugaku, using up to one thousand compute nodes to assess scalability on random stochastic matrices. These large-scale experiments show that the network-on-chip of the A64FX processor acts as an additional level of computation, in between nodes and cores.

1 Introduction

The PageRank algorithm was originally developed to rank Web pages by importance, as the main component powering Web search engines [16]. It generally uses the power iteration method to compute the dominant eigenvector of a stochastic matrix efficiently. In this context, the Web pages and the links between them form a graph that can be represented by a sparse adjacency matrix, on which the PageRank is applied. Beyond its historic roots, this algorithm saw widespread use in applications where data is organized in a graph structure, such as citation networks [15] or traffic grids [17]. More recently, the (personalized) PageRank has been used as a tool to weigh communication between nodes in Graph Neural Networks [12].

The most common method to compute the PageRank exactly is the power iteration, which relies on iterative sparse matrix-vector multiplication (SpMV) as its kernel. In this regard, the PageRank bears a resemblance to other linear algebra methods, such as the conjugate gradient, which is notably used in the HPCG benchmark to measure the performance of supercomputers with a focus on memory and interconnections [6,10]. Indeed, the main component of

D. Groen et al. (Eds.): ICCS 2022, LNCS 13350, pp. 389–402, 2022.
https://doi.org/10.1007/978-3-031-08751-6_28

the conjugate gradient is also sequences of sparse matrix-vector multiplication, and such methods have direct applications for real-life problems. However, the HPCG benchmark uses sparse diagonal matrices, that represent the kind of computational problems arising from physical applications. Such matrices are a best-case scenario for distributed and parallel computing, since the data is organized to minimize cache misses. This benchmark is well-suited to measure the peak optimal performance of HPC systems on graph problems.

In this paper, we focus instead on the PageRank algorithm and its application to data closer to what can be found in networks or Web graphs. These graphs are usually much less structured, and as a result, their underlying adjacency matrices stray quite far from the ideal case of a regular diagonal matrix. Given the very large size of Web graphs, we need to efficienctly scale the PageRank algorithm on graphs up to billions of nodes. To this end, we present a parallel and distributed implementation of the PageRank algorithm, we deploy it on the top-end supercomputer Fugaku, and measure the scalability of the algorithm up to very large graph sizes on hundreds of compute nodes.

The rest of this paper is organized as follows:

- in Sect. 2, we review the relevant existing work on the PageRank algorithm for distributed settings, and put it in perspective with the recent evolution of HPC systems
- in Sect. 3, we present the implementation of the parallel and distributed PageRank algorithm, and the distributed generation of sparse stochastic matrices
- in Sect. 4 we detail the experiments settings on the supercomputer Fugaku, and discuss the results and insights gained from the experiments. We analyze the impact of the network-on-chip on the performance by using different MPI configurations
- in Sect. 5 we summarize the findings and propose further work to do for the future

2   Background and Related Work

The PageRank was designed with scalability as a primary goal. The original study dealt with a database of 75 million URLs and more than 300 million links between them, and the size has grown to billions of elements or more since then. Nevertheless, the number of iterations of the power method remains small compared to the graph size, and the algorithm usually converges in less than 100 iterations with standard parameters [16].

Despite this efficiency, there has been a growing need to design distributed versions of the algorithm to keep up with the size increase and better take advantage of modern HPC systems with many compute nodes. Applying the standard algorithm in a distributed environment implies splitting the matrix in parts, running partial computation independently on each node, then aggregating and distributing the result vector across all nodes between two iterations of the method. A number of previous studies have noted that the overhead induced by the communications between the nodes would be a major problem [11,18]. Indeed, computational applications on real-world graphs suffer from poor locality, since the graph structure is irregular, and high data access to computation ratio, meaning that the performance will be bounded by the data access speed of the system [14]. On this basis, research in this area mainly focused on distributed implementations that did not require this type of broadcast communications.

Regarding the PageRank, an original study outlined the future research directions by first providing a solid definition of the algorithm [4]. It then pointed out that a fully centralised algorithm on a distributed system would incur significant congestion on the communication network, and that it was not suited for graphs that evolve over time due to the synchronization costs. The authors then proposed several versions of distributed algorithms, relying on the parallel execution of random walks to iteratively approximate the steady state distribution of the dominant eigenvector underlying the PageRank. Later studies expanded on these ideas, and notably likened the random walk to a Markov process in order to reduce the amount of information accumulated from walks and communicated to other nodes [18]. In another study, the asynchronous updating process was extended by considering Web pages as agents of a multi-agents system, updating its PageRank value and passing the information to its outbound links [11]. This method uses randomization on the information communicated. Other usually select nodes randomly, from which a walk is started [3].

A related issue is the Personalized PageRank (PPR), where the goal is not to find the importance of a node i among the whole graph, but the importance of a node i relative to another node j. Full naive computation of the PPR on a graph with n nodes requires running the PageRank algorithm n times, which is irrealistic for large graphs, and implies storing a dense matrix of size $n \times n$, which is a major issue as well. Therefore, methods have been proposed that also use Monte Carlo random walks adapted for the PPR [13]. A study used a *Graph Partition Algorithm* to distribute the graph on different compute nodes in order to minimize the links between the subgraphs, and therefore the communications. Each node would compute its local PageRank on the subgraph, which would then be used to build the final result vector. Interestingly, it could also be applied recursively to a hierarchy of subgraphs, allowing for high scalability [7].

The supercomputer Fugaku [5] is #1 in the TOP500 list at the time of the study. Supercomputers have been shifting towards an increase in the number of compute nodes rather than increasing the nominal performance of each node. As a prime example, Fugaku uses more than 150,000 processors. Therefore, particular attention has been brought to the communication network between the nodes, in order to alleviate the limitations that communications may place on computational performance. It uses A64FX processors at 2.2 GHz with 32 GB of HBM2 memory and a memory bandwidth of 1GB/s. Each processor contains 48 compute cores, split into four groups of 12. Each such group is called Core Memory Group (CMG), and has its own L2 cache and memory. The CMGs inside a processor are linked by a network-on-chip that handles the communications between them. The processors are linked by a 6D topology Tofu interconnect [1].

Given the recent developments on these networks, we want to evaluate whether the communications are still a roadblock to scalability in a highly distributed environment, using the PageRank application as a test case.

Recent work has been done about the scalability of the sparse matrix-vector product on the A64FX processor. This operation is the main kernel of the PageRank power iteration method, so it is of prime importance for our study. These past studies showed great performance on regular diagonal matrices, when using an adequate storage format taking advantage of the SIMD capabilities of the processor [2]. The performance decreases quickly as the nonzero elements deviate from the diagonal, due to the aforementioned cost of irregular data access [8]. These studies used only one or two compute nodes. In this study, we aim to extend the experiments using many compute nodes, in order to identify the barriers to scalability for large size problems. We also focus on sequences of SpMV rather than SpMV alone, since these are more directly useful for applications.

3   Parallel and Distributed Implementation

In this Section, we describe the PageRank algorithm in a distributed parallel computing environment, and the sparse matrix data storage formats used.

3.1   PageRank

Regular Algorithm. We focus here on matrices extracted from graphs, where graphs are objects $G = (V, E)$ where a set of vertices V are connected by a set of edges E. A graph can be represented by its adjacency matrix $M \in \mathbb{R}^{n \times n}$, where n is the number of vertices (or nodes) in the graph. In such cases, M is a binary matrix, i.e. all of its elements m_{ij} are either 0 or 1. Given the shape of large graphs, their adjacency matrices are usually very sparse; that is, the number of non-zero values is much smaller than the total number of elements n^2, and usually of the order of n.

The PageRank algorithm outputs a unique score for each node of the graph, based on the graph structure, corresponding to the dominant eigenvector of the normalized adjacency matrix. A higher score means that the node is more important in the graph. The most common way to compute this vector exactly is through the power iteration method, which is essentially a sparse matrix-vector product on the column-normalized transpose of the adjacency matrix, repeated until the result varies by less than a threshold ϵ from one iteration to the next. During each iteration $t + 1$, the SpMV operation is performed on the vector b_t computed in the previous step t, with the initial b_0 being a vector of ones. b_{t+1} is modified to add an uniform probability of teleportation to any node of the graph, using the β parameter, and then normalized. The PageRank algorithm can be seen as a random walk over the graph. In this context, the teleportation corresponds to the probability, at each step, to restart the random walk at another node of the graph. This mechanism is mainly used to avoid getting stuck in sink nodes, i.e. nodes without outbound edges.

Distributed Implementation. The iterated sparse matrix-vector multiplication (SpMV) is the most computationally expensive part of the algorithm, therefore an efficient distributed implementation is required to scale on many compute nodes of modern HPC systems.

We implement the application in C++ with MPI and OpenMP. Each MPI process computes the operation on a part of the matrix, then the results are aggregated and shared between processes at the end of each iteration. OpenMP is used on each process to parallelize the computations for increased efficiency.

3.2   Sparse Matrices

Since each MPI process performs the computations only on a part of the matrix, we need to split the data matrix so that each process has access to the part it uses. More precisely, the matrix is generated directly in a distributed manner, with each process creating and storing its own block of the matrix based on its process rank.

The block distribution consists of both a distribution by rows and a distribution by columns. The $Nc \times Nr$ matrix is split in $Ngc \times Ngr$ sub-matrices. The sub-matrices are stored in a sparse storage format locally. In this case, the input vector is split across the columns of the matrix and the sub-vectors are duplicated on the sub-rows of the same column. The resulting vector has the same size as the number of rows in the sub-matrices of the corresponding row. However, each computing resource contains a part of a sub-vector. To obtain the global result, all the distributed results of the same row have to be summed then each row has to be gathered if the full result vector is needed in one place.

We use three standard storage formats for sparse matrices: CSR, ELLPACK, and SCOO [9]. The computation of the SpMV for these three formats is detailed in the Algorithms 1, 2 and 3 below. CSR and ELLPACK are standard formats used in many applications and frameworks. SCOO is a variant of COO where the matrix is split into blocks, then each block is stored in a COO format. It allows for better locality than COO in distributed settings, and therefore better performance on operations.

4   Experiments

In this Section, we first detail the parameters of the experiments performed to study the scalability of the distributed PageRank algorithm on the supercomputer Fugaku. We present the computing environment and some specificities of the processors. Then, we describe the matrices used for the experiments and present the results.

4.1   Parameters

SVE Implementation. A major feature of the recent ARM-based processors, including the A64FX, is the Scalable Vector Extension (SVE) that enables support for SIMD operations with per-lane prediction, which allows for efficient

Algorithm 1: CSR format data structure and matrix vector product. *idx* is the vector of indexes for the start of each row (of size $n + 1$). *col* and *val* are the vectors of columns and values (of size *nnz* each). *fr* (*fc*) is the index of the first row (column) of the block of data stored on this process, in the context of the whole data matrix

Function *spmv_csr()*
 Data: m : MatrixCSR, v : Vector
 Result: r : Vector
 for $i \leftarrow 0$ to $m.idx.size() - 1$ do
  for $j \leftarrow m.idx[i]$ to $m.idx[i+1] - 1$ do
   r[i] += m.val[j] * v[m.col[j] - m.fc]

Algorithm 2: ELL format data structure and matrix vector product. *col* and *val* are the vectors of columns and values (of size $n \times max\_col$ each)

Function *spmv_ell()*
 Data: m : MatrixELL, v : Vector
 Result: r : Vector
 for $i \leftarrow 0$ to $m.lrs - 1$ do
  for $j \leftarrow 0$ to $m.max\_col - 1$ do
   r[i + m.rpos] += m.val[i * m.max_col + j] * v[m.col[i * m.max_col + j]- m.fc]

Algorithm 3: SCOO format data structure and matrix vector product. *row*, *col* and *val* are the vectors of rows, columns and values (of size *nnz* each)

Function *spmv_scoo()*
 Data: m : MatrixSCOO, v : Vector
 Result: r : Vector
 for $i \leftarrow 0$ to $m.val.size() - 1$ do
  r[m.row[i] - m.fr] += m.val[i] * v[m.col[i] - m.fc]

vectorization [19]. The vector length can be specified as a multiple of 128 bits. We use a vector of 512 bits, which is the default for the processor and allows for the simultaneous computation of 8 double-precision values (64 bits each). The SVE instructions can be generated automatically by the compiler for simple functions. However, the compilers supporting the SVE instructions sometimes fail to vectorize the loops in the SpMV since they require indirect access to store arrays. Instead, we used the ARM SVE intrinsic functions to implement a vectorized version of the SpMV on different matrix formats.[1]

---

[1] Code is available at https://github.com/jgurhem/TBSLA/tree/dev_array.

Computing Environment. The code is implemented in C++ with MPI and OpenMP. The program is compiled using the Fujitsu compiler in Clang mode, with flags '-Kopenmp -fPIC -Ofast -mcpu=native -funroll-loops -fno-builtin -march=armv8.2-a+sve'.

Input Matrices. The PageRank takes as input a stochastic matrix, representing the normalized transpose of the adjacency matrix of a graph, and outputs a single result vector. As described in Sect. 3, each (MPI) process generates a subpart of the matrix, i.e. a block of rows and columns. This block of data is only accessed by the related process, which handles the computation on this part of the matrix. The matrix is therefore never stored in full in a single data structure.

We choose to build the adjacency matrix of graphs where each node has nnz edges linking to other nodes, with these nodes being chosen randomly with uniform distribution (forbidding duplicate edges to ensure a constant number of edges per node). This way, we have a matrix where data access is very irregular, which is often the case in graph-based applications [14]. One could argue that graphs also usually have nodes with different degrees, and regions more or less densely connected. We choose to ignore this parameter here and assign a fixed number of edges originating from each node. This is done in order to simplify the observations and not introduce uncertainty due to the load balancing issues.

4.2 Results

Weak Scaling on the Matrix Density. Since we want to evaluate how the PageRank algorithm scales on Fugaku, it makes sense to perform weak scaling experiments, where we increase the size of the problem along with the amount of computing resources. This also allows us to work on very large matrices, which would not be possible with strong scaling experiments since the instances would have to fit on one single processor in this case. There are two possibilities to increase the problem size: either a) increase the size n of the matrix, or b) increase the density of the matrix by increasing the number of nonzero elements per row. These two can be combined, but since the goal of these weak scaling experiments is to keep the same load per computing resource, it is simpler to do so by only changing either the size n or the density nnz. In this study, we choose the second option and increase the density of the matrix along with the number of compute nodes, while keeping the same matrix size. The first option (increasing the matrix size) implies further limitations that will be discussed at the end of this Section.

As a baseline, we consider a (square) matrix of size $4,000,000 \times 4,000,000$ with either 50 or 100 nonzero elements per row. Therefore, when using 16 compute nodes, the size of the matrix remains $4,000,000 \times 4,000,000$, but the number of nonzero elements per row goes up to 800 and 1600 respectively. As explained in a previous section, the sparse matrix is split in blocks and distributed on the compute nodes. Thus, using additional compute nodes for a matrix of the same size means splitting the matrix in increasingly small blocks. However, since the number of nonzero elements per row increases accordingly, the load per process remains constant. Since the complexity of the PageRank algorithm (using the power iteration method) depends primarily on this number of nonzeroes, this allows us to correctly assess the scalability using this experimental process.

This size was chosen to fit within the memory limits of one processor for all three sparse matrix storage formats. We use two different base densities (50 and 100) to get more insight on how this parameter affects the performance of the application. We increase the number of compute nodes progressively from 1 to 1024, doubling each time. Thus, in the largest case with 1024 compute nodes, the matrix contains $4M$ rows with 102400 nonzero elements each, i.e. more than $400B$ nonzeroes in total. For a given number of processes p, we choose to arrange them on a grid of $p/2$ (horizontally) by 2 (vertically). For example, with 16 processes, the matrix is split into 8 chunks horizontally and 2 chunks vertically, resulting in a sub-matrix of size $500,000$ by $2,000,000$ on each process. Different grid configurations have different impacts on the runtime, but the study of this behavior is outside the range of our present work.

In addition, we experiment on two different MPI/OpenMP configurations:

- 1 MPI process per compute node: each process uses 48 threads, one per processor core. The communications inside the processor are therefore handled by OpenMP, and the communications between nodes by MPI.
- 1 MPI per CMG: each process uses 12 threads, and there are 4 MPI processes per processor. With this setup, the communications between the 4 CMGs inside one processor are also handled by MPI.

The results in Tables 1 and 2 show that in absolute terms, the PageRank only takes a few seconds even for large instances.

Table 1. Median runtime for the PageRank, scaling the *nnz* per row, from a base of *nnz* = 50

Nodes	CSR		ELL		SCOO	
	Node	CMG	Node	CMG	Node	CMG
1	1.89	0.91	1.30	0.91	5.19	2.17
2	2.17	0.76	1.41	0.79	4.24	2.01
4	1.98	0.69	1.33	0.71	3.28	1.84
8	1.58	0.54	1.02	0.55	2.57	1.47
16	1.39	0.47	0.98	0.48	2.24	1.28
32	1.39	0.46	0.88	0.54	2.25	1.33
64	1.40	0.46	0.93	0.47	2.24	1.32
128	1.20	0.43	1.22	0.40	1.89	1.10
256	1.00	0.36	1.02	0.35	1.56	0.95
512	1.00	0.35	1.00	0.35	1.13	0.84
1024	1.00	0.37	1.01	0.37	1.16	0.86

Table 2. Median runtime for the PageRank, scaling the *nnz* per row, from a base of *nnz* = 100

Nodes	CSR		ELL		SCOO	
	Node	CMG	Node	CMG	Node	CMG
1	5.90	1.28	2.59	1.30	5.59	3.53
2	3.92	1.15	2.46	1.19	4.55	3.24
4	3.14	0.90	1.89	0.92	4.37	2.59
8	2.75	0.78	1.70	0.79	3.77	2.27
16	2.78	0.77	1.69	0.81	3.83	2.27
32	2.77	0.77	1.83	0.81	3.81	2.31
64	2.38	0.67	2.20	0.70	3.26	1.99
128	1.98	0.56	2.00	0.58	2.68	1.63
256	1.99	0.56	2.00	0.56	2.68	1.64
512	1.96	0.56	1.96	0.57	2.26	1.55
1024	1.98	0.59	1.97	0.59	2.28	1.58

However, since the convergence of the power iteration method depends on the input matrix, the number of iterations varies from one configuration to another. Therefore, the runtime itself may not be best for assessing the scalability of the application. For more rigorous comparison, we report the median performance in GFlop/s corresponding to the weak scaling results of the above two tables. Figures 1 and 2 show the results with $nnz = 50$ elements per row as basis for matrix density (scaling with the number of compute nodes), using one MPI process per compute node and one MPI process per CMG, respectively.

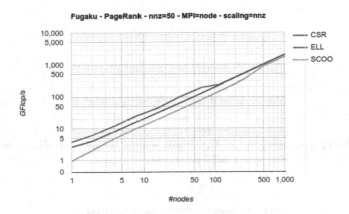

Fig. 1. Median performance for the PageRank, scaling the number of nonzero elements, from a base of $nnz = 50$, with 1 MPI per node

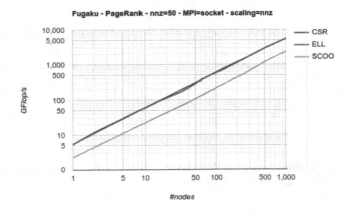

Fig. 2. Median performance for the PageRank, scaling the number of nonzero elements, from a base of $nnz = 50$, with 1 MPI per CMG

Figures 3 and 4 show the results with $nnz = 100$ elements per row as basis for matrix density, using one MPI process per compute node and one MPI process per CMG, respectively.

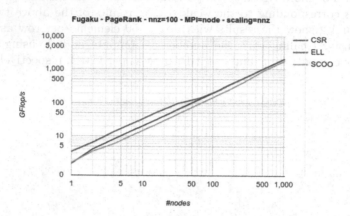

Fig. 3. Median performance for the PageRank, scaling the number of nonzero elements, from a base of $nnz = 100$, with 1 MPI per node

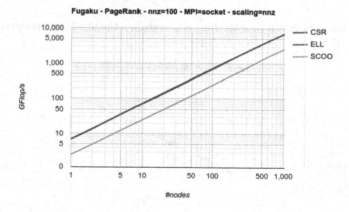

Fig. 4. Median performance for the PageRank, scaling the number of nonzero elements, from a base of $nnz = 100$, with 1 MPI per CMG

Memory Usage. Compared to other supercomputers, Fugaku uses processors with limited memory. Whereas an A64FX processor has 32 GB of HBM2 memory, only 28 GB of allocable RAM can be used, which poses constraints on the size of the data contained in each one. In Fig. 5, we show the memory used per node as the number of nodes increases (using the larger case with base density $nnz = 100$). The patterns are similar with the other MPI configuration (1 MPI process per CMG).

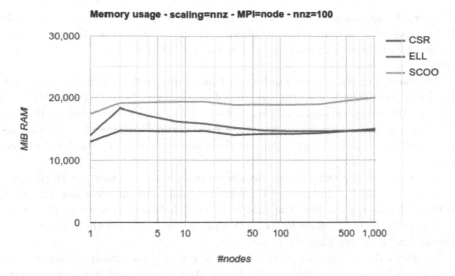

Fig. 5. Memory usage for the PageRank, scaling the number of nonzero elements, from a base of $nnz = 100$, with 1 MPI process per node

4.3 Discussion

Weak Scaling Results. The distributed PageRank shows excellent scalability from 1 to 1024 compute nodes, reaching more than 5TFlop/s on the largest experiments. The runtime per iteration of the power method remains about constant, giving linear speedup on the performance with the number of nodes used, with no signs of faltering. The performance patterns are similar for the two base matrix densities ($nnz = 50$ and $nnz = 100$), although the numbers are higher for the larger case. There are noticeable difference in performance depending on the sparse matrix storage format used, with SCOO performing overall worse than both CSR and ELL. Still, we can observe good scalability with all of these formats.

Another interesting point is the difference between the two MPI configurations; that is, either using 1 MPI process per node, or 1 MPI process per CMG. This parameter has an impact on two aspects. First, using one MPI process per CMG significantly increases performance, as can be observed by comparing Fig. 2 to Fig. 1, and Fig. 4 to Fig. 3. The performance overall is about doubled when using this configuration of 1 MPI per CMG. This indicates that using MPI for communications between the CMGs of a same processor is more efficient than using a pattern of shared memory across the 4 CMGs and using OpenMP for intra-processor communications. Second, the MPI configuration changes the relative behaviors of the different storage formats. When using 1 MPI per node (Figs. 1 and 3), the differences between the storage formats tend to diminish as the number of compute nodes increases. On the contrary, with 1 MPI per CMG (Figs. 2 and 4), there is a larger gap between SCOO on one hand, and CSR and ELL on the other, with these last two being nearly equal.

Memory Usage and Roadblocks. Figure 5 show the memory usage per node of our application, scaling with the number of nodes. We can see that this usage remains nearly constant, and below the limit of 28 GB of allocable RAM per node.

An analysis of the space complexity of our implementation of the PageRank algorithm show that it uses two vectors of n_col elements to store the iterated result of the computation, and two vectors of ln_row elements (all in double-precision), in addition to the matrix storage itself. n_col is the number of columns of the full data matrix, and ln_row is the number of rows in the local matrix block for this process.

In the case of the experiments presented here, where we increase the density of the matrix, the full matrix size remains constant, and the local size may even decrease depending on how the matrix is split between processes. Therefore, the constant memory usage for both the matrix storage and the computation, observed in these two Figures, is normal.

However, these two vectors of size n_col are an issue when scaling on the matrix size, as mentioned previously. At the start of each iteration, each process needs to have the full result vector from the previous iteration in order to perform the computation. At some point, the memory required to store this full result vector on each process becomes larger even than the memory used to store the sparse matrix block. This problem is magnified when using more than 1 MPI process per node, since each process has its copy of the result vector. In our preliminary experiments, we hit the memory limit of 28 GB with 256 nodes when using 1 MPI per node, and with 64 nodes when using 1 MPI per CMG (i.e. 4 MPI per node). This amounts to vectors of more than 1 billion double-precision elements. Consequently, scaling for sparse matrices representing graphs of billions of nodes would require a different implementation of the PageRank algorithm, which does not require the full result vector to be stored on each process.

5   Conclusion

We have presented an implementation of the PageRank algorithm that uses a distributed and parallel sparse matrix vector product as its kernel, in order to scale the exact computation of the PageRank to very large data sets. We have performed weak scaling experiments for this application on the supercomputer Fugaku, increasing the density of a sparse stochastic matrix with the number of compute nodes. The experiments used up to 1024 compute nodes (49152 compute cores), for matrices with hundreds of billions of nonzero elements. The matrix was generated in a distributed setting, with the nonzero elements chosen randomly in order to avoid any structure, and to present the worst possible case for data access patterns. We observed linear scalability for this PageRank algorithm in this context. We compared two MPI configurations and found that using MPI for communications between CMGs on the A64FX processor leads to noticeable improvements in performance, compared to using one MPI per compute node. Whereas the HPCG benchmark uses diagonal matrices, we studied irregular matrices, which are more representative of real-life matrices such as Web graphs or networks commonly used as input for the PageRank algorithm. On these very large irregular matrices, the network-on-chip therefore induces a different programming paradigm, with the communications between CMG being paramount to the performance. The sparse matrix storage formats also exhibit different performance patterns depending on the MPI configuration. Further analysis at a lower level, and profiling of the MPI communications, would be required to investigate these differences in more detail, as well as the discrepancies observed between the different sparse matrix storage formats. Besides that, the main perspective would be to design a PageRank implementation that does not require to store the full result vector on each process, which leads to memory issues when scaling on very large matrices (beyond 1 billion in size). Such an implementation would likely use additional MPI communications to gather the necessary input at each iteration. Therefore, an analysis of the performance tradeoff in this case would be insightful for future applications.

References

1. Ajima, Y., et al.: The tofu interconnect D. In: 2018 IEEE International Conference on Cluster Computing (CLUSTER), pp. 646–654 (2018). https://doi.org/10.1109/CLUSTER.2018.00090
2. Alappat, C.L., et al.: Performance Modeling of Streaming Kernels and Sparse Matrix-Vector Multiplication on A64FX. CoRR abs/2009.13903 (2020). https://arxiv.org/abs/2009.13903
3. Dai, L., Freris, N.M.: Fully distributed pagerank computation with exponential convergence. arXiv preprint arXiv:1705.09927 (2017)
4. De Jager, D.: PageRank: three distributed algorithms. Master's thesis, Imperial College London, London, pubs. doc. ic. ac. uk/pagerank-algorithms (2004)
5. Dongarra, J.: Report on the Fujitsu Fugaku system. University of Tennessee-Knoxville Innovative Computing Laboratory, Technical Report ICLUT-20-06 (2020)
6. Dongarra, J., Heroux, M.A., Luszczek, P.: High-performance conjugate-gradient benchmark: a new metric for ranking high-performance computing systems. Int. J. High Perform. Comput. Appl. **30**(1), 3–10 (2016)
7. Guo, T., Cao, X., Cong, G., Lu, J., Lin, X.: Distributed algorithms on exact personalized PageRank. In: Proceedings of the 2017 ACM International Conference on Management of Data, pp. 479–494 (2017)
8. Gurhem, J., Vandromme, M., Tsuji, M., Petiton, S.G., Sato, M.: Sequences of sparse matrix-vector multiplication on Fugaku's A64FX processors. In: 2021 IEEE International Conference on Cluster Computing (CLUSTER), pp. 751–758. IEEE (2021)
9. Hugues, M.R., Petiton, S.G.: Sparse matrix formats evaluation and optimization on a GPU. In: 2010 IEEE 12th International Conference on High Performance Computing and Communications (HPCC), pp. 122–129. IEEE (2010)
10. Ihde, N., et al.: A survey of big data, high performance computing, and machine learning benchmarks (2021)
11. Ishii, H., Tempo, R., Bai, E.W.: A web aggregation approach for distributed randomized PageRank algorithms. IEEE Trans. Autom. Control **57**(11), 2703–2717 (2012)

12. Klicpera, J., Bojchevski, A., Günnemann, S.: Predict then propagate: graph neural networks meet personalized PageRank. arXiv preprint arXiv:1810.05997 (2018)
13. Lin, W.: Distributed algorithms for fully personalized PageRank on large graphs. In: The World Wide Web Conference, pp. 1084–1094 (2019)
14. Lumsdaine, A., Gregor, D., Hendrickson, B., Berry, J.: Challenges in parallel graph processing. Parallel Process. Lett. **17**(01), 5–20 (2007)
15. Ma, N., Guan, J., Zhao, Y.: Bringing PageRank to the citation analysis. Inf. Process. Manage. **44**(2), 800–810 (2008)
16. Page, L., Brin, S., Motwani, R., Winograd, T.: The pagerank citation ranking: Bringing order to the web. Technical report, Stanford InfoLab (1999)
17. Pop, F., Dobre, C.: An efficient PageRank approach for urban traffic optimization. Mathematical Problems in Engineering 2012 (2012)
18. Das Sarma, A., Molla, A.R., Pandurangan, G., Upfal, E.: Fast distributed PageRank computation. In: Frey, D., Raynal, M., Sarkar, S., Shyamasundar, R.K., Sinha, P. (eds.) ICDCN 2013. LNCS, vol. 7730, pp. 11–26. Springer, Heidelberg (2013). https://doi.org/10.1007/978-3-642-35668-1_2
19. Sato, M., et al.: Co-design for A64FX manycore processor and "Fugaku". In: SC20: International Conference for High Performance Computing, Networking, Storage and Analysis, pp. 1–15 (2020). https://doi.org/10.1109/SC41405.2020.00051

Ultrafast Focus Detection for Automated Microscopy

Maksim Levental[1]([✉]), Ryan Chard[2], Kyle Chard[1,2], Ian Foster[1,2], and Gregg Wildenberg[2]

[1] University of Chicago, Chicago, IL, USA
mlevental@uchicago.edu
[2] Argonne National Lab, Lemont, IL, USA

Abstract. Technological advancements in modern scientific instruments, such as scanning electron microscopes (SEMs), have significantly increased data acquisition rates and image resolutions enabling new questions to be explored; however, the resulting data volumes and velocities, combined with automated experiments, are quickly overwhelming scientists as there remain crucial steps that require human intervention, for example reviewing image focus. We present a fast out-of-focus detection algorithm for electron microscopy images collected serially and demonstrate that it can be used to provide near-real-time quality control for neuroscience workflows. Our technique, *Multi-scale Histologic Feature Detection*, adapts classical computer vision techniques and is based on detecting various fine-grained histologic features. We exploit the inherent parallelism in the technique to employ GPU primitives in order to accelerate characterization. We show that our method can detect out-of-focus conditions within just 20 ms. To make these capabilities generally available, we deploy our feature detector as an on-demand service and show that it can be used to determine the degree of focus in approximately 230 ms, enabling near-real-time use.

1 Introduction

A fundamental goal of neuroscience is to map the anatomical relationships of the brain, an approach broadly called *connectomics*. Electron microscopy, an imaging method traditionally limited to small single 2D images, provides sufficient resolution to directly visualize the connections, or synapses, between neurons. Recently, automated serial electron microscopy (SEM) techniques have been developed where thousands, if not tens of thousands, of individual images are automatically acquired in series and then registered (i.e., aligned) to produce a volumetric dataset. Such datasets allow neuroscientists to follow the tortuous path neurons take through the brain to connect with each other (hence the name connectomics). However, many of the steps that comprise the collection of such datasets for connectomics require manual inspection, causing significant slowdowns in the rate at which datasets can be acquired. Such bottlenecks significantly impact the size of the datasets that can be reasonably acquired and

D. Groen et al. (Eds.): ICCS 2022, LNCS 13350, pp. 403–416, 2022.
https://doi.org/10.1007/978-3-031-08751-6_29

studied. Furthermore, advances in electron microscopes have increased the rate that datasets can be acquired; for example, ~10 Tbs/24hr [6], which, when used to map an entire, mouse brain will result in approximately 1 exabyte of data.

Auto-focus technology is a critical component of many imaging systems; from consumer cameras (for purposes of convenience) to industrial inspection tools to scientific instrumentation [28]. Such technology is typically either *active* or *passive*; active methods exploit some auxiliary device or mechanism to measure the distance of the optics from the scene, while passive methods analyze the definition or sharpness of an image by virtue of a proxy measure called a *criterion function*. Many electron microscopes incorporate auto-focus systems that attempt to focus the microscope before image acquisition. Despite such functionality, out-of-focus (OOF) images still occur at high rates (between 1% and 10%), depending on the quality of the tissue sections being imaged. For instance, it is common to experience occasional staining artifacts, and tears or compression artifacts (i.e., section wrinkles) during ultra-thin serial sectioning. These imperfections can cause auto-focus systems to fail if the microscope centers on them. This results in the system failing to find the correct focal plane, thus necessitating post-acquisition evaluation. These OOF error modes prevent effective automation, since a prerequisite of many downstream transformations is that the images collected all have high degree-of-focus (DOF). Without properly focused images, all downstream computational steps (e.g., 2D tile montaging, 3D alignment, automatic segmentation) will fail.

The DOF of images acquired by an electron microscope is also of critical importance with respect to automation. While seemingly a small step in a potential automation pipeline, focus detection is nevertheless an extremely critical step. In general, imaging tissue sections requires loading and unloading sets of ~100–200 sections at a time. Failure to detect a single OOF image in situ causes significant delays because the affected sample sets need to be reloaded, desired field of view must be reconfigured, and reacquired images need to be realigned into the image stack. All such remediation steps are time and labor intensive, and effectively stops any downstream automation until the problem is remedied. Under ideal conditions, it is estimated that fixing a single image would take several hours of manual intervention, which increases if multiple images in distinct parts of the series have to be manually reacquired and aligned.

In this work we focus on ensuring images acquired by the electron microscope have high DOF, in order to further progress towards to goal of end-to-end automation. To this end, we propose a new technique, *Multi-scale Histologic Feature Detection* (MHFD), that involves a second pass over the collected image, after it has been acquired, using a computer vision system to detect a failure to successfully achieve high DOF. Our technique relies on employing feature detection [16] as a criterion function, in accordance with the hypothesis that the quantity of features detected is positively correlated with DOF. Using this insight, we develop a feature detector based on scale-space representations of images (see Sect. 2.2) but optimized for latency. The design and implementation of our feature detector prioritizes parallelization, specifically in order to target GPU deployments.

Our solution achieves low latency detection of the OOF condition with high accuracy (see Sect. 5). To provide access to these capabilities, we have deployed them as a service that can be consumed on-demand and integrated in automated workflows. The service leverages Argonne National Laboratory's Leadership Computing Facility to provide access to A100 GPUs to rapidly analyze images as they are captured. This allows users to detect low quality images and correct their collection while the sample is still in the microscope, effectively eliminating costly delays in reloading, aligning, and imaging the sample. An important caveat in our work: we explicitly aim to augment existing microscopy equipment without the need for costly and complex retrofitting. This precludes mere improvements to existing auto-focus systems as they are, in essence, proprietary black boxes from the perspective of the end user of an electron microscope.

The rest of this article is organized as follows: Sect. 2 reviews background information on connectomics and scale-space feature detectors. Section 3 describes our focus detection method, in particular optimizations made in order to achieve near-real-time performance. Section 4 describes how we deliver MHFD as a service. Section 5 presents evaluation results. Section 6 discusses related work. Finally, we conclude in Sect. 7.

2   Background

We briefly review a common connectomics workflow and then describe scale-space representations.

2.1   Connectomics

Connectomics is defined as the study of comprehensive maps of connections within an organism's nervous system (called *connectomes*). The data acquisition pipeline for connectomics consists of the following steps:

1. A piece of nervous system (e.g., brain), ranging from $\sim$1mm^3 to $1\,\text{cm}^3$ is stained with heavy metals (e.g., osmium tetroxide, uranyl acetate, lead) in order to provide contrast in resulting images [11];
2. After staining, the section is dehydrated and embedded in a plastic resin to stabilize the tissue for serial sectioning, which is performed with an Automated Serial Sections to Tape (ATUM) device [12] (where ultrathin sections are automatically sectioned and collected on polyimide tape);
3. The sections are mounted to a silicon wafer, with each wafer containing 200–300 sections;
4. The wafer is loaded into a SEM, where the user marks a region of interest (ROI) within the sections for the microscope to image;
5. The SEM initiates a protocol to automatically image the ROI over all the sections at a desired resolution;
6. For each section, the SEM attempts to auto-focus before imaging by sampling different focal planes over a set range of focal depths

The series of collected images are then algorithmically aligned to each other to produce a 3D volumetric image stack where biological features are segmented either manually or by automatic segmentation techniques.

Since the imaging and post-acquisition process (e.g., retakes of blurry images, 3D alignment, segmentation) is slow, connectomics is practically constrained to small volumes ($\sim$100 µm^3), but technologies are rapidly advancing, with near future goals of mapping an entire mouse brain [1]. Even with 100 µm^3 volumes, the scope of the biological problem is large. For instance, a single mouse neuron is estimated to receive $\sim$5000–7000 connections [26] and the cell density of the mouse cortex is $\sim$1.5 × 10^5 cells/mm^3 [10]. A 100 µm^3 volume will therefore contain $\sim$150 neurons receiving 7.5^5 synapses, all of which neuroscientists seek to automatically segment and study. For an entire mouse brain, there are $\sim$7 × 10^6 neurons and 3.5 × 10^{1}0 connections. Ensuring that automatic segmentation algorithms accurately segment neurons depends on having the highest possible quality images and any error in image quality is very likely to produce segmentation errors that propagate in a non-linear fashion. For instance, if the connection between two neurons is improperly assigned, the other neurons that those pair of neurons connect to will also be improperly connected, and so on. Not only is the biological scope of the problem large, but datasets are also large, both in terms of the number of images and data size. Again, using the range of 100 µm^3 to 1 cm^3 datasets, these volumes will equate to 2,500 to 250,000 sections and $\sim$0.7 terabytes or 1 exabyte of data, respectively. Thus, the scope of the data both in terms of the biological goal and data management demands automation in the connectomic pipeline, in order to minimize errors and the need for manual OOF detection and correction.

2.2   Scale-Space Representations

We base our multi-scale histologic feature detection technique on classical scale-space representations of signals and images. We give a brief overview (see [16] for a more comprehensive review). The fundamental principle of scale-space feature detection is that natural images possess structural features at multiple scales and features at a particular scale are isolated from features at other scales. Thus, any image $I(x, y)$ can be transformed into a scale-space representation $L(x, y, t)$, where $L(x', y', t')$ represents the pixel intensity at pixel coordinates (x', y') and *scale* t'. How to construct the representation of the image at each scale is discussed below. More importantly, such a representation lends itself readily to scale sensitive feature detection, owing to the fact that features at a particular scale are decoupled from features at other scales, thereby eliminating confounding detections. Examples of structural features that can be detected and characterized using scale-space representations include edges, corners, ridges, and so called blobs (roughly circular regions of uniform intensity).

A scale-space representation at a particular scale is constructed by convolution of the image with a filter that satisfies the following constraints: non-enhancement of local extrema, scale invariance, and rotational invariance. Other relevant constraints are discussed in [9]. One such filter is the symmetric, mean zero, two dimensional, Gaussian filter [13]:

$$G\left(x,y,\sigma\right) := \frac{1}{2\pi\sigma^2}e^{-\frac{x^2+y^2}{2\sigma^2}}$$

The scale-space representation $L(x,y,t)$ of an image $I(x,y)$ is defined to be the convolution of that image with a mean zero Gaussian filter:

$$L\left(x,y,t\right) := G\left(x,y,t\right) * I\left(x,y\right)$$

where t determines the scale. $L(x,y,t)$ has the interpretation that image structures of scale smaller than $\sqrt{t^2} = t$ have been removed due to blurring. This is due to the fact that the variance of the Gaussian filter is t^2 and features of this scale are therefore "beneath the noise floor" of the filter or, in effect, suppressed by filtering procedure. A corollary is that features with approximate length scale t will have maximal response upon being filtered by $G(x,y,t)$. That is to say, for a t scale feature at pixel coordinates (x,y) and for scales $t' < t < t''$ we have

$$L\left(x,y,t'\right) < L\left(x,y,t\right) < L\left(x,y,t''\right)$$

This is due to the fact that for scales $t' < t$, small scale features will dominate the response and for $t < t''$, as already mentioned, the feature will have been suppressed.

Note that the aforementioned presumes having identified the pixel coordinates (x,y) as the locus of the feature. Hence, in order to detect features across both scale and space dimensions, maximal responses in spatial dimensions (x,y) need to also be characterized. For such characterization one generally employs standard calculus, in order to identify critical points of the second derivatives of $L(x,y,t)$. Hence, we can construct scale-sensitive feature detectors by considering critical points of linear and non-linear combinations of spatial derivatives ∂_x, ∂_y and derivatives in scale ∂_t. For example the scale derivative of the Laplacian

$$\partial_t\nabla^2 L := \partial_t\left(\partial_x^2 + \partial_y^2\right)L \tag{1}$$

effectively detects regions of uniform pixel intensity (i.e., blobs).

Equation (1) permits a discretization called *Difference of Gaussians* (DoG) [19]:

$$t^2\nabla^2 L \approx t \times \left(L\left(x,y,t+\delta t\right) - L\left(x,y,t\right)\right)$$

Therefore, we define the following parameters: n, which determines the granularity of the scales detected; $\min_t$, the minimum scale detected; $\max_t$, the maximum scale detected; $\delta t := \left(\max_t - \min_t\right)/n$; $t_i := \min_t + (i-1)\times\delta t$, the discrete scales detected. We then define the discretized DoG filter:

$$\text{DoG}\left(x,y,i\right) := t_i \times \left(L\left(x,y,t_{i+1}\right) - L\left(x,y,t_i\right)\right) \tag{2}$$

This produces a sequence $\{\text{DoG}\left(x,y,i\right) \mid i=1,\ldots,n\}$ of filtered and scaled images (called a Gaussian pyramid [8]). Note that there are alternative conventions for how each difference in the definition of $\text{DoG}\left(x,y,i\right)$ should be scaled (including partitioning into so called *octaves* [5]); we observe that linear scaling is sufficient, in terms of accuracy and complexity, for the purposes of detecting OOF conditions.

3 Multi-scale Histologic Feature Detection

We propose to use histologic feature detection at multiple scales as a criterion function, reasoning that the absolute quantity of features detected at multiple scales is positively correlated with DOF (see Fig. 1). For our particular use case, this is tantamount to detecting histologic structures ranging from cell walls to whole organelles. The key insight is that the ability to resolve structure across the range of feature scales is highly correlated with a high-definition image. To this end, we develop a feature detector based on Eq. (1) but optimized for latency (rather than accuracy).

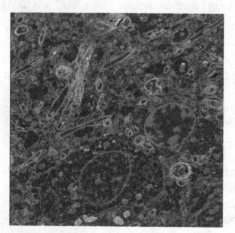
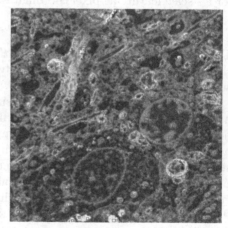

(a) Histologic features of an in-focus section.

(b) Histologic features of an out-of-focus section.

Fig. 1. Comparison of sections with histologic feature recognition as a function of focal depth.

In order to verify our hypothesis, that detecting features across a range of scales is correlated with DOF, we compare the number of histologic features detected as a function of absolute deviation from in-focus ($|f - f'|$ where f' is the correct focal depth) for a series of sections with known focal depth (see Fig. 2a). We observe a strong log-linear relationship (see Fig. 2b). Fitting a log-linear relationship produces a line with $r = -0.9754$, confirming our hypothesis that quantity of histologic features detected is a good proxy measure for DOF. Note that the log-linear relationship corresponds to a roughly quadratic decrease in the number of histologic features detected. This is to be expected since, intuitively, a twice improved DOF of a two dimensional image yields improved detection along both spatial dimensions and thus a four times increased quantity of histologic features detected.

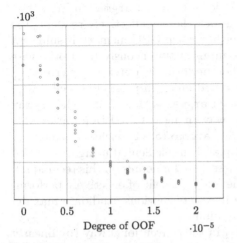 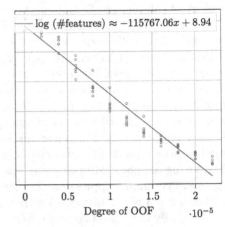

(a) Number of histologic features as a function of absolute deviation from focused ($|f - f'|$ where f' is the correct focal depth).

(b) Log plot and line fit with $r = -0.9754$.

Fig. 2. Comparison of histologic feature recognition as a function of focal depth.

Recall, we aim to achieve near-real-time quality control of SEM images to facilitate error detection and correction while data are being collected. We therefore require low-latency multi-scale histologic feature detection. Here we present the design and implementation of our system that leverages GPUs to rapidly classify images by determining their degree of focus. Our starting point is Eq. (2) for possible optimizations. Computing maxima of DoG (x, y, i) in the scale dimension (equivalently critical points of Eq. (1)) necessarily entails computing maxima in a small pixel neighborhood at every scale. We first make the heuristic assumption that, in each pixel neighborhood that corresponds to a feature, there is a single unique and maximal response at some scale t. This response corresponds to the scale at which the variance of the Gaussian filter G most closely corresponds to the scale of the feature (see Sect. 2.2). We therefore search for *local maxima* in spatial dimensions x, y but *global maxima* in the scale dimension:

$$C := \left\{ \left(\hat{x}_j, \hat{y}_j, \hat{i}_j \right) \right\} := \operatorname*{argmaxlocal}_{x,y} \operatorname*{argmax}_{i} \operatorname{DoG}(x, y, i) \qquad (3)$$

where the subscript j indexes over the features detected. Once all such maxima are identified it suffices to compute and report the cardinality, $|C|$, as the criterion function value.

It is readily apparent that our histologic feature detector is parallelizable: for each scale t_i we can compute $L(x, y, t_i)$ independently of all other $L(x, y, t_j)$ (for $j \neq i$). A further parallelization is possible for the `argmax` operation, since the maxima are computed independently across distinct neighborhoods of pixels. In order to maximally exploit this, we first perform the inner `argmax` in Eq. (3) on a block of columns of $\{\text{DoG}(x, y, i)\}$ in parallel, thereby effectively reducing the Gaussian pyramid to a single image. Note that when GPU memory is sufficient we can compute the `argmax` across all columns simultaneously (and otherwise within a constant number of steps). We then perform the outer `argmaxlocal`$_{x,y}$ on disjoint pixel neighborhoods of the flattened image in parallel as well.

Note that the implementation of the inner `argmax` is "free", since the `argmax` primitive is implemented in exactly this way in most GPGPU libraries [20], and thus our substitution of `argmax`$_i$ for `argmaxlocal`$_i$ yields a moderate latency improvement. The outer `argmaxlocal` is implemented using a comparison against `maxpool_2d`(n, n) (with $n = 3$) (see [15] for details on this technique). Employing `maxpool_2d` in this way has the added benefit of effectively performing non-maximum suppression [21], since it rejects spurious candidate maxima within a 3×3 neighborhood of a true maximum.

Typically one would compute $L(x, y, t_i)$ in the conventional way (by linearly convolving G and I) but prior work has shown that performing the convolution in the Fourier domain is much more efficient [15]; namely

$$L(x, y, t_i) = \mathcal{F}^{-1}\big\{\mathcal{F}\{G(x, y, t_i)\} \cdot \mathcal{F}\{I(x, y)\}\big\}$$

where $\mathcal{F}\{\cdot\}, \mathcal{F}^{-1}\{\cdot\}$ are the Fourier transform and inverse Fourier transform, respectively. This approach has the additional advantage that we can make use of highly optimized Fast Fourier Transform (FFT) routines made available by GPGPU libraries.

One remaining detail is histogram stretching of the images. Due to the dynamic range (i.e., variable bit depth) of the microscope, we need to normalize the histogram of pixel values. We implement this normalization by saturating .175% of the darkest pixels, saturating .175% of the lightest pixels, and mapping the entire range to $[0, 1]$. We find this gives us consistently robust results with respect to noise and anomalous features. This histogram normalization is also parallelized using GPU primitives. We present our technique in Algorithm (1).

4   Histologic Feature Detection as a Service

A key challenge to using our histologic feature detector is that it requires powerful GPUs with large quantities of RAM, something that many commodity GPUs and edge devices lack. To make our detector generally accessible we have deployed it as an on-demand service using the funcX platform [7]. funcX is a high performance function-as-a-service platform designed to provide secure, fire-and-forget remote execution. funcX federates access to remote research cyberinfrastructure via a single, multi-tenant cloud service. Users submit a function

Algorithm 1. Multi-scale Histologic Feature Detection

Input: $I(x, y), n, \min_t, \max_t, M$

1: $I'(x, y) := \text{HistorgramStretch}(I(x, y))$
2: $\text{Broadcast}(I'(x, y), M)$
3: **parfor** $m := 1, \ldots, M$ **do**
4:     **parfor** $i \in I_m$ **do**
5:         $L(x, y, t_i) := \mathcal{F}^{-1}\{\mathcal{F}\{G(x, y, t_i)\} \cdot \mathcal{F}\{I'(x, y)\}\}$
6:     **end**
7: **end**
8: $\text{Gather}(L(x, y, t_i), M)$
9: **parfor** $i := 1, \ldots, n+1$ **do**
10:     $\text{DoG}(x, y, i) := t_i \times (L(x, y, t_{i+1}) - L(x, y, t_i))$
11: **end**
12: $\left\{\left(\hat{x}_j, \hat{y}_j, \hat{i}_j\right)\right\} := \text{argmaxlocal}_{x,y} \text{argmax}_i \text{DoG}(x, y, i)$

Output: $\text{DOF} := \left|\left\{\left(\hat{x}_j, \hat{y}_j, \hat{i}_j\right)\right\}\right|$

---

invocation request to funcX which then routes the request to the desired *endpoint* for execution. Endpoints may be deployed by users on remote computing resources, including clouds, clusters, and edge devices.

We registered our MHFD tool as a funcX function, configuring it such that it requires as input arguments only the location of the input image. The function executes the MHFD tool on an accessible GPU and the resulting feature count and DOF is returned asynchronously via the funcX service. Registration as a funcX function allows others to execute the tool on their own funcX endpoints.

We enable automated invocation of the MHFD via Globus Flows [2]—a research automation platform. funcX is accessible as a Flows Action Provider, enabling users to deploy a flow that detects data creation, transfers data from instrument to analysis cluster, executes the MHFD, and returns results to users.

5 Evaluation

We evaluate our optimized histologic feature detector in terms of runtime performance (in order to assess its fitness a realtime OOF detector). Data used herein were collected using brains prepared in the same manner and as previously described [11]. Using a commercial ultramicrotome (Powertome, RMC), the cured block was trimmed to a ~1.0 mm × 1.5 mm rectangle and ~2,000, 40nm thick sections were collected on polyimide tape (Kapton) using an automated tape collecting device (ATUM, RMC) and assembled on silicon wafers as previously described [12]. Images at different focal distances were acquired using backscattered electron detection with a Gemini 300 scanning electron microscope (Carl Zeiss), equipped with ATLAS software for automated imaging. Dwell times for all datasets were 1.0 microsecond.

We perform runtime experiments across a range of parameters of interest (section resolution, number of feature scales). Our test platform is a NVIDIA

Table 1. Test platform (ALCF ThetaGPU)

CPU	Dual AMD Rome 7742 @ 2.25 GHz
GPU	8x NVIDIA A100-40 GB
HD	4x 3.84 U.2 NVMe SSD
RAM	1TB
Software	CuPy-8.3.0, CUDA-11.0, NVIDIA-450.51.05

DGX A100 (see Table 1). Experiments consist of computing the DOF of a sample section for a given configuration. All experiments are repeated k times (with $k = 21$) and all metrics reported are median statistics, where we discard the first execution as it is an outlier due to various initializations (e.g.,. pinning CUDA memory).

For a section resolution of 1024×1024 pixels we achieve approximately 50 Hz runtime in the single GPU configuration; this is near-real-time. We observe that, as expected, runtime grows linearly with the number of feature scales and quadratically with the resolution of the section; naturally, this is owing to the parallel architecture of the GPU. The principle defect of our technique is that it is highly dependent on the available RAM of the GPU on which it is deployed. In practice, most GPUs available at the edge, i.e., proximal to microscopy instruments, will have insufficient RAM to accommodate large section resolutions and wide feature scale ranges. In fact, even the 40GB of the DGX's A100 is exhausted at resolutions above 4096×4096 for more than approximately 20 feature scales.

Therefore, we further investigate parallelizing MHFD across multiple GPUs. Our implementation parallelizes MHFD in a straightforward fashion: we partition the set of filters across the GPUs, perform the "lighter" FFT-IFFT pair on each constituent GPU, and then gather the results to the root GPU (arbitrarily chosen). That is to say we actually carry out

$$\{L(x, y, t_i) \mid i \in I_m\} = \{\mathcal{F}^{-1}\{\mathcal{F}\{G(x, y, t_i)\} \cdot \mathcal{F}\{I(x, y)\}\} \mid i \in I_m\}$$

where for $m = 1, \ldots, M$ the set I_m indexes the scales allocated to a node m. By partitioning the set of Gaussian filters $\{G(x, y, t_i)\}$ across M nodes, we effectively perform distributed filtering. We use CUDA-aware OpenMPI to implement the distribution. Note that for such multi-GPU configurations the range of feature scales was chosen to be a multiple of the number of GPUs (hence the proportionally increasing sparsity of data in Fig. 3a).

We observe that, as one would expect, runtime is inversely proportional to number of GPUs (see Fig. 3b) but for instances where a single GPU configuration is sufficient it is also optimal. More precise timing reveals that parallelization across multiple GPUs incurs high network copy costs during the gather phase (see Fig. 4). Note that this latency persists even after taking advantage of CUDA IPC [23]. In effect, this is a fairly obvious demonstration of Amdahl's law. Therefore, parallelization across multiple GPUs should be considered in instances where full resolution section images are necessary (e.g., when feature

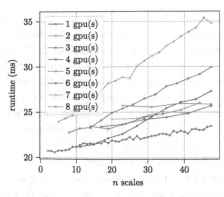
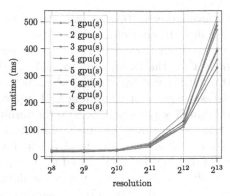

(a) Median runtime as a function of number of feature scales at resolution = 1024 × 1024.

(b) Median runtime as a function of section resolution with 16 feature scales.

Fig. 3. Scaling experiments for runtime with respect to number of GPUs, resolution, and number of feature scales.

scale ranges are very wide, with detection at the lower end of the scale being critical). In all other cases, preprocessing by downsampling, by bilinear interpolation, in order to satisfy GPU RAM constraints yields a more than reasonable tradeoff between accuracy and latency.

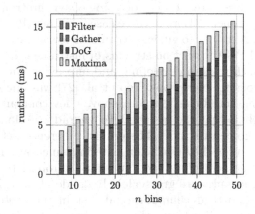

Fig. 4. Breakdown of runtime into the four major phases for two GPUs across feature scales at resolution = 1024 × 1024.

To evaluate the on-demand use of MHFD we deployed a funcX endpoint on the ThetaGPU cluster (Table 1). We registered our MHFD tool as a funcX function and passed the path to the test dataset as input. It is important to note that the SEM imagery is not passed through the funcX platform in these tests but instead is assumed data to be resident on the shared file system. We performed over 1000 invocations of the function after first executing a task to initialize the environment. The tests use a single GPU and a dataset with section resolution of 1024×1024 pixels to be comparable to earlier results. Our findings show that the mean time to perform the feature detection was 25 ms, with a standard deviation of 6 ms, which is similar to the previous result for the same dataset when not using funcX. The mean time taken to submit the request to the funcX service and retrieve the result was 233 ms, with a standard deviation of 44 ms, meaning the funcX platform introduces an overhead of approximately 200 ms. While this overhead is roughly 8× greater than the MHFD analysis itself, the total time required to determine the focus of a dataset is suitable to classify images and report errors as data is collected. In practice, data must also be moved to the ThetaGPU cluster's filesystem before it can be analyzed. When evaluating the time to transfer the single 23MB image from Argonne's Structural Biology Center (where the SEM resides) to the ThetaGPU cluster, we found it was moved at over 200MB/s and took roughly 100 ms. This is due to the two 40Gb/s connections between the SBC and computing facility.

6   Related Work

Automating the control and optimization of scientific instruments is a common area of research that spans a diverse collection of devices and applies an equally diverse set of techniques, including HPC analysis, ML-in-the-loop [22], and edge-accelerated processing [17]. Laszewski et al. [14] and Bicer et al. [4] demonstrate two approaches to perform real-time processing of synchrotron light source data in order to steer experiments. Both of these cases employ HPC systems to rapidly analyze and reconstruct data to guide instruments toward areas of interest. Others have also used FPGAs to act on streams of instrument imagery [25].

There is also much work in developing and improving auto-focus algorithms and their applications to microscopy. Yeo et al. [27] was one of the first investigations of applying auto-focus to microscopy. They compare several criterion functions and conclude that the so-called Tenengrad criterion function is most accurate and most robust to noise. The crucial difference between their evaluation criteria and ours is they select for criterion functions that are suited for optical microscopy, i.e., criterion functions that are robust to staining/coloring (whereas all of our samples are grayscale). Redondo et al. [24] reviews sixteen criterion functions and their computational cost in the context of automated microscopy. Bian et al. [3] address the same issues that motivate us in that they aim to support automated processes in the face of topographic variance in the samples (which leads to comparing OOF rates). Their solutions distinguish themselves in that they employ active devices (such as low-coherence interferometry). Interestingly, seemingly contemporaneously with our project Luo et al. [18] proposed a deep learning architecture that auto-focuses in a "single-shot" manner. Such a solution is appealing given the affinity with our own application of GPGPU to the problem and we intend to experiment with applying it to our data.

7   Conclusions

We presented an OOF detection technique designed to augment existing microscopy instrumentation. Rather than focusing the microscope, as autofocusing algorithms would, our algorithm operates downstream of image acquisition to report out-of-focus events to the user. This enables the user to intervene and initiate reacquisition protocols (on the microscope) before unknowingly proceeding with collecting the next series of images or proceeding with downstream image processing and analysis. Our technique is effective and operates at near-real-time latencies. Thus, this human-in-the-loop remediation protocol already saves the user much wasted collection time triaging defective collection runs.

Acknowledgements. This work was supported by the U.S. Department of Energy, Office of Science, under contract DE-AC02-06CH11357.

References

1. Abbott, L.F., et al.: The mind of a mouse. Cell **182**(6), 1372–1376 (2020)
2. Ananthakrishnan, R., et al.: Globus platform services for data publication. In: Practice and Experience on Advanced Research Computing, pp. 14:1–14:7 (2018)
3. Bian, Z., et al.: Autofocusing technologies for whole slide imaging and automated microscopy. J. Biophoton. **13**(12), e202000227 (2020)
4. Bicer, T., et al.: Real-time data analysis and autonomous steering of synchrotron light source experiments. In: 13th International Conference on e-Science (e-Science), pp. 59–68. IEEE (2017)
5. Burt, P., Adelson, E.: The Laplacian pyramid as a compact image code. IEEE Trans. Commun. **31**(4), 532–540 (1983)
6. Carl Zeiss AG: ZEISS MultiSEM Research Partner Program (10 2018), the World's Fastest Scanning Electron Microscopes, October 2018
7. Chard, R., et al.: funcX: a federated function serving fabric for science. In: Proceedings of the 29th International Symposium on High-Performance Parallel and Distributed Computing, pp. 65–76 (2020)
8. Derpanis, K.G.: The gaussian pyramid (2005)
9. Duits, R., Florack, L., De Graaf, J., ter Haar Romeny, B.: On the axioms of scale space theory. J. Math. Imaging Vis. **20**(3), 267–298 (2004)
10. Herculano-Houzel, S., Watson, C.R., Paxinos, G.: Distribution of neurons in functional areas of the mouse cerebral cortex reveals quantitatively different cortical zones. Front. Neuroanatom. **7** (2013)
11. Hua, Y., Laserstein, P., Helmstaedter, M.: Large-volume en-bloc staining for electron microscopy-based connectomics. Nat. Commun. **6**(1), 1–7 (2015)
12. Kasthuri, N., et al.: Saturated reconstruction of a volume of neocortex. Cell **162**(3), 648–661 (2015)
13. Koenderink, J.J.: The structure of images. Biol. Cybern. **50**(5), 363–370 (1984)
14. von Laszeski, G., et al.: Real-time analysis, visualization, and steering of microtomography experiments at photon sources. Technical report, Argonne National Lab (2000)
15. Levental, M., et al.: Towards online steering of flame spray pyrolysis nanoparticle synthesis. In: IEEE/ACM 2nd Annual Workshop on Extreme-scale Experiment-in-the-Loop Computing (XLOOP), pp. 35–40. IEEE (2020)

16. Lindeberg, T.: Feature detection with automatic scale selection. Int. J. Comput. Vis. **30**, 79–116 (2004)
17. Liu, Z., et al.: Bridging data center AI systems with edge computing for actionable information retrieval. In: 3rd Annual Workshop on Extreme-scale Experiment-in-the-Loop Computing (XLOOP), pp. 15–23. IEEE (2021)
18. Luo, Y., Huang, L., Rivenson, Y., Ozcan, A.: Single-shot autofocusing of microscopy images using deep learning. ACS Photon. **8**(2), 625–638 (2021)
19. Marr, D., Hildreth, E.: Theory of edge detection. Proc. Royal Soc. London. Ser. B. Biol. Sci. **207**(1167), 187–217 (1980)
20. Merrill, D.: Cuda unbound (cub). https://github.com/NVIDIA/cub (2021)
21. Neubeck, A., Van Gool, L.: Efficient non-maximum suppression. In: 18th International Conference on Pattern Recognition, vol. 3, pp. 850–855. IEEE (2006)
22. Pan, J., Libera, J.A., Paulson, N.H., Stan, M.: Flame stability analysis of flame spray pyrolysis by artificial intelligence. Int. J. Adv. Manuf. Technol. 2215–2228 (2021). https://doi.org/10.1007/s00170-021-06884-z
23. Potluri, S., Wang, H., Bureddy, D., Singh, A.K., Rosales, C., Panda, D.K.: Optimizing MPI communication on multi-GPU systems using CUDA inter-process communication. In: 26th International Parallel and Distributed Processing Symposium Workshops PhD Forum, pp. 1848–1857 (2012)
24. Redondo, R., et al.: Autofocus evaluation for brightfield microscopy pathology. J. Biomed. Opt. **17**(3), 1–9 (2012)
25. Stevanovic, U., et al.: A control system and streaming DAQ platform with image-based trigger for x-ray imaging. IEEE Trans. Nucl. Sci. **62**(3), 911–918 (2015)
26. Wildenberg, G.A., Rosen, M.R., Lundell, J., Paukner, D., Freedman, D.J., Kasthuri, N.: Primate neuronal connections are sparse in cortex as compared to mouse. Cell Rep. **36**(11), 109709 (2021)
27. Yeo, T., Ong, S., Jayasooriah, Sinniah, R.: Autofocusing for tissue microscopy. Image Vis. Comput. **11**(10), 629–639 (1993)
28. Sun, Y., Duthaler, S., Nelson, B.J.: Autofocusing algorithm selection in computer microscopy. In: International Conference on Intelligent Robots and Systems, pp. 70–76 (2005)

Models and Metrics for Mining Meaningful Metadata

Tyler J. Skluzacek[1(✉)], Matthew Chen[2], Erica Hsu[3], Kyle Chard[1,4],
and Ian Foster[1,4]

[1] University of Chicago, Chicago, IL, USA
skluzacek@uchicago.edu
[2] University of Illinois at Urbana-Champaign, Champaign, IL, USA
[3] Carnegie Mellon University, Pittsburgh, PA, USA
[4] Argonne National Lab, Lemont, IL, USA

Abstract. The increasing volume and variety of science data has led to
the creation of metadata extraction systems that automatically derive
and synthesize relevant information from files. A critical component of
metadata extraction systems is a mechanism for mapping extractors—
lightweight tools to mine information from a particular file types—to
each file in a repository. However, existing methods do little to address
the heterogeneity and scale of science data, thereby leaving valuable data
unextracted or wasting significant compute resources applying incorrect
extractors to data. We construct an extractor scheduler that leverages file
type identification (FTI) methods. We show that by training lightweight
multi-label, multi-class statistical models on byte samples from files, we
can correctly map 35% more extractors to files than by using libmagic.
Further, we introduce a metadata quality toolkit to automatically assess
the utility of extracted metadata.

Keywords: Metadata quality · Extraction · File type identification

1 Introduction

The many files accumulated within science and engineering organizations may,
both individually and collectively, contain data of great value. However, poor
organization and inadequate documentation frequently make these files difficult for users to navigate. In order to promote repository navigability, metadata
extraction systems [5,11,12,18,23] have been developed to automatically populate rich, searchable data catalogs. Metadata extraction systems generally follow
a common structure, as illustrated in Fig. 1, in which the following steps are performed in order: (A) iterate over all files in a repository; (B) identify the *type(s)*
of each file (e.g., free text, tabular, image); (C) invoke one or more *extractors*
(sometimes called *parsers*) on each file to obtain metadata; and (D) perform an
action with the resulting metadata (e.g., load a search index). However, different metadata extraction systems focus on different use cases, data types, and
communities, and therefore apply different approaches for each stage.

D. Groen et al. (Eds.): ICCS 2022, LNCS 13350, pp. 417–430, 2022.
https://doi.org/10.1007/978-3-031-08751-6_30

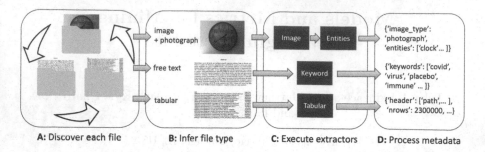

A: Discover each file   B: Infer file type   C: Execute extractors   D: Process metadata

Fig. 1. Automated metadata extraction steps: (A) find all files in repository, (B) infer each file's type such that it can be mapped to applicable extractors, (C) execute one or more extractors, (D) post-process metadata.

In our work we focus on extracting metadata from scientific data, an important step for making these complex data navigable and increasing data utility. While much prior research has focused on metadata extraction from personal and enterprise file collections, there is relatively little focus on scientific data, and in particular on the unique challenges posed by these data. For example, the broad nature of scientific inquiry leads scientists to store data in esoteric formats, without regard for schema or file extension; data are often encoded in multi-dimensional file formats that integrate various data types into single files, or spanning several files; and the rise of IoT and decentralized storage has led to data repositories being spread across disparate compute resources.

The growing volume and velocity of scientific data leads us to closely consider the resources used when extracting metadata. Naively applying all extractors to each file is not only inefficient, but may also lead to incorrect or irrelevant metadata. In Fig. 2, we illustrate execution times when exhaustively invoking a library of eight extractors on every file in the 428 000-file Carbon Dioxide Information Analysis Center (CDIAC) data set [2]. The figure shows that while most extractors fail quickly, significant compute time is wasted; we estimate that successful invocations consume 130 core hours, whereas applying incorrect extractors (e.g., a NetCDF extractor on a Python script) consumes 670 core hours while returning no valid metadata. When mapping files to extractors, even the most advanced extraction systems do little more than map a mimeType, extension, or byte-regex to a single extractor. However, when scientists create data in bespoke formats or store diverse data types within a single file, these modes of mapping extractors to files often fail.

In this paper, we present an intelligent extractor scheduler for the Xtract metadata extraction system [18] that bridges many of the challenges in applying extractors to science data. While our prior work has focused on issues of scale and decentralization [18], we focus here on directly addressing file diversity by leveraging prior research in file type identification (FTI). We construct statistical learning models that, when used as part of our scheduler, can prioritize the application of extractors to collections of files; thereby maximizing the metadata

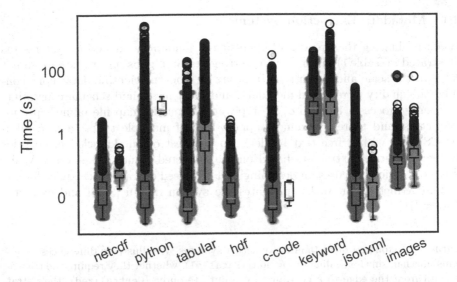

Fig. 2. Box plots (and point distributions) for the invocation of each extractor to each file in CDIAC. The green plots are file-extractor invocations that successfully yield metadata; red are failed invocations (i.e., wasted core hours). (Color figure online)

information obtained. Further, we evaluate the efficacy of these methods via a set of automatically derived metadata quality metrics. The contributions of our work are:

- Parameterization and evaluation of FTI methods for metadata extraction.
- Comparative evaluation that shows that our models outperform a state-of-the-art tool (libmagic [1]) in mapping extractors to files by 35%.
- Application of FTI methods on two large, uniquely-diverse scientific data repositories: the heterogeneous Carbon Dioxide Information Analysis Center (CDIAC) and the homogeneous COVID-19 Open Research Dataset (CORD) [24].
- An automated metadata quality analysis toolkit capable of evaluating extracted metadata.

The remainder of this paper is as follows. Section 2 presents related work in extraction systems and FTI. Section 3 outlines automated metadata quality metrics. Section 4 presents our algorithms, learning models, and quality metrics to be evaluated. Section 5 contains the evaluation of our work on two uniquely diverse scientific data repositories. Finally, Section 6 summarizes our contributions.

2   Related Work

In this section, we review related work in metadata extraction systems and FTI.

2.1 Metadata Extraction Systems

When evaluating the breadth of open-source metadata extraction systems (as illustrated in Table 1), we observe recurring research gaps: most systems do not cater to the scale and decentralized nature of modern scientific data; none consider the quality of returned metadata; and most have rigid schema constraints (i.e., only process a handful of file types) or manually map file mimeTypes for extractors, and therefore cannot support files of multiple types (e.g., a tabular CSV file with a free text header). To the best of our knowledge, no prior system prioritizes extractors based on the expected value of metadata. While this work strictly focuses on designing an FTI-based extractor scheduler for our system Xtract, prior work illuminates the system design [3,18] and extractor library [19].

Table 1. Taxonomy of metadata extraction systems. We illustrate differences in systems' mechanisms for scaling extractions (**Parallel**), whether they require the transfer of data from the edge to a centralized compute resource (**Centralized**), their strategy for mapping extractors to files (**Mapping**), whether they provide quality metrics for automatically extracted metadata (**Quality**), and the supported science domains (**Domains**).

System	Parallel	Central	Mapping	Quality	Domain
Tika [12]	Threads	No	extension, mimeType, byte-matches	None	general
Clowder [11]	Cloud	Yes	mimeType	None	general
BDQC [5]	None	Yes	input schema	None	biomedicine
Constellation [23]	Cloud	Yes	input schema	None	general
ScienceSearch [16]	Cluster	Yes	input schema	None	microscopy
Xtract	Cluster, Cloud	No	FTI	Yes	general

2.2 File Type Identification

File type identification (FTI) aims to automatically classify files from inscribed physical contents and is commonly used in digital forensics [15] and malware detection [21]. FTI methods traditionally rely on easily-attainable features from the file (bytes, extension, size). However, science data creates unique challenges as file creators do not adhere to common file extensions, mimeTypes, or schema [20].

FTI methods are crucial to metadata extraction for two reasons: (1) only extractors yielding metadata should be executed on a file, and (2) the multi-output nature of statistical learning algorithms corresponds well with multi-typed files. We consider significant prior work in evaluating and selecting features and models that may work well for the extractor mapping problem [6,9,13]. We evaluate the performance of our own file type identification methods with metrics from prior FTI work (i.e., train time, precision, recall, F1) and metadata quality metrics, as outlined in the following section.

3   Metadata Quality Determination

Ultimately, the goal of metadata extraction systems is to derive useful metadata; however, current extraction systems do not consider the utility of extracted metadata for either individual files or entire data collections. Metadata quality metrics are thus necessary to illuminate the value of applying a given extractor to a file, and by extension, enables us to evaluate the efficacy of FTI methods and extraction systems. While there is some prior work in metadata quality metrics [8], we specifically seek out metrics to automatically quantify the utility of a metadata corpus. We identify the following metrics that measure various dimensions of utility: yield, completeness, entropy, and readability.

Yield. Metadata yield is the total amount of metadata, measured as the number of bytes of metadata produced. While an imperfect measure, yield is useful for understanding the context of the other metrics, and is easy to obtain. For instance, how do 5 "readable" bytes compare to 1000 that are less readable?

Completeness. Metadata are complete if they contain all possible attributes that could be obtained. In practice, and especially in the presence of diverse schema, some metadata attributes may be left empty. The simplest completeness metric [14] simply divides the number of metadata elements by the total number of elements that could be obtained (i.e., a percentage). We call this metric *simple_completeness* and define it in Eq. (1), where N is the number of attributes and $P(i)$ is 0 if the ith metadata attribute is null, and 1 otherwise:

$$simple\_completeness = \sum_{i=1}^{N} \frac{P(i)}{N} * 100 \tag{1}$$

Other researchers [7] have created weighted versions of completeness to account for some attributes exhibiting higher semantic importance than others. They enable multi-tiered importance by including in their completeness score an "absence" penalty for missing metadata elements, where a higher weight corresponds to subjectively more-relevant attributes. Even further, others have accounted for the weighted importance of values in hierarchical attributes [10]. While the aforementioned measures of completeness arguably better represent a human's subjective view of "completeness", it is difficult to have humans manually provide weights, primarily due to the propensity of users to bias higher weights onto items they personally correlate to value [4]. While we plan to properly address (and de-bias) weighting strategies via user study in future work, we use *simple_completeness* as a sufficient and fair proxy-measure in this paper.

Entropy. Metadata entropy [17] is the degree to which metadata presents information that is different from other metadata. A common approach is to apply Term Frequency-Inverse Document Frequency (TF-IDF) to determine the entropy of a metadata document. TF-IDF for a metadata document provides an importance score for all words in a document, relative to all metadata documents in the corpus. Scientists [14] have proposed a score built on TF-IDF that produces an entropy score for a metadata document as is shown in Eq. (2), where N is the number of text attributes, $attribute_i$ the ith attribute of metadata, and $sum\_tf(attribute_i)$ the sum of TF-IDF scores for a given attribute (in a document):

$$entropy = \log(\sum_{i=1}^{N} sum\_tf(attribute\_i)) \qquad (2)$$

Readability. Readability measures the ability of humans to semantically interpret metadata. In this work, we leverage the Flesch Index [22]—a document score that compounds the complexity of words and sentences onto a 0–100 scale where documents scoring near 0 are unintelligible to most human readers and those scoring near 100 are broadly understood. For metadata documents in a search index, we ideally give higher semantic weight to metadata containing searchable words, thereby penalizing number-dominated metadata. To accomplish this, we weight the Flesch index by the proportion of characters (n_char) that are not numbers (n_num): $W_s = (1 - \frac{n\_num}{n\_char})$. To account for decimal points potentially misrepresenting the ends of sentences in numeric metadata, we remove all mid-numeric decimal points prior to tokenizing. We then define our weighted Flesch index $WFlesch$, where n_word, n_sent, n_syl are the number of words, sentences, and syllables, as follows:

$$WFlesch = \overbrace{(206.835 - 1.015(\frac{n\_word}{n\_sent}) - 84.6(\frac{n\_syl}{n\_word}))}^{\text{original Flesch Index}} * W_s \qquad (3)$$

4   Methodology

We now describe our process for using statistical learning models to identify applicable extractors for each file in a science repository. Specifically, we describe how we label data, generate features, select models, and leverage model outputs as input to the extraction scheduler.

Label Generation. We first create a library of ground-truth labels for all files in both science repositories. To this end, we exhaustively apply each extractor to every file, and record (1) the metadata returned by the extractor, and (2) the time taken to execute the extractor. The file is assigned a label for each extractor that, when applied to it, returns nonempty metadata. If a file receives no labels, we use heuristic methods to determine whether a file might be compressed, a binary executable, or empty—and if none of those—it receives a label of "unknown." For supervised model training, we place each possible type label for a given file $f \in F$ into a label vector $L(f) = [is(f, t_0), is(f, t_1)...is(f, t_m)]$ where $is(f, t_i)$ is 1 if f is of type $t \in T$, and 0 if not.

Feature Selection. We create input feature vectors containing (i) file size and (ii) 16–512 byte samples from the file. As illustrated in Fig. 3, we fetch byte

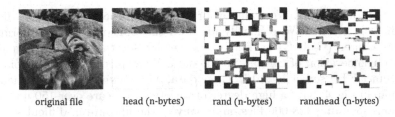

original file   head (n-bytes)   rand (n-bytes)   randhead (n-bytes)

Fig. 3. Feature illustration for head, rand, and randhead.

samples from the following locations in the file: the header (head), randomly throughout (rand), or a combination of both (randhead).

Model Selection. We train models to accomplish the following: given a file $f \in F$, we want to train a model $m \in M$ such that $m(f)$ generates a probability distribution $P(f) = [p(f, e_1), p(f, e_2), ..., p(f, e_n)]$, where $p(f, e)$ is the probability that f should map to extractor $e \in E$. We explore multiple statistical learning models: logistic regression (logit), random forests (rf), and support vector classification (svc). We evaluate model performance primarily in the form of F1 score (which measures overall model performance) and recall (the percentage of correct file-extractor mappings identified) on weighted multi-class probability distributions generated by each model. We also consider model training time. To account for potential overfitting, we evaluate models on both imbalanced (all data) and balanced (subset of the data) classes.

Extraction Scheduler. Our primary goal is to design an extraction system that converts FTI model outputs into a schedule of file/extractor pairs to execute. We first train lightweight regressions that use a file's size to predict metadata yield $Y(e, size(f))$ and extraction time $T(e, size(f))$. We select our regression based on which has the better correlation score between a linear and nonlinear model [25], and fit the corresponding model. Given our probability vector of file-extractor mappings, $P(f) = [p(f, e_1), p(f, e_2), ..., p(f, e_n)]$, the size of a file $size(f)$, and a +1 Laplace smoothing constant, we introduce an objective function to compute predicted metadata yield over time $\alpha(f, e)$:

$$\alpha(f, e) = log(\frac{Y(e, size(f)) * p(f, e) + 1}{T(e, size(f)) + 1}) \qquad (4)$$

We prioritize extractor execution by loading a priority queue in descending order of alpha score (i.e., the system maximizes expected metadata yield over time).

5 Evaluation

We analyze the feature and model performance of the FTI methods and compare with libmagic [1]. We then examine metadata quality when using our scheduler.

Science Repositories. We evaluate our approach in the context of two distinct scientific repositories. We primarily focus on the Carbon Dioxide Information Analysis Center (CDIAC) a climate science dataset that represents a multi-group conglomeration of carbon dioxide data. We copied these data from their now-defunct FTP server in 2017. These data, whose extensions we visualize as a treemap in Fig. 4, have a high degree of variety—there are over 150 unique file extensions spanning 428 000 files, and many of the files are in difficult-to-parse formats (e.g., deprecated Windows installers, Hadoop error logs, and desktop shortcuts) [20]. We include both the unedited file formats consisting of many compressed files (e.g., .Z), and also the decompressed contents. We also examine the more-homogeneous COVID-19 Open Research Dataset (CORD) containing 517 000 JSON-formatted COVID-19 research papers spanning 2019–2021.

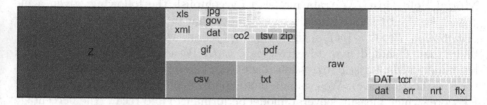

Fig. 4. Treemaps of CDIAC: (left) the unedited repository, (right) all decompressed files. Each box's *area* is the proportion of files of that extension, and *darkness* is the relative total size (darker=bigger). The orange box on the right represents files with no extension. (Color figure online)

Experimental Testbed. We perform our experiments on ALCF Theta, an 11.7-petaflop Cray XC40 supercomputer with second-generation Intel Xeon Phi "Knight's Landing" (KNL) processors. Each node has a 64-core processor and 166 GB MCDRAM, 192 GB DDR4 RAM, with a shared a Lustre file system.

5.1   FTI Modeling

Features. We first want to find the best byte structure (head, rand, randhead) and number of bytes (16–512) to use as features in our analysis. For all experiments, we use a standard 70%/30% train/test split. Figure 5 shows the range of model scores for the different byte structures on each of the three model types: logit, rf, and svc. We see that, for the CDIAC data, the head bytes outperform rand and randhead in every statistical metric. To investigate whether there is significant benefit beyond 512 head bytes, we compare F1 improvements when doubling from 16 to 32, and 256 to 512, respectively. The relative F1 difference when increasing from 16 to 32 bytes for (logit, rf, svc) is (+1.0, +0.6, +0.7), but the difference between 256 and 512 bytes is only (+0.2,+0.1,-1.9). Therefore, we use 512 head bytes, since additional bytes would likely have marginal benefit. As shown in Table 2, CORD can be processed well by any of the feature configurations.

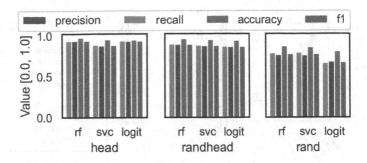

Fig. 5. Model scores for multiple 512-byte feature configurations (CDIAC).

Table 2. Model performance for 16 and 512 bytes for logistic regression (logit), random forests (rf), and support vector classifier (svc) on CDIAC and CORD

Repository	Header bytes	Model	Train time (s)	Precision	Recall	F1 Score
CDIAC		logit	403	0.839	0.836	0.837
	16	rf	2.29	0.890	0.896	0.893
		svc	1010	0.856	0.867	0.861
		logit	1140	0.930	0.936	0.933
	512	rf	4.50	0.939	0.938	0.938
		svc	9240	0.875	0.885	0.880
CORD		logit	17.0	1.00	1.00	1.00
	16	rf	3.56	1.00	1.00	1.00
		svc	418	1.00	1.00	1.00
		logit	183	1.00	1.00	1.00
	512	rf	4.25	1.00	1.00	1.00
		svc	464	1.00	1.00	1.00

Models. To avoid overfitting, we train our models on both imbalanced and balanced classes in CDIAC. The imbalanced class (all-of-CDIAC) experiments shown in Fig. 6 shows that the 512B random forests model can adequately identify most file types in the CDIAC repository. The confusion matrix shows that the model can effectively identify the top hierarchical type for a file, and the multiclass-weighted PR curve shows that, overall, we see high recall, regardless of precision, for each label type. Interestingly, one can see in both diagrams that the most-difficult class for the model to identify is the "unknown" class.

As the imbalanced model is likely overfit to the larger classes, we propose an experiment on balanced classes; we create a balanced subset of CDIAC by randomly selecting 200 files from each class (and omitting those classes with fewer than 200 files). In Fig. 7, the confusion matrix shows that our model can efficiently select the top type of a file, whereas the multiclass-weighted PR curve shows that we see fairly high precision at all levels of recall. Interestingly in the

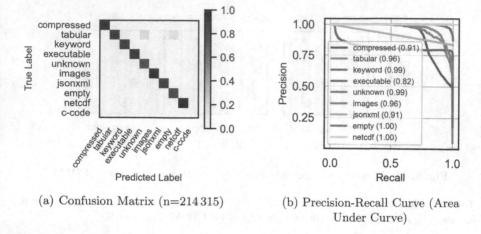

(a) Confusion Matrix (n=214 315)   (b) Precision-Recall Curve (Area Under Curve)

Fig. 6. Imbalanced Classes (CDIAC): confusion matrix (prediction-normalized) and precision-recall curve (multi-class weighted) for random forests model trained on 512 head bytes.

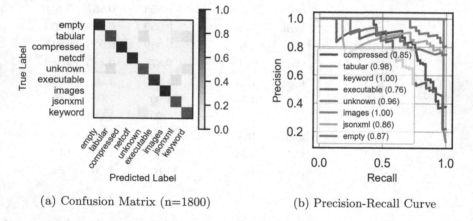

(a) Confusion Matrix (n=1800)   (b) Precision-Recall Curve

Fig. 7. Balanced Classes (CDIAC): confusion matrix (prediction-normalized) and precision-recall curve (multi-class weighted) for random forests model trained on 512 head bytes.

balanced case, we see that the model has added difficulty in selecting tabular files, likely explained by the high overlap of files correctly mapping to multiple extractors.

Finally, to study how the model performs when individual files contain multiple content types, we analyze the output probability distributions for all multi-typed files, and investigate whether each type is represented at the top of the probability distribution. In Table 3, we observe that, for CDIAC, most multi-typed files share a type with the keyword extractor, and the model identifies the keyword type 80% of the time and the other type 96% of the time.

Table 3. Analysis of files of both types Type 1 and Type 2, and how many of each type are included in the top-2 entries of the probability distribution.

Repository	Type 1	Type 2	Count	Type 1 Included	Type 2 Included
CDIAC	keyword	tabular	12 878	10 966	12 415
		jsonxml	3282	1954	3109
		netcdf	252	205	252
		c-code	8	8	3
		python	3	3	3
	tabular	python	7	7	7
CORD	jsonxml	keyword	517 900	517 900	517 900

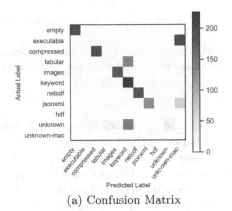

Type	Pr.	Re.	F1
empty	1.00	1.00	1.00
executable	0.00	0.00	0.00
compressed	0.98	1.00	0.99
tabular	0.25	0.06	0.10
images	0.96	0.96	0.96
keyword	0.44	0.91	0.59
netcdf	1.00	0.97	0.98
jsonxml	0.96	0.65	0.78
unknown	0.84	0.13	0.23
unkn.-mac	0.00	0.00	0.00

(a) Confusion Matrix

(b) Precision, Recall, F1

Fig. 8. Libmagic (CDIAC): confusion matrix and performance metrics for mapping extractors to files using the libmagic FTI tool.

Libmagic Comparison. We next compare our approach to the libmagic FTI tool. As libmagic types do not directly map to our extractor library, we manually map libmagic outputs to our types. Some mappings are obvious (e.g., empty:empty, compress'd:compressed) while others required consulting libmagic documentation (e.g., data:unknown). We compare each libmagic output to our extractor labels, and show the result in Fig. 8. Overall, libmagic performs significantly worse than our FTI methods, as it consistently misclassifies tabular and keyword data. Even in this favorable experiment, libmagic only accurately identifies 65% of files (cf. our FTI methods correctly identify 88%).

5.2   Extractor Scheduler and Metadata Analysis

We next evaluate the extraction patterns and metadata output over time when the scheduler uses the predicted probability vectors for each file. Figure 9a shows the extractor executions over all possible file-extractor pairs. Given that the scheduler prioritizes metadata yield over time, we notice that extractors that either succeed quickly (jsonxml, images) or those that produce large quantities of metadata (tabular) have initial spikes within the first 10% of invocations.

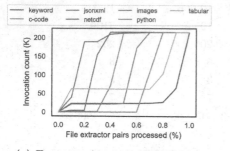

(a) Extractor invocations over time

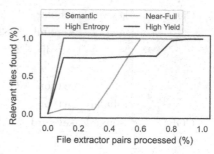

(b) Relevant files found over time

Fig. 9. Scheduler Analysis: (a) extractor invocations over percentage of file-extractor pairs processed; (b) percentage of total quality files discovered over all file-extractor pairs processed.

To measure the relative utility of metadata extraction, we count "useful" metadata documents as defined by the quality metrics: `semantic` metadata are those that contain searchable words (measured as files with nonzero readability scores), `near-full` metadata contain complete data (measured as files with over 50% completeness), `high entropy` metadata add unique information to the corpus (measured as files with nonzero entropy), and `high yield` metadata exceed 500 bytes. Figure 9b shows that maximizing yield over time naturally prioritizes extracting semantically searchable and high entropy metadata. Intuitively this makes sense as semantic metadata are often larger than average (containing many words), and therefore provide high yield. Therefore, we see that this scheduler could add significant value for organizations looking to create a semantically searchable index with high information content on a limited compute budget.

For purposes of illustrating potential quality bias across extractors, we illustrate in Fig. 10 the observed metrics of all CDIAC metadata. We observe that different extractors generate unique quality profiles: tabular metadata exhibit high readability and entropy, keyword metadata exhibit high readability, and hdf and image metadata exhibit high completeness. In future work, we will study how extractor quality profiles can be used in an extraction scheduler.

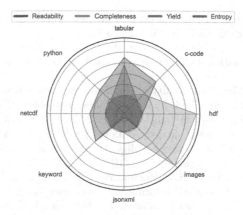

Fig. 10. Spider plot representation of successful metadata extraction metrics on CDIAC. The distance between the center and each extractor is the log-scale range of each metric, and the placement of the colored line represents the median of the corresponding metric. (Color figure online)

6   Conclusion

Accurate and performant metadata extraction is dependent on accurate methods for mapping extractors to files; however, traditional methods are not conducive to the wide, heterogeneous variety of science file formats. We introduce several file type identification methods that use lightweight byte-features from files and various machine learning models to predict scientific file types. We use these models to create a scheduler for the Xtract metadata extraction system, enabling Xtract to prioritize application of extractors to files. Further, we introduce several metrics designed to quantify the utility of metadata, and by extension, the usefulness of the extractor scheduler.

Acknowledgements. We gratefully acknowledge Takuya Kurihana (University of Chicago) for sharing his machine learning expertise. This work is supported in part by the National Science Foundation under Grants No. 2004894 and 1757970, and used resources of the Argonne Leadership Computing Facility.

References

1. Libmagic(3) - linux man page, November 2009. https://linux.die.net/man/3/libmagic
2. Cdiac, March 2018. https://cdiac.ess-dive.lbl.gov/
3. Chard, R., et al.: Funcx: a federated function serving fabric for science. In: Proceedings of the 29th International Symposium on High-Performance Parallel and Distributed Computing, pp. 65–76 (2020)
4. Deng, H., Runger, G., Tuv, E.: Bias of importance measures for multi-valued attributes and solutions. In: International Conference on Artificial Neural Networks, pp. 293–300 (2011)

5. Deutsch, E.W., et al.: BDQC: a general-purpose analytics tool for domain-blind validation of big data. bioRxiv 258822 (2018)
6. Gopal, S., Yang, Y., Salomatin, K., et al.: Statistical learning for file-type identification. In: International Conference on Machine Learning and Applications, pp. 68–73 (2011)
7. Hughes, Baden: Metadata quality evaluation: experience from the open language archives community. In: Chen, Zhaoneng, Chen, Hsinchun, Miao, Qihao, Fu, Yuxi, Fox, Edward, Lim, Ee.-peng (eds.) ICADL 2004. LNCS, vol. 3334, pp. 320–329. Springer, Heidelberg (2004). https://doi.org/10.1007/978-3-540-30544-6_34
8. Király, P.: Measuring Metadata Quality. Ph.D. thesis, Georg-August-Universität Göttingen, June 2019. https://doi.org/10.13140/RG.2.2.33177.77920
9. Li, W.J., Wang, K., Stolfo, S.J., Herzog, B.: Fileprints: identifying file types by n-gram analysis. In: IEEE SMC Information Assurance Workshop, pp. 64–71 (2005)
10. Margaritopoulos, M., Margaritopoulos, T., Mavridis, I., Manitsaris, A.: Quantifying and measuring metadata completeness. J. Am. Soc. Inform. Sci. Technol. **63**(4), 724–737 (2012)
11. Marini, L., Gutierrez-Polo, I., et al.: Clowder: open source data management for long tail data. In: Practice and Experience on Advance Research Computing (2018)
12. Mattmann, C., Zitting, J.: Tika in Action. Manning Publications Co., USA (2011)
13. McDaniel, M., Heydari, M.H.: Content based file type detection algorithms. In: 36th Annual Hawaii Int'l Conference on System Sciences, pp. 10-pp. IEEE (2003)
14. Ochoa, X., Duval, E.: Automatic evaluation of metadata quality in digital repositories. Int. J. Digit. Lib. 67–91 (2009). https://doi.org/10.1007/s00799-009-0054-4
15. Poisel, R., Tjoa, S.: A comprehensive literature review of file carving. In: Int'l Conference on Availability, Reliability and Security, pp. 475–484. IEEE (2013)
16. Rodrigo, G., Henderson, M., et al.: ScienceSearch: enabling search through automatic metadata generation. In: 14th International Conference on e-Science, pp. 93–104 (2018)
17. Shannon, C.E.: A mathematical theory of communication. ACM SIGMOBILE Mob. Comput. Commun. Rev. **5**(1), 3–55 (2001)
18. Skluzacek, T., et al.: A serverless framework for distributed bulk metadata extraction. In: 30th International Symposium on High-Performance Parallel and Distributed Computing (2021)
19. Skluzacek, T.J., : Serverless workflows for indexing large scientific data. In: Int'l Workshop on Serverless Computing, pp. 43–48 (2019)
20. Skluzacek, T.J., et al.: Skluma: an extensible metadata extraction pipeline for disorganized data. In: IEEE 14th International Conference on e-Science, pp. 256–266. IEEE (2018)
21. Tabish, S.M., Shafiq, M.Z., Farooq, M.: Malware detection using statistical analysis of byte-level file content. In: ACM SIGKDD Workshop on CyberSecurity and Intelligence Informatics, pp. 23–31 (2009)
22. Talburt, J.: The Flesch index: An easily programmable readability analysis algorithm. In: International Conference on Systems Documentation, pp. 114–122 (1986)
23. Vazhkudai, S.S., Harney, J., et al.: Constellation: a science graph network for scalable data and knowledge discovery in extreme-scale scientific collaborations. In: IEEE International Conference on Big Data, pp. 3052–3061 (2016)
24. Wang, L.L., Lo, K., et al.: Cord-19: The covid-19 open research dataset. arXiv:2004.10706 (2020). https://doi.org/10.48550/ARXIV.2004.10706
25. Wang, Y., Li, Y., et al.: Efficient test for nonlinear dependence of two continuous variables. BMC Bioinform. **16**(1), 1–8 (2015)

HNOP: Attack Traffic Detection Based on Hierarchical Node Hopping Features of Packets

Jinbu Geng[1,2], Zhenyu Cheng[1(✉)], Zhicheng Liu[1,3], Shuhao Li[1,2], and Rui Qin[1]

[1] Institute of Information Engineering, Chinese Academy of Sciences, Beijing, China
{gengjinbu,chengzhenyu,lishuhao,qinrui}@iie.ac.cn
[2] School of Cyber Security, University of Chinese Academy of Sciences, Beijing, China
[3] National Computer Network Emergency Response Technical Team/Coordination Center of China, Beijing, China
liuzhicheng@cert.org.cn

Abstract. Single packet attack, which is initiated by adding attack information to traffic packets, pose a great threat to cybersecurity. Existing detection methods for single packet attack just learn features directly from single packet but ignore the hierarchical relationship of packet resources, which trends to high false positive rate and poor generalization. In this paper, We conduct an extensive measurement study of the realistic traffic and find that the hierarchical relationship of resources is suitable for identifying single packet attacks. Therefore, we propose HNOP, a deep neural network model equipped with the hierarchical relationship, to detect single packet attacks from raw HTTP packets. Firstly, we construct resource node hopping structure based on the "Referer" field and the "URL" field in HTTP packets. Secondly, hopping features are extracted from the hopping structure of the resource nodes by G_BERT, which are further combined with the lexical features extracted by convolution operation from each node of the structure to form feature vectors. Finally, the extracted features are fed to a classifier, mapping the extracted features to the classification space through a fully connected network, to detect attack traffic. Experiments on the publicly available dataset CICIDS-2017 demonstrate the effectiveness of HNOP with an accuracy of 99.92% and a false positive rate of 0.12%. Furthermore, we perform extensive experiments on dataset IIE_HTTP collected from important service targets at different time. At last, it is verified that the HNOP has the least degraded performance and better generalization compared to the other models.

Keywords: Deep learning · Malicious traffic detection · Hopping features · Hierarchical relationship

1 Introduction

The packet is the basic unit of information transmission in TCP/IP protocol. Single packet attack (also called "atomic attack") is widely abused by attackers for information theft, causing serious losses to companies and individuals. According to the Akamai 2020 Security Report [1], there are 662 million web application layer attacks against the financial services sector and 7 billion against vertical businesses from 2018 to 2019. SQL injection, from all verticals, accounted for more than 72% of all attacks during this period. Therefore, it is crucial to effectively detect single packet attack.

Existing attack traffic detection methods for single packet attack are classified into flow-based and packet-based. Typically, a flow is defined as the set of packets transmitted between two IP addresses on a pair of specific ports using a specific protocol [12]. In the HTTP protocol, flow refers to the set of packets generated during the entire HTTP session. The flow-based detection [14,21] first accumulates multiple packets into a specific flow and then extracts the flow features (inflow and outflow ratio, packet time interval, stream size, number of packets per unit time, etc.), which tends to have a high accuracy. However, the demand for packet accumulation makes it have poor real-time performance. Packet-based detection [17,22] is highly efficient due to extracting traffic features (number of special characters, packet length, URL length, request method, payload information, etc.) directly from the single packet, but tends to have low accuracy due to information loss from inadequate feature extraction. Besides, the use of obfuscation techniques [10,13,15] makes it more difficult to extract lexical features from the attack traffic. Consequently, how to mine more effective features on packets is a crucial topic for packet-based detection.

As deep learning becomes more sophisticated, its effectiveness in extracting features is increasingly recognized by the academic community [7]. More and more attention is being paid to using deep learning methods to improve the effectiveness of feature extraction. Existing deep learning based attack traffic detection methods have made great progress in some cases, however there are two problems that need to be addressed. (1) Previous methods [19,20] almost extract lexical features from packets and neglect the hierarchical relational features, trend to high false positive rate. Hierarchical relational features, as intrinsic properties, expose the process of resource requests from the client to the server and can improve the performance of the detection method. (2) The redundant and invalid information in the packet leads to the lack of good generalization of the extracted features. Therefore, models [5,19] based on these features are also poorly generalized for practical applications and do not meet the needs of realistic network environments.

To address the above issues, we propose a model based on hierarchical node hopping features using deep learning to detect single packet attack effectively. Our insight is built on the fact that the difference in hierarchical relationship and hopping structure due to different attack methods and purposes has great gain in identifying attack traffic. After analyzing the resource request process, we first transform the "URL" field and "Referer" field in the HTTP packet header into a hierarchical node hopping structure. Then G_BERT are utilized to extract their lexical features and hopping features respectively, which are further used for classification after being pooled. Experimental results show that our model achieves effective results in single packet attack detection.

In summary, we make the following contributions.

- We analyze the hierarchical relationship of resources and verify its significance in detecting single packet attack traffic with ablation experiment.
- We transform the packet feature extraction problem for attack traffic into the hierarchical node hopping feature extraction, disregarding flow features and low-level protocols.
- We extract the retained hopping features from the reconstructed hierarchical node hopping structure by BERT and extract the lexical features from each node by convolution operation. Automatically extracted underlying features are difficult to modify and forge, improving the model's resistance to forgery.
- We implement a prototype of HNOP and perform a comprehensive evaluation of HNOP using CICIDS2017 [16] and IIE_HTTP datasets, demonstrating its effectiveness and generalization in detecting single packet attack.

The rest of this paper is outlined as follows. Section 2 reviews related work of HTTP attack detection. Section 3 presents our system scheme and details its different components. Section 4 elaborates the experimental results. Finally, we draw a conclusion in Sect. 5.

2   Related Work

The work in this paper is based on hierarchical node hopping features of packets, using a method based on deep learning for attack traffic detection. In this section, we present the related research work and detection methods.

Attack traffic detection methods based on packet features are explored in many previous works. These methods are primarily based on the idea of machine learning concept of designing a set of traffic features and then modelling and training. Liu *et al.* [11] propose a Bayesian statistical model with a network traffic time-slicing feature to detect attack traffic based on the rule that network traffic changes over time, i.e., network traffic changes at different time slices and some traffic does not occur at specific time slices. Swarnkar *et al.* [17] study a Naive Bayes classifier to detect suspicious payload content in network packets. Gezer *et al.* [6] use the random forest method to monitor banking Trojans and can achieve 99.95% accuracy. Vijayanand *et al.* [18] use a genetic algorithm to select features and detect novel attack traffic by multi-support vector machines. Han *et al.* [8] combine support vector machine (SVM) and cross-entropy to detect controlled network traffic. Kabir *et al.* [9] propose a sampling and least square support vector machine (LS-SVM) for attack traffic detection, which solves the problems that SVM-based methods are ineffective and require lengthy training time when the amount of data is too large. However, such detection methods rely on feature engineering. The uncertainty and limitations of the prior knowledge prevent good robustness and generalization of the detection model.

In order to overcome the long-standing problem of feature dependence, many researchers propose methods based on deep learning to automatically extract the optimal feature set from complex and redundant packets. Since features are automatically abstracted, the method can maintain good generalization as long as the data resources are satisfied [4], which is an advantage over other attack traffic detection methods. Wang *et al.* [20] use Convolutional Neural Network (CNN) and Recurrent Neural Network (RNN) to learn hierarchical spatial-temporal features. Spatial features are learned from each network packet by CNN and temporal features are drawn from the network packet sequence by RNN. Similarly, Wang *et al.* [19] use a one-dimensional convolutional layer to extract the spatial features of the original packets and a Gated Recursive Unit (GRU) structure to extract temporal features. Chiba *et al.* [3] utilize a hybrid optimization framework (IGASAA) based on Improved Genetic Algorithm (IGA) and Simulated Annealing Algorithm (SAA) to automatically build an efficient and effective Deep Neural Network (DNN) based attack traffic detection method. Geng *et al.* [5] transform the HTTP sessions into images and then extract the interactive features from these images by CNN.

To the best of our knowledge, there are little work on detecting HTTP attack traffic by leveraging the HTTP protocol. In addition, almost all previous work involves oversized packet contents and does not combine the attack characteristics of attack traffic to extract more effective features. This makes existing models, while performing well on public datasets, often perform poorly when used on realistic traffic with a time horizon. In this paper, we propose an attack traffic detection model based on hierarchical node hopping features of packets, which can get rid of the reliance on manual feature extraction and can improve the generalization performance of detecting attack traffic.

3   System Scheme

In this section, we build a hybrid structure neural network model, called HNOP. The architecture of the model is shown in Fig. 1, which is composed of hierarchical node hopping structure, G_BERT and classifier. The hierarchical node hopping structure is to transform the resource request process into a hopping problem between hierarchical nodes. G_BERT extracts features from the embedded layer and provides them to the classifier for malicious packet detection. These parts are detailed in the following sections.

3.1   Hierarchical Node Hopping Structure

The HTTP protocol defines the rules for requesting and responding to network resources between the network client and the network server. According to Occam's razor principle [2], we remove the fields in HTTP that are not linked to the resource request and keep only the "URL" field and the "Referer" field. Hence, we propose the HNOP model based on these two fields.

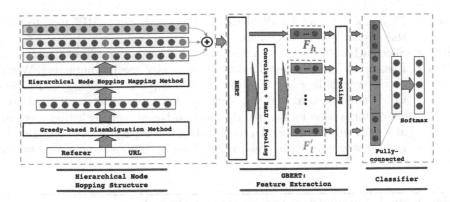

Fig. 1. The architecture of HNOP. (F_l': Lexical Feature; F_h: hopping feature).

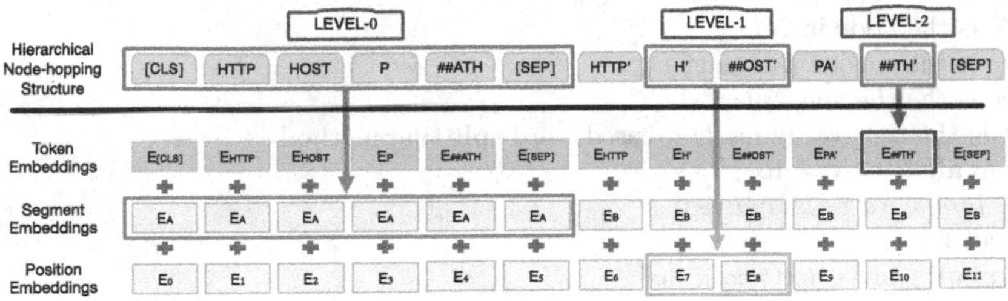

Fig. 2. Hierarchical node hopping mapping.

As shown in Fig. 2, we define the hierarchical nodes based on the entire process of the resource request. First, we define the "Referer" and "URL" fields as level-0 nodes, which characterize the page hopping relationship. Each sub-field of a level-0 node is defined as a level-1 node and is used to characterize the "host-path-resource" resource request process. Finally, the subscripts of the level-1 nodes obtained by the greedy-based disambiguation method are defined as level-2 nodes for representing specific resource information. After defining hierarchical nodes, we use a hierarchical node hopping mapping method to map the hierarchical node hopping structure to the three embedding layers. Greedy-based disambiguation method and hierarchical node hopping mapping method are presented as follows.

Greedy-Based Disambiguation Method. The purpose of greedy-based disambiguation method is to represent the HTTP resource request process more comprehensively and completely. The greedy-based disambiguation method is shown as Algorithm 1 and its steps are followed.

Step 1. Create a fixed-size vocabulary containing individual characters, common words, and sub-words that best fit the structure of the resource request.

Step 2. Combine the level-0 node "Referer" field with the "URL" field to form the original level-0 node hopping structure.

Step 3. Slice each subfield in the level-0 node into a separate level-1 node.

Step 4. Each level-1 node is firstly split into words by special characters. Words in the vocabulary are considered as independent level-2 nodes. For words not in the vocabulary, they are decomposed into sub-words and character tokens by the greedy method-based word splitting method, and each sub-word and character token is considered as a level-2 node.

Step 5. Repeat step 4 until all level-1 nodes are decomposed into level-2 nodes.

Algorithm 1. Greedy-based Disambiguation Method

Input: $ND \{N_1...N_k\}$
Output: $ND \{w_{1_1}...w_{1_m}...w_{k_1}...w_{k_l}\}$
1: N is the node in ND
2: sp is the special character
3: vb is the the vocabulary
4: f is the the greedy method-based word splitting method
5: **for all** $N \in ND$ **do**
6:    $longword \leftarrow N.split(sp)$
7: **end for**
8: **for all** $word \in longword$ **do**
9:    **if** $word \in vb$ **then**
10:       $w \leftarrow word$
11:    **end if**
12:    **if** $word \notin vb$ **then**
13:       $w \leftarrow f(word)[0]$
14:       **for all** $cutword \in f(word)[1:]$ **do**
15:          $w \leftarrow cutword.insert(0,'\#')$
16:       **end for**
17:    **end if**
18: **end for**
19: **return** $ND \{w_{1_1}...w_{1_m}...w_{k_1}...w_{k_l}\}$

With the greedy-based disambiguation method, we get a hierarchical semantics structure. Take the "URL" (http://ngo.mps.gov.cn/ngomh/userfiles/1/_thumbs/images/cms/article/2018/02/1-17-2banner.jpg) as an example. "URL" and the corresponding "Referer" (http://ngo.mps.gov.cn/ngo/portal/index.do?from=singlemessage&isappinstalled=0) are the level-0 nodes, and the relationship between level-0 nodes is used to represent the semantic relationship of the page. The "HOST" (`ngo.mps.gov.cn`) of the "URL" is used as a level-1 node, and the relationship between level-1 nodes is used to represent the semantic relationship of order. Each subfield(''ngo'', ''mps'', ''gov'', ''cn'') obtained by the greedy-based disambiguation method is used as a level-2 node. Different from the parent node, we care about the information of each level-2 node rather than the relationship between nodes.

Hierarchical Node Hopping Mapping Method. Hierarchical node hopping mapping method is based on three embedding operations and is used to map the hierarchical node hopping structure to the three embedding layers. As shown in Fig. 2, the nodes of each level correspond to different embedding layers. The segment embedding layer deals with level-0 nodes, maintaining the page hopping information. The position embedding layer deals with level-1 nodes, maintaining the "host-path-resource" hopping order. The token embedding layer deals with level-2 nodes, vectorizing each specific word.

The hierarchical node hopping structure of length n thus obtains three different vector representations, given as:

Token Embedding. t : (1, n+3, 768), a vector representation of level-2 nodes. Note that two special nodes are inserted at the beginning ([CLS]) of the hierarchical node hopping structure and at the end ([SEP]) of each level-0 node.

Position Embedding. p : (1, n+3, 768), a vector representation of the positions of the level-1 nodes.

Segment Embedding. s : (1, n+3, 768), a vector representation of the positions of the level-0 nodes.

These representations are summed by element to obtain a synthetic representation:

$$X = t + p + s \tag{1}$$

where X(1, n+3, 768) is the vectorized representation of the hierarchical node hopping structure.

3.2 G_BERT

As shown in Fig. 3, the features used in this paper are lexical features and hopping features. Lexical feature refers to the lexical feature of each level-2 node, and hopping feature refers to the association relationship between nodes at all levels. Association relationship include page hopping relationship and resource request process. That is, the hopping feature corresponds to the semantic features of level-1 nodes and level-0 nodes. In order to extract the above two features, G_BERT is divided into three layers. The upper layer uses a BERT network to extract the hopping features. The middle layer uses a convolution operation to extract lexical features. The lower layer is a feature pooling layer used to pool the features extracted from the above two layers into the final hierarchical node hopping features.

Hopping Feature Extraction. Hopping features are extracted by the BERT network, which is composed of l hidden layers with the same structure (same hyperparameters) but different parameters (no shared parameters). The hidden

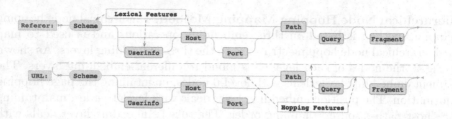

Fig. 3. Location of Lexical Feature and Hopping Feature throughout Resource Request Process. Lexical feature refers to the lexical feature of each level-2 node, and hopping feature refers to the association relationship between nodes at all levels.

layer includes two sub-layers, a multi-head attention function and a feed-forward function. Each sub-layer is accompanied by a residual connection.

The effect of the residual connection is to prevent the neural network gradient from disappearing or exploding in gradient. The residual connection makes the loss surface smoother, which makes the model easier to train and has deeper neural network layers. With residual connectivity, the output of the sub-layer is represented as:

$$X^{i+1} = f_{ln}(X^i + (f_{sn}(X^i))) \tag{2}$$

where X^i is the input vector, X^{i+1} is the output vector, f_{ln} is the layer normalization function for normalization along with the node embedding dimension, and f_{sn} is the current layer's operation function (multi-head attention or feed-forward).

The hopping feature is obtained by the following equation:

$$F_h = X_{CLS}^{2l+1} = X_0^{2l+1} \tag{3}$$

where F_h is the hopping feature, X^{2l+1} is the output vector of the l th hidden layer. F_h is represented as the vector of the [CLS] node in the last layer. X_0^{2l+1} has no obvious semantic information and is more "fairly" to integrate the semantic information of each node in the input than the other X_{OTHER}^{2l+1}. Thus, it is better to represent the overall semantics of the resource request.

Lexical Feature Extraction. For the node vectors N, we perform a convolution operation to extract the text features of the nodes themselves, i.e. lexical features F_l'. In this paper, F_l' are extracted as following:

$$F_l' = \sigma((F_l * K) + b) \tag{4}$$

where K is the convolution kernel. $*$ is convolution operation, b is the bias term and σ is the ReLU activation function. After the above series of operations, the lexical features (F_l') are generated.

Hierarchical Node Hopping Feature. Pooling strategy is added at the end of G_BERT, which is used for the integration of lexical features and hopping features. Three pooling strategies are adopted for comparison in this paper.

Concat Strategy F_{lh} is obtained by splicing lexical features directly after hopping features, for F'_l, F_h:

$$F_{lh} = Concat(F_h, F'_l) \tag{5}$$

Max Strategy. The maximum value of each dimension in F'_{li} and F_{hi} is taken to represent the feature vector F_{lh} by weighting, for F'_{li}, F_{hi}:

$$F_{lhi} = Max\,(F_{hi}, F'_{li}) = \begin{cases} F_{hi}\,, & F_{hi} > F'_{li} \\ F'_{li}\,, & F_{hi} \leqslant F'_{li} \end{cases} \tag{6}$$

Mean Strategy. Calculate the mean of F'_{li} and F_{hi} to represent the feature vector F_{lh}, for F'_{li}, F_{hi}:

$$F_{lhi} = Mean\,(F_{hi}, F'_{li}) = \frac{1}{2}\,(F_{hi} + F'_{li}) \tag{7}$$

3.3 Classifier

Based on the extracted features F_{lh}, we classify the samples using fully connection layer (Linear) and SoftMax. Fully connected layer maps n eigenvectors to K (sample labeling space) eigenvectors by multiplying the weight matrix with the input vectors, adding the bias. SoftMax maps K eigenvectors to K real numbers (probabilities) of $(0, 1)$ and make sure that their sum is 1. The details are as following:

$$Y = \text{SoftMax}(z) = \text{SoftMax}\left(W^T_{n \times K}\hat{X} + b\right) \tag{8}$$

$$\hat{X} = F_{lh} \tag{9}$$

where $\hat{X}$ is the input to the fully connected layer, $W_{n \times K}$ is the weight, b is the bias term. Based on Y, we finally obtain the class probability for each sample.

4   Evaluation

In this section, we evaluate the performance of the proposed model by performing various experiments on CICIDS2017 and IIE_HTTP. In particular, the experiments are intended to satisfy the following demands:

- Analyze the hierarchical relationship of resources and verify its significance with ablation experiment.
- Verify the effectiveness of the hierarchical node hopping feature extraction method by comparing with other methods on the publicly dataset CICIDS2017.
- Verify the generalization of HNOP on a realistic dataset IIE_HTTP by validation comparisons with other methods.

4.1   Data Sets

Under the premise of ensuring security and data privacy, we deploy traffic collection devices at gateways on realistic main protection targets through network operators. We collect about 60 GB of attack traffic data (containing Anti_sequence_attack, SQL_injection, Webshell_attack), as well as a considerable amount of normal data in June and July 2018.

After performing the irreversible desensitization operation to protect user privacy and eliminating the threat of attack data interfering with the online environment, we obtain a long-time dataset, called IIE_HTTP. The distribution of IIE_HTTP is shown in Table 1. In deep learning, each class in the training set maintains the same number by sampling, thus avoiding model training bias.

Table 1. Reprocessing results of the IIE_HTTP.

Categories	All		Train		Test	
	Count	Percentage(%)	Count	Percentage(%)	Count	Percentage(%)
White_sample	198878	19.99	102687	18.96	96191	21.21
Anti_sequence_attack	2496	0.25	1388	0.26	1108	0.24
SQL_injection	466444	46.88	298438	55.12	168006	37.05
Webshell_attack	234195	23.54	83465	15.41	150730	33.24

4.2   The Analysis of Hierarchical Relationship

We number the level-2 nodes in the "URL" from left to right. The hopping from node n to $n+1$ is defined as out_degree of node n and in_degree of node $n+1$. We analyze the hierarchical relationship of three attack types (Anti_sequence_attack, SQL_injection, Webshell_attack).

Table 2. The hierarchical relationship features of three attack types.

Feature	Description	Anti	SQL	Web
Average node	Average value of nodes	1.42	**4.96**	4.17
Maximum node	Maximum value of nodes	5	9	**11**
Key node	Nodes with large degree	0	**10**	8
Average key node	Average value of key nodes	0	**4.36**	3.26
Associated node	Key nodes with large out and in degree	0	0	**3**
Connected graph	Amount of graphs with weak connectivity	0	**10**	3

[a] *Anti: Anti_sequence_attack. SQL: SQL_injection. Web: Webshell_attack.*

As shown in Table 2, the results vary considerably. The explanation for the above phenomenon is that different attacks have different attack methods and attack purposes. SQL_injection occurs before the intrusion and Webshell_attack occurs after the intrusion. Anti_sequence_attack mainly targets reverse sequence vulnerabilities, so its attack "path" is determined.

In our model, the hierarchical relationship are represented as lexical features and hopping features. Therefore, we conduct ablation experiments on these two features to verify the significance of hierarchical relationship in attack traffic identification. In addition, the pooling strategy is to combine lexical features and hopping features as hidden features for hierarchical relationship. We simultaneously incorporate $Concat$, Max, and $Mean$ strategies into the experiments.

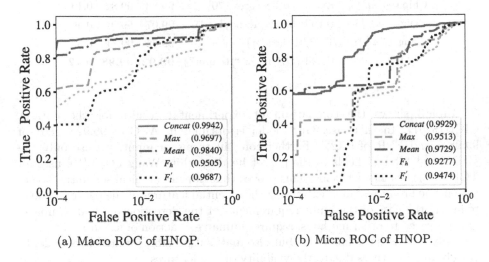

(a) Macro ROC of HNOP.   (b) Micro ROC of HNOP.

Fig. 4. ROC curves of HNOP. $Concat$, Max, and $Mean$ are the results of HNOP under the three pooling strategies. F_h is the result of HNOP with only hopping features, F_l' is the result of HNOP with only lexical features.

Figure 4 show that $Concat$ has the best AUC values for both the macro ROC and the micro ROC. This demonstrates that lexical features and hopping features both play crucial roles in the model. The combination of F_l' and F_h can significantly improve the performance of HNOP compared to others using each feature alone. Note that $Mean$ and Max lose part of the feature information, causing a degradation in the performance of the model.

4.3   Compared with Other Methods

Although deep learning methods are increasingly studied for attack traffic detection in recent years, their relative proportion remains small. HNOP extracts hierarchical node hopping features from the HTTP packet level, and the neural network does this process automatically without human intervention. We expect the features to be more accurate and representative over a certain time span.

Therefore, we compare the experimental results of our method with other published methods. There are three commonly used metrics to evaluate the performance of the model: Accuracy (ACC), Detection Rate (DR), and False Positive Rate (FPR).

Table 3. Performance comparison with other published methods on CICIDS2017 (%).

Method	INPUT	ACC	DR	FPR
Chiba *et al.* [3]	Extracted Features (70)	99.83	99.82	0.13
Geng *et al.* [5]	HTTP(header & payload)	99.07	99.01	0.40
Wang *et al.* [19]	TCP/IP(header)	99.75	99.71	0.32
HNOP	HTTP("URL" & "Referer")	**99.92**	**99.98**	**0.12**

Table 3 shows the comparison of experimental results for the dataset CICIDS2017. In this case, Our model is optimal in ACC of 99.92%, DR of 99.98% and FPR of 0.12%. Furthermore, in terms of input, we use only the "URL" and "Refer" fields of the packet header, while Wang *et al.* [19] uses all TCP/IP packet header fields. This proves that our hierarchical semantics structure in the premise of fully extracting the essential features of the packet, greatly reducing the cost of the input requirements of the detection method. Chiba *et al.*, compared to other methods, requires human extraction of features first. This not only increases the labor cost but also makes it possible for attackers to deliberately forge features due to the visibility of the features.

Table 4. Performance comparison with other published methods on IIE_HTTP (%).

Method	INPUT	ACC	DR	FPR
Chiba *et al.* [3]	Extracted Features (70)	99.22 (↓ 0.61)	98.52 (↓ 1.3)	0.53 (↑ 0.40)
Geng *et al.* [5]	HTTP(header & payload)	98.93 (↓ 0.14)	97.68 (↓ 1.33)	0.74 (↑ 0.34)
Wang *et al.* [19]	TCP/IP(header)	94.12 (↓ 5.63)	89.08 (↓ 10.63)	0.69 (↑ 0.37)
HNOP	HTTP("URL" & "Referer")	**99.84 (↓ 0.08)**	**99.71 (↓ 0.27)**	**0.11 (↓ 0.01)**

[a] *Figures in brackets indicate changes in results compared to CICIDS2017 (Table 3).*
[b] ↓ *indicate decrease,* ↑ *indicate increase.*

Table 4 presents the comparison of the experimental results for the dataset IIE_HTTP. Since the effectiveness of trained models decreases over time, all models perform slightly worse compared to the results of CICIDS2017. However, our model continues to perform well on the test set at different times, achieving ACC of 99.84%, DR of 99.71%, and FPR of 0.11%. DR and ACC drop by at least one order of magnitude less than other models. In addition, in terms of FPR, our model decreases rather than increases. This suggests that HNOP can better identify attack traffic and has higher generalization capability.

5 Conclusion

In this paper, we propose a single packet attack traffic detection method (HNOP) that extracts the hierarchical node hopping features from the hierarchical relationship of resources. Ablation experiments fully demonstrate that hierarchical node hopping features are inherent to attack traffic learning. Extensive experiments on the public dataset demonstrate that our model achieves excellent performance with state-of-the-art work just using less traffic information. In addition, our approach outperforms the others on the realistic network dataset (IIE_HTTP) at low decay rate over time, showing higher generalization capability. In short, we present key insights on how to rereconstruct resource requests, which will shed lights on understanding single packet attack and employing proactive defenses.

Acknowledgement. This work is supported by the National Key Research and Development Program of China (Grant No.2018YFB0804704), and the National Natural Science Foundation of China (Grant No.U1736218).

References

1. Akamai 2020 state of the internet/security. https://www.akamai.com/content/dam/site/en/documents/state-of-the-internet/soti-security-financial-services-hostile-takeover-attempts-report-2020.pdf. Accessed 21 Oct 2021
2. https://en.wikipedia.org/wiki/Occam%27s_razor. Accessed 21 Oct 2021
3. Chiba, Z., Abghour, N., Moussaid, K., Rida, M., et al.: Intelligent approach to build a deep neural network based ids for cloud environment using combination of machine learning algorithms. Comput. Secur. **86**, 291–317 (2019)
4. Dong, B., Wang, X.: Comparison deep learning method to traditional methods using for network intrusion detection. In: 2016 8th IEEE International Conference on Communication Software and Networks (ICCSN), pp. 581–585. IEEE (2016)
5. Geng, J., Li, S., Zhang, Y., Liu, Z., Cheng, Z.: LIFH: learning interactive features from http payload using image reconstruction. In: ICC 2021-IEEE International Conference on Communications, pp. 1–6. IEEE (2021)
6. Gezer, A., Warner, G., Wilson, C., Shrestha, P.: A flow-based approach for trickbot banking trojan detection. Comput. Secur. **84**, 179–192 (2019)
7. Girshick, R.: Fast R-CNN. In: Proceedings of the IEEE International Conference on Computer Vision, pp. 1440–1448 (2015)
8. Han, W., Xue, J., Yan, H.: Detecting anomalous traffic in the controlled network based on cross entropy and support vector machine. IET Inf. Secur. **13**(2), 109–116 (2019)
9. Kabir, E., Hu, J., Wang, H., Zhuo, G.: A novel statistical technique for intrusion detection systems. Futur. Gener. Comput. Syst. **79**, 303–318 (2018)
10. Le, A., Markopoulou, A., Faloutsos, M.: Phishdef: URL names say it all. In: 2011 Proceedings IEEE INFOCOM, pp. 191–195. IEEE (2011)
11. Liu, T., Qi, A., Hou, Y., Chang, X.: Method for network anomaly detection based on bayesian statistical model with time slicing. In: 2008 7th World Congress on Intelligent Control and Automation, pp. 3359–3362 (2008)

12. Moore, A., Zuev, D., Crogan, M.: Discriminators for use in flow-based classification. Technical report (2013)
13. Patil, P., Rane, R., Bhalekar, M.: Detecting spam and phishing mails using SVM and obfuscation URL detection algorithm. In: 2017 International Conference on Inventive Systems and Control (ICISC), pp. 1–4. IEEE (2017)
14. Pontes, C., Souza, M., Gondim, J., Bishop, M., Marotta, M.: A new method for flow-based network intrusion detection using the inverse Potts model. IEEE Trans. Network Serv. Manage. **18**, 1125–1136 (2021)
15. Sahoo, D., Liu, C., Hoi, S.C.: Malicious URL detection using machine learning: a survey. arXiv preprint arXiv:1701.07179 (2017)
16. Stiawan, D., Idris, M.Y.B., Bamhdi, A.M., Budiarto, R., et al.: Cicids-2017 dataset feature analysis with information gain for anomaly detection. IEEE Access **8**, 132911–132921 (2020)
17. Swarnkar, M., Hubballi, N.: OCPAD: One class Naive Bayes classifier for payload based anomaly detection. Expert Syst. Appl. **64**, 330–339 (2016)
18. Vijayanand, R., Devaraj, D., Kannapiran, B.: Intrusion detection system for wireless mesh network using multiple support vector machine classifiers with genetic-algorithm-based feature selection. Comput. Secur. **77**, 304–314 (2018)
19. Wang, B., Su, Y., Zhang, M., Nie, J.: A deep hierarchical network for packet-level malicious traffic detection. IEEE Access **8**, 201728–201740 (2020)
20. Wang, W., et al.: Hast-IDS: learning hierarchical spatial-temporal features using deep neural networks to improve intrusion detection. IEEE Access **6**, 1792–1806 (2017)
21. Xie, J., Li, S., Yun, X., Zhang, Y., Chang, P.: HSTF-model: An http-based trojan detection model via the hierarchical spatio-temporal features of traffics. Comput. Secur. **96**, 101923 (2020)
22. Zand, A., Vigna, G., Yan, X., Kruegel, C.: Extracting probable command and control signatures for detecting botnets. In: Proceedings of the 29th Annual ACM Symposium on Applied Computing, pp. 1657–1662 (2014)

TROPHY: Trust Region Optimization Using a Precision Hierarchy

Richard J. Clancy[1,2]($\boxtimes$), Matt Menickelly[1], Jan Hückelheim[1], Paul Hovland[1], Prani Nalluri[1,3], and Rebecca Gjini[1,4]

[1] Argonne National Laboratory, Lemont 60439, USA
richard.clancy@colorado.edu
[2] University of Colorado, Boulder 80309, USA
[3] Rice University, Houston, TX 77005, USA
[4] University of California, San Diego, CA 92093, USA

Abstract. We present an algorithm to perform trust-region-based optimization for nonlinear unconstrained problems. The method selectively uses function and gradient evaluations at different floating-point precisions to reduce the overall energy consumption, storage, and communication costs; these capabilities are increasingly important in the era of exascale computing. In particular, we are motivated by a desire to improve computational efficiency for massive climate models. We employ our method on two examples: the CUTEst test set and a large-scale data assimilation problem to recover wind fields from radar returns. Although this paper is primarily a proof of concept, we show that if implemented on appropriate hardware, the use of mixed-precision can significantly reduce the computational load compared with fixed-precision solvers.

1 Introduction

Optimization methods are used in many applications, including engineering, science, and machine learning. The memory requirements and run time for different methods have been studied extensively and determine the problem sizes that can be run on existing hardware. Similarly, the energy consumption of each method determines its cost and carbon footprint, which is a growing concern [1].

With the desire to incorporate more data into models and ever-increasing computational power, problem scales have grown as well. To improve efficiency, modern computers tightly integrate graphical processing units (GPUs) and other accelerators. Many of these units natively support data types of differing precision to lessen the storage and computational load. Previous work has found significant differences in the overall energy consumption for double- and single-precision computations [2,3]. Server-level products such as NVIDIA Tensor cores in V100 GPUs show 16× improvement over traditional double precision [4].

Such gains come at a cost, however. Classical algorithms such as the Gram–Schmidt process are well known to suffer from loss of orthogonality and numerical instability due to limited precision [5]. In an effort to ameliorate algorithmic

D. Groen et al. (Eds.): ICCS 2022, LNCS 13350, pp. 445–459, 2022.
https://doi.org/10.1007/978-3-031-08751-6_32

issues with accuracy and stability, there has been a flurry of activity using mixed precision. These methods utilize multiple data types in a principled fashion to reduce the computational burden without sacrificing accuracy. A few of the many applications are tomographic reconstruction [6], seismic modeling [7], and neural network training [8–11]. Mixed-precision methods have been used generically to minimize the cost of linear algebra methods [4], iterative schemes [12], and improved finite element solvers [13].

There are a variety of auto-tuning algorithms that attempt to identify variables within a program that can safely be cast in a lower precision, while satisfying some accuracy constraint [14–18]. Our method proposed does not attempt to identify low precision candidates nor do we try satisfying accuracy constraints. All computation within the objective/gradients are performed in the lowest precision possible and only increase after it is deemed necessary for the solver to proceed.

A recent paper by Gratton and Toint [19] illustrates potential savings in an optimization setting via variable-precision trust region (TR) methods. We investigate the ideas proposed in their work but with an important difference. In particular, their algorithm (TR1DA) requires access to an approximate objective, $\bar{f}(\mathbf{x}_k, \omega_{f,k})$, and gradient, $\bar{\mathbf{g}}(\mathbf{x}_k, \omega_{g,k})$, where $\omega_{f,k}$ and $\omega_{g,k}$ are uncertainty parameters (for the kth iterate $\mathbf{x}_k$) that satisfy

$$|\bar{f}(\mathbf{x}_k, \omega_{f,k}) - f(\mathbf{x}_k)| \leq \omega_{f,k} \quad \text{and} \quad \frac{\|\bar{\mathbf{g}}(\mathbf{x}_k, \omega_{g,k}) - \mathbf{g}(\mathbf{x}_k)\|}{\|\bar{\mathbf{g}}(\mathbf{x}_k, \omega_{g,k})\|} \leq \omega_{g,k}.$$

Their error model requires user specified absolute error bounds on function and gradient values; such bounds are difficult to realize in practice as computational complexity grows for reasons such as catastrophic cancellation and accumulated round-off error. Our focus here is on designing an algorithm that performs well without assumptions on the output error when using lower precision.

In this paper, we introduce TROPHY (**T**rust **R**egion **O**ptimization using a **P**recision **H**ierarch**Y**), a mixed-precision TR method for unconstrained optimization. We provide practically verifiable conditions intended to determine whether the error related to a current precision level may be interfering with the dynamics of the TR algorithm. If the conditions are not satisfied, we increase the precision level until they are. Our goal is to lighten the computational load without sacrificing accuracy of the final solution. By using a limited-memory, symmetric rank-1 update (L-SR1) to the approximate Hessian, the method is suitable for large-scale, high-dimensional problems. We compare the method with a standard TR method—supplied with access to either a single- or double-precision evaluation of the function and gradient—on the Constrained and Unconstrained Testing Environment with safe threads (CUTEst) test problem collection [20] and on a large-scale weather model based on the PyDDA software package [21].

Since computational, storage, and communication savings are based on hardware implementations of different precision types rather than assumed theoretical values, our primary metric for comparison will be adjusted function evaluations rather than time. Simply put, adjusted function evaluations discount computations performed in lower-precision levels. The goal here is to provide a proof of concept for computational gains attainable by exploiting variable precision in TR methods. In practice, improvements in energy consumption, time, communication, and memory must be realized through optimized hardware which is beyond the scope of this paper.

2   Background

Consider the unconstrained minimization of a differentiable function $f : \mathbb{R}^n \to \mathbb{R}$,

$$\min_{\mathbf{x} \in \mathbb{R}^n} f(\mathbf{x}). \tag{1}$$

We are motivated by problems where the objective and its derivatives are expensive to calculate as is typical for large-scale computing. In this paper we focus on the TR framework, but could have studied line-search methods instead such as L-BFGS, which is a popular quasi-Newton method distributed in SciPy [22]. However, it is remarkably simpler to illustrate the effect of error on the quality of models within a TR method; that is likely the reason TR methods were employed in [19]. In the following subsections we give an overview of the general framework for TR methods and describe the model function used in our algorithm.

2.1   Trust Region Methods

Trust region methods are iterative algorithms used for numerical optimization. At each iteration (with the counter denoted by k), a model function $m_k : \mathbb{R}^n \to \mathbb{R}$ is built around the incumbent point or iterate, $\mathbf{x}_k$, such that $m_k(\mathbf{0}) = f(\mathbf{x}_k)$ and $m_k(\mathbf{s}) \approx f(\mathbf{x}_k + \mathbf{s})$. The model, m_k, is intended to be a "good" local model of f on the *trust region*, $\{\mathbf{s} \in \mathbb{R}^n : \|\mathbf{s}\| \le \delta_k\}$ for $\delta_k > 0$. We refer to δ_k as the *trust region radius*. A *trial step*, $\mathbf{s}_k$, is then computed via a(n approximate) solution to the *trust region subproblem*,

$$\mathbf{s}_k = \operatorname*{argmin}_{\|\mathbf{s}\| \le \delta_k} m_k(\mathbf{s}), \tag{2}$$

for $\mathbf{s} \in \mathbb{R}^n$. By an approximate solution to the TR subproblem (2), we mean that one requires the *Cauchy decrease condition* to be satisfied:

$$f(\mathbf{x}_k) - m_k(\mathbf{s}_k) \ge \frac{\mu}{2} \min\left\{\delta_k, \frac{\|\mathbf{g}_k\|}{C}\right\}, \tag{3}$$

where μ and C are constants and $\mathbf{g}_k = \nabla m(\mathbf{x}_k)$. A common choice for m_k is a quadratic Taylor expansion, namely, $m_k(\mathbf{s}) = f(\mathbf{x}_k) + \mathbf{g}_k^T \mathbf{s} + \frac{1}{2}\mathbf{s}^T \nabla^2 f(\mathbf{x}_k)\mathbf{s}$. In practice, $\nabla^2 f(\mathbf{x}_k)$ is typically replaced with a (quasi-Newton) approximation.

Having computed $\mathbf{s}_k$, the standard TR method then compares the true decrease in the function value, $f(\mathbf{x}_k) - f(\mathbf{x}_k + \mathbf{s}_k)$, with the decrease predicted by the model, $m_k(\mathbf{0}) - m_k(\mathbf{s}_k)$. In particular, one computes the quantity

$$\rho_k = \frac{f(\mathbf{x}_k) - f(\mathbf{x}_k + \mathbf{s}_k)}{m_k(\mathbf{0}) - m_k(\mathbf{s}_k)}. \tag{4}$$

If ρ_k is sufficiently positive ($\rho_k > \eta_{good}$ for fixed $\eta_{good} > 0$), then the algorithm accepts $\mathbf{x}_k + \mathbf{s}_k$ as the incumbent point $\mathbf{x}_{k+1}$ and may possibly increase the TR radius $\delta_k < \delta_{k+1}$ (if $\rho_k > \eta_{great}$ for fixed $\eta_{great} \geq \eta_{good}$). This scenario is called a *successful iteration*. On the other hand, if ρ_k is not sufficiently positive or is negative ($\rho_k < \eta_{good}$), then the incumbent point stays the same, $\mathbf{x}_{k+1} = \mathbf{x}_k$, and we set $\delta_{k+1} < \delta_k$. For the experiments below, we chose $\eta_{good} = 10^{-5}$ and $\eta_{great} = 0.10$. This process is iterated until a stopping criterion is met, e.g., when the gradient norm $\|\nabla f(\mathbf{x}_k)\|$ is below a given tolerance. Under mild assumptions, TR methods asymptotically converge to stationary points of $f(\mathbf{x})$ [23].

2.2 Model Function

The model function, m_k, must be specified for a TR algorithm. Popular choices include linear or quadratic approximations of the objective using Taylor series or interpolation methods; the latter are often employed in derivative-free optimization [24]. Since many applications of interest are high-dimensional or have costly objective and derivative functions, it is difficult if not impossible to compute and/or store the Hessian matrix for use in quadratic TR models with memory requirement scaling as $\mathcal{O}(n^2)$. A common technique that exploits derivative information while keeping the cost low is to use *curvature pairs* given by $\mathbf{s}_k$ and $\mathbf{y}_k = \nabla f(\mathbf{x}_k + \mathbf{s}_k) - \nabla f(\mathbf{x}_k)$. After each successful iteration, the curvature pairs are used to update the current approximate Hessian denoted by $\mathbf{H}_k$. These updates employ secant approximations of second derivatives. Common update rules include BFGS, DFP, and SR1 [25].

In this work we use a limited-memory symmetric rank-1 update (L-SR1) to the approximate Hessian. This update rule requires the user to set a memory parameter that specifies a number of secant pairs to use in the approximate Hessian. Since we require only a matrix-vector product and not the explicit Hessian, we can implement a matrix-free version reducing the storage cost to $\mathcal{O}(n)$. Thus, our TR subproblem is

$$\mathbf{s}_k = \underset{\|\mathbf{s}\| \leq \delta_k}{\operatorname{argmin}} \quad \mathbf{s}^T \nabla f(\mathbf{x}_k) + \frac{1}{2}\mathbf{s}^T \mathbf{H}_k \mathbf{s}, \tag{5}$$

which we recast and approximately solve using the Steihaug conjugate gradient method implemented in [26][Appendix B.4].

In the next section we describe the dynamic precision framework and present criteria for when precision should switch. We then are prepared to give a formal statement of TROPHY. In Sect. 4 we describe the problems on which we have tested TROPHY, and in Sect. 5 we discuss the results of our experiments.

3   Method

We assume access to a hierarchy of arithmetic precisions for the evaluation of both $f(\mathbf{x})$ and $\nabla f(\mathbf{x})$, but the direct (infinite-precision) evaluation of

$f(\mathbf{x}), \nabla f(\mathbf{x})$ is unavailable. We formalize this slightly by supposing we are given oracles that compute $f^p(\mathbf{x}), \nabla f^p(\mathbf{x})$ for $p \in \{0, \ldots, P\}$. With very high probability, given a uniform distribution on all possible inputs $\mathbf{x}$, the oracles satisfy the inequalities

$$|f^p(\mathbf{x}) - f(\mathbf{x})| > |f^{p+1}(\mathbf{x}) - f(\mathbf{x})|, \quad \|\nabla f^p(\mathbf{x}) - \nabla f(\mathbf{x})\| > \|\nabla f^{p+1}(\mathbf{x}) - \nabla f(\mathbf{x})\|.$$

For a tangible example, if intermediate calculations involved in the computation of $f(\mathbf{x})$ can be done in half, single, or double precision, then we can denote $f^0(\mathbf{x})$, $f^1(\mathbf{x})$, and $f^2(\mathbf{x})$ as the oracles using only half, single, or double, respectively.

To build on the generic TR method described in Sect. 2, we must specify when and how to switch precision. We can identify two additional difficulties presented in the multiple-precision setting. First, it is currently unclear how to compute ρ_k in (4) since our error model assumes we have no access to an oracle that directly computes $f(\cdot)$. Second, because models m_k typically use function and gradient information provided by $f(\cdot)$ and $\nabla f(\cdot)$, we must specify how to construct models using lower precision oracles.

For the first of these two issues, we make a practical assumption that **the highest level of precision available to us should be treated as if it were infinite precision**. Although this is a theoretically poor assumption, virtually all computational optimization makes it implicitly; algorithms are analyzed over the real numbers but are typically implemented using floating point arithmetic (often double). Thus, in the notation we have developed, the optimization problem we actually aim to solve is not (1) but

$$\min_{\mathbf{x} \in \mathbb{R}^n} f^P(\mathbf{x}), \tag{6}$$

so that the ρ-test in (4) is replaced with

$$\rho_k = \frac{f^P(\mathbf{x}_k) - f^P(\mathbf{x}_k + \mathbf{s}_k)}{m_k(\mathbf{0}) - m_k(\mathbf{s}_k)} = \frac{\text{ared}_k}{\text{pred}_k}. \tag{7}$$

The values ared and pred were introduced to denote "actual reduction" and "predicted reduction", respectively. We note that computing (7) still entails two evaluations of the highest-precision oracle, $f^P(\cdot)$, which is exactly what we hoped to avoid by using mixed-precision. Our algorithm avoids the cost of full-precision evaluations by dynamically adjusting the precision level $p_k \in \{0, \ldots, P\}$ between iterations so that in the kth iteration, ρ_k is approximated by

$$\tilde{\rho}_k = \frac{f^{p_k}(\mathbf{x}_k) - f^{p_k}(\mathbf{x}_k + \mathbf{s}_k)}{m_k(\mathbf{0}) - m_k(\mathbf{s}_k)} = \frac{\text{ered}_k}{\text{pred}_k}, \tag{8}$$

introducing ered to denote "estimated reduction". To update p_k, we are motivated by a strategy similar to one employed in [27] and [28]. We introduce a variable θ_k that is not initialized until the end of the first unsuccessful iteration and set $p_0 = 0$. When the first unsuccessful iteration is encountered, we set

$$\theta_k \leftarrow |\text{ared}_k - \text{ered}_k|. \tag{9}$$

Notice that we must incur the cost of two evaluations of $f^P(\cdot)$ following the first unsuccessful iteration in order to compute ared_k. From that point on, θ_k is involved in a test triggered on every unsuccessful iteration (in which the TR radius is sufficiently small) to determine whether the precision level, p_k, should be increased. We compute θ_k and test for precision when $\delta_k < \Delta_{\text{prec}}$. The value Δ_{prec} is set to be a length scale where numerical imprecision is a concern.

Introducing a predetermined *forcing sequence* $\{r_k\}$ satisfying $r_k \in [0, \infty)$ for all k and $\lim_{k \to \infty} r_k = 0$, and fixing a parameter $\omega \in (0, 1)$, we check on any unsuccessful iteration whether

$$\theta_k^\omega \leq \eta \min \{\text{pred}_k, r_k\}, \tag{10}$$

where $\eta = \min \{\eta_{\text{good}}, 1 - \eta_{\text{great}}\}$. If (10) does not hold, then we increase $p_{k+1} = p_k + 1$ and again update the value of θ_k according to (9) (thus incurring two more evaluations of $f^P(\cdot)$). The reasoning behind the test in (10) is that if (the unknown) ρ_k in (7) satisfies $\rho_k \geq \eta$, then

$$\eta \leq \rho_k = \frac{\text{ared}_k}{\text{pred}_k} \leq \frac{|\text{ared}_k - \text{ered}_k| + \text{ered}_k}{\text{pred}_k} \approx \frac{\theta_k + \text{ered}_k}{\text{pred}_k} = \frac{\theta_k}{\text{pred}_k} + \tilde{\rho}_k. \tag{11}$$

Thus, for the practical test (8) to be meaningful, we need to ensure that $\theta^k / \text{pred}_k < \eta$, which is what (10) attempts to enforce. The use of ω and the forcing sequence in (10) is designed to ensure that we eventually do not tolerate error, since (11) involves an approximation due to the estimate θ_k. The forcing sequence would likely be necessary to guarantee convergence for theoretical analysis, but is not critical to the performance of a practical algorithm, and was not employed in our implementation. For concreteness, if a forcing sequence were employed, one might consider a slowly decaying sequence such as $r_k = 1/\sqrt{k}$.

It remains to describe how we deal with our second identified difficulty, the construction of m_k in the absence of evaluations of $f(\cdot)$ and $\nabla f(\cdot)$. As is frequently done in trust region methods, we will employ quadratic models of the form

$$m_k(\mathbf{s}) = f_k + \mathbf{g}_k^\top \mathbf{s} + \frac{1}{2} \mathbf{s}^\top \mathbf{H}_k \mathbf{s}. \tag{12}$$

Having already defined rules for the update of p_k through the test (10), we take in the kth iteration $f_k = f^{p_k}(\mathbf{x})$ and $\mathbf{g}_k = \nabla f^{p_k}(\mathbf{x})$. In theory, we require $\mathbf{H}_k$ to be any Hessian approximation with a spectrum bounded above and below uniformly for all k. In practice, we update $\mathbf{H}_k$ via L-SR1 updates [29]. By implementing a reduced-memory version, we need not store an explicit approximate Hessian, thus greatly reducing the memory cost and significantly accelerating the matrix-vector products in our model. Pseudocode for TROPHY is provided in Algorithm 1.

4   Test Problems and Implementations

Our initial implementation of TROPHY is written in Python. To validate the algorithm, we focus on a well-known optimization test suite and a problem relating to climate modeling. In all cases, the algorithms terminate when one of the

---

Algorithm 1: TROPHY

Initialize $0 < \eta_{good} \leq \eta_{great} < 1$, $\omega \in (0,1)$, $\gamma_{inc} > 1$, $\gamma_{dec} \in (0,1)$,
$\Delta_{prec} \in (0,1)$, forcing seq. $\{r_k\}$.
Choose initial $\delta_0 > 0$, $\mathbf{x}_0 \in \mathbb{R}^n$.
$\theta_0 \leftarrow 0, p_k \leftarrow 0, k \leftarrow 0, \text{failed} \leftarrow \text{FALSE}$
while *some stopping criterion not satisfied* **do**
    Construct model m_k.
    (Approximately solve) (2) to obtain $\mathbf{s}_k$.
    Compute $\tilde{\rho}_k$ as in (8).
    if $\tilde{\rho}_k > \eta_{good}$ *(successful iteration)* **then**
        $\mathbf{x}_{k+1} \leftarrow \mathbf{x}_k + \mathbf{s}_k$.
        if $\tilde{\rho}_k > \eta_{great}$ *(very successful iteration)* **then**
            $\delta_{k+1} \leftarrow \gamma_{inc}\delta_k$.
        end
    else
        if not failed **then**
            Compute θ_k as in (9).
            failed $\leftarrow$ TRUE.
        end
        if (10) *holds* **or** $\delta_k \geq \Delta_{prec}$ **then**
            $\delta_{k+1} \leftarrow \gamma_{dec}\delta_k$.
        else
            $p_{k+1} \leftarrow p_k + 1$.
            Compute θ_k as in (9).
            $\delta_{k+1} \leftarrow \delta_k$.
        end
        $\mathbf{x}_{k+1} \leftarrow \mathbf{x}_k$.
    end
    $k \leftarrow k + 1$.
end

---

following conditions are met: (1) the first-order condition is satisfied, namely, $\|\nabla f^P(\mathbf{x}_k)\| < \epsilon_{tol}$; (2) the TR radius is smaller than machine precision, namely, $\delta_k < \epsilon_{machine}$; or (3) the first two conditions have not been met after some maximum number of iterations. Condition 1 is a success whereas conditions 2 and 3 are failed attempts. We describe the problem setup and implementation considerations in the current section and then discuss results in Sect. 5.

4.1 CUTEst

Our first example used the CUTEst set [20], which is well known within the optimization community and offers a variety of problems that are challenging to solve. Each problem is given in a Standard Input Format [30] file that is passed to a decoder from which Fortran subroutines are generated. The problems can be built directly by using single or double precision, making the set useful for mixed-precision comparison.

Python Implementation of CUTEst: The PyCUTEst package [31] serves as an interface between Python and CUTEst's Fortran source code. The problems are compiled via the interface; then Python scripts are generated and cached for subsequent function calls. Although CUTEst natively supports both single- and double-precision evaluations, at the time of writing, the single-precision implementations are concealed from the PyCUTEst API.

To access single-precision evaluations, we used PREDUCER [32], a Python script written to compare the effect of round-off errors in scientific computing. PREDUCER parses Fortran source code and downcasts double data types to single. To allow for its use in existing code, all single data types are recast to double after function/gradient evaluation but before returning to the calling program. Because of the overhead associated with casting operations, we do not expect improvements in computational time. However, performance gains in terms of both accuracy and a reduction in the number of adjusted function calls for an iterative algorithm should be realized. We built the double-precision functions, too, and wrapped both functions to pass as a unified handle to TROPHY.

For the subset of unconstrained problems with dimension less than or equal to 100, we ran TROPHY with single/double switching along with the same TR method using only single or double precision. Our first-order stopping criterion was $\|\nabla f^P(\mathbf{x}_k)\| < 10^{-5}$, and the maximum number of allowable iterations was 5,000. We show results for problems solved by at least one TR in Sect. 5.

Julia Implementation of CUTEst: The Julia programming language supports variable-precision floating-point data types. More precisely, it allows users to specify the number of bits used in the mantissa and expands memory for the exponent as necessary to avoid overflow. This is in contrast to the IEEE 754 standard that uses 11 (5), 24 (8), and 53 (11) bits for the mantissa (exponent) of half-, single-, and double-precision floats, respectively. In practice, one can assign enough bits for the exponent to avoid dynamic reallocation.

To exploit variable precision, we hand-coded several of the unconstrained CUTEst objectives in Julia and then computed gradients with forward-mode automatic differentiation (AD) using the ForwardDiff.jl package [33]. We wrote a Julia port that allows us to call this code from Python for use in TROPHY. We opted to use forward-mode AD for its ease of implementation. In all cases used, the hand-coded Julia objective and AD gradient were compared against the Fortran implementation and found to be accurate.

We compared TROPHY with TR methods using a fixed precision of half, single, and double precision (11, 24, and 53 bits, respectively). We used the same first-order condition of $\|\nabla f(\mathbf{x}_k)\| < 10^{-5}$ but allowed this implementation to run only for 1,000 iterations. For TROPHY, we used several precision-switching sets: $\{24, 53\}$, $\{11, 24, 53\}$, $\{8, 11, 17, 24, 53\}$, and $\{8, 13, 18, 23, 28, 33, 38, 43, 48, 53\}$. The third set of precisions was motivated by the number of mantissa bits in bfloat16, fp16, fp24, fp32, and fp64, respectively. The last set increased the number of bits in increments of 5 up to double-precision.

4.2   Multiple Doppler Radar Wind Retrieval:

We also looked at a data assimilation problem for retrieving wind fields for convective storms from Doppler radar returns. Shapiro and Potvin [34, 35] proposed a method for doing so that optimizes a cost functional based on vertical vorticity, mass continuity, field smoothness, and data fidelity, among others. Although the function calls are fairly simple, the wind field must be reconciled on a 3-D grid over space, each with an x, y, and z component. For a $39 \times 121 \times 121$ grid, there are $1,712,997$ variables. Therefore, reducing computational, storage, and communication costs where possible is paramount.

Our work centered on the PyDDA package [21], which was written to solve the aforementioned problem. We amended the code in two significant ways. First, to improve efficiency, we rewrote portions of the code to use JAX, an automatic differentiation package using XLA that exploits efficient computation on GPUs [36]. Since JAX natively supports single precision on CPUs and can be recast to half and double as desired, it nicely serves as a proof of concept on a real application. Second, we modified the solver to use TROPHY rather than the SciPy implementation of L-BFGS.

Once again, we compared TROPHY against single- and double-precision TR methods. TROPHY switched among half, single, and double precision. To avoid overflow initially for half-precision, we warm started the algorithm by providing it with the tenth iterate from the double-precision TR method, i.e., x_{10}. We perturbed this initial iterate 10 times and used the perturbed vectors as the initial guess for each algorithm (including double TR). We measured the average performance when solving each problem to different first-order conditions: $\|\nabla f(x_k)\| < 10^{-3}$ and $\|\nabla f(x_k)\| < 10^{-6}$. The maximum number of allowable iterations was 10,000.

5   Experimental Results

We display results across the CUTEst set using data and performance profiles [37,38]. For a given metric, performance profiles help determine how a set of solvers, S, performs over a set of problems, P. The value $v_{ij} > 0$ denotes a particular metric (say, the final gradient norm) of the jth solver on problem i. We can then consider the performance of each solver in relation to the solver that performed best, that is, the one that achieved the smallest gradient norm. The *performance ratio* is defined as

$$r_{ij} = \frac{v_{ij}}{\min_j \{v_{ij}\}}. \tag{13}$$

Smaller values of r_{ij} are better since they are closer to optimal. The performance ratio was set to ∞ if the solver failed to solve the problem. We can evaluate the performance of a solver by asking what percentage of the problems are solved within a fraction of the best. This is given by the *performance profile*,

$$h_j(\tau) = \frac{\sum_{i=1}^N \mathcal{I}_{\{r_{ij} \leq \tau\}}}{N},$$

where $N = |\mathcal{P}|$ (the cardinality of $\mathcal{P}$) and $\mathcal{I}_{\{A\}}$ is the indicator function such that $\mathcal{I}_{\{A\}} = 1$ if A is true and 0 otherwise. Hence, better solvers have profiles that are above and to the left of the others.

Motivated by the computational models in [2] and [3], we assume that the energy efficiency of single precision is between 2 and 3.6 times higher than double precision [39,40]. The storage and communication have less optimistic savings since we expect the cost of both to scale linearly with the number of bits used in the mantissa. Accordingly, we focus primarily on the model where half- and single-precision evaluations cost $1/4$ and $1/2$ that of a double evaluation, respectively. This gives a conservative estimate for energy cost and a favorable one for execution time. For a given problem and solver, we define *adjusted calls*:

$$\text{Adj. calls} = \sum_{p \in \{0,1,\dots,P\}} \frac{(\#\ \text{bits for prec. } p) \times (\#\ \text{func. calls at prec. } p)}{\#\ \text{bits in prec. } P}. \tag{14}$$

Figures 1 and 2 show performance profiles for the Python and Julia implementations of CUTEst, respectively. All CUTEst problems had their first-order tolerance set to 10^{-5}. Working from right to left in both images, we can see that the first-order condition is steady across methods provided that double-precision evaluations are ultimately available to the solver. When limited to half (11 bits) or single (24 bits), the performance suffers, and a number of problems cannot be solved. For the number of iterations in Python, we see that TROPHY and the double TR method perform comparably. The Julia implementation shows that the iterations count suffers when using low precision or TROPHY with many precision levels available for switching. This behavior is expected for low precision since the solver may never achieve the desired accuracy and hence runs longer, and for TROPHY since each precision switch requires a full iteration. For example, if 10 precision levels are available, TROPHY will take at least 10 iterations to complete. This limits the usefulness of the method on small to medium problems and problems where the initial iterate is close to the final solution. As anticipated, TROPHY shows a distinct advantage for adjusted calls. The

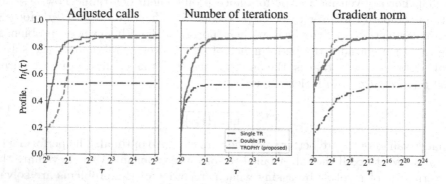

Fig. 1. Performance profiles for Python implementation of unconstrained CUTEst problems of dimension < 100 solved to first-order tolerance of 10^{-5}. Standard single and double TR methods compared against TROPHY using single/double switching.

one exception is when there are many precision levels to cycle through, for the same reason as above. Although the initial iterate might be close to optimal, the algorithm must still visit all precision levels before breaking. The fact is made worse since each time the precision switches, two evaluations at the highest precision are required. Using two or three widely spaces precision levels yields strong results for the CUTEst set.

Table 1. Average performance over ten initializations for single-precision TR, double-precision TR, and TROPHY on PyDDA wind retrieval example. Adjusted calls indicate improved computational efficiency. Half, single, and double costs are 1/4, 1/2, and 1 for linear and 1/16, 1/4, and 1 for quadratic adjustments, respectively. Problem solved to $\|\nabla f(x_{\text{final}})\|_2 < 10^{-3}$ above.

Tolerance $\|\nabla f\| < 10^{-3}$	Half calls	Single calls	Double calls	Adj. calls (linear)	Adj. calls (quad.)	f_{final}	$\|\nabla f_{\text{final}}\|$
Single	–	3411	–	1706	853	5.3×10^{-3}	9.5×10^{-4}
Double	–	–	1877	1877	1877	4.5×10^{-3}	9.1×10^{-4}
TROPHY	465	1898	6	1071	510	4.7×10^{-3}	9.4×10^{-4}

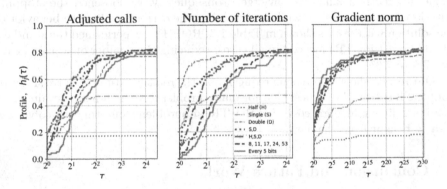

Fig. 2. Performance profiles for Julia implementation of unconstrained CUTEst problems of dimension < 100 solved to first-order tolerance of 10^{-5}. Half, single, and double are standard TR methods using corresponding precision. "S, D", "H, S, D", "8,11,17,24,53", and "Every 5 bits" are TROPHY implementations using different precision regimes. "Every 5 bits" starts at 8 bits and increases to 53 bits in increments of 5 bits. A finer precision hierarchy does not imply better performance.

Table 2. Problem solved to $\|\nabla f(x_{\text{final}})\|_2 < 10^{-6}$ accuracy. The single-precision TR method failed to converge with the TR radius falling below machine precision.

Tolerance $\|\nabla f\| < 10^{-6}$	Half calls	Single calls	Double calls	Adj. calls (linear)	Adj. calls (quad.)	f_{final}	$\|\nabla f_{\text{final}}\|$
Single	–	∞	-	FAIL	FAIL	9.9×10^{-7}	3.9×10^{-6}
Double	–	–	5283	**5283**	5283	1.8×10^{-7}	9.6×10^{-7}
TROPHY	465	7334	601	**4384**	**2464**	1.9×10^{-7}	9.7×10^{-8}

The wind retrieval example shows similar results (Tables 1 and 2). The switching criteria used here differs slightly from the one presented in (10). Specifically, a baseline θ_{p_k} is set after the first failed iteration at the current precision level. The model predicted reduction (pred$_k$) is compared to the baseline θ_{p_k} for successive failures. If pred$_k$ is small compared to the baseline θ_{p_k}, then the precision is increased. We do not expect this to significantly change the qualitative results or behavior of the method for this test case. We included two "adjusted call" columns: one for a linear decay adjustment (memory and communication as above) and the other for quadratic decay (reduction in energy consumption). We originally iterated until $\|\nabla f(\mathbf{x}_k)\| < 10^{-6}$ (Table 2) but observed that the single TR method failed to converge. Consequently, we loosened the stopping criterion as far as possible while maintaining the correct qualitative behavior of the solution with results shown in Table 1. TROPHY outperformed the standard (double-precision TR) method in all cases, reducing the number of adjusted calls by 17% to 73%.

Our results show a promising reduction in the relative cost over naive single or double TR solvers. We expect that for many problems where function evaluations dominate linear algebra costs for the TR subproblem, our time to solve will greatly benefit from the method.

6   Conclusion and Future Work

In this paper we introduced TROPHY, a TR method that exploits variable-precision data types to lighten the computational burden of expensive function/gradient evaluations. We illustrated proof of concept for the algorithm by implementing it on the CUTEst set and PyDDA. The full benefit of our work has not yet been realized. We look forward to implementing similar tests on hardware that can realize the full benefit of lower energy consumption and reduced memory/communication costs and ultimately shorten the time to solution. This will be especially beneficial for large scale climate models.

We would also like to incorporate mixed precision into line-search methods given their popularity in quasi-Newton solvers. By incorporating the same ideas into highly optimized algorithms such as the SciPy implementation of L-BFGS, we could easily deploy mixed precision to a wide population, dramatically reducing computational loads. Although TR methods are, computationally speaking, more appropriate for expensive-to-evaluate objectives, there is no reason the same ideas cannot be extended if practitioners prefer them.

Acknowledgement. We gratefully acknowledge the support by the Applied Mathematics activity within the U.S. Department of Energy, Office of Science, Advanced Scientific Computing Research Program, under contract number DE-AC02-06CH11357, and the computing resources provided on Swing, a high-performance computing cluster operated by the Laboratory Computing Resource Center at Argonne National Laboratory.

References

1. Hao, K.: Training a single AI model can emit as much carbon as five cars in their lifetimes. MIT Technol. Rev. (2019)
2. Molka, D., Hackenberg, D., Schöne, R., Müller, M.S.: Characterizing the energy consumption of data transfers and arithmetic operations on x86–64 processors. In: International Conference on Green Computing, pp. 123–133. IEEE (2010)
3. Kestor, G., Gioiosa, R., Kerbyson, D.J., Hoisie,A.: Quantifying the energy cost of data movement in scientific applications. In: 2013 IEEE International Symposium on Workload Characterization, pp. 56–65. IEEE (2013)
4. Abdelfattah, A., et al.: A survey of numerical linear algebra methods utilizing mixed-precision arithmetic. Int. J. High Perform. Comput. Appl. 10943420211003313 (2021)
5. Golub, G.H., Van Loan, C.F.: Matrix Computations. Johns Hopkins University Press, Baltimore (1996)
6. Doucet, N., Ltaief, H., Gratadour, D., Keyes, D.: Mixed-precision tomographic reconstructor computations on hardware accelerators. In: 2019 IEEE/ACM 9th Workshop on Irregular Applications: Architectures and Algorithms (IA3), pp. 31–38. IEEE (2019)
7. Ichimura, T., et al.: A fast scalable implicit solver for nonlinear time-evolution earthquake city problem on low-ordered unstructured finite elements with artificial intelligence and transprecision computing. In: SC18: International Conference for High Performance Computing, Networking, Storage and Analysis, pp. 627–637. IEEE (2018)
8. Jia, X., et al.: Highly scalable deep learning training system with mixed-precision: training imagenet in four minutes. preprint arXiv:1807.11205 (2018)
9. Micikevicius, P., et al.: Mixed precision training. In: International Conference on Learning Representations (2018)
10. Wang, N., Choi, J., Brand, D., Chen, C.-Y., Gopalakrishnan, K.: Training deep neural networks with 8-bit floating point numbers. In: Proceedings of the 32nd International Conference on Neural Information Processing Systems, pp. 7686–7695 (2018)
11. Carmichael, Z., Langroudi, H.F., Khazanov, C., Lillie, J., Gustafson, J.L., Kudithipudi, D.: Performance-efficiency trade-off of low-precision numerical formats in deep neural networks. In: Proceedings of the Conference for Next Generation Arithmetic 2019, pp. 1–9 (2019)
12. Strzodka, R., Göddeke, D.: Mixed precision methods for convergent iterative schemes. EDGE **6**, 23–24 (2006)

13. Göddeke, D., Strzodka, R., Turek, S.: Performance and accuracy of hardware-oriented native-, emulated-and mixed-precision solvers in FEM simulations. Int. J. Parallel Emergent Distrib. Syst. **22**(4), 221–256 (2007)
14. Chiang, W.-F., Baranowski, M., Briggs, I., Solovyev, A., Gopalakrishnan, G., Rakamarić, Z.: Rigorous floating-point mixed-precision tuning," in Proceedings of the 44th ACM SIGPLAN Symposium on Principles of Programming Languages, POPL 2017, New York, NY, USA, pp. 300–315. Association for Computing Machinery (2017)
15. Graillat, S., Jézéquel, F., Picot, R., Févotte, F., Lathuilière, B.: Auto-tuning for floating-point precision with discrete stochastic arithmetic. J. Comput. Sci. **36**, 101017 (2019)
16. Guo, H., Rubio-González, C.: Exploiting community structure for floating-point precision tuning. In: Proceedings of the 27th ACM SIGSOFT International Symposium on Software Testing and Analysis, pp. 333–343 (2018)
17. Menon, H., et al.: Adapt: Algorithmic differentiation applied to floating-point precision tuning. In: SC18: International Conference for High Performance Computing, Networking, Storage and Analysis, pp. 614–626. IEEE (2018)
18. Rubio-González, C., et al.: Precimonious: tuning assistant for floating-point precision. In: SC 2013: Proceedings of the International Conference on High Performance Computing, Networking, Storage and Analysis, pp. 1–12 (2013)
19. Gratton, S., Toint, P.L.: A note on solving nonlinear optimization problems in variable precision. Comput. Optim. Appl. **76**(3), 917–933 (2020). https://doi.org/10.1007/s10589-020-00190-2
20. Gould, N.I., Orban, D., Toint, P.L.: CUTEst: a constrained and unconstrained testing environment with safe threads for mathematical optimization. Comput. Optim. Appl. **60**(3), 545–557 (2015)
21. Jackson, R., Collis, S., Potvin, C., Munson, T.: PyDDA: a Pythonic direct data assimilation framework for wind retrievals. J. Open Res. Software **8**(1), 1–9 (2020)
22. Virtanen, P., et al.: SciPy 1.0: fundamental algorithms for scientific computing in Python. Nat. Methods **17**(3), 261–272 (2020)
23. Conn, A.R., Gould, N.I., Toint, P.L.: Trust region methods. In: SIAM (2000)
24. Larson, J., Menickelly, M., Wild, S.M.: Derivative-free optimization methods. Acta Numer. **28**, 287–404 (2019)
25. Nocedal, J., Wright, S.: Numerical Optimization. Springer, New York (2006). https://doi.org/10.1007/978-0-387-40065-5
26. Berahas, A.S., Jahani, M., Takác, M.: Quasi-Newton methods for deep learning: forget the past, just sample **16**. preprint arXiv:1901.09997 (2019)
27. Heinkenschloss, M., Vicente, L.: Analysis of inexact trust-region SQP algorithms. SIAM J. Optim. **12**, 283–302 (2002)
28. Kouri, D.P., Heinkenschloss, M., Ridzal, D., van Bloemen Waanders, B.G.: Inexact objective function evaluations in a trust-region algorithm for PDE-constrained optimization under uncertainty. SIAM J. Sci. Comput. 36 (2014)
29. Byrd, R.H., Nocedal, J., Schnabel, R.B.: Representations of quasi-Newton matrices and their use in limited memory methods. Math. Program. **63**(1), 129–156 (1994)
30. Conn, A.R., Gould, G., Toint, P.L.: LANCELOT: a Fortran package for large-scale nonlinear optimization (Release A), vol. 17. Springer (2013). https://doi.org/10.1007/978-3-662-12211-2
31. Fowkes, J., Roberts, L.: PyCUTEst (2018). https://jfowkes.github.io/pycutest
32. Hückelheim, J.: PREDUCER (2019). https://github.com/jhueckelheim/preducer
33. Revels, J., Lubin, M., Papamarkou, T.: Forward-mode automatic differentiation in Julia (2016). arXiv:1607.07892

34. Shapiro, A., Potvin, C.K., Gao, J.: Use of a vertical vorticity equation in variational dual-Doppler wind analysis. J. Atmos. Oceanic Tech. **26**(10), 2089–2106 (2009)
35. Potvin, C.K., Shapiro, A., Xue, M.: Impact of a vertical vorticity constraint in variational dual-doppler wind analysis: tests with real and simulated supercell data. J. Atmos. Oceanic Tech. **29**(1), 32–49 (2012)
36. Bradbury, J., et al.: JAX: composable transformations of Python+NumPy programs (2018). http://github.com/google/jax
37. Dolan, E.D., Moré, J.J.: Benchmarking optimization software with performance profiles. Math. Program. **91**, 201–213 (2002)
38. Moré, J.J., Wild, S.M.: Benchmarking derivative-free optimization algorithms. SIAM J. Optim. **20**(1), 172–191 (2009)
39. Fagan, M., et al.: Overcoming the power wall by exploiting inexactness and emerging COTS architectural features: trading precision for improving application quality. In: 2016 29th IEEE International System-on-Chip Conference (SOCC), pp. 241–246 (2016)
40. Galal, S., Horowitz, M.: Energy-efficient floating-point unit design. IEEE Trans. Comput. **60**(7), 913–922 (2011)

Incremental Mining of Frequent Serial Episodes Considering Multiple Occurrences

Thomas Guyet[1], Wenbin Zhang[2]([✉]), and Albert Bifet[3,4]

[1] Inria, Lyon Center, Villeurbanne, France
thomas.guyet@inria.fr
[2] Carnegie Mellon University, Pittsburgh, USA
wenbinzhang@cmu.edu
[3] University of Waikato, Hamilton, New Zealand
albert.bifet@waikato.ac.nz
[4] LTCI, Telecom Paris, Institut Polytechnique de Paris, Paris, France

Abstract. The need to analyze information from streams arises in a variety of applications. One of its fundamental research directions is to mine sequential patterns over data streams. Current studies mine series of items based on the presence of the pattern in transactions but pay no attention to the series of itemsets and their multiple occurrences. The pattern over a window of itemsets stream and their multiple occurrences, however, provides additional capability to recognize the essential characteristics of the patterns and the inter-relationships among them that are unidentifiable by the existing presence-based studies. In this paper, we study such a new sequential pattern mining problem and propose a corresponding sequential miner with novel strategies to prune the search space efficiently. Experiments on both real and synthetic data show the utility of our approach.

Keywords: Event sequence · Serial episode · Multiple occurrences

1 Introduction

Online mining of frequent patterns over a sliding window is one of the most important tasks in data stream mining with broad applications. In this case, the data stream is made of items or itemsets that arrive continuously. The aim is then to obtain a set of evolving frequent patterns over a sliding window, in which the most recent frequent patterns as well as their evolution are available at any time for information extraction. This motivates work on mining frequent patterns over series of items based on their presence in the stream [2,21]. In this paper, to gain additional information from the stream, we take one step further to extract frequent sequential patterns over a stream of itemsets but also to consider their multiple occurrences in the stream.

Mining frequent sequential patterns from a single long sequence S is better known as serial episode mining [11]. Under this setting, the support of a pattern is the number of times it occurs in S. The way to enumerate the multiple occurrences of a pattern turns out to be important to have the antimonotonicity of the measure. Among the possible enumeration strategies [1], the *minimal occurrences* is the most common [11] with the initial work discussed in [6]. With this property, the classical breadth-first search (like PrefixSpan [13]) or depth-first search algorithms (like GSP [15]) can be adapted to efficiently extract the complete set of frequent sequential patterns occurring in a static sequence. However, applying such algorithms to maintain the recent frequent patterns over the stream would be intractable. In addition, start from scratch each time a new item arrives in the stream is needed, but the computation cost, in practice, is unaffordable.

To address the aforementioned challenges, this paper introduces INCremental SEQuence (INCSEQ), a novel framework to efficiently extract frequent serial episodes over the stream of itemsets. To the best of our knowledge, this is the first work capable of mining series of itemsets incrementally without the need to start from scratch. To summarize, we present the following contributions:

- The formalization of a new incremental sequential pattern mining problem, which counts the exact number of occurrences of sequential patterns.
- A complete algorithm for incremental sequential pattern mining with efficient search space pruning.
- Extensive experiments on both real and synthetic datasets.

2   Basic Concepts and Problem Statement

Suppose that we have a set of items denoted $\mathcal{E}$ and $<$ defines the total order on this set (*e.g.* lexicographic order). An itemset $\beta = (b^i)_{i \in [m]} \subseteq \mathcal{E}$ is a sub-itemset of $\alpha = (a^i)_{i \in [n]} \subseteq \mathcal{E}$, denoted $\beta \sqsubseteq \alpha$, iff there exists a sequence of integers $1 \leq i_1 < i_2 < \cdots < i_m \leq n$ such that $\forall k \in [m], b^k = a^{i_k}$.[1] A *sequence* S is a finite ordered series of itemsets $S = \langle s_1, s_2, \ldots, s_n \rangle$. A *serial episode* (also called *sequential pattern* or pattern for short) is a *sequence*. The length of a sequential pattern S, denoted $|S|$, is the number of itemsets it contains. The total number of items in a pattern S is denoted $\|S\|$. $T = \langle t_1, t_2, \ldots, t_m \rangle$ is a *sub-sequence* of $S = \langle s_1, s_2, \ldots, s_n \rangle$, denoted $T \preceq S$, iff there exists a sequence of integers $1 \leq i_1 < i_2 < \cdots < i_m \leq n$ such that $t_k \sqsubseteq s_{i_k}$ for all $k \in [m]$.

The **minimal occurrences** [11] of a sequential pattern $S = \langle s_1, \ldots, s_n \rangle$ in a sequence $W = \langle w_1, \ldots, w_m \rangle$, denoted $\mathcal{I}_W(S)$, is the list of n-tuple of positions (within W):

$$
\begin{aligned}
\mathcal{I}_W(S) = \{ (i_j)_{j \in [n]} \in [m] \mid \ & \forall j \in [n], s_j \sqsubseteq w_{i_j}, && \text{(a)} \\
& \forall j \in [n-1], \ i_j < i_{j+1}, && \text{(b)} \\
& (w_j)_{j \in [i_1+1, i_n]} \not\preceq S, && \text{(c)} \\
& (w_j)_{j \in [i_1, i_n-1]} \not\preceq S \ \} && \text{(d)}
\end{aligned}
\tag{1}
$$

---

[1] $[n]$ denotes the set of the n first integers $\{1, \ldots, n\}$.

In Eq. 1, condition (a) requires that any itemset of S is a sub-itemset of an itemset of W, while condition (b) specifies the order of itemsets of W needs to respect. In addition, no itemset of W can be a super-itemset of two distinct itemsets of S. This condition does not impose any time constraint between itemsets. Conditions (c) and (d) specify minimal occurrences: if a minimal occurrence of S has been identified in the interval $[i_1, i_n]$, there can not be any minimal occurrence of S in a strict subinterval of $[i_1, i_n]$. For sake of simplification, "occurrence" denotes "minimal occurrence" in the remainder of this paper.

Then, the *support* of a sequential pattern S in sequence W, denoted $supp_W(S)$, is the cardinality of $\mathcal{I}_W(S)$, i.e. $supp_W(S) = card(\mathcal{I}_W(S))$. The support measure $supp_W(\cdot)$ is anti-monotonic on the set of sequential patterns with associated partial order $\preceq$ [16]. Given a threshold σ, we say that a sequential pattern S is *frequent* in a stream window W iff $supp_W(S) \geq \sigma$.

Mining frequent sequential patterns **incrementally** is therefore to extract frequent sequential patterns in a sequence $W = \langle w_1, \ldots, w_m \rangle$ from the ones in $W' = \langle w_0, \ldots, w_{m-1} \rangle$. This recursively mining of frequent sequential patterns enables to mine a stream of itemsets, i.e. to maintain the set of frequent sequential patterns in a window sliding over a stream of itemsets.

3   Incremental Algorithm for Sequential Patterns

Our proposed approach relies on representing the set of frequent sequential patterns (or patterns for short) in a tree structure inspired by the prefixing method of PSP [12]. PSP represents a set of frequent sequential patterns as a tree with two types of edges: the edges representing sequentiality $(\mathcal{S})$ between itemsets and the edges representing the composition $(\mathcal{C})$ of itemsets. Masseglia et al. [12] showed that such representation is memory efficient.

Formally, a tree node N is a 4-tuple $\langle \alpha, \mathcal{I}, \mathcal{S}, \mathcal{C} \rangle$ where:

- $\alpha = (a_i)_{i \in [n]}$ is a sequential pattern of size n,
- $\mathcal{I} = \mathcal{I}_W(\alpha)$, the list of minimal occurrences of α in W,
- $\mathcal{S}$ is the set of descendant nodes which represent patterns $\beta = (b_i)_{i \in [n+1]}$ of size $\|\alpha\| + 1$ such that $\forall i \in [n]$, $a_i = b_i$,
- $\mathcal{C}$ is the set of descendant nodes which represent patterns $\beta = (b_i)_{i \in [n]}$ of size $\|\alpha\| + 1$ such that $\forall i \in [n-1]$, $a_i = b_i$, $a_n \sqsubseteq b_n$ and $\forall j < |a_n|$, $a_n^j < b_n^{|a_n|+1}$, (i.e. itemset b_n extends itemset a_n with the item $b_n^{|a_n|+1}$).

A tree of frequent patterns, denoted $\mathcal{A}_\sigma(W)$, represents all patterns of W having a support greater than σ. The root node of a prefix tree is a node of the form $\langle \{\}, \emptyset, \mathcal{S}, \mathcal{C} \rangle$.

Let N be a node of $\mathcal{A}_\sigma(W)$. The subtree rooted at node N represents the tree composed of all descendants of N (including N). Owing to the anti-monotonicity property, we know that if a node has a support greater than or equal to σ then all its ancestors are frequent sequential patterns in W. In addition, each node – apart from the root – has a single parent. This ensures that a recursive processing

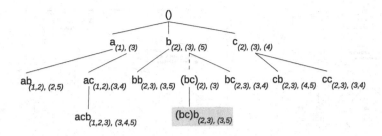

Fig. 1. Example of a tree of frequent sequential patterns $(\sigma = 2)$

of a PSP tree is complete and non-redundant. Figure 1 exemplifies the frequent PSP tree representation followed by its corresponding illustration.

Example 1. *Let $W = \langle a(bc)(abc)cb \rangle$ and $\sigma = 2$. Figure 1 shows the tree $\mathcal{A}_\sigma(W)$. Solid lines indicate membership in the set S (Succession in the sequential pattern), while the dotted lines indicate membership in the set C (Composition with the last itemset). The node $(bc)b$, highlighted in gray, has the pattern node (bc) as parent, since $(bc)b$ is obtained by concatenating b to (bc). The parent node of (bc) is (b) and is obtained by itemset composition (dotted line). At each node of Fig. 1, the list of minimal occurrences is displayed in the index. For example, the pattern $(bc)c$ has two occurrences: $\mathcal{I}(\langle(bc)c\rangle) = \{(2,3), (3,5)\}$.*

3.1 Illustration of the Algorithm

The incremental process aims at updating the tree of frequent patterns with respect to the most recent window of the stream and determining which patterns are frequent. The arrival of a new itemset in the stream triggers two steps: (1) the deletion of occurrences related to the first itemset in the window; (2) the addition of patterns and occurrences related to the new incoming itemset. The addition step incurs the majority of computational load involving three substeps: (i) merging sub-itemsets of the new itemset into the current tree, (ii) completing the lists of occurrences, and (iii) pruning nodes of non-frequent patterns. Our approach therefore performs the deletion step prior to the addition of a new itemset in order to reduce the size of the tree before the computational expensive merging and completion substeps.

Let us consider the window $W = \langle(abc)(ab)(ab)c\rangle$ of length 4, at position 1 of the stream. Assume that $\mathcal{A}_2(W)$, *i.e.* the tree of patterns with support greater than 2, has already been built. The following steps transform the tree of frequent patterns $\mathcal{A}_2(W)$ into the tree $\mathcal{A}_2(W')$ upon the arrival of the new itemset (bc). These steps are illustrated in Fig. 2 and detailed in the following.

1. Deletion of the first itemset: all occurrences starting at the first (oldest) position of the window (orange occurrences at position 1 in the example) are deleted. Then, patterns having a number of occurrences lower than $\sigma = 2$ are deleted from the tree. The result is the tree $\mathcal{A}_2(\langle(ab)(ab)c\rangle)$ where a, (ab), b are

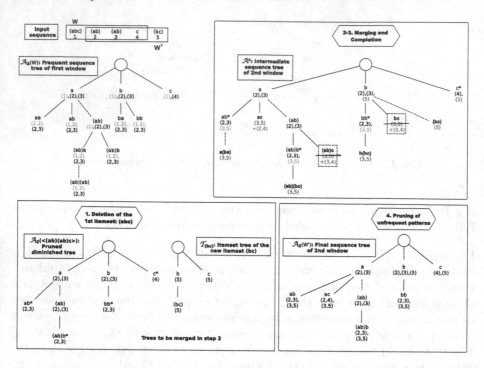

Fig. 2. Successive steps for updating the tree of frequent patterns upon the arrival of itemset (bc) in the window $W = \langle(abc)(ab)(ab)c\rangle$.

frequent. Quasi-frequent patterns (marked with asterisk in the example) are not frequent but may become frequent as they have a support equals to $\sigma - 1$ and they are ended by an item present in the new itemset, *i.e.* (bc). Such nodes are kept in the frequent tree with their occurrences as the following completion step (see below) is not necessary for them.

2. Merging the new current itemset (bc) with every node of the tree of patterns: this step generates all the new candidate patterns of the new window. Intuitively, a pattern is a new candidate (*i.e.* potentially frequent) only if it is the concatenation of a sub-itemset of (bc) to a frequent pattern of $\langle(ab)(ab)c\rangle$. In the tree representation of frequent patterns, this concatenation can be seen as extending each node of $\mathcal{A}_2(\langle(ab)(ab)c\rangle)$ with the itemset tree $\mathcal{T}_{(bc)}$ representing all sub-itemsets of (bc).

In Fig. 2, the tree $\mathcal{T}_{(bc)}$ is merged with the four non-quasi-frequent nodes of $\mathcal{A}_2(\langle(ab)(ab)c\rangle)$:

- with the root node (green occurrences): all subsequences of (bc) become potentially frequent.
- with the nodes a, (ab), b (blue occurrences): all patterns starting with one of these three patterns (frequent in $\langle(ab)(ab)c\rangle$) and followed by a sub-itemset of (bc) become potentially frequent.

We call this procedure "tree merging" because if a node already exists in the tree (*e.g.* node (*b*)), the occurrences related to the new itemset are added to the list of existing occurrences. The list of occurrences of (*b*) becomes $\{(2),(3),(5)\}$. We know that each of these nodes holds all the occurrences of the associated pattern in W'. New nodes are noted in bold face in the frequent tree after the merging step in Fig. 2. Each of these new nodes of $\mathcal{A}^f$, *e.g.* the node (*bc*), has an occurrence list consisting of only one occurrence of a sub-itemset of (*bc*). Quasi-frequent nodes (nodes marked with the asterisk) are not merged with the itemset tree $\mathcal{T}_{(bc)}$. Their occurrence lists are simply updated when needed.

3. Completion of occurrences' lists: Exclusively for new candidate nodes, it is necessary to scan the window W' once again to build the complete list of occurrences of a pattern. For example, the node ab is associated with the list $\{(3,5)\}$. This list must be completed with the occurrences of ab in the previous window $(\{(2,3)\})$. As $\langle ab \rangle$ was unfrequent in W, we must retrieve their occurrences. Red occurrences of the tree $\mathcal{A}^c$ in Fig. 2 show the occurrences added by completion.

4. Pruning non-frequent patterns: $\mathcal{A}^c$, the tree obtained after completion, contains new candidate patterns with complete lists of occurrences. The last step removes patterns with an occurrences' list of size strictly lower than $\sigma = 2$ yielding the tree $\mathcal{A}_2(W')$.

Algorithm 1. MERGING: merging the itemset tree $\mathcal{T}$ with every node of the tree of patterns $\mathcal{A}$.

```
 1: function MERGING(A, T)
 2:     T' ← T
 3:     for  N ∈ A do
 4:         for n ∈ T' do                        ▷ Prefixing T'
 5:             n.α = N.α ⊕ n.α                  ▷ Prefixing the pattern with N.α
 6:             for all I ∈ n.I do               ▷ Prefixing all occurrences
 7:                 I = d ∪ I                    ▷ d is the last element of N.I
 8:             end for
 9:         end for
10:         RECMERGE(T', N)        ▷ Recursive merging of T' with nodes N of A
11:     end for
12:     return A
13: end function
```

3.2   Merging a Tree of an Itemset into a Tree of Frequent Patterns

Now, we detail the merging step which integrates the itemset tree $\mathcal{T}$ into the pattern tree $\mathcal{A}$. Then, we explain the completion of occurrences.

Algorithm 1 describes how the itemset tree $\mathcal{T}$ is merged with every node of the frequent patterns tree $\mathcal{A}$. It consists of two main steps:

- prefixing the itemset tree $\mathcal{T}$ with the pattern of node N,
- recursively merging the prefixed $\mathcal{T}$ with descendants of node N (*cf.* Algorithm 2).

Let $N.\alpha$ be the pattern associated with a node N from the tree of patterns $\mathcal{A}$ and $N.\mathcal{I}$ be the list of occurrences associated with N. For each node N of $\mathcal{A}$, the itemset tree $\mathcal{T}$ is first prefixed by N: on the one hand, the patterns of each node of $\mathcal{T}$ are prefixed by $N.\alpha$; on the other hand, all occurrences of $\mathcal{T}$ are prefixed by the last occurrence of $N.\mathcal{I}$. Using the last occurrence in $N.\mathcal{I}$ enforces the third property (see Eq. 1).

Algorithm 2. RECMERGE: recursively merging the prefixed itemset tree $\mathcal{T}$ with a node of $\mathcal{A}$

Input: n: itemset node tree, N: node of the tree of patterns to be merged with n and
    such that $n.\alpha = N.\alpha$

1: **function** RECMERGE(n, N)
2:     $N.\mathcal{I} \leftarrow N.\mathcal{I} \cup n.\mathcal{I}$ ▷ Merging lists of occurrences
3:     **for** $s_N \in N.\mathcal{S} \cup N.\mathcal{C}$ **do**   ▷ Recursion
4:         **for** $s_n \in n.\mathcal{S} \cup n.\mathcal{C}$ **do**
5:             **if** $s_N.\alpha = s_n.\alpha$ **then**
6:                 $found \leftarrow$ **True**
7:                 RECMERGE(s_n, s_N)
8:             **end if**
9:         **end for**
10:        **if** not $found$ **then**
11:            **if** $s_n \in n.\mathcal{S}$ **then**
12:               $N.\mathcal{S} \leftarrow N.\mathcal{S} \cup \{\text{COPY}(s_n)\}$
13:            **else**
14:               $N.\mathcal{C} \leftarrow N.\mathcal{C} \cup \{\text{COPY}(s_n)\}$
15:            **end if**
16:        **end if**
17:     **end for**
18: **end function**

In a second step, the algorithm recursively merges the root of the itemset tree $\mathcal{T}$ prefixed by N. Algorithm 2 details this merging operation. We first need to make sure that $n.\alpha = N.\alpha$ to verify that the two nodes represent the same pattern. At line 2, occurrences of nodes n and N are merged. By construction of the new occurrence, the conditions of Eq. 1 are satisfied. Then, the descendants of n are processed recursively. For each node of $n.\mathcal{S}$ (resp. $n.\mathcal{C}$), we search a node s_n in $N.\mathcal{S}$ (resp. $N.\mathcal{C}$) such that these nodes represent the same pattern. If such a node is found, then the function RECMERGE is recursively applied. Otherwise, a copy of the entire subtree of s_n is added to $n.\mathcal{S}$ (resp. $n.\mathcal{C}$).

3.3   Completion of a List of Occurrences

When a new pattern is introduced in the tree, it means that it was unfrequent in the previous window, but there might exist occurrences of this pattern. They were simply not stored in the tree (except quasi-frequent patterns). For example, in Fig. 2, the pattern $\langle bc \rangle$ (node surrounded by a dotted line square) is not frequent in W and is not present in the frequent patterns tree $\mathcal{A}_2(W)$. However, after the arrival of itemset (bc) the pattern $\langle bc \rangle$ may become frequent in W. Thus, it is necessary to scan W' to retrieve all occurrences of $\langle bc \rangle$ to compute its frequency.

The completion algorithm is applied exclusively to the nodes newly introduced in the tree. While ensuring the completeness, this method reduces the number of completions. In addition, to make the completion efficient, the occurrences of a pattern β is recursively constructed from the occurrences of its direct parent along the following principles:

- each occurrence $I = (i_1, \ldots, i_{|\delta|})$ of a pattern δ obtained by adding an item e to the last itemset of β (composition) are necessarily occurrences of β, thus the algorithm tests whether e is included in the itemset $w_{i_{|\delta|}}$.
- each occurrence $I = (i_1, \ldots, i_{|\epsilon|})$ of a pattern ϵ, obtained by adding an itemset e to β (succession), are necessarily constructed by adding the element $i_{|\epsilon|}$ to an occurrence of β, thus the algorithm browses a sub-sequence of W' to test the presence of e.

For succession nodes, the completion scans only the sub-sequence of W' composed of the itemsets between $i_{|\beta|} + 1$ and $j_{|\beta|-1}$, where $J = (j_1, \ldots, j_{|\beta|})$ is the occurrence after I in the list of occurrences of β.

As an example, on the tree $\mathcal{A}^c$ in Fig. 2, the occurrences of $\langle bc \rangle$ is $\mathcal{I}(\langle bc \rangle) = \{(3, 5)\}$. This occurrence has been obtained during the merging step by adding the element 5 to the occurrence (3) of pattern $\langle b \rangle$. An occurrence of $\mathcal{I}(\langle bc \rangle)$ is the successor of one of the occurrences of $\langle b \rangle$: $\mathcal{I}(\langle b \rangle) = \{(2), (3), (5)\}$. To complete occurrence (3) from $\mathcal{I}(\langle b \rangle)$, the algorithm looks for one c in W' at a position between 3 $(= 2 + 1)$ and the beginning of the third occurrence of $\mathcal{I}(\langle b \rangle)$, i.e. 5. Here, occurrence (3, 4) is found. But it is a sub-sequence of an existing occurrence $(3, 5)$. Due to the definition of *minimal* occurrences (Eq. 1), $(3, 5)$ is deleted. The same for pattern $(ab)c$ (the other node surrounded by a dotted line square). It is not possible to complete occurrence (2) of $\mathcal{I}(\langle b \rangle)$ because there is no c in the itemset at position 3 (the only possible itemset between the occurrence of $\langle b \rangle$ at position (2) and the next occurrence in $\mathcal{I}(\langle bc \rangle)$).

It is worth mentioning that the proposed algorithm is complete. Specifically, in a streaming context which applies recursively the incremental mining process, it extracts all the frequent sequential patterns for each sliding window of the stream.

4   Experiments and Results

The objective of our experiments is to show that the proposed algorithm is an efficient strategy for mining sequential patterns incrementally. More specifically, we would like to assess the space and time efficiency of the proposal compared to a *Batch* approach, *i.e.* a strategy that does not exploit the incremental changes of the window. The *Batch* algorithm is based on PrefixSpan and uses the PSP tree structure. It rebuilds the entire tree $\mathcal{A}_\sigma(W)$ for each consecutive window of size ws on the data stream. To the best of our knowledge, there is no state-of-the-art competitor for this task.

It is worth noticing that the two approaches are complete and thus extract the exact same sets of patterns. For this reason, we do not discuss the algorithm outputs but only their efficiency.

In a first experiment, we present the result on synthetic data which have been widely used to evaluate the efficiency of sequential pattern mining algorithms. As the purely random nature of this data does not mimic the characteristics of true datasets (with less balanced itemset occurrences or with the presence of significant patterns), we supplement this experiment with an experiment on a real dataset. This dataset also illustrates the practical value and additional information gained by addressing the newly formulated sequential pattern mining problem of this work.

The algorithms were implemented in C++ and ran on a single core. The source code, synthetic datasets and benchmarks scripts are available online[2].

4.1   Experiments on Synthetic Data

In this section, we evaluate INCSEQ against *Batch* on synthetic datasets generated in the same way as the IBM quest data generator. Specifically, at each sequence position, an item is present with a probability of 3%, thus yielding a random sequence of itemsets. The length of the sequence simulating the stream is 1000 times of the windows size, which requires the incremental algorithm to be recursively called 1000 times in a run. The item vocabulary size, $card(\mathcal{E})$, is set to 40. Then, the average number of items per itemset is 1.2. The experiments were conducted by varying the parameters ws (window size from 80 to 300) and σ (minimal support from 3 to 10 occurrences) on 5 different datasets per configuration. The results reported are the average results of all the experiments.

Figure 3-(a) illustrates the execution time with respect to σ. As one can see, the execution time grows exponentially when σ decreases. Note that a timeout is set as 10 min. For more time-consuming mining tasks (with low σ), Batch failed 17 times before a successful completion of the mining process, while INCSEQ failed 16 times. It is also clear that INCSEQ, on average, is an order of magnitude faster than *Batch*. To further assess the superior efficiency of INCSEQ on mining various sizes of window, Fig. 3-(c) and (d) provide the execution time ratio between INCSEQ and *Batch* with respect to σ and ws, respectively. As one can see, INCSEQ dominates *Batch* by 10 to 20 times faster in processing time when σ and ws increase. The different drop for $\sigma = 10$ because the number of frequent patterns is closed to zero. Thus, the computing times are very low for the two approaches.

---

[2] https://gitlab.inria.fr/tguyet/seqstreamminer.

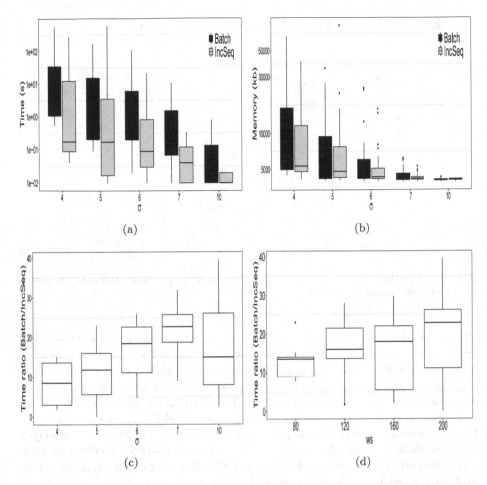

Fig. 3. Comparison of processing time (logarithmic scale) and memory usage with respect to the support threshold σ (with $ws < 25$) and the size of the sliding window ws. (c) and (d) represent the respective computing time ratio of Batch to INCSEQ on the same dataset.

Figure 3-(b) additionally shows the memory usage of the two approaches. As expected, the two approaches are comparable in terms of the memory usage as the required memory is mainly to store the frequent sequential patterns and the two approaches induce identical trees. We also observe that the memory requirement depends upon σ as the lower σ the more frequent patterns. Ensuring memory efficiency is also an essential prerequisite for sequential pattern mining, our proposed method therefore enjoys the advantage of mining sequential patterns with reduced time at no extra memory cost.

4.2   Experiments on Smart Electrical Meter Data

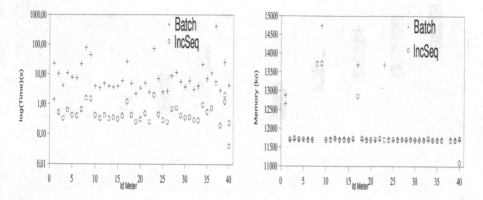

Fig. 4. Comparison of computation time (left) and memory usage (right) when mining the power consumption streams.

We also conducted experiments on real smart electrical meter data. Smart electrical meters record the power consumption of an individual or company in intervals of 30 min and communicate that "instant" information to the electric provider for monitoring and billing purposes. The aim of smart meters is to better anticipate the high consumption of a distribution sector by awarding a consumption profile to each meter, *i.e.* a dynamic model of changes in consumption. As consumption profiles depend on the period of year (seasons, holidays), week (weekdays, weekends) or day and are unpredictable for medium to long-term consumption, we employ INCSEQ and *Batch* to extract the dynamic online profiles of short-term consumption of the meters.

The annual series of instantaneous consumption is a flow of about 18,000 values. We use the SAX algorithm [10] for discretizing the consumption values. A vocabulary size of $|\mathcal{E}| = 14$ and a window aggregation of $PAA = 24$ have been chosen. The consumption profile of a smart meter at time t is the set of frequent consumption patterns during the period $[t - w, t]$ (sliding window of predefined size $w = 28$ itemsets, *i.e.* 2 weeks).

Figure 4 shows the results for 40 m. It is clear that the results obtained on the real data are consistent with those obtained on synthetic data. Specific to the real data, while most of the meters can be processed within seconds, the processing time of some meters are significantly longer (about few minutes). This disparity is attributed to the observed consumption variability. Specifically, the patterns that are more time-consuming to process are relatively constant (*e.g.* industrial consumption) consisting of many repetitions of symbols, thus lead to a large tree depth. It is however clear that the results of real and synthetic datasets conclusively match, which suggests that our proposed method is an efficient sequential pattern miner with manageable memory cost.

5   Related Work

In the field of stream mining, several approaches extended frequent pattern mining in a setting similar to ours. For example, Chang et al. [3] proposed to extract recent frequent patterns in a sliding window over a data stream, while Calders et al. [2] improved such approaches with the adaptive window size. More recently, Giacometti and Soulet [5] proposed a sampling of the pattern to improve the efficiency. Our approach focuses on more complex patterns, i.e., sequential patterns, to extract additional information with a similar streaming setup.

For sequential patterns, less efforts have been made in streaming settings [4]. The incremental or online sequential pattern mining algorithms in the literature address simplified problems of ours: mining frequent sequential patterns in a stream of transactions that are sequences, such as IncSPAM [7], or mining frequent sequential patterns in a collection of itemsets streams, such as PSP-AMS [8]; in both cases, the counting of sequential patterns is based on the number of transactions (resp. number of streams) in which a pattern occurs. However, all these algorithms examine the presence of a pattern in each transaction as the pattern counting method and ignore the multiple occurrences of the pattern in a transaction. Tseng et al. [17] share a similar objective, but their mining algorithms are not incremental. Their framework combines the results of episode mining by batches in a map-reduce architecture without the formal properties of INCSEQ.

Finally, our approach is also different from single-pass serial episodes mining algorithms [9] whose objective is not to maintain the set of frequent serial-episodes, but is to evaluate the support of serial episodes online.

6   Conclusion and Future Works

Although a number of studies have developed approaches to mine sequential patterns over data streams, all of these techniques focus on a stream of items and the number of transactions that contain patterns without considering their multiple occurrences. In this work, we present our incremental algorithm based on counting the minimal occurrences of the sequential patterns over the course of itemsets stream. Experimental studies indicate the superior computational efficiency of our approach compared to the non-incremental method. In the future, we plan to further extend it by considering the condensed representation such as maximum patterns and closed patterns in the context of incremental mining. One immediate future work is to extend these results in conjunction with our previous works [18,19] for fair pattern mining. A relevant avenue is to investigate the ubiquitous graph data representation [14,20] with unique challenges for example the independent and identically distributed (IID) data distribution.

References

1. Achar, A., Laxman, S., Sastry, P.S.: A unified view of automata-based algorithms for frequent episode discovery. CoRR abs/1007.0690 (2010)

2. Calders, T., Dexters, N., Gillis, J.J.M., Goethals, B.: Mining frequent itemsets in a stream. Inf. Syst. **39**, 233–255 (2014)
3. Chang, J.H., Lee, W.S.: A sliding window method for finding recently frequent itemsets over online data streams. J. Inf. Sci. Eng. **20**(4), 753–762 (2004)
4. Fournier-Viger, P., Lin, J.C.-W., Kiran, R.U., Koh, Y.S., Thomas, R.: A survey of sequential pattern mining. Data Sc. Pat. Reco. **1**(1), 54–77 (2017)
5. Giacometti, A., Soulet, A.: Reservoir pattern sampling in data streams. In: Proceedings of the ECML-PKDD, pp. 337–352 (2021)
6. Guyet, T., Quiniou, R.: Incremental mining of frequent sequences from a window sliding over a stream of itemsets, Actes IAF (2012)
7. Ho, C.-C., Li, H.-F., Kuo, F.-F., Lee, S.-Y.: Incremental mining of sequential patterns over a stream sliding window. In: International Conference on Data Mining-Workshops (ICDMW), pp. 677–681 (2006)
8. Jaysawal, B.P., Huang, J.-W.: PSP-AMS: progressive mining of sequential patterns across multiple streams. ACM Trans. Knowl. Discov. Data **13**(1), 1–23 (2018)
9. Li, H., Peng, S., Li, J., Li, J., Cui, J., Ma, J.: Counting the frequency of time-constrained serial episodes in a streaming sequence. Inf. Sci. **505**, 422–439 (2019)
10. Lin, J., Keogh, E., Lonardi, S., Chiu, B.: A symbolic representation of time series, with implications for streaming algorithms. In: Proceedings of the Workshop on Research Issues in Data Mining and Knowledge Discovery (2003)
11. Mannila, H., Toivonen, H., Verkamo, A.I.: Discovering frequent episodes in event sequences. J. Data Min. Knowl. Disc. **1**(3), 210–215 (1997)
12. Masseglia, F., Cathala, F., Poncelet, P.: The PSP approach for mining sequential patterns. In: Żytkow, J.M., Quafafou, M. (eds.) PKDD 1998. LNCS, vol. 1510, pp. 176–184. Springer, Heidelberg (1998). https://doi.org/10.1007/BFb0094818
13. Pei, J., et al.: Mining sequential patterns by pattern-growth: the PrefixSpan approach. Trans. Knowl. Data Eng. **16**(11), 1424–1440 (2004)
14. Le Quy, T., Roy, A., Iosifidis, V., Zhang, W., Ntoutsi, E.: A survey on datasets for fairness-aware machine learning. Data Min. Knowl. Disc. (2022)
15. Srikant, R., Agrawal, R.: Mining sequential patterns: generalizations and performance improvements. In: Apers, P., Bouzeghoub, M., Gardarin, G. (eds.) EDBT 1996. LNCS, vol. 1057, pp. 1–17. Springer, Heidelberg (1996). https://doi.org/10.1007/BFb0014140
16. Tatti, N., Cule, B.: Mining closed strict episodes. Data Min. Knowl. Disc. **25**(1), 34–66 (2012)
17. Tseng, J.C.C., Gu, J.-Y., Wang, P.F., Chen, C.-Y., Li, C.-F., Tseng, V.S.: A scalable complex event analytical system with incremental episode mining over data streams. In: Proceeding of Congress on Evolutionary Computation, pp. 648–655 (2016)
18. Zhang, W., Weiss, J.: Longitudinal fairness with censorship. In: Proceedings of the AAAI Conference on Artificial Intelligence (2022)
19. Zhang, W., Weiss, J.: Rethinking fairness: new definitions and algorithm for fair machine learning under uncertainty. Knowledge and Information Systems (2022)
20. Zhang, W., Weiss, J.C., Zhou, S., Walsh, T.: Fairness amidst non-iid graph data: a literature review. arXiv preprint arXiv:2202.07170 (2022)
21. Zihayat, M., Cheng-Wei, W., An, A., Tseng, V.S., Lin, C.: Efficiently mining high utility sequential patterns in static and streaming data. Proc. Intell. Data Anal. **21**, 103–135 (2017)

MHGEE: Event Extraction via Multi-granularity Heterogeneous Graph

Mingyu Zhang[1,2], Fang Fang[1,2(✉)], Hao Li[1,2], Qingyun Liu[1,2], Yangchun Li[3], and Hailong Wang[3]

[1] Institute of Information Engineering, Chinese Academy of Sciences, Beijing, China
{zhangmingyu,fangfang0703,lihao1998,liuqingyun}@iie.ac.cn
[2] School of Cyber Security, University of Chinese Academy of Sciences, Beijing, China
[3] Chinese Academy of Cyberspace Studies, Beijing, China

Abstract. Event extraction is a key task of information extraction. Existing methods are not effective due to two challenges of this task: 1) Most of previous methods only consider a single granularity information and they are often insufficient to distinguish ambiguity of triggers for some types of events. 2) The correlation among intra-sentence and inter-sentence event is non-trivial to model. Previous methods are weak in modeling interdependency among the correlated events and they have never modeled this problem for the whole event extraction task. In this paper, we propose a novel Multi-granularity Heterogeneous Graph-based event extraction model (**MHGEE**) to solve the two problems simultaneously. For the first challenge, MHGEE constructs multi-granularity nodes, including word, entity and context and captures interactions among nodes by R-GCN. It can strengthen semantic and distinguish ambiguity of triggers. For the second, MHGEE uses heterogeneous graph neural network to aggregating the information of relevant events and hence capture the interdependency among the events. The experiment results on ACE 2005 dataset demonstrate that our proposed MHGEE model achieves competitive results compared with state-of-the-art methods in event extraction. Then we demonstrate the effectiveness of our model in ambiguity of triggers and event interdependency.

Keywords: Event extraction · Heterogeneous graph · R-GCN

1 Introduction

Generally, event extraction(EE) tasks consist of two subtasks. **Event detection** aims to identify and classify event triggers, and **argument extraction** aims to

Supported by the National Key R&D Program of China (2021YFB3101300) and the Strategic Priority Research Program of Chinese Academy of Sciences (Grant No. XDC02030000).

identify arguments of event and label their roles. In Fig. 1, event detection task aims to identify the trigger "dropped" and event type "Conflict: Attack", then event argument extraction task aims to identify the arguments "U.S. planes" and "Iraqis", and their roles "Attacker" and "Victim".

Sentence	Context	Event Extraction		
[S1] A Russian Soyuz capsule (VEH) dropped (*Transport*) the astronauts (PER) off this morning after docking perfectly at the orbiting outpost.	...An American astronaut (PER) and Russian cosmonaut (PER) are settling in at the International Space Station (VEH).	Trigger	dropped	
	... ([S1])	Event type	Transport	
	It was the first manned flight to the space station (VEH) since the February "Columbia" disaster. ...	Arguments	astronauts	capsule
		Roles	Person	Vehicle
[S2] And these pictures show Iraqis (PER) running for cover just before U.S. planes (WEA) dropped (*Attack*) a bomb (WEA) near northern Iraq.	...They say some Iraqis who worked in the field before the war already are asking for their jobs back.	Trigger	dropped	
	... ([S2])	Event type	Attack	
	Just as they disappear over the ridge, you can hear the planes (WEA) and the explosions and then you see the cloud....	Arguments	planes	Iraqis
		Roles	Attacker	Victim
[S3] U.S. President Bush (PER) ordered the firing (*Attack*) on Monday.	...Terrorism threats against the U.S. homeland is growing.	Trigger	firing	
	... ([S3])	Event type	Attack	
	The military action is expected to result in more civilian casualties (PER).	Arguments	Bush	Monday
		Roles	Attacker	Time

Fig. 1. Example documents in ACE 2005 corpus. Triggers and event types are marked in red. Arguments and roles are marked in other colors. The event extraction results of the three sentences are on the far right of the figure. (Color figure online)

There are a lot of research works about sentence-level event extraction, and they still face two critical challenges. 1) **Ambiguity of Triggers**: A word will express different meanings in different sentence so as it will trigger different events. In Fig. 1, both S1 and S2 contain the word "dropped". It would be quite challenging to detect the word "dropped" trigger an "Transport" event in S1 and trigger an "Attack" event in S2 without considering more information with different granularities, such as argument role (Weapon, Place , etc.) and context information. Only the event is detected correctly, can the argument in the event be extracted better. Therefore, how to distinguish different event semantics by a comprehensive understanding of information with different granularities is crucial for improving the accuracy of event extraction. 2) **Event Interdependency**: A sentence may express several correlated events simultaneously. For example, in the event mention "Three people plus the bomber were killed, and at least 30 others were hurt.", a "Die" event is triggered by "killed", and an "Injure" event is triggered by "hurt". This kind of event co-occurrence is also called multiple-event in one sentence. And it is common in ACE 2005 corpus. According to the previous statistics by [9], nearly 27% sentences have more than one events. And these events are often associated with each other, having similar event types with the same roles of arguments. Only modeling the interdependency among them, which is the fundamental to successful extraction, can we extract all events in a sentence correctly, so as to prompt the effect of event extraction. Therefore, how to effective model on such interdependency among the correlated events is one of the key challenges in event extraction.

For **challenge 1)**, most of the existing methods only consider a single granularity information [1,9,11,17,18], especially the inter-sentence information, using entity type or dependency tree. These methods try to make the best of the sentence granularity information to distinguish semantics. Unfortunately, ambiguity can't be solved only by inter-sentence information in many cases. For example, in Fig. 1, it is impossible to distinguish the event type "Attack" or "End_Position" by using entity information in S3. It requires context information "military action" and "casualties". Therefore, we should comprehensively handle the information with different granularities to distinguish different semantic. [2,7] use document granularity information, but bringing a lot of redundant information in the document. Meanwhile, they also neglect sufficient understanding of sentence granularity information. Because of that, there is no obvious improvement in the effect and no method to consider the influence of argument in event, especially roles. For **challenge 2)**, some methods use recurrent neural network to remember previous correlated events [11,13], but they still have the long-distance dependence problem. Another kind of method is graph neural network (GNN), which can effectively model the interdependency among nodes [3,9,12,17]. In order to model the interdependency of events, they construct a graph of word nodes in sentence by dependency tree. But they only consider the words of sentence and are used in the event detection subtask [3,12,17]. That is to say, they not only neglect the key multi-granularity information mentioned above, but also neglect another argument extraction subtask in event extraction and the interaction between the two subtasks.

In order to solve the above problems, inspired by [14], we propose a **Multi-granularity Heterogeneous Graph** model for sentence-level **Event Extraction (MGHEE)** to complete the task of event detection and argument extraction simultaneously. Unlike previous works which only take words as nodes, MHGEE contains another two types of nodes with different granularities: entity and context. Besides, we construct six types of edges. We design three type of nodes by considering the nearest context of the sentence and inter-sentence information to learn multi-granularity semantic information. Then we use R-GCN to enable rich interactions among nodes, so as to distinguish the semantic information of the same trigger word in different events. At the same time, our model also constructs heterogeneous graph to model intra-sentence and inter-sentence event interdependency by aggregating the information of relevant events in the same sentence or context, so as to solve the challenge of multiple events in one sentence. The contributions of this paper can be summarized as follows:

- We propose a novel event extraction model based on Multi-granularity Heterogeneous Graph (MHGEE). Our MGHEE designs multi-granularity nodes and enables rich interactions among nodes by R-GCN, which strengthens the semantic and helps distinguish ambiguity of triggers.
- We are the first to construct heterogeneous graph for whole event extraction. Our MHGEE can model intra-sentence and inter-sentence event interdependency and capture multiple events in one sentence effectively

– Experiments on the ACE 2005 dataset show that our model outperforms the previous SOTA models by nearly 5% F1 on the trigger identification and 2% F1 on the argument identification.

2   Related Work

From the perspective of text, event extraction can be divided into sentence-level event extraction and document-level event extraction. And sentence-level event extraction can be divided into extraction and generation methods. Our work focuses on sentence-level event extraction. We will classify the relevant works based on deep learning from the perspective of the methods used.

Event extraction models based on basic neural network have been widely used to extract features automatically, such as convolutional neural networks (CNN) [1], recurrent neural networks (RNN) [11,13].

Some works operated BERT as the pretrained language model [4,10,16] in recent several years, since BERT has been proven its validity to improve the performance of downstream natural language processing tasks including event extraction.

And with the application of GNN in various fields of natural language processing, some researchers propose to transform the syntactic dependency tree, which contains syntactic information and plays an important role in event extraction, into a graph and employ GCN [5] to conduct event detection through information propagation over the graph [3,12,17]. These works only consider event detection task and ignore argument information.

However, the above existing methods, all focus on the single granularity information in sentence-level, neglecting the multiple granularity information aggregation across sentences. Considering that adjacent sentences from context also store some relevant event information to solve the above challenges, these methods will not integrate multiple granularity information, which would enhance the event signals of the sentence that triggers belong to.

3   Approach of MHGEE Model

Our MHGEE model consists of the following four modules: 1) **Input Layer**: we aim to get initialization vector representations of words, entities, and contexts; 2) **Graph Construction**: we build multi-granularity heterogeneous graph, including three types of nodes and six types of edges; 3) **Information Aggregation over MHG**: we use R-GCN algorithm with gating mechanism to propagate information among multi-granularity information sources, so as to enhance information flow from context and entity nodes for event extraction; 4) **Classification Layer**: we obtain the final embedding representation of words and entities, and get trigger candidates of certain types from trigger labels in BIO form annotation schema, then predict the roles that each entity plays in such events after aggregating word embedding representations to trigger candidate vector t_i and entity vector e_i. Figure 1 gives the architecture of MHGEE model.

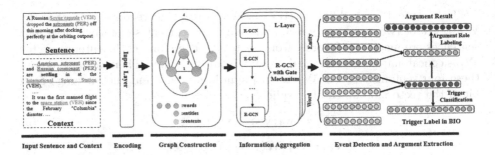

Fig. 2. The architecture of our MHGEE model. Different types of nodes are represented by circles with different colors, and similarity, different types of edges are represented by lines with different numbers in Graph Construction module. Due to space limitations, not all nodes and edges are represented in the graph.

3.1 Input Layer

We need to get the initial embedding representation vector of words, entities and contexts respectively. Let $W = w_1, w_2, \ldots, w_n$ be a sentence of length n where w_i is the i-th word. Similarly, let $E = m_1, m_2, \ldots, m_k$ be the entity in the sentence where m_k is the k-th entity.

The word embedding vector $\mathbf{x_i}$. In order to get the word embedding, each token w_i in the sentence is transformed to a real-valued vector $\mathbf{x_i}$ by looking up in embedding matrices and concatenating the following vectors: 1) **The word embedding vector of $\mathbf{w_i}$:** The word embedding vector is obtained by looking up a pre-trained word embedding matrix GloVe; 2) **The POS-tagging label embedding vector pos$_\mathbf{i}$:** The POS-tagging label embedding is generated by looking up the randomly initialized POS-tagging label embedding table; 3) **The positional embedding vector $\mathbf{p_i}$:** If w_c is the current word in a sentence, then we encode the relative distance $i - c$ from w_i to w_c as a real-valued vector by looking up the randomly initialized position embedding table [11,12]; 4) **The entity type label embedding vector $\mathbf{n_i}$:** Similarly to the POS-tagging label embedding vector of w_i, we annotate the entities in a sentence using BIO annotation schema and transform the entity type labels to real-valued vectors by looking up the embedding table. Thus, the input embedding of w_i can be defined as:

$$\mathbf{x_i} = [\mathbf{w_i}; \mathbf{pos_i}; \mathbf{p_i}; \mathbf{n_i}] \in \mathbb{R}^{d_w + 2 \times d_p + d_{pos} + d_n} \tag{1}$$

where d_w, d_p, d_{pos} and d_n denote the dimension of word embedding, positional embedding, POS-tagging label embedding and entity type embedding respectively.

The entity embedding vector $\mathbf{e_i}$. We calculate the entity embedding vector $\mathbf{e_i}$ with the mean-pooling operation of the vectors of all words, from w_1 to w_n, that make up the entity m_i. Thus, the input embedding of $\mathbf{e_i}$ can be defined as:

$$\mathbf{e_i} = \text{mean-pooling}\left(\sum_{w_i \in m_i}^{n} \mathbf{w_i} \right) \in \mathbb{R}^{d_e} \tag{2}$$

where d_e denotes the dimension of entity.

The context embedding vector c_i. We take the vectors generated by Word2Vec of two sentences above and below the current sentence. These four sentences are made up of words, from w_1 to w_i. Therefore, each sentence vector $\mathbf{W}_j$ is concatenated from all word vectors, from $\mathbf{w}_1$ to $\mathbf{w}_m$. Then we concatenate these four sentence vectors $\mathbf{W}_1$ to $\mathbf{W}_4$ simultaneously, and then average each bit of the vectors obtained after splicing:

$$\mathbf{W}_j = [\mathbf{w}_1; \mathbf{w}_2; ...; \mathbf{w}_m] \in \mathbb{R}^{d_w} \tag{3}$$

$$\mathbf{c}_i = [\mathbf{W}_1; \mathbf{W}_2; \mathbf{W}_3; \mathbf{W}_4] \in \mathbb{R}^{d_c} \tag{4}$$

where d_c denotes the dimension of context.

3.2   Graph Construction

We construct the graph in a multi-granularity way, motivated by the fact that each sentence constituting context contains multiple entities. Then we design three types of nodes to learn multi-granularity semantic information by considering the nearest context, the inter and intra sentence information of sentences: word, entity and context. Here we consider that entities and words are not one-to-one correspondence. Then the MHGEE model is expected to aggregate information from different granularities, as well as model interactions among these nodes for event extraction.

We also define the following six types of edges to reflect the various structural information and intra-sentence and inter-sentence event interdependency in MHGEE. 1) **Word-Word Edge:** via syntactic dependency tree. Then we make these edges exist based on the following assumptions: 2) **Word-Word Edge:** if a word may be the trigger since it has been a trigger ever; 3) **Word-Entity Edge:** if a word belongs to an entity; 4) **Word-Entity Edge:** if an entity and a word ever appeared in a centain event; 5) **Entity-Entity Edge:** if the types of these two entities have been arguments involved in the same event before; 6) **Context-Entity Edge:** if an entity appears in context. These edges enable nodes with different granularities to connect with each other simultaneously with a short path, and enables the MHGEE model to learn node representations specific to different edge types. Different edges are used to learn information of different granularity, which is related to the type of nodes they connect.

3.3   Information Aggregation over MHG

Since previous GNN only considers the node-wise connectivity, ignoring edge types, we employ R-GCN to perform information dissemination over our model. R-GCN performs well at handling high-relational data and distinguishing six different edge types when updating nodes, and the information dissemination over graph nodes can be achieved by aggregation and combination. The update process of the i-th node at the l-th layer can be formally formulated as:

$$\mathbf{n}_i^{(l)} = \frac{1}{|\mathcal{N}_i|} \sum_{j \in \mathcal{N}_i} \sum_{j \in \mathcal{R}_{ij}} f_r \left(\mathbf{h}_j^{(l)} \right) \tag{5}$$

$$\mathbf{u}_i^{(l)} = f_s\left(\mathbf{h}_i^{(l)}\right) + \mathbf{n}_i^{(l)} \tag{6}$$

where $\mathcal{N}_i$ is the set of neighbors of node i, R_{ij} is the set of edge types between i and j, $\mathbf{h}_j^{(l)}$ is the representation of node j in layer l. f_r is a parametrized function specific to an edge type $r \in R$, and both f_r and f_s are implemented with a MLP, $\mathbf{u}_i^{(l)}$ represents the updated representation of node i.

We apply a gate mechanism to provide the way to prevent completely over-writing past information in this module, since it has been shown that GNNs suffer from the smoothing problem when the number of layers is large [5]. Formally:

$$\mathbf{g}_i^{(l)} = \sigma\left(f_g\left(\left[\mathbf{u}_i^{(l)};\mathbf{h}_i^{(l)}\right]\right)\right) \tag{7}$$

where σ is the sigmoid function and f_g is implemented with a MLP. Gating vector $\mathbf{g}_i^{(l)}$ is then applied to control the amount information from neighbor nodes or the original node:

$$\mathbf{h}_i^{(l+1)} = \phi\left(\mathbf{u}_i^{(l)}\right) \odot \mathbf{g}_i^{(l)} + \mathbf{h}_i^{(l)} \odot \left(1 - \mathbf{g}_i^{(l)}\right) \tag{8}$$

where ϕ is the tanh function and $\odot$ denotes element-wise multiplication. After L times of information dissemination, the information of each node will be propagated to L-node distance away, generating L-hop-reasoning relation-aware node representations.

3.4 Classification Layer

We formulate event extraction as a sequence labeling task following previous works [1,8,11,17]. Thus, each word in sentence is assigned a label that contributes to event annotation. We apply the BIO annotation schema to assign trigger label t_i to each token w_i, as there are triggers that consist of multiple tokens, and tag "O" represents the "Other" tag, which means that the corresponding word is irrelevant of the target events. In addition, another two tags "B-type" and "I-type" consist of two parts: the word position in the trigger and any event type.

After aggregating word and entity node embedding representations from R-GCN, we feed the word representation into a fully-connected network, which is followed by a softmax function to compute distribution over all event types:

$$y_{t_i} = \text{softmax}\left(\mathbf{W}_t\mathbf{h} + b_t\right) \tag{9}$$

where $\mathbf{W}_t$ maps the word node representation $\mathbf{h}$ to the feature score for each event type, and b_t is a bias term. We choose event label with the largest probability as the classification result according to the value of y_{t_i}.

After we get trigger candidates of certain types from trigger labels, we then need to predict the roles that each entity e_j plays in such events. We aggregate word embedding representations to trigger candidate vector t_i and entity vector e_j by average pooling along the sequence length dimension. The trigger candidate vector t_i consists of words that combine to form the trigger. Then we concatenate them together and feed into a new fully-connected network to predict the argument role as:

$$y_{a_{ij}} = \text{softmax}\left(\mathbf{W}_a\left[t_i, e_j\right] + b_a\right) \tag{10}$$

where $y_{a_{ij}}$ represents the final output of which role the j-th entity plays in the event triggered by the i-th trigger candidate, and b_a is also a bias term.

3.5   Biased Loss Function

We minimize the joint negative log-likelihood loss function with a bias item as follow:

$$J(\theta) = -\sum_{k=1}^{N}\left(\sum_{i=1}^{n_k} I(O)\log\left(p\left(y_{t_i}\mid\theta\right)\right) + \beta\sum_{i=1}^{t_k}\sum_{j=1}^{e_k}\log\left(p\left(y_{a_{i,j}}\mid\theta\right)\right)\right) \tag{11}$$

where N is the number of sentences in training dataset; n_p, t_p and e_p are the number of words, extracted trigger candidates and entities of the k-th sentence; $I(O)$ represents a switching function to distinguish the loss of tag "O" and event type tags, it outputs number 1 if the tag is "O", otherwise 0; β is a bias weight.

4   Experiments and Results

4.1   Experiment Settings

Dataset and Evaluation Metrics. We conduct our whole experiments on the standard supervised ACE 2005 dataset, which consists of 599 documents annotated with 33 event subtypes, and 34 role classes. Then we add the NONE class and BIO annotation schema to role classes. Therefore, the total number of labels for event detection is 67, and the total number of labels for argument extraction is 37. Tag "O" in both subtasks represents the "Other" tag, which means that the corresponding word is irrelevant of any types. We use the same data split method [1,11,16,17] to compare with the previous works. The data split includes 40 articles with 881 sentences for the test set, 30 other documents with 1087 sentences for the development set and 529 remaining documents with 21,090 sentences for the training set. We follow the traditional evaluation metrics for evaluation: 1) Trigger Identification (TI); 2) Trigger Classification (TC); 3) Argument Identification (AI); 4) Argument Classification (AC). We use the official scorer Precision, Recall and F1-score at the evaluation stage.

Hyper-parameter Setting. The learning rate and batch size we set in our experiments is 2 and 32 respectively. For all experiments below, we use 300 dimensions for word embeddings and 50 dimensions for POS-tagging embedding, positional embedding and entity type embedding. In the R-GCN module, we use a two-layer GCN. The bias parameter in biased loss function β is set to 5.

4.2   Baselines

We compare our proposed MHGEE model with a range of state-of-the-art models in order to comprehensively evaluate performance boost results: 1) **DMCNN** [1], builds a dynamic multi-pooling convolutional model to learn sentence feature; 2) **Cross-Event** [7], uses document level information to improve the performance; 3) **GAIL** [18], bases on an inverse reinforcement learning; 4) **JointBeam** [6], extracts events based on structure prediction by manually designed features; 5) **Joint3EE** [15], bases on the shared hidden representations; 6) **JRNN** [11], employs bidirectional RNN and manually designed features to event extraction jointly; 7) **Embedding+T**: uses word embedding vectors and the traditional sentence-level features; 8) **PSL** [8], uses a probabilistic reasoning model to classify events; 9) **HBTNGMA** [2], models sentence event inter-dependency via a hierarchical and bias tagging model. Some baseline methods operate BERT as the pre-trained language model. 10) **BERT_QA** [4], is a QA-based model which uses machine reading comprehension model for both two subtasks; 11) **TEXT2EVENT** [10], presents a generation-based paradigm; 12) **DMBERT** [16], mainly focuses on the training data augmentation, with external unlabeled data through adversarial mechanism. And some models build a GNN over the dependency tree of a sentence to exploit syntactical information. 13) **GCN-ED** [12], is the first attempt to explore how to effectively use GCN in event detection; 14) **JMEE** [9], enhances GCN with self-attention and highway network to improve the performance of GCN for event detection; 15) **MOGANED** [17], improves GCN with aggregated attention to combine multi-order word representation from different GCN layers.

4.3   Overall Performance and Ablation Analysis

Table 1 shows the overall performance. Our MHGEE model achieves the best F1 scores for event extraction among all the compared methods. There is a significant gain with the trigger identification, which is nearly 5% higher over the best-reported models. There is also a significant gain with the argument identification, which is over 2% higher over the best-reported models. In addition, our MHGEE model still outperforms BERT-based models without using BERT as a pre-trained language model, although the encoder of BERT has been proven its validity to improve the performance of event extraction, which is one of the downstream natural language processing tasks. It demonstrates the effectiveness of aggregating information with different granularities for event extraction tasks. Compared with the previous GNN-based models, our MHGEE model complete the subtask of argument extraction with the consideration of argument information, and the interaction between two subtasks. This information interaction between arguments and trigger has a good effect on improving the performance of event extraction.

Table 1. Overall performance comparing to the SOTA methods

Models	TI			TC			AI			AC		
	P	R	F1	P	R	F1	P	R	F1	P	R	F1
DMCNN	80.4	67.7	73.5	75.6	63.6	69.1	68.8	51.9	59.1	62.2	46.9	53.5
Cross-Event		N/A		68.7	68.9	68.8	50.9	49.7	50.3	45.1	44.1	44.6
GAIL	78.9	66.5	72.2	75.3	63.4	68.9	**69.8**	52.7	60.0	61.6	45.7	52.4
JointBeam	76.9	65.0	70.4	73.7	62.3	67.5	69.8	47.9	56.8	**64.7**	44.4	52.7
Joint3EE	70.5	74.5	72.5	68.0	71.8	69.8	59.9	59.8	59.9	52.1	52.1	52.1
JRNN	68.5	75.7	71.9	66.0	73.0	69.3		N/A			N/A	
Embedding+T	76.9	65.0	70.4	73.7	62.3	67.5	69.8	47.9	56.8	64.7	44.4	52.7
PSL		N/A		75.3	64.4	69.4		N/A			N/A	
HBTNGMA		N/A		77.9	69.1	73.3		N/A			N/A	
BERT-based												
BERT-QA	74.3	77.4	75.8	71.1	73.7	72.4	58.9	52.1	55.3	56.8	50.2	53.3
TEXT2Event		N/A		69.6	74.4	71.9		N/A		52.5	55.2	53.8
DMBERT		N/A		71.6	72.3	70.9		N/A		53.1	54.2	52.8
GNN-based												
JMEE	80.2	72.1	75.9	76.3	71.3	73.7		N/A			N/A	
MOGANED		N/A		**79.5**	72.3	75.7		N/A			N/A	
GCN-ED		N/A		77.9	68.8	73.1		N/A			N/A	
MHGEE	79.8	**81.3**	**80.5**	75.1	**76.0**	**75.7**	63.8	**61.5**	**62.6**	56.6	54.5	**55.5**

Table 2 shows the ablation analysis of our study. We assume that if one type of nodes has been removed, it means the corresponding edges also do not exist in the heterogeneous graph. The F1 score drops more than 2 points regardless of the different edge types, context nodes or entity nodes we remove. If we remove entity nodes, we observe a more significant decline on F1 score than we remove context nodes. It indicates that all kinds of nodes and edges in our MHGEE model play important roles, but entity nodes are more essential. This is because when we alleviate the challenges, we more dependent on the entity information, which means entity nodes can be used as key nodes. If there is no entity information in context to help us determine the triggers, then context is not so necessary.

Table 2. Results of ablation studies on ACE 2005 dataset

Setting	TI	TC	AI	AC
Full (using identified trigger)	80.4	75.6	48.9	44.6
Full (using golden trigger)	**80.5**	**75.7**	**62.6**	**55.5**
- Different Edge Types	77.3	71.5	60.1	52.2
- Context Nodes	77.9	72.3	59.8	54.2
- Entity Nodes	77.1	72.2	57.6	53.4

Additionally, under the condition of using identified trigger, the F1 score of event detection task will not drop significantly. However, the F1 score drops by more than 10% in event argument extraction task. This result shows that we utilize golden trigger other than identified trigger to complete the event argument detection task, since identified trigger can cause the error propagation problem.

Overall, information with different granularities, and all edges with different types, can promote the interaction among nodes by R-GCN, in order to capture the information from the multi-granularity heterogeneous graph to complete event extraction, and finally benefits the performance.

4.4   Effect on Event Interdependency

Following some previous works [1,9,11], we split the test data into two parts: 1/1 and 1/N to evaluate the effect of our model for alleviating the multiple-events phenomena. 1/1 means that one sentence only has a single trigger, and 1/N for all remaining cases. We perform evaluations separately.

Table 3. Performance on single event sentences and multiple event sentences

Setting	Model	1/1	1/N	All
Trigger classification	Embedding+T	68.1	25.5	59.8
	CNN	72.5	43.1	66.3
	DMCNN	74.3	50.9	69.7
	JRNN	75.6	64.8	69.3
	JMEE	75.2	72.7	73.7
	HBTNGMA	**78.4**	59.5	73.3
	MHGEE	74.8	**75.8**	**75.7**
Argument classification	Embedding+T	37.4	15.5	32.6
	CNN	51.6	36.6	48.9
	DMCNN	**54.6**	48.7	53.5
	JRNN	50.0	**55.2**	55.4
	MHGEE	43.7	52.7	**55.5**

We use F1 scores to illustrate the performance of Embedding+T [6], CNN [1], JRNN [11], DMCNN [1], and our model for event extraction in Table 3. CNN is similar to DMCNN, except that it applies the standard max-pooling mechanism. Our MHGEE model significantly outperforms all the other mentioned methods in trigger classification subtask. In the **1/N** data split of triggers, our model is 3.1% better than the JMEE. It demonstrates that our model, with the utilization of multi-granularity heterogeneous graph and the model of intra-sentence and inter-sentence event interdependency, can capture multiple events in one sentence effectively.

Table 4. Performance on single event sentences and multiple event sentences.

Setting	TI	TC	AI	AC
All	80.5	75.7	62.6	55.5
1/1	79.5	74.8	51.8	43.7
1/N	**81.4**	**75.8**	**63.6**	**58.0**

Table 4 shows the event extraction effect of our MHGEE model on both 1/1 and 1/N. Our model performs better on 1/N. It indicates that multiple granularity information has greater gain on distinguishing the semantics between different triggers in one sentence, that is, event interdependency, which sourced from the fact that multiple events are often associated with each other, having similar event types. We model this intra-sentence and inter-sentence event interdependency phenomenon through a heterogeneous graph to not only capture multiple events in one sentence, but also mitigate their similarity.

4.5 Case Study and Effect on Ambiguity of Triggers

In Fig. 3, we show two examples of case study on ambiguity of triggers. In both (a) and (b), both words "discuss" and "fight" trigger two different events respectively. Ambiguity occurs in both cases. However, our MHGEE model can solve this problem in (a), but fails in (b). According to the idea of our model, we need

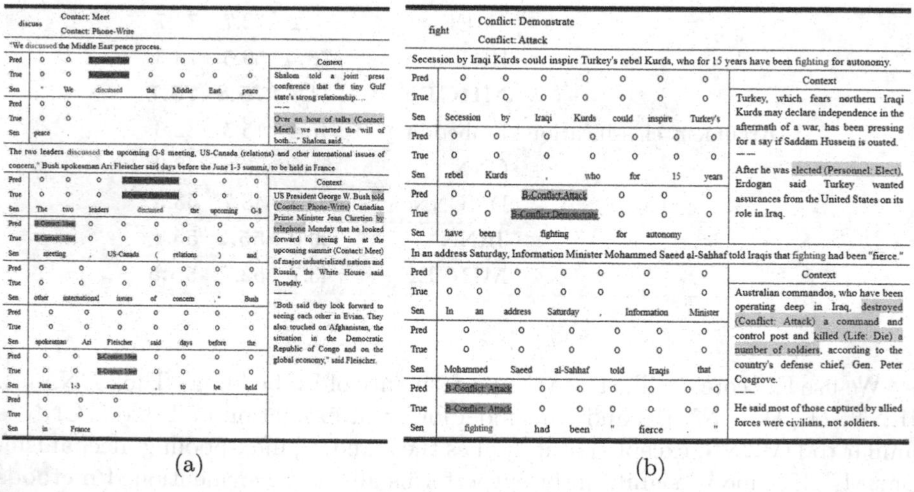

(a)   (b)

Fig. 3. The example of case study. Yellow highlighted content indicates entity information that can be used to solve ambiguity. The context of the sentence where the trigger is located, as well as the true BIO annotation and predicted BIO annotation results of the sentence through other colors are shown. If the colors are the same, our model predicts correctly, otherwise it is different. (Color figure online)

to learn rich different granularity information, including entities and contexts, to solve the ambiguity problem. In (a), there is enough information to help us solve the ambiguity problem, but in (b), there is not enough semantic information.

The two event types triggered by "fight" have certain similarities, and "elected" in context serves as a trigger word for event type "Personnel: elected", which is irrelevant to the two events triggered by "fight", having no evidence to solve the ambiguity problem. The case study shows that multi-granularity information does help to alleviate the problem of ambiguity of triggers. There is indeed rich multi-granularity information around some event mentions, and our MHGEE model can solve the ambiguity problem by aggregating this information.

5   Conclusions and Future Works

In this paper, we propose a novel model MHGEE for event extraction. In order to disambiguate triggers, our MHGEE model aggregate nodes and edges simultaneously into a heterogeneous graph to enable rich information interactions among nodes with different granularities by R-GCN. In addition, we consider the multiple-event phenomenon with modeling intra-sentence and inter-sentence event interdependency. The whole experimental results demonstrate that our MHGEE model can achieve new state-of-the-art performance on the ACE 2005 dataset. In the future, we would like to apply MHGEE to other information extraction tasks, such as aspect extraction and named entity recognition.

References

1. Chen, Y., Xu, L., Liu, K., Zeng, D., Zhao, J.: Event extraction via dynamic multi-pooling convolutional neural networks. In: ACL-IJCNLP 2015–53rd Annual Meeting of the Association for Computational Linguistics and the 7th International Joint Conference on Natural Language Processing of the Asian Federation of Natural Language Processing, Proceedings of the Conference, vol. 1, pp. 167–176 (2015). https://doi.org/10.3115/v1/p15-1017
2. Chen, Y., Yang, H., Liu, K., Zhao, J., Jia, Y.: Collective event detection via a hierarchical and bias tagging networks with gated multi-level attention mechanisms. In: Proceedings of the 2018 Conference on Empirical Methods in Natural Language Processing, EMNLP 2018, pp. 1267–1276 (2020). https://doi.org/10.18653/v1/d18-1158
3. Cui, S., Yu, B., Liu, T., Zhang, Z., Wang, X., Shi, J.: Edge-enhanced graph convolution networks for event detection with syntactic relation. In: Findings of the Association for Computational Linguistics Findings of ACL: EMNLP 2020, pp. 2329–2339 (2020). https://doi.org/10.18653/v1/2020.findings-emnlp.211
4. Du, X., Cardie, C.: Event extraction by answering (almost) natural questions. In: EMNLP 2020–2020 Conference on Empirical Methods in Natural Language Processing, Proceedings of the Conference, pp. 671–683 (2020). https://doi.org/10.18653/v1/2020.emnlp-main.49
5. Kipf, T.N., Welling, M.: Semi-supervised classification with graph convolutional networks. In: 5th International Conference on Learning Representations, ICLR 2017 - Conference Track Proceedings (2017)

6. Li, Q., Ji, H., Huang, L.: Joint event extraction via structured prediction with global features. In: ACL 2013–51st Annual Meeting of the Association for Computational Linguistics, Proceedings of the Conference, vol. 1, pp. 73–82 (2013)
7. Liao, S., Grishman, R.: Using document level cross-event inference to improve event extraction. In: ACL 2010–48th Annual Meeting of the Association for Computational Linguistics, Proceedings of the Conference, pp. 789–797 (2010)
8. Liu, S., Liu, K., He, S., Zhao, J.: A probabilistic soft logic based approach to exploiting latent and global information in event classification. In: 30th AAAI Conference on Artificial Intelligence, AAAI 2016, pp. 2993–2999 (2016)
9. Liu, X., Luo, Z., Huang, H.: Jointly multiple events extraction via attention-based graph information aggregation. In: Proceedings of the 2018 Conference on Empirical Methods in Natural Language Processing, EMNLP 2018, pp. 1247–1256 (2020). https://doi.org/10.18653/v1/d18-1156
10. Lu, Y., et al.: TEXT2EVENT: controllable sequence-to-structure generation for end-to-end event extraction. In: ACL-IJCNLP 2021–59th Annual Meeting of the Association for Computational Linguistics and the 11th International Joint Conference on Natural Language Processing, Proceedings of the Conference, pp. 2795–2806 (2021). https://doi.org/10.18653/v1/2021.acl-long.217
11. Nguyen, T.H., Cho, K., Grishman, R.: Joint event extraction via recurrent neural networks. In: 2016 Conference of the North American Chapter of the Association for Computational Linguistics: Human Language Technologies, NAACL HLT 2016 - Proceedings of the Conference, pp. 300–309 (2016). https://doi.org/10.18653/v1/n16-1034
12. Nguyen, T.H., Grishman, R.: Graph convolutional networks with argument-aware pooling for event detection. In: 32nd AAAI Conference on Artificial Intelligence, AAAI 2018, pp. 5900–5907 (2018)
13. Sha, L., Qian, F., Chang, B., Sui, Z.: Jointly extracting event triggers and arguments by dependency-bridge RNN and tensor-based argument interaction. In: 32nd AAAI Conference on Artificial Intelligence, AAAI 2018, pp. 5916–5923 (2018)
14. Tang, H., Cao, Y., Zhang, Z., Jia, R., Fang, F., Wang, S.: Multi-granularity heterogeneous graph for document-level relation extraction. In: ICASSP, IEEE International Conference on Acoustics, Speech and Signal Processing - Proceedings, vol. 2021-June, pp. 7683–7687 (2021). https://doi.org/10.1109/ICASSP39728.2021.9414755
15. Wadden, D., Wennberg, U., Luan, Y., Hajishirzi, H.: Entity, relation, and event extraction with contextualized span representations. In: EMNLP-IJCNLP 2019–2019 Conference on Empirical Methods in Natural Language Processing and 9th International Joint Conference on Natural Language Processing, Proceedings of the Conference, pp. 5784–5789 (2020). https://doi.org/10.18653/v1/d19-1585
16. Wang, X., Han, X., Liu, Z., Sun, M., Li, P.: Adversarial training for weakly supervised event detection. In: NAACL HLT 2019–2019 Conference of the North American Chapter of the Association for Computational Linguistics: Human Language Technologies - Proceedings of the Conference, vol. 1, pp. 998–1008 (2019). https://doi.org/10.18653/v1/n19-1105

17. Yan, H., Jin, X., Meng, X., Guo, J., Cheng, X.: Event detection with multi-order graph convolution and aggregated attention. In: EMNLP-IJCNLP 2019–2019 Conference on Empirical Methods in Natural Language Processing and 9th International Joint Conference on Natural Language Processing, Proceedings of the Conference, pp. 5766–5770 (2020). https://doi.org/10.18653/v1/d19-1582

18. Zhang, T., Ji, H., Sil, A.: Joint entity and event extraction with generative adversarial imitation learning. Data Intell. 1(2), 99–120 (2019). https://doi.org/10.1162/dint_a_00014

Consistency Fences for Partial Order Delivery to Reduce Latency

Nooshin Eghbal[✉] and Paul Lu

Department of Computing Science, University of Alberta, Edmonton, Canada
{eghbal,paullu}@ualberta.ca

Abstract. For appropriate workloads, partially ordered message delivery can greatly reduce message latency. For example, updates to screens (e.g., remote desktops, VNC) may not have to be totally ordered with respect to different regions of the screen, but ordered with respect to updates to the same region. Similarly, updates to disjoint regions of a file (e.g., bulk-data transfer of sensor data) can be applied in any order, as long as updates (or reads) to the same region of the file are ordered in a consistent way, per data consistency models.

Therefore, we introduce the concept of a *consistency fence* (CF), inspired by a memory fence from data consistency models, as a mechanism to control, specify, and reason about partial orders. If messages are lost on a network, partial ordering via CFs provides a framework to tolerate the latency associated with retransmission, for key workloads.

In a set of simple experiments, based on screen update workloads, we show the latency benefits of partial ordering with CFs. We also show how forward error-correction (FEC) can be combined with CFs and partial ordering to reduce cumulative latency (represented as a cumulative distribution function), as compared to total ordering of messages.

Keywords: Partial order delivery · Latency · TCP · Quality of service · Real-time

1 Introduction

For appropriate workloads (e.g., remote desktops, bulk-data transfers), partially ordered message delivery can greatly reduce message latency. The total ordering of message delivery is more common because total orders are simple and easier to reason about. However, if there is flexibility in the ordering, then retransmissions of lost packets can be overlapped with regular transmissions. For screen updates, different regions can be updated in any order, as long as updates to the same regions have some consistent ordering. For bulk-data transfers, the data for different blocks of the same file can arrive in any order, as long as all updates have arrived by the end of the transfer. The challenge is designing a mechanism and semantics for specifying when ordering matters and when it does not matter.

Therefore, we introduce the concept of a *consistency fence* (CF) as a mechanism to specify when ordering between packets and messages matter. Inspired

© The Author(s), under exclusive license to Springer Nature Switzerland AG 2022
D. Groen et al. (Eds.): ICCS 2022, LNCS 13350, pp. 488–501, 2022.
https://doi.org/10.1007/978-3-031-08751-6_35

by memory fences from data consistency models [11], CFs are like meta-data messages inserted into the stream(s) of messages. Messages emitted *between* a CF can be reordered among themselves, but messages can never be reordered *across* a CF. Furthermore, CFs must be totally ordered between themselves. As a mechanism, a CF can be implemented using well-known sequencing techniques. The semantics of CFs have the beneficial property that inserting extra CFs into the stream can never cause erroneous interleaving, although with some potential loss of performance due to a reduced flexibility of interleaving.

A general way to represent ordering constraints is as a directed acyclic graph (DAG) (Fig. 1). For example, blocks B_1, B_2, and B_3 have no inter-block dependencies and can be delivered to the destination in any order (among those three blocks), but B_9 must be delivered before B_7. Directed edges show the ordering constraints.

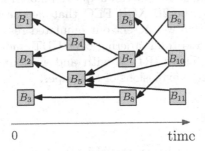

Fig. 1. Dependency graph (Color figure online)

Figure 2 compares total order delivery (such as with TCP) vs. partial order delivery for Fig. 1, if block B_6 is lost. With total order delivery we cannot deliver B_7–B_9 while we are waiting for the retransmission of B_6 (the green block). This source of latency is known as the Head-Of-Line (HOL) blocking problem. However, with partial order delivery we can first deliver B_7–B_9 after B_5, with lower latency for those blocks, and overlap those deliveries with the retransmission of B_6, resulting in lower total latency for delivering B_1–B_{11}.

Workloads that have periodic updates are candidates for partial ordering mechanisms like CFs. In remote desktop applications or online video games, there are periodic screen updates. In each update, different parts of the screen need to be updated and we (usually) need to finish each screen update before starting the next one. Using partial order delivery, we can apply all messages in each update as soon as we receive them in the destination in any order and do not block and wait for the lost ones. There is no need for the total ordering provided by TCP/IP. For bulk-data transfers, which are not discussed any further, the file is not always consumed until the transfer is complete, therefore different blocks of data can arrive in any order, as long as all data arrives before the file is closed.

In Sect. 2, we summarize some of the related papers, and then we describe our CF mechanism in Sect. 3. Then, we explain why adding a forward error

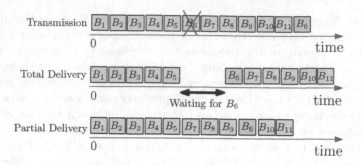

Fig. 2. Total vs. partial order delivery (Color figure online)

correction (FEC) technique to CFs can help even more reduction in the message blocking time and describe 2D XOR FEC that we use for the results in this paper in Sect. 4. In Sect. 5 we present our emulated results to compare partial ordering through CFs and total ordering over UDT under different packet loss rates and Round Trip Times (RTTs) with and without FEC. Section 6 has a summary and some future extensions of the paper.

2 Related Work

The previous work on supporting partial order delivery used some notion of a dependency graph which needs to be sent to the receiver [2,13]. Therefore, the receiver would know in which order the data blocks need to be delivered to the application layer. Using dependency graphs, we can express the ordering precisely but implementing a networking system based on these graphs to support partial ordering is complicated and needs the dependency knowledge in advance.

Also, there are some projects that release the total ordering manner of TCP to solve the HOL blocking problem such as Stream Control Transmission Protocol (SCTP) [15] and Quick UDP Internet Connections (QUIC) [9]. They use multistreaming over a single connection such that only the data blocks within each stream need to be delivered in total order but there is no ordering between the data blocks of different streams. However, there is no strategy for supporting partial ordering over SCTP or QUIC. For example if we want to send the eleven data blocks of Fig. 1 over them, we must use only one stream for the whole graph because we cannot count on the ordering of data of different streams.

UDP-based Data Transfer (UDT) protocol [6] is a user-level, reliable protocol designed for large data transfers over wide-area networks (WANs). UDT has a built-in loss-based Congestion Control Algorithm (CCA) which provides better throughput than TCP CUBIC under WAN settings. UDT has two modes: 1) stream mode and 2) message mode. The stream mode is like TCP and supports total ordering only but in the message mode we can set an order flag for each message to be delivered in total order or arrival order. However, there is no mechanism to express any partial ordering setup inside UDT. As detailed in the next section, our prototype of CFs is layered on top of UDT, using message mode, where the flag of all messages is set to arrival order delivery.

Resource prioritization for HTTP workloads to reduce web page load time has a long history in the literature [10]. The idea is that scheduling web resources based on the file types plays an important role in minimizing the overall page load time. For example, some of the resource files such as CSS and JavaScript need to be downloaded completely before getting used while image files can be rendered incrementally. To communicate the web resource priorities between HTTP client and server, dependency graphs were used for years over different versions of HTTP. However, managing these graphs was complicated and challenging like the concept of dependency graphs in partial ordering setup. Recently a more practical approach was chosen by IETF for further development in HTTP/3, which is based on labeling resources via urgency vs. incremental by the HTTP client [12]. We also aim to introduce a more practical way of handling partial ordering requirements through our consistency fences mechanism.

3   Consistency Fences

We propose the *consistency fence* (CF) mechanism, in which the sender transmits the independent data blocks (or messages) in parallel and inserts a fence before sending any data block that has dependencies to the previously sent blocks. This way, the receiver would know to deliver all the blocks before each fence and then start delivering the blocks after the fence. The CF is analogous to the memory fence/barrier instructions used by CPUs to enforce an ordering constraint on memory operations issued before and after the fence instruction [11].

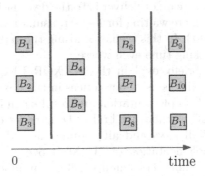

Fig. 3. Consistency fences (Color figure online)

In Fig. 3, we show how the dependency graph in Fig. 1 can be sent using CFs (e.g., vertical red lines). The sender transmits the first three blocks (B_1 to B_3) and then inserts a fence because both B_4 and B_5 are dependent on at least one of the first three blocks, although with no inter-dependencies between B_4 and

B_5 themselves. The same would happen for the third and fourth groups of blocks which are B_6–B_8 and B_9–B_{11}, respectively.

A dependency graph (Fig. 1) can be more precise than consistency fences to express ordering among data blocks and inserting a fence may add more, implied ordering dependencies. For example, in Fig. 3, both B_4 and B_5 seem to be dependent on B_1–B_3, while in the original graph in Fig. 1, B_4 is dependent on B_1 and $B2$, and B_5 is only dependent on B_2. However, communicating the ordering requirements through dependency graphs is complicated and in many situations the sender does not have the complete dependency graph ahead of time. Therefore, we believe that having a more practical mechanism like consistency fences could help encourage application-layer programmers to take advantage of partial order delivery.

4   Combining Partial Order Delivery with Forward Error Correction

Forward Error Correction (FEC) has been discussed in the literature as an effective method to speed up packet loss recovery [1,8]. Using FEC, the sender transmits some redundant packets along with the original packets so the receiver will be able to recover lost packets sooner than the time it needs to wait for receiving the retransmissions especially in the networks with high RTTs like WANs. In our previous work, we showed the benefit of using a specific FEC method called two dimensional XOR-based (2D XOR) FEC on reducing the latency and improving the throughput of data transfers over WANs compared to the state-of-the-art WAN protocols in total order delivery setup [3]. However, we believe that FEC can help reduce latency in partial order delivery in the scenario that we have received some data but we cannot deliver it to the destination because it depends on some lost data and we are waiting for its retransmission (e.g. we have received B_4 and B_5 but B_2 is lost). In this scenario, losing the retransmission of the lost data can make the blocking time even worse.

In this section, we briefly explain the 2D XOR FEC and we evaluate the benefit of combining it with consistency fences in the next section. More details on the the 2D XOR FEC implementation and results can be found elsewhere [3].

In the 2D XOR FEC method, we build 2D matrices of the original packets and send the XOR of all rows and all columns as redundant packets to the destination. In our implementation of 2D XOR over UDT, the user sets the number of rows and columns. For example, if the number of rows is set to 20 and the number of columns is set to 40, we protect each group of 800 packets by building a matrix of packets with 20 rows and 40 columns and send 20 row XORs and 40 column XORs. Therefore, the bandwidth overhead of the mentioned settings would be $\frac{20+40}{800}$ or $\frac{3}{40}$.

The interesting thing about 2D XOR FEC is that we can recover bursty packet loss patterns with the help of column recovery, whereas in 1D XOR FEC we can recover only one packet loss in each row [4,5].

5   Evaluation

We have implemented CFs as a separate mode inside the UDT version 4 system. The user can enter this partial ordering mode by inserting fences using an API call, UDT::sendfence(client_socket), while sending messages. There are two kinds of packets in UDT: 1) data packets, and 2) control packets. UDT allows us to define user-defined control packets. In our implementation, each fence is a control packet which contains three numbers: 1) sequence number, 2) the number of messages from the previous fence to this fence (or the start of the transmission if this is the first fence), and 3) the sequence number of the last message before this fence. Therefore, at the destination we can check if we can deliver each message or we should wait to receive all the messages before the fence to be able to deliver the messages after the fence.

In addition to the default loss-based CCA, UDT has a non-loss-based CCA called UDPBlast which simply gets a fixed sending rate from the user and will not change it in the case of packet loss. We used UDPBlast for all our experiments to be able to better focus and analyse the benefit of supporting partial ordering. Additional experiments with a loss-based CCA remains as a future work.

We evaluate the latency of our CFs implementation over UDT compared to the total ordering setup over UDT in an emulated testbed. We use the Netem-tc Linux tool [7] to set the loss rate and RTT between two nodes with Linux kernel 4.13.0, 2.30 GHz Intel(R) Core(TM) i3-6100U CPU, 2-cores in total, 32 GB of memory, and 1 Gbps bandwidth.

As the workload, we sent 32000 messages with size 1400 bytes. We set the loss rate to {0.1%, 0.5%, 1%}, the RTT to {20 ms, 70 ms}, and the number of messages in each update to {800, 1600}. We inserted a fence after each update in partial ordering mode. Although the workload used is not a full-fledged application, the overall goals of this initial evaluation show:

1. For an appropriate workload (e.g., screen updates) with substantial packet loss (e.g., 1% loss), **the partial ordering capabilities of consistency fences are more effective than FEC** in tolerating the cumulative latencies for each update message (e.g., Fig. 4c)
2. As the RTT increases (e.g., to 70 ms, Fig. 5), FEC can improve on the cumulative latencies for total orderings beyond just partial ordering. However, **partial orderings with consistency fences can also be combined with FEC** (with less FEC overheads) and still have the least cumulative latencies (e.g., Fig. 5c, 6c)
3. When packet losses are lower (e.g., 0.1% or 0.5%), the cumulative latencies for partial orderings are consistently lower than for total orderings, without the overheads of FEC (e.g., Fig. 4b vs. 4c). But, if desired, **FEC can be combined with partial ordering to further reduce cumulative latencies as a trade-off** for FEC overheads (e.g., Fig. 4a vs. 4b).

Figure 4, 5, and 6 show the Cumulative distribution function (CDF) of message blocking time using total vs. partial order delivery with and without FEC for three different packet loss rates. We also report (the number of recovered messages)/(the number of lost messages) for FEC settings in the captions. Lines which are closer to the top-left corner (x = 0 ms, y = 1.0 CDF) show better latency performance (i.e., more messages have lower latency due to blocking). In general, the "Partial" ordering performance lines are better than the "Total" ordering performance lines

The blocking time is the whole time a message was waiting in the receiver buffer (i.e. the time from receiving it from the network to the time being delivered to the application layer). For example, if the RTT is 70 ms, the time from sending a message at the sender application to the time of delivering it to the receiver application (i.e. latency) would be 35 ms plus the blocking time. Therefore, in total order delivery, in the case of losing a message, the latency of the next messages will be increased because we need to block them in the receiver buffer until we receive the retransmission or be able to recover the lost one with FEC. The difference between theses three figures is the RTT and the number of messages in each update so we can study the impact of these two factors on the effectiveness of partial order delivery.

In Fig. 4 we set the RTT to 20 ms and the number of independent messages in each update to 1600. In total ordering setup, all the messages after the lost ones need to wait in the receiver buffer until we receive the retransmissions, which will take about an RTT. Also, if we lose any retransmission, the blocking time will increase by another RTT. However, in partial order delivery, we can still deliver all the messages in each update without waiting for the retransmission of the lost ones. Then, to start delivering the next update's messages we need to wait for receiving the retransmissions.

As we increase the packet loss rate from 0.1% in Fig. 4a to 0.5% in Fig. 4b to 1% in Fig. 4c, the blocking time of the total order setup (i.e. blue) gets worse because for each lost packet we need to wait an RTT to receive the retransmissions and during this time we block all the next messages in the buffer. However, adding FEC to total order setup (i.e. Total+FEC in green and violet) help reduce the blocking time but still cannot reach the performance of partial order delivery (i.e. orange) in Fig. 4b and 4c where we have moderate and high packet loss rates. The reason is that using 2D FEC with 40 rows and 40 columns with total order delivery in Fig. 4b (i.e. green), although we could recover 133 lost messages, we fail recovering 32 messages which result blocking the next messages for an RTT. It gets worse in Fig. 4c when we have a higher loss rate so adding FEC 40 * 40 to total order delivery we can recover 217 lost messages and cannot recover 72 lost messages therefore getting higher blocking time. Also, increasing the number of redundant packets and bandwidth overhead by using FEC 20 * 20 with total ordering setup (i.e. violet) reduces the blocking time compared to total+FEC 40 * 40 but since we still worse than partial ordering setup even without FEC (i.e. orange). However, the combination of FEC 40 * 40 and partial ordering (i.e. red) results close to zero blocking times except for a few number of messages even in Fig. 4c with 1% packet loss rate.

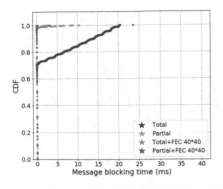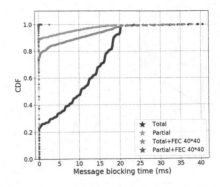

(a) 0.1% loss. Total+FEC 40*40: 34/0. (b) 0.5% loss. Total+FEC 40*40: 133/32.
Partial+FEC 40*40: 30/0.   Partial+FEC 40*40: 151/33.

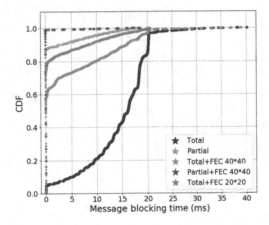

(c) 1% loss. Total+FEC 40*40: 217/72. Partial+FEC 40*40: 237/74. Total+FEC 20*20: 299/22.

Fig. 4. The CDF of message blocking time (ms) for different amount of packet loss when RTT was **20** ms. The number of messages in each update was **1600**. (Color figure online)

As a summary of the results presented in Fig. 4, for all loss rates, partial ordering helps reducing the blocking time compared to total order delivery. Also, adding the same rate of redundant packets with FEC 40 * 40, partial order delivery can get closer to the loss-free scenario compared to total order delivery.

The reason behind choosing 40 for the number of rows and columns in 2D FEC was that we have 1600 messages in each update, so each 2D matrix can include all the messages in each update. Therefore, we can recover the lost messages of each update without needing any messages of the next update, which is important for partial ordering setup.

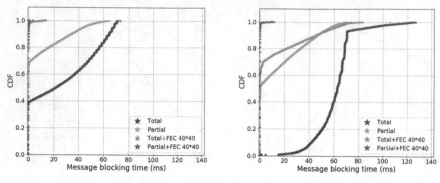

(a) 0.1% loss. Total+FEC 40*40: 34/0. (b) 0.5% loss. Total+FEC 40*40: 146/14.
Partial+FEC 40*40: 29/0.   Partial+FEC 40*40: 149/7.

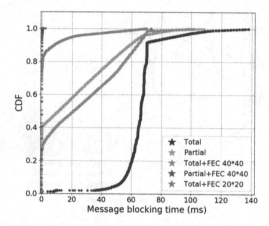

(c) 1% loss. Total+FEC 40*40: 275/45. Par-
tial+FEC 40*40: 304/45. Total+FEC 20*20: 285/2.

Fig. 5. The CDF of message blocking time (ms) for different amount of packet loss when RTT was **70** ms. The number of messages in each update was **1600**. (Color figure online)

Figure 5 is very similar to Fig. 4 except that we have a higher RTT of 70 ms to study the performance of our CFs to support partial ordering in the networks with higher RTTs like WANs. For all three packet loss rates, again partial ordering setup (i.e. orange) helps reduce the blocking time compared to total ordering setup (i.e. blue). However, having a higher RTT compared to Fig. 4 the partial ordering results get worse and move to the right. The reason is that having a higher RTT there would be more number of lost messages for which we have not received a retransmission at the time of receiving the messages of the next update so we need to block them in the buffer in the meantime. In Fig. 5b and 5c we can get close to partial ordering results when we add FEC 40 * 40 to total ordering (i.e. green). Also, adding more overhead and using FEC 20 * 20 with total ordering (violet) in Fig. 5c we can reduce the blocking time more than partial ordering setup. However, using FEC 40 * 40 with partial ordering (i.e. red) is still the best for all loss rates.

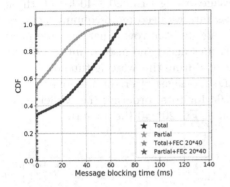 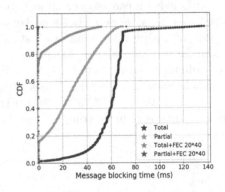

(a) 0.1% loss. Total+FEC 20*40: 33/0. Partial+FEC 20*40: 72/0.

(b) 0.5% loss. Total+FEC 20*40: 158/0. Partial+FEC 20*40: 204/0.

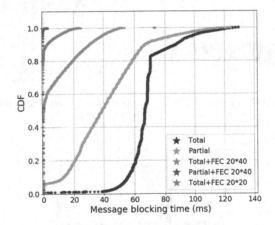

(c) 1% loss. Total+FEC 20*40: 302/0. Partial+FEC 20*40: 304/0. Total+FEC 20*20: 302/0.

Fig. 6. The CDF of message blocking time (ms) for different amount of packet loss when RTT was **70** ms. The number of messages in each update was **800**. (Color figure online)

Figure 6 is very similar to Fig. 5 except that we reduce the number of messages in each update to 800. Having fewer number of messages in each update means having a workload with more ordering requirements. That is why the partial ordering results (i.e. orange) get closer to total ordering results (i.e. blue) but

still better. We set 20 as the number of rows and 40 as the number of columns for
FEC because 20 * 40 gets the number of messages in each update. Addig FEC
20 * 40 to both total and partial ordering setups is beneficial for all packet loss
rates but still partial order delivery can get closer to loss-free scenario compared
to total ordering even with 20 * 20 FEC in Fig. 6c (i.e. violet). The reason in
that although we can recover all lost messages using FEC 20 * 40 for all three
packet loss rates, as we increase packet loss rate we need more column recovery
because of having more than one lost message in each row. Therefore, we need
to wait for receiving all the matrix (i.e. 800 messages) and also column XORs to
start column recovery. In total order delivery, during this waiting time to receive
the whole matrix, we need to block all the messages after the lost ones. That
is why when we use FEC 20 * 20 in Fig. 6c for total ordering (i.e. violet), the
blocking time is reduced compared to total+FEC 20* 40 (i.e. green).

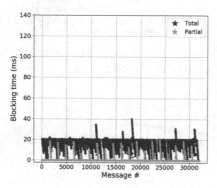

(a) Update size: **1600** messages.
RTT: **20ms**.

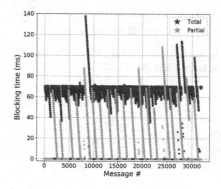

(b) Update size: **1600** messages.
RTT: **70ms**.

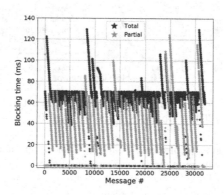

(c) Update size: **800** messages.
RTT: **70ms**.

Fig. 7. The message blocking time (ms) for **1%** packet loss. (Color figure online)

Figure 7 shows the message blocking time (ms) (not the CDF) of the three previous figures but only for 1% loss rate setting. In this figure we can better observe the impact of increasing RTT and decreasing the update size on the message blocking time of partial ordering setup. As we increase RTT from 20 ms in Fig. 7a to 70 ms in Fig. 7b the length of orange vertical lines increase, which means that more number of messages get blocked and for a longer time at the beginning of each update because we are still waiting for receiving the retransmission of the lost messages in the previous update. Also, as we decrease the number messages in each update from 1600 in Fig. 7b to 800 in Fig. 7c, the length of orange vertical lines get shorten because we expect to have about 8 lost messages in each update since the loss rate is 1% while it is 16 lost messages for 1600 messages in each update. Therefore, it would take less time to receive the retransmissions for all lost messages in each update but we have more frequent orange vertical lines since having 32000 messages in total and 800 messages in each update, we have 40 updates where we have 20 updates for 1600 messages in each update. As a summary, both increasing RTT and decreasing the update size would increase the number of blocked messages and the amount of blocking time in partial ordering setup. However, for workloads with periodic updates when there is a pause between updates at the sender side (e.g. remote desktop or online games), that would give the receiver more time after each update to receive the retransmissions and result in fewer blocked messages.

Figure 8 shows the message blocking time (ms) of the combination of either total or partial ordering setups with FEC when the packet loss is 1%. For all three settings, the message blocking time of partial+FEC 40 * 40 (i.e. red) is close to zero except for a few data points as we presented in the CDF graphs. However, for the same rate of redundant packets for total+FEC 40 * 40 in Fig. 8a and Fig. 8b, the vertical green lines show the lost messages that could not be recovered with FEC and we need to wait for the retransmissions. The reason that not-recovered lost messages with FEC do not have the same impact on partial ordering is that with partial order delivery we do not wait for the lost messages in the **middle** of each update and having 1600 messages in each update there is a good chance of getting the retransmissions by the end of each update. Total+FEC 20 * 20 has fewer and shorter vertical violet lines compared to 40 * 40 in green because having more redundant packets we could recover more lost messages and at the same time having smaller matrices we could recover them faster. The interesting and different thing about Fig. 8c is that we can compare the blocking time of total+FEC 20 * 40 and total+FEC 20 * 20 when we could recover all lost messages in both redundant rates that is why there is not green or violet points around 70 ms. Therefore, the difference between total+FEC 20 * 40 and total+FEC 20 * 20 in Fig. 8c is that having smaller matrices in 20 * 20 setting makes the loss recovery faster.

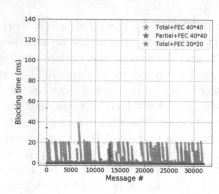
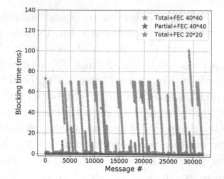

(a) Update size: **1600** messages. RTT: **20ms**.

(b) Update size: **1600** messages. RTT: **70ms**.

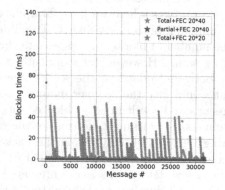

(c) Update size: **800** messages. RTT: **70ms**.

Fig. 8. The message blocking time (ms) for **1%** packet loss. (Color figure online)

6   Concluding Remarks

There are some latency-sensitive workloads (e.g., remote desktops, bulk-data transfers) that do not require total order delivery supported by stream-based protocols such as TCP. When packets are lost in TCP, unnecessary head-of-line blocking can cause extra per-packet latency and lower overall performance. We propose a practical mechanism, called consistency fences, to support partial order delivery and help reduce the unnecessary blocking time and latency. We also show that combining forward error correction with partial order delivery setup makes a better improvement compared to total order delivery setup with the same rate of redundant data. We study the impact of increasing RTT and decreasing the update size on the effectiveness of CFs and show that for the workloads with a pause between updates we will still get noticeable benefit using CFs under high RTT and low update size scenarios.

Studying the performance of our CFs implementation with loss-based CCAs remains as a future work. Also, we want to expand our evaluation of the combination of CFs and FEC to other more efficient FEC methods like rate-less codes such as Raptor codes [14].

Acknowledgments. Thank you to Steve Sutphen, the Natural Sciences and Engineering Research Council (NSERC) of Canada, and Huawei.

References

1. Balakrishnan, M., Marian, T., Birman, K.P., Weatherspoon, H., Ganesh, L.: Maelstrom: transparent error correction for communication between data centers. IEEE/ACM Trans. Netw. **19**(3), 617–629 (2011)
2. Connolly, T., Amer, P., Conrad, P.: An extension to TCP: partial order service. Internet RFC1693 (1994)
3. Eghbal, N., Lu, P.: Low-variance latency through forward error correction on wide-area networks. In: 2021 IEEE 46th Conference on Local Computer Networks (LCN), pp. 90–98. IEEE (2021)
4. Ferlin, S., Kucera, S., Claussen, H., Alay, Ö.: MPTCP meets FEC?: supporting latency-sensitive applications over heterogeneous networks. IEEE/ACM Trans. Netw. **26**(5), 2005–2018 (2018)
5. Flach, T., Dukkipati, N., Cheng, Y., Raghavan, B.: TCP instant recovery: Incorporating forward error correction in TCP. Working Draft, IETF Secretariat, Internet-Draft draft-flach-tcpm-fec-00 (2013)
6. Gu, Y., Grossman, R.L.: UDT: UDP-based data transfer for high-speed wide area networks. Comput. Netw. **51**(7), 1777–1799 (2007)
7. Hemminger, S., et al.: Network emulation with NetEm. In: Linux conf au, vol. 5, p. 2005. Citeseer (2005)
8. Kim, M., Cloud, J., ParandehGheibi, A., Urbina, L., Fouli, K., Leith, D., Médard, M.: Network coded TCP (CTCP). arXiv preprint arXiv:1212.2291 (2012)
9. Langley, A., et al.: The quic transport protocol: Design and internet-scale deployment. In: Proceedings of the Conference of the ACM Special Interest Group on Data Communication, pp. 183–196 (2017)
10. Marx, R., De Decker, T., Quax, P., Lamotte, W.: Resource multiplexing and prioritization in HTTP/2 over TCP versus HTTP/3 over QUIC. In: Bozzon, A., Domínguez Mayo, F.J., Filipe, J. (eds.) WEBIST 2019. LNBIP, vol. 399, pp. 96–126. Springer, Cham (2020). https://doi.org/10.1007/978-3-030-61750-9_5
11. Mosberger, D.: Memory consistency models. ACM SIGOPS Oper. Syst. Rev. **27**(1), 18–26 (1993)
12. Oku, K., Pardue, L.: Extensible prioritization scheme for http. Work in Progress, Internet-Draft, draft-ietfhttpbis-priority-02 1 (2020)
13. Pooya, S., Lu, P., MacGregor, M.H.: Structured message transport. In: 2012 IEEE 31st International Performance Computing and Communications Conference (IPCCC), pp. 432–439. IEEE (2012)
14. Shokrollahi, A.: Raptor codes. IEEE Trans. Inf. Theory **52**(6), 2551–2567 (2006)
15. Stewart, R., Metz, C.: SCTP: new transport protocol for TCP/IP. IEEE Internet Comput. **5**(6), 64–69 (2001)

Dynamic Classification of Bank Clients by the Predictability of Their Transactional Behavior

Alexandra Bezbochina, Elizaveta Stavinova[✉], Anton Kovantsev, and Petr Chunaev

National Center for Cognitive Research, ITMO University,
Saint Petersburg 199034, Russia
{aabezbochina,stavinova,ankovantcev,chunaev}@itmo.ru

Abstract. We propose a method for dynamic classification of bank clients by the predictability of their transactional behavior (with respect to the chosen prediction model, quality metric, and predictability measure). The method adopts incremental learning to perform client segmentation based on their predictability profiles and can be used by banks not only for determining predictable (and thus profitable, in a sense) clients currently but also for analyzing their dynamics during economical periods of different types. Our experiments show that (1) bank clients can be effectively divided into predictability classes dynamically, (2) the quality of prediction and classification models is significantly higher with the proposed incremental approach than without it, (3) clients have different transactional behavior in terms of predictability before and during the COVID-19 pandemics.

Keywords: Predictability · Incremental learning · Transactional data

1 Introduction

Analyzing client's (especially transactional) behavior is highly demanded nowadays [3,10,23,24,26] and is related to different tasks—from the prediction of client's next purchase [24,26] to that of future client's location [16,23]. These studies are particularly motivated by the purposes of the company's marketing efficiency and risk management. For example, companies are interested in determining profitable clients [27,28] and understanding the general client's behavior with respect e.g. to demographics. This is known as client segmentation [1,4,17] and aims at increasing company's profitability. Furthermore, the COVID-19 pandemics and the related economical processes have emphasized once again the necessity to analyze client's transactional behavior in dynamics [3,10] so that a company can use the results to withstand future possible crises.

This research is financially supported by the Russian Science Foundation, Agreement 17-71-30029, with co-financing of Bank Saint Petersburg, Russia.

In this paper, we face the task of analyzing bank clients' transactional behavior in a dynamic manner in terms of *predictability*. This means that we study how the behavior can be predicted by a chosen *prediction model, quality metric* and *predictability measure*, see [24]. However, we do not evaluate the predictability on the whole client set (as is usually performed) but distinguish and analyze *classes* of clients by means of their *predictability profiles*. One can consider predictable clients more profitable as they have stable patterns of behavior and thus a bank can make marketing activities for such clients less risky. The essential novelty of our study consists in that we adopt an *incremental approach* that allows solving the problem of such classification of clients dynamically and can be used by banks not only for determining predictable clients currently but also for analyzing their dynamics during economical periods of different types. To reduce computational time for predictability analysis when the number of bank clients is large, we also use an incremental *classifier* to divide clients by their predictability. One can use the classifier to estimate the client's predictability already without the necessity to perform actual prediction by the prediction model. To summarize, the impact of this study is as follows:

- we propose a new method for dynamic classification of bank clients by the predictability of their transactional behavior (with respect to the bank's chosen prediction model, quality metric, and predictability measure);
- we use the prediction and classification models based on Long Short-Term Memory (LSTM) network [20] that are adaptive in the sense that they exploit the incremental learning principle;
- we show in our experimental study of the proposed method that it is efficient in dividing bank clients into predictability profile classes dynamically and moreover the quality of prediction and classification models is significantly higher with the proposed incremental approach than without it (as e.g. in [24]) for economical periods of different types (in particular, before and during the COVID-19 pandemics).

The code, public datasets, and experimental results are given on GitHub[1].

2 Related Work

The task considered in this paper is connected with several scientific topics such as predictability analysis, classification, clients' segmentation and incremental learning. That is why a review covering several related topics is provided below.

Predictability Analysis. The topic of predictability analysis can be divided into two sub-topics: the topic of *intrinsic* [7] predictability analysis, which is aimed at evaluating the prediction quality using only the data characteristics regardless of the prediction model, and *realized* [12] predictability analysis, that estimates the chosen model's predictive quality on a certain data.

---

[1] https://github.com/AlgoMathITMO/Dynamic-classifier.

The existing papers on the topic of predictability estimation can be classified according to the object of the research: *univariate time series, multivariate time series, event sequences* and *network links*. However, there is no single methodological approach even inside these groups as different researchers look at the predictability phenomenon from the positions of different science areas (information theory, dynamical systems theory, etc.).

Many works in the field are devoted to measuring the predictability of a univariate time series before the actual forecasting. The first realized predictability measure is proposed in [12], and it is further used in [18] as a base for a new measure of similar type. There is a significant group of intrinsic predictability estimation methods that exploit entropy evaluation: in [2] predictability measure based on permutation entropy is proposed, in [5] its weighted modification is presented, in [22] the Shannon entropy is used to estimate the categorical time series predictability, in [14] several time series features representing its predictability is analyzed, and, finally, in [13] time series predictability is explored via its transition graph analysis. As for the event sequences predictability estimation, to the best of our knowledge, there are only three works on this topic. In [8,11] the intrinsic predictability is estimated, while in [24] the realized one.

Client's Predictability Class Identification. To verify the novelty of the proposed method two papers should be regarded.

The first one is [24], where a method for client's predictability classification is proposed. There are several similarities between [24] and our paper: the experiments are conducted on transactional data, the object of the researches is categorical time series (or event sequences), the method for assessing the client's predictability class in the current is a generalization of the method from [24] on dynamic settings. Nevertheless, our paper considers the problem of estimating the clients predictability classes from a new perspective. Firstly, in our work the information about client's belonging to a predictability class in the sense of a single financial event is supplemented with the similar information corresponding to several events described by different financial categories. Secondly, our method uses incremental learning to dynamically estimate arriving data.

The second paper is [25], where a classification method for network links' predictability is introduced. The idea is that by using link features one can determine link predictability classes, i.e. [25] is rather close ideologically to [24].

Clients' Segmentation. There is also a group of works aimed at clients' segmentation (an unsupervised task, in opposite to classification considered in our paper) that is usually performed to obtain similar client groups according to some features [1,4,17]. Applying personalized marketing company for these groups makes it possible to better satisfy individual needs of clients. The important type of the clients' segmentation task is determining valuable clients. For example, [28] proposes a method for identification of valuable travellers. Also, purchasing behavior of valuable clients is considered in [27].

Incremental Learning. In classical scenarios of Machine Learning it is assumed that all the training data is available at the beginning of training but it is not always true. In turn, incremental learning is used when the training data becomes available gradually. The goal of incremental learning is to adapt the learning model to new data (and not to forget the existing knowledge) without entire retraining the model.

The following groups can be distinguished among the neural network approaches to incremental learning. Regularization-based approaches [15] are usually based on imposing additional restrictions on changing the weights of the neural network when training on new data. Approaches based on dynamic architecture [19] are based on "freezing" the already trained neural network weights and adding new neurons for training on new data. Finally, learning with replay [21] involves saving old data in order to use not only new data, but also samples of old data for future training.

3   The Description of the Proposed Method

Pipeline. Suppose there is a set of bank clients $\{c_i\}_{i=1}^n$. Every client is represented by a set of transactions. Every transaction is given a MCC, or a Merchant Category Code, which is a 4-digit code representing a certain category of transactions. All MCC are grouped into N categories. Therefore, the spending of a client c_i during a period t can be defined as a vector of dimension N: $S_{i,t} = (m_{1,t}, \ldots, m_{N,t})_i$, where $m_{k,t} \in \{0,1\}$ is the indicator of spending in category k. In this way we can define the history of ith client's transactions as the sequence of vectors $\{S_{i,t}\}_{t=1}^T$, where T is the total number of periods.

Having a set of bank clients $\{c_i\}_{i=1}^n$ presented by event sequences $\{S_{i,t}\}_{t=1}^T$, our task is to divide this set into predictability classes according to values of a chosen *quality metric* for a chosen *prediction model* and a certain category of transactions (at each time step). In this work, for simplicity we distinguish two classes of client's transactional predictability: namely, high (*Class A*, green plots in figures) and low (*Class B*, red plots in figures) predictability classes. The presence of a client in Class A means that his/her transactional behavior is predicted with higher quality than the behavior of any client from Class B.

Fig. 1. The pipeline of the proposed method. (Color figure online)

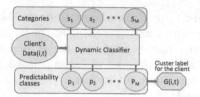

Fig. 2. The formation of client's predictability profiles.

We use the pipeline presented in Fig. 1 to perform the above-mentioned task. At the first stage, the input data is processed. We do some transformations on the history of transactions and extract vectors that act as an input of the prediction model. At the second stage, the prediction model is trained on the processed data and the probability of making a transaction in a particular category is predicted. Then we calculate the values of the quality metric at the next stage. Based on these values two classes are determined. Next, we train a *classification model* that identifies the classes without making the actual predictions.

The proposed pipeline is constructed in a dynamical manner and serves as a tool for maintaining arriving transactional data according to the principle of incremental learning. In this work we use a simple but efficient approach based on updating the model every time after new data arrival. Our approach can be classified as learning with replay because updating is performed using not only the new data, but also a sample of the already used data.

In the first step in Fig. 1, the prediction as well as the classification models are trained on the data that we have at a certain point in time. The model trained with the initial data is called a *base model* below while the dynamic one is called an *incremental model*. After training the base model, it is saved. At the second step, the saved model is loaded and updated using the new data that has arrived. The test and training samples are shifted by some interval every step.

Predictability Profile Analysis. For a set of clients $\{c_i\}_{i=1}^n$, we further define a set of M categories $s_1, \ldots, s_M$, in terms of which we want to analyze the clients' behavior. Having the history of the ith client's transactions $\{S_{i,t}\}_{t=1}^T$, we are able to obtain the labels of predictability classes $(p_1, \ldots, p_M)_t$, $p_m \in \{0,1\}$ for this client according to chosen categories using the predictability classifier.

We call the vector of predictability class labels $(p_1, \ldots, p_M)_t$, where $p_m \in \{0,1\}$, a *predictability profile*. In fact, the vector can be interpreted as a predictability cluster label $G(i,t)$ for ith client in time period t, see Fig. 2. These clusters, their population and clients' transitions from one cluster to another characterize the clients' preferences along the time in a quite explainable way.

4   Experimental Study

Now we aim to show the efficiency of the proposed method. It is important to check that the quality of prediction and classification models is higher within the method than for non-updated models (as in [24]). We use transactional data for economical periods of different nature for this purpose.

Data Description and Processing. For our experimental study we use two datasets: the first one (called $D1$ below) is public and provides the reproducibility of our results, while the second one (called $D2$ below) cannot be made public as provided by our commercial partner. Recall that the code, the public dataset $D1$ and the results for both $D1$ and $D2$ are available on GitHub, see Sect. 1.

The dataset $D1$ is from the Kaggle competition by the Raiffeisen bank (was publicly available at the time of competition). It represents the transaction history of 10,000 bank clients from January to December 2017. The data contains information for one year about the country, city, transaction address, date, amount of money spent and other values. MCC codes are categorized into $N = 87$ categories (in order to reduce the number of categories in the data).

The other dataset $D2$ contains client transactions for the period of one year from October 2019 till September 2020. Thus it contains the data of the first period caused by COVID-19 restrictions which were proclaimed between the 30th of March and the 11th of May 2020 in Russia. There are 11,166,746 records about 7,287 clients. The structure of $D2$ is the same as for $D1$.

Data preprocessing is identical for both datasets. First of all, corrupted or missing values are removed. Then the data are aggregated into categories by MCC codes and resampled to the weekly frequency with sum of transaction numbers for each client per week. Zero value is set if a client has no transaction during a certain week. After that the clients with less than one transaction in the period are excluded from the set. Thus we obtain a table with the columns of client's identifier, week number and amount of transaction in each category.

Prediction and Classification Models and Their Quality Metrics. The first model in the proposed pipeline (Fig. 1) is the prediction model which is aimed at the predicting the fact of transaction at a certain category. In this paper LSTM network [9] is used as the prediction model because of its advantage in remembering time dependencies [6,23,24]. Our predictive model consists of two layers of 64 LSTM-cells and one dense layer for output with the dimension of the number of categories. It takes the data according to the length of the training period and the number of categories in batches of 64 and returns the predicted probabilities of transaction for every client in each category in a certain period.

For this task, we form the following input vectors from the processed data: $A_{i,t} = (b_{1,t}, \ldots, b_{N,t})$, where $b_{k,t}$ is the number of transactions made by a client c_i during a period t $(t \in \{1, \ldots, T\})$. Here we choose the time step for t equal to one week. In this terms, the input for the prediction model can be formulated as $\{A_{i,t}\}_{t=1}^{\widetilde{T}}$, where $\widetilde{T}$ is a chosen length of the input sequence. The desired output for the prediction model is the indicator of the transaction made by a client c_i in category k at the next time step $(\widetilde{T}+1)$: $Q_{k,\widetilde{T}+1}^i$. The LSTM network used as a prediction model outputs the estimated probability of this event $\hat{Q}_{k,\widetilde{T}+1}^i$. For

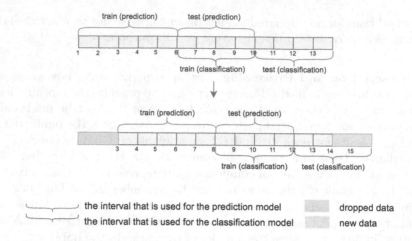

Fig. 3. Intervals for the chosen prediction and classification models.

our purposes, the category called Restaurants is selected as a target category in our experiments. This category represents visits to catering places (restaurants, cafes, fast food places, coffee houses, etc.)

Then the data is split into train and test periods according to the chosen threshold corresponding to a certain week. The data splitting scheme is presented in Fig. 3. Note that the prediction model allows to update its weights dynamically on data streams. With the new data arrival, the train and test periods are shifted to the number of weeks presented in the new data and the LSTM's weights are updated using the remaining old data and the new data. To estimate the quality of the proposed prediction model we use the Precision-Recall curves.

In our experiments we use a Bidirectional LSTM network [20] as a classification model. The *input* of used BiLSTM network is a set of categorical sequences consisting of the event indicators corresponding to a client c_i with the step of one week: $D_i = (d_1, \ldots, d_K)_i$, where d_l is the number of transactions in a chosen category, K is a chosen window length. The network consists of two layers: bidirectional LSTM of 20 cells and one output dense with two output cells for two predictability classes. The network *outputs* the predictability class p for each sequence from the input. The train period for the classification model coincides with the test period for the prediction model, and the test period for the classification model is the next K months. The details of the data splitting can be found in Fig. 3. The classification model also has the ability of dynamical learning on data streams, whose mechanism is the same as the prediction model has. To evaluate the quality of the classification model we use the ROC-AUC curves.

Predictability Measure. We use the sample *predictability rate* of an event in the next period from [24] as the *predictability measure*:

$$C(L, Q, m, i, k) = 1 - \frac{1}{L} \sum_{j=1}^{L} |Q_{k,j}^i - \hat{Q}_{k,j}^i| \in [0, 1], \tag{1}$$

where L is the test period size, $Q^i_{k,j}$ is the actual event indicator, $\hat{Q}^i_{k,j}$ is the predicted probability of the event, m is a forecasting model, i corresponds to the ith client in a set $\{c_i\}_{i=1}^n$, k is an index of a chosen transaction category.

We compute the values of (1) for every client in the set $\{c_i\}_{i=1}^n$ to distinguish clients by predictability, thus, forming the set $\{C(L, Q, m, i, k)\}_{i=1}^n$. Namely, those clients for whom the values of C belong to $[0, median$ $(\{C(L, Q, m, i, k)\}_{i=1}^n)]$ are in the class with low predictability (Class B), and vice versa (Class A).

Using the chosen prediction model, we perform the prediction of a transaction in the Restaurants category. Then, using the chosen quality metric we divide all bank clients into the predictability classes. Finally, using the trained classification model, we solve the problem of identifying the client's predictability class skipping the stage of using the prediction model. In the Fig. 4 (a)–(b) one can see Precision-Recall curves for predictability classes after the first step of training: classes obtained after the prediction model and classes obtained after the classification. In the case of the classification model's perfect quality, the left and right figures will be the same. But we can see that the current predictability classes obtained by the classification model have more similar quality between each other than in the case of the division by the prediction model. But still the division by the classification model saves the classes quality hierarchy.

Dynamic Classifier Analysis. The length of our first dataset allows us to simulate the appearance of the new data and to train the model in nine steps. At every step the test and training samples are shifted by 2 weeks. Using this data, we can update the model's weights and the predictability classes labels.

Since the proposed method assumes constant updating, we have the opportunity to evaluate the forecasting accuracy within several steps. Figure 5 (a) shows the median of coefficient (1) dynamics of prediction model. When the base model is applied to new data, the prediction error stays approximately the same. On the contrary, with dynamic relearning, the error reduces sharply, then changes insignificantly. This shows that the dynamic classifier can distinguish the changes of arriving data distribution better than the model trained once.

In Fig. 4 (c)–(d) Precision-Recall curves for different predictability classes after the ninth step of training are presented: classes obtained after prediction model and classes obtained after classification. Comparing this figure with Fig. 4 (a)–(b), one can note that the quality of the prediction model (the left figures) has increased (from 0.73 to 0.77 in terms of Precision-Recall AUC). After the ninth training step the division into the predictability classes obtained by the prediction model is more contrast than it was earlier: two classes are further from each other. As for the classification model, it catches the division better after the whole training process as seen from the plots.

Figure 5 (c) shows the comparison of the base and incremental models. While the accuracy of the base model decreases, incremental training allows to achieve a higher level of accuracy by constant updating.

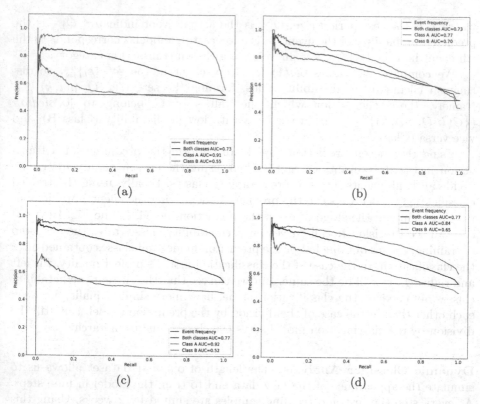

Fig. 4. Precision-recall curves for predictability classes on $D1$: (a) true classes (by the prediction model) after the first training, (b) the classes obtained by the classifier after the first training, (c) true classes (by the prediction model) after the ninth training, (d) the classes obtained by the classifier after the ninth training.

We now apply our method to the dataset $D2$ that contains data for the period of COVID-19 restrictions. The quality scores for incremental learning and base model are calculated according to (1). Their median values for each step are shown in Fig. 5 (b). One can notice that decreasing prediction quality happens after a time delay of three of four weeks after a critical transition has occurred. It was the last week of March 2020 when the restrictions were proclaimed, while the predictive quality fell by the end of April. But nevertheless, the incremental training model not only provides higher quality for each step of the process, but it can recover faster when the crisis is over. Furthermore, it is shown in Fig. 5 (d) how our models manage to overcome the difficulties caused by data volatility during Christmas holidays and COVID-19 restrictions.

Client Predictability Profile Analysis. In order to have the most versatile evaluation of client's predictability in our study we choose five transaction categories in different spheres of interest: restaurants and cafes, food stores,

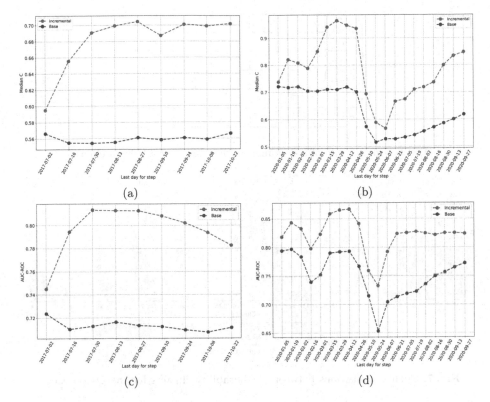

Fig. 5. The median C values for the prediction models: (a) $D1$, (b) $D2$; the ROC AUC values for the classification models: (c) $D1$, (d) $D2$.

hairdressers and beauty salons, cosmetic stores, medical care. Figure 6 shows an example of a predictability profile for one bank client. The first column of the predictability profile indicates the step of model training, while the other five columns represent the five chosen transaction categories. The values inside these five columns are binary; they indicate the predictability class at which a client belongs to (at a certain step of model training and a certain transaction category). For example, at the first step of training the model (Fig. 6) we can say that for two categories out of five, we can define the client's behaviour as "predictable". Over time, the predictability profile changes and by tracking it we can analyze client's behavior.

Let us say that the five client predictability classes together represent a binary number. When this number is converted to a decimal number system for each of the clients, we get 32 segments, or predictability clusters from 0 which means "00000" to 31 which means "11111". Then we can trace the changes in clients' behavior and their transit from one cluster to the other from step to step during the incremental learning process.

step number	# 1	# 2	# 3	# 4	# 5
step 1	1	0	1	0	0
step 2	1	0	1	0	0
step 3	1	1	1	1	0
			...		
step 9	1	1	1	1	0

Fig. 6. An example of a dynamic predictability profile for a client.

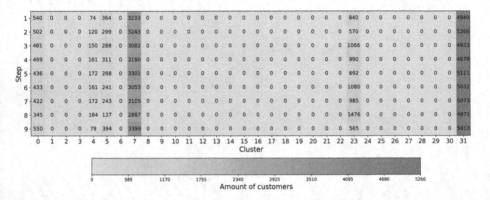

Fig. 7. Clients' transitions between predictability clusters for the $D1$ dataset.

So, we can analyze the dynamic of group clients' behavior along the time and see how its predictability changes from step to step of our incremental process. In Fig. 7 we can see what was happening with predictability clusters in $D1$ with more or less stationary data. The x-axis represents the number of cluster and the y-axis shows the course of time in terms of the model training step. The value inside the cells indicate the number of clients belonging to a cluster at a certain time step. Moreover, the cell colour highlights the most represented clusters. Analyzing Fig. 7, we can conclude that all clients are concentrated in clusters 0 ("00000"), 4 ("00100"), 5 ("00101"), 7 ("00111"), 23 ("10111"), 31 ("11111"). The most populated is the cluster where clients have good predictability, because it includes those who use their cards very rarely. Obviously, lack of transactions for a long time causes good predictions of no transaction in future.

The experiment with $D2$ set shows a bit different distribution of clients in predictability profiles. Most of them perform good predictability in every trade category so they belong to cluster 31. Many clusters are empty. Nonetheless, transitions between clusters happen on each step and the population of clusters never stays unaltered, as is seen in Fig. 8.

How the number of transitions from one cluster to the other changes in unstable situation we can see in Fig. 9 (b). It is on the increase when the critical period begins, then it falls down and recovers on the lower level after this period is finished. During the stationary period this amount keeps on a more or less sustainable level as Fig. 9 (a) illustrates. The analysis conducted in this section can be used as a tool for client profiling.

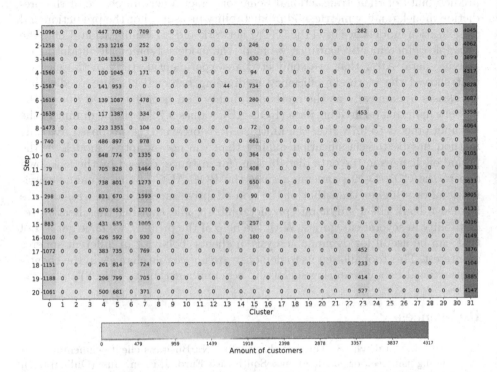

Fig. 8. Clients' transitions between predictability clusters for the $D2$ dataset.

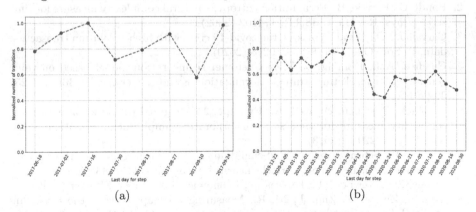

Fig. 9. Normalized number of transitions between classes: (a) $D1$, (b) $D2$.

5 Conclusions

We have proposed a method for dynamic classification of bank clients by the predictability of their transactional behavior under a certain choice of the prediction model, quality metric, and predictability measure. For the prediction and classification models, we used the LSTM network and its modification. Using the principle of incremental learning, we made the models dynamically updating on arriving data. After that, we conducted an experimental study of the method that showed the method's efficiency in dividing bank clients into predictability classes dynamically as the method's quality had been increased due to the model's updating on arriving data (ROC-AUC values 0.74 at the beginning of the learning process to 0.78 after the ninth step). Moreover, the proposed dynamic method has better classification quality than the non-adaptive models in the period of changes in the data distribution (for instance, caused by the lockdown due to the COVID-19 pandemic), since the ROC-AUC of the dynamic classifier is always higher than that of the non-adaptive model (Fig. 5). Finally, we formed a bank client's dynamic predictability profile showing the client's predictability in several categories. With the help of the profiles, we have got a tool that can be useful for dynamic analysis of clients' behavior in different spheres of interest. This tool allowed us to demonstrate noticeable changes in the client's transactional behavior during social and economic instability (Fig. 8).

References

1. Bach, M.P., Jukovic, S., Dumicic, K., Sarlija, N.: Business client segmentation in banking using self-organizing maps. South East Euro. J. Econ. Bus. (Online) **8**(2), 32 (2013)
2. Bandt, C., Pompe, B.: Permutation entropy: a natural complexity measure for time series. Phys. Rev. Lett. **88**(17), 174102 (2002)
3. Carvalho, V.M., et al.: Tracking the covid-19 crisis with high-resolution transaction data. R. Soc. Open Sci. **8**(8), 210218 (2020)
4. Cuadros, A.J., Domínguez, V.E.: Customer segmentation model based on value generation for marketing strategies formulation. Estudios Gerenciales **30**(130), 25–30 (2014)
5. Fadlallah, B., Chen, B., Keil, A., Principe, J.: Weighted-permutation entropy: a complexity measure for time series incorporating amplitude information. Phys. Rev. **87**(2), 022911 (2013)
6. Gers, F.A., Schraudolph, N.N., Schmidhuber, J.: Learning precise timing with LSTM recurrent networks. J. Mach. Learn. Res. **3**(Aug), 115–143 (2002)
7. Granger, C., Newbold, P.: Forecasting Economic Time Series. Elsevier (1986)
8. Guo, J., Zhang, S., Zhu, J., Ni, R.: Measuring the gap between the maximum predictability and prediction accuracy of human mobility. IEEE Access **8**, 131859–131869 (2020)
9. Hochreiter, S., Schmidhuber, J.: Long short-term memory. Neural Comput. **9**(8), 1735–1780 (1997)
10. Horvath, A., Kay, B.S., Wix, C.: The covid-19 shock and consumer credit: Evidence from credit card data. Available at SSRN 3613408 (2021)

11. Järv, P.: Predictability limits in session-based next item recommendation. In: Proceedings of the 13th ACM Conference on Recommender Systems, pp. 146–150 (2019)
12. Kaboudan, M.: A measure of time series' predictability using genetic programming applied to stock returns. J. Forecast. 18(5), 345–357 (1999)
13. Kovantsev, A., Chunaev, P., Bochenina, K.: Evaluating time series predictability via transition graph analysis. In: 2021 International Conference on Data Mining Workshops (ICDMW), pp. 1039–1046 (2021)
14. Kovantsev, A., Gladilin, P.: Analysis of multivariate time series predictability based on their features. In: 2020 International Conference on Data Mining Workshops (ICDMW), pp. 348–355. IEEE (2020)
15. Li, Z., Hoiem, D.: Learning without forgetting. IEEE Trans. Pattern Anal. Mach. Intell. 40(12), 2935–2947 (2017)
16. Moon, G., Hamm, J.: A large-scale study in predictability of daily activities and places. In: MobiCASE, pp. 86–97 (2016)
17. Panuš, J., Jonášová, H., Kantorová, K., Doležalová, M., Horáčková, K.: Customer segmentation utilization for differentiated approach. In: 2016 International Conference on Information and Digital Technologies (IDT), pp. 227–233. IEEE (2016)
18. Prelipcean, G., Popoviciu, N., Boscoianu, M.: The role of predictability of financial series in emerging market applications. In: Proceedings of the 9th WSEAS International Conference on Mathematics & Computers in Business and Economics (MCBE'80), pp. 203–208 (2008)
19. Rusu, A.A., et al.: Progressive neural networks. arXiv preprint arXiv:1606.04671 (2016)
20. Schuster, M., Paliwal, K.K.: Bidirectional recurrent neural networks. IEEE Trans. Sign. Process. 45(11), 2673–2681 (1997)
21. Shin, H., Lee, J.K., Kim, J., Kim, J.: Continual learning with deep generative replay. arXiv preprint arXiv:1705.08690 (2017)
22. Song, C., Qu, Z., Blumm, N., Barabási, A.L.: Limits of predictability in human mobility. Science 327(5968), 1018–1021 (2010)
23. Stavinova, E., Bochenina, K.: Forecasting of foreign trips by transactional data: a comparative study. Procedia Comput. Sci. 156, 225–234 (2019)
24. Stavinova, E., Bochenina, K., Chunaev, P.: Predictability classes for forecasting clients behavior by transactional data. In: Paszynski, M., Kranzlmüller, D., Krzhizhanovskaya, V.V., Dongarra, J.J., Sloot, P.M.A. (eds.) ICCS 2021. LNCS, vol. 12744, pp. 187–199. Springer, Cham (2021). https://doi.org/10.1007/978-3-030-77967-2_16
25. Stavinova, E., Evmenova, E., Antonov, A., Chunaev, P.: Link predictability classes in complex networks. In: International Conference on Complex Networks and Their Applications, pp. 376–387. Springer (2021)
26. Vaganov, Danila, Funkner, Anastasia, Kovalchuk, Sergey, Guleva, Valentina, Bochenina, Klavdiya: Forecasting Purchase Categories with Transition Graphs Using Financial and Social Data. In: Staab, Steffen, Koltsova, Olessia, Ignatov, Dmitry I.. (eds.) SocInfo 2018. LNCS, vol. 11185, pp. 439–454. Springer, Cham (2018). https://doi.org/10.1007/978-3-030-01129-1_27
27. Wong, E., Wei, Y.: Customer online shopping experience data analytics: integrated customer segmentation and customised services prediction model. Int. J. Retail Distrib. Manag. (2018)
28. Wong, J.Y., Chen, H.J., Chung, P.H., Kao, N.C.: Identifying valuable travelers and their next foreign destination by the application of data mining techniques. Asia Pacific J. Tourism Res. 11(4), 355–373 (2006)

Unveiling User Behavior on Summit Login Nodes as a User

Sean R. Wilkinson[(✉)] [iD], Ketan Maheshwari [iD], and Rafael Ferreira da Silva [iD]

Oak Ridge National Laboratory, Oak Ridge, TN, USA
{wilkinsonsr,maheshwarikc,silvarf}@ornl.gov

Abstract. We observe and analyze usage of the login nodes of the leadership class Summit supercomputer from the perspective of an ordinary user—not a system administrator—by periodically sampling user activities (job queues, running processes, etc.) for two full years (2020–2021). Our findings unveil key usage patterns that evidence misuse of the system, including gaming the policies, impairing I/O performance, and using login nodes as a sole computing resource. Our analysis highlights observed patterns for the execution of complex computations (workflows), which are key for processing large-scale applications.

Keywords: High Performance Computing · Workload characterization · User behavior

1 Introduction

HPC systems have been designed to address computing, storage, and networking needs for complex, high-profile applications. Specifically, leadership class supercomputers [16] meet the needs of applications that require high-speed interconnects, low latency, high I/O throughput, and fast processing capabilities (currently petascale, and soon exascale) [6]. Understanding the performance of these systems and applications is a cornerstone for the design and development of efficient, reliable, and scalable systems. To this end, several works have focused on the development of system- and application-level monitoring and profiling tools that can provide fine-grained characterizations of systems' and applications' performance.

The current landscape of HPC systems performance research is mostly focused on the system's performance—which is utterly valuable for systems design [4,9,15]. However, the *user perception* of the system is often disregarded, and there is a common misconception that application execution performance is the only consideration for user satisfaction. Although application performance

This manuscript has been authored by UT-Battelle, LLC, under contract DE-AC05-00OR22725 with the US Department of Energy (DOE). The publisher acknowledges the US government license to provide public access under the DOE Public Access Plan (http://energy.gov/downloads/doe-public-access-plan).

is one of the chief goals of HPC, there are several additional factors that impact user experience. More specifically, before experiencing the capabilities of the HPC nodes, users' first interactions are with the *login nodes*, where users share resources like CPU, memory, storage, and network bandwidth while performing basic tasks like compiling code, designing experiments, and orchestrating services. The login nodes on an HPC system represent a gateway to the system which is often overlooked when considering the capabilities and performance of the overall system. We argue that the experience on the login nodes may impact a user's perception of and behavior on the system, thus influencing whether and how the user continues to utilize that system in future work.

In this paper, we attempt to identify long-term usage patterns by collecting observational data on the login nodes from the Summit leadership class supercomputer hosted at the Oak Ridge Leadership Computing Facility (OLCF) at Oak Ridge National Laboratory (ORNL). Every hour for two years (2020–2021), we have collected data about the login nodes' performance with respect to CPU, memory, and disk usage, and we also collected data about their activity with respect to logged-in users, the programs they were running on the login nodes, and the status of all user jobs. Figure 1 shows the number of users per login node within this time window. In addition to examining traditional metrics like system usage and distribution of jobs and users across the nodes, we also seek to (i) highlight atypical usage and user misconducts, (ii) relate these behaviors to potential performance issues, (iii) identify usage patterns of a complex class of applications such as scientific workflows, and (iv) establish relationships between users' sessions length and system load.

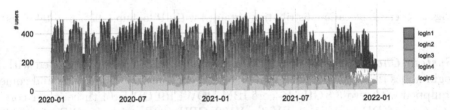

Fig. 1. Distribution of users over 5 Summit login nodes (Jan 2020–Dec 2021). Gaps in the time series indicate outages or system downtime.

2   Characteristics of the Summit Login Node Data

Table 1 summarizes the main characteristics of the collected data. The dataset represents activity from 1,967 unique users, who connected using 9,841 unique IPs and submitted 1,783,867 jobs, of which 1,073,754 completed successfully while 705,103 had a non-zero exit code. Figure 2 shows the distribution of users' geolocations, which were resolved through an IP geolocation tool [1]. For the sake of privacy, any user-specific data had been previously anonymized and not retained.

Table 1. Characteristics of Summit login node data for a period of two years (Jan 2020–Dec 2021). Totals for "Unique Users" and "Unique IPs" do not sum additively due to Summit users whose use spanned both years. Additionally, the total number of jobs may not coincide with the sum of individual years because jobs may be carried over from one year to the next.

Year	# Unique users	# Unique IPs	# Jobs			
			Completed	Suspended	Exited	Total
2020	1,509	5,094	480,550	1,869	313,257	795,676
2021	1,514	5,467	668,580	3,264	410,493	1,082,337
Total	1,967	9,841	1,073,754	5,010	705,103	1,783,867

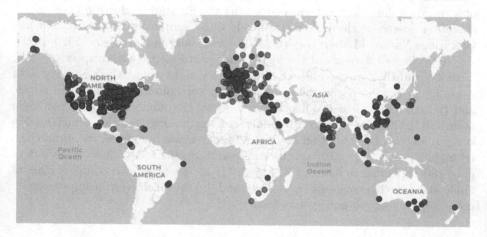

Fig. 2. Users' geolocation distribution obtained with IP lookup (~93% of total users).

System Characteristics and Data Collection. Summit is equipped with 5 login nodes [18]. Each login node runs Red Hat Enterprise Linux v8.2 and comes equipped with two 3.8 GHz 16-core IBM POWER9 CPUs (4 threads per core), 512 GiB of DDR4-2667, 4 NVidia V100 GPUs each with 16 GiB of HBM2, and connection to a 250 PB GPFS scratch filesystem. Users usually log into Summit via SSH to the load-balanced `summit.olcf.ornl.gov` hostname, but they can optionally connect to a specific login node. Data were collected hourly, starting January 1, 2020, on all five login nodes. A shell script ran in the user space as a `while` loop within a Linux `tmux` session because user cronjobs are not allowed, and it collected traditional system usage performance metrics as well as user behavior (e.g., running processes and jobs). One caveat is that the hourly sampling frequency may have failed to capture fine grained behavior, as many things can happen between samples. Nevertheless, we believe that the large volume of samples sufficiently captures most of the representative system and user activity. More precisely, each sample collects the following data:

– List of currently logged-in users using the `w` command;

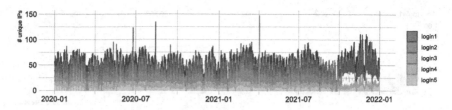

Fig. 3. Distribution of unique IPs across summit login nodes (Jan 2020–Dec 2021).

- CPU and memory usage using the `top` and `ps` commands (which also provides the list of running processes), and statistics from `meminfo` and `vmstat` in the `/proc` filesystem;
- Status of users' batch jobs via the `bjobs` command;
- Disk usage statistics using the `df -h` command and disk throughput by measuring the timespan for writing a 1GB data file to GPFS.

Data Preparation. Real-world data may be incomplete, noisy, and inconsistent, which can obscure useful patterns [20]. Data preparation techniques cannot be fully automated; it is necessary to apply them with knowledge of their effect on the data being prepared. We used our prior knowledge about the execution of scientific applications on HPC to extract and combine relevant information from each source of data. We have then pre-processed the dataset by removing redundancies and missing data (e.g., due to outages and system downtimes), sanitizing lists of programs and users for long-running processes and jobs, and resolving IP addresses for filtering and identifying individual users and their locations, among other things.

3   System Metrics

In this section, we examine overall characteristics and performance metrics from Summit. The assessed set of metrics are restricted to an ordinary user's perspective of the system, as viewed from a login node. Although these metrics are often reported and analyzed in-depth from the system's perspective by using system-wide monitoring and profiling tools, here we have used a subset of these metrics to support our claims regarding user experience and behavior.

3.1   Users Access

Figure 1 shows the distribution of user sessions per login node. The average percentage of user distribution is 21.9% ($\pm$9.2%), 20.7% ($\pm$9.2%), 17.9% ($\pm$7.8%), 20.2% ($\pm$8.8%), and 19.3% ($\pm$8.1%) for login nodes 1–5, respectively. Although this distribution is relatively balanced among login nodes, by inspecting the distribution of unique IPs per login session (Fig. 3) we observe that there is an imbalance on the disposition of individual users among the nodes. Specifically, the average percentage of unique IPs distribution is 20.5% ($\pm$8.8%), 20.8%

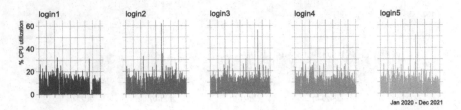

Fig. 4. CPU utilization on summit login nodes (Jan 2020–Dec 2021).

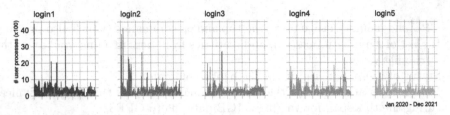

Fig. 5. User processes on summit login nodes (Jan 2020–Dec 2021).

($\pm$9.2%), 16.7% ($\pm$8.6%), 23.5% ($\pm$10.1%), and 18.5% ($\pm$8.1%) for login nodes 1–5, respectively. This indicates that a subset of users may be (involuntarily) benefitting from an increased number of concurrent login sessions; thus, their perceived experience of the system may be more favorable when compared with users who share resources with a larger number of individual users.

To evaluate the above claim, we examined CPU utilization and the number of user processes per login node (Figs. 4 and 5). Overall, CPU utilization is relatively balanced among nodes (around 15% in average across nodes) with some spikes on login nodes 2, 3, and 5. Unsurprisingly, the number of user processes follows similar trends as for the distribution of unique IPs. Both of these results corroborate the claim that a small subset of users have been benefited from lower concurrency. More precisely, the balanced distribution of CPU utilization on login nodes 3 and 5 indicates that this small set of users consumes as many resources on these nodes as the larger set of users on the other nodes.

3.2   I/O Throughput

Every hour, we have measured the I/O throughput of Summit's GPFS for writing a 1 GB randomly generated binary data file to a shared folder. Notice that we do not aim to assess peak write speeds; instead our goal is to identify potential low performance caused by user-related I/O operations within the login nodes. Figure 6 shows the distribution of the number of user processes running per login node in relation to the I/O throughput for writing a 1 GB file. Note that the performance of the GPFS filesystem may also be affected by I/O operations occurring on the compute nodes; thus a weak correlation is expected with processes running on the login nodes. That said, we can observe that a low performance is highly correlated with an increased number of user processes

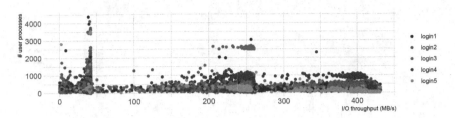

Fig. 6. I/O throughput of Summit's GPFS for writing a 1 GB file in relation to the number of user processes.

running on the login nodes. Specifically, throughput values as low as 42 MB/s are reported when more than 3,000 user processes are running for more than 3 consecutive hours. For the same set of datapoints, user processes running on login5 run for more than 5 consecutive hours, which coincides with the timespan in which the filesystem yields low performance (recall that login5 has, on average, a reduced number of concurrent unique users). Analogously, impaired performance (around 250 MB/s) is observed for a very small group of users who run more than 2,500 processes on login3 for more than 7 h.

3.3 Computational Jobs

The fundamental purpose of leadership-class supercomputers is to improve science by running the largest-scale computational jobs. It is expected that user satisfaction is mostly dictated by the ability to execute batch jobs successfully with good performance and without long waits in the queue. Figure 7-*top* shows the percentage distribution of jobs based on their status. The workload average jobs submitted, running, and completed, as shown by the LSF scheduler, are 529 (±271), 81 (±24), and 90 (±69), respectively. Given that the number of individual users (see Table 1) is orders of magnitude higher than the average number of running jobs, the variation in the number of running jobs seems relatively low.

Figure 7-*bottom* shows the distribution of node-hours consumed per job. Intriguingly, the shape of the distributions are alike across years and months. More precisely, the average root mean square error (RMSE) is below 6 for every month comparison between the two years, with most jobs consuming between 1,000 and 10,000 node-hours. This result suggests that jobs are mostly submitted by a small set of users running similar, yet large, workloads. Indeed, by examining the number of jobs submitted per user, we observe that *only 29 users (−1.4% of total number of users in the dataset), submitted more than 50% of all Summit jobs over the measured period of time.* These jobs represent more than 82% of the total consumed node-hours in the dataset (Fig. 8). As expected, most individual jobs consume between 1,000 and 10,000 node-hours, which corroborates the findings asserted from Fig. 7-*bottom*. Most users submitted a very small number of jobs, though they span a wide range of node-hours consumption, with a few

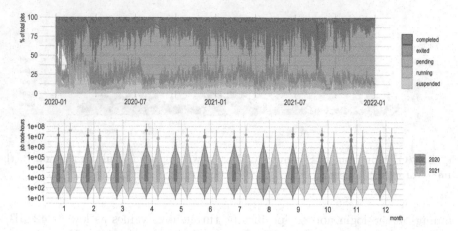

Fig. 7. Distribution of jobs' statuses (*top*) and jobs' sizes (*bottom*). Each "violin" represents the distribution of jobs' sizes in a given month in terms of node-hours as a rotated kernel density plot on top of a box-plot that shows the first and third quartiles of the distribution; the width of the violin corresponds to the number of jobs, and the dots indicate outliers in the tails.

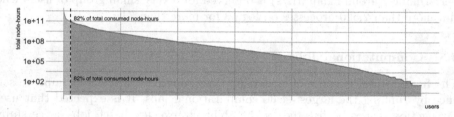

Fig. 8. Distribution of total node-hours consumed per user. (The dashed vertical line delineates the total node-hours consumed by 29 users, which represents more than 82% of the entire dataset.)

jobs consuming nearly all available compute resources. We can also observe that specific users submitted sets of individual jobs with a wide range of node-hours (e.g., from 96 up to 193,537), but also submitted more than 1,100 jobs with the same size (e.g., ~4,300 node-hours).

4   User Behavior

HPC performance metrics are traditionally associated with success metrics such as high system utilization and large number of users and jobs, which correlate to wide system adoption by the community and fulfillment of scientific goals. Understanding and modeling user behavior in HPC environments is key to exhibiting usage patterns that may help improve the design of the system, relate performance bottlenecks to specific behaviors, and ascertain violations of policies and best practices, among other things. Previous studies have mainly focused on

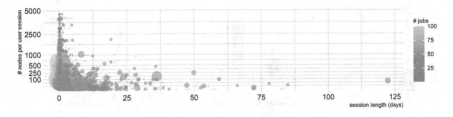

Fig. 9. Users' session lengths (in days) in relation to the number of jobs submitted and maximum number of nodes requested.

job characteristics (performance metrics as presented in Sect. 3.3) and scheduling (queuing time, wall time, etc.) [12,13,19]. In this section, we examine user behavior from the standpoint of (i) the average user session length, (ii) misuses, and (iii) usage patterns of a complex class of applications such as scientific workflows.

4.1 User Sessions

In this section, we investigate the length of user sessions in an attempt to characterize user behavior by relating the time users spend logged into the system with the number and size (in terms of number of nodes) of jobs submitted. We define a *session* as a time interval indicated by activity which begins and ends with inactivity. We use batch job submission as the indicator of activity, and for inactivity, we leverage *think time* [7], which quantifies the time between the completion of a job and the submission of the next job by the same user. Thus, a session is the time period that complements two subsequent think times for the same user. In this work, we assume that a think time is characterized by an interval of more than 24 h. We do not consider weekends, holidays, or system downtimes or outages as think times.

We identified 27,789 sessions, the longest of which spans 123 days and runs 68 jobs over a maximum of 64 nodes. Most of the users (about 92%) established more than one session, and most of the user sessions (about 84%) span less than one day; also, more than 50% of these sessions request only 1 or 2 nodes per job. Sessions with large-scale jobs that use nearly all of Summit's compute nodes span only a few hours, with only 3 spanning slightly more than one day. This supports the idea that user experience on login nodes significantly impacts user satisfaction, because users spend most of their time testing and debugging while using the login nodes. Figure 9 shows the distribution of user sessions' lengths in relation to the total number of nodes used by all jobs within a session.

4.2 Misuse

Typically, HPC systems balance users across the set of login nodes to improve the overall user experience and limit any potential performance impact due to

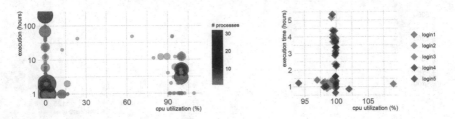

Fig. 10. *Left:* Execution of `mpirun` and `mpiexec` on login nodes (1,172 instances from 74 users). *Right:* 56 executions of GROMACS (`gmx_mpi`) on login nodes by two users. Each user runs instances of GROMACS on every login node, which occupy the available GPUs for several hours.

heavy user processes (see Sect. 3.1). To prevent low quality of service, most HPC systems provide guidance and best practices for operations that should not be performed on login nodes because they are shared resources. For instance, it is discouraged to run long-term and/or heavy services (e.g., databases) on such nodes. In this section, we examine whether users run processes that could harm the overall performance of these shared resources. To this end, we mined the dataset for processes that did not represent typical, system-related tasks, that consumed a substantial amount of resources (CPU/GPU/memory), or that ran for a long period of time. We limit our discussion in this section to two representative use cases: (i) execution of tightly coupled applications using `mpirun` and `mpiexec`, and (ii) execution of high-throughput applications.

Tightly Coupled Applications. We have identified 1,172 uses of `mpirun` and `mpiexec` by 74 users for running tightly coupled applications in the login node. (Our filtering process removed mentions to compiling operations and flags, environment variables, etc.) In further investigation, we noticed that 816 out of the 1,172 instances of `mpirun` and `mpiexec` spawned only a single process for less than one hour – which suggests that those executions were simple tests. Figure 10-*left* shows execution times for the `mpirun` and `mpiexec` instances, their associated CPU utilization, and the number of processes spawned. The longest execution runs for 204 h and spawns 16 processes, followed by a dozen of executions that run for about 100 h. There is also a cluster of instances that consume more than 90% of CPU for an average of 12 h, with two instances running for 47 and 49 h each. A detailed look at these instances unveiled that they use up to 4 cores from the login nodes and up to 4 GB of RAM each, which could then considerably impact the performance of sound processes (compilation, (de)compression, file synchronization, etc.) from other users.

Figure 10-*right* shows a subset of the executions shown in Fig. 10-*left*, which corresponds to executions of GROMACS [17], a widely used molecular dynamics package, on GPUs in the login nodes. Specifically, we highlight a use case in which two users attempt to "game the system" by launching concurrent executions of the GPU-enabled version of GROMACS (`gmx_mpi`), configured to spawn one CPU process and as many GPU processes as available in the system. To prevent such behaviors, Summit enforces limits on the login nodes to ensure resource availability by leveraging the Linux kernel feature `cgroups`: each user is limited to 16 hardware threads, 16 GB of memory, and 1 GPU; and after 4 h of CPU-time all login sessions are limited to 0.5 hardware threads; after 8 h, the process is automatically killed. These limits are reset as new login sessions are started. These two users consumed 50% of all GPU resources across login nodes for about 84 consecutive hours, however, through a synchronized process in which each of them re-initiated a session periodically, so the limits would be reset. This behavior is not only substantially harmful to other users by preventing a fair share of resources, but also it conflicts with best practices of not running scientific applications within login nodes.

High-Throughput Applications. We have identified a substantial number of executions of high-throughput applications on the login nodes. Here, we focus on a subset of these executions that consumes more than 90% of CPU per process, which comprises 8,014 instances executed by 549 users (27.9% of total users). Figure 11-*left* shows the distribution of user processes *vs.* their length, in hours, that run user codes (i.e., scientific applications) on the login nodes. As for the tightly-coupled applications above, we have filtered out all instances related to sound processes (compilation, (de)compression, file synchronization, etc.). Users ran a wide range of codes—495 unique programs—in which –78% of them run for less than an hour; thus, we consider them as execution tests. Some instances span 16 threads (`cgroups` limit) and run up to 7 h, while others (about 7% of the dataset) use more than 8 threads and run between 3 and 8 h. We then consider these instances as misuse of the login nodes. Due to the limits imposed by Summit, we do not observe any attempt to "game the system"; these processes are mostly evenly distributed across login nodes, with a slightly higher number for login3 (405 instead of 330 on average for the other login nodes).

In spite of the large variation of user programs, we have identified that 4,478 instances (from 329 users) are running Python programs (Fig. 11-*right*). These instances represent 72.6% of the instances shown in Fig. 11-*left*, which run for more than an hour. This result indicates that some users may tend to use these login nodes as additional computing resources, or even as their sole computing node. In order to assert the latter, we attempted to isolate the list of users that ran any of these codes without ever submitting a single job to the batch queue. Astonishingly, we identified 41 users that fall into this category, which comprises 1,012 instances, i.e., 12.6% of the original dataset (Fig. 11-*bottom*). Although running user programs on login nodes as an extension of computing resources is against best practices, using a leadership-class HPC system for running user-based codes uniquely on login nodes must be prevented—strict policies and processes should then be defined to impede similar misuse of resources.

While the `cgroups` mechanism protects the overall login node resources, it falls short in "low key" and "gaming the system" misuse, as shown above. Several measures may be taken to mitigate these issues. For example, the data collected by this work can be used to identify misusers, either to educate them about best practices or perhaps to introduce punitive actions. We will not conjecture about potential new policies here, however.

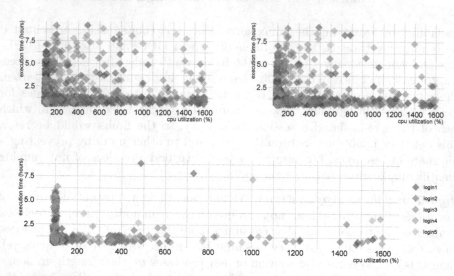

Fig. 11. *Left:* Execution of user processes (high-throughput applications) on login nodes. *Right:* Execution of Python programs on login nodes. *Bottom:* Execution of Python programs by users that have never submitted a batch job to the system. (Note that 1600% CPU utilization means that a process comprising 16 threads consumed 100% CPU utilization each.)

4.3 Scientific Workflows

Scientific workflows are used almost universally across scientific domains for solving complex and large-scale computing and data analysis problems. The importance of workflows is highlighted by the fact that they have underpinned some of the most significant discoveries of the past few decades [3]. Many of these workflows have significant demands for computation, storage, and communcation, and thus they have been increasingly executed on large-scale computer systems [14]. In this section, we seek to identify how and to what extent workflows have been used on Summit. Typically, workflow systems run a coordinator process that manages workflow tasks' dependencies, launches jobs to the batch queue as their dependencies are satisfied, monitors their jobs' execution, and performs data movement operations on behalf of the user. Table 2 shows the total number of processes run by workflow systems in Summit login nodes. In total, 71 users utilized workflow technologies for automating the execution of their scientific applications. These processes often refer to agents that manage the workflow execution and they can take several formats: from single orchestration components (e.g., Swift/T) to the management of ensembles (e.g., RADICAL/EnTK). The former leverages batch jobs for defining workflows within a parallel, tightly coupled application (thus the lower number of processes), while the latter manages sets of tasks as high-throughput applications, i.e. the so-called *pilot jobs* [2].

Figure 12 shows the cumulative number of workflow-related processes across Summit login nodes for our dataset. Overall, workflow technology adoption has

Table 2. Total number of workflow management systems' processes observed across Summit login nodes (Jan 2020–Dec 2021).

	Parsl	Swift/t	Pegasus	Fireworks	Dask	Maestro	Cylc	Dagman	Snakemake	Radical
Processes	3,807	88	5,399	319	40,875	2,225	106	4	15,797	2,113,192
Users	7	3	3	5	27	6	1	1	5	13

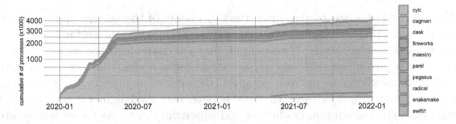

Fig. 12. Cumulative number of workflow management systems' processes observed across Summit login nodes (Jan 2020–Dec 2021), shown with square root scale.

gradually increased throughout these past two years. A notable growth in workflow usage is observed in the first two quarters of 2020, which coincides with research conducted to understand the COVID-19 pandemic through the use of HPC. Specifically, this research leveraged the RADICAL/EnTK framework for investigating spike dynamics in a variety of complex environments, including within a complete SARS-CoV-2 viral envelope simulation [5]. This research has been awarded the 2020 ACM Gordon Bell Special Prize for High Performance Computing-Based COVID-19 Research.

5   Related Work

Analyzing and characterizing HPC workloads is a common practice for measuring system and application performance metrics and thus identifying potential bottlenecks and atypical behaviors [7]. For example, the National Energy Research Scientific Computing Center (NERSC) has profiled and characterized three generations of their supercomputing systems [11]. In these studies, HPC benchmarks are used to obtain performance measurements, which are then used for the procurement process of machines. Similarly, a characterization of the workload of Tianhe-1A at the National Supercomputer Center in Tianjin presents equivalent system-level metrics [8]. In [10], a characterization of a parallel filesystem unveils I/O bottlenecks for different classes of applications. Conversely, our analyses in this paper target users' experience and behavior on login nodes—the interface to HPC systems.

In [13], user behavior is studied with regards to think time, the time between the completion of a job and the submission of the next job by the same user. Although this work leverages this same concept for defining user sessions, the study conducted in [13] attempted to understand and characterize patterns of job submissions. Our work, instead, seeks to understand user behavior on login nodes and relate their actions to misuses of the system or performance issues. To the best of our knowledge, this is the first work that conducts such a study.

6   Conclusion and Future Work

We examined observation data from the login nodes of the leadership-class Summit supercomputer at OLCF. We analyzed traditional system performance metrics such as user access, I/O throughput, and job characteristics, as well as user behavior regarding session lengths, misuse of login nodes, and how users have leveraged workflows to perform complex, distributed computing. Our findings identified key usage patterns that we believe will shed light on the usage of login nodes on contemporary clusters and supercomputers. As immediate future work, we will continue to collect this observation data for the rest of the life of Summit, and we will start data collection for the upcoming exascale Frontier supercomputer at OLCF. We also intend to analyze the data further into other dimensions, including resource usage balancing and correlation of external events (e.g., conference deadlines, call for proposals deadlines, etc.), as well as the impact of the COVID-19 pandemic on the user behavior.

Acknowledgments. This research used resources of the Oak Ridge Leadership Computing Facility at the Oak Ridge National Laboratory, which is supported by the Office of Science of the U.S. Department of Energy under Contract No. DE-AC05-00OR22725. We acknowledge Suzanne Parete-Koon for early brainstorming of some of the ideas presented here. We thank Scott Atchley, Bronson Messer, and Sarp Oral for their thorough revision of this paper.

References

1. IP Geolocation API (2022). https://www.abstractapi.com/ip-geolocation-api
2. Ananthraj, V., et al.: Towards exascale computing for high energy physics: the atlas experience at ornl. In: 2018 IEEE 14th International Conference on e-Science (e-Science), pp. 341–342 (2018). https://doi.org/10.1109/eScience.2018.00086
3. Badia Sala, R.M., Ayguadé Parra, E., Labarta Mancho, J.J.: Workflows for science: a challenge when facing the convergence of HPC and big data. Supercomput. Front. Innov. **4**(1), 27–47 (2017). https://doi.org/10.14529/jsfi170102
4. Bang, J., et al.: HPC workload characterization using feature selection and clustering. In: Proceedings of the 3rd International Workshop on Systems and Network Telemetry and Analytics, pp. 33–40 (2020). https://doi.org/10.1145/3391812.3396270
5. Casalino, L., et al.: AI-driven multiscale simulations illuminate mechanisms of SARS-CoV-2 spike dynamics. Int. J. High Perform. Comput. Appl. (2021). https://doi.org/10.1177/10943420211006452
6. Dongarra, J., Gottlieb, S., Kramer, W.T.: Race to exascale. Comput. Sci. Eng. **21**(1) (2019). https://doi.org/10.1109/MCSE.2018.2882574

7. Feitelson, D.G.: Looking at data. In: 2008 IEEE International Symposium on Parallel and Distributed Processing, pp. 1–9 (2008). https://doi.org/10.1109/IPDPS. 2008.4536092

8. Feng, J., Liu, G., Zhang, J., Zhang, Z., Yu, J., Zhang, Z.: Workload characterization and evolutionary analyses of Tianhe-1A supercomputer. In: Shi, Y., et al. (eds.) ICCS 2018. LNCS, vol. 10860, pp. 578–585. Springer, Cham (2018). https://doi.org/10.1007/978-3-319-93698-7_44

9. Liu, Z., et al.: Characterization and identification of HPC applications at leadership computing facility. In: Proceedings of the 34th ACM International Conference on Supercomputing, pp. 1–12 (2020). https://doi.org/10.1145/3392717.3392774

10. Lockwood, G.K., Snyder, S., Wang, T., Byna, S., Carns, P., Wright, N.J.: A year in the life of a parallel file system. In: Proceedings of the International Conference for High Performance Computing, Networking, Storage, and Analysis, pp. 931–943. IEEE Press (2018). https://doi.org/10.1109/SC.2018.00077

11. Nersc benchmarking and workload characterization (2021). https://www.nersc.gov/research-and-development/benchmarking-and-workload-characterization

12. Rodrigo, G.P., Östberg, P.O., Elmroth, E., Antypas, K., Gerber, R., Ramakrishnan, L.: Towards understanding HPC users and systems: a NERSC case study. J. Parallel Distrub. Comput. **111**, 206–221 (2018). https://doi.org/10.1016/j.jpdc. 2017.09.002

13. Schlagkamp, S., Ferreira da Silva, R., Deelman, E., Schwiegelshohn, U.: Understanding user behavior: from HPC to HTC. Procedia Comput. Sci. **80**, 2241–2245 (2016). https://doi.org/10.1016/j.procs.2016.05.397. International Conference on Computational Science 2016, ICCS 2016

14. Ferreira da Silva, R., Filgueira, R., Pietri, I., Jiang, M., Sakellariou, R., Deelman, E.: A characterization of workflow management systems for extreme-scale applications. Fut. Gen. Comput. Syst. **75**, 228–238 (2017). https://doi.org/10.1016/j.future.2017.02.026

15. Ferreira da Silva, R., et al.: Characterizing a high throughput computing workload: the compact muon solenoid (CMS) experiment at LHC. Procedia Comput. Sci. **51**, 39–48 (2015). https://doi.org/10.1016/j.procs.2015.05.190, International Conference On Computational Science, ICCS 2015 Computational Science at the Gates of Nature

16. Top 500 (2021). https://www.top500.org

17. Van Der Spoel, D., Lindahl, E., Hess, B., Groenhof, G., Mark, A.E., Berendsen, H.J.: Gromacs: fast, flexible, and free. J. Comput. Chem. **26**(16) (2005). GROMACS: fast, flexible, and free

18. Vazhkudai, S.S., et al.: The design, deployment, and evaluation of the coral pre-exascale systems. In: SC18: International Conference for High Performance Computing, Networking, Storage and Analysis, pp. 661–672. IEEE (2018)

19. Wolter, N., McCracken, M.O., Snavely, A., Hochstein, L., Nakamura, T., Basili, V.: What's working in HPC: investigating HPC user behavior and productivity. CTWatch Q. **2**(4A), 9–17 (2006)

20. Zhang, S., Zhang, C., Yang, Q.: Data preparation for data mining. Appl. Artif. Intell. **17**(5–6) (2003). https://doi.org/10.1080/713827180

n-type B-N Co-doping and N Doping in Diamond from First Principles

Delun Zhou[1], Lin Tang[1], Jinyu Zhang[1], Ruifeng Yue[1,2(✉)], and Yan Wang[1,2,3(✉)]

[1] School of Integrated Circuit, Tsinghua University, Beijing, China
yuerf@tsinghua.edu.cn, wangy46@mail.tsinghua.edu.cn
[2] Beijing National Research Center for Information Science and Technology, Beijing, China
[3] Beijing Innovation Center for Future Chips (ICFC), Beijing, China

Abstract. The boron-nitrogen (B-N) co-doped diamond with different structures have been studied by the first-principle calculations to find possible defect structures to achieve effective n-type doping. Nitrogen doped diamond itself shows the characteristics of direct bandgap, however its big gap between donor level and conduction band minimum (CBM) may contribute to its undesirable ionization energy. We found for the first time B-N co-doping as a promising method to overcome the disadvantages of N doping in diamond. B-N co-doped diamond, especially the B-N_3 defect, retains the characteristics of direct band gap, and has the advantages of low ionization energy and low formation energy. The effective mass of electron/ hole of B-N co-doped diamond is less than that of pure diamond, indicating better conductivity in diamond. The N-2p states play vital role in the conduction band edge of B-N_3 co-doped diamond. Hence, the B-N_3 has outstanding performance and is expected to become a promising option for N-type doping in diamond.

Keywords: B-N co-doping · n-type diamond · Band structure · Effective mass

1 Introduction

As an ultra-wide bandgap semiconductor material, diamond has extraordinary physical and chemical properties and is expected to be utilized in high-power electronic devices. The wide bandgap (-5.50 eV) and extreme thermomechanical properties of diamond make it a potentially vital material for future electronic devices, especially for high-frequency, high-power, high irradiation tolerant and high-temperature applications. Besides, its carrier mobility, critical breakdown field and thermal conductivity make diamond stand out from other wide-bandgap semiconductor materials.

Diamond doping has been the focus of continuous attention in the past few decades and various scientific experiments have been conducted. P-type doping in the diamond can be realized by boron doping, whose acceptor level is 0.37 eV. Wide doping range (10^{14} to 10^{21} cm^{-3}) has been realized as well [1]. However, achieving an appropriate n-type diamond is still challenging. Elements from Group I, V and VI have been studied as dopant for diamonds in recent years. Substitutional nitrogen may provide an extra

D. Groen et al. (Eds.): ICCS 2022, LNCS 13350, pp. 530–540, 2022.
https://doi.org/10.1007/978-3-031-08751-6_38

electron, but its deep donor level (1.4 eV below the edge of the conduction band) limits its applications on room temperature semiconductor devices [2].Though the ionization energy of P doping is only 0.43 eV, its low carrier mobility (−23 cm/V.s) makes it unsuitable for room temperature applications as well [3]. Shallow levels cited for sulfur doping turned out to be incorrect due to inadvertent contamination with boron, and S donor level is currently believed to be closer to 1.4 eV [4]. In addition, the formation energy of sulfur is far greater than that of phosphorus (4.2 eV), indicating sulfur is more difficult to incorporate into diamond than phosphorus, and its solubility in bulk diamond will also be lower, which directly affects doping efficiency. Prins [5] obtained a shallow n-type diamond(donor level:0.32eV) through oxygen implantation. Yet, the oxygen donors would be deactivated after being annealed above 600 °C. Interstitial Li and Na in diamond have been calculated: the donor level of interstitial Li is 0.1 eV [6] and the donor level of interstitial Na is 0.3 eV [7]. However, their solubility in diamond is extremely low and are likely to combine with other impurities or defects in diamond, leading to electrical inactivity.

Up to now, no suitable single dopant has been found to achieve the n-type shallow donor doping of diamond, especially at ionization energy and formation energy angles.

Researchers have begun to use co-doping method to study diamond impurity doping. Some experiments have achieved n-type diamond through B-S co-doping [8]. Theoretical explanations for B-S co-doping have been published [9].

However, the performance of B-N co-doped diamond is far from ideal and better explanation of co-doped diamond system is still needed to be explored. Previous work [11] has mainly studied the bond length and ionization energy of B-N clusters in diamond. There is still a lack of calculation and research on key parameters such as DOS, band structure, and carrier effective mass of B-N co-doped diamond. Hence, it is necessary to explore the conductivity mechanism of B-N co-doped diamond with shallow donor level.

2 Calculation Methods

This paper follows the calculation method in our previously published work [9, 10]. All calculations were performed based on density functional theory (DFT) to optimize the geometric structure and compute the band structure and density of state. The exchange-correlation function is implemented by VASP through the Perdew–Burke–Ernzerhof (PBE) within generalized gradient approximation (GGA) [1] Projector augmented plane-wave (PAW) potentials is chosen to describe the core-valence interaction. In this research, the cutoff energy of plane-wave was set to 500 eV using convergence verification [12], the convergence criterion of electronic structure relaxation calculation was set as 1×10^{-5} eV, and the convergence criterion of the inter-atomic forces was set as $1 \times 10-4$ eV/Å. A 216-atom supercell (3 × 3 × 3) of diamond, with a 9 × 9 × 9 mesh (Monkhorst-Pack) of KPOINTS was adopted for the calculation. A large cell size was essential to ensure the reliability of calculation results, especially for shallow doping studies. The bulk diamond's lattice constant was converged to 3.573 Å, in line with previous experiments, and was served as the original building block for supercell construction.

3 Results and Discussion

3.1 Impurity Formation Energy(Ef)

The impurity formation energy is calculated to judge the possibility and stability of impurity doping into the material. The lower the value, the easier and more stable the impurity doping is, which is conducive to effective doping.

The formation energy of impurity X in charge state q is defined as:

$$E_f[Xq] = E_{tot}[Xq] - E_{tot}[C, bulk] - \sum nX \, \mu X - q(E_F + \Delta V) \qquad (1)$$

where Etot[Xq] denotes the total energy of the whole structure and Etot[C,bulk] denotes the total energy of the perfect diamond bulk without any impurity. nX describes the number of doping/removing impurity atoms of atom X (C atoms or doping atoms), μX denotes the chemical potential of impurity X. In this study, the chemical potential of nitrogen was calculated from N_2 and the chemical potential of boron was from B_2H_6 [13]. E_F denotes the Fermi level referenced to the VBM (Ev), and ΔV is an alignment of electrostatic potentials between the defect supercell and the bulk. In this study, we primarily concentrate on the neutral charge state, which means that the value of the last term in the expression is basically zero.

The solubility of impurity is strongly related to $N\exp(-E_f /kT)$, where k is Boltzmann's constant, E_f represents the formation energy of the impurity, T denotes the temperature, and N represents the density of sites in the bulk where the impurity may incorporate [14]. Hence, the impurity solubility is strongly associated with its formation energy. A smaller formation energy usually indicates better doping efficiency.

First, we investigated the formation energy of a nitrogen atom or boron atom in diamond at substitutional site or interstitial site. As shown in Table 1, the formation energy of substitutional boron atom (or substitutional nitrogen atom) doped diamond is significantly lower than that of interstitial nitrogen/boron atom doped diamond, suggesting that substitutional sites of nitrogen and boron atoms are easier to be incorporated. Therefore, we focus on the substitutional doping when studying B-N co-doped diamond (Fig. 1).

Table 1. Impurity formation energy (E_f) of interstitial B (B_i), substitutional B (B_s), interstitial N (N_i), and substitutional N (N_s) doping in diamond

Compound	Position	E_f /eV
$C_{215}N$	N_s	2.127
$C_{216}N$	N_i	8.869
$C_{215}B$	B_s	−0.257
$C_{216}B$	B_i	5.675

The covalent radius of nitrogen (0.734 Å) and boron (0.82 Å) are close to that of carbon (0.77 Å), indicating much lighter lattice distortion than other substitutional

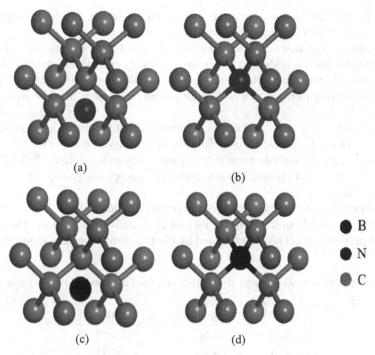

Fig. 1. Doping structure of interstitial nitrogen(a), substitutional nitrogen(b), interstitial boron(c), substitutional boron(d).

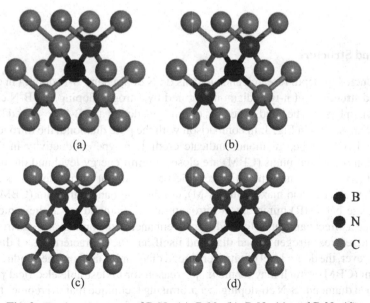

Fig. 2. Doping structures of B-N$_1$ (a), B-N$_2$ (b), B-N$_3$ (c), and B-N$_4$ (d).

dopants. Smaller lattice distortion results in less internal strain, which contributes to the construction of doped structures with low formation energy.

Multiple nitrogen atoms were added around the substitutional boron atom to replace those original nearest neighbor carbon atoms. Figure 2 demonstrates the four structures mainly studied in this paper, which are a substitutional boron atom connected with one to four neighboring nitrogen atoms. The formation energies of the four structures are shown in Table 2.

Calculation results in Table 2 distinctly illustrate that the formation energies of the four structures are low, which are -1.662 eV, -1.257 eV, -0.875 eV, and 0.523 eV. The formation energies of other common dopants, such as S (11.1eV), P (10.4 eV) and B-S (6.2 eV) are much greater. The low formation energy suggests great potential for effective doping.

The low formation energies of these impurities indicate that nitrogen and boron atoms exist as a whole in the diamond instead of as individual dopants. Therefore, its electrical properties shall not be explained by the traditional compensation effect.

Table 2. Impurity formation energy (Ef) of B-N co-doped diamond with different structures.

Compound	Structure	Ef/eV
$C_{214}B_1N_1$	B-N_1	-1.662
$C_{213}B_1N_2$	B-N_2	-1.257
$C_{212}B_1N_3$	B-N_3	-0.875
$C_{211}B_1N_4$	B-N_4	0.523

3.2 Band Structure

We conducted in-depth research and analysis on N doping and B-N doping in diamond. The band structures of n-type diamond formed by nitrogen doping and B-N co-doping are shown in Fig. 3. The band structures of nitrogen doped and B-N co-doped led n-type diamond are shown in Fig. 3. In comparison with the pure diamond, the nitrogen, B-N_2, B-N_3, and B-N_4 doped diamonds indicate credible n-type conductivity in that those conduction band minimums (CBM) are close to fermi energy level and the intermediate band (IB) in the bandgap can be regarded as the donor energy level. As shown in Fig. 3, the valence band maximum (VBM), conduction band minimum (CBM) and the intermediate band (IB) minimum of nitrogen doped diamond are at the same k-point, suggesting a direct bandgap is formed, hence, enhancing the optical and electronic properties. Therefore, nitrogen doped diamond itself has the characteristics of direct band gap. However, the donor level of nitrogen doped diamond is far from the conduction band minimum (CBM), which may be the major reason for undesirable electrical properties of N doped diamond. B-N co-doping is a promising technique that overcomes the disadvantages of N doped diamond, especially in the field of ionization energy. B-N co-doped diamond preserves the characteristics of the direct band gap and significantly reduces

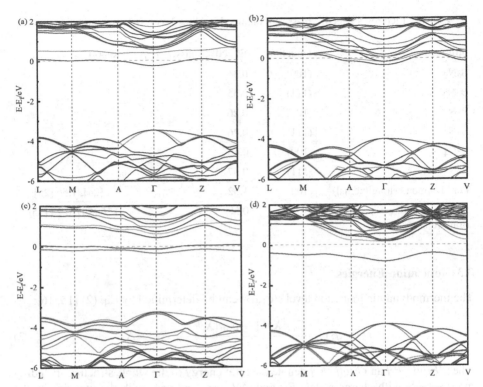

Fig. 3. Band structures of doped diamond with B-N$_2$ (a), B-N$_3$ (b), (c) B-N$_4$, and (d) Nitrogen

the gap between the donor level and the conduction band minimum (CBM). B-N$_3$ doped diamond seems to be the most promising defect structure, since there is no obvious gap between the donor level and the CBM, in other words, there is no apparent intermediate level. The novel characteristic may indicate great potential for better ionization energy performance. As far as we know, this is the first time these results have been reported. B-N co-doped diamond is expected to have outstanding electrical and optoelectronic properties.

Furthermore, we calculated the electron effective masses of diamond with B-N$_2$, B-N$_3$, and B-N$_4$ doped. The calculation results of electron effective masses are presented in Table 3. In order to ensure the accuracy of the calculation results, we also calculated the effective mass of pure diamond (longitudinal and transverse), and the calculated results are in good agreement with the experimental results. The electron's effective masses of B-N co-doped diamond is basically smaller than pure diamond.

Table 3. Electron's effective masses of diamond co-doped with B-N and pure diamond.

Structure	Direction	Effective masses (m_0)	Reports
B-N$_2$	(100)	0.34	
B-N$_2$	(111)	0.35	
B-N$_3$	(100)	0.26	
B-N$_3$	(111)	0.26	
B-N$_4$	(100)	0.45	
B-N$_4$	(111)	0.57	
Pure Diamond (longitudinal)	(100)	1.52	1.4 [24], 1.56 [25]
Pure Diamond (transverse)	(100)	0.38	0.36 [24], 0.28 [25]

3.3 Ionization Energies

The thermodynamic transition level $\varepsilon(q_1, q_2)$ can be determined by Eq. (2) [15, 16]:

$$\varepsilon(q_1, q_2) = \frac{E_{tot}[X^{q_1}] - E_{tot}[X^{q_2}]}{q_2 - q_1} - E_V - \Delta V \tag{2}$$

where $Etot[X^{q_1}]$ and $Etot[X^{q_2}]$ denote the total energy of the whole structure including the defect X with charge states. E_V and ΔV are consistent with the meaning in the previous formula (1). The acceptor ionization energy E_A is equal to the transition level $\varepsilon(0/-)$, and the donor ionization energy E_D equals Eg minus $\varepsilon(0/+)$ [17], where Eg equals the bandgap of pure diamond calculated by VASP(4.1 eV).

To ensure the correctness of our calculation results, we calculated some doping results already verified in experiments and other theoretical calculations. Our calculation results are shown in Table 4, consistent with previous theoretical and experimental results.

Table 4. Ionization energies (E_D, eV) of some n-type dopants in diamond.

Dopant	E_D/eV(Our work)	E_D/eV(Ref.)
P	0.59	0.43[b]-0.56[a][3, 19]
S	1.44	1.4 [b] [4]
O	0.45	0.32[b] [5]
N	1.43	1.4[a,b] [2, 21]
Li(interstitial)	0.04	0.1[a] [7]
BS	0.55	0.39[a]–0.52[b] [22, 23]
Li-N$_4$	0.232	0.271[a] [20]

[a] Theoretical values [b] Experimental values

Table 5. Ionization energies (E_D, eV) and formation energy(E_f, e) of n-type B-N co-doped defects in diamond

Dopant	E_D/eV	E_f/eV
B-N$_2$	0.63	−1.257
B-N$_3$	0.23	−0.875
B-N$_4$	0.54	0.523

The donor ionization energies (E_D) of B-N co-doping dopants are listed in Table 5. It is apparent that BN$_2$, BN$_3$, and BN$_4$ demonstrated n-type diamond characteristics with low donor levels. The donor ionization energy of BN$_3$ is 0.23 eV, which is the best performer in our research.

Our ionization energy calculation results are not quite consistent with previous research [11], in which BN$_2$, BN$_3$, and BN$_4$ have larger ionization energies(greater than 1eV). This may be due to different calculation methods. In the previous article, the ionization energy was obtained by the difference between the CBM and the impurity energy level, while we calculated the ionization energy by formula (2). We believe that the calculation method of the previous work is not accurate enough, and the method we adopt is a widely used calculation method, thus, the calculation results are more credible.

3.4 Electronic Structure

We calculated the total density of states (TDOS) of B-N$_2$, B-N$_3$, and B-N$_4$ co-doped diamond, and the results are displayed in Fig. 4. As shown in Fig. 4, the impurity level and fermi level of B-N$_2$, B-N$_3$, and B-N$_4$ are near the conduction band minimum (CBM). According to the calculation results of the band structure, it is evident that the B-N co-doping can achieve n-type diamond doping, which are in line with the previous discussion of ionization energy. Together, the calculation results of these two parts verify that B-N$_2$, B-N$_3$, B-N$_4$ co-doped diamonds are n-type diamonds.

Combining the performances of TDOS, ionization energy, and formation energy, we assume B-N$_3$ co-doping as a very promising defect structure to realize shallow n-type doping in diamond. We calculated the partial densities of state (PDOS) of B-N$_3$ for in-depth knowledge of its donor characteristics. As shown in Fig. 5, N-2p states play decisive role in the conduction band edge.

Based on previous calculations and discussions, the performance of B-N$_3$ doping is extremely outstanding, and it has negative formation energy and extremely low ionization energy. Other B-N co-doped structures can also achieve N-type diamond doping. B-N$_3$ has strong stability in diamond, is easy to form effective doping, and has a shallow donor energy level. Therefore, it can be concluded that B-N$_3$ co-doping may be a promising alternative to achieve n-type doping with shallow donor levels in diamond.

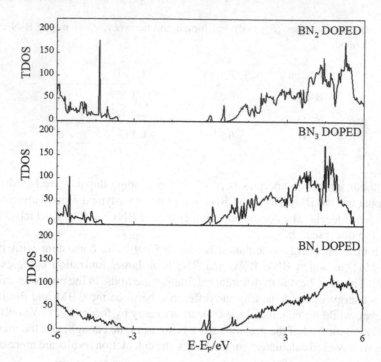

Fig. 4. TDOS of the B-N$_2$, B-N$_3$, B-N$_4$ doped diamond.

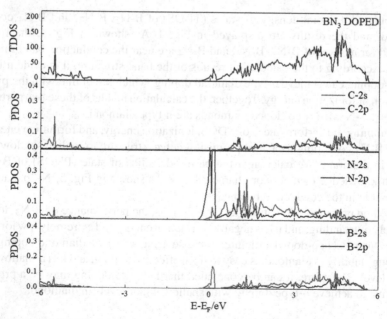

Fig. 5. TDOS and PDOS of the B-N$_3$-doped diamond. (a) TDOS of the B-N$_3$-doped diamond; PDOS of the B-N$_3$-doped diamond for (b) C atoms, (c) N atoms, (d) B atom.

4   Conclusions

In conclusion, B-N co-doped can be expected to realize desirable n-type diamond. B-N3 co-doped diamond has the advantage of low formation energy (-0.875eV) and low ionization energy (0.23eV), indicating great potential for effective doping and shallow n-type doping. Besides, we found that the valence band maximum (VBM), the intermediate band (IB) minimum and covalent band maximum (CBM) of boron and nitrogen co-doping diamond are at the same k-point, which contrasts with pure diamond. N-2p states mainly determine the conduction band edge of B-N3 doped diamond, suggesting the incorporated nitrogen atoms play crucial role in conductivity besides improvement in effective doping. Our study demonstrates that the B-N3 structure has the merit of shallow donor level and high solubility, thus, it can be seen as a promising alternative for n-type diamonds. Our calculation results need to be further verified by experiments.

References

1. Czelej, K., Spiewak, P., Kurzydowski, K.J.: Electronic structure and N-type doping in diamond from first principles. Mrs Adv. 1(16), 1093–1098 (2016)
2. Shah, Z.M., Mainwood, A.: A theoretical study of the effect of nitrogen, boron and phosphorus impurities on the growth and morphology of diamond surfaces. Diam. Relat. Mater. 17(7–10), 1307–1310 (2008)
3. Kato, H., et al.: Diamond bipolar junction transistor device with phosphorus-doped diamond base layer. Diam. Relat. Mater. 27(28), 19–22 (2012)
4. Sque, S.J., Jones, R., Goss, J.P., et al.: Shallow donors in diamond: chalcogens, pnictogens, and their hydrogen complexes. Phys. Rev. Lett. 92(1), 017402 (2004)
5. Prins, J.F.: n-type semiconducting diamond by means of oxygen-ion implantation. Phys. Rev. 61(11), 7191–7194 (2000)
6. Kajihara, S.A., et al.: Nitrogen and potentialn-type dopants in diamond. Phys. Rev. Lett. 66(15), 2010–2013 (1991)
7. Goss, J.P., Briddon, P.R.: Theoretical study of Li and Na as n-type dopants for diamond. Phys. Rev. 75(7), 2978–2984 (2007)
8. Eaton, S.C., Anderson, A.B., Angus, J.C., et al.: Co-doping of diamond with boron and sulfur. Electrochem. Solid-State Lett. 5(8), G65 (2002)
9. Tang, L., Yue, R., et al.: N-type B-S co-doping and S doping in diamond from first principles. Carbon Int. J. Sponsored Am. Carbon Soc. 130, 458–465 (2018)
10. Zhou, D., Tang, L., Geng, Y., et al.: First-principles calculation to N-type Li N Co-doping and Li doping in diamond. Diam. Related Mater. 110, 108070 (2020)
11. Croot, A., Othman, M.Z., Conejeros, S., et al.: A theoretical study of substitutional boron-nitrogen clusters in diamond. J. Phys. Condens. Matt. 30(42) (2018)
12. Chadi, D.J.: Special points for Brillouin-zone integrations. Phys. Rev. 16(4), 1746–1747 (1977)
13. Spiewak, P., Kurzydlowski, K., et al.: Electronic structure of substitutionally doped diamond: Spin-polarized, hybrid density functional theory analysis. diamond & related materials, 2017.Walle V D , Chris G . First-principles calculations for defects and impurities: applications to III-nitrides. J. Appl. Phys. 95(8), 3851–3879 (2004)
14. Ullah, M., Ahmed, E., Hussain, F., Rana, A.M. , Raza, R., Ullah, H.: Electronic structure calculations of oxygen-doped diamond using DFT technique. Microelectr. Eng. 146(1), 26–31 (2015)

15. Freysoldt, C., Grabowski, B., Hickel, T., et al.: First-principles calculations for point defects in solids. Rev. Mod. Phys. **86**(1) (2014)
16. Walle, V.D., Chris, G.: First-principles calculations for defects and impurities: applications to III-nitrides. J. Appl. Phys. **95**(8), 3851–3879 (2004)
17. Goss, J.P., Briddon, P.R., Eyre, R.J.: Donor levels for selected n-type dopants in diamond: a computational study of the effect of supercell size. Phys. Rev. Condens. Matt. Mater. Phys. **74**(24), 245217.1–245217.7 (2006)
18. Koizumi, S., Kamo, M., Sato, Y., et al.: Growth and characterization of phosphorous doped {111 homoepitaxial diamond thin film. FASEB J. **9**(8), 651–658 (1997)
19. Miyazaki, T., Okushi, H.A.: A theoretical study of a sulfur impurity in diamond. Diam. Relat. Mater. **10**(3–7), 449–452 (2001)
20. Moussa, J.E., Marom, N., Sai, N., Chelikowsky, J.R.: Theoretical design of a shallow donor in diamond by lithium-nitrogen codoping. Phys. Rev. Lett. **108**(22), 226404.1–226404.5 (2012)
21. Schwingenschloegl, U., Chroneos, A., Schuster, C., et al.: Doping and cluster formation in diamond. J. Appl. Phys. **110**(V110N5), 162 (2011)
22. Jing, Z., Li, R., Wang, X., et al.: Study on the microstructure and electrical properties of boron and sulfur codoped diamond films deposited using chemical vapor deposition. J. Nanomater **2014**(21), 4338–4346 (2014)
23. Eaton, S.C., Anderson, A.B., Angus, J.C., et al.: Diamond growth in the presence of boron and sulfur. Diam. Relat. Mater. **12**(10–11), 1627–1632 (2003)
24. Nava, F., Canali, C., Jacoboni, C., et al.: Electron effective masses and lattice scattering in natural diamond. Solid State Commun. **33**(4), 475–477 (1980)
25. Naka, N., Fukai, K., Handa, Y., et al.: Direct measurement via cyclotron resonance of the carrier effective masses in pristine diamond. Phys. Rev. **88**(3), 035205 (2013)

Stock Predictor with Graph Laplacian-Based Multi-task Learning

Jiayu He[1(✉)], Nguyen H. Tran[1], and Matloob Khushi[1,2]

[1] The University of Sydney, Sydney, Australia
jihe5893@uni.sydney.edu.au, nguyen.tran@sydney.edu.au
[2] University of Suffolk, Ipswich, UK
matloob.khushi@sydney.edu.au

Abstract. The stock market is a complex network that consists of individual stocks exhibiting various financial properties and different data distribution. For stock prediction, it is natural to build separate models for each stock but also consider the complex hidden correlation among a set of stocks. We propose a federated multi-task stock predictor with financial graph Laplacian regularization (FMSP-FGL). Specifically, we first introduce a federated multi-task framework with graph Laplacian regularization to fit separate but related stock predictors simultaneously. Then, we investigate the problem of graph Laplacian learning, which represents the association of the dynamic stock. We show that the proposed optimization problem with financial Laplacian constraints captures both the inter-series correlation between each pair of stocks and the relationship within the same stock cluster, which helps improve the predictive performance. Empirical results on two popular stock indexes demonstrate that the proposed method outperforms baseline approaches. To the best of our knowledge, this is the first work to utilize the advantage of graph Laplacian in multi-task learning for financial data to predict multiple stocks in parallel.

Keywords: Federated learning · Multi-task learning · Graph learning · Stock prediction

1 Introduction

Deep learning based stock prediction modeling has been intensively studied in recent years [17,27]. From the point of view of market analysis, stocks exhibit highly different properties [9]. It is natural to build separate models for a group of stocks and select portfolios based on the prediction [19]. In fact, most researches build prediction models independently, which ignore the dynamic relationship among different stocks, instead of learning models simultaneously.

It is well known that the stock market is a complex network [20]. The price movement of an individual stock is correlated to its historical behavior and also highly depends on other stocks, namely the *inter-series relationship*. For example, investors often assess the performance of an individual stock by exploring the relative impact of each company in its supply-chain network. The financial market shows a hierarchical structure [18], where stocks in different groups (clusters) respond to the same economic factor in a different manner; on the other hand, stocks in the same cluster always demonstrate strong similarities when responding to information, which is referred to as the *intra-cluster relationship*.

Therefore, the main objective of stock prediction is to build prediction models for different stocks and utilize both inter-series and intra-cluster relationships among stocks. The first challenge is building prediction models to extract the temporal dependencies of time series. To address the challenge, we introduce a federated multi-task learning method [11,25] to learn separate models for each stock simultaneously. Federated multi-task learning is able to handle the diversity of different tasks and build the best model for each task in parallel. On the other hand, in order to model both inter-series and intra-cluster relationships, we present a graph Laplacian [22] learning optimization problem with stylized financial constraints. Here, we consider an undirected weighted graph to represent the stocks network, where each stock task is treated as one node, and the edges represent the dependencies between pairs of stocks. To capture the intra-cluster relationships, we introduce a k-cluster Laplacian constraint to learn a graph with exact k connected groups. The learned graph is added to the multi-task learning framework as a regularization term to control the relationship between tasks. Then, we learn the graph and stock prediction models in an alternating fashion.

The main contribution of this work can be summarized as follows:

- We propose the federated multi-task learning with estimates from stock market data to predict a set of stocks simultaneously.
- We propose the formulation to learn a $k-$cluster graph with rank constraints to captures inter-series and intra-cluster dependencies between stocks.
- We propose the first stock prediction framework that utilizes the advantage of graph Laplacian learning.

To show the effectiveness of our proposed methods, we compare the prediction results with baseline approaches over two popular stock indexes.

2   Related Works

This work is highly related to multi-task learning and graph learning. Multi-task learning algorithms have been intensively studied and have a wide range of applications, such as healthcare, wireless networks [5,8]. In Federated multi-task learning (FMTL) [25], given datasets that are distributed over multiple clients, the goal is to learn separate models for each clients. Each model works best for each client. Stratified Model [26] is introduced in a similar manner while the objective function is minimized by the alternating direction methods of multipliers [4]. However, the critical limit of both FMTL and Stratified Model is that they are unable to solve non-convex objective functions. Recently, a unified framework for FMTL (FedU) [11] had success in solving multi-task learning applications in both convex and non-convex objective functions. It should be noted that FedU treats Laplacian in the regularization term as prior information; however, the relationships between tasks are often unknown in the field of the stock market. Thus, considering the natural diversity of stocks distribution, the objective of our work is to fit separate tasks for each stock and estimate the Laplacian simultaneously.

From the perspective of financial engineering, graph learning [20] is an increasingly important problem that carries out graph signal processing and machine learning tasks. Given the observations of each node, the goal of graph learning is to learn the optimal matrix, which represents the relationship between each pair of nodes. The graph structure is usually embedded by a Laplacian matrix [22]. Among all methods, learning graphs under smoothness assumption [16] has gained popularity. [16] assumes that observations change smoothly between connected nodes and shows that the optimal Laplacian matrix can be found by minimizing the Dirichlet energy. Motivated by that, [12] adds a variable to approximate the observations, which allows some noise in the observations. Although the above methods have achieved promising results on graph learning tasks, they are not designed to learn graphs with clustering sub-class; therefore, they can not be directly used for financial tasks. Recently, [21] proposes a graph-based clustering method to perform clustering on nodes by adding a constrained Laplacian rank to the objective function. However, the above method works in a two-stages process. Specifically, an initial estimate of graph is needed in the first stage, then it projects the initial estimate onto a rank-constrained Laplacian. A disadvantage of the two-stage process is that the final Laplacian estimate is not directly learned from the data. Furthermore, the results depend on the initial graph's estimate. Recent work [6] investigate the graph Laplacian as a candidate to capture the relationship of stocks from a probabilistic perspective. However, these methods have not been studied for stock prediction, and the advantages of graph learning in financial markets remain open for further research.

3 Methods

3.1 Problem Formulation

Suppose there are N stocks, each of them consists of $m+1$ time series. We specify one time series as the target series, while the other series are used as exogenous series. We use $Y = \{Y^1, Y^2, \ldots, Y^N\} \in R^{N \times T}$ to denote the observations of all target series, where T is the length of window size. For example, given a dataset with N stocks, we use the closing price of each stock as the target series. Then, for each stock, we use $X = (x_1, x_2, \ldots, x_T) \in R^{m \times T}$ to denote the observations of its exogenous series, such as hand-crafted technical indicators. Given the previous values of the target series, i.e., $Y^i = (y_1^i, y_2^i, \ldots, y_T^i)$, and the historical observations of the exogenous series, i.e., $X^i = (x_1^i, x_2^i, \ldots, x_T^i)$, the problem is to build a stock predictor, $F_i(w_i)$, which can predict the price movement $y_{T+p}^{i,\text{binary}}$ of each stock in the next p time step:

$$\hat{y}_{T+p}^{i,\text{binary}} = F_i(w_i | X^i, Y^i),$$

where $y_{T+p}^{i,\text{binary}} = \text{sign}(y_{T+p}^i - y_T^i)$.

Also, we consider a undirected weighted graph $\mathcal{G} = \{\mathcal{N}, \mathcal{E}, \boldsymbol{A}\}$, where $\mathcal{V} = \{1, 2, \ldots, N\}$ is the node set representing stocks, $\mathcal{E} \subseteq \{u, v \in \mathcal{V}\}$ is the edge set representing all possible connections between pairs of nodes. $\boldsymbol{A} \in R_+^{N \times N}$ is a symmetric weighted adjacency matrix that satisfies $A_{ii} = 0, A_{ij} > 0$ if $\{i, j\} \in \mathcal{E}$ and $A_{ij} = 0$, otherwise. The graph Laplacian matrix is defined as $\boldsymbol{L} \triangleq \boldsymbol{D} - \boldsymbol{A}$. $\boldsymbol{D} \triangleq \mathrm{diag}(\boldsymbol{A1})$ is the degree matrix, $\boldsymbol{1} \in R^N$ is the all-ones vector.

3.2   Federated Multi-task Stocks Predictor

From the point of view of market analysis, a market predictor is used to predict a market movement by extracting the dynamic historical temporal information and utilizing the related economic or public events. However, different stocks are not necessarily influenced by the same events or information. Different assets exhibit various properties and data distribution [9]. Thus, it is natural to fit personalized models to each stock. On the other hand, the price movement of an individual stock is usually related to other stocks besides its own information [15]. In this work, to fit separated models for each stock and consider the connection between stocks, we introduce the Federated Multi-task Learning with Financial Graph Laplacian Regularization entitled "Federated Multi-task Stocks Predictor with Financial Graph learning (FMSP-FGL)".

The introduced FMSP-FGL follows the module of FMTL [11,25]. In this work, we fit separate models to each stock (node) to capture the temporal dynamics for prediction and use the Laplacian matrix as a regularization term to consider the inter-series structure by using the following formulation:

$$
\begin{aligned}
\min_{\boldsymbol{W}, \boldsymbol{L}} \quad & \sum_{i=1}^N F_i(\boldsymbol{w}_i) + \alpha \, \mathrm{Tr}(\boldsymbol{W}^T \boldsymbol{L} \boldsymbol{W}) + \beta \|\boldsymbol{L}\|_F^2 \\
\text{s.t.} \quad & \boldsymbol{L1} = \boldsymbol{0}, L_{ij} = L_{ji} \le 0, \forall i \ne j, \\
& \mathrm{diag}(\boldsymbol{L}) = \boldsymbol{1}, \\
& \mathrm{rank}(\boldsymbol{L}) = N - k.
\end{aligned}
\tag{1}
$$

where $\boldsymbol{W} = [\boldsymbol{w}_1, \ldots, \boldsymbol{w}_N]^T \in R^{N \times d}$ is a collective matrix whose i-th column is the weight vector for the i-th stock predictor (task), α and β are two positive regularization parameters, and k represents number of stocks clusters (groups). In addition, $\mathrm{Tr}(\cdot)$ and $\| \cdot \|_F$ denote the trace and Frobenius norm of a matrix, respectively; $\mathrm{diag}(\cdot)$ and $\mathrm{rank}(\cdot)$ denote the diagonal vector and rank of a matrix, respectively. Here, $F_i(\boldsymbol{w}_i) = E[l(\boldsymbol{w}_i | X^i, Y^i)]$ represents the expected negative log-likelihood loss corresponding to i-th task's sample and weights. Specifically, we fit separated models for different stocks because different stocks have non-i.i.d. distribution, which means that $F_i(\boldsymbol{w}_i)$ and $F_j(\boldsymbol{w}_j)$ should be personalized when $i \ne j$.

It should be noted that we use the Laplacian to measure the hidden structure of the given weights matrix $\boldsymbol{W}$, and minimize the Dirichlet energy [16] to find the optimal Laplacian as follow,

$$\mathrm{Tr}(\boldsymbol{W}^T \boldsymbol{L} \boldsymbol{W}) = \frac{1}{2} \sum_{i,j} A_{ij} \|\boldsymbol{w}_i - \boldsymbol{w}_j\|^2,$$

where $\|\boldsymbol{w}_i - \boldsymbol{w}_j\|^2$ is the squared Euclidean distances between two stock predictors. The learned Laplacian matrix contains the correlated structures among all stock tasks, i.e., $-L_{ij} = A_{ij} \geq 0$ measures the conditional dependency between two tasks, and $L_{ij} = A_{ij} = 0$ iif $\boldsymbol{w}_i$ and $\boldsymbol{w}_j$ are independent. The Frobenius norm is added as a penalty term in the objective function to control the edge weights of the Laplacian matrix. Moreover, we enforce a k-cluster graph Laplacian to capture the intra-cluster relationship between stocks in the same group. The above constraints of the Laplacian matrix are designed considering the stylized facts of financial tasks, as discussed in Sect. 3.3.

The proposed optimization problem (1) is not jointly convex in $\boldsymbol{L}$ and $\boldsymbol{W}$. Therefore, we adopt an alternating optimization method [25,28], in which $\boldsymbol{L}$ or $\boldsymbol{W}$ is optimized alternatively with the other variable fixed until it converges. The whole scheme of the algorithm is shown in Algorithm 1. The algorithm consists of the model weights updating part (lines 2–12) and the graph Laplacian updating part (lines 13–14). First, we fix $\boldsymbol{L}$ and solve the following optimization problem with respect to $\boldsymbol{W}$:

$$\min_{\boldsymbol{W}} \quad \sum_{i=1}^{N} F_i(\boldsymbol{w}_i) + \alpha \,\mathrm{Tr}(\boldsymbol{W}^T \boldsymbol{L} \boldsymbol{W}). \tag{2}$$

At the second part, we solve the graph learning optimization problem with respect to $\boldsymbol{L}$ given $\boldsymbol{W}$ as:

$$\begin{aligned}
\min_{\boldsymbol{L}} \quad & \alpha \,\mathrm{Tr}(\boldsymbol{W}^T \boldsymbol{L} \boldsymbol{W}) + \beta \|\boldsymbol{L}\|_F^2 \\
\text{s.t.} \quad & \boldsymbol{L}\mathbf{1} = \mathbf{0}, L_{ij} = L_{ji} \leq 0, \forall i \neq j, \\
& \mathrm{diag}(\boldsymbol{L}) = \mathbf{1}, \\
& \mathrm{rank}(\boldsymbol{L}) = N - k.
\end{aligned} \tag{3}$$

Specifically, when $\boldsymbol{L}$ is fixed, in each global iteration (line 1), each task performs R local updates first. The updated models are sent to their related task to perform the Laplacian regularization, which determines the correlated structure. For each local stock predictor, $F_i(\boldsymbol{w}_i)$, the goal is to learn the temporal dynamics of time series and predict future values of each stock. To achieve this goal, we adopt N modules of DARNN [24]. We choose DARNN due to its capability of selecting the most relevant exogenous input features and exploit the temporal dependencies in predicting target series. We follow the network structure of DARNN with some modifications for stock time series prediction as discussed in Sect. 4.1. It should be noted that the proposed problem (1) reduces to a decentralized version of FedU [11] when $\boldsymbol{L}$ is fixed. However, FedU does not consider the variation of the correlated structure among different tasks.

Algorithm 1. FMSP-FGL: Federated Multi-task Stocks Predictor with Financial Graph learning

---

Input: Data $\{X^i, Y^i\}$, initial $\boldsymbol{w}_i^{(0)}$, for $i = 1, \ldots, N$ tasks, initial matrix $\boldsymbol{L}$, learning rate μ, regularization parameter α, and number of local and global iteration R and $T_{\max}$.

1: **for** $t = 1, 2, \ldots, T_{\max}$ **do**
2:     **Step to update W:**
3:         **for** task $i \in \{1, \ldots, N\}$ in parallel **do**
4:             initialize local model $\boldsymbol{w}_{i,0}^{(t)} := \boldsymbol{w}_i^{(t)}$
5:             **for** $r = 0, \ldots, R-1$ **do**
6:                 compute mini-batch gradient $\nabla F_i(\boldsymbol{w}_{i,r}^{(t)})$
7:                     update local task weights $\boldsymbol{w}_{i,r+1}^{(t)} := \boldsymbol{w}_{i,r}^{(t)} - \mu \nabla F_i(\boldsymbol{w}_{i,r}^{(t)})$
8:             **end for**
9:         **end for**
10:         **for** task $i \in \{1, \ldots, N\}$ in parallel **do**
11:             $\boldsymbol{w}_i^{(t+1)} := \boldsymbol{w}_{i,R}^{(t)} + \alpha \mu R \sum_{j \neq i} L_{ij}(\boldsymbol{w}_{i,R}^{(t)} - \boldsymbol{w}_{j,R}^{(t)})$
12:         **end for**
13:     **Step to update Graph Laplacian L:**
14:         Solve the problem (3) to update $\boldsymbol{L}$
15: **end for**

When $\boldsymbol{W}$ is fixed, the problem (3) is non-convex and non-differentiable due to the constraint $\text{rank}(\boldsymbol{L}) = N - k$, which enforce a k-cluster graph Laplacian. We solve the problem by using optimization relaxation and alternating optimization methods as presented in Sect. 3.3.

3.3   Graph Laplacian Interpretation for Financial Tasks

Learning graphs from data is a fundamental problem to capture the hidden relationship between different assets [6]. To uncover the conditional dependencies between stock prediction tasks, we propose the problem (3) with constraints considering the stylized financial facts.

The first constraint in (3), $\boldsymbol{L1} = \boldsymbol{0}, L_{ij} = L_{ji} \leq 0, \forall i \neq j$, follows the definition of a positive semidefinite Laplacian matrix. This constraint implies that $\boldsymbol{L}$ only represents non-negative relationships, which meets the assumption that assets are always positively dependent [9]. The second constraint, $\text{diag}(\boldsymbol{L}) = \boldsymbol{1}$, controls the degree of the nodes to avoid isolated nodes. This constraint meets the fact that there is no independent asset in the financial market as all assets are treated as a complex network [20]. Thus, in our setting, the correlation between two nodes (tasks) can be measure as $-(L_{ij}/\sqrt{L_{ii}L_{jj}}) = -L_{ij}, \forall i \neq j$. In practice, the stock market is a well-defined complex network [18] where hierarchical structures can be detected, assets in different clusters have different reaction to market information. More interestingly, the intra-cluster assets demonstrate much more lead-lag correlations than inter-group components [15]. Thus, we add the rank constraint, $\text{rank}(L) = N - k$, to enforce the graph to have k connected clusters to learn the intra-cluster relationship.

The proposed problem (3) is non-convex due to the rank constraint, $\text{rank}(L) = N - k$, which is that the sum of the k smallest eigenvalues of L is equal to zero, i.e., $\sum_{i=1}^{k} \sigma_i(L) = 0$. According to Fan's theorem [13],

$$\sum_{i=1}^{k} \sigma_i(L) = \min_{V \in R^{N \times k}, V^T V = I} \text{Tr}(V^T L V),$$

thus we have the problem (3) equivalent to the relaxed version as following,

$$\min_{L, V \in R^{N \times k}} \alpha \text{Tr}(W^T L W) + \beta \|L\|_F^2 + \lambda \text{Tr}(V^T L V)$$
$$\text{s.t.} \quad L\mathbf{1} = 0, L_{ij} = L_{ji} \le 0, \forall i \ne j, \tag{4}$$
$$\text{diag}(L) = 1, V^T V = I,$$

where λ is a regularization parameter. Note that when λ is large enough, the optimal solution will enforce the sum of k smallest eigenvalues of L to be zero. Then we rewrite the problem (4) into two convex sub-problems in an alternating fashion. When L is fixed, we have the sub-problem for V:

$$\min_{F \in R^{N \times k}, V^T V = I} \text{Tr}(V^T L V), \tag{5}$$

whose solution is given by the k collective eigenvectors of L corresponding to the k smallest eigenvalues according to Fan's theorem [13].

For a fixed V, we have the following sup-problem for L:

$$\min_{L} \alpha \text{Tr}(W^T L W) + \beta \|L\|_F^2 + \lambda \text{Tr}(V^T L V)$$
$$\text{s.t.} \quad L\mathbf{1} = 0, L_{ij} = L_{ji} \le 0, \text{diag}(L) = 1. \tag{6}$$

We can rewrite the sub-problem (6) as a quadratic program by using *half-vec operator* and *duplication matrix* [2]. Specifically, given the vectorization of L, denoted as $\text{vec}(L) \in R^{N^2}$, we introduce the *half-vec operator* $\text{vech}(\cdot)$. Then $\text{vech}(L) \in R^{N(N+1)/2}$ denotes the vector obtained from $\text{vec}(L)$ by eliminating all superdiagnoal elements of L. Now, notice that L is symmetric, there exists a unique constant matrix $D_N \in R^{N^2 \times N(N+1)/2}$, called the duplication matrix [2], that transforms, for symmetric L, $\text{vech}(L)$ into $\text{vec}(L)$, that is,

$$D_n \text{vech}(L) = \text{vec}(L) \quad (L = L^T). \tag{7}$$

Now, together with the facts that, $\text{Tr}(W^T L W) = \text{vec}(WX^T)^T \text{vec}(L)$, and $\|L\|_F^2 = \text{vec}(L)^T \text{vec}(L)$, we can rewrite the sup-problem (6) as:

$$\min_{\text{vech}(L)} \left[\alpha \text{vec}(WW^T) + \lambda \text{vec}(VV^T) \right] D_N \text{vech}(L)$$
$$+ \beta \text{vech}(L)^T D_N^T D_N \text{vech}(L) \tag{8}$$
$$\text{s.t.} \quad G \text{vech}(L) \le h,$$
$$A \text{vech}(L) = b,$$

Algorithm 2. Graph learning algorithm to solve L in (3)

    Input: Model weights W, initial $L, \alpha, \beta, \lambda$.

1: **while** not converge **do**
2:     Update $V^{(l+1)}$ by solving the problem (5) fixing L at $L^{(l)}$
3:     Update $L^{(l+1)}$ by solving the problem (8) fixing V at $V^{(l)}$
4: **end while**
    Output: Graph Laplacian L.

---

where the constraints in the problem (8) handle the inequality and equality constraints in the problem (6). Problem (8) is convex and can be solved efficiently by convex programming languages [3], e.g. *CVXPY* [10]. Algorithm 2 summarized the implementation to solve the problem (3).

4 Experiments

In this section, we first introduce the stock dataset, parameters setting, and performance evaluation metrics. In order to show the effectiveness of the proposed model, we then compare FMSP-FGL with several cutting-edge approaches. Finally, we then use a step-by-step justification to demonstrate its capability of capturing the inter-series association as well as the intra-cluster relationship.

4.1 Experiment Settings

We choose two prominent stock indexes, namely DJIA, and SP500, which contain 30, 500 constituent stocks, respectively. We collect the constituent stocks' time series data from Jan-3-2017 to Dec-31-2020 from Yahoo! Finance.[1] For SP500, we collect the data of 55 stocks among 11 GICS [1] sectors. For each sector, stocks with top 5 market capitalization are selected. The frequency of the data collection is day-by-day. Each data sample contains 5 features: the opening price, highest price, lowest price, closing price, and volume (OHLCV). As previous works [14,17], we select OHLV as well as 8 popular technical indicators[2] as the exogenous series, and use closing price as the target series. We pre-processed the collected time series data by calculating the relative percentage change of each stocks on each day with respect to its observations 5 days ago. We aim to predict the next trading day price direction (up/down) of stocks given the percentage changes of stocks over 5 consecutive days.

In our experiment, the first 90% data are used for training, and the following 10% are used as the test set. We select four metrics in evaluation, i.e., Accuracy (Acc), Precision (Pre), Recall and F1 scores.

---

[1] In total, we collect 29 stocks from DJIA, because the stock, DOW, was listed after 2017.
[2] Technical indicators: Moving Average Convergence Divergence, Average Directional Movement Index, Awesome Oscillator, Money Flow Index, Upper Bollinger Bands, Lower Bollinger Bands, Chaikin Money Flow, On-balance Volume Mean Range.

There are two important parameters in the proposed algorithm, i.e., the graph learning regularization parameters $\{\alpha, \beta\}$ and the number of clusters k in (1). In practice, the regularization parameters setting are carried out on different ratio $\frac{\beta}{\alpha}$ to maximize the prediction performance. We fix α as 1, the ratio was determined by conducting a grid search to achieves the highest test accuracy. To determine the number of cluster k of all tasks, we use a pre-defined sector classification list, Global Industry Classification Standard (GICS) [1], as an advantage of prior domain knowledge. We denote the number of unique sectors of a dataset as k_{max}, then we choose k from the finite set $k = \{1, 2, ..., k_{max}\}$.

FMSP-FGL contains N modules of DARNNs [24], where N is the number of constituent stocks of each dataset. DARNN is used as our basic predictive module due to its capability of capturing dynamic temporal dependencies. We modify the output of DARRN as a single scalar value to perform movement prediction, with a negative log-likelihood loss function. The dimension of the hidden state and cell state are fixed as 256. All DARNN have the same window size $T = 5$, and prediction step $p = 1$. We treat the prediction of one stock as one single task. Each task shares the same hyper-parameters setting. The size of the minibatch is 128. The number of global iteration is 50, and the number of local iteration is 2. The learning rate is 0.001. Considering that λ should be large enough as we discuss in Sect. 3, we start λ with a small value 2, and double the value if k is larger than the number of zero eigenvalues of L. All tasks are trained with stochastic gradient descent. All experiments are repeated five times, and the average performance is reported. A Tesla V100 GPU is used for training. All experiments are implemented by PyTorch [23] version 1.7.

4.2   Results

To show the effectiveness of our proposed method, we compare our algorithm with the following baselines methods: Local Model (training one separate model for each stock), Federated multi-task learning framework (FedU) [11], also, we compare to FMSP with state-of-the-art graph learning methods, i,e., SigRep [12], CLR [21]. Considering the fact that FedU requires a prior graph for training, we use the sector classification list, GICS, as the prior information, where stocks in the same sector share the same weight connection. We denote Local as single task learning (STL) and the others as multi-task learning algorithms (MTL).

The price movement prediction results of FMSP and baseline methods are summarized in Table 1. The results show that federated multi-task learning algorithms have better performance over all evaluation metrics than single task algorithm, Local, which confirms the effectiveness of multi-task learning. The reason is that multi-task learning with Laplacian regularization can fit personalized but related models for each stock. The algorithm, FedU, with graph pre-defined setting outperforms Local, which confirms that the prior domain knowledge can be used to increase predictive performance. We conclude that federated learning algorithms can improve the overall prediction performance.

We observe that the proposed FMSP-FGL has the best performance across two datasets. The accuracy and precision of FMSP-FGL are constantly higher

Table 1. Prediction results. All models predict price direction (up/down) on the next day.

Algorithm		DJIA				SP500			
		Acc	Pre	Recall	F1	Acc	Pre	Recall	F1
STL	Local	70.61	78.91	71.76	75.16	69.05	69.45	77.96	73.46
MTL	FedU	71.61	72.88	78.64	75.65	70.82	70.78	79.84	75.04
	FMSP-SigRep	75.45	74.58	**85.50**	**79.67**	72.45	71.73	82.26	76.64
	FMSP-CLR	74.76	74.65	83.45	78.80	74.13	73.20	**83.49**	78.00
	FMSP-FGL*	**75.84**	**77.92**	79.56	78.73	**74.70**	**74.09**	82.94	**78.27**

than others, which is preferred for stock prediction. It shows evidence that the proposed graph learning algorithm can help with learning the correlated structure of different tasks thus improving overall prediction performance. One interesting point to note is that the recall rate of FMSP-FGL is lower than the other baseline, which can be seen as a trade off between accuracy and recall. In stock markets, the goal of predictor is to maintain a higher level of precision. To demonstrate the effectiveness of our graph learning algorithm, we visualize the learned graph of the percentage change of the closing price with $k = 8$ (number of sectors) in Fig. 1. The figure shows the SigRep [12] is unable to learn financial data, which leads to many possibly fake connections and fails to capture a meaningful network. The CLR [21] learns a k-cluster graph; however, the graph contains isolated nodes, namely, *DIS, INTC*, which contradicts the financial network theory in the real-world [7]. The proposed algorithm returns a k-cluster graph meaningful representation with much less fake connections. Precisely, the graph captures the prior GICS sector classification information (node-colored edges) as well as the inter-cluster connections (grey-colored edges) learned from the data without a two-stage process. Together with the prediction results, it shows the financial graph learning algorithm can help with stocks prediction.

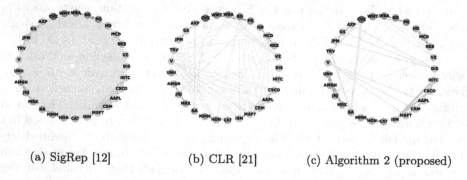

| (a) SigRep [12] | (b) CLR [21] | (c) Algorithm 2 (proposed) |

Fig. 1. Graph learning visualization. The graph Laplacian is learned directly from the percentage change of each stocks in the DJIA dataset.

We study the sensitivity of FMSP-FGL with respect to the ratio of regularization parameters $\frac{\beta}{\alpha}$. We plot the number of edges of the learned graph and the prediction performance of the proposed algorithm versus different ratios $\frac{\beta}{\alpha}$ over the DJIA dataset in Fig. 2. As the ratio increase, we observe that both the number of edges and the prediction accuracy tends to increase. The intuitions behind this observation are as follows. When the ratio increases, the Frobenius norm of the learned Laplacian matrix in (2) tends to be small. Because we set the diagonal of Laplacian to be 1 in the constraint, the number of edges tends to increase with small values to enforce the Frobenius norm to be small. Next, we see that the proposed algorithm outperforms FedU when the number of edges of the learned graph exceeds the one of the prior graph provided by GICS. These results show that FMSP-FGL is able to learn a graph that captures the complex connection between different stocks, which contains more useful and unobserved information for prediction compared to the prior graph.

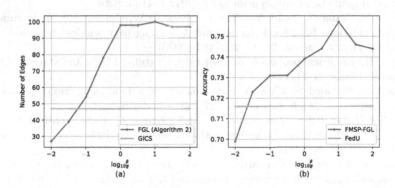

Fig. 2. (a) Number of edges in the graphs learned by FGL (Algorithm 2), and (b) prediction performance of FMSP-FGL, for different ratios $\frac{\beta}{\alpha}$ over the DJIA dataset

5   Conclusion

In this paper, we first design a multi-task stock predictor to fit separated but related tasks simultaneously. Then a graph-based clustering optimization problem with rank constraint and degree control constraint is presented to capture both the inter-series and intra-cluster relationship between stock tasks. We transform the proposed non-convex optimization problem into a relaxed convex problem and solve it alternatively. Empirical results on two stocks dataset show our methods outperform competing approaches. To our best knowledge, the proposed technique is the first to apply graph Laplacian learning with meaningful financial interpretations on multi-task stock prediction.

References

1. Gics. https://www.msci.com/our-solutions/indexes/gics. Accessed 7 Oct 2021
2. Abadir, K.M., Magnus, J.R.: Matrix Algebra, vol. 1. Cambridge University Press, Cambridge (2005)
3. Boyd, S., Boyd, S.P., Vandenberghe, L.: Convex Optimization. Cambridge University Press, Cambridge (2004)
4. Boyd, S., Parikh, N., Chu, E.: Distributed Optimization and Statistical Learning via the Alternating Direction Method of Multipliers. Now Publishers Inc., Delft (2011)
5. Brisimi, T.S., Chen, R., Mela, T., Olshevsky, A., Paschalidis, I.C., Shi, W.: Federated learning of predictive models from federated electronic health records. Int. J. Med. Inform. **112**, 59–67 (2018)
6. Cardoso, J.V.d.M., Palomar, D.P.: Learning undirected graphs in financial markets. arXiv preprint arXiv:2005.09958 (2020)
7. Cardoso, J.V.d.M., Ying, J., Palomar, D.P.: Algorithms for learning graphs in financial markets. arXiv preprint arXiv:2012.15410 (2020)
8. Chen, M., Yang, Z., Saad, W., Yin, C., Poor, H.V., Cui, S.: A joint learning and communications framework for federated learning over wireless networks. IEEE Trans. Wirel. Commun. **20**(1), 269–283 (2020)
9. Cont, R.: Empirical properties of asset returns: stylized facts and statistical issues. Quanti. Financ. **1**(2), 223 (2001)
10. Diamond, S., Boyd, S.: CVXPY: a Python-embedded modeling language for convex optimization. J. Mach. Learn. Res. **17**(83), 1–5 (2016)
11. Dinh, C.T., Vu, T.T., Tran, N.H., Dao, M.N., Zhang, H.: Fedu: a unified framework for federated multi-task learning with Laplacian regularization. arXiv preprint arXiv:2102.07148 (2021)
12. Dong, X., Thanou, D., Frossard, P., Vandergheynst, P.: Learning Laplacian matrix in smooth graph signal representations. IEEE Trans. Sig. Process. **64**(23), 6160–6173 (2016)
13. Fan, K.: On a theorem of Weyl concerning eigenvalues of linear transformations I. Proc. Natl. Acad. Sci. U.S.A. **35**(11), 652 (1949)
14. He, J., Khushi, M., Tran, N.H., Liu, T.: Robust dual recurrent neural networks for financial time series prediction. In: Proceedings of the 2021 SIAM International Conference on Data Mining (SDM), pp. 747–755. SIAM (2021)
15. Hou, K.: Industry information diffusion and the lead-lag effect in stock returns. Rev. Financ. Stud. **20**(4), 1113–1138 (2007)
16. Kalofolias, V.: How to learn a graph from smooth signals. In: Artificial Intelligence and Statistics, pp. 920–929. PMLR (2016)
17. Li, C., Song, D., Tao, D.: Multi-task recurrent neural networks and higher-order Markov random fields for stock price movement prediction: Multi-task RNN and higer-order MRFs for stock price classification. In: Proceedings of the 25th ACM SIGKDD International Conference on Knowledge Discovery and Data Mining, pp. 1141–1151 (2019)
18. Mantegna, R.N.: Hierarchical structure in financial markets. Eur. Phys. J. B Condens. Matter Compl. Syst. **11**(1), 193–197 (1999)
19. Markowitz, H.M.: Portfolio Selection. Yale University Press, London (1968)
20. Marti, G., Nielsen, F., Bińkowski, M., Donnat, P.: A review of two decades of correlations, hierarchies, networks and clustering in financial markets. Prog. Inf. Geom. 245–274 (2021)

21. Nie, F., Wang, X., Jordan, M.I., Huang, H.: The constrained Laplacian rank algorithm for graph-based clustering. In: Thirtieth AAAI Conference on Artificial Intelligence (2016)
22. Oellermann, O.R., Schwenk, A.J.: The Laplacian spectrum of graphs (1991)
23. Paszke, A., et al.: Automatic differentiation in pytorch (2017)
24. Qin, Y., Song, D., Chen, H., Cheng, W., Jiang, G., Cottrell, G.: A dual-stage attention-based recurrent neural network for time series prediction. arXiv preprint arXiv:1704.02971 (2017)
25. Smith, V., Chiang, C.K., Sanjabi, M., Talwalkar, A.: Federated multi-task learning. In: Proceedings of the 31st International Conference on Neural Information Processing Systems, pp. 4427–4437. Curran Associates Inc. (2017)
26. Tuck, J., Barratt, S., Boyd, S.: A distributed method for fitting Laplacian regularized stratified models. arXiv preprint arXiv:1904.12017 (2019)
27. Yoo, J., Soun, Y., Park, Y.c., Kang, U.: Accurate multivariate stock movement prediction via data-axis transformer with multi-level contexts. In: Proceedings of the 27th ACM SIGKDD Conference on Knowledge Discovery and Data Mining, pp. 2037–2045 (2021)
28. Zhang, Y., Yeung, D.Y.: A convex formulation for learning task relationships in multi-task learning. arXiv preprint arXiv:1203.3536 (2012)

Identification of MEEK-Based TOR Hidden Service Access Using the Key Packet Sequence

Xuebin Wang[1,2], Zhipeng Chen[4], Zeyu Li[3(✉)], Wentao Huang[3], Meiqi Wang[1,2], Shengli Pan[3], and Jinqiao Shi[3(✉)]

[1] Institute of Information Engineering, Chinese Academy of Sciences, Beijing, China
[2] School of Cyber Security, University of Chinese Academy of Sciences, Beijing, China
[3] School of Cyberspace Security, Beijing University of Posts and Telecommunications, Beijing, China
{lizeyu,shijinqiao}@bupt.edu.cn
[4] National Internet Emergency Center(CNCERT/CC), Beijing, China

Abstract. Tor enables end user the desirable cyber anonymity with obfuscation technologies like MEEK. However, it has also manifested itself a wide shield for various illegal hidden services involved cyber criminals, motivating the urgent need of deanonymization technologies. In this paper, we propose a novel communication fingerprint abstracted from key packet sequences, and attempt to efficiently identify end users MEEK-based access to Tor hidden services. Specifically, we investigate the communication fingerprint during the early connection stage of MEEK-based Tor rendezvous establishment, and make use of deep neural network to automatically learn and form a key packet sequence. Unlike most of existing approaches that rely on the entire long communication packet sequence, experiments demonstrate that our key packet sequence enabled scheme can significantly reduce both the time and hardware resource consumption for the identification task by 23%–37% and 80%–86%, respectively, while being able to keep a slightly better accuracy.

Keywords: Tor · Hidden service · Traffic analysis · MEEK

1 Introduction

Tor [17], used by more than two million users daily [1], is one of the most popular anonymous communication systems, aiming to protect users' online privacy. Tor also provides hidden services (HSs), the so-called darknet, to protect server-side anonymity. Therefore, some people use this mechanism to publish sensitive contents on hidden services, making the deep-dark cyberspace a hotbed of crime [4]. Hence, it is necessary to identify Tor hidden service traffic.

In order to enhance the availability of the network in censorship countries, Tor proposes many obfuscation methodologies to bypass censorship, such as Tor bridge mechanisms, MEEK-based [5] obfuscation and Obfs4-based [3] obfuscation. According to Tor project statistics, MEEK is one of the most popular bridges currently. The obfuscation introduced by MEEK protocol brings difficulties for detecting and identifying Tor hidden service traffic.

Many previous work has shown that it is possible to identify whether a user is accessing a hidden service [8,9,11] and even distinguish the specific hidden service the user is accessing [7,10,12,15,19,20]. However, the detection granularity of prior work is in the unit of entire access trace, leading to a relatively lower detection timeliness and much more memory resource consumption, which is not suitable for online identification. Moreover, the obfuscation brings by MEEK protocol has not been taken into consideration, need to be thoroughly analyzed.

In this paper, we present a novel approach to identify Tor hidden service access activity with key sequences under MEEK-based obfuscation scenario, which only uses the specific TCP packet sequence as input and identify the Tor hidden service access behavior in the early stage of the access procedure, effectively improving the identification timeliness as well as reducing the cost of hardware resources. The contributions of this paper are listed as follows:

- verify that there does exist a TCP package sequence contributes significantly to identify Tor hidden service access behavior under MEEK-based obfuscation scenario.
- propose a novel method to identify Tor hidden service access activity under MEEK-based obfuscation scenario only based on key TCP package sequence, which can effectively improve the identification timeliness as well as reduce the cost of hardware resources.
- capture a large and practical dataset on four different MEEK-based obfuscation scenarios to validate our method. Besides, we make the dataset public[1], allowing researchers to replicate our results and evaluate new approaches in the future.
- Based on the collected dataset, compared with the existing method using all data as input, only with the key TCP packet sequence as input, the identification effect is improved by 3%–4%, the timeliness is improved by 23%–37%, and the hardware resource consumption is reduced by 80%–86%.

Organization. The rest of the paper is organized as follows. In Sect. 2, we illustrate the background on Tor hidden service, MEEK protocol as well as the related work. We present the key sequence based identification methodology in Sect. 3. In Sect. 4, we describe the data collection and processing methodology. We next present, in Sect. 5, our observations and experimental results under four MEEK-based obfuscation scenarios. Finally, we draw the conclusion in Sect. 6.

---

[1] The dataset can be found on the following URL: https://github.com/Meiqiw/meek-mingan/.

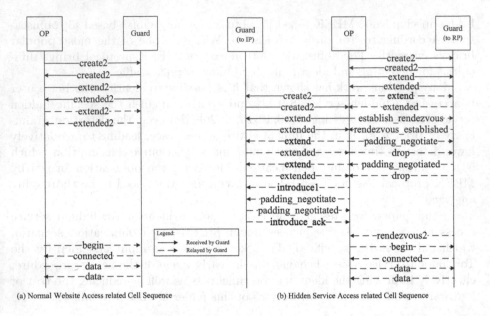

(a) Normal Website Access related Cell Sequence   (b) Hidden Service Access related Cell Sequence

Fig. 1. Different cell sequences between hidden service and normal website access behaviors

2   Background and Related Work

In this section, we introduce the necessary background for our work, including Tor hidden service mechanism and MEEK obfuscation protocol. Besides, we describe the related work about MEEK detection as well as WFP (Website Fingerprint Attack) on Tor hidden service.

2.1   Tor Hidden Service

Hidden service was launched in 2004, it can hide the location information for the service provider, thus can anonymous the responder. According to the specification [2], there are obvious differences between hidden service and normal website access activities. Figure 1 shows the detail different cell sequences exchanged when accessing normal website and hidden service from the OP(Onion Proxy) side. There are less control cells when accessing the normal website than that of accessing the hidden service, like *relay-establish-rendezvous* and *relay-rendezvous2*. Standing between the OP and the entry relay, we can see that the cell sequences is significantly different when accessing the hidden service and accessing the normal website. Accordingly, the different cell sequences will cause the difference of traffic sequences between two accessing behaviors as Tor uses TLS as the transport layer protocol which is based on TCP, and that provides the basis for our work.

2.2 Meek

MEEK is one of the most popular pluggable transports commonly used by users in censorship areas. MEEK uses domain fronting [5] technology to avoid censorship, so as to access a domain name prohibited by the censor. The structure of MEEK mechanism mainly include three different parts: MEEK client bounded with Tor client, fronted server with an allowed domain name provide by cloud service provider and MEEK server bounder with Tor router. Firstly, MEEK client encapsulates the data from Tor client into the payload of an HTTP post request and sends it to the fronted server within a HTTPS request. The domain name in the DNS query and the SNI field of the TLS in the request are both **allowed.example**, and the host header field in the HTTP part of the request is **forbidden.example**. Because the HTTP part has been encrypted by TLS, the censor can only get **allowed.example**, but can't resolve the domain name(forbidden.example) that user really wants to access. Secondly, after receiving the request, the fronted server extracts the internal HTTP request and sends it to the MEEK server. What's more, as the MEEK server is not allowed to push data to the MEEK client actively, the MEEK client needs to continuously poll the fronted server to check whether the MEEK server forwards data back and finally extracts the response content.

2.3 Related Work

Many researches use machine learning methods to detect tor pluggable transport traffic [6,13,16], and have achieved good results. MEEK is one of Tor's pluggable transports that has a wide audience, more and more work has been done on MEEK traffic recognition [14,18,21,22], which all got state-of-the-art detection performance.

Another research area related to our work is website fingerprinting attacks. WFP needs to collect client-side traffic, and then use traffic classification method to train the classifier. The difference is that WFP needs to answer the question of which website the user has visited, while we need to answer whether the user has visited Tor hidden service. Many outstanding works [7,10,12,15,19,20] in the WFP field have proposed classification algorithms with great reference value.

At present, there are few researchers who pay attention to distinguishing whether users are accessing a hidden service. In the existing researches on distinguishing hidden services, attackers can be divided into node level attackers and network level attackers according to their capabilities and locations. **Node level attackers** control relays of Tor network, passively record cell sequences and traffic data, and can infer whether users are accessing hidden services by analyzing the traffic of the collected information. Kwon et al. [9] showed that by controlling the entry nodes, the traffic accessing hidden services can be distinguished from the traffic accessing normal services with an accuracy rate of over 90%. Recently, Jansen et al. [8] control the middle nodes to execute the circuit fingerprinting attach, which proves that the attack from the middle nodes is as effective as the attack from the entry nodes. However, the effectiveness of node level attacks largely depends on the number of controlled nodes. While **network level attackers** are located in the network path between users and entry nodes, and they can only collect some widely known traffic data. Hayes and Danezis [7] detected the onion site in 100,000 sites, and distinguished the onion site from other conventional web pages. The true positive rate was 85%, and the false positive rate was only 0.02%. Panchenko et al. [11] used machine learning to identify hidden service related traffic, with an precision rate of over 90% and a recall rate of over 80%. These works are all about taking the whole traffic flow as the input of machine learning or deep learning, and then extracting features for recognition, which is not suitable for online classification scenario.

Besides, the obfuscation brings by MEEK protocol has not been taken into consideration, need to be thoroughly analyzed. In this paper, we present a novel approach to identify Tor hidden service access activity with **key sequence** under MEEK-based obfuscation scenario, which only uses the specific TCP packet sequence as input and identify the Tor hidden service access behavior in the early stage of the access procedure, effectively improving the identification timeliness as well as reducing the cost of hardware resources.

3   Methodology

3.1   Threat Model

In this work, we assume a network-level adversary, which means the adversary can not process traffic by modifying, dropping or delaying packets. Besides, we assume that the adversary knows the client's identify so that the position of adversary is between the OP and MEEK Server, aiming at distinguishing normal website or hidden service access behavior without decrypting packets. The adversary scenario can be shown in Fig. 2.

3.2   Key Sequence Definition

According to the Tor hidden service protocol, there exists some particular cells representing the start point and ready status for accessing Tor hidden service. The **establish_rendezvous** cell which is send by the OP along the OP-RP circuit indicating the start point of accessing Tor hidden service, while the **rendezvous2** cell which is received from HS along the OP-RP circuit indicating the ready status of accessing Tor hidden service. In this paper, we define the TCP sequences between sending **establish_rendezvous** and receiving **rendezvous2** as the **key sequence** for distinguishing Tor hidden service access behavior from public website access activity.

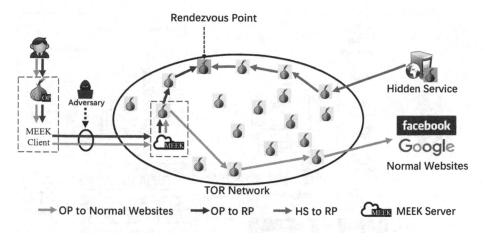

Fig. 2. The adversary scenario of this approach

3.3   Key Sequence Based Identification Methodology

According to the Tor specification, many circuits with different purposes are multiplexed in one single Tor TLS connection, resulting that a network level attacker can not distinguish each circuit from others. What's more, according to the MEEK protocol, MEEK client only persists one HTTPS connection with MEEK server simultaneously, indicating that all data communicating between OP and HS are encapsulated and transferred in one single HTTPS connection. Additionally, the particular polling mechanism of MEEK protocol generates numerous heartbeat packages obfuscating the location distribution and sequence length of the **key sequence** defined above.

In this paper, from the perspective of a network level attacker, we propose a lightweight method to identify Tor hidden service access behavior based on key sequence under MEEK-based obfuscation scenario. As shown in Fig. 3, the identification framework contains Data Collection, Data Preprocess and Prediction three phrases. In data collection phrase, we collect traffic record and cell information as our raw data, we describe the data collection process in next section in detail. While in data preprocess phrase, we extract TCP package sequence as well as cell sequence and build our dataset which we will describe in next section. At last, in the prediction phrase, we find the Key Sequence and use deep learning networks to perform identification task. Firstly, we perform statistics on all the cell sequences, finding the start time point and end time point of the key sequence which is shown in the green box. In detail, we count the absolute position of establish_rendezvous and rendezvous2 in the unit of one connection which is shown in the yellow box. Then, we find the corresponding TCP package index and window size with the help of start time point and end time point respectively, which is shown in the red box. Moreover, we extract the **key sequence** as input of identification model to distinguish Tor hidden service access activity from public website access activity.

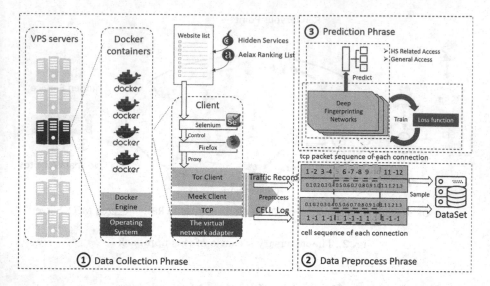

Fig. 3. KEY sequence based hidden service access identification framework (Color figure online)

4   Data Collection and Processing

4.1   MEEK Access Scenarios Configuration

Despite of the public MEEK bridge node extracted from the Tor browser, user can setup private MEEK bridge on various CDNs(cloud service providers). The forwarding mechanism of different CDNs may have impact on the location and window size of key sequences used for Tor hidden service access identification. Thus, in order to simulate users using different CDNs for Tor network access, four scenarios are defined: **public, cdn77, fastly, stackpath**. In all four scenarios, users use the MEEK plugin to browse the web in the Tor network, where public refers to the public MEEK bridge node setup on azure cloud service by Tor official, and the other three are private MEEK bridges built on different cloud platforms respectively.

4.2   Data Collection

To increase the diversity of traffic, we use different country networks for different scenarios: **public** and **cdn77** within China, which use Virtual Machines (VMs) of 8-core CPU, 31G of RAM and 1T of disk; **fastly** scenario leases two VPS nodes in the US and **stackpath** scenario leases two VPS nodes in the UK, each VPS is provided by Vultr, with a 4-core CPU, 8G of RAM and 160G of disk. In each VPS or VMs, multiple dockers are used for distributed data capture which is shown in the data collection phrase of Fig. 3. In each docker, we use Selenium (version 3.12.0) to control headless browser Firefox (version 60.0.2)

Table 1. Data collection of four scenarios

Scenario	Position	Time	WebsiteTrace	OnionTrace	Total
Public	China	2021.4–2021.5	15071	10622	25693
cdn77	China	2021.12–2022.1	15370	20984	36354
Fastly	US	2021.12–2022.1	31174	27494	58688
Stackpath	UK	2021.12–2022.1	32294	28129	60423

to access the Tor network, utilizing a SOCKS5 proxy listened by Tor. At the same time, we require the client to access the Tor network via the MEEK plugin with corresponding configuration in the torrc file.

For each access activity, we record traffic trace of web pages leveraging tcpdump. Each Web page is given 120 s to load, and upon loading the page, it is left open for an additional 10 s, after which the browser is closed and the Tor process is killed. Next, tcpdump and Tor process are restarted. A script to monitor the bootstrap status of the Tor process is deployed, ensuring Tor is ready before each visit. With this setup, new connections and circuits are established each time as the client visits a website, ensuring that we never used the same circuit to download more than one instance of a single page. Besides, in order to find out the location and window size of key sequence, we modify the source code of Tor OP and record the connection creation, circuit construction, stream info, cell sequences into the notice log file, aiming at showing light on the real activity Tor instance occurs during each access trace.

Following our data collection method, we use Alexa Top websites and Tor hidden services[2] as our target website for four scenarios, each with 10,000 websites. After data collection, we filter out invalid traces and outliers, which caused by timeout or crash of the browser or Selenium driver. Eventually, we obtain huge amount of traces as shown in Table 1, each trace accomplishes with one corresponding notice log file.

4.3 Data Extraction and Processing

Our dataset contains two type of data: **traffic traces** and **cell records**. With the cell records, we split the cell sequences according to different connectionID, generating cell sequences of one specific connection, which commonly multiplexed with multiple circuits. For the traffic traces, we split each traffic trace into different flows. And then transfer each flow into TCP packet sequences. The detail processing procedure described as follows respectively.

Cell Record Processing. By parsing the notice log file, much basic information about the connection, circuit and cell are extracted, including connection creation time, connectionID, circuitID, cell command and direction etc. Firstly,

---

[2] As the prior work, we chose hidden services based on the list provided by the .onion search engine http://www.ahmia.fi/.

we order cells of each circuit with timestamp and tag the circuit with different flags according to the circuit purpose. We divide circuit into two categories: **clientdata**, meaning that this circuit is built to access non-hidden service related data, **clientip, clientrp, clienthsdir**, those three are hidden service related. Secondly, we select circuits belongs to the same connection and put corresponding cells together, generating the cell sequence of one specific connection. At last, we tag each connection according to the circuits categories multiplexed in the same connection.

Traffic Trace Processing. As for collected traffic traces, our process performs as follows: Firstly, we tag each connection the same category as the connection recorded in the notice log file which processed in the prior subsection. What's more, we parse the single flow pcap files into TCP packet sequences, with the help of Tshark (version 1.12.1), an industrial grade widely used tool for network traffic analysis.

DATASET MEEK21. After completing the above operation, we eventually obtain *MEEK21*, consisting of two subsets: 10,000 instances of hidden service related and general website related for each scenario, each instance contains packet size, direction and time information of each packet.

5   Evaluation and Discussion

In this section, we first select the state-of-the-art classification model used in this paper. Next, we verify the existence of the key sequence. Then, we analyze the location distribution and window size of the key sequence under four MEEK scenarios. At last, we perform iterative experiments to learn the best choice of the start point and window size to perform our attack.

5.1   Select Classification Model

To find the best performing model for the method proposed in this paper, we compare the best performing CUMUL [10] method for machine learning and the best performing DF [15] method for deep learning. To check the robustness and accuracy of the models, we divide the dataset into a training set, a validation set and a test set according to 1:1:2. In particular, as deep learning methods require uniform inputs, we sort the length of each trace in the dataset and select trace length at 95% loci as model inputs to ensure that the vast majority of trace information can be fed into the model. As shown in Table 2, in each scenario, the DF method achieves better identification performance compared to the CUMUL method, which indicates that the deep learning approach works better in identifying the access behavior of Tor hidden service under MEEK obfuscation scenario. Hence, we subsequently validate our proposed approach using the DF model.

Table 2. Results of DF VS Cumul

	Input_size	DF				CUMUL			
		acc	pre	recall	f1	acc	pre	recall	f1
Public	857	0.9684	0.9578	0.98	0.9687	0.9036	0.9042	0.9036	0.9035
cdn77	1893	0.986	0.9742	0.9984	0.9862	0.9234	0.9244	0.9234	0.9234
Fastly	793	0.9845	0.9797	0.9894	0.9845	0.9192	0.9195	0.9192	0.9191
Stackpath	3217	0.9872	0.9792	0.9956	0.9873	0.9295	0.931	0.9295	0.9294

5.2   Key Sequence Verification

To verify the existence of key sequence in Tor hidden service identification task, we put the whole trace as input with DF model and calculate the gradients value of each packet, aiming at finding out which part of the trace contributes the most to the identification task.

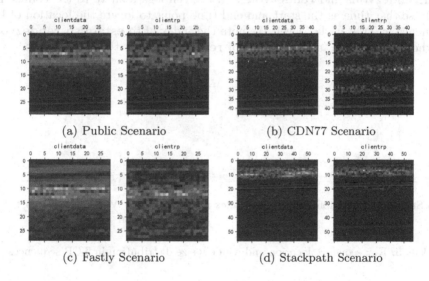

(a) Public Scenario   (b) CDN77 Scenario

(c) Fastly Scenario   (d) Stackpath Scenario

Fig. 4. Gradients value of each packet to the identification task in four scenarios

As shown in Fig. 4, each square in the diagram represents the gradient value of the packet for distinguishing public website access activity(**clientdata**) and Tor hidden service access activity(**clientrp**). Brighter colors indicate that the packet at that position has more weight to identification task. From the result, we draw the conclusion that there does exists a **key sequence** which contributes significantly for identifying Tor hidden service in each MEEK scenario. But the location and window size for key sequence varies for each scenario.

Table 3. Identification results based on key sequence

Scenario	Start_index	Window_size	Accuracy	Precision	Recall	f1
Public	186	101	0.9379	0.9052	0.9608	0.9321
cdn77	309	156	0.9666	0.9568	0.9648	0.9607
Fastly	278	114	0.9593	0.9461	0.9558	0.9509
Stackpath	338	181	0.9652	0.9368	0.9756	0.9558

5.3 Key Sequence Location Distribution Observation

To find the location distribution of key sequence, for each scenario, we count the absolute position of the **establish_rendezvous** and **rendezvous2** cells, the time relative to the start of the trace(we use seconds as the unit), and the number of packets between the two cells. Then, we count the time of **establish_rendezvous** and **rendezvous2** in the cell logs relative to the connection and use that time as the start and end time point to locate the location of key sequence in TCP packet level. We note the index of to **establish_rendezvous** as the start position, and the index of **rendezvous2** as the end position.

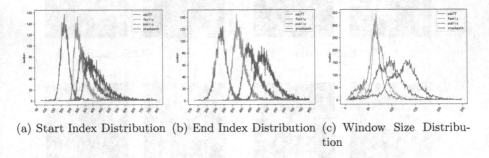

(a) Start Index Distribution (b) End Index Distribution (c) Window Size Distribution

Fig. 5. Key sequence location and window size distribution in TCP sequences

As shown in Fig. 5, the location of the key sequence varies from each scenario while all accord with normal distribution. In the **public** scenario, the key sequence is the earliest, starting at the 184th TCP packet and ending at the 352ed TCP packet, with an interval of about 90 TCP packets; followed by the **fastly** scenario, where the key sequence starts at about the 277th TCP packet and ends at the 491st TCP packet, with an interval of about 104 TCP packets; followed by the **cdn77** scenario starts at about the 306th TCP packet and ends at the 629th TCP packet, with an interval of about 146 TCP packets, followed by the **stackpath** scenario, which starts at the 343rd TCP packet and ends at the 713th TCP packet, with an interval of about 176 TCP packets.

5.4 Classification with Key Sequence

In this section, we try to search the best value of the start point and window size for the DF classification method for four scenarios. We denote the search space as S*W, which S indicates the space of start TCP index and W indicates the window size. According to the observation described above, we set S belongs to [start index-5, start index+6] and W belongs to [window size-10, window size+11]. Then, by setting the radio of training, validation and testing as 1:1:2, we perform experiments with DF classification method iteratively by increase the S and W parameter with a step by 1 for four MEEK scenarios.

Table 4. Results of full TCP sequence without key sequence

Scenario	Input_size	Accuracy	Precision	Recall	f1
Public	756	0.8883	0.8592	0.9288	0.8926
cdn77	1737	0.9181	0.872	0.98	0.9228
Fastly	679	0.8871	0.8517	0.9374	0.8924
Stackpath	3036	0.8612	0.7866	0.9914	0.8772

Table 5. Method effectiveness in timeliness and resource consumption

Scenario	ADKSF	ANPKSF	ADC	ANPC	Time saving	Memory saving
Public	92.74	101	137.74	511	0.33	0.8
cdn77	76.76	156	120.89	971	0.37	0.84
Fastly	68.09	114	88.57	566	0.23	0.8
Stackpath	70.15	181	105.13	1304	0.33	0.86

As shown in Table 3, in each scenario, only with key sequence can achieve 3%-4% better performance compared with the results of existing research work CUMUL [10] using full TCP sequence which are shown in Table 2, indicating that with the key TCP package sequences as input, a network level attacker can distinguish whether a user is accessing Tor hidden service with a high accuracy without decrypting the packets.

Besides, we remove the key sequence in each trace and use DF model for classification again, the results are shown in Table 4. Compared the identification results of using only key sequence, without key sequence, full TCP packet sequence as input, which results are shown in Table 3, Table 4 and Table 2 respectively, indicating that the key sequence proposed in this paper contributes tremendously in Tor hidden service access identification task. The advantage of using key sequence is that only a portion of the TCP sequence needs to be observed, rather than the entire trace, effectively improving the identification timeliness as well as reducing the cost of hardware resources.

Moreover, we calculate the TCP index at the end of the key sequence fragment and the average time duration, and compare it with the average number and time duration of entire access trace to obtain the percentage of time and memory consumption saved by our method. In the Table 5, **ADKSF** means average time duration of the key sequence, **ANPKSF** means average number of packets of the key sequence, **ADC** means average time duration of one access trace, **ANPC** means average number of packets of one access trace. As shown in Table 5, compared with the existing methods using all data as input, the timeliness is improved by 23%–37%, and the hardware resource consumption is reduced by 80%–86%, indicating that our method has high feasibility and good universality, which is much suitable for online identification scenario.

6   Conclusion

In this paper, we verify that there does exist a TCP package sequence contributes significantly to identify Tor hidden service access behavior under MEEK scenario. Moreover, we present a novel approach to identify Tor hidden service access activity with key sequence under MEEK scenario. What's more, we perform comprehensive experiments and thorough analysis in both public and private MEEK scenarios, the results show that our method can identify the Tor hidden service access behavior in the early stage of the access procedure, effectively improving the identification timeliness as well as reducing the cost of hardware resources, indicating that our method has high feasibility and good universality, which is much suitable for online identification scenario.

The approach we present is to identify Tor hidden service access behavior under MEEK scenario, and is useful for supervisor to monitor the violation of criminal activity. However, it is true that the approach can identify which user uses Tor to access hidden service, but it is not clear which hidden service the user accesses. Therefore, the approach is a pre-work for recognizing hidden service for fine-grainedness.

The MEEK establish one connection from OP to MEEK server to transmit packets. In some cases, the user will access both hidden services and normal website at the same time. Whether the approach can identify the access behaviors which users access multiple hidden services or access boss hidden services and normal websites by using multiple tabs or not, and whether this approach can apply to the different status of network, needs further study.

Acknowledgements. This work is supported by the Fundamental Research Program(JCKY2019211B001), the Key Research and Development Program for Guangdong Province under grant(No.2019B010137003) and the Strategic Priority Research Program of the Chinese Academy of Sciences with No. XDC02030000.

References

1. Tor project, users - tor metrics. https://metrics.torproject.org/userstats-relay-country.html?start=2021-11-20&end=2022-01-20&country=all&events=off. Accessed Jan 2022
2. Tor specification. https://gitweb.torproject.org/torspec.git/tree/rend-spec-v2.txt
3. Angel, Y., Winter, P.: obfs4 (the obfourscator), May 2014. https://gitweb. torproject.org/pluggable-transports/obfs4.git/tree/doc/obfs4-spec.txt
4. Christin, N.: Traveling the silk road: a measurement analysis of a large anonymous online marketplace. Arch. Neurol. **2**(3), 293 (2012)
5. Fifield, D., Lan, C., Hynes, R., Wegmann, P., Paxson, V.: Blocking-resistant communication through domain fronting. Proc. Priv. Enhancing Technol. **2015**(2), 46–64 (2015)
6. Guan, Z., Gou, G., Guan, Y., Wang, B.: An empirical analysis of plugin-based tor traffic over SSH tunnel. In: MILCOM 2019–2019 IEEE Military Communications Conference (MILCOM), pp. 616–621. IEEE (2019)
7. Hayes, J., Danezis, G.: k-fingerprinting: a robust scalable website fingerprinting technique. In: USENIX Security Symposium, pp. 1187–1203 (2016)
8. Jansen, R., Juárez, M., Galvez, R., Elahi, T., Díaz, C.: Inside job: applying traffic analysis to measure tor from within. In: 25th Annual Network and Distributed System Security Symposium, NDSS 2018, San Diego, California, USA, 18–21 February 2018. The Internet Society (2018)
9. Kwon, A., Alsabah, M., Lazar, D., Dacier, M., Devadas, S.: Circuit fingerprinting attacks: passive deanonymization of tor hidden services. In: Usenix Security Symposium, pp. 287–302 (2015)
10. Panchenko, A., et al.: Website fingerprinting at internet scale. In: NDSS (2016)
11. Panchenko, A., Mitseva, A., Henze, M., Lanze, F., Wehrle, K., Engel, T.: Analysis of fingerprinting techniques for tor hidden services. In: Proceedings of the 2017 on Workshop on Privacy in the Electronic Society, Dallas, TX, USA, 30 October - 3 November 2017, pp. 165–175. ACM (2017)
12. Rimmer, V., Preuveneers, D., Juarez, M., Van Goethem, T., Joosen, W.: Automated website fingerprinting through deep learning. In: Network & Distributed System Security Symposium (NDSS) (2018)
13. Shahbar, K., Zincir-Heywood, A.N.: An analysis of tor pluggable transports under adversarial conditions. In: 2017 IEEE Symposium Series on Computational Intelligence (SSCI), pp. 1–7. IEEE (2017)
14. Sheffey, S., Aderholdt, F.: Improving meek with adversarial techniques. In: 9th {USENIX} Workshop on Free and Open Communications on the Internet ({FOCI} 19) (2019)
15. Sirinam, P., Imani, M., Juarez, M., Wright, M.: Deep fingerprinting: undermining website fingerprinting defenses with deep learning. In: Proceedings of the 2018 ACM SIGSAC Conference on Computer and Communications Security, pp. 1928–1943 (2018)
16. Soleimani, M.H.M., Mansoorizadeh, M., Nassiri, M.: Real-time identification of three tor pluggable transports using machine learning techniques. J. Supercomput. **74**(10), 4910–4927 (2018)
17. Syverson, P., Dingledine, R., Mathewson, N.: Tor: the second generation onion router. In: Usenix Security, pp. 303–320 (2004)
18. Wang, L., Dyer, K.P., Akella, A., Ristenpart, T., Shrimpton, T.: Seeing through network-protocol obfuscation. In: Proceedings of the 22nd ACM SIGSAC Conference on Computer and Communications Security, pp. 57–69 (2015)

19. Wang, M., Li, Y., Wang, X., Liu, T., Shi, J., Chen, M.: 2ch-TCN: a website finger-printing attack over tor using 2-channel temporal convolutional networks. In: 2020 IEEE Symposium on Computers and Communications (ISCC), pp. 1–7 (2020). https://doi.org/10.1109/ISCC50000.2020.9219717
20. Wang, T., Goldberg, I.: Improved website fingerprinting on tor. In: Proceedings of the 12th ACM Workshop on Workshop on Privacy in the Electronic Society, pp. 201–212. ACM (2013)
21. Xie, H., Wang, L., Yin, S., Zhao, H., Shentu, H.: Adaptive meek technology for anti-traffic analysis. In: 2020 International Conference on Networking and Network Applications (NaNA), pp. 102–107. IEEE (2020)
22. Yao, Z., et al.: Meek-based tor traffic identification with hidden Markov model. In: 2018 IEEE 20th International Conference on High Performance Computing and Communications; IEEE 16th International Conference on Smart City; IEEE 4th International Conference on Data Science and Systems (HPCC/SmartCity/DSS), pp. 335–340. IEEE (2018)

Towards a Scalable Set Similarity Join Using MapReduce and LSH

Sébastien Rivault[(✉)], Mostafa Bamha, Sébastien Limet, and Sophie Robert

Université Orléans, INSA Centre Val de Loire, LIFO EA 4022, Orléans, France
{Sebastien.Rivault,Mostafa.Bamha,Sebastien.Limet,
Sophie.Robert}@univ-orleans.fr

Abstract. Set similarity joins consists in computing all pairs of similar sets from two collections of sets. In this paper, we introduce an algorithm called *MRSS-join*, an extended version of our previous MRS-Join algorithm for the treatment of similarity in the trajectories. *MRSS-join* algorithm is based on the MapReduce computation model and a randomized redistribution approach guaranteeing perfect load balancing properties during all similarity join calculation steps while significantly reducing communication costs and the number of sets comparisons with regard to the best known algorithms based on prefix filtering. All our claims are supported by theoretical guarantees and a series of experiments that show the effectiveness of our approach in handling large datasets collections on large-scale systems.

Keywords: Similarity join operations · Local Sensitive Hashing (LSH) · MapReduce model · Data skew · Hadoop framework

1 Introduction

The Set Similarity Join (SSJ) consists in finding all the pairs of sets having a distance smaller than a given threshold. SSJ has a large amount of applications including data cleaning [4], entity resolution [7], similar text detection [19, 22], and collaborative filtering [2]. The pruning power of the SSJ is also used to reduce the number of candidate pairs for edit-based string similarity joins [1].

Formally, the R-S join for two collections R and S of sets from the universe $\mathcal{U}$ is $R \bowtie_\lambda S = \{(u, v) \in R \times S \mid Dist(u, v) \leq \lambda\}$ where $Dist(u, v)$ is a distance between u and v, and λ is the threshold parameter. Throughout this paper, a set $u \in R \cup S$ is called a record and the elements of u are called tokens.

We restrict the scope in this paper to one of the most popular distances in the literature, namely the similarity function Jaccard. It is defined as follows: $Jaccard(u, v) = \|u \cap v\| / \|u \cup v\|$ where $\|\cdot\|$ is the cardinality of a set. Note that by design $0 \leq Jaccard(u, v) \leq 1$, and $Jaccard(u, v) = 1$ if and only if u and v are equals. To satisfy the metric space properties, we define the corresponding Jaccard distance as $Dist_J(u, v) = 1 - Jaccard(u, v)$.

© The Author(s), under exclusive license to Springer Nature Switzerland AG 2022
D. Groen et al. (Eds.): ICCS 2022, LNCS 13350, pp. 569–583, 2022.
https://doi.org/10.1007/978-3-031-08751-6_41

Naively, the SSJ computations can be performed by comparing all the data pairs which requires a Cartesian product computation. This may have a disastrous effect on performance and limits their scalability to process large datasets. In a BigData context, the amount of data easily exceeds the storage capacity and the processing capability of a single machine. Accordingly, a cluster of machines and scalable distributed algorithms are required.

In the literature, we distinguish two classes of SSJ algorithms according to their result completeness. We refer to algorithms that produce the full similarity join result as exact and others as approximate. Exact SSJ has received much attention and an experimental survey has been conducted on the most recent algorithms [8]. It concludes that none of the evaluated algorithms scale for large datasets processing.

Approximate SSJ is usually based on Locality Sensitive Hashing (LSH) that is a randomized method for generating candidate pairs. In the massively parallel computation model, [11,12] present an algorithm relying on LSH that achieves guarantees on the result completeness and balanced load of the processing nodes. However, it assumes that the dataset is not skewed. This issue is solved by the *MRS-join* algorithm [17] that guarantees perfect balancing properties among the processing nodes while reducing communication costs. *MRS-join* has been described to perform similarity joins on trajectories using MapReduce [6]. The MapReduce paradigm has received a lot of attention for being a scalable parallel shared-nothing data-processing platform. In order to generalize our framework, we present *MRSS-join* that performs set similarity joins using the Jaccard distance. Furthermore, additional filtering steps and new communication templates are introduced in the self join case.

The *MRSS-join* is compared to VernicaJoin (VJ) [21] which is the state-of-the-art algorithm in terms of runtime and robustness in an exact computation according to [8]. VJ is a multistep algorithm based on prefix and length filtering. By sorting records according to the global frequency of tokens, the w-prefix of a record corresponds to its w first tokens. By determining the prefix sizes depending on the Jaccard similarity threshold and the record length, a candidate pair of records can be pruned if their prefixes have no common token. In the experiments, we compare the performance and the quality of the LSH filtering. The quality is measured in terms of *recall* and *precision*. The *recall* is the fraction of the number of pairs of similar records correctly produced over the exact number of similar records, whereas *precision* corresponds to the fraction of the number of pairs of similar records correctly produced over the number of candidates.

The remaining of this paper is organized as follows: Sect. 2 presents requirements for the understanding of the *MRSS-join* algorithm. Section 3 describes the *MRSS-join* algorithm. Experimental results presented in Sect. 4 confirms the efficiency of our approach. We then conclude in Sect. 5.

2   Preliminaries

This section is organized as follows: Sect. 2.1 introduces LSH and its associated algorithm to perform set similarity joins using LSH; Sect. 2.2 explains distributed histograms and randomized communication templates; Sect. 2.3 presents the *MRS-join* algorithms based on LSH, distributed histograms and randomized communication templates to guarantee a perfect balancing of the load and computation among the processing nodes while reducing communication and computation to only relevant data.

2.1 Locality Sensitive Hashing (LSH)

Indyk and Motwani introduced a randomized hashing framework [9,13] that solves efficiently the (λ, c)-near neighbor problem even in high dimensional spaces. It is based on a hashing scheme that ensures that close data points are more likely to collide than distant ones. More formally, it is characterized by the following definition.

Let u, v be two records from a common universe $\mathcal{U}$, *Dist* a distance and λ the threshold distance parameter. Given an approximation factor $c > 1$ and two probabilities p_1 and p_2 such that $0 \leq p_2 < p_1 \leq 1$, $\mathcal{H}$ is a family of LSH functions, if it satisfies the following conditions for any hash function $h \in \mathcal{H}$ chosen uniformly:

1. If $Dist(u, v) \leq \lambda$ then $\mathsf{P}[h(u) = h(v)] \geq p_1$
2. If $Dist(u, v) \geq c * \lambda$ then $\mathsf{P}[h(u) = h(v)] \leq p_2$

Subsequently, we focus on the set similarity join using Jaccard similarity function, however the approach may be generalized using any LSH family that has constant probabilities. MinHash [3] is a family of LSH function that estimates the Jaccard distance. It is defined from a random permutation π of the universe $\mathcal{U}$. For any element e of $\mathcal{U}$, let note $\pi(e)$ be the position of e in the permutation of $\mathcal{U}$. The hashing function h is then defined by $h(u) = \min_{e \in u} \pi(e)$. It is easy to prove that $\mathsf{P}[h(u) = h(v)] = Jaccard(u, v)$.

It is common to concatenate several independent hash functions to improve *precision* and use many independent repetitions to improve *recall*. In a formal way, let $\mathcal{H}_K$ be the LSH family in which a hash function is obtained by concatenating $K \geq 1$ hash functions uniformly and independently selected from $\mathcal{H}$. Accordingly, it holds that $\mathsf{P}_{g^K \in \mathcal{H}_K}[g^K(u) = g^K(v)] = \mathsf{P}_{h \in \mathcal{H}}[h(u) = h(v)]^K = p_1^K$.

K-partition [15,18] is a variant of MinHash which is efficient to compute several independent hash functions using a single permutation. The idea is to partition the permutation into K bins $B_1, \ldots, B_K$ and $h_i(u) = \min_{e \in u \cap B_i} \pi(e)$. If $u \cap B_i$ is empty, $h_i(u)$ is the set to the first on right (circular) $h_j(u)$ where $u \cap B_j$ is not empty. We refer the reader to [18] for more detailed information. In the rest of the paper MinHash refers to K-partition one-permutation MinHash.

For now, we have left K unspecified, we primarily review Hu et al.'s algorithm [11,12] that provides the following load bounds in the massively parallel computation model.

Theorem 1 ([11,12]). *There is a randomized similarity join algorithm that runs in $O(1)$ rounds on P processors that reports each join result with at least a constant probability and the following expected load:*

$$\tilde{O}\left(\sqrt{\frac{\|R \bowtie_\lambda S\|}{P^{1/(1+\rho)}}} + \sqrt{\frac{\|R \bowtie_\sigma S\|}{P}} + \frac{N}{P^{1/(1+\rho)}}\right).$$

where $\sigma = c * \lambda$, $\rho = \frac{\log(p_1)}{\log(p_2)} < 1$ and N is the number of inputs.

The corresponding algorithm is given by the three following steps by setting $K = \left\lceil \log(p_1, 1/P^{\frac{\rho}{1+\rho}}) \right\rceil$:

1. Randomly and independently select Q hash functions $g_1^K, g_2^K, ..., g_Q^K$ from $\mathcal{H}_K$,
2. For each record u, emit a key/value pair $<(i, g_i^K(u)), u>$ for all $i \in 1, ..., Q$,
3. Perform a join by treating $(i, g_i^K(u))$ as the join attribute value, i.e., two records u, v join if $g_i^K(u) = g_i^K(v)$ for all i. For a pair of records (u, v), output them if $Dist_J(u, v) \leq \lambda$.

The number of repetitions is given by $Q = \lceil p_1^{-K} \rceil$ in [11] that gives an optimal output sensitive algorithm. Although result of the similarity join is reported with at least a constant probability, users may want to generate the full similarity join result. By setting $Q = \lceil 3 * p_1^{-K} * \ln(N) \rceil$, the probability to report all join results is $1 - 1/N$ [12]. Thereafter, we prefer to let users specify the desired result expectation by setting $Q = \lceil E * p_1^{-K} \rceil$ with $1 \leq E \leq 3 * \ln(N)$. This algorithm forms the basic building blocks of the similarity join using LSH.

For sake of clarity, we introduced the similarity join using LSH between two collections R and S but in the following, we will consider self joins of a dataset Γ which aim is to compute $\Gamma \bowtie_\lambda \Gamma$. Self joins can be handled as R-S joins from the LSH side.

2.2   Distributed Histograms and Communication Templates

We first review the notion of distributed histograms introduced in [10] to reduce communication costs while guaranteeing perfect balancing properties among all processing nodes. Then, we explain the histogram distribution used in [17]. We implement the memory extensions following the ideas of [17] to guarantee that a distributed histogram always fits in processing nodes' memory. At last, we introduce communication templates for the self join case in the same spirit as the ones introduced for the general case [10,17] to which we refer the reader for further information.

The histogram of a join is defined as the association between a join attribute value and its frequency. It is used to generate communication templates, allowing to transmit only relevant data fairly during the join phase. More formally, for a dataset Γ where $L(\Gamma)$ denotes the set of its LSH join attribute values, the histogram Hist(Γ) is the list of the pairs $<x, \mathbf{f_x}>$ where $x \in L(\Gamma)$ and $\mathbf{f_x}$ is its frequency.

In order to reduce communication costs to relevant data, only join attribute values which might appear in the join result are present in the histogram. Join attribute values that produce a result imply that their frequencies are greater than or equal to two. Thus, the histogram for the similarity join $\Gamma \bowtie_\lambda \Gamma$ which contains only relevant data is defined as follows.

Definition 1. $Hist(\Gamma \bowtie_\lambda \Gamma) = \{<x, f_x>, \forall x \in L(\Gamma) \mid f_x > 1\}$

For large datasets, we expect that the corresponding histogram does not fit in memory. Therefore, [17] introduced a way to distribute it for multiple join attribute values. The distribution principle is based on the appearance of join attribute values in different *splits* where a *split* is the portion of data that a **map** function processes. When constructing the histogram, for all join attribute values, the *split* identifiers are also stored. The distribution job requires as many **reduce** tasks as the total number of *splits*. Each join attribute value and its corresponding entry in the histogram are transmitted for each split identifier. At the end, each **reduce** task output corresponds to the distributed histogram required by a *split*.

In order to guarantee that a distributed histogram always fits in memory even in the case of large *splits*, an additional parameter noted t_{max} is used. This parameter is chosen in such a way that each *Mapper* can buffer a distributed histogram of size at most $Q * t_{max}$. The method consists in constructing groups of consecutive records in a *split* so that a group is composed of at most t_{max} records. Handling distributed histogram on groups requires no new algorithm. Essentially, instead of storing only the splits identifiers, the set of pairs (*splitId*, *groupId*) is stored. Later, a pair (*splitId*, *groupId*) is called a *chunk* identifier. A *chunk* is the corresponding portion of data. At the end, each **reduce** task output corresponds to the sorted list of *chunk*'s histogram required by a *split*.

Distributed histograms are then used to reduce communication costs while guaranteeing perfect balancing properties among all processing nodes. It also avoids the effects of data skew in large datasets processing. To this end, communication templates use a parameter denoted by f_{max}. This parameter defines the number of records that a *Reducer* will have to store and process during the similarity join step. Owing to this parameter, the records having a common join attribute value will be divided into several buckets (blocks), so that each bucket can be stored in memory. This makes the *MRS-join* algorithm scalable and insensitive to the effects of data distribution skew.

For a given join attribute value x, communication templates will distribute all buckets according to two cases:

a. $f_x < f_{max}$: the records corresponding to the join attribute value are transmitted to a single *Reducer*, without special processing, using a hashing approach.
b. $f_{max} \leqslant f_x$: The join attribute value is highly frequent and as illustrated Fig. 1. In order to balance the computations, the join attribute value x is divided into several blocks (5 on our example).

In Fig. 1, the generated communication templates are arranged in rows and columns. Each cell corresponds to a bucket. Each column corresponds to data transmitted to a **reduce** task. These tasks are identified starting from i_0 that is a random integer which can be stored in the histogram. For the sake of the clarity, it will not be mentioned in the following.

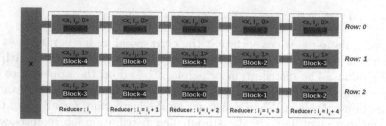

Fig. 1. Communication templates for a highly frequent join attribute value in a self join case. Distributed buckets (red) are stored to compute the similarity join with replicated buckets (black). (Color figure online)

To ensure that the buckets are sorted in the correct order appropriate MapReduce <Key, Value> pairs are used. The keys are composed of the join attribute value, the column and the row identifier. Pairs are then redirected by the MapReduce **partition** function by means of the **reduce** task identifier. For a *Reducer* task, the join is computed using the following algorithm for a highly frequent join attribute value:

- Store in memory the distributed buckets (i.e., the row identifier is at zero),
- Compute the join within the stored buckets,
- Compute the join with replicated buckets.

Later, we introduce an additional filtering step to this algorithm in the case of a highly frequent join attribute value to reduce the number of comparisons.

2.3   MRS-join

MRS-join [17] is an algorithm built on top of the MapReduce framework that uses LSH, distributed histograms and randomized communication templates to guarantee balanced load and computation among the processing nodes. It is a multi-steps algorithm with time and space guarantees for all the join computation steps.

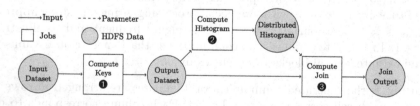

Fig. 2. MapReduce similarity join computation steps.

MRS-join proceeds in 3 steps, each step including one or two MapReduce jobs. The Fig. 2 represents the interactions between the different steps of the algorithm:

❶ Compute the LSH join attribute values,

❷ The histogram of the join is computed and distributed to guarantee balanced communication patterns regardless the data distribution,

❸ By using distributed histograms, efficient and scalable communication templates are generated and the distance between the pairs identified as similar, is computed to produce the similarity join output.

The step ❶ computes the Q LSH join attribute values, in our implementation it is performed before steps ❷ and ❸. The step ❷ is composed of two MapReduce *jobs*, the first one is used to compute the histogram of the join and the second to distribute it. The main difference is that the histogram is constructed for a self-join and distributed by *chunks* instead of *splits* as explained previously. Using distributed histograms, the step ❸ computes the similarity join. To reduce the number of comparisons, we introduce additional filters during this step in the *MRSS-join* algorithm.

For space reason, we do not include a theoretical analysis for each computation step in the following. However, the load of each *Reducer* during the last step of *MRSS-join* is bounded by Theorem 1. In addition, the *MRS-join* algorithm has an asymptotic optimal complexity when the maximum cost to read all distributed histograms corresponding to a *Mapper* is less than the maximum cost to transmit all data and to perform the similarity join in a processing node. We recall that the sizes of distributed histograms are very small compared to input datasets sizes, and we refer the reader to [17] for more details regarding the cost model. Using *chunks* instead of *splits* to distribute the histogram only adds to the cost model a constant factor that depends on t_{max} and the number of records in a *split*. By maximizing t_{max} while ensuring that the corresponding distributed histograms fit in memory, we expect that using *chunks* instead of *splits* has a limited impact on the size of distributed histograms. Of course, this is only an intuitive argument that would deserve a detailed cost model.

3   *MRSS-join*: A Scalable Set Similarity Join Algorithm Using LSH and MapReduce

We assume that, before the start of the **map** phase, the Q MinHash functions are randomly and uniformly selected and stored in the HDFS. The MinHash function is implemented using Zobrist hashing [23]. Zobrist hashing has been shown theoretically and practically to have strong MinHash properties while being fast in practice [5,20].

The histogram computation *job* is described in the Algorithm A. To compute frequency of each join attribute value, the **map** phase emits, for each record, and for each join attribute value x, two key/value pairs, allowing to handle the frequency and the set of *chunkId* apart. To ensure that the *groups* within distributed histograms are sorted in the correct order, a *chunkId* is implemented using a 64 bits integer. The first 32 bits correspond to the *splitId* and the last ones to the *groupId*.

Algorithm A: Histogram computation step

> **Map:** $<id,\ u> \rightarrow <(x,\ 0/1),\ (1/,\ /chunkId)>$
> > **init:**
> > | Read from HDFS the Q MinHash functions.
> > $chunkId \leftarrow$ getCurrentChunkId();
> > Compute the Q LSH join attribute values of the record u.
> > For each join attribute value x emit the pairs:
> > - $<(x,\ 0),\ (1,\ \emptyset)>$
> > - $<(x,\ 1),\ (\varepsilon,\ chunkId)>$
>
> **Combine:** $<(x,\ 0),\ (1,\ \emptyset)^*> \rightarrow <x,\ (lf_x,\ \emptyset)>$
> > Compute local frequency using the size of input values.
> > Emit a pair $<(x,\ 0),\ (lf_x,\ \emptyset)>$.
>
> **Combine:** $<(x,\ 1),\ (\varepsilon,\ chunkId)^*> \rightarrow <x,\ (\varepsilon,\ chunkId^*)>$
> > Compute the set of $chunkId$.
> > Emit a pair $<(x,\ 1),\ (\varepsilon,\ chunkId^*)>$.
>
> **Reduce:** $<(x,\ 0),\ (lf_x,\ \emptyset)^*>$
> > Compute the global frequency of the join attribute value.
>
> **Reduce:** $<(x,\ 1),\ (\varepsilon,\ chunkId^*)^*> \rightarrow <x,\ (\mathbf{f_x},\ chunkId^*)>$
> > Compute the set of $chunkId$; whenever the set size exceeds a given limit, the
> > following pair is emitted and the set cleared.
> > Emit the pair $<x,\ (\mathbf{f_x},\ chunkId^*)>$ if $\mathbf{f_x} > 1$.

Algorithm B: Histogram distribution step

> **Map:** $<x,\ (\mathbf{f_x},\ chunkId^*)> \rightarrow <chunkId,\ (x,\ \mathbf{f_x})>$
> > | Emit a pair $<id,\ (x,\ \mathbf{f_x})>$ for all $id \in chunkId^*$.
>
> **Partition:** $<chunkId,\ (x,\ \mathbf{f_x})> \rightarrow Integer$
> > | Return the $splitId$ from the $chunkId$.
>
> **Reduce:** $<chunkId,\ (x,\ \mathbf{f_x})^*> \rightarrow <chunkId,\ (x,\ \mathbf{f_x})^*>$
> > Compute the set of values received to eliminate duplicates.
> > Emit a pair $<x,\ \mathbf{f_x}>$ for all remaining values.

The *Combiner* computes the local frequencies and the set of *chunkIds* of the current *split*. *Reducers* sum up the received local frequencies and filter out join attribute values that produce no results. The set of *chunkIds* where the join attribute value appears is stored to be able to distribute the histogram. Since the set of *chunkIds* is computed after the frequency, it does not have to fit in memory. Duplicate entries may be produced, which will be eliminated during the distribution step according to Algorithm B. This algorithm relies on the **partition** function to distribute the histogram based on the stored *split* identifiers. Owing to the parameter $\mathbf{t_{max}}$, duplicates can be eliminated in memory during the **reduce** phase. The similarity join *job*, described in the Algorithm C, uses distributed histograms to generate efficient communication templates. The similarity join is computed by filtering out false positive pairs. It should be noted that the computation of the Jaccard distance is performed only once for a pair of records. We introduce additional filtering steps to reduce the number of

---

Algorithm C: Similarity join computation step

Map: $<id,\ u> \rightarrow\ <(x,\ reducerId,\ rowId),\ (u,\ LSHKeys)>$
  init:
    | Read from HDFS the Q MinHash functions.
  Read and store the corresponding distributed histogram of the current
  chunkId.
  Compute the Q LSH join attribute values of the record u.
  Retain only the join attribute values that appear in the distributed
  histogram, which we will denote by *LSHKeys*.
  Emit pairs according to communication templates described previously 2.2.
Partition: $<(x,\ reducerId,\ rowId),\ (u,\ LSHKeys)> \rightarrow\ Integer$
  | Redirect each pair according to communication templates.
Reduce: $<(x,\ reducerId,\ rowId),\ (u,\ LSHKeys)> \rightarrow\ (id_1,\ id_2)$
  Compute the similarity join using the communication templates.
  For each pair of different record:
  - Compute the intersection of the sets *LSHKeys*.
  - If the current join attribute value is the minimal among the intersection,
    compute the Jaccard distance and output them if the distance is lower
    than λ.

---

comparisons. We distinguish two cases depending on whether the similarity join computations of a join attribute value are performed by one or several **reduce** tasks.

In the case where the similarity join computations are performed by a single *Reducer* task, the length filter can be applied [1]. This filter allows only the Jaccard distance between pairs of records of similar size to be computed. For this purpose, the length of the records is appended to the transmitted keys during the join step, allowing the records to be processed in a sorted manner.

In the case of a highly frequent join attribute value, the filtering step relies on LSH. For a **reduce** task and a join attribute value, instead of storing the distributed block, each record is partitioned into F_Q buckets using MinHash. The number of concatenate MinHash function F_K is based on F_Q and the desired expectation F_E, i.e., formally, $F_K = \log_{p_1^{-1}}(F_Q/F_E)$.

When $K < F_K$, we expect the number of comparisons to be drastically reduced, resulting in a time saving. However, this time saving is only guaranteed if the cost of filtering and comparing remaining records is less than the cost of comparing all records. The filtering cost is increasing with F_Q, we set $F_Q = 32$ and $F_E = 4$ in our experiments.

4   Experiments

In this section, we discuss the efficiency and the strength of our theoretical analysis by experimenting the *MRSS-join* algorithm on real world and synthetic datasets. We measured the *recall* and *precision* as well as the efficiency of the *MRSS-join* compared to the state-of-the-art algorithm. The experiments

Table 1. Characteristics of the experimental datasets.

Dataset	N $\cdot 10^5$	Record length max	Record length avg	Universe $\cdot 10^3$	Size (B)
AOL	100	245	3	3900	396 MB
ENRO	2.5	3162	135	1100	254 MB
LIVE	31	300	36	7500	873 MB
NETF	4.8	18000	210	18	576 MB
ORKUT	2	40000	120	8700	2.5 GB
WDC	41	17000	15	184644	5.8 GB
UNIFORM	1	25	10	0.21	4.5 MB
ZIPF (1.0)	4.4	84	50	100	33 MB
ALL	182	40000	21	205962	10.3 GB

were performed using Hadoop 3.2.1 framework on a cluster of 11 machines. Each machine has the following characteristics: Intel(R) Xeon(R) CPU E5-2650 @2.60 GHz, 16 Gb of memory, 300 Gb of HDD disk and 6 Gb as a value for Heap memory of Map/Reduce tasks. The nodes are connected by a 1 Gb/s network. The map output compression is enabled as well as the output compression. For the following experiments, the duplicates in the input dataset are removed since duplicate removal is a different problem from similarity joins. In addition, this makes our results comparable to the existing surveys [8, 16].

4.1   Performance and Results Quality

To analyze the performance and the quality of the *MRSS-join* algorithm, we used 6 real-world and 2 synthetic datasets. The datasets mainly come from the survey [16] and the distributed survey [8] on the exact set similarity join. The entire English relational subset of the WDC Web Table Corpus 2015 [14] has been added to test the scalability of the algorithms. Textual datasets are preprocessed to translate original strings to integers using the tools from [16]. Since we focus on the set similarity join, this pre-processing step is not measured in our experiments. Table 1 presents characteristics of the selected datasets. The characteristics of the datasets are given by the number of records, the maximum, and average size of a record and the size of the universe. Records in AOL, LIVE and WDC datasets are short with a large universe which favors the *VJ* algorithm. The token frequencies follow a Zipfian token distribution for most of real-world dataset, that means a large part of the universe is infrequent. An exception is the NETF dataset which has few infrequent tokens. The two synthetic datasets are UNIFORM and ZIPF which are generated to follow a uniform and a Zipfian token distribution using Zipf factor 1.0 and the generators from [16]. The dataset ALL is simply the union of the datasets presented above and is used to test scalability. We used the state-of-the-art algorithm *VJ* to compare the performance of *MRSS-join*. We performed a self join on all previous datasets by varying the Jaccard similarity threshold in $\{0.6, 0.7, 0.8, 0.9, 0.95\}$. Table 2 shows the average join processing time in seconds over three independent runs. For each run and practical reasons, we set a timeout of 2 h, which is more than three times the runtime of *MRSS-join* for all dataset. The letter "T" denotes that this timeout has been reached for all runs. For the additional parameters, the desired expectation of generating any output pair is set to $E = 2$, and we set $\rho = 0.5$. For space reasons, we omit the experiments varying these parameters. However, we have selected parameters that give the best trade-off between time and quality in our setup.

The processing time for *MRSS-join* is always lower than the corresponding processing time of *VJ* except for AOL. However, in this setting the runtimes are of the same magnitude. In the remaining cases, *MRSS-join* achieves a speedup that can exceed an order of magnitude. Especially, for datasets such as ALL, NETF, ORKUT and WDC, where *VJ* fails to compute the similarity join within the time allocated for one or more similarity thresholds.

More in-depth analysis is given Table 3 that compares the transmitted data during the communication phase of the joining step between *MRSS-join* and *VJ*. It shows that *VJ* is inefficient in terms of transmitted data for dataset containing long records on average as ENRO, NETF and ORKUT. This is due to the fact that *VJ* uses prefixes to compute the similarity join which makes it very sensitive to long records and low similarity join thresholds. This is not the case in *MRSS-join* because it is based on LSH framework which is independent of dimensionality. In addition, we recall that only relevant data is transmitted in *MRSS-join* which drastically reduces the transmitted data during the communication phase of the similarity join step.

In Table 3, the size of transmitted data for ALL and WDC datasets are not reported for *VJ* since it failed before the similarity join step. Indeed, to compute prefixes according to the lowest frequency, *VJ* constructs the histogram of the universe. This histogram is then buffered in memory. For datasets with a very large universe, such as ALL and WDC, memory overloads may occur which limits its efficiency and scalability. This cannot happen in *MRSS-join* because the histogram of LSH join attribute values is distributed and read by *chunks* that fit in memory.

Finally, *VJ* groups intermediate data by prefix token during the joining step. The size of a group depends on the token frequency thus for large datasets, a group may hit the memory limit of a Reducer which limits its scalability. Even in the case of small datasets with few infrequent tokens as NETF, this limits significantly its efficiency because the join similarity computations are not well distributed among all the processing nodes. This cannot happen in *MRSS-join* because the join computations for a highly frequent join attribute value, are partitioned into buckets and transmitted to distinct **reduce** tasks in a randomized manner. This makes *MRSS-join* scalable and insensitive to the data distribution. To achieve these performances, *MRSS-join* is based on LSH that produces almost all the results of the similarity join. Table 4 shows the quality of the LSH filtering by *MRSS-join*. ZIPF is omitted because the similarity join produces no result on the queried thresholds since a large part of the tokens are unique. Each cell reports the *recall* and the *precision* values. The full similarity

Table 2. Time in seconds of *MRSS-join* algorithm compared to *VJ* algorithm.

Threshold	0.6		0.7		0.8		0.9		0.95	
Dataset	VJ	MRSS	VJ	MRSS	VJ	MRSS	VJ	MRSS	VJ	MRSS
AOL	**260**	383	**198**	249	**172**	220	**168**	195	**165**	191
ENRO	481	**142**	341	**137**	295	**136**	231	**131**	196	**137**
LIVE	480	**202**	349	**177**	306	**157**	227	**149**	221	**147**
NETF	T	**217**	2809	**172**	909	**148**	460	**139**	299	**143**
ORKUT	5176	**172**	3654	**167**	2717	**154**	1183	**156**	695	**153**
WDC	T	**1798**	T	**831**	T	**620**	T	**458**	T	**401**
UNIFORM	175	**135**	153	**129**	149	**134**	145	**130**	147	**134**
ZIPF	151	**132**	152	**131**	147	**125**	152	**127**	156	**127**
ALL	T	**2086**	T	**1043**	T	**766**	T	**606**	T	**549**

Table 3. Transmitted data of *MRSS-join* algorithm compared to *VJ* algorithm.

Threshold	0.6		0.7		0.8		0.9		0.95	
Dataset	VJ	MRSS	VJ	MRSS	VJ	MRSS	VJ	MRSS	VJ	MRSS
AOL	**455 MB**	2 GB	**409 MB**	1 GB	**333 MB**	811 MB	**255 MB**	327 MB	209MB	**133 MB**
ENRO	16 GB	**741 MB**	12 GB	**403 MB**	8 GB	**247 MB**	4 GB	**146 MB**	2 GB	**78 MB**
LIVE	13 GB	**1 GB**	10 GB	**762 MB**	7 GB	**222 MB**	3 GB	**96 MB**	2 GB	**45 MB**
NETF	79 GB	**3 GB**	59 GB	**1 GB**	40 GB	**742 MB**	20 GB	**9MB**	10GB	**1MB**
ORKUT	234 GB	**1 GB**	176 GB	**780 MB**	118 GB	**37 MB**	59 GB	**17 MB**	30 GB	**12 MB**
WDC	T	**14 GB**	T	**7 GB**	T	**5 GB**	T	**2 GB**	T	**1 GB**
UNIFORM	**13 MB**	87 MB	**10 MB**	37 MB	**8 MB**	28 MB	5 MB	**1 MB**	4 MB	**110 KB**
ZIPF	306 MB	**51 MB**	234 MB	**29 MB**	161 MB	**2MB**	87 MB	**20 KB**	52 MB	**20 KB**
ALL	T	**25 GB**	T	**13 GB**	T	**8 GB**	T	**4 GB**	T	**2 GB**

Table 4. Results quality of the *MRSS-join* algorithm.

Dataset	Threshold				
	0.6	0.7	0.8	0.9	0.95
AOL	0.93 | 0.010	0.96 | 0.003	0.96 | 0.003	0.96 | 0.000	0.98 | 0.000
ENRO	0.99 | 0.086	0.99 | 0.124	0.99 | 0.402	0.98 | 0.202	0.97 | 0.355
LIVE	0.93 | 0.027	0.97 | 0.010	0.97 | 0.014	0.98 | 0.033	0.98 | 0.077
NETF	0.96 | 0.001	0.97 | 0.001	0.97 | 0.001	0.96 | 0.015	1.00 | 0.024
ORKUT	0.97 | 0.005	0.98 | 0.003	0.97 | 0.015	0.97 | 0.019	0.97 | 0.017
WDC	0.98 | 0.488	0.98 | 0.032	0.98 | 0.024	0.97 | 0.052	0.98 | 0.063
UNIFORM	0.92 | 0.001	0.96 | 0.001	0.98 | 0.001	1.0 | 0.0	1.0 | 0.0
ALL	0.97 | 0.460	0.98 | 0.030	0.98 | 0.026	0.98 | 0.045	0.98 | 0.049

join is computed using *VJ* except for ALL and WDC, where we used *MRSS-join* with the expectation set to $E = 3 * \ln(N)$. We observe that, *MRSS-join* achieves at least 90% *recall* for all datasets. One can notice, the low values of the *precision* in the experiments. These values are not bad due to the fact that, there is no hashing or sorting techniques allowing to find similar pairs and even for these low precision values, *MRSS-join* reduces drastically the number of set comparisons compared to *VJ* as shown in the Table 5.

Table 5. Computed distances of the *MRSS-join* algorithm compared to the *VJ* algorithm.

Threshold	0.6		0.7		0.8		0.9		0.95	
Dataset	VJ	MRSS	VJ	MRSS	VJ	MRSS	VJ	MRSS	VJ	MRSS
AOL	1.3E+10	1.2E+09	5.5E+09	5.1E+08	1.6E+09	1.4E+08	2.4E+08	3.0E+07	1.2E+08	7.8E+06
ENRO	2.3E+09	2.4E+07	6.2E+08	9.6E+06	1.1E+08	1.8E+06	1.2E+07	5.4E+05	1.8E+06	1.4E+05
LIVE	7.6E+09	1.5E+08	2.1E+09	7.6E+07	4.1E+08	1.1E+07	5.3E+07	7.7E+05	8.9E+06	1.3E+05
NETF	6.4E+10	1.1E+08	2.2E+10	3.9E+07	5.0E+09	3.0E+06	4.6E+08	4.9E+03	4.9E+07	2.6E+02
ORKUT	4.6E+09	4.9E+06	1.1E+09	2.0E+06	1.7E+08	1.4E+05	1.2E+07	2.6E+04	1.5E+06	6.4E+03
WDC	T	8.2E+09	T	5.1E+09	T	1.4E+09	T	1.1E+08	T	2.0E+07
UNIFORM	2.4E+09	3.6E+07	1.3E+09	1.9E+07	5.4E+08	5.4E+05	1.1E+08	2.5E+03	3.2E+07	2.4E+02
ALL	T	8.2E+09	T	5.5E+09	T	1.4E+09	T	1.3E+08	T	2.7E+07

5   Conclusion

In this article, we have introduced *MRSS-join* an efficient and scalable MapReduce set similarity join algorithm, using LSH and randomized communication templates approach allowing to reduce drastically the number of sets comparisons and communication costs while guaranteeing perfect balancing properties during all the steps of large datasets similarity join computation.

MRSS-join theoretical guarantees and experiments, using real world and synthetic benchmarks datasets, show that the overhead related to the use both Min-Hash and our communication templates remains very small compared to the gain in performance by reducing communication and data processing to almost all relevant data (this avoids sets pairwise comparisons). We showed that, *MRSS-join* avoids memory overflows by controlling the size of generated buckets. This makes the algorithm scalable and insensitive to the data distribution. It also solves the limitations of existing approaches to handle large datasets whenever data associated to a MapReduce key cannot fit in the available reducer's local memory.

Future work will be devoted to extend *MRSS-join* algorithm to a more general purpose framework for most similarity joins operations by using LSH techniques. We also plan to compute sequences similarity processing in large datasets using similar techniques based on our randomized MapReduce data redistribution to balance load among processing nodes while guaranteeing the scalability of the proposed solutions in large scale systems.

References

1. Arasu, A., Ganti, V., Kaushik, R.: Efficient exact set-similarity joins. In: Proceedings of the 32nd International Conference on Very Large Data Bases, pp. 918–929 (2006)
2. Bayardo, R.J., Ma, Y., Srikant, R.: Scaling up all pairs similarity search. In: Proceedings of the 16th International Conference on World Wide Web, pp. 131–140 (2007)
3. Broder, A.Z., Glassman, S.C., Manasse, M.S., Zweig, G.: Syntactic clustering of the Web. Comput. Netw. ISDN Syst. **29**(8), 1157–1166 (1997)

4. Chaudhuri, S., Ganti, V., Kaushik, R.: A primitive operator for similarity joins in data cleaning. In: 22nd International Conference on Data Engineering, p. 5 (2006)
5. Dahlgaard, S., Knudsen, M.B.T., Thorup, M.: Practical hash functions for similarity estimation and dimensionality reduction. In: Proceedings of the 31st International Conference on Neural Information Processing Systems, pp. 6618–6628 (2017)
6. Dean, J., Ghemawat, S.: MapReduce: simplified data processing on large clusters. Commun. ACM **51**(1), 107–113 (2008)
7. Dey, D., Sarkar, S., De, P.: A distance-based approach to entity reconciliation in heterogeneous databases. IEEE Trans. Knowl. Data Eng. **14**(3), 567–582 (2002)
8. Fier, F., Augsten, N., Bouros, P., Leser, U., Freytag, J.C.: Set similarity joins on mapreduce: an experimental survey. Proc. VLDB Endow. **11**(10), 1110–1122 (2018)
9. Gionis, A., Indyk, P., Motwani, R.: Similarity search in high dimensions via hashing. In: Proceedings of the 25th International Conference on Very Large Data Bases, pp. 518–529 (1999)
10. Hassan, M.A.H., Bamha, M., Loulergue, F.: Handling data-skew effects in join operations using mapreduce. Procedia Comput. Sci. **29**, 145–158 (2014)
11. Hu, X., Tao, Y., Yi, K.: Output-optimal parallel algorithms for similarity joins. In: Proceedings of the 36th ACM SIGMOD-SIGACT-SIGAI Symposium on Principles of Database Systems, pp. 79–90. ACM (2017)
12. Hu, X., Yi, K., Tao, Y.: Output-optimal massively parallel algorithms for similarity joins. ACM Trans. Datab. Syst. **44**(2), 1–36 (2019)
13. Indyk, P., Motwani, R.: Approximate nearest neighbors: towards removing the curse of dimensionality. In: Proceedings of the Thirtieth Annual ACM Symposium on Theory of Computing, pp. 604–613 (1998)
14. Lehmberg, O., Ritze, D., Meusel, R., Bizer, C.: A large public corpus of web tables containing time and context metadata. In: Proceedings of the 25th International Conference Companion on World Wide Web, pp. 75–76 (2016)
15. Li, P., Owen, A., Zhang, C.H.: One permutation hashing. In: Pereira, F., Burges, C.J.C., Bottou, L., Weinberger, K.Q. (eds.) Advances in Neural Information Processing Systems. vol. 25. Curran Associates, Inc. (2012)
16. Mann, W., Augsten, N., Bouros, P.: An empirical evaluation of set similarity join techniques. Proc. VLDB Endow. **9**(9), 636–647 (2016)
17. Rivault, S., Bamha, M., Limet, S., Robert, S.: A Scalable MapReduce Similarity Join Algorithm Using LSH. Université d'Orléans. To appear to IJPP int. Journal, Research report, LIFO (2021)
18. Shrivastava, A., Li, P.: Densifying one permutation hashing via rotation for fast near neighbor search. In: Proceedings of the 31st International Conference on Machine Learning, pp. 557–565 (2014)
19. Theobald, M., Siddharth, J., Paepcke, A.: Spotsigs: robust and efficient near duplicate detection in large web collections. In: Proceedings of the 31st Annual International ACM SIGIR Conference on Research and Development in Information Retrieval, pp. 563–570 (2008)
20. Thorup, M.: Fast and powerful hashing using tabulation. Commun. ACM **60**(7), 94–101 (2017)
21. Vernica, R., Carey, M.J., Li, C.: Efficient parallel set-similarity joins using MapReduce. In: Proceedings of the 2010 ACM SIGMOD International Conference on Management of Data, pp. 495–506 (2010)

22. Wandelt, S., et al.: State-of-the-art in string similarity search and join. SIGMOD Rec. **43**(1), 64–76 (2014)
23. Zobrist, A.L.: A new hashing method with application for game playing. ICGA J. **13**, 69–73 (1990)

Cyberbullying Detection with Side Information: A Real-World Application of COVID-19 News Comment in Chinese Language

Jian Xing[1,2,3(✉)], Xiaoyu Zhang[1,2], Lin Chen[1,2], Yu Ding[1,2(✉)], Yaru Zhang[1,2], Wei Hu[3], Zhicheng Jin[3], Jingya Wang[3], Yaowei Chen[3], and Yi Hong[3]

[1] Institute of Information Engineering, Chinese Academy of Sciences, Beijing, China
{xingjian,zhangxiaoyu,chenlin,dingyu,zhangyaru}@iie.ac.cn
[2] School of Cyber Security, University of Chinese Academy of Sciences, Beijing, China
[3] National Computer Network Emergency Response Technical Team/Coordination Center of China Xinjiang Branch, Urumqi, China

Abstract. Cyberbullying is an aggressive and intentional behavior committed by groups or individuals, and its main manifestation is to make offensive or hurtful comments on social media. The existing researches on cyberbullying detection underuse natural language processing technology, and is only limited to extracting the features of comment content. Meanwhile, the existing datasets for cyberbullying detection are non-standard, unbalanced, and the data content of datasets is relatively outdated. In this paper, we propose a novel Hybrid deep Model based on Multi-feature Fusion (HMMF), which can model the content of news comments and the side information related to net users and comments simultaneously, to improve the performance of cyberbullying detection. In addition, we present the JRTT: a new, publicly available benchmark dataset for cyberbullying detection. All the data are collected from social media platforms which contains Chinese comments on COVID-19 news. To evaluate the effectiveness of HMMF, we conduct extensive experiments on JRTT dataset with five existing pre-trained language models. Experimental results and analyses show that HMMF achieves state-of-the-art performances on cyberbullying detection. To facilitate research in this direction, we release the dataset and the project code at https://github.com/xingjian215/HMMF.

Keywords: Cyberbullying detection · Side information · New benchmark dataset · COVID-19 · Nature language processing · Chinese language processing

1 Introduction

Cyberbullying, which means to bully or harass others by online comments on social media, has become a widely discussed problem in recent years [1,2] . The

D. Groen et al. (Eds.): ICCS 2022, LNCS 13350, pp. 584–598, 2022.
https://doi.org/10.1007/978-3-031-08751-6_42

anonymity and concealment of the mobile Internet have accelerated cyberbullying into a widespread social phenomenon. Meanwhile, the global pandemic of COVID-19 in recent years has exacerbated the anomie of such online comments [3]. The anomie phenomenon of online comments is mainly manifested in the use of insulting and discriminatory language, such as abuse, slander, contempt and ridicule without the constraints of moral norms and laws. It makes others suffer mental and psychological violations and damage by language violence. In order to create a harmonious network atmosphere and purify the language environment for comments, it is necessary to effectively detect and analyze cyberbullying on social media.

As is known to all, detecting cyberbullying on social media is a difficult and challenging problem that needs much efforts to be devoted [4]. The reasons are two-fold.

Firstly, the task of extracting and identifying such language is generally attributed to the field of natural language processing(NLP) [5,6]. However, with the flexibility and irregularity of language comments, it is difficult to directly find and deal with cyberbullying in time. For instance, comments that do not directly contain malicious words, sarcastically asked comments, and comments that quote questionable statements. Furthermore, most existing detection methods of cyberbullying mainly focus on the modeling of comment content, while ignoring the rich side information in social comments [4]. Significantly, the popularity of social media enables us to collect relevant side information from the perspective of net users, which helps us capture rich information except the content of comments. Another is that, with the explosive growth of Internet information in modern society, manually checking users' comments by the administrators of social media platforms is completely inadequate for cyberbullying detection. Therefore, the application of machine learning technology has become a practically feasible approach for automated cyberbullying detection [7–9].

Secondly, most existing researches mainly focus on English social media platforms, such as twitter and Instagram [10–13]. But at the same time, various Chinese pre-trained models provide a basis for us to conduct in-depth research on automated detection technology of cyberbullying based on Chinese [14–18], and creating a baseline dataset is the priority of this research [1].

In this paper, we study the problem of cyberbullying detection with side information. Particularly, the goal is to focus on how to effectively use the rich side information generated by social media for performance improvement of cyberbullying detection. To cope with this, we propose a novel Hybrid deep Model based on Multi-feature Fusion framework, called HMMF, for the problem of cyberbullying detection. The main contributions of our work are summarized as follows:

(1) Model-oriented: We propose a novel framework namely HMMF, which is the first to integrate diverse Chinese pre-trained models with Transformer for automated cyberbullying detection. The HMMF can perform the mutual fusion of multiple features and improve the performance of cyberbullying detection.

(2) Feature-oriented: We mine more effective features from the side information related to net users and comments. The HMMF can effectively extract sentence embeddings of the comment content through the comprehensive application of existing Chinese pre-training models. Meanwhile, it can learn more useful social features, attribute features and interaction features from side information through Transformer model.

(3) Data-oriented: We publish a new benchmark dataset based on Chinese social media for cyberbullying detection, called JRTT. It is the first publicly available dataset based on Toutiao with side information. The data content of JRTT set is closely related to COVID-19 and can better reflects the characteristics of the current era of cyberbullying.

(4) We conduct in-depth research in real-world applications of COVID-19 news comments. Experimental results on real-world dataset show that the HMMF achieves better performance than previous methods in cyberbullying detection.

The rest of the paper is organized as follows. Section 2 introduces the related work. Section 3 gives a formal definition of cyberbullying detection. Section 4 describes the proposed model in detail. Section 5 introduces the dataset and the experimental settings, present the contrast models and shows the experimental results. Section 6 concludes the paper with discussion.

2 Related Work

Cyberbullying detection on social media is a new research field. The existing researches on cyberbullying detection still have some limitations.

In Model-oriented, the existing researches mainly utilized traditional machine learning methods and deep learning algorithms to detect cyberbullying. Algorithms such as Support Vector Machine [19,20], Naive Bayes [21–23], Random Forest [20,24], Decision Tree [25], and Logistic Regression [19,21] were used to construct cyberbullying detection models. Models such as CNN [26], BiLSTM [27], and C-LSTM [28] were used to detect cyberbullying. However, few researches used knowledge-based advanced NLP models to perform automated cyberbullying detection [29]. In recent years, a large number of researches [14–18] had shown that the pre-trained models based on large corpus can learn the general language representation, which is conducive to the downstream NLP tasks. It can also avoid training the model from scratch.

In Feature-oriented, the existing researches mainly focused on identifying aggressive language through text analysis. They used feature representation methods, such as sentiment analysis [30], TF-IDF [10,13,31], and word embedding vector [2]. The existing researches invested a lot of energy to model text content generated by net users [32]. Both binary classification [12] and fine-grained classification [33,34] methods mainly focused on the features based on text content, while ignoring the side information. Therefore, it is a new research direction to improve the performance of cyberbullying detection through side information [35–37].

In Data-oriented, at present, there is no standard dataset for cyberbullying detection. Most studies [10,11,28,38,39] independently created datasets from the social media platforms (such as twitter and Youtube) by using public APIs . However, these datasets could not be compared with each other and were quite unbalanced. Less than 20% of the available samples were classified as cyberbullying. Most existing datasets are in English and a few of them are in other languages. For example, Dutch [20], Arabic [40] and Bangla [41]. Due to some problems in Chinese processing, such as polysemy and word vector pre-training, there are relatively few researches using Chinese datasets [42]. The existing publicly available datasets mainly include: Ask.fm [28], Formspring [38], Myspace [10], Twitter [11], and Toxicity [39]. The above datasets were produced before 2019, and they could not reflect the situation and characteristics of cyberbullying incidents in the current network society.

The above analysis shows that automated cyberbullying detection on social media is still an extremely challenging task. Aiming at the limitations of existing researches, we first construct a manually labeled Chinese cyberbullying detection dataset, and then propose a hybrid deep model with side information based on multi-feature fusion for cyberbullying detection.

3   Problem Definition

A Definition of Cyberbullying: In the past few years, many social and psychological researches have attempted to give an accurate definition of cyberbullying. However, there is no consensus on the definition of cyberbullying. Based on various concepts proposed in existing literature [36,40], we define cyberbullying as aggressive and intentional behavior committed by groups or individuals. The typical feature is to make offensive or hurtful remarks on social media, which are specifically manifested in abuse, slander, threat and ridicule. In this paper, we focus on cyberbullying, i.e. language violence.

Cyberbullying Detection: Generally, cyberbullying detection can be defined as automated detecting cyberbullying through deep learning technology using text features or other higher-order information features in social media data [36,37]. In this paper, cyberbullying detection on social media is defined as a binary classification problem, because our work focuses on providing a general detection benchmark rather than a multi-class classification.

In this section, we detail the mathematical definition of cyberbullying detection on social media. First, we introduce the mathematical definitions of the main components of comment and side information. Second, we give a formal definition of cyberbullying detection by referring to the mathematical definitions given by existing researches. We define the basic notations as follows.

- Let w denotes the *comment content*. It only consists of a major component: user's comments. In general, it refers to Chinese short news comments on social media.
- Let s denotes the *side information*. It contains two main components: side information of comments and side information of users. Side information of

comment s_c consists of a series of properties that contain the comment interaction, such as the number of people who agree with comments and the
degree of comments interaction. Side information of user s_u contains a list of
key characteristics that describe net users, such as user name, user personal
description, users' social history behavior statistics, and user interaction data.
– Let $E = \{e_1, e_2, ..., e_T\}$ denotes a corpus of T social media sessions. Each
social media session contains one comment $C = \{w_1, w_2, ..., w_m\}$ and the corresponding side information $S = \{s_1, s_2, ..., s_n\}$, where m and n represent the
number of words in the comment and the number of types of side information
respectively. We treat the cyberbullying detection as the binary classification problem. Each social media session e_i is associated with a binary label
$y \in \{0, 1\}$, where 0 represents non-bullying session, and 1 represents bullying
session.

Definition. *Given the side information of session s and corresponding comment
w, the target of cyberbullying detection is to automatically predict y for unlabeled
comment w, i.e., $F : s\&w \to y$ such that,*

$$F(s, w) = y \tag{1}$$

where F is the target model we want to obtain.

4   The Proposed Model

In this section, we fully illustrate the proposed **H**ybrid deep **M**odel based on
Multi-feature **F**usion (HMMF) for cyberbullying detection, which is composed
of three modules: (1) a *semantic context encoding* module that encodes text
content of comments and text category side information of session, (2) a *digital
data encoding* module for encoding digital side information of session, (3) a
session prediction module that integrates semantic context information features
with side information features into the final session representation, and uses it to
predict whether the session is bullying or non-bullying. The overall framework
of HMMF is shown in Fig. 1. Specifically, HMMF consists of the following three
parts:

(1) Semantic Context Encoding: Pre-trained model of BERT or its variants are
used to extract sentence embeddings of comment content and word embeddings of text category side information. The above sentence embeddings
and word embeddings are concatenated as the representation vector t.
(2) Digital Data Encoding: A fully connected layer followed by a Transformer
model are applied to encode digital side information. Finally, Max-Pooling
operation is applied to output a high-level representation vector d of digital
data.
(3) Session Prediction: We directly connect the output vectors of above two
modules to form the final feature vector v, and make the final prediction
through a fully connected layer.

Next, we introduce the detailed construction procedure of each module in
HMMF, and then present the training process for the whole framework.

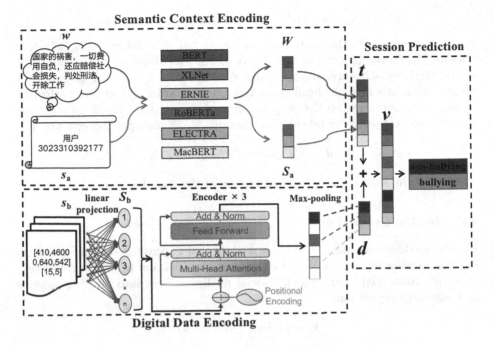

Fig. 1. The framework of HMMF.

4.1 Semantic Context Encoding

In the first module, the comment content w and text category side information s_a are used as input respectively, and then their pre-computed feature vectors are encoded by BERT or its variants. We get the feature vectors of dimension 768. The detailed process is:

$$t = Concat(M(w) \rightarrow W, M(s_a) \rightarrow S_a) \tag{2}$$

where W is a matrix of sentence embeddings representing comment content and S_a is a matrix of word embeddings representing text category side information. M is the pre-trained model.

4.2 Digital Data Encoding

In the second module, linear transformation is used to map digital side information s_b to higher dimensions. Then, we encode digital side information using Transformer, which uses a stack encoder consisting of two identical layers. Each layer includes a multi-head self-attention mechanism followed by a position-wise fully connected feedforward network. The formulation of multi-head self-attention mechanism with s_b can be defined as [43]:

$$Attention(Q, K, V) = softmax\left(\frac{QK^T}{\sqrt{d_k}}\right) V \tag{3}$$

$$MultiHead(S_b) = Concat(Attention_1, ..., Attention_h) \qquad (4)$$

where S_b, represents the matrix vector of digital side information, which is obtained by linear transformation. Q, K, V are its different subspace matrices respectively. d_k represents the dimension of vector Q and h represents the number of heads in multi-head self-attention mechanism. Next, we perform a max-pooling operation on the output vector of the encoder. Finally, we get the final representation of digital side information d by a max-pooling operation:

$$d = Maxpooling(r_1, r_2, ..., r_j) \qquad (5)$$

where r_i is the i-th column of output matrix of the encoder.

4.3   Session Prediction

Finally, we design a session prediction module that concatenates t and d to form the final session representation v for the final classification task. Then, v is sent to a fully connected layer to divide social media sessions into bullying sessions and non-bullying sessions.

$$Y = softmax(W^T v + b) \qquad (6)$$

where, $Y \in \mathbb{R}^n$, n is the number of classes equals 2, respectively represent bullying and non-bullying.

4.4   Training Process

As shown in Algorithm 1, we describe the HMMF training process. In each iteration of the algorithm, w and s_a are concatenated, and then the sentence embeddings W and word embeddings S_a are extracted by the pre-trained model of BERT or its variants (line 2 and 3, Eq. 2). After that, S_b is given by linear transformation (line 5). Representation vector d is computed through encoders (Eq. 3, Eq. 4, and Eq. 5). Representation vector t and d are concatenated to build the final session representation v (line 12). The final prediction Y is computed through fully connected layer (Eq. 6). Finally, once the training converges, the target model F is returned, which can be used for session prediction (line 16).

5   Experiments

In this section, we evaluate HMMF on a new publicly available benchmark dataset. We compare HMMF with a set of representative models. First, we introduce the real-world dataset JRTT, which is first published by us. Second, we describe the experimental settings including the contrast models and evaluation metrics. Finally, we present the experimental results and analyze them in detail from both macroscopic and microscopic perspectives.

Algorithm 1: Training procedure of HMMF.

Input: comment content w and side information s, and labels $y = \{0, 1\}$
Output: target model F

1 **for** *number of epoch* **do**
2 $\quad$ $w \to W$ through Pre-trained model of BERT or its variants;
3 $\quad$ $s_a \to S_a$ through Pre-trained model of BERT or its variants;
4 $\quad$ compute t according to Eq. 2: $t = Concat(M(w) \to W, M(s_a) \to S_a)$;
5 $\quad$ $s_b \to S_b$ linear transformation;
6 $\quad$ **foreach** *encoder* **do**
7 $\quad\quad$ compute Q, K, V according to Eq. 3 and Eq. 4:
8 $\quad\quad$ $Attention(Q, K, V) = softmax(\frac{QK^T}{\sqrt{d_k}})V$;
9 $\quad\quad$ $MultiHead(S_b) = Concat(Attention_1, ..., Attention_h)$;
10 $\quad$ **end**
11 $\quad$ compute d according to Eq. 5: $d = Maxpooling(r_1, r_2, ..., r_j)$;
12 $\quad$ integrate t and d into v: $v = t + d$;
13 $\quad$ compute Y according to Eq. 6: $Y = softmax(W^T v + b)$;
14 **end**
15 **if** *the training converges* **then**
16 $\quad$ $Calculate - Centers(s \& w : F)$;
17 **end**
18 **return** F;

5.1 JRTT: A New Benchmark Dataset

We collect, process and publish a new benchmark dataset, called JRTT. It is the first publicly available Chinese dataset based on Toutiao, which is the largest information platform in China. It ranks second in news Apps and has the most downloads. Therefore, it can be well used for cyberbullying detection research. The JRTT dataset includes 4016 manually annotated news comments from Toutiao COVID-19 special column. These comments come from 3875 different net users. There are two types of comments labels: bullying and non-bullying. The distribution of labels in the JRTT dataset is relatively well-balanced: 1833 bullying sessions and 2183 non-bullying sessions. The comments dates are from January 2021 to December 2021. Following the standard setting, we divide the dataset into three subsets, i.e., training set, validation set, and test set. They respectively account for 80%, 10%, and 10% of the entire dataset. The data of JRTT are described in detail as follows.

- **Comment content.** It includes only one data field: comment. The comments are mainly short sentences from net users, and they only involve one topic: COVID-19. We model the comments using the architecture described in Sect. 4.1.
- **Side information of comment.** It consists of two data fields: the number of likes on comments and the number of replies on comments. They represent whether other users agree with the comments and the level of interactions on

comment. It is digital side information about the comment. We combine and model them using the architecture described in Sect. 4.2.

- **Side information of user.** It consists of five data fields: username, number of Toutiao posted by user, the total number of likes, Number of fans and Number of concerns. They represent user's characteristics and profile information. It is primarily the digital side information about user. We model username using the architecture described in Sect. 4.1. We combine other four data fields and model them using the architecture described in Sect. 4.2.

To make a fair comparison, we run a series of comparative experiments on the real-world dataset JRTT, which is a new publicly available benchmark dataset for cyberbullying detection.

5.2 Experimental Settings

We use several Chinese pre-trained models, including BERT [14], XLNet [15], ERINE [16], RoBERTa[17], and MacBERT [18], to initialize sentence embeddings and word embeddings. The padding size is 256 and batch size is 10. The dropout rate is 30% and the learning rate is 0.00005. The optimizer selects Adam optimizer. The epoch parameter is set to 30. In Digital Data Encoding, the number of layers and the number of heads are set to 3.

Experimentally we compared HMMF with five representative models:

BERT. A new language representation model. The full name is Bidirectional Encoder Representations from Transformers. BERT is a language transformation model introduced by Google. It is a general language model trained on a very large corpus, and then used for NLP tasks. Therefore, using BERT requires two steps: pre-training and fine tuning. The pre-trained BERT model can create the most advanced model for eleven tasks. It uses the same architecture in different tasks.

XLNet. A generalized autoregressive pre-trained method for NLP that significantly. It improves upon BERT on 20 tasks and achieves the current state-of-the-art on 18 tasks. XLNet is a variant of BERT released by CMU and Google brain team in 2019. It learns the bidirectional contexts over all permutations of the factorization order and integrates the idea of the current optimal autoregressive model Transformer-XL.

ERNIE. Enhanced Representation through Knowledge Integration. ERNIE is the model first released by Baidu in 2019. It surpasses BERT and XLNet in 16 Chinese and English tasks and achieves state-of-the-art effect, especially in Chinese NLP tasks. ERNIE's advantage is that it learns the semantic representation of complete concepts in the real world through the learning of entity concept knowledge, and it extends training corpus, especially the introduction of forum dialogue corpus, enhances the semantic representation ability of the model.

RoBERTa. A Robustly Optimized BERT Pretraining Approach. Roberta has mainly improved BERT in three aspects: one is to improve the optimization function; the other is to use the dynamic mask to train the model, which proves the shortcomings of NSP (next sense prediction) training strategy and adopts a larger batch size; the third is to use a larger dataset for training and BPE (byte pair encoding) to process text data.

MacBERT. A new pre-trained language model, which replaces the original MLM task into MLM as correction (Mac) task and mitigates the discrepancy of the pretraining and fine-tuning stage. MacBERT is the model released by Harbin Institute of Technology SCIR Laboratory. It achieves state-of-the-art performances on many NLP tasks.

Performance Metrics: To evaluate the model, we utilize the standard metrics in classification, i.e., accuracy, precision, recall and F1-score. Existing researches [35,37,40] mainly use F1-score as an evaluation metrics to evaluate cyberbullying detection models. For binary classification problems, such metrics are easy to obtain. In order to compare the models with each other, we also use AUC score (Area Under Curve) to evaluate the models. Most researches use AUC scores [28], which is the criteria specified in many classification challenges.

5.3   Experimental Results and Analysis

In this section, we empirically evaluate the proposed model HMMF on JRTT, a new publicly available benchmark dataset. Through a series of experiments, we demonstrate the effectiveness of the model based on multi-feature fusion. Particularly, we answer the following research questions.

- *Q1*. How does the proposed model perform on cyberbullying detection?
- *Q2*. Can the various features generated by side information improve the detection performance?

To answer questions *Q1* and *Q2*, we compare HMMF with above five representative models. The comparison results on JRTT are shown in Table 1, 2, 3, 4 and Table 5. Among them, data sources D1 is comment content, D2 is user name, and D3 includes all other side information. For a fair experimental comparision, HMMF and the corresponding comparison model use the same pre-trained model.

For question *Q1*, compared with five representative models, HMMF achieves the best detection performance in accuracy, precision, recall, F1-score and AUC. Among them, HMMF based on MacBERT and ERNIE pre-trained model perform best. For instance" the F1-score of HMMF increased by an average of 2.64%, among which the HMMF based on RoBERTa pre-trained model increased the most, reaching 6.50%.

For question *Q2*, the experimental results fully demonstrate that the fusion of comment content features and side information features can improve the performance of cyberbullying detection. Specifically, when we fuse D2 and D3 based on D1, the detection performance of all models becomes better. The reasons for the better performance of HMMF are as follows: Firstly, HMMF can extract sentence embeddings of comment content and word embeddings of text category side information. Secondly, using Transformer model, we can make better use of digital side information. Thirdly, besides news comment, the side information of session contains effective features, which can be used to improve the performance of cyberbullying detection. Next, we study a specific case to illustrate the importance of side information features.

A Case Study: RoBERTa mispredicts some comments as non-bullying, which are: "为人师表, 只是口头上的, 实际做法大相径庭！", "公布信息吧, 让大家看看这两位老师是怎么为人师表的", and "现在的老师有几个有师德的". However, HMMF correctly predicts them as bullying through the features generated by side information. The above comments do not contain malicious words, but are comments in the form of irony or rhetorical questions. Therefore, it is difficult to predict cyberbullying only by comment content. At this time, side information is needed to aid detection. Moreover, through this case, we can find that these bullying comments do not concern COVID-19, but express dissatisfaction with specific identity groups.

Table 1. Performance comparison based on BERT pre-trained model.

Model	BERT		HMMF	
Data source	D1	D1+D2	D1+D3	D1+D2+D3
Accuracy	91.78%	95.47%	94.62%	**95.75%**
Precision	91.79%	95.55%	94.62%	**95.77%**
Recall	91.78%	95.47%	94.62%	**95.75%**
F1-score	91.78%	95.46%	94.61%	**95.75%**
AUC	91.73%	95.36%	94.62%	**95.78%**

Table 2. Performance comparison based on XLNet pre-trained model.

Model	XLNet		HMMF	
Data source	D1	D1+D2	D1+D3	D1+D2+D3
Accuracy	92.35%	93.20%	92.63%	**93.77%**
Precision	92.38%	93.24%	92.66%	**93.77%**
Recall	92.35%	93.20%	92.63%	**93.77%**
F1-score	92.35%	93.19%	92.63%	**93.77%**
AUC	92.28%	93.12%	92.67%	**93.73%**

Table 3. Performance comparison based on ERNIE pre-trained model.

Model	ERNIE		HMMF	
Data source	D1	D1+D2	D1+D3	D1+D2+D3
Accuracy	96.32%	**97.17%**	96.60%	**97.17%**
Precision	96.35%	**97.19%**	96.63%	97.17%
Recall	96.32%	**97.17%**	96.60%	**97.17%**
F1-score	96.31%	**97.17%**	96.60%	**97.17%**
AUC	96.25%	**97.21%**	96.64%	97.16%

Table 4. Performance comparison based on RoBERTa pre-trained model.

Model	RoBERTa		HMMF	
Data source	D1	D1+D2	D1+D3	D1+D2+D3
Accuracy	89.80%	94.05%	95.75%	**96.32%**
Precision	89.95%	94.07%	95.88%	**96.35%**
Recall	89.80%	94.05%	95.75%	**96.32%**
F1-score	89.81%	94.05%	95.75%	**96.31%**
AUC	89.90%	94.08%	95.85%	**96.25%**

Table 5. Performance comparison based on MacBERT pre-trained model.

Model	MacBERT		HMMF	
Data source	D1	D1+D2	D1+D3	D1+D2+D3
Accuracy	95.47%	95.47%	97.17%	**97.45%**
Precision	95.50%	95.53%	97.17%	**97.49%**
Recall	95.47%	95.47%	97.17%	**97.45%**
F1-score	95.47%	95.47%	97.17%	**97.45%**
AUC	95.51%	95.53%	97.16%	**97.39%**

5.4   Repeatability

The experimental equipment is configured as 128 GB memory and a GeForce RTX 2080 GPU. Source code is available at https://github.com/xingjian215/HMMF.

6 Conclusion

In this paper, we present a new hybrid deep model HMMF for cyberbullying detection, which can learn useful representations from both comment contents and side information to boost the detection performance. Inspired by the widely used pre-trained model in NLP technology, we utilize several Chinese pre-trained models for encoding the semantic context and utilize Transformer for encoding the digital data synthetically to mining more effective features contained in side information. We introduce JRTT, a new benchmark dataset for automated cyberbullying detection. JRTT's authentic, real-world comments on COVID-19 from diverse net users can fully reflect the characteristics of cyberbullying and promote the further development of research. Experimental results further demonstrate that HMMF achieves new SOTA performance in the real-world application of COVID-19 news comments. In the future, we can mine more side information features, such as net user profile, gender, historical comment information, etc.

Acknowledgement. This work was supported by the National Natural Science Foundation of China (Grant U2003111, 61871378 and Grant U1803263).

References

1. Hee, C., et al.: Automatic detection of cyberbullying in social media text. PLoS One **13**(10), e0203794 (2018)
2. Rosa, H., et al.: Automatic cyberbullying detection: a systematic review. Comput. Hum. Behav. **93**, 333–345 (2019)
3. Lin, Y., et al.: Psychological intervention for three COVID patients who suffered online violence. Chin. J. Psychiat. 239–242 (2021)
4. Emmery, C., et al.: Current limitations in cyberbullying detection: on evaluation criteria, reproducibility, and data scarcity. Lang. Res. Eval. **55**(3), 597–633 (2020). https://doi.org/10.1007/s10579-020-09509-1
5. Parma, N., et al.: How bullying is this message: a psychometric thermometer for bullying. In: COLING, pp. 695–706 (2016)
6. Walisa R., et al. Automated cyberbullying detection using clustering appearance patterns. In: IEEE International Conference on Knowledge and Smart Technology, pp. 242–247 (2017)
7. Gutiérrez-Esparza, et al. Classification of cyber-aggression cases applying machine learning. Appl. Sci. **9**(9), 1828 (2019)
8. Ducharme, D.N.: Machine Learning for the Automated Identification of Cyberbullying and Cyberharassment. Open Access Dissertations (2017)
9. Haidar, B., et al.: A multilingual system for cyberbullying detection: Arabic content detection using machine learning. Adv. Sci. Technol. Eng. Syst. J. **2**(6), 275–284 (2017)
10. Sugandhi, R., et al.: Automatic monitoring and prevention of cyberbullying. Int. J. Comput. Appl. **8**, 17–19 (2016)
11. Zhao, R., et al.: Cyberbullying detection based on semantic-enhanced marginalized denoising auto-encoder. IEEE Trans. Affect. Comput. **8**(3), 328–339 (2016)

12. Zhao, R., et al.: Automatic detection of cyberbullying on social networks based on bullying features. International Conference on Distributed Computing and Networking, pp. 1–6 (2016)
13. Hosseinmardi, H., et al.: Prediction of cyberbullying incidents in a media-based social network. In: International Conference on Advances in Social Networks Analysis and Mining, pp. 186–192 (2016)
14. Devlin, J., et al.: BERT: Pre-training of Deep Bidirectional Transformers for Language Understanding. arXiv:1810.04805 (2018)
15. Yang, Z., et al.: XLNet: Generalized Autoregressive Pretraining for Language Understanding. arXiv:1906.08237 (2020)
16. Zhang, Z., et al.: ERNIE: enhanced language representation with informative entities. In: Proceedings of the 57th Annual Meeting of the Association for Computational Linguistics (2019)
17. Liu, Y., et al.: RoBERTa: A Robustly Optimized BERT Pretraining Approach. arXiv:1907.11692 (2019)
18. Cui, Y., et al.: Revisiting pre-trained models for Chinese natural language processing. In: EMNLP (2020)
19. Chavan, V.S., et al.: Machine learning approach for detection of cyber-aggressive comments by peers on social media network. In: ICACCI, pp. 2354–2358 (2015)
20. Van Hee, C., et al.: Detection and fine-grained classification of cyberbullying events. In: RANLP, pp. 672–680 (2015)
21. Mangaonkar, A., et al.: Collaborative detection of cyberbullying behavior in Twitter data. In: EIT, pp. 611–616 (2015)
22. Sanchez, H., et al.: Twitter bullying detection. J. Jpn. Assoc. Periodontol. (2011)
23. Dinakar, K., et al.: Modeling the detection of textual cyberbullying. In: ICWSM, vol. 5, no. 3, pp. 11–17 (2011)
24. García-Recuero, Á.: Discouraging abusive behavior in privacy preserving online social networking applications. In: Proceedings of the 25th International Conference Companion on World Wide Web, pp. 305–309 (2016)
25. Reynolds, K., et al.: Using machine learning to detect cyberbullying. In: ICMLA, vol. 2, pp. 241–244 (2011)
26. Akhter, M.P., et al.: Abusive language detection from social media comments using conventional machine learning and deep learning approaches. Multimedia Syst. 1–16 (2021)
27. Agrawal, S., Awekar, A.: Deep learning for detecting cyberbullying across multiple social media platforms. In: Pasi, G., Piwowarski, B., Azzopardi, L., Hanbury, A. (eds.) ECIR 2018. LNCS, vol. 10772, pp. 141–153. Springer, Cham (2018). https://doi.org/10.1007/978-3-319-76941-7_11
28. Rosa, H., et al.: A deeper look at detecting cyberbullying in social networks. In: IJCNN, pp. 1–8 (2018)
29. Hong, F., et al.: Social media toxicity classification using deep learning: real-world application UK Brexit. Electron. 10(11), 1332 (2021)
30. Walaa M., et al. Sentiment analysis algorithms and applications: a survey. Ain Shams Eng. J. 5(4), 1093–1113 (2014)
31. Perera, A., et al.: Accurate cyberbullying detection and prevention on social media. Procedia Comput. Sci. 181, 605–611 (2021)
32. Dani, H., Li, J., Liu, H.: Sentiment informed cyberbullying detection in social media. In: Ceci, M., Hollmén, J., Todorovski, L., Vens, C., Džeroski, S. (eds.) ECML PKDD 2017. LNCS (LNAI), vol. 10534, pp. 52–67. Springer, Cham (2017). https://doi.org/10.1007/978-3-319-71249-9_4

33. Hee, C., et al.: Detection and fine-grained classifification of cyberbullying events. In: International Conference Recent Advances in Natural Language. RANLP, pp. 672–680 (2015)
34. Honnibal, M., et al. spaCy 2: natural language understanding with Bloom embeddings, convolutional neural networks and incremental parsing. Appear, **7**(1), 411–420 (2017)
35. Dadvar, M., Trieschnigg, D., Ordelman, R., de Jong, F.: Improving cyberbullying detection with user context. In: Serdyukov, P., Braslavski, P., Kuznetsov, S.O., Kamps, J., Rüger, S., Agichtein, E., Segalovich, I., Yilmaz, E. (eds.) ECIR 2013. LNCS, vol. 7814, pp. 693–696. Springer, Heidelberg (2013). https://doi.org/10.1007/978-3-642-36973-5_62
36. Ge, S., et al. Improving Cyberbully Detection with User Interaction. arXiv:2011.00449 (2020)
37. Chen, H.Y., et al.: HENIN: learning heterogeneous neural interaction networks for explainable cyberbullying detection on social media. In: EMNLP (2020)
38. Zhang, X., et al.: Cyberbullying detection with a pronunciation based convolutional neural network. In: ICMLA, pp. 740–745 (2017)
39. Thain, N., et al.: Wikipedia Talk Labels: Toxicity
40. Khairy, M.M, Tarek, M., Abd-El-Hafeez, T.: Automatic detection of cyberbullying and abusive language in Arabic content on social networks: a survey - scienceDirect. Procedia Comput. Sci. **189**, 156–166 (2021)
41. Ahmed, M.F., et al.: Cyberbullying Detection Using Deep Neural Network from Social Media Comments in Bangla Language (2021)
42. Liu, Z., et al.: Detection and analysis of cybernetics bullying language on common Chinese social network platforms. J. Southwest China Normal Univ. **46**(8), 86–94 (2021)
43. Ashish, V., et al.: Attention Is All You Need. arXiv:1706.03762 (2017)

Designing a Training Set for Musical Instruments Identification

Daniel Kostrzewa(✉)(iD), Blazej Koza, and Pawel Benecki(iD)

Department of Applied Informatics, Silesian University of Technology,
Gliwice, Poland
{daniel.kostrzewa,pawel.benecki}@polsl.pl

Abstract. This paper presents research on one of the most challenging branches of music information retrieval – musical instruments identification. Millions of songs are available online, so recognizing instruments and tagging them by a human being is nearly impossible. Therefore, it is crucial to develop methods that can automatically assign the instrument to the given sound sample. Unfortunately, the number of well-prepared datasets for training such algorithms is very limited. Here, a series of experiments have been carried out to examine how the mentioned methods' training data should be composed. The tests were focused on assessing the decision confidence, the impact of sound characteristics (different dynamics and articulation), the influence of training data volume, and the impact of data type (real instruments and digitally created sound samples). The outcomes of the tests described in the paper can help make new training datasets and boost research on accurate classifying instruments that are audible in the given recordings.

Keywords: Music information retrieval · Musical instruments identification · Dataset design · Training data · Analog sound · Digital sound

1 Introduction

One of the most challenging parts of music information retrieval is identifying the musical instrument. In order to fulfill this goal, many algorithms for automatic instrument recognition were developed. Different machine learning strategies can be used for automatic instruments classification. They can be entirely computational [4,5] or perceptual [17,18]. The task can be carried out with classic classifiers [6,7], by analyzing fundamental frequency [14], the usage of hidden Markov models [10], and based on deep neural networks [12,21]. In this research, the simple comparison of the distance of mel-frequency cepstral coefficient (MFCC) vectors was used [11].

This work was supported by research funds for young researchers of the Department of Applied Informatics, Silesian University of Technology, Gliwice, Poland (grant no. 02/0100/BKM22/0021 – DK and grant no. 02/0100/BKM22/0023 – PB).

D. Groen et al. (Eds.): ICCS 2022, LNCS 13350, pp. 599–610, 2022.
https://doi.org/10.1007/978-3-031-08751-6_43

The task of automatic musical instruments recognition has a strong practical justification. People often like listening to similar pieces of music. This is dictated by their musical taste and current mood. As the Internet allows access to millions of tracks, automatic instrument recognition has become a necessity. However, creating high-quality classification methods is not enough, it is crucial to provide appropriate training data to achieve the highest results.

1.1   Related Work

As the possibilities and popularity of deep learning methods have grown in recent years, the need to create a well-prepared dataset for training such algorithms has become urgent. Unfortunately, the task of instrument classification is undertaken by researchers exceptionally rarely due to its difficulty. Therefore, the conducted studies are performed on various, far from being perfect, databases.

There is no single recognized dataset for the task of identifying musical instruments. However, there are several collections worthy of attention. The NSynth Dataset [9], TinySOL [8], and Good-sounds [2,19] contain 305979, 2913, 8750 single annotated notes, respectively, which is not enough (because of having separate notes) for train meaningful model. These collections provide some limited information about the dynamics, type of play, and quality. Medley DB [3] collects 122 songs with information about instruments, while Medley-solos-DB [1,15] contains 21571 audio clips of 8 instruments with a fixed duration of 2972 ms, that is, 65536 discrete-time samples. However, it does not offer information about the different playing techniques, dynamics, and articulation. Last but not least, the OpenMIC-2018 dataset [13] contains 2000 excerpts (10s each) from 20 instruments with no additional information about dynamics and articulation.

Moreover, according to our best knowledge, there is no research on the impact of different aspects of designing training dataset for musical instruments recognition methods. We believe that broadening training datasets with recordings with different dynamics, articulation, data sources (real instruments and digitally created sound samples) will allow for significant improvement in the quality of musical instruments recognition.

1.2   Contribution and Paper Structure

The main goal of the research is to determine the best approach for designing a training set for the task of musical instruments recognition from an audio signal. Several key aspects have been thoroughly examined, i.e., the base confidence of the classification, the impact of the different sound characteristics in the training set, the influence of training data volume, and the impact of data type (analog and digital instruments). Moreover, a simple method for automatic identification of musical instruments was developed. This strategy consists of frequency detection, parameterizing the sound, and classifying it. The analysis is based on samples of five real musical instruments: piano, clarinet, alto saxophone, trumpet, and accordion. All sound samples come from the authors' private recordings. To ensure reliable results, the sound samples have been recorded with different articulation and dynamics to reproduce the instrument's pattern as accurately as possible.

The proposed method created for musical instruments identification is presented in Sect. 2. Section 3 describes the used dataset, the conducted research, and the obtained results, while the summary and the final conclusions have been included in Sect. 4.

2 Identification of Musical Instruments

2.1 Extraction of Sound Features

The critical issue in designing a method that enables automatic recognition of the instrument's timbre is the digital representation of sound characteristics, which can be appropriately processed and classified. The way the sound recordings of musical instruments are parameterized has a great influence on the final effectiveness of the algorithms that classify the sound. It is necessary to determine the vector of sound characteristics, whose size, i.e., the number of parameters determining the sound, should be as small as possible. This will allow the representation of strongly correlated parameters to reduce the computational complexity and more accurate ordering of information. This is important because the recordings are easy to be compared with each other.

In digital tone analysis, one of the most frequently used sound parameterization methods is the usage of Mel-frequency cepstral coefficients (MFCCs) [20]. MFCCs are computed in a following steps:

1. Calculate the Fourier transform of the windowed part of the signal.
2. Map the powers of the spectrum obtained to the mel scale using overlapping triangular windows or cosine windows.
3. Perform logarithms of the powers for each mel frequency.
4. Perform a discrete cosine transform of the list of logarithmic mel powers as if it were a signal.
5. The MFCC values are the amplitudes of the resulting spectrum.

In the developed method, the input audio signal's conversion into the MFCC feature vector is done using the Accord.NET library. The resulting vector (i.e., a vector representing an unknown musical instrument) is compared with the vectors obtained from the reference sounds (i.e., all reference vectors of known musical instruments). Distances calculated between the vector of the unknown instrument and reference vectors for each musical instrument are averaged, and this average value is the final distance between the vector representing the unknown instrument and instruments stored in the dataset.

In order to visualize the results, a set of graphs from MFCC vectors was created. For each graph, the horizontal axis determines the frequencies, and the vertical axis indicates the signal power of a particular frequency. The values on the axes do not correspond to the real values of frequencies and decibels. The scaling of the graph is done automatically by the Accord.NET library. Therefore, the graphs are only used to illustrate the sound vector of the instrument (Fig. 1).

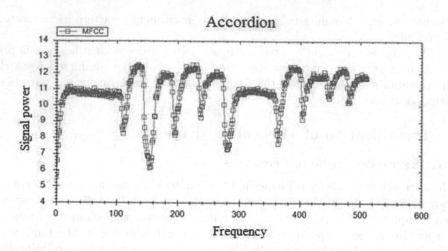

Fig. 1. Sample reference vector for instrument used in experiments.

2.2   Classification Method

Appropriate data classification is another element necessary for automatic recognition of the timbre of musical instruments. Artificial neural networks [12, 16, 21], minimal-distance algorithms including the closest neighbor method [4, 6], are the most often used for this purpose. Since the main goal of this research is to determine the best approach for designing a training set for the task of musical instruments recognition from an audio signal, only the simple closest neighbor algorithm is employed for the task of classification. The comparison consists of finding the absolute distance between the input sound vector and individual pattern vectors (reference sound vectors). The lowest average distance between the input sound vector and patterns of the given instrument means matching the input sample to this particular instrument. The distance between two n-dimensional vectors is calculated from the Euclidean distance:

$$|u - v| = \sqrt{(u_1 - v_1)^2 + (u_2 - v_2)^2 + ... + (u_n - v_n)^2} \tag{1}$$

where u and v are vectors belonging to the space R^n and represent two different sounds in the MFCC domain.

3   Experiments

3.1   Dataset and Hardware Setup

It was decided to analyze various sounds with different properties in the conducted research. It is also important to compare effectiveness of the classification method on instruments with similar sounds (belonging to the same group, e.g., woodwind) and instruments belonging to different groups. Therefore, it was decided to use five instruments, which will allow us to evaluate the method and designing training set procedures by different aspects: alto saxophone, clarinet, trumpet, accordion, and piano. Clarinet and saxophone, belonging to the same woodwind group, allow testing the classification of similarly sounding instruments (which can be difficult to distinguish). The trumpet is another instrument of the wind group, but the way it produces sound is different from the clarinet or saxophone and therefore belongs to the brass family. The next two instruments are the piano and the accordion. The piano is an instrument belonging to the chordophones group, where the sound is produced by hitting a hammer on a string. In the accordion, the sound, as in the wind instruments, is produced by air flowing from bellows through reeds activated using a keyboard. The possibility of playing chords determined the choice of the last two instruments. This will make it possible to test the effectiveness of polyphonic and monophonic sounds. Also, various possible sounds of accordions (they have so-called *registers*, which allow changing the timbre of the sound) were recorded to create a robust instrument reference vector.

The first phase of data collection was the recording of instrument sound samples. This was done using the RODE NT2-A microphone, the Focusrite Scarlett 2i2 analog-to-digital converter, and the free software for multi-track sound recording – Audacity. All samples were recorded in a 44.1 kHz .wav format with a resolution of 16 bits. All instruments' sounds were recorded in one room and on an identical hardware configuration to ensure the highest possible reliability of the experiment. To reflect the sample sound, identical sequences were recorded for each of the five instruments: ranges, passages, melodies, and long sounds. In the case of accordion and piano, apart from single sounds, chords were used. Each recording was made with different dynamics and articulation. A total of 70 recordings were made for each instrument, 60 were used as the training set, and 10 – as the test set.

The dataset consisting of instrumental sound recordings was additionally enriched with recordings from digital sound synthesizers (Logic Pro X application on Mac OS) in order to be able to classify both types of sound sources for each instrument correctly. This allows generating sounds of many instruments with different dynamics and articulation. For each of the five instruments tested, twenty longer samples were recorded, which were divided into smaller fragments for each experiment at the testing phase. As a result, a total of 450 recordings were prepared (real instruments and synthesizers together).

3.2   Assessment of the Decision Confidence

The first part of the experiments was to determine the overall assessment of the proposed method's confidence. This was necessary to determine whether the algorithms were implemented correctly and whether further research makes sense. It was assumed that the tested sound sample belongs to a given instrument if the distance between the reference vector, i.e., averaged MFCC vector for the given instrument, and the tested instrument vector is the smallest. The result of the analysis process is the distance between reference and sample vectors.

The determination of the decision confidence required a relative percentage of confidence based on the distance obtained. For this purpose, all the distances between vectors were collected, and the smallest and the largest were selected. The lowest value was determined as 100% confidence, and the highest value corresponds to 0%. Since simple accuracy shows only classification results, we introduced a more insightful metric. The idea behind creating such a metric shows how instruments can be differentiated from one another.

In this experiment, all fifty test recordings were used (ten from each instrument), and an average of 60% of the decisions confidence was achieved. The highest confidence was obtained for the clarinet (79%). The piano, accordion, and trumpet were recognized with confidence of 65%, 61%, and 57%, respectively. The saxophone has lowest confidence – 38%.

3.3   Impact of Sound Characteristics

The next part of the research was to find the optimal method configuration and training samples to build a universal sound pattern for a given instrument. The detection accuracy was tested depending on the constants defining the dimension of a single vector of sound signal features and the number of vectors in a two-dimensional MFCC array. In addition, the influence of the characteristics of the training recordings on the results was analyzed. The sound samples were sorted into groups with different dynamics, articulation, and pitch. The values of constants determining the dimensions of the MFCC matrix were set accordingly: 512 – the dimension of a single vector of sound signal features and 13 – the number of vectors in the MFCC object.

The sound emitted by the instrument depends on the musician's style of playing, habits, or the genre of music performed. For example, the characteristics of a ballad melody are very different from rock or pop music. This is due to various dynamics and articulations in sound. In this experiment, the influence of differentiation of the dataset in terms of sound characteristics on the proposed method's accuracy was observed. The data were divided into three groups: X1 – long sounds with similar dynamics and articulation, X2–X1 set enriched with samples with different dynamics, and X3–X1 and X2 sets with added recordings with different articulation. Training sets contained 20 samples each (for each instrument). Importantly, all three groups (i.e., X1, X2, and X3) have 20 samples of each instrument, however, their composition is different. 10 separate recordings (for each instrument) were used for testing. The influence of data differentiation on detection accuracy was observed, and the results of this experiment are presented in Table 1.

After enriching the dataset with samples of different dynamics, the average accuracy increased by 2.2%. The most significant improvement was achieved for the accordion and trumpet. In wind instruments, where a stronger airflow causes higher dynamics, the timbre changes, it is sharper and clearer. In the case of the piano, the change of dynamics does not significantly affect the timbre, and the accuracy of the proposed method has not increased.

Table 1. The impact of dataset diversity on the proposed method accuracy.

Instrument	X1	X2	X3
Clarinet	58%	59%	65%
Saxophone	30%	32%	33%
Trumpet	49%	53%	53%
Piano	63%	63%	65%
Accordion	55%	59%	59%
Average	**51%**	**53.2%**	**55%**

The enrichment of the training set with various articulation recordings allowed to increase the average accuracy by 1.8%. The recordings were made in the following techniques: staccato (separate sounds, with shortened values), legato (sounds played smoothly, without any breaks between them), glissando (smooth transition from one sound to another) and, in the case of accordion and piano, arpeggio (chord broken into a sequence of notes). The highest accuracy increase (6%) was achieved for the clarinet, while the trumpet and accordion were recognized with the same accuracy as for the X2 set.

3.4 The Influence of Training Data Volume

The variety of instrument sounds is practically endless. The sound of one instrument may vary in many ways and for many reasons. These include the previously described sound characteristics, the musician's playing style, and the materials used to make the instrument. The developed method compares the MFCC signal vector with a previously prepared pattern that is a MFCC vector of individual recordings with different articulation and dynamics of a given instrument. For this reason, the sound patterns of instruments should be as varied as possible, which requires much training data.

In this experiment, the number of training samples on the effectiveness of method accuracy was examined. While maintaining the diversity of sound characteristics, six datasets of sizes 1, 5, 10, 20, 40, and 60 samples were prepared (for each instrument). For the tests, 10 separate recordings (for each instrument) were used. The results obtained in this experiment are presented in Table 2.

Table 2. The influence of the amount of training data on the effectiveness of the proposed method.

Instrument	Y1 – 1	Y2 – 5	Y3 – 10	Y4 – 20	Y5 – 40	Y6 – 60
Clarinet	25%	40%	50%	65%	70%	75%
Saxophone	15%	20%	30%	35%	45%	50%
Trumpet	30%	40%	40%	50%	60%	60%
Piano	20%	40%	45%	65%	75%	75%
Accordion	55%	55%	55%	60%	75%	80%
Average	**29%**	**39%**	**44%**	**55%**	**65%**	**68%**

The obtained results of the proposed method accuracy proved to be strongly dependent on the size of the training dataset. The lowest accuracy was achieved for the Y1 set, where for each instrument, the dataset contained only one sample. The obtained results ranged from 15–55%, and the average recognition was 29%. The lowest 15% result was achieved for the saxophone, which was most often mistaken with a clarinet. This may be since these are instruments from the same family (i.e., woodwind).

The gradual enrichment of the dataset allowed for a considerable improvement in decision-making quality. For the last dataset (Y6), the method recognized instruments with an average accuracy of 68%. The improvement varies from 25% for the accordion up to 55% for the piano. The lowest average performance difference (3%) was observed between the Y5 and Y6 datasets, it can be concluded that the size of the Y6 dataset allows for an impeccable reproduction of the timbre pattern of the instruments and high accuracy sound recognition. A small difference in accuracy may also indicate the upper limit of the amount of data needed to reproduce the timbre pattern.

3.5   The Impact of Data Type

Modern technology allows for sound production through analog musical instruments and the use of software for music production, sound generators, and synthesizers. Training data for timbre recognition cannot be limited to analog instruments only because the pace of digital technology development and the development of electronic music genres can make such a method less and less useful over time.

As a result, it was decided to enrich the set of training data with digital instruments samples and investigate the effectiveness of recognizing such sounds. In the first part of the study, the accuracy of recognizing digital instrument sounds on a set of training data recorded on analog instruments was examined. For this purpose, the Y6 set from the previous experiment with 60 samples size was used. The results obtained during the conducted research are shown in Table 3 (column a), which compares the obtained results with those from the previous experiment (column b).

The average recognition accuracy of digital instruments is 16% lower. This may indicate poor timbre reproduction of instruments by synthesizers. The most significant timbre deviation was observed for the accordion, where the difference was 50%. In the case of saxophone and piano, the accuracy of the method did not change and remained at 50% and 75%, respectively.

Another aim of the experiment was to add samples of digital instruments to the training data and retest the effectiveness of the classification. The dataset was enriched with 10 samples of digital instrument sounds, and the results are presented in Table 3 (column c). The outcomes obtained show that training data from various sources increase the effectiveness of the developed method significantly. After extending the training dataset with samples of digital instruments, an increase of 17% in average accuracy was noted. The final average accuracy (69%) is higher than the average accuracy obtained during the previous experiment, where only analog instruments were used.

Table 3. Performance of proposed method for sounds of different types.

Instrument	Analog training set		Analog and digital training data
	a) Digital sounds	b) Analog sounds	c) Digital sounds
Clarinet	65%	75%	75%
Saxophone	50%	50%	65%
Trumpet	40%	60%	60%
Piano	75%	75%	75%
Accordion	30%	80%	70%
Average	**52%**	**68%**	**69%**

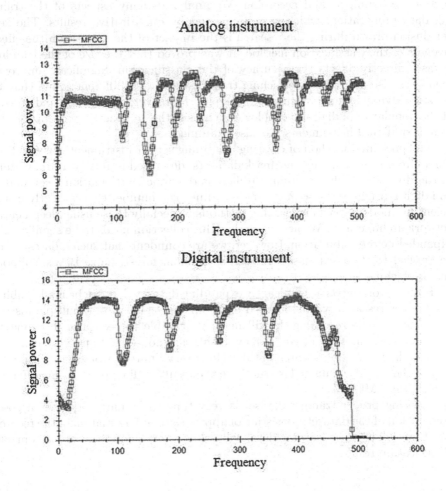

Fig. 2. The comparison of analogue and digital accordion MFCC vectors.

This research proved the difference between traditional musical instruments and electronic sounds, which is visible in Fig. 2. Ultimately, it can be stated that the diversity of the dataset of samples of sounds from different sources allows achieving greater accuracy of the developed method.

4   Conclusions

In this paper, we developed a strategy and gave insights how to build a training dataset to identify musical instruments automatically. This strategy can be treated as the pathway to creating an extensive and robust dataset that could be used in a data-centric AI approach. Moreover, the simple classification method was proposed based on comparing the MFCC values. Experiments were carried out for the real instruments belonging to different groups: piano, clarinet, alto saxophone, trumpet, and accordion. We analyzed many aspects of the training data preparation, which were supported by quantitative results. The key conclusion drawn during the research is the impact of the dataset on the effectiveness of the classification method. It was proved that the size of the training dataset directly affects the accuracy of the classification. Significant improvement was also achieved by enriching training data with different sound characteristics (dynamics and articulation) samples. Moreover, due to the rapid growth of the popularity of digital technologies in music, the accuracy in recognizing the sounds of digital instruments was also examined.

The presented method of creating a training set for instrument classification differs from the available benchmark datasets, described in related work, mainly in the diversity of the recordings (different dynamics, articulation, and analog and digital origin of the sound). 450 recordings is a number that is clearly insufficient for modern machine learning methods, especially those using deep neural network architectures. We are aware that this collection needs to be significantly expanded before publication. However, we are confident that such a dataset can be applied to standard classifiers such as random forest, naive Bayes, support vector machine, etc.

Besides working on a significantly expanding dataset that can be made public, our further research will be focused on unmixing signals from multiple instruments in a single recording. It will make it possible to recognize instruments from authentic musical pieces. However, this is a non-trivial and very complex task, so the number of research in this field is very limited. Moreover, it is necessary to thoroughly examine how sensitive the results will be to choosing different metrics than MFCCs.

Creating proper training dataset is very time-consuming. However, a good preparation of the dataset, consisting of appropriate differentiation of the recordings that will serve as learning data, will achieve very good musical instruments identification results.

References

1. Andén, J., Lostanlen, V., Mallat, S.: Joint time-frequency scattering. IEEE Trans. Sig. Process. **67**(14), 3704–3718 (2019)
2. Bandiera, G., Romani Picas, O., Tokuda, H., Hariya, W., Oishi, K., Serra, X.: Good-sounds.org: a framework to explore goodness in instrumental sounds. In: International Society for Music Information Retrieval Conference, pp. 414–419 (2016)
3. Bittner, R.M., Wilkins, J., Yip, H., Bello, J.P.: Medleydb 2.0: New data and a system for sustainable data collection. ISMIR Late Breaking and Demo Papers (2016)
4. Brown, J.C.: Computer identification of musical instruments using pattern recognition with cepstral coefficients as features. J. Acoust. Soc. Am. **105**(3), 1933–1941 (1999)
5. Brown, J.C., Houix, O., McAdams, S.: Feature dependence in the automatic identification of musical woodwind instruments. J. Acoust. Soc. Am. **109**(3), 1064–1072 (2001)
6. Chakraborty, S.S., Parekh, R.: Improved musical instrument classification using cepstral coefficients and neural networks. In: Mandal, J.K., Mukhopadhyay, S., Dutta, P., Dasgupta, K. (eds.) Methodologies and Application Issues of Contemporary Computing Framework, pp. 123–138. Springer, Singapore (2018). https://doi.org/10.1007/978-981-13-2345-4_10
7. Chandwadkar, D., Sutaone, M.: Role of features and classifiers on accuracy of identification of musical instruments. In: National Conference on Computational Intelligence and Signal Processing, pp. 66–70. IEEE (2012)
8. Emanuele, C., Ghisi, D., Lostanlen, V., Lévy, F., Fineberg, J., Maresz, Y.: TinySOL: an audio dataset of isolated musical notes (2020). https://zenodo.org/record/3685367
9. Engel, J., et al.: Neural audio synthesis of musical notes with wavenet autoencoders (2017)
10. Eronen, A.: Musical instrument recognition using ICA-based transform of features and discriminatively trained HMMs. In: International Symposium on Signal Processing and Its Applications, vol. 2, pp. 133–136. IEEE (2003)
11. Gulhane, S.R., Shirbahadurkar Suresh, D., Badhe Sanjay, S.: Identification of musical instruments using MFCC features. In: Smys, S., Iliyasu, A.M., Bestak, R., Shi, F. (eds.) ICCVBIC 2018, pp. 957–968. Springer, Cham (2020). https://doi.org/10.1007/978-3-030-41862-5_97
12. Han, Y., Kim, J., Lee, K.: Deep convolutional neural networks for predominant instrument recognition in polyphonic music. IEEE/ACM Trans. Audio Speech Lang. Process. **25**(1), 208–221 (2016)
13. Humphrey, E., Durand, S., McFee, B.: Openmic-2018: an open data-set for multiple instrument recognition. In: ISMIR, pp. 438–444 (2018)
14. Kitahara, T., Goto, M., Okuno, H.G.: Pitch-dependent identification of musical instrument sounds. Appl. Intell. **23**(3), 267–275 (2005)
15. Lostanlen, V., Cella, C.E.: Deep convolutional networks on the pitch spiral for musical instrument recognition. In: International Society for Music Information Retrieval Conference, pp. 1–7 (2016)
16. Loughran, R., Walker, J., O'Neill, M., O'Farrell, M.: The use of Mel-frequency cepstral coefficients in musical instrument identification. In: International Computer Music Conference (2008)

17. McAdams, S.: Recognition of sound sources and events. In: Thinking in Sound: The Cognitive Psychology of Human Audition, pp. 146–198 (1993)
18. McAdams, S.: Musical timbre perception. In: The Psychology of Music, pp. 35–67 (2013)
19. Romani Picas, O., et al.: A real-time system for measuring sound goodness in instrumental sounds. In: Audio Engineering Society Convention, vol. 138 (2015)
20. Sahidullah, M., Saha, G.: Design, analysis and experimental evaluation of block based transformation in MFCC computation for speaker recognition. Speech Commun. **54**(4), 543–565 (2012)
21. Solanki, A., Pandey, S.: Music instrument recognition using deep convolutional neural networks. Int. J. Inf. Technol. **14**, 1–10 (2019). https://doi.org/10.1007/s41870-019-00285-y

Characterizing Wildfire Perimeter Polygons from QUIC-Fire

Li Tan[1], Raymond A. de Callafon[1(✉)], and Ilkay Altıntaş[2]

[1] Department of Mechanical and Aerospace Engineering,
University of California San Diego, La Jolla, CA, USA
`{ltan,callafon}@eng.ucsd.edu`
[2] San Diego Supercomputer Center, University of California San Diego,
La Jolla, CA, USA
`altintas@ucsd.edu`

Abstract. QUIC-Fire is a modern fire simulation tool that can simulate the progression of three-dimensional fuel consumption over a landscape, modeling the interaction of a wildfire with weather such as wind conditions around the wildfire. The resulting simulation gives a detailed progression of the consumed three-dimensional fuel that can be eloquently mapped to an image of a burn area in the landscape as the wildfire progresses over time. Although an image of burned vegetation over a landscape gives detailed information of the activity and coverage area of a wildfire, a numerical characterization of the boundary of the burn area can be used for a variety of computations. The boundary of the burn area, also labeled as the wildfire perimeter, can be parametrized with a closed polygon. The set of ordered vertices of the closed polygon provide a compact numerical representation of the location of the wildfire and can be used for computations related to fire coverage area and modern wildfire assimilation techniques to improve the prediction of wildfire progression. Designing a robust algorithm to create a wildfire perimeter in the form of a set of ordered vertices of a closed polygon around the image of consumed vegetation in a landscape is not a trivial task. This paper discusses the properties of two such algorithms: the iterative minimum distance algorithm (IMDA) and quadriculation algorithm (QA) to obtain a closed polygon for a wildfire perimeter. To illustrate the effectiveness, these two algorithms are applied to multiple image (raster) data of a burn area in the landscape of a wildfire created by QUIC-Fire simulations. It is shown that both algorithms are robust in computing wildfire perimeters, and computational time are less than one second for each image created by QUIC-Fire. As such, this work contributes to the development of computational methods to automate the process of characterizing the closed polygon of a wildfire perimeter based on burn area images.

Keywords: Wildland fire · QUIC-Fire · Polygons · Automation

Work is supported by WIFIRE Commons and funded by NSF 2040676 and NSF 2134904 under the Convergence Accelerator program.

1   Introduction

Vegetation dispersed over a landscape is the main fuel component that drives many wildfires. As a wildfire consumes this fuel under the influence of external wind and other weather conditions, it creates a 'burn area' or 'burn scar' of consumed fuel in the landscape that can cause significant damage, economic loss and environmental impacts. Clearly, understanding the wildland fire behavior and reducing the effects of wildfires by either controlling vegetation via prescribed burns or improving predicting the progression of a wildland fire are desirable.

Improving the prediction of wildfire progression has been an active area of research [7,11,12]. Data assimilation by combining wildfire modeling and ensemble Kalman filter is applied in [4,15,17], while several studies on the influence of wind condition and fuel have been conducted[2,18,19]. Many fire behavior models have been developed to improve the prediction of the wildfire progression [1,5,9,10,14], and the focus on controlling wildfires by prescribed burns is driven by QUIC-Fire [10]. QUIC-Fire can serve as a modern fire simulation tool to simulate the progression of three-dimensional (3D) fuel consumption over a landscape, while also approximating the dynamic interaction of a fire with weather including wind conditions in the atmosphere around the fire. QUIC-Fire can also takes into account the interactions between multiple fires and can compute fire progression at the resolution of one meter.

A wildfire perimeter, defined as a closed polygon around the burn area of a wildfire or prescribed burn, is an important numerical characterization of the impact of the fire and can be used for a variety of computations. Most wildfire perimeters are obtained from 2D images [3,13,21], while the consumed 3D fuel created by QUIC-Fire simulation is mapped to a 2D image of a burn area in the landscape as the wildfire progresses over time. So even for the output of QUIC-Fire, it is desirable to create an algorithm to compute the closed polygon of the wildfire perimeter.

Designing a robust algorithm to create a wildfire perimeter in the form of a set of ordered vertices of a closed polygon around the image of consumed vegetation in a landscape is not a trivial task. Edge detection methods have been applied to wildfire images [16,20], but only find a set of unordered boundary points that is not suitable to produce a closed polygon. In addition, a wildfire perimeter may include one main closed polygon and multiple additional closed polygons due to sporadic fire spread caused by embers and no assumption can be made on the shapes of the polygons. Due to this complexity of multiple wildfire perimeters, traditional pattern recognition algorithms [6,8] are not directly applicable.

This paper discusses the properties of two algorithms: the iterative minimum distance algorithm (IMDA) and quadriculation algorithm (QA) to create a closed polygon of a wildfire perimeter. The IMDA is based on continually connecting two closest points in the set of unordered boundary points determined by conventional image edge detection. A threshold value is set up to assist in determining whether two points in a cluster are closely located. If one cluster is far away from other clusters, then it is regarded as a new isolated polygon representing a separate fire perimeter due to embers. From a completely different point of view, the QA creates a polygon by recursively dividing the raster image into indivisible rectangles, where all the internal pixels of the rectangles have the same color, and then merging adjacent rectangles that have the same color. In general, QA avoids the process of ordering the unordered boundary points, but takes a longer time when merging the different polygons.

2   QUIC-Fire Output Data

As mentioned in the introduction, the focus of this paper is to discuss the properties of the iterative minimum distance algorithm (IMDA) and quadriculation algorithm (QA) to obtain closed polygons of wildfire perimeters based on images of consumed vegetation in a landscape. This section summarizes the QUIC-Fire output data used for the evaluation of the IMDA and QA. The QUIC-Fire output data consist of images of the fuel densities over a landscape at ground level (below 10 m) at different time stamps (100 s, 300 s, 500 s, 700 s, 900 s, 1100 s) as a prescribed burn or wildfire progresses. The images of the QUIC-Fire output data are given in Fig. 1.

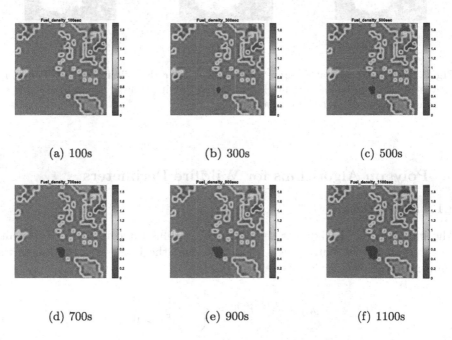

(a) 100s   (b) 300s   (c) 500s

(d) 700s   (e) 900s   (f) 1100s

Fig. 1. Fuel densities at different time stamps after the wildfire begins. (Color figure online)

From Fig. 1, it can be observed that as time increases, the dark blue area with near zero fuel density becomes larger, which means the fuels are consumed and the wildfire is spreading. The burn area of the wildfire at different time stamps can then be detected by comparing the difference of the corresponding fuel densities and the fuel density before the wildfire starts, leading to the black/white images given in Fig. 2.

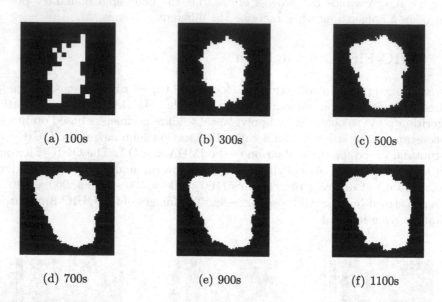

| (a) 100s | (b) 300s | (c) 500s |

| (d) 700s | (e) 900s | (f) 1100s |

Fig. 2. Burn area data at different time stamps after the wildfire begins. The white area is the burn area, and the black area is the unburned area. The scales of the six plots are selected differently for a better view.

3   Polygon Algorithms for Wildfire Perimeters

3.1   Image Data

The burn area at six different time stamps are illustrated in Fig. 2. The white area represents the burn area with $y = 1$, and the black area represents the unburned area with $y = 0$, where

$$y_{i,j} = f([i,j], b) = \begin{cases} 0, & \text{if } b = 0 \\ 1, & \text{if } b > 0 \end{cases} \qquad (1)$$

In (1), $[i,j]$ is a vector providing the position information of the target pixel in the image, and b is the absolute difference value between the fuel densities at the current time stamp and before the wildfire starts. The variable y is used to describe the area (burn area or unburned area) the pixel (at $[i,j]$) belongs to.

To better cover all possible situations of wildfire and illustrate the performances of IMDA and QA, one of the burn area outputs of Fig. 2 has been

Fig. 3. Modified output of the QUIC-Fire with extra rectangular burn area. The white area represents the burn area, and the black area represents the unburned area.

increased in complexity by adding a separate (rectangular) burn area, and removing part of the original burn area, as indicated in Fig. 3. The additional burn area is added to verify if both the IMDA and QA can recognize multiple wildfire perimeters within the data of Fig. 3.

3.2 Quadriculation Algorithm

The first method of finding an ordered set of vertices of a closed polygon around a burn area is the quadriculation algorithm (QA). Inspired by fire simulation tool FARSITE [5], the QA solves the problem in two main steps of division and union. Due to the fact that the minimum unit of a rasterized burn area image is a pixel, QA quadriculates the target image into four squares or rectangles recursively until all pixels in one square or rectangle have same value y. The process is illustrated on a simple example in Fig. 4(a). It can be observed that after the first division, only the pixels in the right-top square of the image have the same value ($y = 0$). Therefore, another quadriculation is needed. The second division should be applied on left-top, left-bottom, right-bottom squares because pixels have different values $y = 1$ and $y = 0$ in these three squares, and no division should be applied on the right-top square.

After the recursive division, the adjacent squares or rectangles with same value y should be joined, and the perimeter of the polygon in Fig. 4(a) can be obtained in Fig. 4(b). With the precision of one meter for QUIC-Fire, the size of each cell is one meter times one meter. Therefore, the polygon obtained by QA can be accurate enough to describe the wildfire perimeter. The process of QA is summarized in Algorithm 1.

Algorithm 1. QA

Input: Fire image
Output: Polygons representing wildfire perimeters
  1: Recursively quadriculate the image into four squares or rectangles until all pixels in one square or rectangle have same value y.
  2: Join the adjacent squares or rectangles until the pixels in adjacent squares or rectangles have different value y.

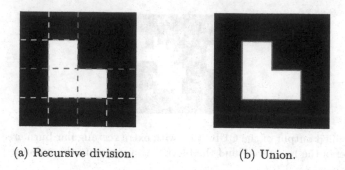

(a) Recursive division.   (b) Union.

Fig. 4. Division and union in QA. The dashed red and green lines represents the first and second division respectively. The red solid line represents the polygon of the wildfire perimeter. The white and black area are with $y = 1$ and $y = 0$ respectively. (Color figure online)

The QA is known to take quite some computation time due to two main steps of division and union that scales up as the image size increases. It would be beneficial to have an algorithm that can also handle large images with multiple burn areas. The proposed algorithm is presented in the next section.

3.3 Iterative Minimum Distance Algorithm (IMDA)

Preparatory Work. The IMDA solves the problem of finding an ordered set of vertices of a closed polygon around a burn area by selecting and ordering the set of unordered boundary points. First, a standard image edge detection algorithm is applied to Fig. 3 to acquire the boundary points. The boundary points are detected by comparing the value $y_{i,j}$ of the target pixel with its surroundings. An abrupt change in the y value of the pixel expressed by

$$|y_{i-1,j} - y_{i,j}| \neq 0 \text{ and } y_{i-1,j} = 1,$$
$$\text{or} |y_{i,j} - y_{i,j-1}| \neq 0 \text{ and } y_{i,j-1} = 1,$$
$$\text{or} |y_{i+1,j} - y_{i,j}| \neq 0 \text{ and } y_{i+1,j} = 1,$$
$$\text{or} |y_{i,j} - y_{i,j+1}| \neq 0 \text{ and } y_{i,j+1} = 1,$$

and the pixel at i, j can be regarded as a boundary point. Due to the fact that the precision of the QUIC-Fire data can be as small as one meter, the edge detection achieves a resolution of one meter.

Naive Minimum Distance. To motivate the IMDA, first consider the simplest method for the rearrangement of the unordered vertices or boundary points: choosing an arbitrary starting point and find the closest point to the previous selected point. In this native minimum distance (NMD) check, an important requirement is to avoid a self-intersection of the polygon.

With the set of the unordered boundary points B, the starting point b_1 is first selected arbitrarily. Then, remove b_1 from the set B, and find a new point

b_v $(v > 1)$ with the minimum distance to b_1 in B. If the distance between the last two selected points b_{v-1} and b_v, where $v \geq 3$, is larger than the distance from b_{v-1} to b_1, b_{v-1} is connected to b_1 directly to produce a closed polygon. To ensure there is no problem of self-intersection, the NMD checks whether the line segment $b_{v-1}b_v$ intersects with any previous created line segments. If there exists an intersection, the point b_{v-1} is deleted and connect $b_{v-2}b_v$. This process iterates until no intersection exists.

During the process of finding b_v, two or more points can be found with same distance to the previous selected point (multi-choice situation). To solve this problem, each choice will be stored and the corresponding closed polygon is recorded. The polygon with the largest number of vertices is picked as an optimal choice because more vertices means more detailed information. If there are multiple polygons with same number of vertices, more constraints such as the area of the polygon, can be added to select the optimal polygon.

The main problem for the NMD check is that for each multi-choice situation, two or more complete polygons that are generated also need to be stored for comparison purpose. Storing and comparing polygons may be an computationally expensive process, especially when the numbers of boundary points and multi-choice situations increase. This problem is illustrated in a simple case of Fig. 5. It can be observed that the red polygon better describes the burn area than the cyan dashed polygon, and the only difference between these two polygons is located inside the green dashed rectangle in the figure.

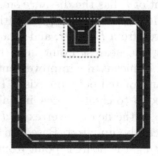

Fig. 5. Two possible polygons after removing self-intersections (red line and cyan dashed line). The green dashed rectangle shows the two-choice difference. (Color figure online)

Next to storing and comparing multiple polygons, the NMD check cannot deal with the case when a wildfire has multiple disjoint burn areas to create multiple wildfire perimeters. These problems lead to a modification of the NMD check and result in the actual IMDA.

Computation of Ordered Vertices of the Closed Polygon. In the computation of ordered vertices of the closed polygon in IMDA, one initial main polygon is first obtained by arbitrarily choosing a point in the multi-choice situation. All left points are used to modify the initial main polygon or create a new isolated polygon. It is still assumed that all the unordered boundary points can be used only once, but with one more constraint: the largest distance between two adjacent boundary points should be smaller than $d = \sqrt{2}$ due to point-to-point pixel distances. Following this distance observation, there are two main steps in IMDA: the first step is to obtain an initial main polygon, and the second step is to modify the obtained polygon and decide whether there is an extra isolated polygon. The logic of each step is described as follows.

For the first step, an arbitrary starting point b_1 is selected from the set B of the unordered boundary points to be the first point of the set P that is used to restore the ordered vertices of the polygon of a wildfire perimeter. The point with the minimum distance to the previously selected point in P is chosen from B and added to P one by one. If there are multiple points with the same minimum distance to the previously selected point, the first point in order is selected. During the selection, if no other points in set B have the distance smaller than d with respect to the last point in P, the distance from the last point in P to the starting point b_1 is checked. If the distance is smaller than d, a closed polygon is created. On the contrary, if the distance is larger than d, it means the current trajectory is not correct. Therefore, the last point in P needs to be deleted and moved to a different set B_c so that this point can be reused again and ordered correctly. Repeat deleting the last point in P and move it to set B_c until a point with a distance smaller than d to the updated last point of P can be found in B, or the updated last point of P has the distance smaller than d to the starting point b_1. The first step is finished by creating an initial closed polygon.

With all the points moved from B_c to B, and clearing the set B_c, the second step is initiated by finding the nearest point in B to any vertex in P, if the distance is larger than d, it means no improvement can be achieved by the initial main polygon, and an isolated polygon exists. Hence, the first step should be repeated for the updated B to create a new initial polygon. If there exists a point in B with a distance to the nearest vertex in P smaller than d, it means the initial closed polygon can be updated. Based on closest vertex in P as the first point of the trajectory P_c, the nearest point from set B to the last point in P_c is found. If the distance from the newly detected point in B to the last point in P_c is smaller than d, then add the newly detected point to the set P_c. If no more points in B has the distance smaller than d to the last point of P_c, find the closest point in P to the last point in P_c. If the distance from the closest point in P to the last point in P_c is smaller than d, add the detected closest point in P to the set P_c. If the distance is larger than d, delete the last point in P_c, and add it to B_c until the distance from the closest point in P to the last point in P_c is smaller than d. Then add the detected closest point in P to the set P_c.

One important thing to note here is that the first point and the last point in P_c should be different from each other. Based on the first point and the last point of P_c, add P_c to the initial created polygon. If the previously created polygon has other vertex between the first point and the last point of P_c, which means connecting P_c to P will lead to the deletion of previously selected vertices. Then, whether connecting P_c to P depends on whether connecting P_c will increase the area of the polygon. If connecting P_c to P can increase the area of the polygon, P_c is connected to P and replace the corresponding part selected in the first step. Otherwise, keep P as it is. Iterate this process until there is no point left in B and B_c. The logic process of the IMDA is summarized in Algorithm 2.

Algorithm 2. IMDA

Input: Unordered boundary points B and threshold value d.
Output: Polygons representing wildfire perimeters
1: Pick the arbitrary starting point b_1 in B, and delete b_1 in B.
2: Find a closed polygon P based on finding the point with the minimum distance that is smaller than d to the previously selected point.
3: Find a trajectory P_c when the distance from any point in B to P is smaller than d.
4: If adding P_c to P will not result in the deletion of the previously selected point in P, add P_c to P.
5: If adding P_c to P will result in the deletion of the previously selected point in P, P_c is added to P when it increases the area of the polygon.
6: Iterate steps 3-5 until no points in B have distance smaller than d to P.
7: Repeat the above steps if there are multiple polygons.

---

4 Numerical Results

IMDA and QA are applied to the modified burn area data of Fig. 3 to verify the detection of multiple fire perimeters. The resulting closed polygons created by IMDA and QA are shown in Fig. 6. It is clear that both IMDA and QA produce the two distinct fire perimeters, but it can also be observed that IMDA provides slightly tighter polygons around the burn area as the polygons are not restricted to horizontal and vertical lines as in QA.

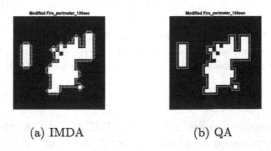

(a) IMDA   (b) QA

Fig. 6. Polygons of the wildfire perimeter (red lines) of the modified burn area. Burn area (white), unburned area (black), and detected boundary points (yellow circles). (Color figure online)

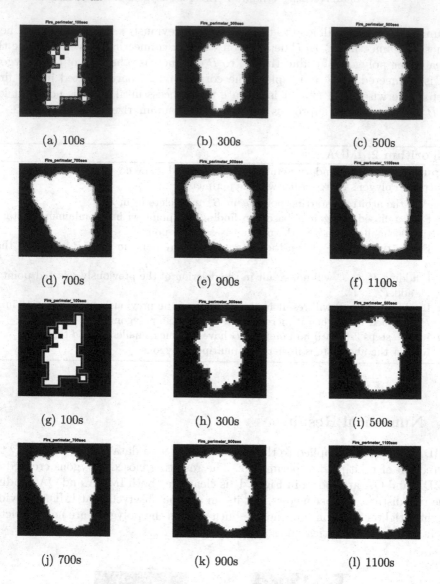

Fig. 7. Polygons of the wildfire perimeter (red lines) of the burn area at six different time stamps. (a)-(f) are obtained by IMDA, and (g)-(l) are obtained by QA. Burn area (white), unburned area (black) and detected boundary points (yellow circles). (Color figure online)

To further compare the performance of IMDA and QA, the algorithms are applied to the burn area data of at the six different time stamps of Fig. 2. The visual results are summarized in Fig. 7 with the same conclusion: both IMDA and QA produce correct results, but IMDA provides slightly tighter polygons. The more telling observations come from Table 1, where it can be seen that the computation time of IMDA scaled favorably compared to QA as the image size and the burn area of the wildfire perimeter increases. As reference for the computation time, all calculations were performed on an Intel Core i7-7500U CPU with 16 GB RAM.

Table 1. Computation time of IMDA and QA

	100 s	300 s	500 s	700 s	900 s	1100 s
IMDA	0.019 s	0.019 s	0.021 s	0.020 s	0.023 s	0.024 s
QA	0.041 s	0.132 s	0.185 s	0.205 s	0.362 s	0.352 s

5 Conclusions

This paper compares two algorithms, iterative minimum distance algorithm (IMDA) and quadriculation algorithm (QA), to obtain the closed polygons that parametrize wildfire perimeters. The IMDA is based on continually connecting two closest points in the set of unordered boundary points determined by conventional image edge detection. A threshold value is set up to assist in determining whether two points are closely located. From a completely different point of view, the QA creates a polygon by recursively dividing the raster image into indivisible rectangles, where all the internal pixels of the rectangles have the same color. Then, the QA merges adjacent rectangles that have the same color. Using simulation data produced by QUIC-Fire that consist of raster images of consumed vegetation in a landscape, the performance of IMDA and QA is compared. Although the logic of IMDA is more complicated, IMDA produces slightly tighter polygons around the burn area compared to QA, as the polygons are not restricted to horizontal and vertical lines in the image resolution. Moreover, the computation time of IMDA scaled favorably compared to QA as the image size and the burn area of the wildfire perimeter increase.

References

1. Achtemeier, G.L.: Field validation of a free-agent cellular automata model of fire spread with fire-atmosphere coupling. Int. J. Wildland Fire 22(2), 148–156 (2012)
2. Cheney, N., Gould, J., Catchpole, W.: The influence of fuel, weather and fire shape variables on fire-spread in grasslands. Int. J. Wildland Fire 3(1), 31–44 (1993)
3. Dickinson, M.B., et al.: Measuring radiant emissions from entire prescribed fires with ground, airborne and satellite sensors-RxCADRE 2012. Int. J. Wildland Fire 25(1), 48–61 (2015)
4. Fang, H., Srivas, T., de Callafon, R.A., Haile, M.A.: Ensemble-based simultaneous input and state estimation for nonlinear dynamic systems with application to wildfire data assimilation. Control Eng. Pract. 63, 104–115 (2017)

5. Finney, M.A.: FARSITE, Fire Area Simulator - model development and evaluation, vol. 4. US Department of Agriculture, Forest Service, Rocky Mountain Research Station (1998)
6. García, N.L.F., Martínez, L.D.M., Poyato, Á.C., Cuevas, F.J.M., Carnicer, R.M.: Unsupervised generation of polygonal approximations based on the convex hull. Pattern Recogn. Lett. **135**, 138–145 (2020)
7. Gollner, M., et al.: Towards data-driven operational wildfire spread modeling: A report of the NSF-funded WIFIRE workshop. Technical Report (2015)
8. Graham, R.L., Yao, F.F.: Finding the convex hull of a simple polygon. J. Algorithms **4**(4), 324–331 (1983)
9. Linn, R., Reisner, J., Colman, J.J., Winterkamp, J.: Studying wildfire behavior using FIRETEC. Int. J. Wildland Fire **11**(4), 233–246 (2002)
10. Linn, R.R., et al.: QUIC-fire: a fast-running simulation tool for prescribed fire planning. Environ. Model. Softw. **125**, 104616 (2020)
11. Mandel, J., et al.: Towards a dynamic data driven application system for wildfire simulation. In: Sunderam, V.S., van Albada, G.D., Sloot, P.M.A., Dongarra, J.J. (eds.) ICCS 2005. LNCS, vol. 3515, pp. 632–639. Springer, Heidelberg (2005). https://doi.org/10.1007/11428848_82
12. Mandel, J., et al.: A note on dynamic data driven wildfire modeling. In: Bubak, M., van Albada, G.D., Sloot, P.M.A., Dongarra, J. (eds.) ICCS 2004. LNCS, vol. 3038, pp. 725–731. Springer, Heidelberg (2004). https://doi.org/10.1007/978-3-540-24688-6_94
13. Manzano-Agugliaro, F., Pérez-Aranda, J., De La Cruz, J.: Methodology to obtain isochrones from large wildfires. Int. J. Wildland Fire **23**(3), 338–349 (2014)
14. Rothermel, R.C.: A mathematical model for predicting fire spread in wildland fuels, vol. 115. Intermountain Forest and Range Experiment Station, Forest Service, United ... (1972)
15. Srivas, T., Artés, T., De Callafon, R.A., Altintas, I.: Wildfire spread prediction and assimilation for FARSITE using ensemble Kalman filtering. Procedia Comput. Sci. **80**, 897–908 (2016)
16. Stow, D.A., Riggan, P.J., Storey, E.J., Coulter, L.L.: Measuring fire spread rates from repeat pass airborne thermal infrared imagery. Remote Sens. Lett. **5**(9), 803–812 (2014)
17. Subramanian, A., Tan, L., de Callafon, R.A., Crawl, D., Altintas, I.: Recursive updates of wildfire perimeters using barrier points and ensemble Kalman filtering. In: Krzhizhanovskaya, V.V., et al. (eds.) ICCS 2020. LNCS, vol. 12142, pp. 225–236. Springer, Cham (2020). https://doi.org/10.1007/978-3-030-50433-5_18
18. Tan, L., de Callafon, R.A., Block, J., Crawl, D., Altıntaş, I.: Improving wildfire simulations by estimation of wildfire wind conditions from fire perimeter measurements. In: Paszynski, M., Kranzlmüller, D., Krzhizhanovskaya, V.V., Dongarra, J.J., Sloot, P.M.A. (eds.) ICCS 2021. LNCS, vol. 12746, pp. 231–244. Springer, Cham (2021). https://doi.org/10.1007/978-3-030-77977-1_18
19. Tan, L., de Callafon, R.A., Block, J., Crawl, D., Çağlar, T., Altıntaş, I.: Estimation of wildfire wind conditions via perimeter and surface area optimization. J. Comput. Sci. **61**, 101633 (2022)
20. Valero, M., Rios, O., Pastor, E., Planas, E.: Automated location of active fire perimeters in aerial infrared imaging using unsupervised edge detectors. Int. J. Wildland Fire **27**(4), 241–256 (2018)
21. Zajkowski, T.J., et al.: Evaluation and use of remotely piloted aircraft systems for operations and research-RxCADRE 2012. Int. J. Wildland Fire **25**(1), 114–128 (2015)

On a New Generalised Iteration Method in the PSO-Based Newton-Like Method

Ireneusz Gościniak$^{(\boxtimes)}$⑩ and Krzysztof Gdawiec⑩

Institute of Computer Science, University of Silesia,
Będzińska 39, 41-200Sosnowiec, Poland
{ireneusz.gosciniak,krzysztof.gdawiec}@us.edu.pl

Abstract. The root-finding problem is very important in many applications and has become an extensive research field. One of the directions in this field is the use of various iteration schemes. In this paper, we propose a new generalised iteration scheme. The schemes like Mann, Ishikawa, Das–Debata schemes are special cases of the proposed iteration. Moreover, we use the proposed iteration with the PSO-based Newton-like method in two tasks. In the first task, we search for the roots, whereas in the second one for patterns with aesthetic features. The obtained results show that the proposed iteration can decrease the average number of iterations needed to find the roots and that we can generate patterns with potential artistic applications.

Keywords: Root-finding · Dynamics · Iterations · Visualisation

1 Introduction

Let $f_1, f_2, \ldots, f_D : \mathbb{R}^D \to \mathbb{R}$ and let

$$
\mathbf{F}(z^1, z^2, \ldots, z^D) = \begin{bmatrix} f_1(z^1, z^2, \ldots, z^D) \\ f_2(z^1, z^2, \ldots, z^D) \\ \vdots \\ f_D(z^1, z^2, \ldots, z^D) \end{bmatrix} = \begin{bmatrix} 0 \\ 0 \\ \vdots \\ 0 \end{bmatrix} = \mathbf{0}. \tag{1}
$$

Moreover, let us assume that $\mathbf{F} : \mathbb{R}^D \to \mathbb{R}^D$ is a continuous function and has continuous partial derivatives to appropriate order. To find the roots of $\mathbf{F}$, i.e., solve $\mathbf{F}(\mathbf{z}) = \mathbf{0}$, where $\mathbf{z} = [z^1, z^2, \ldots, z^D]$, we can use the Newton's method [3].

The root-finding problem (1) is very important in many applications [3,9]. Therefore, in the literature, we can find many root-finding methods. One of the methods is the mentioned earlier Newton's method which was the base for many other methods. For instance, in [6] the authors combined Newton's method with the Particle Swarm Optimisation (PSO) approach obtaining a method in which we can control the dynamics of the method very easily. Many of the root-finding methods are based on the finding of fixed points, so the use of the iteration

D. Groen et al. (Eds.): ICCS 2022, LNCS 13350, pp. 623–636, 2022.
https://doi.org/10.1007/978-3-031-08751-6_45

methods for approximate finding such point was proposed in recent years [2,5], and became a very popular study direction.

In this paper, we propose a new iteration scheme allowing an extension of the number of mappings. Moreover, we use the proposed scheme in the root-finding problem using the PSO-based Newton-like methods proposed in [6], and to generate patterns with aesthetic features.

The paper is organised as follows. In Sect. 2 we introduce the PSO-based Newton-like root-finding method. Then, in Sect. 3, we review some of the iteration schemes known in the literature. The new generalised iteration scheme is proposed in Sect. 4. In Sect. 5, we present the experimental setup, and in Sect. 6, we discuss the obtained results. Finally, in Sect. 7, we give the concluding remarks.

2   PSO-Based Newton-Like Method

In [6], the authors introduced a PSO-based Newton-like method to solve (1). The method combines the well-known Newton's method with the idea of PSO, namely it implements the best position of particle similarly to the PSO.

The method is defined using the following formula:

$$\mathbf{z}_{n+1} = \mathbf{z}_n + \mathbf{v}_{n+1}, \tag{2}$$

where $\mathbf{z}_0 \in \mathbb{R}^D$ is the starting point, $\mathbf{v}_0 = [0, 0, \ldots, 0]$ is the starting velocity, $\mathbf{v}_{n+1}$ is the current velocity ($\mathbf{v}_{n+1} = [v_{n+1}^1, v_{n+1}^2, \ldots, v_{n+1}^D]$), $\mathbf{z}_n$ is the previous position point ($\mathbf{z}_n = [z_n^1, z_n^2, \ldots, z_n^D]$). The algorithm sums the position of the particle $\mathbf{z}_n$ with its current velocity $\mathbf{v}_{n+1}$, which is given by the formula:

$$\mathbf{v}_{n+1} = \omega \mathbf{v}_n + \eta \mathbf{N}(\mathbf{z}_n), \tag{3}$$

where $\mathbf{v}_n$ – the previous velocity of the particle, $\omega \in [0, 1)$ – inertia weight, $\eta \in (0, 1]$ – acceleration constant. Moreover, $\mathbf{N}$ represents the Newton's method part, and it is defined by

$$\mathbf{N}(\mathbf{z}) = -\mathbf{J}^{-1}(\mathbf{z})\mathbf{F}(\mathbf{z}) \tag{4}$$

where $\mathbf{J}^{-1}$ is the inverse of the Jacobian matrix of $\mathbf{F}$.

Method (2) reduces to the classical Newton's method if $\omega = 0$ and $\eta = 1$, and to the relaxed Newton's method if $\omega = 0$ and $\eta \neq 1$. For $\omega < 1$ the algorithm has good convergence and can reach the solution. The ω and η should be selected by tuning—it is a kind of art [6].

3   Iteration Processes

If we take $\mathbf{T}(\mathbf{z}_n) = \mathbf{z}_n + \mathbf{v}_{n+1}$, then (2) takes the following form:

$$\mathbf{z}_{n+1} = \mathbf{T}(\mathbf{z}_n). \tag{5}$$

This type of iteration is known as the Picard iteration [13], and is widely used among others in fixed point theory.

In fixed point theory, we can find many other iteration processes. For example,

1. The Mann iteration [11]:

$$\mathbf{z}_{n+1} = (1 - \alpha_n)\mathbf{z}_n + \alpha_n \mathbf{T}(\mathbf{z}_n), \ n = 0, 1, 2, \ldots, \tag{6}$$

where $\alpha_n \in (0, 1]$ for all $n \in \mathbb{N}$, and for $\alpha_n = 1$ it reduces to the Picard iteration.

2. The Ishikawa iteration [8]:

$$\mathbf{u}_n = (1 - \beta_n)\mathbf{z}_n + \beta_n \mathbf{T}(\mathbf{z}_n),$$
$$\mathbf{z}_{n+1} = (1 - \alpha_n)\mathbf{z}_n + \alpha_n \mathbf{T}(\mathbf{u}_n), \ n = 0, 1, 2, \ldots, \tag{7}$$

where $\alpha_n \in (0, 1]$ and $\beta_n \in [0, 1]$ for all $n \in \mathbb{N}$, and we obtain the Mann iteration when $\beta_n = 0$, and the Picard iteration when $\alpha_n = 1$ and $\beta_n = 0$.

3. The Agarwal iteration [1] (or S-iteration):

$$\mathbf{u}_n = (1 - \beta_n)\mathbf{z}_n + \beta_n \mathbf{T}(\mathbf{z}_n),$$
$$\mathbf{z}_{n+1} = (1 - \alpha_n)\mathbf{T}(\mathbf{z}_n) + \alpha_n \mathbf{T}(\mathbf{u}_n), \ n = 0, 1, 2, \ldots, \tag{8}$$

where $\alpha_n \in [0, 1]$ and $\beta_n \in [0, 1]$ for all $n \in \mathbb{N}$, Moreover, when $\alpha_n = 0$, or $\alpha_n = 1$ and $\beta_n = 0$ the S-iteration reduces to the Picard iteration.

For an overview of some other iteration processes and their dependencies, see [5], where 17 different iterations are reviewed.

All the iterations shown so far use one mapping, but in fixed point theory, there are iterations that use several mappings. Examples of this type of iterations are the following:

1. The Das–Debata iteration [4]:

$$\mathbf{u}_n = (1 - \beta_n)\mathbf{z}_n + \beta_n \mathbf{T}_1(\mathbf{z}_n),$$
$$\mathbf{z}_{n+1} = (1 - \alpha_n)\mathbf{z}_n + \alpha_n \mathbf{T}_2(\mathbf{u}_n), \ n = 0, 1, 2, \ldots, \tag{9}$$

where $\alpha_n \in (0, 1]$ and $\beta_n \in [0, 1]$ for all $n \in \mathbb{N}$. For $\mathbf{T}_1 = \mathbf{T}_2$ the Das–Debata iteration reduces to the Ishikawa iteration.

2. The Khan–Cho–Abbas iteration [10]:

$$\mathbf{u}_n = (1 - \beta_n)\mathbf{z}_n + \beta_n \mathbf{T}_1(\mathbf{z}_n),$$
$$\mathbf{z}_{n+1} = (1 - \alpha_n)\mathbf{T}_1(\mathbf{z}_n) + \alpha_n \mathbf{T}_2(\mathbf{u}_n), \ n = 0, 1, 2, \ldots, \tag{10}$$

where $\alpha_n \in (0, 1]$ and $\beta_n \in [0, 1]$ for all $n \in \mathbb{N}$, when $\mathbf{T}_1 = \mathbf{T}_2$ the equation reduces to the Agarwal iteration.

3. The generalised Agarwal's iteration [10]:

$$\mathbf{u}_n = (1 - \beta_n)\mathbf{z}_n + \beta_n \mathbf{T}_1(\mathbf{z}_n),$$
$$\mathbf{z}_{n+1} = (1 - \alpha_n)\mathbf{T}_3(\mathbf{z}_n) + \alpha_n \mathbf{T}_2(\mathbf{u}_n), \ n = 0, 1, 2, \ldots, \tag{11}$$

where $\alpha_n \in (0, 1]$ and $\beta_n \in [0, 1]$ for all $n \in \mathbb{N}$, moreover when $\mathbf{T}_1 = \mathbf{T}_3$ the equation reduces to the Khan–Cho–Abbas iteration, and the Agarwal iteration is obtained when $\mathbf{T}_1 = \mathbf{T}_2 = \mathbf{T}_3$.

4 The Generalised Iteration

Let us consider the following general iteration scheme:

$$
\begin{aligned}
\mathbf{z}_{0,n+1} &= p_{0,0}\mathbf{z}_{i,n} + p_{0,1}\mathbf{T}_{0,0}(\mathbf{z}_{i,n}),\\
\mathbf{z}_{1,n+1} &= p_{1,0}\mathbf{z}_{i,n} + p_{1,1}\mathbf{T}_{1,0}(\mathbf{z}_{i,n}) + p_{1,2}\mathbf{z}_{0,n+1} + p_{1,3}\mathbf{T}_{1,1}(\mathbf{z}_{0,n+1}),\\
\mathbf{z}_{2,n+1} &= p_{2,0}\mathbf{z}_{i,n} + p_{2,1}\mathbf{T}_{2,0}(\mathbf{z}_{i,n}) + p_{2,2}\mathbf{z}_{0,n+1} + p_{2,3}\mathbf{T}_{2,1}(\mathbf{z}_{0,n+1}) +\\
&\quad + p_{2,4}\mathbf{z}_{1,n+1} + p_{2,5}\mathbf{T}_{2,2}(\mathbf{z}_{1,n+1}),\\
&\;\;\vdots\\
\mathbf{z}_{i,n+1} &= p_{i,0}\mathbf{z}_{i,n} + p_{i,1}\mathbf{T}_{i,0}(\mathbf{z}_{i,n}) + p_{i,2}\mathbf{z}_{0,n+1} + ... +\\
&\quad + p_{i,2i}\mathbf{z}_{i-1,n+1} + p_{i,2i+1}\mathbf{T}_{i,i}(\mathbf{z}_{i-1,n+1}),\ n = 0,1,2,\ldots,
\end{aligned}
\tag{12}
$$

where $p_{i,k} \in [0,1]$ for $k \in \{0,1,\ldots,2i+1\}$ for all $i \in \mathbb{N}$ and $\sum_{k=0}^{2i+1} p_{i,k} = 1$ for the given $i \in \mathbb{N}$. The $\mathbf{z}_{i,n}$ is the previous position of the particle and the $\mathbf{z}_{i,n+1}$ is the next position of the particle. The $\{\mathbf{z}_{0,n+1}, \mathbf{z}_{1,n+1}, \ldots, \mathbf{z}_{i-1,n+1}\}$ create a set of reference points – it plays a role similar to a swarm in PSO.

Iteration (12) is a general form of the iterations presented in Sect. 3, when the sequences of the parameters are constant:

1. The Mann iteration is obtained for $i = 0$ and $p_{0,0} + p_{0,1} = 1$:

$$
\mathbf{z}_{0,n+1} = (1 - p_{0,1})\mathbf{z}_{0,n} + p_{0,1}\mathbf{T}_{0,0}(\mathbf{z}_{0,n}),\ n = 0,1,2,\ldots,
\tag{13}
$$

where $p_{0,1} \in (0,1]$.

2. The Ishikawa iteration is obtained for $i = 1$ and $p_{0,0}+p_{0,1} = 1$, $p_{1,0}+p_{1,3} = 1$, $p_{1,1} = p_{1,2} = 0$, and $\mathbf{T}_{0,0} = \mathbf{T}_{1,1}$:

$$
\begin{aligned}
\mathbf{z}_{0,n+1} &= (1 - p_{0,1})\mathbf{z}_{1,n} + p_{0,1}\mathbf{T}_{0,0}(\mathbf{z}_{1,n}),\\
\mathbf{z}_{1,n+1} &= (1 - p_{1,3})\mathbf{z}_{1,n} + p_{1,3}\mathbf{T}_{1,1}(\mathbf{z}_{0,n+1}),\ n = 0,1,2,\ldots,
\end{aligned}
\tag{14}
$$

where $p_{0,1} \in [0,1]$ and $p_{1,3} \in (0,1]$.

3. The Agarwal iteration is obtained for $i = 1$ and $p_{0,0}+p_{0,1} = 1$, $p_{1,1}+p_{1,3} = 1$, $p_{1,0} = p_{1,2} = 0$, and $\mathbf{T}_{0,0} = \mathbf{T}_{1,0} = \mathbf{T}_{1,1}$:

$$
\begin{aligned}
\mathbf{z}_{0,n+1} &= (1 - p_{0,1})\mathbf{z}_{i,n} + p_{0,1}\mathbf{T}_{0,0}(\mathbf{z}_{1,n}),\\
\mathbf{z}_{1,n+1} &= (1 - p_{1,3})\mathbf{T}_{1,0}(\mathbf{z}_{1,n}) + p_{1,3}\mathbf{T}_{1,1}(\mathbf{z}_{0,n+1}),\ n = 0,1,2,\ldots,
\end{aligned}
\tag{15}
$$

where $p_{0,1} \in [0,1]$ and $p_{1,3} \in (0,1]$.

4. The Das–Debata iteration is obtained for $i = 1$ and $p_{0,0}+p_{0,1} = 1$, $p_{1,0}+p_{1,3} = 1$, $p_{1,1} = p_{1,2} = 0$, and $\mathbf{T}_{0,0} \neq \mathbf{T}_{1,1}$:

$$
\begin{aligned}
\mathbf{z}_{0,n+1} &= (1 - p_{0,1})\mathbf{z}_{1,n} + p_{0,1}\mathbf{T}_{0,0}(\mathbf{z}_{1,n}),\\
\mathbf{z}_{1,n+1} &= (1 - p_{1,3})\mathbf{z}_{1,n} + p_{1,3}\mathbf{T}_{1,1}(\mathbf{z}_{0,n+1}),\ n = 0,1,2,\ldots,
\end{aligned}
\tag{16}
$$

where $p_{0,1} \in [0,1]$ and $p_{1,3} \in (0,1]$.

5. For $i = 1$ and $p_{0,0} + p_{0,1} = 1$, $p_{1,1} + p_{1,3} = 1$, $p_{1,0} = p_{1,2} = 0$, and $\mathbf{T}_{0,0} = \mathbf{T}_{1,0} \neq \mathbf{T}_{1,1}$ we obtain the Khan–Cho–Abbas iteration:

$$\begin{aligned}
\mathbf{z}_{0,n+1} &= (1 - p_{0,1})\mathbf{z}_{i,n} + p_{0,1}\mathbf{T}_{0,0}(\mathbf{z}_{1,n}), \\
\mathbf{z}_{1,n+1} &= (1 - p_{1,3})\mathbf{T}_{1,0}(\mathbf{z}_{1,n}) + p_{1,3}\mathbf{T}_{1,1}(\mathbf{z}_{0,n+1}), \quad n = 0, 1, 2, \ldots,
\end{aligned} \tag{17}$$

where $p_{0,1} \in [0, 1]$ and $p_{1,3} \in (0, 1]$.

6. The generalised Agarwal's iteration is obtained for $i = 1$ and $p_{0,0} + p_{0,1} = 1$, $p_{1,1} + p_{1,3} = 1$, $p_{1,0} = p_{1,2} = 0$, and $\mathbf{T}_{0,0} \neq \mathbf{T}_{1,0} \neq \mathbf{T}_{1,1}$:

$$\begin{aligned}
\mathbf{z}_{0,n+1} &= (1 - p_{0,1})\mathbf{z}_{i,n} + p_{0,1}\mathbf{T}_{0,0}(\mathbf{z}_{1,n}), \\
\mathbf{z}_{1,n+1} &= (1 - p_{1,3})\mathbf{T}_{1,0}(\mathbf{z}_{1,n}) + p_{1,3}\mathbf{T}_{1,1}(\mathbf{z}_{0,n+1}), \quad n = 0, 1, 2, \ldots,
\end{aligned} \tag{18}$$

where $p_{0,1} \in [0, 1]$ and $p_{1,3} \in (0, 1]$.

Iteration (12) not only reduces to the existing iteration methods, but it can be used to obtain completely new ones. For instance, we can consider the following iteration scheme:

$$\begin{aligned}
\mathbf{z}_{0,n+1} &= (1 - p_{0,1})\mathbf{z}_{i,n} + p_{0,1}\mathbf{T}_{0,0}(\mathbf{z}_{2,n}), \\
\mathbf{z}_{1,n+1} &= (1 - p_{1,3})\mathbf{T}_{1,0}(\mathbf{z}_{2,n}) + p_{1,3}\mathbf{T}_{1,1}(\mathbf{z}_{0,n+1}), \\
\mathbf{z}_{2,n+1} &= (1 - p_{2,5})\mathbf{T}_{2,1}(\mathbf{z}_{0,n+1}) + p_{2,5}\mathbf{T}_{2,2}(\mathbf{z}_{1,n+1}), \quad n = 0, 1, 2, \ldots,
\end{aligned} \tag{19}$$

Similar to the Agarwal's iterations, we can use one or several mappings in this iteration. For one mapping, i.e., $\mathbf{T}_{0,0} = \mathbf{T}_{1,0} = \mathbf{T}_{1,1} = \mathbf{T}_{2,1} = \mathbf{T}_{2,2}$ we will name this scheme as New', and for several mappings, i.e., $\mathbf{T}_{0,0} = \mathbf{T}_{1,0} \neq \mathbf{T}_{1,1} = \mathbf{T}_{2,1} \neq \mathbf{T}_{2,2}$, as New". The computational cost of iterations can be calculated as the number of different mappings of a point in the scheme. The Mann iteration has the same computational cost as Picard's iteration – only one mapping. The Ishikawa, Agarwal, Das-Debata, Khan-Cho-Abbas iterations have two different mappings. The Ishikawa, Agarwal, Das-Debata, Khan-Cho-Abbas have mappings for two different points, so we can consider that the computational cost is the same. The generalized Agarwal iteration and the proposed New 'and New" schemes have three different mappings, so they have the same computational cost.

5   Experimental Setup

In the experiments, we study the case in which $D = 2$, and the PSO-based Newton-like method from Sect. 2 is taken as the operators $T_{i,j}$ in the iteration process from Sect. 4. The experiments relied on the optimisation of the iteration's parameters for two tasks. In the first task, we minimise the number of iterations needed to find the root, whereas, in the second task, we are interested in obtaining graphical patterns with aesthetic features.

Let $\mathbb{C}$ be the field of complex numbers with a complex number $c = x + iy$ where $i = \sqrt{-1}$ and $x, y \in \mathbb{R}$. In the experiments, we want to solve the following non-linear equation

$$p(c) = 0 \tag{20}$$

where $p(c) = c^3 - 1$. This equation can be written in the following form:

$$0 = c^3 - 1 = (x + iy)^3 - 1 = x^3 - 3xy^2 - 1 + (3x^2y - y^3)i. \tag{21}$$

Expression (21) can be transformed into a system of two equations with two variables:

$$\mathbf{F}(x, y) = \begin{bmatrix} f_1(x, y) \\ f_2(x, y) \end{bmatrix} = \begin{bmatrix} 0 \\ 0 \end{bmatrix} = \mathbf{0}, \tag{22}$$

where $f_1(x, y) = x^3 - 3xy^2 - 1$, $f_2(x, y) = 3x^2y - y^3$. The set of solutions of this system is the following: $[1, 0]$, $[-0.5, -0.866025]$, $[-0.5, 0.866025]$.

In a similar way, we can transform other commonly used, in the literature, complex polynomial equations into systems of non-linear equations:

$$0 = c^4 - 10c^2 + 9 = x^4 - 6x^2y^2 + y^4 - 10x^2 + 10y^2 + 9 + (4x^3y - 4xy^3 - 20xy)i, \tag{23}$$

where $f_1(x, y) = x^4 - 6x^2y^2 + y^4 - 10x^2 + 10y^2 + 9$, $f_2(x, y) = 4x^3y - 4xy^3 - 20xy$ and the set of solutions of this system is the following: $[-3.0, 0.0]$, $[-1.0, 0.0]$, $[1.0, 0.0]$, $[3.0, 0.0]$;

$$0 = c^5 - c = x^5 - 10x^3y^2 + 5xy^4 - x + (5x^4y - 10x^2y^3 + y^5 - y)i, \tag{24}$$

where $f_1(x, y) = x^5 - 10x^3y^2 + 5xy^4 - x$, $f_2(x, y) = 5x^4y - 10x^2y^3 + y^5 - y$ and the set of solutions of this system is the following: $[-1.0, 0.0]$, $[0.0, -1.0]$, $[0.0, 0.0]$, $[0.0, 1.0]$, $[1.0, 0.0]$;

$$0 = c^6 + 10c^3 - 8 = x^6 - 15x^4y^2 + 15x^2y^4 - y^6 + 10x^3 - 30xy^2 - 8 +$$
$$+ (6x^5y - 20x^3y^3 + 6xy^5 + 30x^2y - 10y^3)i, \tag{25}$$

where $f_1(x, y) = x^6 - 15x^4y^2 + 15x^2y^4 - y^6 + 10x^3 - 30xy^2 - 8$, $f_2(x, y) = 6x^5y - 20x^3y^3 + 6xy^5 + 30x^2y - 10y^3$ and the set of solutions of this system is the following (approximately): $[-2.207, 0]$, $[-0.453, -0.785]$, $[-0.453, 0.785]$, $[0.906, 0]$, $[1.103, -1.911]$, $[1.103, 1.911]$.

Nowadays, visual analysis is an essential part of modern analysis of the quality of the root-finding methods [12]. To visualise the dynamics of the algorithm's operations, we use Algorithm 1, which is a standard algorithm used in polynomiography [9].

Algorithm 1: Visualisation of algorithm's operations dynamics.

Input: F – function; $\mathbf{A} \subset \mathbb{R}^D$ – solution space; m – the maximum
number of iterations; I_q – iteration (12) with the parameters q; C
– colouring function; ε – accuracy

Output: Visualisation of the dynamics

1 **foreach** $\mathbf{z}_0 \in \mathbf{A}$ **do**
2 $\quad$ $i = 0$
3 $\quad$ $\mathbf{v}_0 = [0, 0, \ldots, 0]$
4 $\quad$ **while** $i \le m$ **do**
5 $\quad\quad$ $\mathbf{z}_{n+1} = I_q(\mathbf{z}_n)$
6 $\quad\quad$ **if** $\|\mathbf{z}_{n+1} - \mathbf{z}_n\| < \varepsilon$ **then**
7 $\quad\quad\quad$ $\lfloor$ **break**
8 $\quad\quad$ $i = i + 1$
9 $\quad$ colour $\mathbf{z}_0$ with $C(i)$

The algorithms used in the experiments were implemented in the C++ programming language. The experiments were conducted on a computer with the Intel Core i7-8750H CPU 2.20 GHz processor, 16 GB RAM, NVIDIA GeForce GTX 1060 Mobile and Linux Ubuntu 20.04 LTS. The iterations minimisation is implemented using the simple genetic algorithm from GAlib [14]. Each of the optimised parameters has a 63 bit representation mapped to the decimal phenotype with a given range: $[0, 1]$ or $[0, 1.5]$. The simple genetic algorithm operation parameters are: the probability of the mutation is 1.5%, the cross-over probability is 80%, the population size is 50, and the number of generations is 400. The fitness function is based on Algorithm 1 implemented using OpenCL 2.0 [7]. Implementing the genetic algorithm does not require any further discussion because the GAlib library is elementary to use. Optimisation of the algorithm's operation to obtain aesthetic patterns is based on expert knowledge. The selection of the iteration parameter values is based on the authors' experience, and assessing the aesthetics of the pattern is subjective. The influence of parameters on the algorithm's operation will be discussed in the next section.

To generate the images using Algorithm 1, we used a colour map with $m = 256$ levels (Fig. 1), $\varepsilon = 1.0e-2$, image resolution is 800×800 pixels, and the area **A** for the considered systems were the following: $[-2, 2]^2$, $[-4, 4] \times [-2, 2]$, $[-2, 2]^2$, $[-2.3, 1.7] \times [-2, 2]$, respectively.

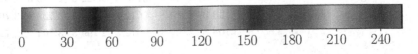

Fig. 1. Colour map used in the experiments. (Color figure online)

6   Discussion on the Obtained Results

As we mentioned in the previous section, the selection of iteration parameters is analysed for two purposes: to minimise the number of iterations and to obtain patterns with aesthetic features.

In the tables with the results we use the following abbreviations: A – Eq. (21), B – Eq. (23), C – Eq. (24), D – Eq. (25), and the letter I denotes the simulation without inertia weight (which means that $\omega = 0$) and the simulation in which it is used by II.

6.1   Minimising the Number of Iterations

The simple genetic algorithm minimises the value of the average number of iterations by selecting appropriate coefficients. Table 1 contains the average iteration values obtained for the analysed algorithms without inertia weight (I) and with their use (II). The values of the optimised coefficients for the algorithms without inertia weight are presented in Table 2, and taking into account the inertia weight in Table 3. Unambiguous conclusions can be drawn from the data analysis. The inertia weight is of marginal importance. Only the transformation of the particle and the reference point significantly influence the operation of the algorithm. Algorithms are transformed into a sequence of such transformations. The iterations proposed in the article (New', New") increase the number of such transformations to obtain the best results – the lowest average iteration value.

Table 1. The average numbers of iterations of the analysed algorithms for the considered approach minimizing the number of iterations.

Iteration	Test							
	A		B		C		D	
	I	II	I	II	I	II	I	II
Picard	5:822 7		5.654 8		5.986 3		8.168 1	
Mann	5.820 6	5.792 9	5.654 8	5.617 0	5.985 8	5.676 0	8.115 9	8.116 7
Ishikawa	3.668 6	3.657 2	3.719 2	3.704 9	3.747 8	3.600 4	4.800 4	4.801 5
Agarwal	3.668 7	3.654 5	3.719 3	3.705 6	3.748 2	3.524 3	4.799 8	4.804 8
Das–Debata	3.668 5	3.656 3	3.719 2	3.705 2	3.748 3	3.524 7	4.800 5	4.803 2
Khan–Cho–Abbas	3.669 0	3.654 9	3.719 3	3.705 6	3.748 3	3.523 2	4.800 2	4.804 7
Generalised Agarwal	3.668 9	3.655 2	3.719 3	3.707 0	3.748 3	3.522 9	4.799 3	4.817 2
New'		2.914 0	2.869 6	3.108 3	3.061 8	3.012 2	2.795 1	3.703 4	3.704 9
New"		2.914 3	2.850 9	3.108 8	3.013 4	3.012 3	2.799 3	3.704 0	3.727 2

Moreover, taking into account the values of the coefficients selected in the optimization process it can be concluded that the Mann iteration transforms into Picard's iteration. The Ishikawa, Agarwal, Das–Debata, Khan–Cho–Abbas and

Table 2. Values of the optimised coefficients for algorithms without inertia weight for the considered approach minimizing the number of iterations.

Test environment	Coefficients	A Coefficients		B Coefficients		C Coefficients		D Coefficients	
Picard	$p_{0,1}$ $\eta_{0,0}$	1.000 0	1.000 0	1.000 0	1.000 0	1.000 0	1.000 0	1.000 0	1.000 0
Mann	$p_{0,1}$ $\eta_{0,0}$	0.999 9	0.999 7	1.000 0	1.000 0	1.000 0	1.000 0	0.998 7	0.968 0
Ishikawa	$p_{0,1}$ $\eta_{0,0}$	0.999 8	0.999 8	0.996 3	1.000 0	1.000 0	1.000 0	0.960 6	0.984 8
	$p_{1,3}$	1.000 0		1.000 0		1.000 0		1.000 0	
Agarwal	$p_{0,1}$ $\eta_{0,0}$	0:998 5	0.999 7	0.998 9	1.000 0	0.999 6	0.999 7	0.955 5	0.989 7
	$p_{1,3}$	0.999 7		1.000 0		0.999 5		0.995 0	
Das–Debata	$p_{0,1}$ $\eta_{0,0}$	0.999 4	0.998 3	0.997 5	1.000 0	0.999 6	0.999 5	0.998 7	0.991 4
	$p_{1,3}$ $\eta_{1,1}$	0.999 8	0.999 7	1.000 0	0.998 7	1.000 0	0.999 6	0.999 9	0.935 3
Khan–Cho–Abbas	$p_{0,1}$ $\eta_{0,0}$	0.998 3	1.000 0	0.999 4	0.999 9	0.999 6	0.999 7	0.964 4	0.991 3
	$p_{1,3}$ $\eta_{1,1}$	0.999 1	0.999 9	1.000 0	0.996 1	0.999 8	0.999 5	0.993 4	0.962 6
Generalised Agarwal	$p_{0,1}$ $\eta_{0,0}$	0.999 7	0.999 0	0.990 1	1.000 0	0.999 5	0.999 6	0.984 3	0.990 4
	$p_{1,3}$ $\eta_{1,1}$	0.999 7	0.999 2	1.000 0	0.998 7	0.999 9	0.999 6	0.997 0	0.941 2
	$\eta_{2,2}$		0.583 7		0.559 6		0.705 2		0.890 4
New'	$p_{0,1}$ $\eta_{0,0}$	0.999 9	1.000 0	0.999 9	1.000 0	0.999 7	0.999 9	0.951 8	0.988 5
	$p_{1,3}$	1.000 0		0.999 7		0.999 7		0.9740	
	$p_{2,5}$	0.995 6		1.000 0		0.998 5		0.987 1	
New"	$p_{0,1}$ $\eta_{0,0}$	0.999 5	0.999 3	0.999 4	0.999 8	1.000 0	0.999 4	0.977 9	0.954 4
	$p_{1,3}$ $\eta_{1,1}$	0.999 7	0.999 7	0.999 9	0.999 9	0.999 7	0.999 9	0.985 0	0.969 6
	$p_{2,5}$ $\eta_{2,2}$	0.998 4	0.999 8	0.999 8	1.000 0	0.999 2	0.999 3	0.986 1	0.993 7

generalised Agarwal iterations are also transformed into the same form. And, the New' iteration is transformed to the New". These observations confirm the polynomiographs presented in Fig. 2 for iteration with transformations without inertia weight and in Fig. 3 with inertia weight. The acceleration constant may take values greater than 1.0. It is possible due to the properties of the analysed test environments. Polynomiographs for the proposed schemes look smoother compared to other methods, this is primarily caused by the improved convergence behaviour.

6.2 Obtaining an Aesthetic Pattern

Table 4 presents the average iteration values for algorithms without inertia weight (I) and with their use (II). The values of the coefficients for the algorithms without inertia weight are presented in Table 5, and with the inertia weight in Table 6. One way to obtain aesthetic patterns is to vary the number of iterations creating the polynomiograph significantly. It is associated with a considerable extension of the average iteration value. The number of iterations can be differentiated by increasing the inertia weight and decreasing the acceleration constant. Changes in the p parameters allow the control of the particle dynamics in different areas of the image. We can conclude – the greater the number of coefficients, the greater the ability to control particle dynamics.

Table 3. Values of the optimised coefficients for algorithms with inertia weight for the considered approach minimizing the number of iterations.

Test environment	Iteration	Coefficients	A Coefficients		B Coefficients		C Coefficients		D Coefficients	
Mann	$p_{0,1}$	$\omega_{0,0}$	0.919 0	0.000 3	0.739 8	0.000 4	0.902 6	0.028 8	0.834 0	0.000 3
		$\eta_{0,0}$		1.150 8		1.381 7		1.386 9		1.158 8
Ishikawa	$p_{0,1}$	$\omega_{0,0}$	0.981 9	0.001 4	0.990 4	0.000 2	0.999 6	0.002 4	0.964 3	0.000 1
	$p_{1,3}$	$\eta_{0,0}$	0.996 1	1.079 2	0.999 8	1.019 9	0.971 5	1.172 7	0.999 3	0.981 1
Agarwal	$p_{0,1}$	$\omega_{0,0}$	0.985 2	0.000 0	0.951 3	0.000 1	0.997 4	0.013 0	0.928 5	0.001 0
	$p_{1,3}$	$\eta_{0,0}$	0.916 8	1.105 9	0.988 8	1.031 3	0.740 7	1.499 2	0.986 8	0.997 1
Das–Debata	$p_{0,1}$	$\omega_{0,0}$	0.830 3	0.002 3	0.902 6	0.000 4	0.999 9	0.011 2	0.688 0	0.000 2
	$p_{1,3}$	$\eta_{0,0}$	0.998 7	1.061 5	0.999 8	1.019 1	0.999 7	1.095 1	0.997 9	0.989 6
		$\omega_{1,1}$		0.000 3		0.000 3		0.009 6		0.000 7
		$\eta_{1,1}$		1.275 0		1.097 0		1.499 0		1.348 6
Khan–Cho–Abbas	$p_{0,1}$	$\omega_{0,0}$	0.998 9	0.001 5	0.996 9	0.000 2	0.999 3	0.013 8	0.996 6	0.000 1
	$p_{1,3}$	$\eta_{0,0}$	0.784 5	1.321 9	0.781 9	1.307 3	0.908 8	1.222 9	0.737 4	1.341 7
		$\omega_{1,1}$		0.000 6		0.000 3		0.006 2		0.002 6
		$\eta_{1,1}$		1.075 8		0.995 8		1.499 3		0.950 0
Generalised Agarwal	$p_{0,1}$	$\omega_{0,0}$	0.712 3	0.001 3	0.748 3	0.000 3	1.000 0	0.005 4	0.651 5	0.000 4
	$p_{1,3}$	$\eta_{0,0}$	0.972 4	1.070 7	0.980 1	1.039 2	0.972 5	1.134 3	0.901 6	1.100 6
		$\omega_{1,1}$		.000 0		0.003 2		0.011 9		0.013 4
		$\eta_{1,1}$		1.491 2		1.351 7		1.498 3		1.455 1
		$\omega_{2,2}$		0.024 2		0.004 6		0.197 8		0.007 1
		$\eta_{2,2}$		1.069 4		0.987 1		1.483 5		0.972 6
New'	$p_{0,1}$	$\omega_{0,0}$	0.994 6	0.000 0	0.999 9	0.000 0	0.998 9	0.001 1	0.950 9	0.000 1
	$p_{1,3}$	$\eta_{0,0}$	0.997 5	1.191 6	0.990 3	1.162 1	0.994 1	1.496 3	0.948 9	0.997 0
	$p_{2,5}$		0.749 4		0.875 4		0.638 8		0.988 6	
New"	$p_{0,1}$	$\omega_{0,0}$	0.993 7	0.006 9	0.996 8	0.032 4	0.975 1	0.000 5	0.999 2	0.026 5
	$p_{1,3}$	$\eta_{0,0}$	0.850 4	1.144 7	0.999 0	0.980 9	0.993 7	1.499 0	0.985 2	0.992 8
	$p_{2,5}$	$\omega_{1,1}$	0.939 0	0.009 7	0.692 5	0.054 7	0.692 3	0.005 0	0.866 8	0.001 5
		$\eta_{1,1}$		1.429 2		1.474 6		1.497 8		0.929 9
		$\omega_{2,2}$		0.003 2		0.000 8		0.003 8		0.000 9
		$\eta_{2,2}$		0.966 6		1.468 0		1.464 6		1.144 1

Table 4. The average number of iterations of the analysed algorithms for the considered approach generating aesthetic patterns.

Iteration	Test							
	A		B		C		D	
	I	II	I	II	I	II	I	II
New'	21.26	14.50	45.87	122.41	9.30	9.60	19.83	15.33
New"	6.44	26.42	29.44	58.04	13.19	17.88	10.88	27.07

The generalised form of iteration proposed in the article gives such possibilities. The possibilities of creating patterns with aesthetic features are presented in

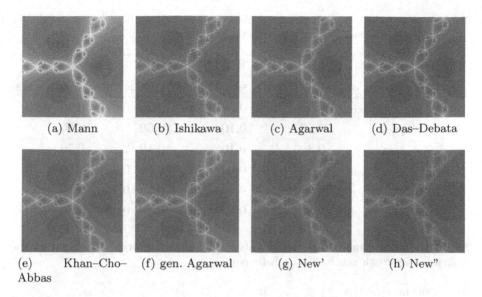

(a) Mann   (b) Ishikawa   (c) Agarwal   (d) Das–Debata

(e)   Khan–Cho–   (f) gen. Agarwal   (g) New'   (h) New"
Abbas

Fig. 2. Polynomiographs of the iterations without inertia weights.

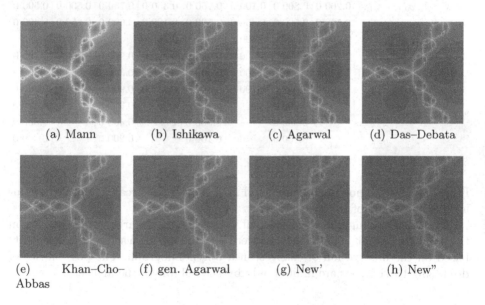

(a) Mann   (b) Ishikawa   (c) Agarwal   (d) Das–Debata

(e)   Khan–Cho–   (f) gen. Agarwal   (g) New'   (h) New"
Abbas

Fig. 3. Polynomiographs of the iterations with inertia weights.

Table 5. Values of the selected coefficients for algorithms without inertia weight for the considered approach generating aesthetic patterns.

Test environment			A		B		C		D	
Iteration	Coefficients		Coefficients		Coefficients		Coefficients		Coefficients	
New'	$p_{0,1}$	$\eta_{0,0}$	0.50	0.10	0.20	0.10	0.20	0.40	0.10	0.20
	$p_{1,3}$		0.80		0.10		0.50		0.20	
	$p_{2,5}$		0.30		0.10		0.10		0.50	
New"	$p_{0,1}$	$\eta_{0,0}$	0.10	0.10	0.04	0.20	0.50	0.20	0.20	0.20
	$p_{1,3}$	$\eta_{1,1}$	0.20	0.70	0.10	0.15	0.20	0.15	0.10	0.50
	$p_{2,5}$	$\eta_{2,2}$	0.10	0.20	0.10	0.40	0.30	0.40	0.30	0.40

Table 6. Values of the selected coefficients for algorithms with inertia weight for the considered approach generating aesthetic patterns.

Test environment			A		B		C		D	
Iteration	Coefficients		Coefficients		Coefficients		Coefficients		Coefficients	
New'	$p_{0,1}$	$\omega_{0,0}$	0.200 0	0.800 0	0.100 0	0.750 0	0.350 0	0.750 0	0.500 0	0.500 0
	$p_{1,3}$	$\eta_{0,0}$	0.500 0	0.500 0	0.450 0	0.600 0	0.100 0	0.600 0	0.250 0	0.400 0
	$p_{2,5}$		0.400 0		0.400 0		0.350 0		0.100 0	
New"	$p_{0,1}$	$\omega_{0,0}$	0.200 0	0.900 0	0.200 0	0.900 0	0.200 0	0.900 0	0.300 0	0.700 0
	$p_{1,3}$	$\eta_{0,0}$	0.100 0	0.200 0	0.300 0	0.300 0	0.100 0	0.300 0	0.300 0	0.300 0
	$p_{2,5}$	$\omega_{1,1}$	0.100 0	0.900 0	0.100 0	0.900 0	0.200 0	0.900 0	0.400 0	0.800 0
	$\eta_{1,1}$			0.200 0		0.200 0		0.300 0		0.400 0
	$\omega_{2,2}$			0.900 0		0.500 0		0.400 0		0.800 0
	$\eta_{2,2}$			0.100 0		0.100 0		0.200 0		0.300 0

Figs. 4 and 5. The polynomiographs in Fig. 5 show much greater particle dynamics due to the use of the inertia weight.

Even slight changes in the coefficient values can cause large visual changes on the polynomiograph. The coefficients were selected during a series of experiments based on the observation of changes on a polynomiograph. The possibilities of discussing these issues are limited only by the volume of the article.

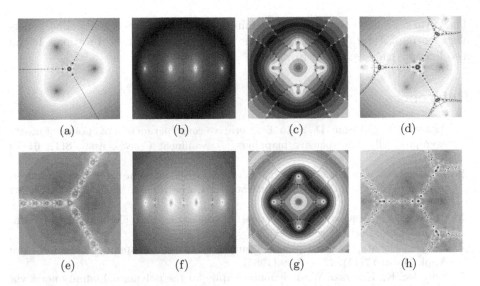

Fig. 4. Polynomyographs with aesthetic features obtained using the iterations without inertia weights.

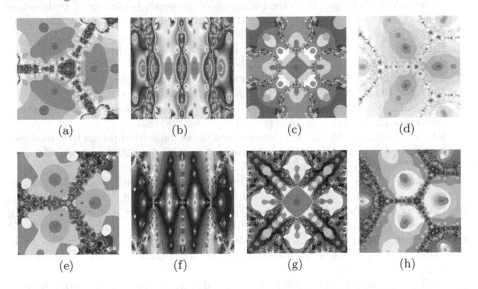

Fig. 5. Polynomyographs with aesthetic features obtained using the iterations with inertia weights.

7   Conclusions

The article proposes a generalised form of iteration. Based on it, a new iteration for one and several mappings is proposed. It was shown that the proposed algorithm enables fast root-finding and creating patterns with aesthetic features.

These tasks, due to the algorithm tuning, are contradictory goals. The generalised form of iteration will allow for the creation of many new iterations. It can also become the basis for hybridisation with other algorithms.

References

1. Agarwal, R., O'Regan, D., Sahu, D.: Iterative construction of fixed points of nearly asymptotically nonexpansive mappings. J. Nonlinear Convex Anal. **8**(1), 61–79 (2007)
2. Ardelean, G., Cosma, O., Balog, L.: A comparison of some fixed point iteration procedures by using the basins of attraction. Carpathian J. Math. **32**(3), 277–284 (2016)
3. Cheney, W., Kincaid, D.: Numerical Mathematics and Computing, 6th edn. Brooks/Cole, Pacific Groove (2007)
4. Das, G., Debata, J.: Fixed points of quasinonexpansive mappings. Indian J. Pure Appl. Math. **17**(11), 1263–1269 (1986)
5. Gdawiec, K., Kotarski, W.: Polynomiography for the polynomial infinity norm via Kalantari's formula and nonstandard iterations. Appl. Math. Comput. **307**, 17–30 (2017). https://doi.org/10.1016/j.amc.2017.02.038
6. Gościniak, I., Gdawiec, K.: Visual analysis of dynamics behaviour of an iterative method depending on selected parameters and modifications. Entropy **22**(7), 734 (2020). https://doi.org/10.3390/e22070734
7. Howes, L., Munshi, A.: The OpenCL specification (2015). http://www.khronos.org/registry/OpenCL/specs/opencl-2.0.pdf
8. Ishikawa, S.: Fixed points by a new iteration method. Proc. Am. Math. Soc. **44**(1), 147–150 (1974). https://doi.org/10.1090/S0002-9939-1974-0336469-5
9. Kalantari, B.: Polynomial Root-Finding and Polynomiography. World Scientific, Singapore (2009). https://doi.org/10.1142/9789812811837
10. Khan, S., Cho, Y., Abbas, M.: Convergence to common fixed points by a modified iteration process. J. Appl. Math. Comput. **35**(1), 607–616 (2011). https://doi.org/10.1007/s12190-010-0381-z
11. Mann, W.: Mean value methods in iteration. Proc. Am. Math. Soc. **4**(3), 506–510 (1953). https://doi.org/10.1090/S0002-9939-1953-0054846-3
12. Petković, I., Rančić, L.: Computational geometry as a tool for studying root-finding methods. Filomat **33**(4), 1019–1027 (2019). https://doi.org/10.2298/FIL1904019P
13. Picard, E.: Mémoire sur la théorie des équations aux dérivées partielles et la méthode des approximations successives. J. de Mathématiques Pures et Appliquées **6**(4), 145–210 (1890)
14. Wall, M.: GAlib, a C++ library of genetic algorithm components (2008). http://lancet.mit.edu/ga/

Peridynamic Damage Model Based on Absolute Bond Elongation

Shangyuan Zhang and Yufeng Nie[(✉)]

Research Center for Computational Science, Northwestern Polytechnical University,
Xi'an 710129, People's Republic of China
zhangshangyuan@mail.nwpu.edu.cn, yfnie@nwpu.edu.cn

Abstract. A bond-based peridynamic damage model is proposed to incorporate the deformation and the damage process into a unified framework. This new model is established based on absolute bond elongation, and both the elastic and damage parameters of the material are embedded in the constitutive relationship, which makes the model better characterize the process of material damage. Finally, different phenomenons for various damage patterns is observed by numerical experiments, rich damage patterns will make this model better suitable for damage simulation.

Keywords: Damage · Peridynamic · Bond-based · Absolute bond elongation

1 Introduction

Peridynamic [19] provides an alternative theory to classical continuum mechanics in modeling complex crack problems. Different from the classical continuum mechanics, the mechanical behavior of the material is characterized by nonlocal interactions between material points. The spatial derivative of the displacement in the model is replaced by the integration. It is this feature makes it an advantage in dealing with crack propagation problems. Its effectiveness in modeling material damage has been shown in numerical simulation of crack nucleation [21], crack propagation [15] and branching [4,13], phase transformations in solids [6], impact damage [3] and so on. Mathematical analysis and numerical approximation of the peridynamic model have been studied in [5,8–11].

Silling and Askari introduced a peridynamic damage model in [20]. However, in this model, the damage is a function of the bond stretch (the rate of elongation), which is not continuous about bond stretch. This brings a lot of trouble to the well-posedness of the model in mathematics. It is impossible to describe the process of the bond from elastic deformation to damage and finally failure. Yang et al. [7] proposed a damage model to investigate mode-I crack propagation in concrete by constructing a trilinear softening curve of the bond stretch. For the first time, Emmrich give a well-posed result for a nonlinear peridynamic model

D. Groen et al. (Eds.): ICCS 2022, LNCS 13350, pp. 637–650, 2022.
https://doi.org/10.1007/978-3-031-08751-6_46

with Lipschitz continuous pairwise force function [9], and extended it to inherit irreversible damage [10]. To make the model accurately describe the fracture phenomenon, there are two main problems need to be considered. One is how to involve damage in the constitutive relationship of the material. The other is the mechanism of crack nucleation in fracture. The absolute bond elongation contains higher order deformation than the stretch, it can describe deformation better. It is essential for the peridynamic model to integrate these components into the model.

Based on the above viewpoints, a peridynamic damage model is established based on absolute bond elongation, and the fracture criterion based on absolute bond elongation is also given. By setting the damage term to be continuous, the well-posedness could be ensured. The ability to treat both the deformation and damage within the same mathematical framework will make the peridynamic model a practical tool to simulate the whole process of the real material deformation.

This paper is organized as follows. The peridynamic damage model for PMB (Prototype Microelastic Brittle) material is presented in Sect. 2. Then, the damage model based on bond elongation are presented in Sect. 3. Finally, numerical experiments are then presented in Sect. 4, showing that the effects of different damage patterns on material fracture behavior and further discussing the peridynamic modeling.

2   The Peridynamic Theory

The equation of motion for the bond-based peridynamic model is

$$\rho \ddot{\mathbf{u}}(\mathbf{x}, t) = \int_{\mathcal{H}_\mathbf{x}} \mathbf{f}\left(\mathbf{u}\left(\mathbf{x}', t\right) - \mathbf{u}(\mathbf{x}, t), \mathbf{x}' - \mathbf{x}\right) dV_{\mathbf{x}'} + \mathbf{b}(\mathbf{x}, t), \ (x, t) \in \Omega \times (0, T).$$

(1)

where ρ denotes the mass density, $\mathcal{H}_x$ is the peridynamic neighborhood of $\mathbf{x} \in \Omega$, $\mathbf{f}$ is the pairwise force function, $\mathbf{b}$ is the external force density, the vector $\boldsymbol{\xi} = \mathbf{x}' - \mathbf{x}$ denotes the relative position vector between the two material points, which we call bond. $\boldsymbol{\eta} = \mathbf{u}\left(\mathbf{x}', t\right) - \mathbf{u}(\mathbf{x}, t)$ represents the relative displacement. The interaction between the material points will decrease with the distance increasing, once the distance between the two material points beyond the horizon δ, the interaction $\mathbf{f}(\boldsymbol{\eta}, \boldsymbol{\xi}) = \mathbf{0}(|\boldsymbol{\xi}| > \delta)$. The bond stretch is the relative change of the bond, which is defined as follows

$$s(\boldsymbol{\xi}, \boldsymbol{\eta}) = \frac{|\boldsymbol{\eta} + \boldsymbol{\xi}| - |\boldsymbol{\xi}|}{|\boldsymbol{\xi}|}.$$

(2)

For PMB material, the pairwise force function is proportional to the bond stretch. In order to describe the damage phenomenon, a damage indicator is multiplied in original constitutive relation [20], and the constitutive relation can be written as

$$\mathbf{f}(\boldsymbol{\eta}, \boldsymbol{\xi}) = cs(\boldsymbol{\xi}, \boldsymbol{\eta})\mu(\boldsymbol{\xi}, \boldsymbol{\eta})\frac{\boldsymbol{\eta} + \boldsymbol{\xi}}{|\boldsymbol{\eta} + \boldsymbol{\xi}|},$$

(3)

Then the spring constant c can be expressed with the known material constants in the classical theory [2]. $\mu(\boldsymbol{\xi}, \boldsymbol{\eta})$ is the bond damage indicator, and its expression is

$$\mu(\boldsymbol{\xi}, \boldsymbol{\eta}) = \begin{cases} 1, & \text{otherwise} . \\ 0, & \exists\, t' \in [0, t], \ s.t. \ s\left(\boldsymbol{\xi}, \boldsymbol{\eta}(t')\right) > s_0. \end{cases} \tag{4}$$

Here, s_0 is critical bond stretch that might be determined from experimental data. However, since there exists a jump break point at the critical stretch, which brings the difficulty in the proof of the well-posedness.

3   The Peridynamic with Damage

In our model, both the elastic and damage parameters of the material are embedded in the constitutive relationship. Besides, the absolute bond elongation contains more deformation information than the bond stretch. The damage model based on absolute bond elongation is given as follows

$$\mathbf{f}(\boldsymbol{\eta}, \boldsymbol{\xi}) = \omega(|\boldsymbol{\xi}|) e(|\boldsymbol{\eta} + \boldsymbol{\xi}|, |\boldsymbol{\xi}|) \mu(\boldsymbol{\xi}, e) \frac{\boldsymbol{\eta} + \boldsymbol{\xi}}{|\boldsymbol{\eta} + \boldsymbol{\xi}|}. \tag{5}$$

The ω is the influence function, which reflects that the different bond have effect on its own force and the material property, $e = |\boldsymbol{\eta} + \boldsymbol{\xi}| - |\boldsymbol{\xi}|$ is the absolute bond elongation. Two common forms of influence function in the literature [4, 20] are given as follows.

(a) PMB material: $\omega(|\boldsymbol{\xi}|) = \frac{c}{|\boldsymbol{\xi}|}$.   (b) Soda-Lime Glass: $\omega(|\boldsymbol{\xi}|) = \frac{c}{|\boldsymbol{\xi}|} \left(1 - \frac{|\boldsymbol{\xi}|}{\delta}\right)$.

In constitutive relation (5), μ is a function of absolute bond elongation e, and its forms were constructed by using Hermite interpolation, such as

(i) The function itself is continuous:

$$\mu(e) = \begin{cases} 1, & \text{if } e < \lambda e_c(|\boldsymbol{\xi}|), \\ \frac{e_c - e}{e_c - \lambda e_c}, & \text{if } \lambda e_c(|\boldsymbol{\xi}|) \le e \le e_c(|\boldsymbol{\xi}|), \\ 0, & \text{if } e > e_c(|\boldsymbol{\xi}|). \end{cases} \tag{6}$$

(ii) First-order derivative function is continuous:

$$\mu(e) = \begin{cases} 1, & \text{if } e < \lambda e_c(|\boldsymbol{\xi}|), \\ 1 - \frac{3(e - \lambda e_c)^2}{(e_c - \lambda e_c)^2} + \frac{2(e - \lambda e_c)^3}{(e_c - \lambda e_c)^3}, & \text{if } \lambda e_c(|\boldsymbol{\xi}|) \le e \le e_c(|\boldsymbol{\xi}|), \\ 0, & \text{if } e > e_c(|\boldsymbol{\xi}|). \end{cases} \tag{7}$$

(iii) Second-order derivative function is continuous:

$$\mu(e) = \begin{cases} 1, & \text{if } e < \lambda e_c(|\boldsymbol{\xi}|), \\ 1 - \frac{10(e - \lambda e_c)^3}{(e_c - \lambda e_c)^3} + \frac{15(e - \lambda e_c)^4}{(e_c - \lambda e_c)^4} - \frac{6(e - \lambda e_c)^5}{(e_c - \lambda e_c)^5}, & \text{if } \lambda e_c(|\boldsymbol{\xi}|) \le e \le e_c(|\boldsymbol{\xi}|), \\ 0, & \text{if } e > e_c(|\boldsymbol{\xi}|). \end{cases} \tag{8}$$

Where λ is the parameter given in the computing process, and $e_c(|\boldsymbol{\xi}|)$ is the critical elongation of each bond. If we assume that $\omega(|\boldsymbol{\xi}|)$ has form (a) or (b), and $e_c(|\boldsymbol{\xi}|) = d|\boldsymbol{\xi}|^s$, then the peridynamic damage model based on absolute bond elongation is obtained. We now turn to determine the model parameters. It should be noticed that the determination of c is before the damage ($\mu = 1$).

3.1 Determination of Elastic Constants

Influence Function in Form (a). Assumed that $\omega(|\boldsymbol{\xi}|) = \frac{c}{|\boldsymbol{\xi}|}$ and the body undergo a homogeneous deformation, that is $\boldsymbol{\eta} = \epsilon\boldsymbol{\xi}$, then the bond elongation $e = \epsilon|\boldsymbol{\xi}|$. Therefore, the energy in bond $\boldsymbol{\xi}$ is

$$W_{bond} = \int_0^{\epsilon|\boldsymbol{\xi}|} \frac{c}{|\boldsymbol{\xi}|}e\, de = \frac{c\epsilon^2|\boldsymbol{\xi}|}{2}. \tag{9}$$

The energy density at x is

$$W_{nonlocal} = \frac{1}{2}\int_{\mathcal{H}_x} W_{bond}dVx' = \begin{cases} \int_0^\delta \frac{c\epsilon^2\xi}{4}4\pi\xi^2 d\xi = \frac{\pi c\epsilon^2\delta^4}{4}, & D = 3, \\[2mm] h_2\int_0^\delta \frac{c\epsilon^2\xi}{4}2\pi\xi d\xi = \frac{\pi ch_2\epsilon^2\delta^3}{6}, & D = 2, \\[2mm] h_1\int_0^\delta \frac{c\epsilon^2\xi}{4}2d\xi = \frac{ch_1\epsilon^2\delta^2}{4}, & D = 1. \end{cases} \tag{10}$$

Undergo the same deformation, the strain energy density of the classical theory is

$$W^{3D}_{classical} = \frac{9k\epsilon^2}{2}, \quad W^{2D}_{classical} = 2k'\epsilon^2, \quad W^{1D}_{classical}\frac{E\epsilon^2}{2}. \tag{11}$$

then we obtain that

$$c^{3D} = \frac{18k}{\pi\delta^4}, \quad c^{2D} = \frac{12k'}{\pi h_2\delta^3}, \quad c^{1D} = \frac{2E}{h_1\delta^2}. \tag{12}$$

where k and k' is the bulk modulus in 3D and 2D respectively, E is the Young's modulus, h_1 and h_2 is the rod cross-sectional area and plate thickness.

Influence Function in Form (b). Assumed that $\omega(|\boldsymbol{\xi}|) = \frac{c}{|\boldsymbol{\xi}|}\left(1 - \frac{|\boldsymbol{\xi}|}{\delta}\right)$, then the energy in bond $\boldsymbol{\xi}$ is

$$W_{bond} = \int_0^{\epsilon|\boldsymbol{\xi}|} \frac{c}{|\boldsymbol{\xi}|}\left(1 - \frac{|\boldsymbol{\xi}|}{\delta}\right)e\, de = \frac{c\epsilon^2|\boldsymbol{\xi}|}{2}\left(1 - \frac{|\boldsymbol{\xi}|}{\delta}\right) \tag{13}$$

The energy density at x is

$$W_{nonlocal} = \frac{1}{2}\int_{\mathcal{H}_x} W_{bond} dV x' = \begin{cases} \int_0^\delta \frac{ce^2|\xi|}{4}\left(1 - \frac{|\xi|}{\delta}\right) 4\pi\xi^2 d\xi = \frac{\pi ce^2\delta^4}{20}, & D = 3, \\ h_2\int_0^\delta \frac{ce^2|\xi|}{4}\left(1 - \frac{|\xi|}{\delta}\right) 2\pi\xi d\xi = \frac{\pi ch_2\epsilon^2\delta^3}{24}, & D = 2, \\ h_1\int_0^\delta \frac{ce^2|\xi|}{4}\left(1 - \frac{|\xi|}{\delta}\right) 2d\xi = \frac{ch_1\epsilon^2\delta^2}{12}, & D = 1. \end{cases}$$

(14)

Undergo the same deformation, the strain energy density of the classical theory is

$$W^{3D}_{classical} = \frac{9k\epsilon^2}{2}, \quad W^{2D}_{classical} = 2k'\epsilon^2, \quad W^{1D}_{classical} = \frac{E\epsilon^2}{2}, \quad (15)$$

then we obtain that

$$c^{3D} = \frac{90k}{\pi\delta^4}, \quad c^{2D} = \frac{48k'}{\pi h_2\delta^3}, \quad c^{1D} = \frac{6E}{h_1\delta^2}. \quad (16)$$

where k and k' is the bulk modulus in 3D and 2D respectively, E is the Young's modulus, h_1 and h_2 is the rod cross-sectional area and plate thickness.

3.2 Determination of Damage Constants

If we assume that $e_c(|\xi|) = d|\xi|^s$, then we need to determine the parameter d for bond damage. Assume the bond undergo a elongation such that the bond broken, The superscripts below are used to represent the energy under different influence functions and damage patterns. Using following formulation

$$W_{bondbroken} = \int_0^{e_c} \omega(|\xi|)e\mu(e)\, de. \quad (17)$$

Then the energy in a single bond under different patterns can be written as

- $W^{ai}_{bondbroken} = \dfrac{cd^2|\xi|^{2s-1}(\lambda^2 + \lambda + 1)}{6}$,

- $W^{aii}_{bondbroken} = \dfrac{cd^2|\xi|^{2s-1}(3\lambda^2 + 4\lambda + 3)}{20}$,

- $W^{aiii}_{bondbroken} = \dfrac{cd^2|\xi|^{2s-1}(2\lambda^2 + 3\lambda + 2)}{14}$,

- $W^{bi}_{bondbroken} = cd^2|\xi|^{2s-1}\left(1 - \dfrac{|\xi|}{\delta}\right)\dfrac{\lambda^2 + \lambda + 1}{6}$,

- $W^{bii}_{bondbroken} = cd^2|\xi|^{2s-1}\left(1 - \dfrac{|\xi|}{\delta}\right)\dfrac{3\lambda^2 + 4\lambda + 3}{20}$,

- $W^{biii}_{bondbroken} = cd^2|\xi|^{2s-1}\left(1 - \dfrac{|\xi|}{\delta}\right)\dfrac{2\lambda^2 + 3\lambda + 2}{14}$.

Next, we will obtain the parameters in one, two and three dimensions.

3D Case. The critical energy release rate in 3D case can be expressed as

$$G_0 = \int_0^\delta \int_0^{2\pi} \int_z^\delta \int_0^{cos^{-1}z/\xi} W_{bondbroken}\xi^2 sin\phi d\phi d\xi d\theta dz. \tag{18}$$

Then the damage constant in different influence function and damage patterns will be calculated as

$$d^{ai} = \sqrt{\frac{6(2s+3)G_0}{\pi c\delta^{2s+3}(1+\lambda+\lambda^2)}}, \qquad d^{bi} = \sqrt{\frac{6(2s+3)(2s+4)G_0}{\pi c\delta^{2s+3}(1+\lambda+\lambda^2)}},$$

$$d^{aii} = \sqrt{\frac{20(2s+3)G_0}{\pi c\delta^{2s+3}(3+4\lambda+3\lambda^2)}}, \qquad d^{bii} = \sqrt{\frac{20(2s+3)(2s+4)G_0}{\pi c\delta^{2s+3}(3+4\lambda+3\lambda^2)}}, \tag{19}$$

$$d^{aiii} = \sqrt{\frac{14(2s+3)G_0}{\pi c\delta^{2s+3}(2+3\lambda+2\lambda^2)}}, \qquad d^{biii} = \sqrt{\frac{14(2s+3)(2s+4)G_0}{\pi c\delta^{2s+3}(2+3\lambda+2\lambda^2)}}.$$

2D Case. The critical energy release rate in 2D case can be expressed as

$$G_0 = 2h \int_0^\delta \int_z^\delta \int_0^{cos^{-1}z/\xi} W_{bondbroken}\xi d\phi d\xi dz. \tag{20}$$

Then the damage constant in different influence function and damage patterns in 2D case will obtained as follows.

$$d^{ai} = \sqrt{\frac{6G_0}{chF_a(1+\lambda+\lambda^2)}}, \qquad d^{bi} = \sqrt{\frac{6G_0}{chF_b(1+\lambda+\lambda^2)}},$$

$$d^{aii} = \sqrt{\frac{20G_0}{chF_a(3+4\lambda+3\lambda^2)}}, \qquad d^{bii} = \sqrt{\frac{20G_0}{chF_b(3+4\lambda+3\lambda^2)}}, \tag{21}$$

$$d^{aiii} = \sqrt{\frac{14G_0}{chF_a(2+3\lambda+2\lambda^2)}}, \qquad d^{biii} = \sqrt{\frac{14G_0}{chF_b(2+3\lambda+2\lambda^2)}},$$

where

$$F_a = \frac{\delta^{2+2s}(4s+\frac{1}{1+s}+\frac{\sqrt{\pi}\Gamma(-1-s)}{\Gamma(-1/2-s)})+\frac{2\delta^{2+2s}\sqrt{\pi}\Gamma(-s)}{(2+2s)\Gamma(-1/2-s)}}{(1+2s)^2}, \tag{22}$$

$$F_b = \frac{\delta^{2+2s}}{3+5s+2s^2}, \tag{23}$$

are the functions used in (21), h is the thickness of the plate. However, there is *Gamma* function in (22), which is not exist in some case, we can use three-dimensional damage parameters as an alternative way.

1D Case. The critical energy release rate in 1D case can be expressed as

$$G_0 = h \int_0^\delta \int_z^\delta W_{bondbroken} d\xi dz. \tag{24}$$

Then the damage constant in different influence function and damage patterns in 1D case can be calculated as

$$d^{ai} = \sqrt{\frac{6(2s+1)G_0}{ch\delta^{2s+1}(1+\lambda+\lambda^2)}}, \qquad d^{bi} = \sqrt{\frac{6(2s+2)(2s+1)G_0}{ch\delta^{2s+1}(1+\lambda+\lambda^2)}},$$

$$d^{aii} = \sqrt{\frac{20(2s+1)G_0}{ch\delta^{2s+1}(3+4\lambda+3\lambda^2)}}, \qquad d^{bii} = \sqrt{\frac{20(2s+2)(2s+1)G_0}{ch\delta^{2s+1}(3+4\lambda+3\lambda^2)}}, \tag{25}$$

$$d^{aiii} = \sqrt{\frac{14(2s+1)G_0}{ch\delta^{2s+1}(2+3\lambda+2\lambda^2)}}, \qquad d^{biii} = \sqrt{\frac{14(2s+2)(2s+1)G_0}{ch\delta^{2s+1}(2+3\lambda+2\lambda^2)}},$$

where h is the cross sectional area of the bar.

3.3 Conservation of Energy and Energy Decay

The peridynamic damage model based on absolute bond elongation is given as follows:

$$\begin{cases} \rho\ddot{u}(t,\mathbf{x}) = \int_{\Omega\cup\Omega_c} \omega(|\boldsymbol{x}'-\boldsymbol{x}|)f(c(t,\boldsymbol{x},\boldsymbol{x}',u)\mu(e,e^*)\mathbf{M}(t,\boldsymbol{x},\boldsymbol{x}',u)d\mathbf{x}' \\ \qquad + \mathbf{b}(t,\mathbf{x}), \\ \mathbf{u}(0,\mathbf{x}) = \mathbf{w}(\mathbf{x}), \quad \dot{\mathbf{u}}(0,\mathbf{x}) = \mathbf{v}(\mathbf{x}), \quad \mathbf{u}(t,\cdot)|_{\Omega_c} = 0. \end{cases} \tag{26}$$

where ρ is the mass density, the term $\omega_\delta(\boldsymbol{x}'-\boldsymbol{x})$ is the influence function, $f(e(t,\boldsymbol{x},\boldsymbol{x}',\boldsymbol{u})) = ce(t,\boldsymbol{x},\boldsymbol{x}',\boldsymbol{u})$ reflects the relationship between the magnitude of the bond force and the bond elongation, e is bond elongation, e^* is the largest bond elongation in historical time, $\mu(e,e^*)$ denotes the damage, $\mathbf{M}(t,\boldsymbol{x},\boldsymbol{x}',\boldsymbol{u})$ indicates the direction of the bond force, $\mathbf{b}(t,\mathbf{x})$ is the body force. Assume that external forces don't change over time, the total energy of the system is

$$E(t) = \frac{1}{2} \int_{\Omega\cup\Omega_I} \int_{\Omega\cup\Omega_I} \omega(|\mathbf{x}'-\mathbf{x}|)p(e(t,\mathbf{x},\mathbf{x}',\mathbf{u}))d\mathbf{x}'d\mathbf{x}$$
$$+ \frac{1}{2} \int_\Omega \rho|\dot{\mathbf{u}}(t,\mathbf{x})|^2 d\mathbf{x} - \int_\Omega \mathbf{u}(t,\mathbf{x}) \cdot \mathbf{b}(\mathbf{x})d\mathbf{x}. \tag{27}$$

where $\omega_\delta(|\mathbf{x}'-\mathbf{x}|)p(e)$ is the energy produced by the deformation of a single bond.

$$p(e) = \int_0^e f(e')\mu(e',e^*)de' \tag{28}$$

Theorem 1. *(Conservation of Energy and Energy Decay) If the bond never broken, the total energy of system* (26) *is conserved. If the bond broken, the total energy of system* (26) *is nonincreasing in time.*

Proof.

$$\frac{dE(t)}{dt} = \int_\Omega \rho \dot{\mathbf{u}}(t,\mathbf{x}) \cdot \ddot{\mathbf{u}}(t,\mathbf{x}) dx - \int_\Omega \dot{\mathbf{u}}(t,\mathbf{x}) \cdot \mathbf{b}(\mathbf{x}) dx$$

$$+ \frac{1}{2} \int_{\Omega \cup \Omega_I} \int_{\Omega \cup \Omega_I} \omega_\delta(|\mathbf{x}' - \mathbf{x}|) f(e) \mu(e,e^*) \dot{e}(t,\mathbf{x},\mathbf{x}',\mathbf{u}) dx' dx$$

$$+ \frac{1}{2} \int_{\Omega \cup \Omega_I} \int_{\Omega \cup \Omega_I} \omega_\delta(|\mathbf{x}' - \mathbf{x}|) \left(\int_0^e f(e') \frac{d\mu(e,e^*)}{de^*} de' \right) \dot{e}^*(t,\mathbf{x},\mathbf{x}',\mathbf{u}) dx' dx$$

$$= \int_\Omega \rho \dot{\mathbf{u}}(t,\mathbf{x}) \cdot \ddot{\mathbf{u}}(t,\mathbf{x}) dx - \int_\Omega \dot{\mathbf{u}}(t,\mathbf{x}) \cdot \mathbf{b}(\mathbf{x}) dx$$

$$+ \frac{1}{2} \int_{\Omega \cup \Omega_I} \int_{\Omega \cup \Omega_I} \omega_\delta(|\mathbf{x}' - \mathbf{x}|) f(e) \mu(e) M(t,\mathbf{x},\mathbf{x}',\mathbf{u})(\dot{u}(t,x') - \dot{u}(t,x)) dx' dx$$

$$+ \frac{1}{2} \int_{\Omega \cup \Omega_I} \int_{\Omega \cup \Omega_I} \omega_\delta(|\mathbf{x}' - \mathbf{x}|) \left(\int_0^e f(e') \frac{d\mu(e,e^*)}{de^*} de' \right) \dot{e}^*(t,\mathbf{x},\mathbf{x}',\mathbf{u}) dx' dx$$

$$= \int_\Omega \rho \dot{\mathbf{u}}(t,\mathbf{x}) \cdot \ddot{\mathbf{u}}(t,\mathbf{x}) dx - \int_\Omega \dot{\mathbf{u}}(t,\mathbf{x}) \cdot \mathbf{b}(\mathbf{x}) dx$$

$$- \int_{\Omega \cup \Omega_I} \int_{\Omega \cup \Omega_I} \omega_\delta(|\mathbf{x}' - \mathbf{x}|) f(e) \mu(e) M(t,\mathbf{x},\mathbf{x}',\mathbf{u}) \dot{u}(t,x) dx' dx$$

$$+ \frac{1}{2} \int_{\Omega \cup \Omega_I} \int_{\Omega \cup \Omega_I} \omega_\delta(|\mathbf{x}' - \mathbf{x}|) \left(\int_0^e f(e') \frac{d\mu(e,e^*)}{de^*} de' \right) \dot{e}^*(t,\mathbf{x},\mathbf{x}',\mathbf{u}) dx' dx$$

$$= \frac{1}{2} \int_{\Omega \cup \Omega_I} \int_{\Omega \cup \Omega_I} \omega_\delta(|\mathbf{x}' - \mathbf{x}|) \left(\int_0^e f(e') \frac{d\mu(e,e^*)}{de^*} de' \right) \dot{e}^*(t,\mathbf{x},\mathbf{x}',\mathbf{u}) dx' dx.$$

In the above process, we need the integrad is continuous. If bond never never broken, then $\dot{e}^*(t,\mathbf{x},\mathbf{x}',\mathbf{u}) = 0$, and $\frac{dE(t)}{dt} = 0$, the total energy of system (26) is conserved. When bond broking, because the damage is a nonincreasing function and the bond can't recover, $\frac{d\mu}{de^*} \leq 0$, then $\frac{dE(t)}{dt} \leq 0$, the total energy of the system is decreasing. So when bondbroken happening, the system will have the energy decay property, that is said the system consistent with the laws of thermodynamics.

4   Numerical Examples

In this section, numerical examples are given to demonstrate the effectiveness of the model, and the meshfree method [16] is used to solve the model equation. First, the body is discretized into material points, and the equation have the form

$$\rho \ddot{\mathbf{u}}_i = \sum_{p \in \mathcal{H}_\delta(x_i)} \mathbf{f}(\mathbf{u_p} - \mathbf{u_i}, \mathbf{x_p} - \mathbf{x_i}) V_p + b_i, i = 1, 2, ..., N. \tag{29}$$

Then, the equation in the time direction can be discretized as follows.

$$\rho \left(\frac{\mathbf{u}_i^{n+1} - 2\mathbf{u}_i^n + \mathbf{u}_i^{n-1}}{\Delta t^2} \right) = \sum_p \mathbf{f}(\mathbf{u_p^n} - \mathbf{u_i^n}, \mathbf{x_p} - \mathbf{x_i}) V_p + b_i^n, i = 1, 2, ..., N. \tag{30}$$

that is,

$$\rho(\frac{\dot{\mathbf{u}}_i^{n+1} - \dot{\mathbf{u}}_i^n}{\Delta t}) = \sum_p \mathbf{f}(\mathbf{u}_p^n - \mathbf{u}_i^n, \mathbf{x}_p - \mathbf{x}_i)V_p + b_i^n, i = 1, 2, ..., N. \tag{31}$$

where

$$\dot{\mathbf{u}}_i^{n+1} = \frac{\mathbf{u}_i^{n+1} - \mathbf{u}_i^n}{\Delta t}. \tag{32}$$

Therefore, rewriting the above procedure in the following format.

$$\mathbf{u}_i^{n+1} = \dot{\mathbf{u}}_i^{n+1}\Delta t + \mathbf{u}_i^n, i = 1, 2, ..., N. \tag{33}$$

$$\dot{\mathbf{u}}_i^{n+1} = \ddot{\mathbf{u}}_i^{n+1}\Delta t + \dot{\mathbf{u}}_i^n, i = 1, 2, ..., N. \tag{34}$$

In following numerical experiments, numerical results under different damage relations are given, and the damage areas are highlighted to show the impact of different damage on numerical simulation. The damage index ϕ whose definition is given by

$$\phi(\mathbf{x}, t) = 1 - \frac{\int_{B_\delta(\mathbf{x})} \mu dV_{x'}}{\int_{B_\delta(\mathbf{x})} dV_{x'}}, \tag{35}$$

where μ is the bond damage factor.

4.1   Example 1

A simple benchmark problem in dynamic fracture is performed using the peri-dynamic damage model based on bond elongation to investigate the influence of different influence functions and continuity on numerical results. A thin square plate with a pre-existing crack which subjected to a velocity boundary condition is given below. To verify the validity of the model and the influence of different damage models. Numerical simulation results in several modes are presented. The geometry of the plate can be seen in Fig. 1, the thickness of the plate is 0.1 mm, pre-existing crack length is 10 mm at the center. In the process of computing, grid spacing is 0.1 mm. The mechanical properties of the material is presented in Table 1. Where E is Young modulus, ν is the Poisson ratio, ρ is

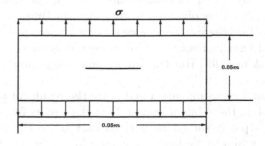

Fig. 1. Geometry of the plate with pre-existing crack in the center

Table 1. Material parameters for the plate

ρ	$E(Gpa)$	ν	$G_0(J/m^2)$
2450	32	1/3	3.0

density, and G_0 is the critical energy release rate. The uniform normal stress is applied to the top and bottom edges of the plate perpendicular to the crack. The figure is the result at 20.05 μs, similar experiments can be seen in [1,18].

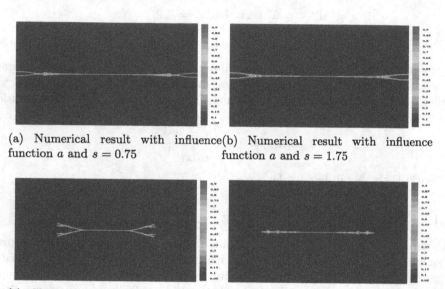

(a) Numerical result with influence function a and $s = 0.75$

(b) Numerical result with influence function a and $s = 1.75$

(c) Numerical result with influence function b and $s = 0.75$

(d) Numerical result with influence function b and $s = 1.75$

Fig. 2. Damage is continuous about the bond elongation

Figures 2, 3 and 4 show the damage contour plots in the plate when the damage is zero, first and second-order continuous about the bond elongation, respectively. This may reveal some differences that might arise due to the continuity of the damage. It is observed from the plots that as the continuity of the damage increasing, there was little change in the damage contour plots. It can be understood that increasing the continuity doesn't change the energy in a single bond breaking a lot. But this might preserve some underlying physical properties.

In Figs. 2, 3 and 4, the subfigure a and b are the result for influence function (a) with stress 5 Mpa, the subfigure c and d are the results for influence function (b) with stress 5 Mpa. It should be noticed that when the influence function changed, the energy in a single bond changed a lot, different crack propagation phenomenon can be observed. When s changed, the energy of bond-breaking

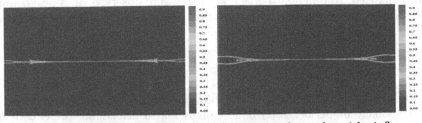

(a) Numerical result with influence function a and $s = 0.75$ (b) Numerical result with influence function a and $s = 1.75$

(c) Numerical result with influence function b and $s = 0.75$ (d) Numerical result with influence function b and $s = 1.75$

Fig. 3. Damage is first order continuous about the bond elongation

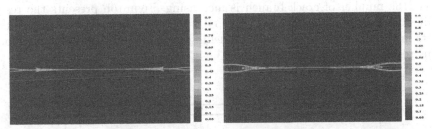

(a) Numerical result with influence function a and $s = 0.75$ (b) Numerical result with influence function a and $s = 1.75$

(c) Numerical result with influence function b and $s = 0.75$ (d) Numerical result with influence function b and $s = 1.75$

Fig. 4. Damage is second order continuous about the bond elongation

is also changed, so different damage patterns were observed. It is an advantage that for different physical materials, the real description of physical phenomena can be achieved by adjusting parameters.

4.2 Example 2

A simple benchmark problem in dynamic fracture is performed using the peridynamic damage model based on bond elongation to investigate the effectiveness of our model. The dynamic crack propagation and branching can be observed in the numerical simulation. The problem considered here is that of crack branching in a plate made of glass subjected to sudden stress loading conditions. This problem has been simulated with peridynamic damage model [12, 14, 17], and the experiment result can also be seen in [13]. As shown in Fig. 5, the setup consists of a plate with a pre-crack from the left edge to the center of the plate. The material properties considered are in Table 2, and the influence function a is used in the experiment. The plate is loaded dynamically at the top and bottom surfaces with a sustained stress of 1.2 Mpa. A regular lattice of material particles is used for the discretization, the lattice spacing of 0.1 mm is used, the horizon chosen is three times the lattice spacing. Figure 6a shows the damage contour in the plate at 1.2 Mpa. It becomes very clear from this figure that the crack branching happened. The peridynamic damage model predicts symmetrical crack path with simple branching. As it can be seen in Fig. 6a, this result coincides with the experiments [13] reported by Ha and Bobaru. With increasing the applied force, the number of crack branch is increasing. Figure 6b presents the results while the applied load is increased up to 2.4 Mpa. As observed from the damage plots in Fig. 6b, a secondary crack branching is observed by peridynamic model. This result confirms by experiments [13].

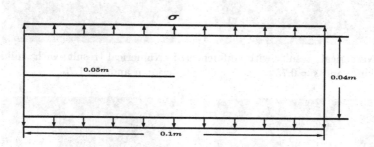

Fig. 5. Configuration of the plate

Table 2. Material parameters for the plate

ρ	$E(Gpa)$	ν	$G_0(J/m^2)$
2440	72	1/3	3.8

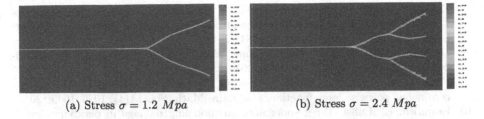

(a) Stress $\sigma = 1.2\ Mpa$ (b) Stress $\sigma = 2.4\ Mpa$

Fig. 6. Crack propagation in the plate with stress $\sigma = 1.2$ Mpa, 2.4 Mpa

5   Conclusions

A new peridynamic damage model is proposed which is based on absolute bond elongation. In this model, the elastic parameters and damage parameters are embedded into the constitutive relation, so that the whole process of the material from elastic to the damage can be modeled. Moreover, the effects of different influence functions and different damage patterns are investigated numerically. In particular, a pre-cracked plate under applied traction is simulated to assess the accuracy of the model and the numerical results agree better with the crack branching experiment. However, the mechanism of fracture and damage is very complex, the parameter in this model is depending on the classical model, it is not accurate for describing the fracture problem. In the future, extending the model based on experimental and molecular dynamics will enable the model to effectively describe the practical problem.

Acknowledgements. This research was supported by National Natural Science Foundation of China (No. 11971386) and the National Key R&D Program of China (No. 2020YFA0713603).

References

1. Abbasiniyan, L., Hoseini, S.H., Faroughi, S.: Fracture analysis of pre-cracked and notched thin plates using peridynamic theory. J. Comput. Appl. Res. Mech. Eng. 11(2), 329–338 (2019). https://doi.org/10.22061/jcarme.2019.5136.1628
2. Bobaru, F., Foster, J.T., Geubelle, P.H., Silling, S.A.: Handbook of Peridynamic Modeling. CRC Press, Boca Raton (2016)
3. Bobaru, F., Ha, Y.D., Hu, W.: Damage progression from impact in layered glass modeled with peridynamics. Cent. Eur. J. Eng. 2(4), 551–561 (2012)
4. Bobaru, F., Zhang, G.: Why do cracks branch? A peridynamic investigation of dynamic brittle fracture. Int. J. Fract. 196(1–2), 59–98 (2015)
5. Chen, X., Gunzburger, M.: Continuous and discontinuous finite element methods for a peridynamics model of mechanics. Comput. Methods Appl. Mech. Eng. 200(9–12), 1237–1250 (2011)
6. Dayal, K., Bhattacharya, K.: Kinetics of phase transformations in the peridynamic formulation of continuum mechanics. J. Mech. Phys. Solids 54(9), 1811–1842 (2006)

7. Dong, Y., Wei, D., Xuefeng, L., Shenghui, Y., Xiaoqiao, H.: Investigation on mode-I crack propagation in concrete using bond-based peridynamics with a new damage model. Eng. Fract. Mech. **199**, 567–581 (2018)
8. Du, Q., Tao, Y., Tian, X.: A peridynamic model of fracture mechanics with bond-breaking. J. Elast. **132**(2), 197–218 (2018)
9. Emmrich, E., Puhst, D.: Well-posedness of the peridynamic model with lipschitz continuous pairwise force function. Commun. Math. Sci. **11**(4), 1039–1049 (2013)
10. Emmrich, E., Puhst, D.: A short note on modeling damage in peridynamics. J. Elast. **123**(2), 245–252 (2016)
11. Emmrich, E., Weckner, O.: The peridynamic equation and its spatial discretisation. Math. Model. Anal. **12**(1), 17–27 (2007)
12. Ha, Y.D., Bobaru, F.: Studies of dynamic crack propagation and crack branching with peridynamics. Int. J. Fract. **162**(1–2), 229–244 (2010)
13. Ha, Y.D., Bobaru, F.: Characteristics of dynamic brittle fracture captured with peridynamics. Eng. Fract. Mech. **78**(6), 1156–1168 (2011)
14. Huang, D., Lu, G., Wang, C., Qiao, P.: An extended peridynamic approach for deformation and fracture analysis. Eng. Fract. Mech. **141**, 196–211 (2015)
15. Kilic, B., Madenci, E.: Prediction of crack paths in a quenched glass plate by using peridynamic theory. Int. J. Fract. **156**(2), 165–177 (2009)
16. Madenci, E., Oterkus, E.: Peridynamic theory. In: Peridynamic Theory and Its Applications. Springer, New York (2014). https://doi.org/10.1007/978-1-4614-8465-3_2
17. Prakash, N., Seidel, G.D.: A novel two-parameter linear elastic constitutive model for bond based peridynamics. In: 56th AIAA/ASCE/AHS/ASC Structures, Structural Dynamics, and Materials Conference (2015)
18. Ramulu, M., Kobayashi, A.S.: Mechanics of crack curving and branching - a dynamic fracture analysis. Int. J. Fract. **27**(C5), 187–201 (1985)
19. Silling, S.: Reformulation of elasticity theory for discontinuities and long-range forces. J. Mech. Phys. Solids **48**(1), 175–209 (2000)
20. Silling, S.A., Askari, E.: A meshfree method based on the peridynamic model of solid mechanics. Comput. Struct. **83**(17–18), 1526–1535 (2005)
21. Silling, S.A., Weckner, O., Askari, E., Bobaru, F.: Crack nucleation in a peridynamic solid. Int. J. Fract. **162**(1–2), 219–227 (2010)

Privacy Paradox in Social Media: A System Dynamics Analysis

Ektor Arzoglou[⊠], Yki Kortesniemi, Sampsa Ruutu, and Tommi Elo

Department of Communications and Networking, Aalto University, Helsinki, Finland
{ektor.arzoglou,yki.kortesniemi,sampsa.ruutu,tommi.elo}@aalto.fi

Abstract. The term 'privacy paradox' refers to the apparent inconsistency between people's concerns about their privacy and their actual privacy behaviour. Although several possible explanations for this phenomenon have been provided so far, these assume that (1) all people share the same privacy concerns and (2) a snapshot at a given point in time is enough to explain the phenomenon. To overcome these limitations, this article presents a system dynamics simulation model that considers the diversity of privacy concerns during the process of social media adoption and identifies the types of situations in which the privacy paradox emerges. The results show that (1) the least concerned minority can induce the more concerned majority to adopt social media and (2) even the most concerned minority can be hindered by the less concerned majority from discarding social media. Both (1) and (2) are types of situations that reflect the privacy paradox.

Keywords: Digital platforms · Privacy · Privacy paradox · Social media · System dynamics

1 Introduction

Social media, such as Facebook and Instagram, are platforms that have changed how people interact and share experiences by acting as *mediators* between users and content [8]. Users actively construct their online identities, engage in active data sharing, and therefore satisfy various personal and professional needs. As a result, the influence exerted by current users on potential users to also adopt social media is reinforced, and as it evolves into a *social norm*, it becomes harder to resist regardless of privacy preferences and concerns [1]. At the same time, surveys show that people who use social media every day are highly concerned about the data collected about them on the Internet [12]. This inconsistency of privacy concerns and actual behaviour is often referred to as the *privacy paradox* [4,9].

Over the last couple of decades, privacy researchers have provided several possible explanations for the privacy paradox, with studies assuming that (1) all people share the same privacy concerns [9] and (2) a snapshot at a given point in time is enough to explain the phenomenon. However, privacy concerns are not

D. Groen et al. (Eds.): ICCS 2022, LNCS 13350, pp. 651–666, 2022.
https://doi.org/10.1007/978-3-031-08751-6_47

of the same degree for all people [15]. Moreover, the cross-sectional approaches, such as surveys and experiments, used in previous privacy paradox studies do not explain the changes in privacy concerns over time [9]. These limitations can be addressed with a process theory, hence motivating the development of a simulation model using *system dynamics* [13] in this article.

The research question guiding this article is: *In what types of situations can a social norm outweigh privacy concerns, thus resulting in social media adoption, and how does this help understand the privacy paradox?* The results of the developed system dynamics simulation model show that (1) the least concerned minority can induce the more concerned majority to adopt social media and (2) even the most concerned minority can be hindered by the less concerned majority from discarding social media. Both (1) and (2) are types of situations that reflect the privacy paradox. Finally, the contributions of this article also include demonstrating the potential of system dynamics as a tool for analysing privacy behaviour.

The rest of the article is organised as follows. Section 2 reviews literature on informational privacy and privacy paradox. Section 3 describes the applicability of the methodology used, namely system dynamics modelling, to the privacy paradox. Section 4 presents the model of the social media platform. The simulation results are discussed in Sect. 5. Finally, Sect. 6 concludes the article.

2   Theoretical Background

The concept of privacy has three main aspects: (1) *territorial privacy*, protecting the close physical area surrounding a person, (2) *privacy of the person*, protecting a person against undue interference, and (3) *informational privacy*, controlling whether and how personal data can be gathered, stored, processed, or selectively disseminated [9,10]. This article focuses exclusively on the third aspect.

2.1   Informational Privacy

One of the most influential privacy theories is that developed by Alan Westin, who defines privacy as "the claim of individuals, groups, or institutions to determine for themselves when, how, and to what extent information about them is communicated to others" [14]. In addition, Westin discusses privacy as a *dynamic* process (i.e. over time) of interpersonal boundary control and argues that not all people share the same privacy preferences. Westin's privacy segmentation divides the public into three (empirically- and not theoretically-derived) groups: (1) *privacy fundamentalists*, who see privacy as paramount, (2) *privacy unconcerned*, who see no need for privacy, and (3) *privacy pragmatists*, who weigh potential personal or societal benefits of information disclosure, assess privacy risks, and then decide whether they will agree or disagree with specific information activities [15].

Since the focus of this article is on informational privacy, Westin's (informational) privacy segmentation acts as a key driver for the model development.

2.2 Privacy Paradox

The term 'privacy paradox' emerged from studying privacy in the context of consumer behaviour. In 2001, Brown "uncovered something of a privacy paradox" through a series of interviews with online shoppers; despite expressing high privacy concerns, consumers were still willing to give their personal details to online retailers as long as they had something to gain in return [4,9]. Some of the most important explanations for the privacy paradox are based on: (1) privacy calculus, (2) incomplete information, bounded rationality, and decision biases, and (3) social influence [9]. The explanations are summarised in Table 1.

Table 1. Privacy paradox explanations

Explanation	Description
Privacy calculus	People perform a perfectly informed and rational cost-benefit analysis and decide to share their data only when benefits outweigh costs. However, they might still express concerns about the privacy of their shared data, resulting in the inconsistency between expressed privacy concerns (or attitude) and actual behaviour
Incomplete information, bounded rationality, and decision biases	People use heuristics, which compensate for limitations in information, time, and cognitive capabilities, in order to make decisions. However, these heuristics often result in unexpected outcomes
Social influence	People's behaviour is influenced by social factors and therefore might not match their unbiased attitude

3 A System Dynamics Model of the Privacy Paradox

System dynamics is a methodology that uses feedback loops, accumulations, and time delays to understand the *behaviour of complex systems over time* [13]. One of the primary strengths of system dynamics is that it allows for the inclusion of both social and technical elements into the same model and therefore the study of complex sociotechnical systems, such as social media.

Over the last decade, researchers have started to conceptualise digital platforms and multi-sided markets as dynamic systems and use system dynamics to study related phenomena, such as platform adoption, over time [5,11]. However, these studies neglect the fact that privacy concerns are a significant factor in the adoption of online services, such as digital platforms [6], and therefore the concept of privacy concerns is missing from existing platform adoption models. To overcome this limitation, this article presents a system dynamics simulation model that considers the diversity of privacy concerns during the process of social media adoption and identifies the types of situations in which the privacy paradox emerges.

In system dynamics, the model development is preceded by (1) *reference modes*, which are graphs illustrating the problem (e.g. the privacy paradox) as a pattern of behaviour over time, and (2) a *dynamic hypothesis*, which aims to explain the problematic behaviour shown in the reference modes in terms of the underlying *feedback and stock-flow structure* (see Sect. 4) of the system [13].

3.1   Problem Articulation

In order to illustrate the privacy paradox in the context of social media, this article uses four reference modes that are most relevant to platform adoption: (1) there is a growth in platform adoption that ultimately stabilises (i.e. S-shaped growth) (e.g. a social media platform can be steadily adopted by highly concerned users, who are influenced by less concerned users), (2) the S-shaped growth in platform adoption is followed by a minor decline, which ultimately stabilises (e.g. a social media platform can maintain a large fraction of highly concerned users, who are hindered by less concerned users from discarding), (3) an initial period of growth in platform adoption is followed by a decline, but platform adoption subsequently overtakes this decline and continues to grow until it ultimately stabilises (e.g. a social media platform can experience only a transient loss of highly concerned users, who discard but are eventually influenced by less concerned users to re-adopt), and (4) an initial period of growth in platform adoption is followed by a collapse (i.e. overshoot and collapse) (e.g. a social media platform can be discarded by highly concerned users, who also influence less concerned users). Reference modes (1)–(3) illustrate situations in which privacy concerns are inconsistent with platform adoption, thus reflecting the privacy paradox, whereas in (4) privacy concerns are consistent with platform discard, thus not reflecting the privacy paradox.

The purpose of the model is to explain the types of situations in which a social norm can outweigh privacy concerns using these four modes of dynamic behaviour. As these dynamic behaviours can occur in different settings, the model was built as a generic representation of social media without focusing on any specific platform. Finally, the time horizon of the model is in the order of multiple years, so that the entire platform adoption phase is included in the simulation results.

3.2   Dynamic Hypothesis

The dynamic hypothesis guiding the model development is that an extended feedback structure of the Bass model of *innovation diffusion*, which describes the adoption of new products or services (over time) [3], can produce the four modes of dynamic behaviour. As such, platform adoption can be influenced by different factors, such as privacy concerns, that have an effect on the feedback loops of the model.

The model includes a social media platform of potential and current users that is modelled *endogenously*. That is, every platform variable is affected by one or more other platform variables. Conversely, since privacy concerns can be described as a merely negative concept not bound to any specific context [7, 9], they are modelled *exogenously*. That is, privacy concerns affect but are not affected by the platform.

4   Model Development

In system dynamics, (1) *stocks*, shown as rectangles, represent accumulations of either matter or information, (2) *flows*, shown as pipes and valves, regulate the rate of change of the stocks, and (3) *auxiliaries*, shown as intermediate variables between stocks and flows, clarify the sequence of events that cause the flows to change the stocks.

Variables at the tail of the causal links are independent, indicating a cause, while variables at the head of the causal links are dependent, indicating an effect. All causal links indicate that a change in the independent variable causes the dependent variable to change in the *same* direction, except for these labelled with a minus sign (−) that indicate a change in the *opposite* direction.

Finally, *feedback* is the process whereby an initial cause gradually spreads through a chain of causal links to ultimately re-affect itself, thereby forming a loop that can be either *reinforcing* (R) (i.e. amplifying change) or *balancing* (B) (i.e. counteracting and opposing change). In this case, variables constituting feedback loops are at the same time both causes and effects.

4.1   Model Structure

The social media platform is modelled by extending the Bass model of innovation diffusion, which considers adoption through exogenous efforts, such as advertising, and adoption through word-of-mouth [3] (Fig. 1). In addition, the model utilises several equations from Ruutu et al. [11].

When the platform is launched, the initial number of users is zero, so the only source of adoption are external influences, such as advertising (B1: "Market Saturation"). When the first users enter the platform, the adoption rate increases through word-of-mouth (R1: "WoM"). As the stock of users grows, platform value increases, and the norm related to platform adoption becomes stronger and consequently harder to deviate from. As a result, more potential users conform and adopt the platform (R2: "Social Norm"). The advertising and word-of-mouth effects are largest at the start of the platform diffusion process and steadily diminish as the stock of potential users is depleted (B1, B2: "Market Saturation"). Finally, current users may decide to discard the platform and re-enter the stock of potential users (since they may be persuaded to adopt again in the future). In this case, the discard rate depends on the number of current users and the decrease, caused by privacy concerns, in platform value (B3: "Discard").

The behaviour of potential and current users is modelled using rules of bounded rationality, which depend on the information available to users at a given point in time. In other words, potential and current users are not assumed

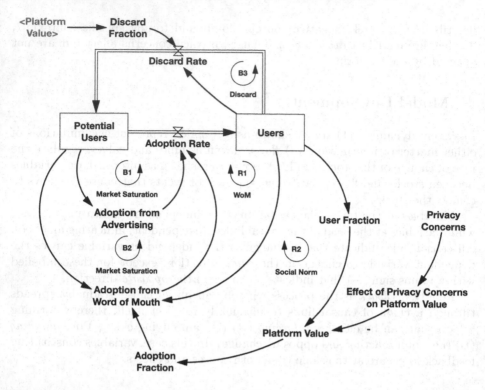

Fig. 1. Social media adoption affected by social norm and privacy concerns

to have perfect foresight of how platform adoption will progress, and they make their decisions regarding platform adoption and discard based on their perception of platform value to them.

4.2   Model Parameters

The total population (N) considered in the model is 1000 users, divided into 550 Pragmatists, 250 Fundamentalists, and 200 Unconcerned as per Westin's first privacy segmentation [15]. The degree of privacy concerns for Pragmatists is modelled with parameter P^*, and the additional degree of privacy concerns for Fundamentalists compared to Pragmatists is modelled with parameter F^*, which is a multiplier of P^*. Finally, parameter U^* determines the degree of privacy concerns for Unconcerned. This can range from the extreme condition of zero ($U^* = 1$) to matching the degree of privacy concerns for Pragmatists ($U^* = 0$).

Furthermore, the time at which privacy concerns start is modelled with parameter $T^0 PC$, and the initial value of privacy concerns is modelled with parameter $PC(0)$. The effect of privacy concerns on platform value is modelled using exponential smoothing. Here, parameter τPC is used to determine the erosion of privacy concerns. This essentially indicates the time for users to develop feelings of exhaustion, resignation, and even cynicism towards privacy (i.e. privacy fatigue) [6]. As such, privacy concerns are assumed to be boundedly rational.

In addition to the parameters determining privacy concerns, the model includes eight further parameters that have an effect on platform adoption. First, an external advertising effort (a), starting at time T^0 and ending at time T, is initially required to bring the first users in the platform. Thereafter, platform adoption continues only with word-of-mouth, which depends on the contact rate (c) between potential and current users. Conversely, platform discard depends on parameter τ, which is used to determine the time for users to process the decrease, caused by privacy concerns, in platform value and react by discarding the platform. Moreover, the reference value of platform competitors is modelled with parameter V^*, and the reference user fraction is modelled with parameter uf^*. High values of V^* imply that users receive high value from competitive platforms, and high values of uf^* imply that more users are needed in order to obtain the same level of benefits. Therefore, high values of these two parameters are making platform adoption harder. Finally, the equation of platform value contains an exponent (γ) determining the strength of social norm. High values of γ imply that platform value is strongly dependent on the number of current users. Hence, in the beginning, when there is a lack of users, high values of γ make platform adoption harder. The model equations and parameter values are listed in the Appendix.

4.3 Model Testing and Validation

The model was built using Vensim DSS for Mac Version 9.0.0 (Double Precision), and the simulation experiments were performed using time step 0.0625 and Euler numerical integration. Several validation tests have been successfully passed to gradually build confidence in the soundness and usefulness of the model. The validation tests assess the validity of the model structure with respect to the purpose presented in Sect. 3 and are grouped into *direct structure tests*, which do not involve simulation experiments, and *structure-oriented behaviour tests*, in which simulation experiments are used [2]. The results are presented in Table 2.

5 Simulation Results

Using the model, it is possible to simulate the four modes of dynamic behaviour presented in Sect. 3.1 and therefore identify the types of situations in which the privacy paradox emerges.

Table 2. Validation tests applied to the model

Test	Result
Direct structure tests	
Structure confirmation	The feedback structures of the model have been formulated and extended based on the Bass model of innovation diffusion [3]
Parameter confirmation	All parameters in the model have clear and meaningful counterparts in the real world. In addition, all parameters were set to limited ranges with minimum and maximum values. Since the model was built as a generic representation of social media, the exact parameter values are not significant, and the parameters have not been estimated based on any specific platform
Direct extreme condition	The model includes formulations to ensure that stock variables remain valid at all times. For example, the sum of potential and current users stays constant to ensure that conservation laws are met, and the rate of adoption is adjusted with a conditional function (i.e. min) to ensure that the stocks of potential and current users stay non-negative
Dimensional consistency	The units of all variables and parameters have been specified, and the model passes Vensim's dimensional consistency test
Structure-oriented behaviour tests	
Indirect extreme condition	The model behaves as expected when individual variables are subjected to extreme conditions. For example, setting the number of users to zero results in zero platform value
Behaviour sensitivity	The model behaves plausibly when individual parameters are set to the limits of their meaningful ranges of variation as well as when several parameters are varied simultaneously in a Monte Carlo experiment (see Sect. 5.5)

5.1   Simulation Experiment 1

For the first simulation experiment (Fig. 2), the degree of privacy concerns is the same for Pragmatists and Fundamentalists ($P^* = 0.5$, $F^* = 1$), on the assumption that the two user groups share the same privacy preferences. In addition, a zero degree of privacy concerns is considered for Unconcerned ($U^* = 1$), assuming that this user group sees no need for privacy.

Initially, platform adoption takes place through advertising (B1) and word-of-mouth (B2, R1) until $Time = 4$. At this point, advertising efforts (B1) end,

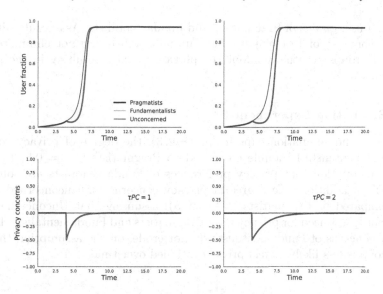

Fig. 2. The social norm created by Unconcerned results in platform adoption also for Pragmatists and Fundamentalists, although privacy concerns of the last two user groups are not eliminated.

and platform adoption continues only with word-of-mouth (B2, R1). Moreover, privacy concerns start for Pragmatists and Fundamentalists ($T^0 PC = 4$). In the beginning, the effect of privacy concerns on platform value outweighs social norm (R2), causing the number of these two user groups to decline. At the same time, the number of Unconcerned continues to grow, since there is no effect of privacy concerns to decrease platform value for this user group. In other words, during this phase, discards (B3) dominate adoptions (B2, R1) for Pragmatists and Fundamentalists, while adoptions (B2, R1) continue to dominate discards (B3) for Unconcerned. However, as privacy concerns of Pragmatists and Fundamentalists erode, the effect of privacy concerns on platform value is also falling back. Therefore, as the number of Unconcerned grows, social norm (R2) outweighs privacy concerns of Pragmatists and Fundamentalists, hence recovering platform value for the last two user groups as well. As a result, adoptions (B2, R1) dominate discards (B3) once more for Pragmatists and Fundamentalists, allowing the number of these two user groups to grow again ($Time = 5$). Thus, on one hand, if privacy concerns erode faster ($\tau PC = 1$, left side of Fig. 2), platform adoption is easier. On the other hand, if privacy concerns erode slower ($\tau PC = 2$, right side of Fig. 2), platform adoption becomes harder. In other words, the slower the erosion of privacy concerns, the longer it takes for social norm to outweigh privacy concerns and therefore the longer the delay in platform adoption.

Both runs of the first simulation experiment illustrate an example of the so called *minority rule*, where the smallest user group of Unconcerned influences the larger user groups of Pragmatists and Fundamentalists. As a result, although privacy concerns of Pragmatists and Fundamentalists are not eliminated, the two user groups eventually adopt the platform, hence resulting in the privacy paradox.

5.2   Simulation Experiment 2

For the second simulation experiment (Fig. 3), the degree of privacy concerns for Fundamentalists is double compared to Pragmatists ($P^* = 0.5$, $F^* = 2$), on the assumption that privacy preferences of Fundamentalists are somewhat stronger. In addition, the degree of privacy concerns for Unconcerned is one-fifth compared to Pragmatists ($U^* = 0.8$), assuming that Unconcerned have significantly less need for privacy than Pragmatists and Fundamentalists. Finally, privacy concerns of Fundamentalists do not erode, on the assumption that this user group is less likely to feel privacy fatigued over time.

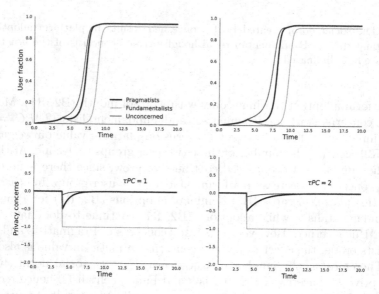

Fig. 3. The social norm created by Pragmatists and Unconcerned results in platform adoption also for Fundamentalists, although privacy concerns of the last user group remain constant.

As before, advertising efforts (B1) end at $Time = 4$, and platform adoption continues only with word-of-mouth (B2, R1). In addition, privacy concerns start for all three user groups ($T^0PC = 4$). At this point, the number of Unconcerned continues to grow as in the previous simulation experiment. The reason is that privacy concerns of this user group are low and have a trivial effect on platform value. At the same time, the number of Pragmatists and Fundamentalists is again starting to decline. However, privacy concerns of Pragmatists erode and are eventually outweighed by social norm (R2), which recovers platform value for this user group, and therefore the number of Pragmatists is once more starting to grow ($Time = 5$). On the other hand, the number of Fundamentalists continues to decline, since privacy concerns of this user group remain constant. Nevertheless, as the number of Pragmatists and Unconcerned grows, social norm (R2) outweighs privacy concerns of Fundamentalists too, hence recovering platform value for the last user group as well. As a result, the number of Fundamentalists starts to grow again ($Time = 7$), although privacy concerns of this user group do not erode. Thus, as before, the faster the erosion of privacy concerns ($\tau PC = 1$, left side of Fig. 3), the shorter the delay in platform adoption. Conversely, the slower the erosion of privacy concerns ($\tau PC = 2$, right side of Fig. 3), the longer the delay in platform adoption and therefore the more likely the platform adoption is to collapse.

Both runs of the second simulation experiment illustrate once more the minority rule, since the smallest user group of Unconcerned is initially influencing the largest user group of Pragmatists, before both eventually influence the user group of Fundamentalists. As a result, the privacy paradox for Pragmatists is similar to the previous simulation experiment. In addition, although privacy concerns of Fundamentalists remain constant, this user group eventually adopts the platform too, thus exhibiting a more severe privacy paradox.

5.3   Simulation Experiment 3

For the third simulation experiment (Fig. 4), the degree of privacy concerns is the same for Pragmatists and Fundamentalists ($P^* = 0.5$, $F^* = 1$), on the assumption that the two user groups share the same privacy preferences. In addition, the degree of privacy concerns for Unconcerned is half compared to the first two user groups $U^* = 0.5$), assuming that Unconcerned have somewhat less need for privacy than Pragmatists and Fundamentalists. Finally, privacy concerns erode slower for all three user groups ($\tau PC = 5$), on the assumption that users are less willing to give up their privacy. As a result, platform adoption becomes dependent more on the strength of social norm and less on the erosion of privacy concerns.

Once more, platform adoption takes place through advertising (B1) and word-of-mouth (B2, R1) until $Time = 4$. This is when advertising efforts (B1) end, and platform adoption continues only with word-of-mouth (B2, R1). Here, if the start time of privacy concerns is the same with the previous two simulation experiments ($T^0 PC = 4$, left side of Fig. 4), the platform's installed user base is small and social norm (R2) weak relative to privacy concerns. Hence, discards (B3) dominate adoptions (B2, R1), causing the stock of users to deplete. In other words, the platform is not able to gather critical mass sufficient to recover platform value from the effect of privacy concerns. Conversely, if privacy concerns start later ($T^0 PC = 5$, right side of Fig. 4), the user stock is large and social norm (R2) strong relative to privacy concerns. Hence, adoptions (B2, R1) continue to dominate discards (B3), allowing the number of users to grow. In this

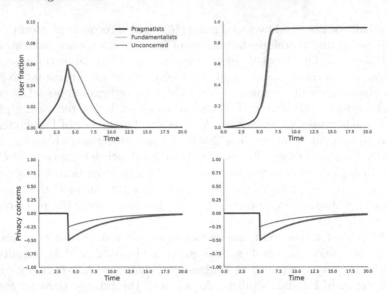

Fig. 4. Weak social norm results in platform adoption collapse because of privacy concerns, whereas strong social norm renders platform adoption nearly unaffected by privacy concerns.

case, the platform has enough time to gather critical mass sufficient to render platform value nearly unaffected by privacy concerns.

In the first run of the third simulation experiment, no privacy paradox emerges, since privacy concerns are consistent with platform adoption for all three user groups. That is, platform adoption increases when no privacy concerns exist, decreases when privacy concerns start, and collapses while privacy concerns are not eliminated. In the second run, privacy concerns of Pragmatists and Fundamentalists have nearly zero effect on platform adoption of these two user groups, hence resulting in the privacy paradox. In addition, Unconcerned start to exhibit also some extent of the privacy paradox, since this user group is somewhat more concerned here compared to the previous two simulation experiments.

5.4   Simulation Experiment 4

For the fourth simulation experiment (Fig. 5), the degree of privacy concerns for Fundamentalists is triple compared to Pragmatists ($P^* = 0.5$, $F^* = 3$), on the assumption that privacy preferences of Fundamentalists are significantly stronger. In addition, the degree of privacy concerns for Unconcerned is half compared to Pragmatists $U^* = 0.5$), assuming that Unconcerned have somewhat less need for privacy than Pragmatists and significantly less need for privacy than Fundamentalists. Finally, erosion of privacy concerns applies only to Pragmatists and Unconcerned ($\tau PC = 5$), once more on the assumption that Fundamentalists are less likely to feel privacy fatigued over time.

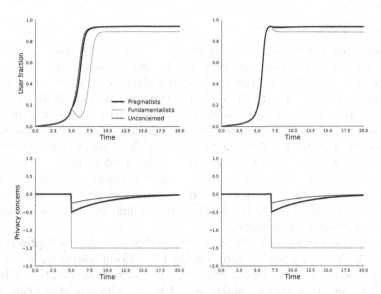

Fig. 5. The social norm created by Pragmatists and Unconcerned results in platform adoption also for a large fraction of Fundamentalists, although privacy concerns of the last user group remain constant.

Here, if the start time of privacy concerns is only one year later compared to the first two simulation experiments ($T^0PC = 5$, left side of Fig. 5), social norm (R2) outweighs privacy concerns of Pragmatists and Unconcerned but is outweighed by privacy concerns of Fundamentalists, thus preserving platform value and adoption only for the first two user groups. By contrast, the number of Fundamentalists is starting to decline. However, as the number of Pragmatists and Unconcerned grows, social norm (R2) outweighs privacy concerns of Fundamentalists too, hence recovering platform value for the last user group as well. As a result, the number of Fundamentalists starts to grow again ($Time = 6$), although privacy concerns of this user group do not erode. Similarly, if privacy concerns start even later ($T^0PC = 7$, right side of Fig. 5), they are once more initially causing the number of Fundamentalists to decline ($Time = 7$), despite the fact that the platform has enough time to gather critical mass. However, social norm (R2) is strong to eventually counterbalance privacy concerns of Fundamentalists, who are therefore hindered from discarding, hence allowing for a large fraction of this user group to remain in the platform ($Time = 8$).

In both runs of the fourth simulation experiment, the privacy paradox for Pragmatists and Unconcerned is similar to the previous simulation experiment. In addition, although privacy concerns of Fundamentalists remain constant, a large fraction of this user group eventually adopts the platform too, thus exhibiting a more severe privacy paradox.

5.5 Sensitivity Analysis

The parameters determining the degree (P^*, F^*, and U^*), erosion (τPC), and start time (T^0PC) of privacy concerns are the key factors for the privacy paradox. Low values of τPC and privacy concerns allow for an easier platform adoption, whereas high values of τPC and privacy concerns make platform adoption harder. In other words, higher and more persistent privacy concerns prevent the reinforcement of social norm, hence causing a longer delay in platform adoption and possibly a platform adoption collapse. In this case, the longer the delay in platform adoption, the more likely the platform adoption is to collapse and therefore the less likely the privacy paradox is to emerge. On the other hand, lower and less persistent privacy concerns are more easily outweighed by social norm, hence allowing platform adoption to continue with a shorter delay and resulting in the privacy paradox.

Moreover, low values of T^0PC may cause either a platform adoption collapse, in case of high privacy concerns, or a delay in platform adoption, in case of low privacy concerns. In other words, high privacy concerns raised in the early stages of platform adoption may eliminate social norm, which is still weak, causing possibly a platform adoption collapse. On the other hand, early and low privacy concerns may outweigh social norm temporarily, causing only a delay in platform adoption. In addition, for high values of T^0PC, relatively high values of privacy concerns are required to have an impact on social norm, which becomes stronger as platform adoption takes place. Thus, the earlier the privacy concerns start, the easier it is for privacy concerns to outweigh social norm, hence making the paradox less likely, and vice versa.

6 Concluding Discussion

This article presents a system dynamics simulation model that considers the diversity of privacy concerns during the process of social media adoption and identifies the types of situations in which the privacy paradox emerges. The model illustrates that (1) the least concerned minority can induce the more concerned majority to adopt social media and (2) even the most concerned minority can be hindered by the less concerned majority from discarding social media. Both (1) and (2) are types of situations that reflect the privacy paradox.

Since the model was built as a generic representation of social media, a limitation of the simulation results is that they do not apply exactly to every platform and context. As such, a fruitful topic for future research would be to empirically test and validate the simulation results and thus support the usefulness and applicability of the model both in the context of specific social media platforms and in additional contexts, such as peer-to-peer (P2P) platforms like Airbnb and Uber. Finally, the model could be developed further to study the privacy paradox in the context of platform competition. For example, the model could include two or more platforms and therefore identify the types of situations in which people may continue to use platforms that were launched early but face privacy issues (e.g. WhatsApp) or decide to switch to privacy alternatives that were launched later (e.g. Signal).

Acknowledgements. The authors would like to thank Pekka Nikander for his insightful comments.

Appendix: Model Equations and Parameter Values

The model equations and parameter values are shown in Table 3. In the equations, subscript w refers to the user group (p: Pragmatists, f: Fundamentalists, u: Unconcerned). The model includes formulations to ensure that users cannot be added or removed spontaneously (i.e. mass balance) and that stock variables stay non-negative. For clarity, these have been omitted from the equations shown in

Table 3. Model equations and parameter values

Name	Equation/parameter value		Unit	#
Potential users	$\dot{P}_w$	$= DR_w - AR_w$	User	1
	$P_w(0)$	$= 1000$		
Users	$\dot{U}_w$	$= AR_w - DR_w$	User	2
	$U_w(0)$	$= 0$		
Adoption rate	AR_w	$= P_w \cdot (a + c \cdot af_w \cdot U_w/N_w)$	User/Year	3
Discard rate	DR_w	$= U_w \cdot df_w/\tau$	User/Year	4
Adoption fraction	af_w	$= V_w/(V_w + V^*)$	-	5
Discard fraction	df_w	$= V^*/(V^* + V_w)$	-	6
Total population	N_w	1000 (divided into 550 Pragmatists, 250 Fundamentalists, and 200 Unconcerned)	User	
Advertising start time	T^0	0	Year	
Advertising end time	T	4	Year	
Advertising effectiveness	a	0.01	1/Year	
Contact rate	c	10	1/Year	
User reaction time	τ	1.5	Year	
User fraction	uf_w	$= U_w/N_w$	-	7
Reference user fraction	uf^*	0.5	-	
Platform value	V_w	$= (\frac{\sum_w uf_w}{uf^*})^\gamma + E_w$	-	8
Reference value	V^*	3	-	
Effect of users on platform value	γ	0.7	-	
Privacy concerns (Pragmatists)	PC_p	$= PC(0) - \text{Step}(P^*, T^0PC)$ Step input function	-	9a
Privacy concerns (Fundamentalists)	PC_f	$= PC(0) - \text{Step}(P^* \cdot F^*, T^0PC)$ Step input function	-	9b
Privacy concerns (Unconcerned)	PC_u	$= PC(0) - \text{Step}(P^* - (P^* \cdot U^*), T^0PC)$ Step input function	-	9c
Reference privacy concerns	PC_w^*	$= \text{Smoothi}(PC_w, \tau PC, PC(0))$ Exponential smoothing function	-	10
		$= PC(0)$ (no erosion of privacy concerns)		10'
Privacy concerns initial value	$PC(0)$	0	-	
Privacy concerns start time	T^0PC	5	Year	
Privacy concerns erosion time	τPC	5	Year	
Reference pragmatism	P^*	0.5	-	
Reference fundamentalism	F^*	3	-	
Reference unconcern	U^*	0.5	-	
Effect of privacy concerns on platform value	E_w	$= PC_w - PC_w^*$	-	11

Table 3. For details of the formulations and to ensure the replicability of the simulation results, the simulation model Vensim file is openly available upon request.

References

1. Arzoglou, E., Kortesniemi, Y., Ruutu, S., Elo, T.: The role of privacy obstacles in privacy paradox: a system dynamics analysis (Submitted) (2022)
2. Barlas, Y.: Formal aspects of model validity and validation in system dynamics. Syst. Dyn. Rev. **12**(3), 183–210 (1996)
3. Bass, F.M.: A new product growth for model consumer durables. Manage. Sci. **15**(5), 215–227 (1969)
4. Brown, B.: Studying the internet experience. HP Laboratories Technical Report 49 (2001)
5. Casey, T.R., Töyli, J.: Dynamics of two-sided platform success and failure: an analysis of public wireless local area access. Technovation **32**(12), 703–716 (2012)
6. Choi, H., Park, J., Jung, Y.: The role of privacy fatigue in online privacy behavior. Comput. Hum. Behav. **81**, 42–51 (2018)
7. Dienlin, T., Trepte, S.: Is the privacy paradox a relic of the past? An in-depth analysis of privacy attitudes and privacy behaviors. Eur. J. Soc. Psychol. **45**(3), 285–297 (2015)
8. van Dijck, J.: Facebook and the engineering of connectivity: a multi-layered approach to social media platforms. Convergence **19**(2), 141–155 (2013)
9. Kokolakis, S.: Privacy attitudes and privacy behaviour: a review of current research on the privacy paradox phenomenon. Comput. Secur. **64**, 122–134 (2017)
10. Rosenberg, R.S.: The Social Impact of Computers. Academic Press Inc., Cambridge (1992)
11. Ruutu, S., Casey, T., Kotovirta, V.: Development and competition of digital service platforms: a system dynamics approach. Technol. Forecast. Soc. Change **117**, 119–130 (2017)
12. Statista: Special Eurobarometer 447 - Online Platforms (2016). https://www.statista.com/study/37575/social-networks-search-engines-and-online-marketplaces-in-the-eu-28/
13. Sterman, J.D.: Business Dynamics: Systems Thinking and Modeling for a Complex World. McGraw-Hill, New York (2000)
14. Westin, A.F.: Privacy and Freedom. Atheneum, Berlin (1967)
15. Westin, A.F.: Social and political dimensions of privacy. J. Soc. Issues **59**(2), 431–453 (2003)

GPU Power Capping
for Energy-Performance Trade-Offs
in Training of Deep Convolutional Neural
Networks for Image Recognition

Adam Krzywaniak⬛, Pawel Czarnul⬛, and Jerzy Proficz$^{(\boxtimes)}$⬛

Faculty of Electronics, Telecommunications and Informatics Centre of Informatics —
Tricity Academic Supercomputer and networK (CI TASK), Gdansk University
of Technology, Narutowicza 11/12, 80-233 Gdańsk, Poland
adam.krzywaniak@pg.edu.pl, pczarnul@eti.pg.edu.pl, j.proficz@task.gda.pl

Abstract. In the paper we present performance-energy trade-off investigation of training Deep Convolutional Neural Networks for image recognition. Several representative and widely adopted network models, such as Alexnet, VGG-19, Inception V3, Inception V4, Resnet50 and Resnet152 were tested using systems with Nvidia Quadro RTX 6000 as well as Nvidia V100 GPUs. Using GPU power capping we found other than default configurations minimizing three various metrics: energy (E), energy-delay product (EDP) as well as energy-delay sum (EDS) which resulted in considerable energy savings, with a low to medium performance loss for EDP and EDS. Specifically, for Quadro 6000 and minimization of E we obtained energy savings of 28.5%–32.5%, for EDP 25%–28% of energy was saved with average 4.5%–15.4% performance loss, for EDS ($k = 2$) 22%–27% of energy was saved with 4.5%–13.8% performance loss. For V100 we found average energy savings of 24%–33%, for EDP energy savings of 23%–27% with corresponding performance loss of 13%–21% and for EDS ($k = 2$) 23.5%–27.3% of energy was saved with performance loss of 4.5%–13.8%.

Keywords: Energy-aware computing · High performance computing ·
Green computing · Machine learning

1 Introduction

Nowadays, energy consumption has become one of the key aspects, apart from execution time and scalability, for practically all types of parallel compute intensive applications, not only traditional high performance computing (HPC) applications executed in clusters [4] but also various workloads run in a cloud environment [7,11]. This is also becoming a very interesting and important factor for the currently very popular and time consuming training of AI models. These are mostly executed on workstations and systems that feature powerful GPUs.

© The Author(s), under exclusive license to Springer Nature Switzerland AG 2022
D. Groen et al. (Eds.): ICCS 2022, LNCS 13350, pp. 667–681, 2022.
https://doi.org/10.1007/978-3-031-08751-6_48

In this paper we tackle investigation of performance-energy configurations that stem from using GPU power capping for several popular Deep Convolutional Neural Networks used for image recognition, typically trained on both GPUs and multi- and many-core CPUs [2,9,14]. This power limiting method can be used for different purposes, such as performance maximization under a defined power budget or plain energy saving. Former research confirmed the method to be reliable and provide more predictable results in terms of performance losses in comparison with a DVFS technique [17]. The main contributions of this paper in the aforementioned context include:

- deriving configurations showing larger energy savings than performance loss (percentage wise) results using GPU power capping for specific models of deep Convolutional Neural Networks for image recognition, and
- investigation of differences in impact of power capping on various network models using various high performance GPUs including Nvidia V100 and Nvidia Quadro RTX 6000 GPUs.

The following section presents the related works in the GPU power capping subject, including general approach and energy-aware machine learning issues. Then, our testbed systems and used benchmarks are described. Afterwards, in Sect. 5, we described the performed experiments, monitoring methods and the results. Finally, we provided some conclusions and future works.

2   Related Works

2.1   Power Capping for GPU Servers

GPU power capping, or limiting, related works are mainly grouped into power capping design and implementation techniques or methods of management of heterogeneous systems under a defined power limit. The former is dedicated for achieving defined power limitation (usually in Watts). The latter generalizes the hardware/software power limiting methods designed for one component into a system wide limitation of power or energy, with additional constraints usually related to the overall performance.

A typical example of the power capping solution is a GPU-CAPP micro-architectural technique [20] dedicated for power capping limitation over GPU, with an additional objective to accelerate computations of parallel workloads. The solution assumes a direct use of hardware on-chip and global voltage regulators, which led to speed up the computations in comparison to two different fixed frequency sets, using a Nvidia GTX480 GPU system.

Another approach to imply power limitation into GPU is using a dynamic voltage and frequency scaling (DVFS) technology. A solution [8] proposed by Huang et al. uses a global-based neural network for modeling Nvidia GPU behavior under DVFS limitations, based on task characteristics, for the presented test cases, the solution enables decreasing energy along with performance improvements.

In [15] Mclaughlin et al. proposed another power capping solution dedicated for a graph traversal breadth-first algorithm. It was based on two techniques: DVFS and core scaling, features provided by an AMD A10-5800K GPU card. The described and evaluated power management algorithm, designed for maximizing efficiency under a given power cap, showed promising results for a wide range of graph data.

In [21] Tsuzuku et al. proposed power capping method for CPU-GPU heterogeneous systems. They presented performance and power consumption models used for a static solution setting initial frequencies (with Nvidia DVFS support), and then during runtime their exact settings are tuned using dynamic approach.

In [1] Ahmed et al. presented a power cap solution for CPU-GPU heterogeneous systems by defining a power cap allocation model, and implementing its simulator using discreet-event simulation engine. They performed trace-based experiments and crosschecked them with real parallel application executions on Nvidia based system. The results showed capability of the solution to decrease energy consumption of the performed computations.

In [3] Ciesielczyk et al. presented an approach limiting power usage for heterogeneous systems including GPU servers. They proposed an optimization model along with heuristic and exact solutions, where the test results using application benchmarks showed energy consumption reduction and minimal, negative impact on the performance.

In [16] Mishra et al. described how DVFS and other power-aware techniques, such as load balancing and task mapping, can influence the performance and power usage of CPU-GPU systems. They presented a collection of the up-to-date solutions and their comparative analysis. However, there was no power capping technique considered.

In [13] we performed an analysis of the GPU-based power capping for a collection of typical HPC benchmarks. We found out that using such a technique provides many possibilities of trade-off between performance and energy consumption, with different aspects of the overall evaluation, including solely energy and some in-between metrics. However, the presented results did not cover any machine learning related applications, which are addressed in this paper.

Power capping can be considered as a higher level mechanism than DVFS as it can employ core frequency scaling but potentially also shorter/longer term limits with peaks exceeding the limits and thus is of interest to be explored.

2.2 Energy Aware Machine Learning

Firstly, authors of paper [5] introduce several approaches and models of estimation of energy consumption, specifically as a key element to consider energy-aware processing in machine learning. Subsequently, they consider activities related to deep learning where energy estimation was adopted in the literature i.e. training and inference, as well as CPU and GPU based ones. Then energy estimation is applied to two real use cases. For data stream mining, HAT and VFDT algorithms obtain accuracy of over 97% for non-concept drift estimations and 56–65% for concept drift datasets. For energy estimation of inference using a regression model they obtained accuracy of 73.7% for Inception-V3, 63% for MobileNet, 70% for DenseNet.

Furthermore, authors of [14] investigate power and energy of using CNN models on CPU and GPU systems. Specifically, thorough analysis includes tests on Xeon CPU, K20 GPU, Titan X GPU based systems, for various frameworks on GPUs (Caffee, Torch, MXNet,), various libraries on CPU (Atlas, OpenBLAS and MKL). Additionally, breakdown of energy and power per layers of a model has been provided, for various batch sizes, considering HyperThreading as well as various memory and core frequencies. Results could be used for setting configurations for minimization of energy on CPU and GPU systems. However, further exploration considering training times and accuracies for time limited training could be extensions of that work. On the other hand, in paper [2] the author proposed NeuralPower—a framework based on sparse polynomial regression aimed at prediction of power, runtime, and energy consumption of CNNs used on a GPU. Specifically, the author was able to obtain average accuracy of over 88% for runtime and power for tested CNN architectures.

In paper [22], authors proposed GPOEO, a new online energy optimization framework for iterative machine learning applications run on GPUs. The tool is able to detect iterative phases, measures performance and energy used. Firstly, in an offline stage, it collects performance metrics and energy for various frequencies of SM and memory and then in the online stage applies the selected frequencies to optimize a function of energy and time. Similarly, in paper [24], Zou et al. proposed a power limiting solution, using resource utilization as an indicator of iteration bounds. The proposed method used GPU DVFS mechanisms to set up a defined power limit with performance maximization or performance degradation level with minimizing the energy consumption. Our work is focused on research of the power limiting techniques introduced in the tested GPU cards. We demonstrated and evaluated these mechanisms against the ML applications, however, they are more general and are useful for other types of workloads.

In paper [10] authors optimize deep learning at a higher level i.e. provide an allocation method for a GPU cluster for deep learning jobs (training and inference) minimizing energy consumption and meeting performance requirements. The approach uses a mixed-integer nonlinear problem (MINLP) formulation. The solution is able to turn off the nodes that have no DL jobs. Experiments have been conducted using GTX 1000 and RTX 2000 series GPUs showing e.g. energy savings of 43% compared to PA-MBT and 15% compared to EPRONS approaches.

Another approach, related to optimization of selected steps of training CNNs meant for energy consumption reduction was proposed in [23]. Specifically, three types of optimizations were proposed: stochastic skipping mini-batches with 0.5 probability, selection of a different subset of CNN layers for updates, and computing the sign of a gradient without computing the full gradient. The authors have demonstrated results from an FPGA board e.g. for ResNet-74 trained using CIFAR-10, energy was saved of over 90% and 60%, with a top-1 accuracy loss of about 2% and 1.2%.

In paper [18], authors explored energy-aware optimization of running deep learning workloads using GPU power capping as well as frequency capping for reduction of Energy-to-Solution (ETS) and Energy-Delay-Product (EDP) on a system with POWER8 and one P100 GPU. The authors demonstrated savings of ETS up to 27% and half of the examples decreasing EDP.

3 Testbed Systems

Table 1 presents testbed systems used for the research in this paper. We have performed the tests on two systems with modern Nvidia GPUs. Testbed 1 has 8 Nvidia Quadro RTX 6000 (Turing architecture) cards which were released in August 2018. Testbed 2 has 8 Nvidia Tesla V100-SXM2-16GB (Volta architecture) cards which were firstly released in June 2017. Both systems run on Ubuntu OS (20.4 and 18.4 respectively), both have CUDA 11 installed (11.5 and 11.2 respectively) and both has python 3.8 installed.

Table 1. Testbed configurations

Testbed system	Testbed 1: Quadro 6000	Testbed 2: V100
CPU model	2 x Intel® Xeon® Silver 4210 CPU @ 2.20 GHz	2 x Intel® Xeon® CPU E5-2686 v4 @ 2.30 GHz
CPU cores [physical/logical]	2 x [10/20]	2 x [16/32]
System memory size (RAM)	376 GB RAM	480 GB RAM
GPU model	Nvidia Quadro RTX 6000 (Turing)	Nvidia Tesla V100-SXM2-16GB (Volta)
GPU memory	24 GB GDDR6	16 GB HBM2
GPU default power limit	260 W	300 W
GPU available power limit range	100 W–260 W	150 W–300 W
Cuda cores	4608	5120
Core clock speed	1440 MHz	1370 MHz
Operating System	Ubuntu 20.04.3 LTS	Ubuntu 18.04.6 LTS
Python version	3.8.10	3.8.12
Cuda version	V11.5.119	V11.2.152
Tensorflow version	2.8.0	2.4.0

4 Deep CNN Benchmarks Used for Experiments

In the experiments we have used six popular Convolutional Neural Networks (CNN) designed for image recognition. The CNNs we have chosen are: Alexnet (presented in 2012, $63e6$ parameters), VGG-19 (2014, $143e6$ parameters), Inception V3 (2015, $24e6$ parameters), Inception V4 (2016, $43e6$ parameters), Resnet50 (2015, $26e6$ parameters) and Resnet152 (2015, $60e6$ parameters). The models used by us as a representative set of CNN benchmarks for the experiments were trained with synthetic ImageNet dataset. We have reduced the number of synthetic data samples to 100,000 for V100 system and to 32,000 for Quadro 6000 system. Each benchmark was executed on a single GPU, with either the batch size 128 (for Alexnet, VGG-19, Inception V3, Resnet50) or batch size 64 (for Inception V4 and Resnet152) where the model reached the GPU memory limit. The code of the benchmarks were taken from the official Tensorflow benchmarks github available online[1].

---

[1] https://github.com/tensorflow/benchmarks/.

5 Experiments and Results

5.1 Power Monitoring and Controlling Methodology

For the Nvidia GPUs power monitoring and controlling we have used the Application Programmable Interface (API) exposed in Nvidia Management Library (NVML). The API allows to read the current power usage in milliwatts using `nvmlDeviceGetPowerUsage` and – according to Nvidia's documentation – it is accurate to within $\pm5\%$ of current power draw. The GPU power limit is set via `nvmlDeviceSetPowerManagementLimit` API call. The power consumption is read with the fixed $T = 0.5\,s$ period. The energy consumption E is calculated as a definite integral of current power P and the sampling period T products for the given interval identical with application total execution time. The application execution time is measured using C++ `std::chrono::high_resolution_clock`. The whole process of power limits exploration and evaluation of the performance-energy trade-offs was fully automated and enclosed within a software tool we call StaticEnergyProfiler (StEP). The tool was originally introduced as EnergyProfiler in [12] as an automatic tool for exploration of software power caps in Intel CPUs. Since then we have extended the tool with Nvidia GPU support and published the code with new naming convention in an open source repository with Software Power Limiting Tools (SPLiT) suite available online[2].

All the results presented in Sect. 5.3 and 5.4 were obtained with the StEP automatic tool. For each system we have evaluated the full available power limits range (100 W–260 W for Quadro 6000 system and 150 W–300 W for V100 system) with a 5 W step. For each power limit we have executed 5 test runs and we present an average result. Each of six benchmark CNNs was executed in a training mode for just a one epoch with reduced ImageNet synthetic dataset. We assume the potential real life use case of proposed StEP tool as a fast way of power limits exploration with performance-energy trade-offs evaluation as an initial step before launching the full CNN training for the target number of epochs and full dataset with a power limit selected by user.

It should be noted that our approach extends benchmarking of various models for various hyperparameters and compute devices such as in [9] by generating more configurations by power capping. However, this does not result in changes to the important model accuracy, precision or recall values.

5.2 Target Metrics for Bi-Objective Energy-Performance Optimization

Exploration of available power limits results in a series of the energy-performance results which may form a Pareto-optimal front. If we target for a bi-objective optimization which is considering both energy savings and performance loss any result included in a Pareto-optimal front might be chosen as a desired solution. In order to evaluate the results within Pareto-optimal front we may want to evaluate

---

[2] https://projects.task.gda.pl/akrz/split/.

each result using various target metrics. One of the simplest is consideration of only the total energy consumption where we seek for energy minimum:

$$M_E(E, t) = E \tag{1}$$

Another metric that we would like to evaluate is already used by us in [13] total energy and total execution time product which aims for bi-objective optimization. The energy-performance product is known in the literature as energy-delay product (EDP)[6]:

$$M_{EDP}(E, t) = Et \tag{2}$$

Finally, the third bi-objective oriented metric for energy aware optimization which we will evaluate in this paper is energy-delay sum (EDS) proposed by Roberts et al. in [19]:

$$M_{EDS}(E, t) = \alpha E + \beta t \tag{3}$$

EDS is a weighted sum of energy E and total execution time t. The metric assumes that we adjust the proposed weights α and β the way that we arbitrarily choose a theoretical acceptable time increase for the abstract scenario when the energy consumption would be 0. Since we need to know the theoretical time increase represented by k parameter we also need to know the reference time and energy result. Thus, the EDS metric may be only evaluated as a relative value based on some reference result. Taking the two points with a reference result and a theoretical result of $k \cdot t_{ref}$ time and zero energy we may construct an equation considering the metric as a linear function based on two points: $E - E_{ref} = \frac{0 - E_{ref}}{k \cdot t_{ref} - t_{ref}} \cdot (t - t_{ref})$ out of which after transformation we obtain the formula:

$$1 = \frac{k-1}{k \cdot E_{ref}} \cdot E + \frac{1}{k \cdot t_{ref}} \cdot t \tag{4}$$

and we may read the $\alpha = \frac{k-1}{k \cdot E_{ref}}$ and $\beta = \frac{1}{k \cdot t_{ref}}$ weights where k is a theoretical accepted time increase.

Any result for which the Eq. 4 is true will indicate that the performance loss with given energy savings is proportionally equivalent to the reference result with respect to proportion defined by k parameter. Any result for which the EDS metric value is lower than 1 will be considered as better than the reference result in terms of EDS metric evaluation for particular k parameter value.

Figure 1 presents graphical visualization of three target metrics aforementioned above. The axes represent normalized energy and normalized execution time. The value of 1 on each axis represent the reference result obtained for the default system setup. Each data point represents a single energy-performance result obtained with some power limit applied. We present the E metric as a horizontal line with the value of 1 on the normalized energy axis. The EDP metric is a hyperbole which pass through point with (1,1) values. The EDS metric is a linear function which is presented by us for two values of k parameter ($k = 1.5$ and $k = 2.0$). Any result which is below the metric line will be accepted by this metric as a better than default result. The result point which euclidean distance from the metric line is the biggest will be considered as a result with minimal value for this metric.

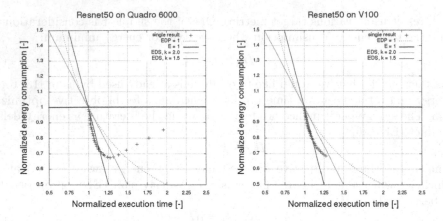

Fig. 1. Graphical visualization of selected target metrics for exemplary series of energy-performance results obtained for 1-epoch training of Resnet50 with different power limits executed on both Quadro 6000 (on the left) and V100 (on the right) systems.

5.3 Results Obtained for Quadro 6000 System

Fig. 2 presents the relative results obtained for all six CNN benchmarks executed on Quadro 6000 system. Table 2 presents specific absolute results with all the target metrics optimal results and their corresponding power limits obtained for the Quadro 6000 system.

For all of the tested CNN benchmarks any power limit lower than default one results in a lower energy (E) value. The minimum of energy for each of six tested CNN benchmarks on Quadro 6000 system can be found somewhere within a range of power limits of 120 W–130 W. The typical obtained minimum of energy results in 28.5%–32.5% of energy saved.

The EDP metric, which represents the energy and time product, has a value less than 1 for any of tested CNN benchmarks typically in the range of power limits of 120 W–255 W. That means that for any power limit within that range we obtain the EDP value better than the default one. Typically the EDP minimum can be found within the power limits range of 140 W–170 W on the Quadro 6000 system. The aforementioned EDP minimum results mostly in 25%–28% of energy saved with an average 4.5%–15.4% of performance loss. Considering the EDP metric minimum we get the trade-off of much higher energy savings than the cost of performance loss we bear.

The EDS metric was evaluated for two values of the k parameter. For $k = 1.5$ the values of a metric which are better than the reference result are mostly found when power limit is set in the range of 160 W–255 W. For $k = 2.0$ the range is wider and starts at 140 W. Typically, the minimum of EDS($k = 1.5$) is found in the power limits range of 190 W–215 W and results on average in 12%–26% of energy savings with only 3.9%–6.8% of performance loss. For EDS($k = 2.0$) the typical power limits range for finding a minimum of this metric is within 140 W–180 W what results on average in 22%–27% of energy savings with 4.5%–13.8%

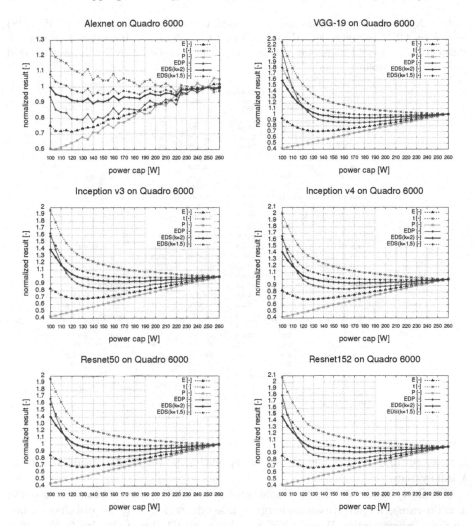

Fig. 2. Normalized results (average power, total energy consumption, total execution time, EDP and EDS metric) for all six CNN benchmarks (Alexnet, VGG-19, Inception 3, Inception 4, Resnet50, Resnet152) obtained for different power caps in range 100 W–260 W applied to Quadro 6000 system.

of performance loss. Minimization of EDS metric for both k parameter values allows then for significant energy savings with minor or negligible performance loss.

5.4   Results Obtained for V100 System

Fig. 3 presents relative results obtained for all six CNN benchmarks executed on V100 system. Table 3 presents specific absolute results with all the target

Table 2. Results of minimization of selected three target metrics (E, EDP, EDS) for all six CNN benchmarks (Alexnet, VGG-19, Inception 3, Inception 4, Resnet50, Resnet152) obtained for Quadro 6000 system with synthetic ImageNet dataset reduced to 32,000 of samples.

CNN benchmark	Target Metric	Power cap [W]	Average Power [W]	Total Energy [kJ]	Total Energy vs default [%]	Total Time [s]	Total Time vs default [%]
Alexnet	default	260	116.459	3.288	–	28.2	–
	min E	120	71.9	2.336	−28.5	32.5	+11.5
	min EDP	140	79.6	2.427	−25.7	30.5	+4.5
	min EDS(k=2.0)	140	79.6	2.427	−25.7	30.5	+4.5
	min EDS(k=1.5)	140	79.6	2.427	−25.7	30.5	+4.5
VGG-19	default	260	233.6	51.516	–	220.5	–
	min E	135	126.9	36.342	−29.9	286.4	+29.9
	min EDP	165	153.8	38.646	−25.0	251.2	+13.9
	min EDS(k=2.0)	175	162.5	39.734	−22.9	244.5	+10.9
	min EDS(k=1.5)	205	188.5	43.686	−15.2	231.8	+5.1
Inception3	default	260	224.4	41.382	–	184.4	–
	min E	125	115.0	28.157	−32.0	244.9	+32.8
	min EDP	155	139.8	29.720	−28.2	212.6	+15.3
	min EDS(k=2.0)	175	156.6	31.668	−23.5	202.2	+9.7
	min EDS(k=1.5)	190	169.3	33.346	−19.4	197.0	+6.8
Inception4	default	260	235.9	82.316	–	348.9	–
	min E	125	119.1	56.450	−31.4	474.0	+35.8
	min EDP	160	149.9	60.360	−26.7	402.6	+15.4
	min EDS(k=2.0)	165	154.1	61.202	−25.7	397.3	+13.8
	min EDS(k=1.5)	215	198.4	71.921	−12.6	362.4	+3.9
Resnet50	default	260	215.5	27.068	–	125.4	–
	min E	130	114.8	18.264	−32.5	159.1	+26.8
	min EDP	155	134.5	19.323	−28.6	143.7	+14.5
	min EDS(k=2.0)	160	139.1	19.683	−27.3	141.2	+12.8
	min EDS(k=1.5)	200	169.9	22.445	−17.1	132.1	+5.3
Resnet152	default	260	230.4	64.522	–	280.2	–
	min E	130	121.1	43.839	−32.1	362.0	+29.2
	min EDP	170	155.1	47.958	−25.7	309.2	+10.4
	min EDS(k=2.0)	180	163.6	49.471	−23.4	302.4	+7.9
	min EDS(k=1.5)	205	183.3	53.645	−16.9	292.7	+4.5

metrics optimal results and their corresponding power limits obtained for the V100 system.

Similarly to the Quadro 6000 system, on the V100 system we can also observe a wide range of performance-energy trade-offs which may be obtained when applying power limits. What is different between V100 and Quadro 6000 is that V100 has much higher lowest power limit value available and therefore in most of tested CNN benchmark cases the minimal value for the energy (E) metric is obtained for the lowest available power limit which is 150 W. The minimal energy (E) value for the V100 system allows for average energy savings of 24%–33%.

The EDP bi-objective metric which represents the energy and time product has its minimum for the most of tested CNN benchmarks when the power limit is set within range 170 W–180 W. We observe for EDP metric minimum the average energy savings of 23%–27% with corresponding performance loss of 13%–21%. This shows that for the V100 system the EDP metric minimum allows again for obtaining interesting and satisfactory trade-offs when proportionally more energy may be saved than we loose on execution time increase.

The EDS metric minimum for the V100 system was also evaluated for two values of the k parameter. For $k = 1.5$ most of the tests with selected CNN

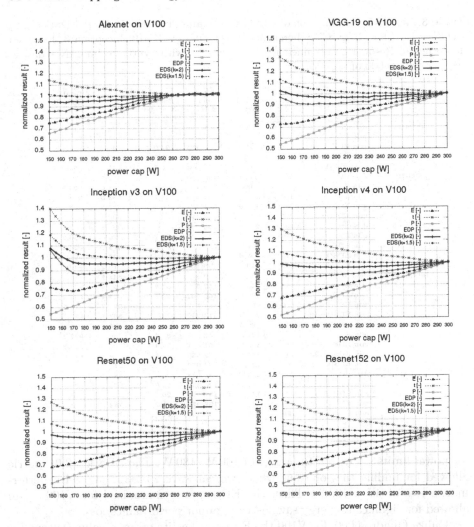

Fig. 3. Normalized results (average power, total energy consumption, total execution time, EDP and EDS metric) for all six CNN benchmarks (Alexnet, VGG-19, Inception 3, Inception 4, Resnet50, Resnet152) obtained for different power caps in range 100 W–260 W applied to V100 system.

benchmarks show that the results better than the reference one which means results with the value of EDS metric less than 1 were found for the V100 system mostly when power limit was set within 210 W–295 W. The optimal solutions (min of EDS) for $k = 1.5$ were found for power limits set within the range of 190 W–215 W. We observed significant energy savings of 12.6%–25.7% with only a minor performance loss of 3.9%–6.8%. For $k = 2.0$ the range of power limits with better than default results was covered by the range of 150 W–295 W. The minimal EDS($k = 2.0$) metric values for most of the tested CNNs were obtained

Table 3. Results of optimization of selected four target metrics (E, EDP, EDS(k = 2.0), EDS(k = 1.5)) for all six CNN benchmarks (Alexnet, VGG-19, Inception 3, Inception 4, Resnet50, Resnet152) obtained for V100 system with synthetic ImageNet dataset reduced to 100,000 of samples.

CNN benchmark	Target Metric	Power cap [W]	Average Power [W]	Total Energy [kJ]	Total Energy vs default [%]	Total Time [s]	Total Time vs default [%]
Alexnet	default	300	181.4	7.206	–	39.7	–
	min E	150	119.4	5.425	−24.7	45.4	+14.4
	min EDP	155	121.9	5.477	−24.0	44.9	+13.1
	min EDS(k=2.0)	165	127.5	5.603	−22.2	43.9	+10.6
	min EDS(k=1.5)	195	145.5	6.091	−15.5	41.9	+5.4
VGG-19	default	300	264.9	130.369	–	492.5	–
	min E	150	143.8	94.703	−27.4	658.6	+33.8
	min EDP	180	170.5	99.494	−23.7	583.5	+18.6
	min EDS(k=2.0)	200	189.9	105.181	−19.3	553.8	+12.5
	min EDS(k=1.5)	275	242.9	122.194	−6.3	503.1	+2.2
Inception3	default	300	265.8	104.941	–	394.8	–
	min E	170	160.9	75.608	−26.7	470.0	+20.2
	min EDP	170	160.9	75.608	-26.7	470.0	+20.2
	min EDS(k=2.0)	210	195.0	83.257	−19.4	427.0	+9.2
	min EDS(k=1.5)	245	219.7	90.106	−12.7	410.2	+4.9
Inception4	default	300	273.1	230.8	–	845.1	–
	min E	150	143.4	157.5	−31.8	1098.5	+30.0
	min EDP	170	162.6	166.602	−27.8	1024.4	+21.2
	min EDS(k=2.0)	200	190.9	181.532	−21.4	950.7	+12.5
	min EDS(k=1.5)	270	245.9	213.249	−7.6	867.4	+2.6
Resnet50	default	300	262.3	69.886	–	266.4	–
	min E	150	141.2	47.866	−31.5	339.0	+27.3
	min EDP	180	166.7	51.431	−26.4	308.4	+15.8
	min EDS(k=2.0)	190	177.6	53.430	−23.5	300.8	+12.9
	min EDS(k=1.5)	260	229.5	63.002	−9.9	274.5	+3.0
Resnet152	default	300	273.7	186.948	–	683.0	–
	min E	150	143.2	125.352	−33.0	875.3	+28.2
	min EDP	185	176.3	137.879	−26.2	782.1	+14.5
	min EDS(k=2.0)	185	176.3	137.879	−26.2	782.1	+14.5
	min EDS(k=1.5)	255	233.7	165.032	−11.7	706.2	+3.4

for power limits set within the range of 160 W–180 W. The EDS(k = 2.0) metric allowed for finding the solutions where we can save 23.5%–27.3% with performance loss of 4.5%–13.8%. Thus, the EDS metric minimized for V100 again allowed for significant energy savings with minor performance loss.

Unlike Quadro 6000, for V100 the E metric has its optimal value (except for Inception v3) for minimal power limit. The reason for that seem to be much higher minimal power limit available on V100 (150 W) compared to Quadro 6000 (100 W). If the V100 offered a wider range of available power limits, with the lower possible minimum, we could probably observe slightly lower minimal energy consumption with a bigger performance penalty. Then, the E characteristics would probably have their own clear minimal values for the corresponding power limits, that could be different from the imposed by the manufacturer minimal power caps.

6   Conclusions and Future Work

In this paper, we investigated performance-energy trade-offs for a collection of popular machine learning architectures, supported by power limiting (a.k.a.

capping) features of two Nvidia GPU cards: V100 and Quadro 6000. The performed benchmarks cover training of Deep Convolutional Neural Networks, such as Alexnet, Resnet or Inception, which were used used for image recognition.

The results proved that power capping and its underlying implementation limiting computational use of GPU resources can result in limited performance and consequently prolonged execution time but for some configurations allows even larger, percentage wise, reduction of energy used. Power limiting can imply much lower energy consumption (up to 33% for V100 and Quadro 6000), along with a low to medium performance penalty. Moreover, usage of well defined metrics enabled this bi-objective optimization to support the selection of the desired configuration.

The future works will cover the following areas:

- automation of the selection of a power-related configuration, based on a chosen metric, supporting the dynamic adaptation to the application behavior,
- CPU/GPU resource allocation, depending on the performance and energy requirements defined for a whole HPC system,
- modeling and simulation of the GPU/CPU behavior for selection of a static power-related configuration.

Basing on our research results, we are convinced that the power-performance optimization can significantly decrease the carbon footprint and energy cost of the machine learning solutions.

References

1. Ahmed, K., Tasnim, S., Yoshii, K.: Energy-efficient heterogeneous computing of parallel applications via power capping. In: 2020 International Conference on Computational Science and Computational Intelligence (CSCI), pp. 1237–1242, IEEE, December 2020
2. Cai, E., Juan, D.-C., Stamoulis, D., Marculescu, D.: Learning-based Power and Runtime Modeling for Convolutional Neural Networks (2019). https://personal.utdallas.edu/~fxc190007/courses/20S-7301/example-report-2.pdf. Accessed 8 June 2022
3. Ciesielczyk, T., et al.: An approach to reduce energy consumption and performance losses on heterogeneous servers using power capping. J. Sched. **24**(5), 489–505 (2021)
4. Czarnul, P., Proficz, J., Krzywaniak, A.: Energy-aware high-performance computing: survey of state-of-the-art tools, techniques, and environments. Sci. Program. **2019**, 8348791:1–8348791:19 (2019)
5. García-Martín, E., Rodrigues, C.F., Riley, G., Grahn, H.: Estimation of energy consumption in machine learning. J. Parallel Distrib. Comput. **134**, 75–88 (2019)
6. Gonzalez, R., Horowitz, M.: Energy dissipation in general purpose microprocessors. IEEE J. Solid-State Circuits **31**(9), 1277–1284 (1996)
7. Hogade, N., Pasricha, S., Siegel, H.J.: Energy and network aware workload management for geographically distributed data centers. CoRR abs/2106.00066 (2021)
8. Huang, Y., Guo, B., Shen, Y.: GPU energy consumption optimization with a global-based neural network method. IEEE Access **7**, 64303–64314 (2019)

9. Jabłońska, K., Czarnul, P.: Benchmarking deep neural network training using multi- and many-core processors. In: Saeed, K., Dvorský, J. (eds.) CISIM 2020. LNCS, vol. 12133, pp. 230–242. Springer, Cham (2020). https://doi.org/10.1007/978-3-030-47679-3_20

10. Kang, D.-K., Lee, K.-B., Kim, Y.-C.: Cost Efficient GPU Cluster Management for Training and Inference of Deep Learning. Energies **15**, 474 (2022). https://doi.org/10.3390/en15020474

11. Khan, T., Tian, W., Ilager, S., Buyya, R.: Workload forecasting and energy state estimation in cloud data centres: Ml-centric approach. Futur. Gener. Comput. Syst. **128**, 320–332 (2022)

12. Krzywaniak, A., Czarnul, P., Proficz, J.: Extended investigation of performance-energy trade-offs under power capping in HPC environments. In: 2019 International Conference on High Performance Computing and Simulation (HPCS), pp. 440–447 (2019)

13. Krzywaniak, A., Czarnul, P.: Performance/energy aware optimization of parallel applications on GPUs under power capping. In: Wyrzykowski, R., Deelman, E., Dongarra, J., Karczewski, K. (eds.) PPAM 2019. LNCS, vol. 12044, pp. 123–133. Springer, Cham (2020). https://doi.org/10.1007/978-3-030-43222-5_11

14. Li, D., Chen, X., Becchi, M., Zong, Z.: Evaluating the energy efficiency of deep convolutional neural networks on CPUs and GPUs. In: 2016 IEEE International Conferences on Big Data and Cloud Computing (BDCloud), Social Computing and Networking (SocialCom), Sustainable Computing and Communications (SustainCom) (BDCloud-SocialCom-SustainCom), pp. 477–484 (2016)

15. Mclaughlin, A., Paul, I., Greathouse, J.L.: A power characterization and management of GPU graph traversal (2014)

16. Mishra, A., Khare, N.: Analysis of DVFS techniques for improving the GPU energy efficiency. Open J. Energy Effic. **04**(04), 77–86 (2015)

17. Patki, T., et al.: Comparing GPU power and frequency capping: a case study with the MuMMI workflow. In: 2019 IEEE/ACM Workflows in Support of Large-Scale Science (WORKS). pp. 31–39, IEEE, November 2019. https://doi.org/10.1109/WORKS49585.2019.00009, https://ieeexplore.ieee.org/document/8943552/

18. Mazuecos Pérez, M.D., Seiler, N.G., Bederián, C.S., Wolovick, N., Vega, A.J.: Power efficiency analysis of a deep learning workload on an IBM Minsky platform. In: Meneses, E., Castro, H., Barrios Hernández, C.J., Ramos-Pollan, R. (eds.) CARLA 2018. CCIS, vol. 979, pp. 255–262. Springer, Cham (2019). https://doi.org/10.1007/978-3-030-16205-4_19

19. Roberts, S.I., Wright, S.A., Fahmy, S.A., Jarvis, S.A.: Metrics for energy-aware software optimisation. In: Kunkel, J.M., Yokota, R., Balaji, P., Keyes, D. (eds.) ISC High Performance 2017. LNCS, vol. 10266, pp. 413–430. Springer, Cham (2017). https://doi.org/10.1007/978-3-319-58667-0_22

20. Straube, K., Lowe-Power, J., Nitta, C., Farrens, M., Akella, V.: Improving provisioned power efficiency in HPC systems with GPU-CAPP. In: 2018 IEEE 25th International Conference on High Performance Computing (HiPC), pp. 112–122, IEEE, December 2018

21. Tsuzuku, K., Endo, T.: Power capping of CPU-GPU heterogeneous systems using power and performance models. In: Proceedings of the 4th International Conference on Smart Cities and Green ICT Systems, pp. 226–233. SCITEPRESS - Science and and Technology Publications (2015)

22. Wang, F., Zhang, W., Lai, S., Hao, M., Wang, Z.: Dynamic GPU energy optimization for machine learning training workloads. IEEE Trans. Parallel Distrib. Syst. **33**(11), 2943–2954 (2022). https://doi.org/10.1109/TPDS.2021.3137867

23. Wang, Y., et al.: E2-train: energy-efficient deep network training with data-, model-, and algorithm-level saving. CoRR abs/1910.13349 (2019)
24. Zou, P., Li, A., Barker, K., Ge, R.: Indicator-directed dynamic power management for iterative workloads on GPU-accelerated systems. In: Proceedings-20th IEEE/ACM International Symposium on Cluster, Cloud and Internet Computing, CCGRID 2020, pp. 559–568 (2020). https://doi.org/10.1109/CCGrid49817.2020.00-37

Forecasting Bank Default with the Merton Model: The Case of US Banks

Kihwan Jo[1], Gahyun Choi[1,2], Jongwook Jeong[3(✉)], and Kwangwon Ahn[1,2(✉)]

[1] Center for Finance and Technology, Yonsei University, Seoul, South Korea
k.ahn@yonsei.ac.kr
[2] Department of Industrial Engineering, Yonsei University, Seoul, South Korea
[3] CIO2, SK Square, Seoul, South Korea
jongwook.jeong@sk.com

Abstract. This paper examines whether the probability of default (Merton 1974) can be applied to banks' default predictions. Using the case of US banks in the post-crisis period (2010–2014), we estimate several Cox proportional hazard models as well as their out-of-sample performance. As a result, we find that the Merton measure, that is, the probability of default, is not a sufficient statistic for predicting bank default, while, with the 6-month forecasting horizon, it is an extremely significant predictor and its functional form is a useful construct for predicting bank default. Findings suggest that (i) predicting banks' defaults over a mid- to long-term horizon can be done more effectively by adding the inverse of equity volatility and the value of net income over total assets, and (ii) the role of the capital adequacy ratio is doubtful even in short-run default prediction.

Keywords: Bank default · Prediction · Probability of default

1 Introduction

The sound operation of the banking sector underpins the safety of the market economy due to its role in offering liquidity to the marketplace in which industry players commonly trade goods and services in the physical market. Moreover, the default of a particular bank can quickly spread to other banks through the creditor–debtor network, resulting in a significant impact on the economy as a whole along with the globalization of the financial market. Since the global financial crisis, the need for the preemptive management of the banking sector and early warning indicators have gained much attention; the bankruptcy of individual banks has much more economic ramifications and costs than that of a corporation. As a banking crisis imposes significant social costs, it is important for regulators to establish a system that can detect prevailing risks in advance and implement an immediate response. Although there have been many prior studies on predictions of default, they are mostly focused on general enterprises rather than banks. Due to the unique nature of the banking industry, there are not many prior studies that thoroughly investigated bank default predictions.

Merton [1] proposed a structural model for assessing the credit risk of a corporation by presuming the firm's equity as a call option on its assets. Specifically, Merton's "distance to default" (hereafter, DD) expresses the distance at which corporate values fall into debt levels in Z-score. Merton's model has been widely used; for example, Moody's KMV commercialized a corporate default prediction model. However, there are a number of opinions on the use of Merton's model for default prediction. Some studies have argued that the model outperforms Altman's Z-score and Ohlson's O-score [2], while others have provided counter-evidence that Merton's DD is not a sufficient statistic for measuring credit quality and the predictability of the reduced-form model, particularly with market value, is much more accurate [3, 4]. In addition, some studies have reported evidence that the predictability of Merton's DD increases when it is used together with volatility and leverage [5]. As discussed, the assessment of Merton's DD model and the ways to improve its use are still inconclusive, even in corporate default.

However, there are still limited discussions about whether or not Merton's DD model can be useful for predicting bank default. Some studies have reported that information on credit ratings with Merton's DD model has predictability for bank default when used with bond spreads; particularly for downgrading banks [6]. Another study provided evidence that Merton's DD and its spreads are a superior measure than accounting data in default prediction in a case study of Japanese banks [7]. These studies provided an explanation of why and how Merton's DD has difficulty being used for banks' default prediction in two-fold: (i) market and funding risk stemming from high-leverage assets by short-term procurement and (ii) regulation and policy intervention in the market. Accordingly, some studies proposed the concept of "distance to capital" (hereafter, DC) by introducing the capital adequacy ratio (CAR) into Merton's DD to predict bank default [8–10],[1] and others revised the assumptions of Merton's model, such as asset value following a lognormal distribution, as it underestimates bank default risk [11]. As such, several attempts have been made to improve bank default prediction by including the CAR or relaxing the assumptions of the Merton model.

This study examines whether the probability of default (hereafter, POD) calculated using Merton's DD measure can be used for bank default prediction. Specifically, we test the three hypotheses as follows: (i) the POD is a sufficient statistic for forecasting bank default; (ii) the functional form of the Merton DD model creates useful information, like the case of a corporate, for forecasting bank default; and (iii) the POD has predictability for bank default. For this purpose, we estimate several Cox proportional hazard models and examine their out-of-sample performance for US banks in the post-crisis period (2010–2014). As a result, we find that the POD is not a sufficient statistic for forecasting default. Yet, over a 6-month forecasting horizon, the functional form of the Merton DD model is useful and Merton DD probability has significant default predictability. The findings suggest that the CAR, a unique characteristic of each bank, fails to predict bank default, even in the short-run. Yet, in over the mid- to long-term horizon, bank default predictions can be made more effectively by adding the inverse of equity volatility and the value of net income over total assets for in-sample and out-of-sample forecasts, respectively.

---

[1] In particular, Ji et al. [9] employed the time-varying volatility when calculating Merton's DD and its extension, DC, by sampling the posterior distribution and proposed an early warning indicator using the difference between DD and DC.

In this paper, Sect. 2 explains data and methodology, and Sect. 3 presents the results and discussion. Section 4 concludes.

2 Data and Methodology

2.1 Data

Quarterly data of 322 US banks, including 60 failed banks, were retrieved from the COMPUSTAT database for the 2010–2014 period. The term "failed bank" is used following the definition of Fahlenbrach et al. [12]. Failed banks are categorized by their type of failure as follows: (i) banks that are on the list of failed banks maintained by the Federal Deposit Insurance Corporation (FDIC); (ii) banks that have filed under Chapter 11 and are not on the FDIC list; (iii) banks that are merged at a discount; (iv) banks that have been forced to de-list from the stock exchange; and (v) banks that have voluntarily de-listed. During the sample period, most failed banks belong to the "FDIC" and "Forced delisting" categories, and the number of failed banks decreased over time as the effects of the global financial crisis diminished. For the period from 2010 to 2011, the number of failed banks was 26 and 15, and FDIC cases were the main type of failed banks. Table 1 summarizes the annual status of failed banks by type.

Table 1. Sample construction of failed banks

Year	2010	2011	2012	2013	2014	Total
FDIC	13	6	1	2	2	24
Chapter 11	4	1	1	2	0	8
Merged at a discount	3	1	0	1	0	5
Forced delisting	3	6	5	0	0	14
Voluntary delisting	3	1	0	4	1	9
Total	26	15	7	9	3	60

Note: A case filed under Chapter 11 is frequently referred to as a "reorganization" or "rehabilitation" bankruptcy.

2.2 Hypotheses

This paper aims to investigate whether the probability of default [1] can be used for banks' default prediction. For this purpose, we estimate several Cox proportional hazard models to test the following three hypotheses:

- The first hypothesis is that the POD is a sufficient statistic for forecasting bank default, implying that any other variable in a hazard model should not be a statistically significant covariate other than the POD.

- The second hypothesis is that the functional form of the Merton DD model is useful for forecasting bank default, implying that the POD should remain statistically significant in a hazard model that includes all of the variables used to calculate the POD.
- The third hypothesis is that the POD has bank default predictability, implying that the POD should remain as a statistically significant default predictor in our hazard model, regardless of the other variables that we include in the models.

As a robustness check, we further examine the out-of-sample performance of our models.

2.3 Merton's DD Probability and Its Extension

The Merton model makes two crucial assumptions. The first is that the asset value of a firm follows geometric Brownian motion,

$$dV_t = \mu V_t dt + \sigma V_t dW_t$$

where V_t is the asset value of the firm, μ is the continuously compounded average return on V_t, σ is the volatility of firm value, and dW_t is a standard Wiener process. The second assumption is that the firm has issued only one discount bond maturing after T periods. Under these two assumptions, the equity of the firm E_t is regarded as a European call option on the underlying asset value of the firm with the strike price being the obligated debt payment L at maturity T,

$$E_T = \max[V_T - L, 0].$$

in the end, the resulting Z-score, namely the POD, is the probability that a borrower cannot fulfill its promised payment at maturity,

$$POD = P(V_T < L) = N(-DD)$$

where $N(\cdot)$ is the cumulative density function (CDF) of the standard normal distribution and DD is the distance to default, namely Merton's DD.

In contrast to the probability of default for each firm, the insolvency risk of a financial institution can be measured by estimating the DC considering the minimum capital requirement [9]. The probability of undercapitalization (hereafter, POU) is conceptually similar to the well-known POD. For simplicity, we assume that capital consists completely of equity [9] and that the statutory minimum capital adequacy ratio is $c = 0.08$ as the threshold for undercapitalization (Basel I). Accordingly, once $V_T - L < c \cdot V_T$ holds at time T after a debt payment, the bank is presumed to be undercapitalized. In particular, the POU is modified from Merton's model [8] and measures insolvency risk rather than default risk, as follows:

$$POU = P(V_T - L < c \cdot V_T) = P(V_T < \lambda L) = N(-DC)$$

where DC is the distance to capital and λ is the correction factor for DC, as follows:

$$\lambda = \frac{1}{1-c}.$$

For the estimation strategy, we follow Ji et al. [9], presenting the sampling procedure from the posterior distribution through state filtering and parameter learning.

2.4 Hazard Model

To assess the Merton DD model's accuracy, we need a method to compare the POD to alternative predictor variables. Thus, we employ a Cox proportional hazard model to test our hypotheses [13]. Proportional hazard models assume that the hazard rate $\lambda(t)$, that is, the probability of default at time t conditional on survival until time t, is as follows:

$$\lambda(t) = \lambda_0(t) \exp[Z(t)'\beta],$$

where $\lambda_0(t)$ is the baseline hazard rate and the term $Z(t)'\beta$ allows the expected time to default to vary across banks according to their covariates, $Z(t)$. The baseline hazard rate is common to all banks.

The Cox proportional hazard model does not impose any structure on the baseline hazard rate $\lambda_0(t)$. Cox's partial likelihood estimator provides a way of estimating β without requiring estimate of $\lambda_0(t)$. It can also handle censoring of observations, which is one of the features of the data. Details about estimating the proportional hazard model can be found in many sources, including [13].

3 Results and Discussion

3.1 Hazard Model Results

Panels A and B of Table 2 summarize the results of estimating several Cox proportional hazard models with the 6- and 9-month forecasting horizons. Models 1 and 2 in both panels are univariate hazard models, which explain time to default as a function of the Merton DD probability and its extension incorporating the CAR, that is, the POD and the POU, respectively. These are simple univariate models. Yet, the fact that their explanatory variables vary over time implies that it is more complicated than it might appear. Models 1 and 2 confirm that the POD and the POU are both extremely significant predictors of default.

Table 2. Hazard model estimates

Panel A: Time to default – 6 months							
Variable	Model 1	Model 2	Model 3	Model 4	Model 5	Model 6	Model 7
POD	2.901***		1.628***		1.562***		1.12
	(0.295)		(0.470)		(0.723)		(0.688)
POU		3.039***	1.671***			1.630***	1.495***
		(0.383)	(0.565)			(0.634)	(0.681)
ln E				−0.115	0.090	0.058	0.190
				(0.125)	(0.159)	(0.151)	(0.171)
ln F				0.041	−0.182	−0.129	−0.273
				(0.130)	(0.164)	(0.155)	(0.175)

(*continued*)

Table 2. (*continued*)

Panel A: Time to default – 6 months

Variable	Model 1	Model 2	Model 3	Model 4	Model 5	Model 6	Model 7
$1/\sigma(E)$				-0.828^{***}	-0.604^{***}	-0.644^{***}	-0.484^{**}
				(0.181)	(0.207)	(0.188)	(0.212)
$r_{i,t-1} - r_{m,t-1}$				-0.001	0.003	0.004	0.001
				(0.001)	(0.001)	(0.001)	(0.001)
NI/TA				-0.018	-0.024	-0.038	-0.039
				(0.031)	(0.031)	(0.0.32)	(0.033)
Pseudo R^2	0.085	0.077	0.095	0.124	0.128	0.128	0.130

Panel B: Time to default – 9 months

Variable	Model 1	Model 2	Model 3	Model 4	Model 5	Model 6	Model 7
POD	2.685^{***}		1.564^{***}		0.911		0.667
	(0.284)		(0.463)		(0.609)		(0.621)
POU		2.704^{***}	1.446^{***}			0.903	0.736
		(0.350)	(0.530)			(0.580)	(0.630)
$\ln E$				-0.061	0.073	0.000	0.190
				(0.120)	(0.153)	(0.152)	(0.166)
$\ln F$				0.029	-0.121	-0.097	-0.183
				(0.122)	(0.156)	(0.151)	(0.169)
$1/\sigma(E)$				-0.905^{***}	-0.779^{***}	-0.812^{***}	-0.727^{***}
				(0.156)	(0.181)	(0.170)	(0.191)
$r_{i,t-1} - r_{m,t-1}$				-0.003	-0.0009	-0.0003	0.0008
				(0.001)	(0.001)	(0.001)	(0.001)
NI/TA				-0.016	-0.012	-0.017	-0.013
				(0.022)	(0.023)	(0.0.22)	(0.023)
Pseudo R^2	0.083	0.076	0.089	0.103	0.109	0.108	0.111

Note: There are 322 banks in total and 60 defaults in the sample. The PODs and POUs are expressed in percentage. $\ln E$ and $\ln F$ are the natural logarithms of equity (in millions of dollars) and the face value of debt (in millions of dollars), respectively. $1/\sigma(E)$ is the inverse of equity volatility measured using daily data from the previous year, $r_{i,t-1} - r_{m,t-1}$ is the stocks return over the previous year minus the market return over the same period, and NI/TA is the ratio of net income to total assets. A positive coefficient on a particular variable implies that the hazard rate is increasing in that value. Standard errors are in parentheses. *** and ** denote significance at the 1% and 5% levels, respectively.

Model 3 in Panels A and B combines the POD and POU in one hazard model. Both covariates are statistically significant, indicating that the Merton DD probability is not a sufficient statistic for bank default prediction resulting in the rejection of our first hypothesis for the 6- as well as 9-month forecasting horizons. While the coefficients of two covariates have similar magnitudes and statistical significance, their magnitudes are much smaller in Model 3 than in the first two models, Models 1 and 2 for both forecasting horizons. This suggests that the POU shares the information content of the POD related to predicting bank default.

Model 4 in both panels is a simple reduced-form model that employs the same inputs as the Merton DD model: the log of the bank's equity value, the log of the bank's debt, its returns over the past year, and the inverse of the bank's equity volatility. Unlike Bharath and Shumway [4], only the covariate of the inverse of the bank's equity volatility is strongly statistically significant in both forecasting horizons, implying that volatility is a strong predictor of bank default unlike corporate bankruptcy. For example, low volatility over a prolonged period leads to higher risk-taking [14].

Models 5 and 6 in both panels include all the covariates of Model 4 and also include the POD and POU, respectively. Comparing the estimates of Models 4, 5, and 6 in Panel A, we find that the POD and POU are significant predictors, even when all of the quantities used to calculate the Merton DD probability are included in the hazard model. Accordingly, Model 5 in Panel A indicates that the functional form of the POD is a useful construct for bank default forecasting in the 6-month forecasting horizon, providing strong evidence in favor of our second hypothesis. Yet, only $1/\sigma(E)$ in Model 5 of Panel B shows significant predictability in the 9-month prediction. Thus, the functional form of the Merton DD probability fails to create any meaningful information for bank default prediction when the prediction horizon is extended to 9 months. Moreover, unlike in the 6-month default prediction, the role of the CAR also disappears for bank default prediction in the 9-month forecasting horizon shown in Model 6 of Panel B.

In Panels A and B, Model 7 adds the POU to Model 5; the POU includes information about the Merton DD probability and the CAR into a single risk metric. From the estimates of Model 7 of Panel A, we find that none of the predictors, other than the POU and $1/\sigma(E)$, is statistically significant. Yet, this cannot force to reject our third hypothesis in the 6-month prediction horizon since the POU inherits information content from the POD, in terms of its inputs as well as functional structure, and continues to be significant. This evidence suggests that the POU, that is, the POD incorporating the CAR, could deliver more meaningful information in regard to bank default prediction than the POD, in a 6-month forecasting window. In the 9-month predictions (Models 5, 6, and 7 in Panel B of Table 2), Merton's DD probability does not have predictability for bank default and including the CAR into the POD also fails to provide any meaningful information content as a default predictor.

Overall, Table 2 shows that the Merton DD probability is an extremely significant predictor but is not a sufficient statistic for predicting bank default. Moreover, at least in the 6-month prediction, it indicates that the functional form of the Merton DD model is useful for forecasting bank default and the unique characteristics of each bank could be used in bank default prediction for short-run default predictions: the POU is more important than the Merton DD probability for forecasting bank default. However, both the functional form of the Merton DD probability and even adding the CAR as an additional input for calculating the Z-score fail to produce any meaningful information for bank default prediction in the 9-month prediction.

3.2 Out-of-Sample Results

Table 3 documents our assessment of the out-of-sample predictability of several variables. To create the table, banks are sorted into deciles of a particular forecasting variable. Then, the number of defaults that occurred up to each decile group is tabulated in terms of the percentage. One advantage of this approach is that the default predictability of a specific variable can be summarized without estimating actual default probabilities. Even if our model for translating the distance to default (and distance to capital) into the POD (and the POU) is slightly misspecified, our out-of-sample results will remain unaffected. Specifically, the normal CDF is not the most appropriate choice.

Table 3. Out-of-sample forecasts

Panel A: Out-of-sample forecast – 6 months

Decile	POD	POU	E	$r_{i,t-1} - r_{m,t-1}$	NI/TA
	60 failures, 322 firm–months (6 months)				
1	53.3	53.3	46.7	51.7	51.7
2	81.7	71.7	75.0	73.3	80.0
3	90.0	85.0	85.0	81.7	88.3
4	100	90.0	90.0	81.7	88.3
5	100	91.7	93.3	86.7	96.7
6–10	100	100	100	100	100

Panel B: Out-of-sample forecast – 9 months

Decile	POD	POU	E	$r_{i,t-1} - r_{m,t-1}$	NI/TA
1	51.7	40.0	43.3	50.0	53.3
2	76.7	68.3	65.0	61.7	78.3
3	83.3	80.0	81.7	68.3	85.0
4	88.3	86.7	85.0	71.7	90.0
5	100	90.0	90.0	71.7	90.0
6–10	100	100	100	100	100

Note: Panel A examines the accuracy over a 6-month forecasting horizon. There are 322 firms in our sample with 60 defaults. The POD (in percentage) is the Merton DD probability, the POU (in percentage) is the extension of POD incorporating the CAR information, E (in millions of dollars) is market equity, $r_{i,t-1} - r_{m,t-1}$ is the stock's return over the previous year minus the market's return over the same period, and NI/TA is the bank's ratio of net income to total assets. Panel B considers defaults in the 9-month forecasting horizon, and it includes the same forecast variables as Panel A.

Panel A compares the predictions of the Merton DD model to its extension with the CAR, market equity, past returns, and the ratio of net income to total assets in the 6-month forecasting horizon. While both the Merton DD model probability and its extension (namely, the POD and POU, respectively) are able to classify 53.3% of defaulting banks in the highest probability decile at the beginning of the quarter in which they default, the POU underperforms in terms of classifying defaulting banks in the other deciles. In particular, the POU is an even a worse predictor than E and NI/TA from the 2nd decile, indicating that the inclusion of the CAR in Merton's DD probability, in the form of the POU, even hurts the POD's out-of-sample performance, unlike the hazard models.

In the 6-month forecasting horizon, the out-of-sample performance of the POD is much better than simply sorting firms on their market equity in the entire deciles. This is consistent with the results of corporate default prediction, implying that the success of the POD does not simply reflect the predictive value of market equity [4, 15]. Apparently, it is useful to form a probability measure by creating a Z-score and using a cumulative distribution to calculate its corresponding probability. Given that the POU does not perform better than the POD in out-of-sample forecasts contrary to hazard models, the functional form of the probability measure suggested by the Merton DD model appears to be a more valuable innovation than the incorporation of the CAR into the POD.

Panel B reports similar forecasting results with hazard models for a longer prediction horizon, that is, 9 months. Remarkably, the POD and POU, both of which exploit the same Z-score functional form suggested by theory, are not better at default prediction than NI/TA. Yet, both models perform quite well in classifying low-risk banks. In particular, the misclassification of risky banks into the lower risk deciles (deciles 5–10) is obviously the lowest for the POD, implying that using the functional form suggested by theory produces fewer low-risk misclassifications than any of the others considered. These results again confirm that we should reject our first hypothesis in the 9-month forecasting horizon, and the role of the CAR as a default predictor disappears regardless of the forecasting horizon.

Simply sorting banks by the value of their net income over total assets has surprisingly strong forecasting power, greater than any of the other indicators, including the POD, at least for 9-month forecasting. This is in contrast to the economic and statistical significance of the POD and NI/TA in the hazard model, in which NI/TA is not a significant predictor, as well as to the results of the out-of-sample forecasts reported in [4] in the case of firms. Thus, since the Merton DD model has no simple way to capture the innovations in the value of net income over total assets, it is difficult to believe that the POD is a sufficient statistic for default prediction.

4   Conclusion

For US banks in the post-crisis period (2010–2014), we find that the Merton DD measure is not a sufficient statistic for predicting bank default, that is, similar to the prediction of corporate default. However, particularly in the 6-month forecasting horizon, the Z-score calculated by Merton's DD is an extremely significant predictor and its functional form is a useful construct for forecasting bank default. The findings suggest that, over the mid- to long-term horizon, bank default prediction can be improved by adding the inverse of equity volatility (in-sample forecast) and the value of net income over total assets (out-of-sample forecast) in addition to the POD. Yet, the role of the capital adequacy ratio is doubtful even in short-run default prediction.

For corporate default prediction, the forecasting horizon is typically a year. However, the POD has bank default predictability only for a 6-month forecasting horizon, and it is difficult to predict bank default over a longer time horizon, even for 9 months. This is due to the fact that financial institutions are sensitive to investor confidence and the progress of default proceeds much more rapidly than with firms. In conclusion, social costs can be minimized by early diagnosis and rapid response to banks' default by paying attention to the POD. The role of equity volatility and the value of net income over total assets should not be overlooked in bank default prediction. In addition, we could (i) extend Merton's framework by adopting stochastic volatility and (ii) propose an early warning indicator for banks' credit risk for future studies.

Acknowledgements. This research was supported by the Future-leading Research Initiative at Yonsei University (Grant Number: 2021–22-0306; K.A.).

References

1. Merton, R.C.: On the pricing of corporate debt: The risk structure of interest rates. J. Financ. **29**(2), 449–470 (1974)
2. Hillegeist, S.A., Keating, E.K., Cram, D.P., Lundstedt, K.G.: Assessing the probability of bankruptcy. Rev. Acc. Stud. **9**(1), 5–34 (2004)
3. Du, Y., Suo, W.: Assessing credit quality from equity markets: can a structural approach forecast credit ratings? Can. J. Adm. Sci. **24**, 212–228 (2007)
4. Bharath, S.T., Shumway, T.: Forecasting default with the Merton distance to default model. Rev. Financ. Stud. **21**(3), 1339–1369 (2008)
5. Campbell, J.Y., Hilscher, J., Szilagyi, J.: In search of distress risk. J. Financ. **63**(6), 2899–2939 (2008)
6. Gropp, R., Vesala, J., Vulpes, G.: Equity and bond market signals as leading indicators of bank fragility. J. Money Credit Bank **38**(2), 399–428 (2006)
7. Harada, K., Ito, T., Takahashi, S.: Is the distance to default a good measure in predicting bank failures? a case study of Japanese major banks. Jpn. World Econ. **27**, 70–82 (2013)
8. Chan-Lau, J.A., Sy, A.N.: Distance–to–default in banking: a bridge too far? J. Bank. Regul. **9**(1), 14–24 (2007)
9. Ji, G., Kim, D.S., Ahn, K.: Financial structure and systemic risk of banks: evidence from Chinese reform. Sustainability **11**, 3721 (2019)
10. Expósito, R.R., Veiga, J., Touriño, J.: Enabling Hardware Affinity in JVM-Based Applications: A Case Study for Big Data. In: Krzhizhanovskaya, V.V., Závodszky, G., Lees, M.H., Dongarra, J.J., Sloot, P.M.A., Brissos, S., Teixeira, J. (eds.) ICCS 2020. LNCS, vol. 12137, pp. 31–44. Springer, Cham (2020). https://doi.org/10.1007/978-3-030-50371-0_3
11. Nagel, S., Purnanandam, A.: Banks' risk dynamics and distance to default. Rev. Financ. Stud. **33**(6), 2421–2467 (2020)
12. Fahlenbrach, R., Prilmeier, R., Stulz, R.M.: This time is the same: using bank performance in 1998 to explain bank performance during the recent financial crisis. J. Financ. **67**(6), 2139–2185 (2012)
13. Cox, D.R., Oakes, D.: Analysis of Survival Data (Vol. 21). CRC Press. (1984)
14. Danielsson, J., Valenzuela, M., Zer, I.: Learning from history: Volatility and financial crises. LSE SRC Discussion Paper No 57 (2018)
15. Vassalou, M., Xing, Y.: Default risk in equity returns. J. Financ. **59**(2), 831–868 (2004)

Deep Neural Sequence to Sequence Lexical Substitution for the Polish Language

Michał Pogoda[✉] [ID], Karol Gawron[ID], Norbert Ropiak[ID],
Michał Swędrowski[ID], and Jan Kocoń[ID]

Wrocław University of Science and Technology, 50-370 Wrocław, Poland
{michal.pogoda,karol.gawron,norbert.ropiak,michal.swedrowski,
jan.kocon}@pwr.edu.pl

Abstract. The aim of this paper is to investigate the applicability of language models to the problem of lexical substitution in a strongly inflected language. For this purpose, we focus on pre-trained models based on transformer architectures, in particular BERT and BART. We present a solution in the form of the BART-based sequence-to-sequence model. Then we propose and explore a number of approaches to generate an artificial dataset for lexical substitution, using the adapted PLEWiC dataset as a reference. During this study we focus on Polish as an example of a strongly inflected language.

Keywords: Natual lanugage processing · Lexical substitution · Sequence to sequence

1 Introduction

Lexical substitution is the task of finding alternative word substitutions that preserve a statement's meaning in context. Its applications include text simplification and paraphrasing. The main challenge is finding a word substitute that not only preserves the meaning of the original sentence but is also grammatically correct and fits the context.

To approach near-human quality in this task, models must perfectly match the relationships between synonymous expressions and their meaning in the context. Often expressions, that by definition are not synonyms can become ones through contextual clarification. For example, a context can restrict the meaning of a hyperonym so that the meaning becomes equivalent to its hyponym).

Her pet was barking loudly

Here we can - with reasonable probability - replace the word *"pet"* with a word *"dog"* even despite the fact that they do not share a synonymy relationship.

Large pre-trained models such as BERT [3] embed a large amount of information regarding word placement in a context. It has been shown [18,22] that, despite not being trained strictly in a lexical substitution task, they exhibit state-of-the-art results in similar problems. In this paper, we evaluate a solution analogous to the one proposed in [18] in a strongly inflected language setting (the Polish Language) and compare the approach to the supervised learning of the sequence-to-sequence model on synthetic and natural sets.

This paper contains the following contributions:

- Adaptation of the PLEWiC dataset as a lexical substitution benchmark.
- Comparison of different methods for generating synthetic lexical substitution datasets.
- Evaluation of BERT-based lexical substitution for the Polish language and comparison with fine-tuned BART [12] model.

2 Related Work

One of the earliest approaches to finding word alternatives were based on lexical networks (WordNets) using synonyms. As an example, one can consider the baseline method proposed by D. McCarthy et al. [14]. This approach, however, has some limitations: they rely only on manually annotated relations and omit all words that are not marked as synonyms, but can still be good alternatives to the selected word; they do not use the context of the word being replaced.

In a highly influential work, T. Mikolov at al. [16] proposed the idea of unsupervised word embeddings becoming a source of inspiration for approaches like the usage of skip-gram-based word representations [15] to perform lexical substitution based on vector distance.

Modern approaches are mostly based on deep language models. N. Arefyev et al. [1] showed that large pretrained models can give better performance than previous unsupervised and supervised methods of lexical substitution. They also noted that for BERT-type models whose tokenization is subword, a multi-token prediction could give better results.

According to our knowledge, for SemEval [14] and CoinCo [10] benchmarks, the best models for English are based on transformer architectures. W. Zhou et al. [22] present methods for target masking: full, partial, none. They also show a positive effect on the model proposal sorting method's performance, which is based on the cosine similarity of the embeddings of the proposed sentences to the original sentences.

To address the high computational complexity of sorting, instead of using the original model, distilled versions of models can be used, additionally trained to evaluate the sentence semantic similarity (STS). N. Reimers et al. [20] show the advantage of such a trained model in the STS task and make the distilUSE model available as part of the Sentence Transformers framework[1].

In a paper on a similar lexical simplification task, J. Qiang et al. [18] present a method that does not hide part of the information from the model (as in partial masking, which can make it challenging to find suitable alternatives) but

---

[1] https://www.sbert.net/.

combines the original sentences with the sentence where the selected (target) word is masked.

Encoder-only architectures like BERT pose a significant computational complexity problem for multi-token prediction. To get around it, a Seq2Seq BART model can be used. L. Martin [13] applied it to the sentence simplification task, which leads us to believe that it would also work well for the lexical substitution task. In the following subsections, we summarize how substitutions are obtained in the models trained with the masked language modeling objective. In our work, applying ideas from the presented papers, we compare the model used in the current SOTA i.e. BERT, and an approach using the BART sequence-to-sequence model for the task of lexical substitution in the Polish Language.

2.1 Full Masking

One of the simplest approaches involves replacing a piece of text with a mask token (or multiple tokens) [22]. The problem with this approach is that the semantic meaning of the masked text fragment is completely lost. For example, if we replace the word *cat* in the sentence *She has a cat* with a mask, the linguistically correct responses will be tokens such as *car* or *computer* (see Fig. 1).

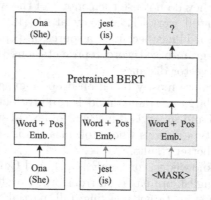

Fig. 1. Full masking approach. We replace a word with a `<MASK>` token (or tokens), and the model finds a substitution based purely on context. It does not have any information on what word was initially used, and substitutions will likely be returned based only on the frequency of occurrence in a given context.

2.2 Partial Masking

The second approach, that partially address the issue with full masking, is the so-called partial masking [22]. Instead of completely replacing the selected fragment with a mask, dropout can be applied after the initial embedding of the input tokens. The assumption is that with this approach, some semantic information is preserved, while at the same time, the model has such little information about the initial token that it can propose insertions other than the trivial copy of the input token (see Fig. 2).

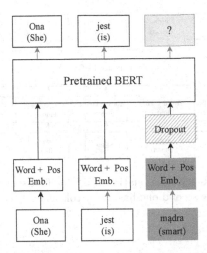

Fig. 2. Partial masking approach. Dropout is applied after initial embedding. That makes it harder for the model to just copy the input with nearly 100% confidence.

2.3 Cross-sentence Relationship

To our best knowledge, currently, the most promising approach is to use language models that have been trained on the Next Sentence-Prediction-task [18]. In particular, the original BERT models, as later approaches, often dropped this task from pretraining. J. Qiang et al. [18] exploited the fact that to perform next-sentence predictions, the model must learn some semantical relationship between the sentences. The authors empirically showed that if we prepare the input to the model in a manner analogous to the examples in the next-sentence-prediction task, but instead of two consecutive sentences, we repeat the same sentence once in its entirety and once with the target word masked, the model will suggest semantically similar words in place of the mask. This approach is shown in Fig. 3. Experiments we performed in Sect. 8 prove, that this kind of approach works noticeably better than raw MLM and dropout methods for generating semantically similar words.

3 Multi-token Prediction

The problem of converting single-token expressions into multi-token ones (and vice versa) is critical in strongly inflected languages (such as the Polish Language) because the limited number of tokens is not able to cover all inflectional varieties even for relatively popular words and, as a result, multi-token words are a frequent phenomenon. An issue with using encoder-only models (like BERT) is that the number of predicted tokens in the output always matches the number of tokens masked in the input. This means that the model, knowing how many tokens are masked, will only predict the words that fully fit into space. As a result, when converting a single-token expression to a multi-token expression (or

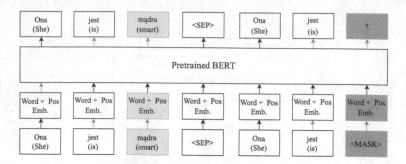

Fig. 3. BERT using cross-sentence relationship. The first sentence is provided without any changes, whereas the second one has mask tokens in place of the desired substitution.

vice versa), multiple passes would have to be performed, at least one for each possible length of the target expression (see Fig. 4).

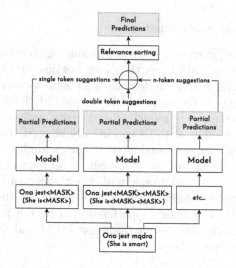

Fig. 4. Multi-token prediction with an encoder only. If token-length of possible substitution is not known in advance, we would have to perform inference for every possible substitution length.

For this reason, we also investigated the BART model, which does not cause this problem. The reasons for choosing BART are:

- Seq2Seq architecture: During inference, the model is not restricted to fill n-tokens space only. Because of that, we can perform a single beam-search to get predictions for multiple-tokens lengths.

- One of the tasks on which the model is pre-trained is multi-token infilling. This task is similar to the problem on which we want to perform model fine-tuning.
- Availability for the Polish language: There exists a model pre-trained only on the Polish corpora (which can be a significant advantage over multilingual models).

4 Datasets

To perform fine-tuning and model testing, we selected the PlEWi Corpus – PLEWiC [5]. It is a collection of language errors for the Polish language, based on the editing history of the Polish version of Wikipedia. From the perspective of this work, the subset of stylistic errors is essential, especially edits based on synonyms. The error type is automatically annotated using a set of hand-crafted heuristics. The authors manually evaluated 200 samples from each category, and in the case of synonyms replacement, they claimed that their heuristics achieved 99% precision. We used a subset of examples from the dataset whose substitution is marked as a synonymous replacement. Unlike other datasets in this field, this one has, in general, only one correct substitution for each element. For training on this dataset, we used 18635 samples for training and 6212 samples for validation.

We also generated a synthetic training set based on the KPWr Corpus [17] and excerpts from the Polish version of the book *Sherlock Holmes* manually annotated with wordnet senses [6]. The exact process of generating the collection is described in Sect. 6.

5 Sequence-to-Sequence Substitution

The BART[12] model is a complete encoder-decoder transformer, and both the input and output are text. Originally, BART was pre-trained on a variety of denoising tasks, including multi-token masking. For example, from the text:

John really likes Anna.

Randomly selected tokens (both their number and position) were replaced by a single mask token, e.g.:

John<mask>Anna.

The purpose of this task is to reproduce the original sentence. In this way, a conditional language model is produced. Unfortunately, there is no information about the semantics of the masked part of the sentence in the original problem formulation. Therefore, the predicted substitutions' order depends only on the frequency of use in a given context. For this reason, we designed a new task that is as close as possible to the original pre-training task but includes the necessary information by prepending the masked sentence with the original content as follows:

[really likes] John<mask>Anna.

This gives the model (at least in theory) easy access to the masked fragment's semantic content. As we wanted the problem to be as similar as possible to the pre-trained definition (due to the scarcity of training data), we defined the model's output to be the whole sentence (like in a masking pre-training task) rather than just the substituted part. To extract the substituted part for evaluation, we truncate the output from the beginning and the end of the sentence by the number of non-masked characters in the input sentence. Unfortunately, here the most considerable advantage, which is the Seq2Seq architecture, is at the same time the most significant problem. We have no guarantee that the model will not change anything except the masked fragment. For example, the output of the model for the presented example could theoretically be:

John adore Kristina

Furthermore, the original BART model was pre-trained to remove words, and we could still observe such behavior after fine-tuning, making the substitution extraction a problematic task.

6   Synthetic Dataset Generation

The PLEWiC collection - despite its relatively good quality - includes edits of mainly single words. Its language domain is also limited due to the presence of Wikipedia texts and the fact that there are only examples classified as synonyms by specific heuristics. For those reasons, in our work, we explore the possibility of using synthetically generated datasets. The idea behind this approach is to take any corpus, then perform part-of-speech tagging using WCRFT2 [19], run a word sense disambiguation WoSeDon tool [7], use plWordNet to find potential synonyms, and finally use morphosyntactic dictionary Morfeusz 2 [8] to apply original word form to a synonym (see Fig. 5). We formulated three methods to generate the training set.

6.1   Single-word Replacement (SWR)

After performing disambiguation, we check the synset size of each word in the sentence. If the synset's size to which a word belongs is greater than one, we generate input and output sentences for each combination of two lexical units from the synset (see Fig. 6). To preserve the correct form of the words, for each combination of lexical units, we use the synthetic dataset generation pipeline described above (Fig. 5). The dataset was split into training (88182 entries) and validation (33766 entries) part.

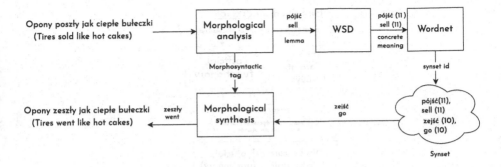

Fig. 5. General synthetic dataset generation schema.

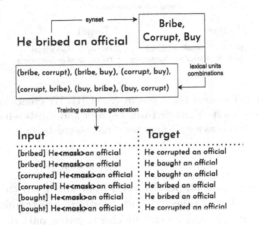

Fig. 6. Single-word replacement dataset generation. After finding a word suitable for substitution, we create all possible ordered combinations of lexical units from the synset to generate training pairs.

6.2 Single Word and Context Replacement (SWCR)

In the SWR method, there are many examples with identical contexts in the dataset (e.g., if we have a synset of size 3 for a single word in a given sentence, there are six training examples with identical contexts). We apply a second approach to increase the context diversity and model robustness against irrelevant context changes, where words that are not masked are also replaced when generating the dataset. The words substituted outside of the masked area are the same in the input sequence and the output sequence. The process is illustrated in Fig. 7. For the SCWR dataset, the training part had 3205239 samples, and the validation part had 1359584 samples.

6.3 Substitution of a Longer Sentence Fragment (SLSF)

In a sentence corpus, we randomly select the beginning and end of the fragment we are replacing. We then search for words with synset greater than one in

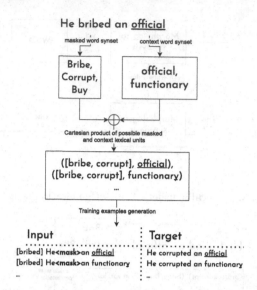

Fig. 7. Single word and context replacement (SWCR). We create an ordered combination of possible substitutions of masked-word synset and words outside of synset. Then we combine both sets (also as a cartesian product) to create a set of training examples.

this fragment and generate all possible combinations from them. The sentences formed from a pair of two different combinations are training examples. The process is depicted in Fig. 8. The size of the training part of the SLSF dataset has 97562 entries and the validation had 35086 entries.

6.4   Wordnet-based Dataset Generation

During the manual evaluation of the synthetic substitution examples, we noticed that, despite using WoSeDon, words are still empirically often assigned to incorrect meanings - e.g. sentence"Plane is a flat, two-dimensional surface." could lead to generation of "Aircraft is a flat, two-dimensional surface.". Thus, especially when several words in one sentence are replaced, it is sometimes difficult to guess the original semantic meaning of the sentence. To overcome the limitations of WoSeDon, we took usage example word usage sentences from plWordNet (Polish Wordnet, [4]) as base sentences. In this case, there was no problem with the word sense disambiguation task because we can assume with high probability that the occurrence of the lemma of the analyzed word in the usage example can be considered as the meaning of the lexical unit for which the example was created (see Fig. 9). For this dataset, the training part consists of 37436 examples and the validation part consists of 13012 examples.

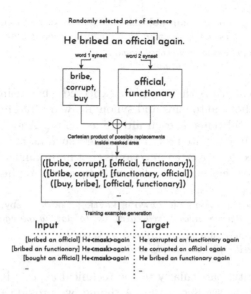

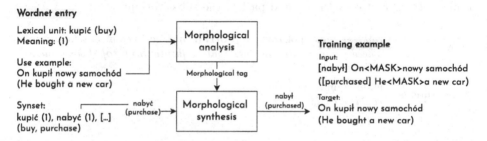

Fig. 8. Substitution of a longer sentence fragment (SLSF). Multiple words are replaced at once so that the model can learn multi-word substitutions.

Fig. 9. Generating artificial dataset from plWordNet. The approach is similar to SWR approach. However, WSD stage is omitted as the example is already assigned to the concrete meaning.

7 Experiment Setup and Initial Results

We used pre-trained Polish BERT [11] and pretrained Polish BART *"A Repository of Polish NLP Resources"* [2] models compatible with the Huggingface Transformers framework [21]. We trained BART model with Adam [9] optimizer for three epochs, with a learning rate 3×10^{-5}, batch size of 2 with two gradient accumulation steps.

Initial tests of the model show that it can perform well with replacing one or more words while preserving the sense of the sentence, at least in a few propositions. However, it could be noticed that the proposed alternatives are quite often antonyms. For example:

Input:	Było dziś bardzo [zimno], mimo że pora roku na to nie wskazuje
	It was very [cold] today, even though the time of year doesn't indicate it.
Substitutions:	ciepło, chłodne, wilgotno, gorąco
	warm, chilly, humid, hot

This is a group of errors that could be expected when using models based on MLM training, as both antonyms and synonyms will often have similar context and this behavior still remains even after fine-tuning. Another problem among the errors spotted in generated sets are phrases, such as *red card* or *broke a leg*, which are not correctly converted into words that maintain the sense of the whole phrase, but only the sense of the individual words of the phrase. For example:

Input:	Zawodnik otrzymał [czerwoną kartkę] z powodu zbyt agresywnej gry.
	The player received a [red card] due to playing too aggressively.
Substitutions:	czerwoną/żółtą/brązową/fioletową kartkę
	red/yellow/brown/purple card

The last difficulty, particularly in the technical aspect, is a problem arising from the nature of the seq2seq model. Although we provide a mask in the place where the model's suggestion should appear, the model can replace the rest of the sentence as well. If words outside the masked area are changed, it can be difficult to extract only the desired part of the substitution:

Input:	Reżyser chciał pokazać, że [potrafi] nakręcić coś nieoczekiwanego.
	The director wanted to show that he [knew how to] shoot something
	unexpected.
Substitutions:	umie, e umie, mie
	could, e could, coul

8 Evaluation

To examine the substitutions' semantic quality and grammatical correctness, we used a test sample from the PLEWiC dataset. To evaluate the quality of the models, we selected a quality measure of Recall at 10 and Recall at 5. There is always only 1 value assigned as correct in the PLEWiC dataset, so Recall at K for a single sample can only reach values of 0 and 1. In the rest of this paper, by Recall at K we mean the Recall at k averaged over all samples in the PLEWiC test set, so it tells us how often the target phrase appeared in the K best predictions of the model. We used the pre-trained (i.e.: without any fine-tuning) BERT as the base model, where only one token is masked, and only one token can be substituted. Other models used for the evaluation are BART-based models fine-tuned using: PLEWiC dataset (model: B-PLEWiC), SWR, SWCR, SLSF generated synthetic datasets described in Sect. 6 (models: B-SWR, B-SWCR, B-SLSF) and wordnet-based dataset described in Sect. 6.4 (model: B-WN).

Table 1 shows the final results on the PLEWiC test set of various models, differing mainly in the set on which they were trained. The evaluation consisted of comparing the model proposal with the word that was actually used. The first 500 records from the PLEWiC collection were used for testing. The results show the percentage of cases in which the correct word appeared in the first N model proposals. Prediction is counted as correct only if both lemma and form were correct (i.e., only "perfect" match counted).

The best models turned out to be those learned on mixed datasets, and among these, the one trained without adding the Wordnet dataset turned out to be better according to Recall at 1. However, for Recall at 10, the better one was the model trained on the mixed dataset with all synthetic sets: Wordnet and PLEWiC.

Table 1. Result [Recall@N] on PLEWiC test set of BERT baseline and BART models with different fine-tuning datasets. *The BERT model in most cases gives the original occurring word as the first proposition, hence the low Recall@1 score. In comparison, the Recall@2 for BERT is 28.8%.

Model/Dataset	Recall@1	Recall@5	Recall@10
Baseline w/o fine-tuning			
BERT	2.9*	42.3	58.9
PLEWiC only			
B-PLEWiC	56.4	73.6	76.6
Synthetic only			
B-SLSF	11.4	33.8	43.8
B-SWR	13.6	35.4	44.6
B-SWCR	14.6	35.6	43.8
B-WN	7.2	36.2	47.4
Mixed datasests			
B-PLEWiC+SLSF	**56.8**	**74.4**	77.6
B-PLEWiC+SLSF+WN	55.0	**74.4**	**78.0**

9   Results and Future Work

The models based on synthetically generated datasets were inferior to the dataset developed from actual Wikipedia editions (PLEWiC dataset) and the BERT-based baseline. The following causes may have contributed to this result:

- Introduction of grammatical structure errors. In the presented algorithm, we did not take context editing into account. Thus, when examining exact substitutions, a reduction in the quality of inflectional forms could have significantly reduced the results obtained.
- A difference in the domain of texts. BERT was pre-trained on a dataset consisting of Wikipedia (which is the base for PLEWiC). On the other hand, BART was fine-tuned on KPWr and Sherlock Holmes novel.

Fine-tuning on the PLEWiC set significantly improved the results relative to baseline in the form of pre-trained BERT. The already obtained Recall at 1 value was higher than the baseline Recall at 10. Additionally, the use of the Seq2Seq model significantly simplified the problem in terms of computational complexity, as variable-length output is supported. Adding a synthetic dataset did improve overall results, however the impact was not significant.

Further research will include, in particular, study of the impact of sequence-to-sequence query definition on the final performance of the system and explore the possibility of incorporating active learning into the process of fine-tuning deep lexical substitution models

Acknowledgements. This work was financed by (1) the National Science Centre, Poland, project no. 2019/33/B/HS2/02814; (2) the Polish Ministry of Education and Science, CLARIN-PL; (3) the European Regional Development Fund as a part of the 2014-2020 Smart Growth Operational Programme, CLARIN – Common Language Resources and Technology Infrastructure, project no. POIR.04.02.00-00C002/19; (4) the statutory funds of the Department of Artificial Intelligence, Wrocław University of Science and Technology.

References

1. Arefyev, N., Sheludko, B., Podolskiy, A., Panchenko, A.: Always keep your target in mind: studying semantics and improving performance of neural lexical substitution. In: Proceedings of the 28th International Conference on Computational Linguistics, pp. 1242–1255 (2020)
2. Dadas, S.: A repository of polish NLP resources. Github (2019). https://github.com/sdadas/polish-nlp-resources/
3. Devlin, J., Chang, M.W., Lee, K., Toutanova, K.: Bert: pre-training of deep bidirectional transformers for language understanding. arXiv preprint arXiv:1810.04805 (2018)
4. Dziob, A., Piasecki, M., Rudnicka, E.: plWordNet 4.1-a linguistically motivated, corpus-based bilingual resource. In: Fellbaum, C., Vossen, P., Rudnicka, E., Maziarz, M., Piasecki, M. (eds.) Proceedings of the 10th Global WordNet Conference, 23–27 July 2019, Wroclaw (Poland), pp. 353–362. Oficyna Wydawnicza Politechniki Wrocławskiej, Wrocław (2019)
5. Grundkiewicz, R.: Automatic extraction of polish language errors from text edition history. In: Habernal, I., Matoušek, V. (eds.) TSD 2013. LNCS (LNAI), vol. 8082, pp. 129–136. Springer, Heidelberg (2013). https://doi.org/10.1007/978-3-642-40585-3_17
6. Janz, A., Chlebus, J., Dziob, A., Piasecki, M.: Results of the PolEval 2020 shared task 3: word sense disambiguation. In: Proceedings of the PolEval 2020 Workshop. Institute of Computer Science, Polish Academy of Sciences, Warsaw, Poland (2020). http://poleval.pl/files/poleval2020.pdf
7. Kedzia, P., Piasecki, M., Orlinska, M.: WoSeDon (2016). CLARIN-PL digital repository, http://hdl.handle.net/11321/290
8. Kieraś, W., Woliński, M.: Morfeusz 2 - analizator i generator fleksyjny dla języka polskiego. Język Polski XCVI **I**(1), 75–83 (2017)
9. Kingma, D.P., Ba, J.: Adam: a method for stochastic optimization. CoRR abs/1412.6980 (2015)

10. Kremer, G., Erk, K., Padó, S., Thater, S.: What substitutes tell us-analysis of an "all-words" lexical substitution corpus. In: Proceedings of the 14th Conference of the European Chapter of the Association for Computational Linguistics, pp. 540–549 (2014)

11. Kłeczek, D.: Polbert: attacking polish NLP tasks with transformers. In: Ogrodniczuk, M., Łukasz Kobyliński (eds.) Proceedings of the PolEval 2020 Workshop. Institute of Computer Science, Polish Academy of Sciences (2020)

12. Lewis, M., et al.: Bart: denoising sequence-to-sequence pre-training for natural language generation, translation, and comprehension, pp. 7871–7880 (2020). https://doi.org/10.18653/v1/2020.acl-main.703

13. Martin, L., Fan, A., de la Clergerie, É., Bordes, A., Sagot, B.: Multilingual unsupervised sentence simplification. arXiv preprint arXiv:2005.00352 (2020)

14. McCarthy, D., Navigli, R.: The English lexical substitution task. Lang. Resour. Eval. **43**(2), 139–159 (2009)

15. Melamud, O., Levy, O., Dagan, I.: A simple word embedding model for lexical substitution. In: Proceedings of the 1st Workshop on Vector Space Modeling for Natural Language Processing, pp. 1–7 (2015)

16. Mikolov, T., Chen, K., Corrado, G., Dean, J.: Efficient estimation of word representations in vector space (2013)

17. Oleksy, M., et al.: Polish corpus of wrocław university of technology 1.3 (2019). CLARIN-PL digital repository, http://hdl.handle.net/11321/722

18. Qiang, J., Li, Y., Zhu, Y., Yuan, Y., Wu, X.: Lsbert: a simple framework for lexical simplification. arXiv preprint arXiv:2006.14939 (2020)

19. Radziszewski, A.: A tiered CRF tagger for polish. In: Bembenik, R., Skonieczny, L., Rybinski, H., Kryszkiewicz, M., Niezgodka, M. (eds.) Intelligent Tools for Building a Scientific Information Platform. Studies in Computational Intelligence, vol. 467, pp. 215–230. Springer, Berlin, Heidelberg (2013). https://doi.org/10.1007/978-3-642-35647-6_16

20. Reimers, N., Gurevych, I.: Making monolingual sentence embeddings multilingual using knowledge distillation. In: Proceedings of the 2020 Conference on Empirical Methods in Natural Language Processing (EMNLP), pp. 4512–4525 (2020)

21. Wolf, T., et al.: Transformers: state-of-the-art natural language processing. In: Proceedings of the 2020 Conference on Empirical Methods in Natural Language Processing: System Demonstrations, pp. 38–45. Association for Computational Linguistics, Online, October 2020. https://doi.org/10.18653/v1/2020.emnlp-demos.6, https://aclanthology.org/2020.emnlp-demos.6

22. Zhou, W., Ge, T., Xu, K., Wei, F., Zhou, M.: Bert-based lexical substitution. In: Proceedings of the 57th Annual Meeting of the Association for Computational Linguistics, pp. 3368–3373 (2019)

Facial Mask Impact on Human Age and Gender Classification

Krzysztof Małecki[1][(✉)], Adam Nowosielski[1], and Mateusz Krzak[2]

[1] Faculty of Computer Science and Information Technology,
West Pomeranian University of Technology, Żołnierska 52 Str.,
71-210 Szczecin, Poland
{kmalecki,anowosielski}@wi.zut.edu.pl
[2] Netcompany Poland Sp.z o.o., Puławska 182 Str., 02-670 Warsaw, Poland

Abstract. The human face contains important information enabling the social identification of the owner about the age and gender. In technical systems, the face contains a number of important information that enables the identification of a person. The COVID-19 pandemic made it necessary to cover the face with a mask and thus hide a significant part of information content in the face, important for social or technical purposes. The paper analyses how covering the face with a mask makes it difficult to identify a person in terms of age and gender determination. Analyzes with the employment of state of the art models based on deep neural networks are performed. Their effectiveness is investigated in the context of the limited information available, as with the case of the face covered with a mask.

Keywords: Face recognition · Deep learning · COVID-19

1 Introduction

The human face contains a lot of essential information. The basic information that can be read from the face is age and gender. From the face, it is also possible to read emotions. In everyday life a person can be identified by others on the basis of the face. All the information contained in the face is also more and more often used in technical systems [1–4], especially those based on multispectral data [5]. Face ID allows to unlock the device or provide access to a security restricted area. Face can be utilized as the medium in touchless human-computer interfaces [6]. Novel HCI interfaces can read emotions of its user and adapt their behaviour and react appropriately to the emotional state of a human. Finally, from facial expressions the recognition of student emotions during a lecture is possible [7], providing instant feedback to the lecturer and contributing to the increase in the quality of education.

COVID-19 pandemic has changed how the image of the face functions in a society. Obligatory wearing of a mask to stop the spread of the virus made a significant influence on how people engage in social activities. Significant part of the information contained in the face is hidden behind a mask. The areas around the mouth is highly informative providing clues for emotional state, and also premises for determining age and sex. By covering the face with a mask these hints become unavailable. Face-oriented algorithms also assume a particular level of visibility of facial features. Subjected to faces covered with masks might fail to operate.

The problem of facial mask impact on reading information contained in the human face has been already addressed in the literature. One of the first publications were dedicated to detecting a face covered with a mask, e.g. [8–10]. Wearing a mask incorrectly makes the prevention method pointless therefore algorithms for detection when facemasks are being worn incorrectly emerge [11]. The National Institute of Standards and Technology (NIST) started examining the performance of face recognition algorithms on faces occluded by masks [12], run under the ongoing Face Recognition Vendor Test (FRVT). In the augmented report [13] (after the COVID-19 pandemic was declared) NIST observed that developers have adapted their algorithms to support face recognition on subjects potentially wearing face masks [13]. Despite this, the performance of the evaluated recognition algorithms is, in general, deteriorated. A detailed overview of the progress in masked face recognition can be found in [14].

In the paper, we analyze the impact of the face mask on reading age and gender information from a face. The problem alone is challenging. The appearance of the human face, apart from the congenital determinants, is influenced by a lifestyle, environmental conditions, health, emotional state and others. A facial mask, by concealing a large part of the information, makes the task even more difficult. In our research, we focus on deep learning algorithms. With the help of Tensorflow and Keras libraries, different models were utilized and trained. Their effectiveness was evaluated on prepared database.

2   Idea

In the model training process photos of computer-generated faces based on the artificial intelligence algorithm StyleGAN2 [15] developed by NVIDIA were used. For the purposes of the article, a computer program has been prepared that automatically downloads photos from the website [16] with the age and gender information. Then, the Face-Recognition-Model-with-Gender-Age-and-Emotions-Estimations computer program [17] available on GitHub was modified and used, and the photos were divided into appropriate age and gender categories, saving this data to a CSV file. The computer-generated photos present the faces of people aged 9 to 74, both male and female.

The photo database consists of 8,000 photos divided into gender categories: female and male. Then, of these 8,000 images, 4 age groups (0–12, 13–20, 21–50, 51+) of 1500 images each were separated. Groups of equal size were chosen to guarantee the same training and testing conditions for each case. Furthermore, both sex categories have been split into 3000 training images, 500 validation, and 500 pictures for testing. Similar division was applied to each of the age groups with distribution values of: 1000, 250, and 250 appropriately.

(a)   (b)

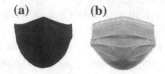

Fig. 1. Two types of masks used: (a) a textile, (b) a surgical.

In the next step of preparing the benchmark database, the MaskTheFace program was used to apply face protective masks [18] available on the GitHub [19] platform, under the MIT license. In order to differentiate the research material, two types of masks were used: a textile mask and a surgical mask (Fig. 1). It must be noticed, that procedure of adding artificial masks to face images, instead of acquiring real data, is found in the literature (e.g. [12,13]). The process of obtaining real photos is expensive and time-consuming, the face databases already exist, and the mask only hides the information already available. Some researchers use the Internet community to obtain data (e.g. [11] use citizens by asking to take different selfies through an app and placing the mask in different positions).

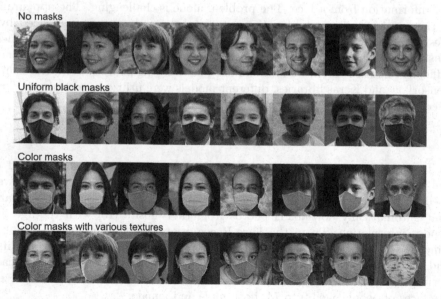

Fig. 2. Sample images from benchmark database containing faces without masks and with different categories of masks.

To illustrate the established and diversified benchmark database, the authors presented selected images of the faces without protective masks, with textile protective masks in a uniform black color, with different types of masks and different colors and shades, and faces of people with masks of different colors and textures (Fig. 2).

Four different training models have been utilized in this paper: a deep neural network (DNN) model and three convolutional neural network models - containing one (CNN1), three (CNN3) and four convolutional layers (CNN4). The DNN model consisted of seven layers, including four Dense layers (fully-connected layers): 2nd, 4th, 6th and 7th. The CNN1 model contained only one Conv2D layer (layer 1) and one MaxPooling2D layer (layer 2) and one flatten layer and three dense layers. The CNN3 model consisted of three alternating layers Conv2D (layers 1, 3, 5) and MaxPooling2D (layers 2, 4, 6) and four final layers: flatten (layer 7), dense (layers: 8, 10), dropout (last layer). Finally, the CNN4 model consisted of four alternating Conv2D layers (layers: 1, 3, 5, 7) and MaxPooling2D (layers: 2, 4, 6, 8), one flatten layer and four dense layers.

All models have been executed with the same parameters. The number of epochs was 150 and the number of samples taken from the generator before the end of the epoch was 75. Uploaded images were scaled so that the input object size was $64 \times 64 \times 3$. The output layer of each network was activated with the softmax function. For classifiers containing convolutional layers, the Adam optimizer with a value of 0.0001 was selected. The deep neural network (DNN) model had the RMSprop optimizer applied.

3 Baseline Validation

For comparison purposes, the previously mentioned Face-Recognition-Model-with-Gender-Age-and-Emotions-Estimations trained model [17] was executed on faces without applied masks.The original model obtained values above 90% for both gender and age groups. The detailed results of this experiment when no masks were worn are shown in Fig. 3 for age groups and in Fig. 4 for gender.

For age groups, all models performed well in correct assignment for two age groups: 0–12 and 51+. For the 0–12 age group, the percentage results of correct assignment are presented as DNN: 82%, CNN1: 91.2%, CNN3: 89.2%, CNN4: 90%, respectively. In the 51+ group, the models obtained similar values, i.e. DNN: 84.8%, CNN1: 93.2%, CNN3: 87.2%, CNN4: 90.8%. For groups 13–20 and 21–50, the classifiers performed significantly less well in correctly assigning categories. In the 13–20 group, of all the models, CNN4 performed the best with an efficiency of 46.8%. In the 21–50 group, CNN1 had the best efficiency with a value of 72.8%.

The correctness of gender assignment is high. This means that all classifiers performed well with the following percentages. The DNN model achieved an efficiency of 83.2%, CNN1: 81.4%, CNN3: 87.6%, CNN4: 90.2% in correctly assigning women. For men, the values were from DNN: 85.8%, CNN1: 88.6%, CNN3: 88.8%, CNN4: 89%.

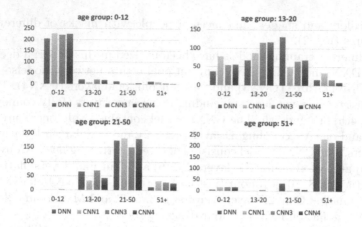

Fig. 3. Assigned categories for age groups with no masks applied

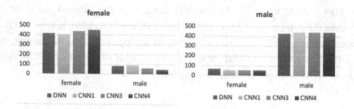

Fig. 4. Assigned categories for gender with no masks applied

The results obtained above will be considered as the baseline and the results of subsequent experiments will be compared to them. In the next step, the prepared models are evaluated for the correct assignment of people's faces with masks to the appropriate class categories.

4   The Results of Age Groups and Gender Classification

This section presents the results of the study on the recognition performance of faces obscured by black cloth masks (Sect. 4.1), cloth and surgical masks with different colors (Sect. 4.2), and cloth and surgical masks with different colors and textures (Sect. 4.3), by individual classifiers (DNN, CNN1, CNN3, CNN4).

4.1   Black Fabric Masks

In the task of classifying face images obscured by uniform black cloth masks (Fig. 5), models based on convolutional layers achieved the highest performance: 80.8% (CNN3) and 98.4% (CNN4). The DNN model was least effective in the task of classifying into the appropriate age group, generating the following results: 59.6% (for 0–12 group), 41.2% (for 13–20), 5.2% (for 21–50), 90.8% (for 51+).

In the gender classification task (Fig. 6), correct assignment for women equals: 96.4% (DNN), 98% (CNN1), 95.8% (CNN3) and 98% (CNN4). An interesting result is the score of 78% obtained by the DNN classifier in the task of classifying a face image into a group of men. The other classifiers obtained very close results as those of the female group.

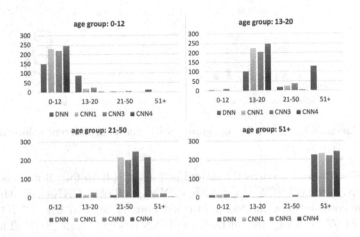

Fig. 5. Assigned categories for age groups with black masks applied.

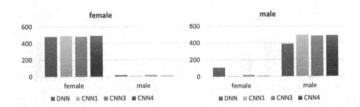

Fig. 6. Assigned categories for gender with black masks applied.

4.2   Masks in Different Colors

In the classification of face images covered by different types of masks of different colors, the best results were obtained for two age groups: 0–12 and 51+. These were respectively: 78% (DNN), 92% (CNN1), 89.6% (CNN3), 90.4% (CNN4) and 91.2% (DNN), 91.2% (CNN1), 90.4% (CNN3), 86.4% (CNN4). The effectiveness in the 21–50 age group for the DNN model was only 28.8%, while the other classifiers gave results ranging from 61.2% to 76%. The lowest recognition efficiency occurred in the group of 13–20. The DNN model obtained: 41.6%, CNN1: 28.4%, CNN3: 30.4%, CNN4: 44% (Fig. 7).

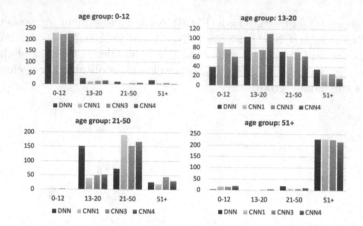

Fig. 7. Assigned categories for age groups with masks in different colors applied.

The effectiveness of correctly assigning men was high for all models (Fig. 8): 95.8% (DNN), 82.6% (CNN1), 92.2% (CNN3), 89.6% (CNN4). In the female group, the convolutional models performed equally well: 87.6% (CNN1), 83.6% (CNN3), 84.6% (CNN4). In contrast, the DNN model scored 48.4%. This classifier may not have learned to recognize certain features such as short hair, which can occur in both men and women.

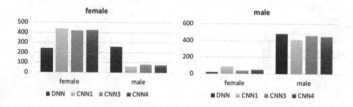

Fig. 8. Assigned categories for gender with masks in different colors applied.

4.3   Masks in Different Colors and Textures

In the task of classifying images of faces covered by different types of masks of different colors and textures, the highest performance was observed for the age groups 0–12 and 51+ (Fig. 9). In detail, DNN: 97.6%, CNN1: 89.6%, CNN3: 91.6%, CNN4: 92.4%. The experiment highlighted the problem of image classification for the age group 13–20, for which the highest efficiency is 44.4% (CNN4) and for the age group 21–50, for which the effectiveness oscillates between 56.4% and 70%. It can be concluded that the patterns on the masks disrupted the correct learning process of the models.

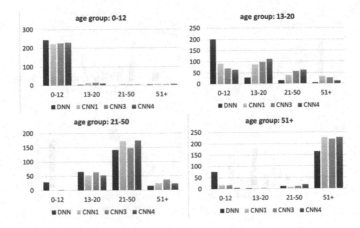

Fig. 9. Assigned categories for age groups with masks in different colors and textures applied.

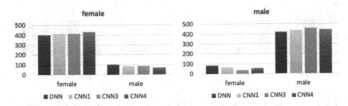

Fig. 10. Assigned categories for gender with masks in different colors and textures applicd.

In the gender classification task (Fig. 10), for the women's group, the results were 80.2% (DNN), 83% (CNN1), 82.4% (CNN3), 86.4% (CNN4). The results for the men's group, on the other hand, are as follows: 83.6% (DNN), 87.4% (CNN1), 92% (CNN3), 89.2% (CNN4).

5   The Performance Results Targeted at Each Model Separately

In the following subsections we analyse the classification performance of different models separately.

5.1   Deep Neural Network Model (DNN)

For the DNN classifier, the results for age recognition are shown in Fig. 11 and for gender recognition in Fig. 12.

In the 51+ age group, the classifier achieved face recognition performance: 84.8% (no masks), 90.8% (uniform black masks), 91.2% (color masks) and 65.6% (color masks with different textures). For the 0–12 age group, the following

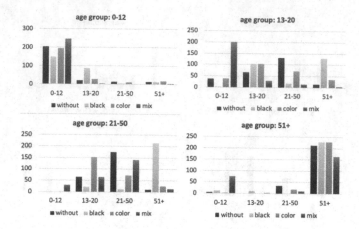

Fig. 11. The results of the DNN model for various age groups and masks.

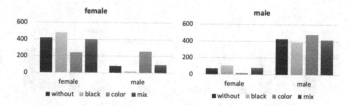

Fig. 12. The results of the DNN - gender classification task.

results were achieved, respectively: 82%, 59.6%, 78% i 97.6%. In the case of the other age groups, the model achieved relatively low efficiency. In the male classification task, the DNN model achieved an efficiency of 85.8% (no masks), 78% (solid black masks), 95.8% (color masks), and 83.6% (color masks with different textures). In the assignment to the group of women, the results were as follows: for no masks - 83.2%, 96.4% for uniform black masks, 48.4% for color masks, 80.2% for color masks with different textures.

5.2 One Convolutional Layer Model (CNN1)

For the 0–12 and 51+ groups, the type of mask applied, or lack thereof, had no significant effect on the correctness of imputation the assignment (Fig. 13). In the 0–12 age group the results are respectively: 91.2% (no masks), 91.6% (uniform black masks), 92% (color masks), 89.6% (color masks with different textures). Whereas for the group 51+ it will be respectively: 93.2%, 93.6%, 91.2%, 90.8%. Lower classification efficiency was observed for people in the age group of 21–50. Complication of the mask appearance by applying different colors, shades and textures causes a decrease in accuracy: 86.4% for black masks, 76% for color masks and finally 69.2% different texture masks. In the 13–20 age group, only one case (face images with uniform black masks) had the highest classification of 90%. For the remaining cases, the results did not exceed 40%.

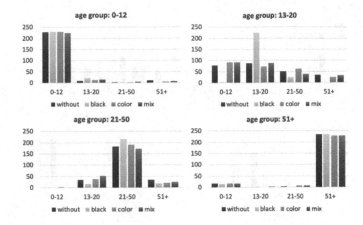

Fig. 13. The results of the CNN1 model for various age groups and masks.

In the case of gender classification (Fig. 14), in a group of women, varied appearance of the mask causes a slight decrease in accuracy: from 98% for black masks, through 87.6% for color masks, and ending on 83% for different textures masks. In the male classification, the matching to group is highest for face images masked with black masks and equal to 98.2%.

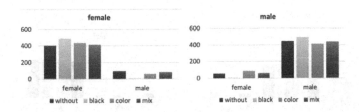

Fig. 14. The results of the CNN1 - gender classification task.

5.3   Three Convolutional Layers Model (CNN3)

Face classification for all types of masks for groups: 0–12 and 51+ indicated that the type of mask applied had no significant effect on the CNN3 model. In the 0–12 group for no masks case, the model achieved an efficiency of 89.2%, for uniform black masks - 88%, color masks - 89.6%, and for mixed masks - 91.6%. For the 51+ age group, similar results were achieved. In the 13–20 group, the highest efficacy was achieved for black masks (80.8%). Classification of the remaining cases ended with efficiencies ranging from 30.4% for color masks, through 39.2% for mixed masks, and obtaining 46.4% for the no masks case. In the 21–50 group, recognition of faces covered by uniform black masks ended with an efficiency of 80.8%. Classification efficiency for the remaining cases ranged from 59.2% to 61.2% (Fig. 15).

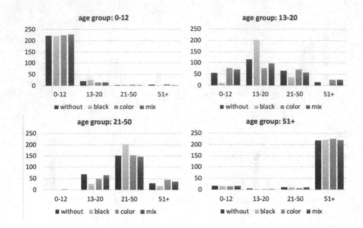

Fig. 15. The results of the CNN3 model for various age groups and masks.

Figure 16 shows the results of correct and incorrect face assignment to the group of women and men. For the women's group the values are respectively: without mask applied - 87.6%, uniform black masks - 95.8%, color masks - 89.6% and mixed masks - 91.6%. For the group of men, the results are as follows: for photos of the face without the mask on - 88.8%, with black masks - 96.2%, for color masks - 92.2% and 92% for different textures masks.

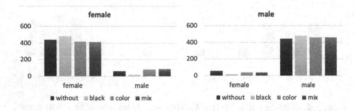

Fig. 16. The results of the CNN3 - gender classification task.

5.4   Four Convolutional Layers Model (CNN4)

In the experiments for the classifier referred to as CNN4 results are presented in (Fig. 17) for age groups, and in in (Fig. 18) for gender.

For face classification to the 0–12 age group, all test cases showed high model performance: 90% (no masks), 98.4% (uniform black masks), 90.4% (color masks) and 92.4% (mixed masks). Almost identical results were obtained for the 51+ age group. For the other age groups, the CNN4 model achieved the best classification results for black masks. A large difference in classification effectiveness can be observed in the 13–20 age group: 98.4% (black masks), 46.8% (no masks),

44% (color masks), 44.4% (mixed masks). For the 21–50 group, and black masks the accuracy equals 98.4%, 72.4% for no masks case, 66.8% for color masks, and 70% for mixed masks case (Fig. 17).

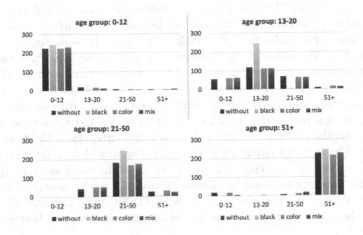

Fig. 17. The results of the CNN4 model for various age groups and masks.

Gender classification task (Fig. 18), for both groups ended with a high success rate. For a group of women: 90.2% (no masks), 98% (black masks), 84.6% (color masks), and 86.4% (mixed masks). For the male group, the classification results are similar and are as follows: for facial images without masks - 89%, black masks - 98%, color masks - 89.6%, and 89.2% for different textures masks.

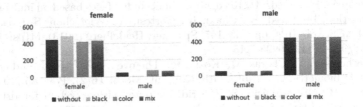

Fig. 18. The results of the CNN4 - gender classification task.

6 Conclusion

The purpose of this study was to investigate the effect of facial masking on automatic estimation of a person's age and gender. The effectiveness of four deep learning models has been examined. From the study, some conclusions can be drawn regarding the human features contained in faces.

The main, rather expected, result is an overall deterioration in the age and gender classification of masked face subjects. Masks obscure facial hair, which is a special feature for men, making it easy to quickly categorize them. Long hair, characteristic of women but found in men, influences misclassification into the wrong group. And in reverse, short hair in women provides a valid rationale for classification into the male group, especially when other facial features are obscured by a mask. The differences that occur between the different variants of the masks applied indicate the limited learning capabilities of the existing models. This is especially true for masks with miscellaneous textures. This is important from the point of view of the increasingly popular face monitoring and recognition systems. Nowadays, when it is common to mask parts of the face with a protective mask, it seems that detecting the mask itself and removing it programmatically, especially for different colored masks with complex textures, could improve the results of person identification or age and gender classification.

As for age, the situation is similar. For some age categories (13–20 and 21–50) all prepared models did not learn to recognize them, which could be due to the fact that the protective mask obscures large facial features such as lips, nose, and wrinkles making the amount of information the model processes limited as well. Interestingly, for convolutional models (CNNs) with different numbers of layers, in the vast majority of cases, the black masks performed better than the others.

It can be concluded from our study that DNN model was found to be more susceptible to face masks, especially those which additionally contained textures. Convolutional networks proved high performance in the task of classifying face images for the 0–12 and 51+ age groups, despite covering these faces with different masks.

References

1. Forczmański, P., Łabędź, P.: Recognition of occluded faces based on multi-subspace classification. In: Saeed, K., Chaki, R., Cortesi, A., Wierzchoń, S. (eds.) CISIM 2013. LNCS, vol. 8104, pp. 148–157. Springer, Heidelberg (2013). https://doi.org/10.1007/978-3-642-40925-7_15
2. Wojciechowska, A., Choraś, M., Kozik, R.: The overview of trends and challenges in mobile biometrics. J. Appl. Math. Comput. Mech. 16(2), 173–185 (2017)
3. Cyganek, B., Gruszczyński, S.: Hybrid computer vision system for drivers' eye recognition and fatigue monitoring. Neurocomputing 126, 78–94 (2014)
4. Małecki, K., Forczmański, P., Nowosielski, A., Smoliński, A., Ozga, D.: A new benchmark collection for driver fatigue research based on thermal, depth map and visible light imagery. In: Burduk, R., Kurzynski, M., Wozniak, M. (eds.) CORES 2019. AISC, vol. 977, pp. 295–304. Springer, Cham (2020). https://doi.org/10.1007/978-3-030-19738-4_30
5. Małecki, K., Nowosielski, A., Forczmański, P.: Multispectral data acquisition in the assessment of driver's fatigue. In: Mikulski, J. (ed.) TST 2017. CCIS, vol. 715, pp. 320–332. Springer, Cham (2017). https://doi.org/10.1007/978-3-319-66251-0_26
6. Nowosielski, A., Forczmański, P.: Touchless typing with head movements captured in thermal spectrum. Pattern Anal. Appl. 22(3), 841–855 (2018). https://doi.org/10.1007/s10044-018-0741-0

7. Tonguç, G., Ozkara, B.O.: Automatic recognition of student emotions from facial expressions during a lecture. Comput. Educ. **148**, 103797 (2020)
8. Sethi, S., Kathuria, M., Kaushik, T.: Face mask detection using deep learning: an approach to reduce risk of coronavirus spread. J. Biomed. Inform. **120**, 103848 (2021)
9. Kansal, I., Popli, R., Singla, C.: Comparative analysis of various machine and deep learning models for face mask detection using digital images. In: 9th International Conference on Reliability, Infocom Technologies and Optimization (Trends and Future Directions) (ICRITO), pp. 1–5 (2021)
10. Joshi, A.S., Joshi, S.S., Kanahasabai, G., Kapil, R., Gupta, S.: Deep learning framework to detect face masks from video footage. In: 12th International Conference on Computational Intelligence and Communication Networks (CICN), pp. 435–440 (2020)
11. Tomás, J., Rego, A., Viciano-Tudela, S., Lloret, J.: Incorrect facemask-wearing detection using convolutional neural networks with transfer learning. Healthcare **9**, 1050 (2021)
12. Ngan, M., Grother, P., Hanaoka, K.: Ongoing face recognition vendor test (FRVT) Part 6A: face recognition accuracy with masks using pre-COVID-19 algorithms, NIST Interagency/Internal Report (NISTIR), National Institute of Standards and Technology, Gaithersburg, MD (2020). https://doi.org/10.6028/NIST.IR.8311. Accessed 10 Jan 2022
13. Ngan, M., Grother, P., Hanaoka, K.: Ongoing face recognition vendor test (FRVT) Part 6B: face recognition accuracy with face masks using post-COVID-19 algorithms, NIST Interagency/Internal Report (NISTIR), National Institute of Standards and Technology, Gaithersburg, MD (2020). https://doi.org/10.6028/NIST.IR.8331. Accessed 11 Jan 2022
14. Alzu'bi, A., Albalas, F., AL-Hadhrami, T., Younis, L.B., Bashayreh, A.: Masked face recognition using deep learning: a review. Electronics **10**, 2666 (2021)
15. Karras, T., Laine, S., Aittala, M., Hellsten, J., Lehtinen, J., Aila, T.: Analyzing and improving the image quality of StyleGAN. In: Proceedings of the IEEE/CVF Conference on Computer Vision and Pattern Recognition, pp. 8110–8119 (2020)
16. Benchmark. https://www.thispersondoesnotexist.com. Accessed 8 Jan 2022
17. GitHub Face Recognition. https://www.github.com/rileykwok/Face-Recognition-Model-with-Gender-Age-and-Emotions-Estimations. Accessed 8 Jan 2022
18. Anwar, A., and Raychowdhury, A.: Masked Face Recognition for Secure Authentication. arXiv preprint arXiv:2008.11104 (2020)
19. MaskTheFace. https://www.github.com/aqeelanwar/MaskTheFace. Accessed 8 Jan 2022

Simple and Efficient Acceleration of the Smallest Enclosing Ball for Large Data Sets in E^2: Analysis and Comparative Results

Vaclav Skala(✉), Matej Cerny, and Josef Yassin Saleh

Faculty of Applied Sciences, Department of Computer Science and Engineering,
University of West Bohemia, Pilsen, CZ 301 00, Czech Republic
{skala,matcerny,salehj}@kiv.zcu.cz
http://www.VaclavSkala.eu

Abstract. Finding the smallest enclosing circle of the given points in E^2 is a seemingly simple problem. However, already proposed algorithms have high memory requirements or require special solutions due to the great recursion depth or high computational complexity unacceptable for large data sets, etc. This paper presents a simple and efficient method with speed-up over 100 times based on processed data reduction. It is based on efficient preprocessing, which significantly reduces points used in the final processing. It also significantly reduces the depth of recursion and memory requirements, which is a limiting factor for large data processing. The proposed algorithm is easy to implement and it is extensible to the E^3 case, too. The proposed algorithm was tested for up to 10^9 of points using the Halton's and "Salt and Pepper" distributions.

Keywords: Smallest enclosing circle · Smallest enclosing ball · Algorithm complexity · Preprocessing · Convex hull · Convex hull diameter

1 Introduction

Algorithms for finding the smallest enclosing circle in the E^2 case, or the enclosing ball in the E^k general case, have been studied for a long time and many algorithms have been published with many modifications. Sylvester [62] made the first problem formulation in 1857 and later by others, see Elzinga [10]. Several algorithms have been published, e.g. Megiddo's algorithm [31] with an overview of some other interesting solutions, Ritter [46], etc. A brief introduction to the problem is available at WiKi [71,72].

An interesting approach was published by Welzl [67] in 1991. It is a "brute force" recursive algorithm with a random selection of points. It leads to a

Research supported by the University of West Bohemia - Institutional research support.

significant speed-up due to random point selection. However, it is not directly usable for large data sets due to the very deep recursion calls.

Unfortunately, the originally proposed Welzl's algorithm is partially incorrect and Matoušek, Sharir, Welzl's published the corrected version, as the MSW algorithm [30] (the code available on WiKi [72]), see [70, 72].

Algorithm 1. MSW - Matousek, Sharir, Welzl's algorithm

Require: Finite sets P and R of points in the plane $|R| = 3$
Ensure: Minimal disk enclosing $P \cup R$
   if P is empty **then**
      return trivial(R)
   end if
   choose p in P ▷ randomly and uniformly
   D := msw($P - \{p\}, R$)
   if p is in P **then**
      return(D)
   end if
   $q = $ nonbase($R \cup \{p\}$)
      ▷ Welzl's algorithm for 4 points could be used to find what would not be in R
   return MSW ($P - \{p\} \cup \{q\}, R \cup \{p\} - \{q\}$)

It should be noted, that there is no significant difference between the original Welzl's and MSW algorithms as far as the timing is concerned.

2 Proposed Preprocessing

The smallest enclosing center problem is closely related to the diameter of the convex hull problem. A simple algorithm with the $O_{exp}(N)$ complexity for finding a diameter of the convex hull of points using preprocessing was published by Skala [53, 55, 58]. The algorithm based on polar space subdivision was introduced in Skala, Smolik, Majdisova [59] and extended in Skala, Majdisova, Smolik [57] for the E^3 case.

2.1 Convex Hull Diameter Estimation

It is based on a simple idea. The AABB points and points closest to the AABB corners form a convex hull. The maximum distance d defines an estimation of the convex hull diameter, i.e. radius r. Then the given points Ω are split to $\Omega_0, \ldots, \Omega_4$, see Fig. 1.

It can be seen that the Ω_0 points cannot contribute to the convex hull diameter. Then points of Ω_1 and Ω_3 are processed and the value r is updated. Similarly, updates of the radius r are made after Ω_2 and Ω_4, Ω_1 and Ω_2, Ω_2 and Ω_3, Ω_3 and Ω_4, Ω_4 and Ω_1. This leads to significant reduction of points that could form the final convex hull. Then the final diameter of the convex hull is computed; see Skala [53, 56, 58] for details.

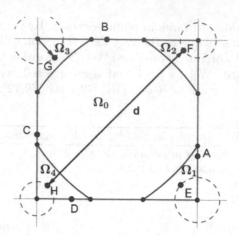

Fig. 1. Maximum distance estimation

Other efficient algorithms for finding the convex hull of points in the E^2 case were published in Skala [55,59] and for the convex hull in E^3 case was described in Skala [57]. This preprocessing leads to significant speed up, see Skala [58] for details. However, the polar subdivision used in Smolik [61] is quite complex to implement.

The Welzl's recursive algorithm is based on the "brute force" approach actually, but the randomized point heuristic selection use leads to the $O_{exp}(N)$ expected complexity, where N is a number of points. Unfortunately, it leads to deep recursive calls, which is a very limiting factor for large data processing.

2.2   Theoretical Analysis

The simplest acceleration of the smallest enclosing circle algorithm is to find points that form the Axis Aligned Bounding Box (AABB), i.e. points A, B, C, D. The worst case is when the AABB is a square and the points A, B, C, D are at the middle of the edges, see Fig. 2a. In this case, all points inside of the area Ω_0 can be removed from the future processing. However, if points closest to the AABB corners are found, i.e. points E, F, G, H, all points inside of the convex polygon $A, \ldots, H$ can be removed. The points E, F, G, H are on an expected distance r from corners and the radius r decreases with the number N of the given points.

The Fig. 2a presents a general case with a rectangular area. The closest point to a corner of the AABB lies on a circle with the expected radius r. It should be noted that only $\frac{1}{4}$ of the circle area is inside of the AABB. The radius r depends on the number of the given points N. If the regular orthogonal distribution of points in the E^2 case forms a mesh of $\sqrt{N} \times \sqrt{N}$ points. However, in the following a uniform distribution of points is expected and the "Salt and Pepper" Chen [5] and Halton's [71] distributions were used in experiments.

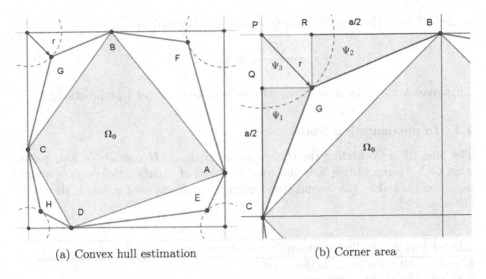

(a) Convex hull estimation   (b) Corner area

Fig. 2. Axis Aligned Bounding Box and convex polygon

Let P_0 is the area of the square with edges of the length a. The P_1 is the area of the circular sector of the radius r, see Fig. 2a, containing just one point.

$$\frac{1}{N} = \frac{P_1}{P_0} \qquad pN = 1 \qquad p = \frac{P_1}{P_0} \qquad N\frac{\frac{1}{4}\pi r^2}{a^2} = 1 \qquad r^2 = \frac{4a^2}{\pi N} \qquad (1)$$

Then in the E^2 case the expected radius r can be estimated as:

$$r = \frac{2a}{\sqrt{\pi}\sqrt{N}} = \frac{2}{\sqrt{\pi}}\frac{a}{\sqrt{N}} \approx 1.1284\frac{a}{\sqrt{N}} \qquad (2)$$

The sizes of the corner's areas $\Psi_1 + \Psi_2 + \Psi_3$ can be estimated as:

$$P_2 = \frac{a}{2}\frac{r}{\sqrt{N}} \qquad (3)$$

It means, that the P_2 area decreases with the value N significantly. If the total number of points is $N = 10^8$, i.e. $\sqrt{N} = 10^4$, then

$$r \approx 1.1284\frac{a}{10^4} = 0.00011284\,a \qquad P_2 \approx \frac{0.00011284}{2\sqrt{N}}\,a^2 \qquad (4)$$

It can be seen, that the estimated point reduction is very high. In the E^3 case, the number of points in the given set is N and only $\frac{1}{8}$ of the corner ball volume are inside of the AABB. The expected radius r can be estimated as:

$$pN = 1 \qquad p = \frac{V_1}{V_0} \qquad N\frac{\frac{1}{8}\frac{4}{3}\pi r^3}{a^3} = 1 \qquad r^3 = \frac{6a^3}{\pi\sqrt[3]{N}} \qquad (5)$$

In the E^3 case, the expected radius r can be estimated as:

$$r = \frac{\sqrt[3]{6}\,a}{\sqrt[3]{\pi}\sqrt[3]{N}} \approx 1.2407\frac{a}{\sqrt[3]{N}} \tag{6}$$

This gives some estimation of the efficiency of the proposed preprocesing.

2.3   Implementation Notes

The algorithm for finding the convex polygon $A,\ldots,H$ consists of four passes with $O(N)$ complexity (N is the total number of points). However, it should be noted that after the second step a significant fraction of points is discarded already.

Algorithm 2. Smallest enclosing circle algorithm

Require: All given points in the Ω set
Ensure: The smallest enclosing circle
  FIND the points forming the AABB, i.e. $A,\ldots,D$.
  SPLIT the points into four areas half-planes based on their position relative to the line segments formed by the points $A,\ldots,D$.
                    ▷ points that do not fit into any of the half-planes, i.e. quad
                    ▷ points inside of $ABCD$ quad, are promptly discarded
  FIND points $E,\ldots,H$.   ▷ distance of the points is measured only to the corner
                    ▷ of their respective half-plane
  REMOVE the remaining points inside of the convex polygon $A,\ldots,H$
              ▷ information about point's half-plane is used to reduce further testing
      ▷ the points in Ω_0 are removed as they cannot influence the final smallest ball
  CALL the Welzl's algorithm [MSW] for the remaining points   ▷ see Fig.5a

---

This approach of discarding the significant fraction of points at the beginning has proven to be superior to the simple point-in-polygon test, as such test needs to find the whole polygon first, which, among other things, requires measuring distance to all four AABB corners per point.[1]

Further reduction of read/write operations can be achieved by using separate data structure for storing the index of a region Ω_i for a given point.

3   Experimental Results

The proposed modification of the Welzl's algorithm was tested for a large number of points (up to $N > 4 * 10^8$ points) in the E^2 case. The Halton's and "Salt and Pepper" distributions were used for experiments. Experiments proved the following expected properties:

---

[1] Also, instead of computing the distance between points d, the $\sqrt{d^2}$ should not be used and d^2 can be used for distance comparisons Skala [52,54]. Same idea can be applied to the radius of a circumscribed circle in Welzl's algorithm.

- the proposed smallest enclosing circle with preprocessing algorithm is of the $O_{exp}(N)$ time complexity even for large data sets (the Halton's and "Salt and Pepper" distributions used),
- timing and significant speed up due to preprocessing, see Fig. 3a and Fig. 3b respectively,

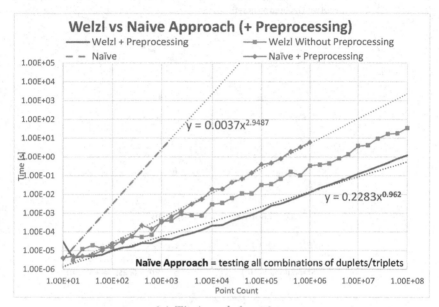

(a) Timing of algorithms

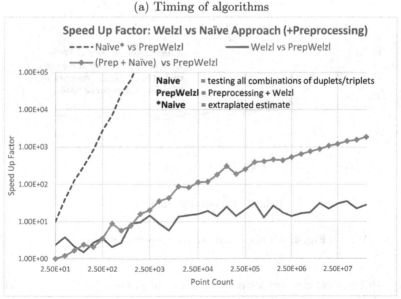

(b) The corner area influence

Fig. 3. Timing and corner areas influence

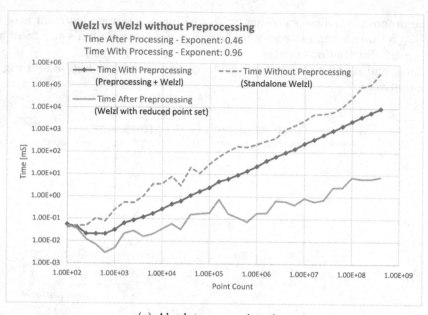

(a) Absolute processing time

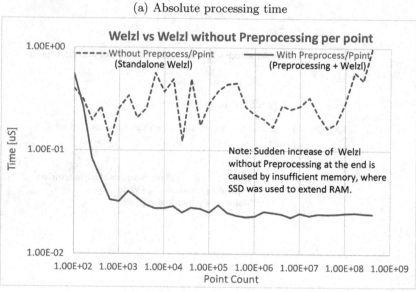

(b) Relative processing time, i.e. $\frac{time}{N}$

Fig. 4. Absolute and relative processing times

- the reduction ratio grows with the number of points $O\sqrt{N}$, see Fig. 5a,
- the relative processing time $time/N$ is nearly constant for $N \geq 10^4$, see Fig. 4b,

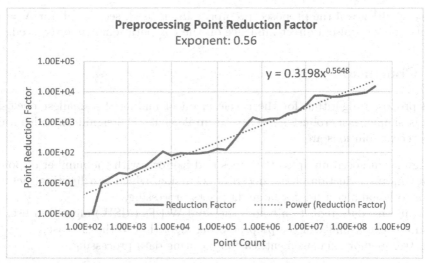

(a) Preprocessing point reduction factor

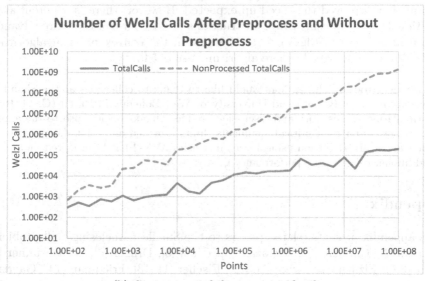

(b) Comparison of the recursive depth

Fig. 5. Reduction and recursive depth

– significant decrease in the recursion depth as the result of preprocessing, which in turn leads to higher memory efficiency.

Implementation was done in C++, x64, compiler MSVC, Windows 10, 16GB RAM, Intel i7-10750H, 2.60GHz, 6 Cores CPU.
One of the main advantages of the proposed preprocessing is the significant reduction of the recursion depth. It can be seen that the depth recursion required

is nearly 10^4 less if the proposed preprocessing is used, see Fig. 5b for $N = 10^8$ points. There is also a direct influence on the computational time required.

4   Conclusion

The proposed algorithm for the acceleration of finding the smallest enclosing ball is simple, fast, robust and easy to implement. The significant advantages over recent solutions are:

- significant speed-up up to 10^2 times and it grows with the number of points,
- significant reduction of the depth of recursion, which is a limiting factor of the original algorithm for large data sets processing,
- significant data reduction before the final Welzl's (MSW) algorithm use,
- simple extensibility of the preprocessing algorithm to the E^3 case,
- better memory management, caching, during data processing,

The Halton's and "Salt and Pepper" distributions were used and the experimental results proved the speed-up expected. However, there is a potential for additional speed-up using SIMD instructions or GPU, use of the more advanced algorithms, e.g. the $O(\lg N)$ or $O(1)$ point in the convex polygon algorithms Skala [51], circumscribed sphere algorithm Skala [54]. [2]

Acknowledgment. The authors would like to thank to colleagues at the Shandong University in Jinan (China) and University of West Bohemia in Pilsen (Czech Rep.) for stimulating of this work and discussions made. Thanks belong also to the anonymous reviewers for their critical comments, hints provided and for unknown relevant references. [3]([3]SIMD version using Intel's intrinsics AVX-2 has been tested and led to an additional roughly 20% performance gain.)

Appendix

This appendix presents related papers to the Smallest Enclosing Ball problem: Agarwal [1], Cavaleiro [2,3], Cazals [4], Chen [5], Drager [6], Edelsbrunner [7], Efrat [8,9], Elzinga [10], wiki: [70–72], Fischer [11–13], Friedman [14], Gaertner [15], Gao [16], Goaoc [17], Har-Peled [18], Jiang [19,20], Kallberg [21], Karmakar [22,23], Krivosija [24], Larsson [25], Li [26], Liu [27], Martinetz [28], Martyn [29], Matousek [30], Megiddo [31], Mordukhovich [32], Mukherjee [33], Munteanu [34], Nam [35], Nielsen [38–40], Nielsen [36,37,41,42], Nock [43], Pan [44], Pronzato [45], Ritter [46], Saha [47], Shen [48], Shenmaier [49], Shi [50], Skyum [60], Smolik [61], Sylvester [62], Tao [63], Wang [64,65], Wei [66], Welzl [67–69], Xu [73,74], Yildirim [75], Zhou [76–78].

---

[2] Responsibilities: Algorithm design and analysis, manuscript preparation - V.Skala, Experimental implementation and verification - M. Cerny and J.Y. Saleh.

References

1. Agarwal, P., Ben Avraham, R., Sharir, M.: The 2-center problem in three dimensions. Comput. Geometry Theory Appl. **46**(6), 734–746 (2013). https://doi.org/10.1016/j.comgeo.2012.11.005
2. Cavaleiro, M., Alizadeh, F.: A faster dual algorithm for the euclidean minimum covering ball problem. Ann. Oper. Res. (2020). https://doi.org/10.1007/s10479-018-3123-5
3. Cavaleiro, M., Alizadeh, F.: A dual simplex-type algorithm for the smallest enclosing ball of balls. Comput. Optim. Appl. **79**(3), 767–787 (2021). https://doi.org/10.1007/s10589-021-00283-6
4. Cazals, F., Dreyfus, T., Sachdeva, S., Shah, N.: Greedy geometric algorithms for collection of balls, with applications to geometric approximation and molecular coarse-graining. Comput. Graph. Forum **33**(6), 1–17 (2014). https://doi.org/10.1111/cgf.12270
5. Chen, Q.Q., Hung, M.H., Zou, F.: Effective and adaptive algorithm for pepper-and-salt noise removal. IET Image Proc. **11**(9), 709–716 (2017). https://doi.org/10.1049/iet-ipr.2016.0692
6. Drager, L., Lee, J., Martin, C.: On the geometry of the smallest circle enclosing a finite set of points. J. Franklin Inst. **344**(7), 929–940 (2007). https://doi.org/10.1016/j.jfranklin.2007.01.003
7. Edelsbrunner, H., Virk, Z., Wagner, H.: Smallest enclosing spheres and chernoff points in bregman geometry. Leibniz International Proceedings in Informatics, LIPIcs **99**, 351–3513 (2018). https://doi.org/10.4230/LIPIcs.SoCG.2018.35
8. Efrat, A., Sharir, M., Ziv, A.: Computing the smallest k-enclosing circle and related problems. In: Dehne, F., Sack, J.-R., Santoro, N., Whitesides, S. (eds.) WADS 1993. LNCS, vol. 709, pp. 325–336. Springer, Heidelberg (1993). https://doi.org/10.1007/3-540-57155-8_259
9. Efrat, A., Sharir, M., Ziv, A.: Computing the smallest k-enclosing circle and related problems. Comput. Geometry Theory Appl. **4**(3), 119–136 (1994). https://doi.org/10.1016/0925-7721(94)90003-5
10. Elzinga, D.J., Hearn, D.W.: The minimum covering sphere problem. Manage. Sci. **19**(1), 96–104 (1972). https://doi.org/10.1287/mnsc.19.1.96
11. Fischer, K., Gartner, B.: The smallest enclosing ball of balls: combinatorial structure and algorithms. Int. J. Comput. Geom. Appl. **14**(4–5), 341–378 (2004). https://doi.org/10.1142/s0218195904001500
12. Fischer, K., Gärtner, B.: The smallest enclosing ball of balls: combinatorial structure and algorithms. In: Proceedings of the Annual Symposium on Computational Geometry, pp. 292–301 (2003)
13. Fischer, K., Gärtner, B., Kutz, M.: Fast smallest-enclosing-ball computation in high dimensions. In: Di Battista, G., Zwick, U. (eds.) ESA 2003. LNCS, vol. 2832, pp. 630–641. Springer, Heidelberg (2003). https://doi.org/10.1007/978-3-540-39658-1_57
14. Friedman, F., Stroudsburg, E.: Minimal enclosing circle and two and three point partitions of a plane. In: Proceedings of International Conference on Scientific Computing (2006)
15. Gärtner, B.: Fast and robust smallest enclosing balls. In: Nešetřil, J. (ed.) ESA 1999. LNCS, vol. 1643, pp. 325–338. Springer, Heidelberg (1999). https://doi.org/10.1007/3-540-48481-7_29

16. Gao, S., Wang, C.: A new algorithm for the smallest enclosing circle. In: Proceedings of the 2018 8th International Conference on Management, Education and Information (MEICI 2018), pp. 562–567. Atlantis Press, December 2018. https://doi.org/10.2991/meici-18.2018.111
17. Goaoc, X., Welzl, E.: Convex hulls of random order types. In: Leibniz International Proceedings in Informatics, LIPIcs 164 (2020). https://doi.org/10.4230/LIPIcs.SoCG.2020.49
18. Har-Peled, S., Mazumdar, S.: Fast algorithms for computing the smallest k-enclosing circle. Algorithmica (New York) **41**(3), 147–157 (2005). https://doi.org/10.1007/s00453-004-1123-0
19. Jiang, Y., Cai, Y.: A reformulation-linearization based algorithm for the smallest enclosing circle problem. J. Ind. Manage. Optim. **17**(6), 3633–3644 (2021). https://doi.org/10.3934/jimo.2020136
20. Jiang, Y., Luo, C., Ling, S.: An efficient cutting plane algorithm for the smallest enclosing circle problem. J. Ind. Manage. Optim. **13**(1), 147–153 (2017). https://doi.org/10.3934/jimo.2016009
21. Kallberg, L., Shellshear, E., Larsson, T.: An external memory algorithm for the minimum enclosing ball problem. VISIGRAPP 2016 - Proceedings of the 11th Joint Conference on Computer Vision, Imaging and Computer Graphics Theory and Applications, pp. 83–90 (2016). https://doi.org/10.5220/0005675600810088
22. Karmakar, A., Roy, S., Das, S.: Fast computation of smallest enclosing circle with center on a query line segment. In: CCCG 2007–19th Canadian Conference on Computational Geometry, pp. 273–276 (2007)
23. Karmakar, A., Roy, S., Das, S.: Fast computation of smallest enclosing circle with center on a query line segment. Inf. Process. Lett. **108**(6), 343–346 (2008). https://doi.org/10.1016/j.ipl.2008.07.002
24. Krivosija, A., Munteanu, A.: Probabilistic smallest enclosing ball in high dimensions via subgradient sampling. Leibniz International Proceedings in Informatics, LIPIcs 129 (2019). https://doi.org/10.4230/LIPIcs.SoCG.2019.47
25. Larsson, T., Kallberg, L.: Fast and robust approximation of smallest enclosing balls in arbitrary dimensions. Comput. Graph. Forum **32**(5), 93–101 (2013). https://doi.org/10.1111/cgf.12176
26. Li, X., Ercan, M.F.: An algorithm for smallest enclosing circle problem of planar point sets. In: Gervasi, O., Murgante, B., Misra, S., Rocha, A.M.A.C., Torre, C., Taniar, D., Apduhan, B.O., Stankova, E., Wang, S. (eds.) ICCSA 2016. LNCS, vol. 9786, pp. 309–318. Springer, Cham (2016). https://doi.org/10.1007/978-3-319-42085-1_24
27. Liu, Y.-F., Diao, R., Ye, F., Liu, H.-W.: An efficient inexact newton-CG algorithm for the smallest enclosing ball problem of large dimensions. J. Oper. Res. Soc. China **4**(2), 167–191 (2015). https://doi.org/10.1007/s40305-015-0097-8
28. Martinetz, T., Mamlouk, A., Mota, C.: Fast and easy computation of approximate smallest enclosing balls. In: Brazilian Symposium of Computer Graphic and Image Processing, pp. 163–168 (2006). https://doi.org/10.1109/SIBGRAPI.2006.20
29. Martyn, T.: Tight bounding ball for affine IFS attractor. Comput. Graph. (Pergamon) **27**(4), 535–552 (2003). https://doi.org/10.1016/S0097-8493(03)00089-X
30. Matoušek, J., Sharir, M., Welzl, E.: A subexponential bound for linear programming. Algorithmica (New York) **16**(4–5), 498–516 (1996). https://doi.org/10.1007/bf01940877, the code available at https://news.ycombinator.com/item?id=14475832
31. Megiddo, N.: On the ball spanned by balls. Discrete Comput. Geometry **4**(6), 605–610 (1989). https://doi.org/10.1007/BF02187750

32. Mordukhovich, B., Nam, N., Villalobos, C.: The smallest enclosing ball problem and the smallest intersecting ball problem: existence and uniqueness of solutions. Optim. Lett. **7**(5), 839–853 (2013). https://doi.org/10.1007/s11590-012-0483-7

33. Mukherjee, D.: Reduction of two-dimensional data for speeding up convex hull computation (2022). arxiv:2201.11412, https://arxiv.org/pdf/2201.11412.pdf

34. Munteanu, A., Sohler, C., Feldman, D.: Smallest enclosing ball for probabilistic data. In: Proceedings of the Annual Symposium on Computational Geometry, pp. 214–223 (2014). https://doi.org/10.1145/2582112.2582114

35. Nam, N., Nguyen, T., Salinas, J.: Applications of convex analysis to the smallest intersecting ball problem. J. Convex Anal. **19**(2), 497–518 (2012)

36. Nielsen, F.: The siegel-klein disk: Hilbert geometry of the siegel disk domain. Entropy **22**(9) (2020). https://doi.org/10.3390/e22091019

37. Nielsen, F., Hadjeres, G.: Approximating covering and minimum enclosing balls in hyperbolic geometry. In: Nielsen, F., Barbaresco, F. (eds.) GSI 2015. LNCS, vol. 9389, pp. 586–594. Springer, Cham (2015). https://doi.org/10.1007/978-3-319-25040-3_63

38. Nielsen, F., Nock, R.: Approximating smallest enclosing balls. In: Laganá, A., Gavrilova, M.L., Kumar, V., Mun, Y., Tan, C.J.K., Gervasi, O. (eds.) ICCSA 2004. LNCS, vol. 3045, pp. 147–157. Springer, Heidelberg (2004). https://doi.org/10.1007/978-3-540-24767-8_16

39. Nielsen, F., Nock, R.: A fast deterministic smallest enclosing disk approximation algorithm. Inf. Process. Lett. **93**(6), 263–268 (2005). https://doi.org/10.1016/j.ipl.2004.12.006

40. Nielsen, F., Nock, R.: On approximating the smallest enclosing bregman balls. In: Proceedings of the Annual Symposium on Computational Geometry 2006, pp. 485–486 (2006). https://doi.org/10.1145/1137856.1137931

41. Nielsen, F., Nock, R.: On the smallest enclosing information disk. Inf. Process. Lett. **105**(3), 93–97 (2008). https://doi.org/10.1016/j.ipl.2007.08.007

42. Nielsen, F., Nock, R.: Approximating smallest enclosing balls with applications to machine learning. Int. J. Comput. Geom. Appl. **19**(5), 389–414 (2009). https://doi.org/10.1142/S0218195909003039

43. Nock, R., Nielsen, F.: Fitting the smallest enclosing bregman ball. In: Gama, J., Camacho, R., Brazdil, P.B., Jorge, A.M., Torgo, L. (eds.) ECML 2005. LNCS (LNAI), vol. 3720, pp. 649–656. Springer, Heidelberg (2005). https://doi.org/10.1007/11564096_65

44. Pan, S., Li, X.: An efficient algorithm for the smallest enclosing ball problem in high dimensions. Appl. Math. Comput. **172**(1), 49–61 (2006). https://doi.org/10.1016/j.amc.2005.01.127

45. Pronzato, L.: On the elimination of inessential points in the smallest enclosing ball problem. Optim. Methods Softw. **34**(2), 225–247 (2019). https://doi.org/10.1080/10556788.2017.1359266

46. Ritter, J.: An efficient bounding sphere, pp. 301–303. Graphics Gems, Academic Press Professional, Inc (1990)

47. Saha, A., Vishwanathan, S., Zhang, X.: New approximation algorithms for minimum enclosing convex shapes. Proceedings of the Annual ACM-SIAM Symposium on Discrete Algorithms pp. 1146–1160 (2011). https://doi.org/10.1137/1.9781611973082.86

48. Shen, K.W., Wang, X.K., Wang, J.Q.: Multi-criteria decision-making method based on smallest enclosing circle in incompletely reliable information environment. Comput. Ind. Eng. **130**, 1–13 (2019). https://doi.org/10.1016/j.cie.2019.02.011

49. Shenmaier, V.: Complexity and approximation of the smallest k-enclosing ball problem. Eur. J. Comb. **48**, 81–87 (2015). https://doi.org/10.1016/j.ejc.2015.02.011

50. Shi, Y.Z., Wang, S.T., Wang, J., Deng, Z.H.: Fast classification for nonstationary large scale data sets using minimal enclosing ball. Kongzhi yu Juece/Control and Decision **28**(7), 1065–1072 (2013)

51. Skala, V.: Trading time for space: an O(1) average time algorithm for point-in-polygon location problem: theoretical fiction or practical usage? Mach. Graph. Vision **5**(3), 483–494 (1996)

52. Skala, V.: Barycentric coordinates computation in homogeneous coordinates. Comput. Graph. (Pergamon) **32**(1), 120–127 (2008). https://doi.org/10.1016/j.cag.2007.09.007

53. Skala, V.: Fast $o_{expected}(n)$ algorithm for finding exact maximum distance in E2 instead of $O(N^2)$ or $O(N\ lgN)$. AIP Conf. Proc. **1558**, 2496–2499 (2013). https://doi.org/10.1063/1.4826047

54. Skala, V.: A new robust algorithm for computation of a triangle circumscribed sphere in E3 and a hypersphere simplex. In: AIP Conference Proceedings 1738 (2016). https://doi.org/10.1063/1.4952269

55. Skala, V.: Diameter and convex hull of points using space subdivision in E^2 and E^3. In: Gervasi, O., Murgante, B., Misra, S., Garau, C., Blečić, I., Taniar, D., Apduhan, B.O., Rocha, A.M.A.C., Tarantino, E., Torre, C.M., Karaca, Y. (eds.) ICCSA 2020. LNCS, vol. 12249, pp. 286–295. Springer, Cham (2020). https://doi.org/10.1007/978-3-030-58799-4_21

56. Skala, V., Majdisova, Z.: Fast algorithm for finding maximum distance with space subdivision in E2. LNCS **9218**, 261–274 (2015). https://doi.org/10.1007/978-3-319-21963-9_24

57. Skala, V., Majdisova, Z., Smolik, M.: Space subdivision to speed-up convex hull construction in E3. Adv. Eng. Softw. **91**, 12–22 (2016). https://doi.org/10.1016/j.advengsoft.2015.09.002

58. Skala, V.: Fast $o_{expected}(n)$ algorithm for finding exact maximum distance in E2 instead of $O(N^2)$ or $O(N\ lgN)$. In: Misra, S., Gervasi, O., Murgante, B., Stankova, E., Korkhov, V., Torre, C., Rocha, A.M.A.C., Taniar, D., Apduhan, B.O., Tarantino, E. (eds.) ICCSA 2019. LNCS, vol. 11619, pp. 367–380. Springer, Cham (2019). https://doi.org/10.1007/978-3-030-24289-3_27

59. Skala, V., Smolik, M., Majdisova, Z.: Reducing the number of points on the convex hull calculation using the polar space subdivision in E2. SIBGRAPI **2016**, 40–47 (2017). https://doi.org/10.1109/SIBGRAPI.2016.015

60. Skyum, S.: A simple algorithm for computing the smallest enclosing circle. Inf. Process. Lett. **37**(3), 121–125 (1991). https://doi.org/10.1016/0020-0190(91)90030-L

61. Smolik, Z.M., Skala, V.: Efficient speed-up of the smallest enclosing circle algorithm. Informatica, pp. 1–11 (2022). https://doi.org/10.15388/22-INFOR477, accepted for publication, online 2022-03-23

62. Sylvester, J.: A question in the geometry of situation. Quarterly J. Pure Appl. Math. **1**, 79 (1857). https://doi.org/10.1049/iet-ipr.2016.0692

63. Tao, J.W., Wang, S.T.: Large margin and minimal reduced enclosing ball learning machine. Ruan Jian Xue Bao/Journal of Software **23**(6), 1458–1471 (2012). https://doi.org/10.3724/SP.J.1001.2012.04071

64. Wang, Y., Li, Y., Chang, L.: Approximate minimum enclosing ball algorithm with smaller core sets for binary support vector machine. In: 2010 Chinese Control and Decision Conference, CCDC 2010, pp. 3404–3408 (2010). https://doi.org/10.1109/CCDC.2010.5498584

65. Wang, Y., Li, Y., Tan, K.L.: Coresets for minimum enclosing balls over sliding windows. Proceedings of the ACM SIGKDD International Conference on Knowledge Discovery and Data Mining, pp. 314–323 (2019). https://doi.org/10.1145/3292500.3330826

66. Wei, L.Y., Anand, A., Kumar, S., Beri, T.: Simple methods to represent shapes with sample spheres. In: SIGGRAPH Asia 2020 Technical Communications, SA 2020 (2020). https://doi.org/10.1145/3410700.3425424

67. Welzl, E.: Smallest enclosing disks (balls and ellipsoids). In: Maurer, H. (ed.) New Results and New Trends in Computer Science. LNCS, vol. 555, pp. 359–370. Springer, Heidelberg (1991). https://doi.org/10.1007/BFb0038202

68. Welzl, E.: Geometric optimization and unique sink orientations of cubes. In: Fiala, J., Koubek, V., Kratochvíl, J. (eds.) MFCS 2004. LNCS, vol. 3153, pp. 176–176. Springer, Heidelberg (2004). https://doi.org/10.1007/978-3-540-28629-5_9

69. Welzl, E.: The smallest enclosing circle - a contribution to democracy from switzerland? Algorithms Unplugged, pp. 357–360 (2011). https://doi.org/10.1007/978-3-642-15328-0_36

70. Wikipedia contributors: Article: Smallest-circle problem - Wikipedia, the free encyclopedia (2021). https://en.wikipedia.org/wiki/Smallest-circle_problem. Accessed 29 Jan 2022

71. Wikipedia contributors: Halton sequence - Wikipedia, the free encyclopedia (2021). en.wikipedia.org/wiki/Halton_sequence. Accessed 27 Jan 2022

72. Wikipedia contributors: Talk: Smallest-circle problem - Wikipedia, the free encyclopedia (2021). https://en.wikipedia.org/wiki/Talk:Smallest-circle_problem. Accessed 29 Jan 2022

73. Xu, J., Bu, F., Si, W., Qiu, Y., Chen, Z.: An algorithm of weighted Monte Carlo localization based on smallest enclosing circle. In: Proceedings - 2011 IEEE International Conference on Internet of Things and Cyber, Physical and Social Computing, iThings/CPSCom 2011, pp. 157–161 (2011). https://doi.org/10.1109/iThings/CPSCom.2011.67

74. Xu, S., Freund, R., Sun, J.: Solution methodologies for the smallest enclosing circle problem. Comput. Optim. Appl. 25(1–3), 283–292 (2003). https://doi.org/10.1023/A:1022977709811

75. Yildirim, E.: Two algorithms for the minimum enclosing ball problem. SIAM J. Optim. 19(3), 1368–1391 (2008). https://doi.org/10.1137/070690419

76. Zhou, G., Tohemail, K.C., Sun, J.: Efficient algorithms for the smallest enclosing ball problem. Comput. Optim. Appl. 30(2), 147–160 (2005). https://doi.org/10.1007/s10589-005-4565-7

77. Zhou, Q., Zhu, H.S., Xu, Y.J., Li, X.W.: Smallest enclosing circle based localization approach for wireless sensor networks. Tongxin Xuebao/J. Commun. 29(11), 84–90 (2008)

78. Zhou, Y., Yang, B., Wang, J., Zhu, J., Tian, G.: A scaling-free minimum enclosing ball method to detect differentially expressed genes for RNA-SEQ data. BMC Genomics 22(1), 479 (2021). https://doi.org/10.1186/s12864-021-07790-0

Elastic Resource Allocation Based on Dynamic Perception of Operator Influence Domain in Distributed Stream Processing

Fan Liu[1,2], Weilin Zhu[1,2], Weimin Mu[1(✉)], Yun Zhang[1], Mingyang Li[1], Ziyuan Zhu[1], and Weiping Wang[1]

[1] Institute of Information Engineering, Chinese Academy of Sciences, Beijing, China
{liufan,zhuweilin,muweimin,zhangyun,limingyang,zhuziyuan,
wangweiping}@iie.ac.cn
[2] School of Cyber Security, University of Chinese Academy of Sciences,
Beijing, China

Abstract. With the development of distributed stream processing systems, elastic resource allocation has become a powerful means to deal with the fluctuating data stream. The existing methods either focus on a single operator or only consider the static correlation between operators to perform elastic scaling. However, they ignore the dynamic correlation between operators in data stream processing applications, which leads to lagging and inaccuracy resource allocation, increasing processing latency. To address these issues, we propose an elastic resource allocation method, which is based on the dynamic perception of operator influence domain, to perform resource allocation dynamically and in advance. The experimental results show that compared with the existing methods, our method not only guarantees that the end-to-end latency meets QoS requirements but also reduces resource utilization.

Keywords: Data stream processing · Dynamic correlation · Adaptive partition · Meta-learning

1 Introduction

Distributed stream processing systems (DSPSs) can quickly analyze and mine the real-time value of data, which are powerful means to process continuous and massive data. In DSPSs, resource allocation determines the operator parallelism of data stream processing applications (DSPAs), which is the key to ensure the QoS of DSPAs. Because data streams exhibit the characteristics of dynamic fluctuation and mutation, elastic resource allocation has become a dominant method.

Many researchers have proposed elastic resource allocation methods. Zhang et al. [1] monitors the actual processing time of input data load for each operator, which is compared with the required time to identify the bottlenecks, and increases the parallelism of those operators to maximize the throughput of DSPAs in shared memory multi-core architectures. Mu et al. [2] predicts the load of each operator and compares it with the processing performance to allocate resources quantitatively. These methods focus on a single operator to perform elastic scaling and do not consider the correlation between operators. Actually, the changes in the upstream operators may create ripple effects throughout DSPAs [3,4], and some researchers utilize the correlation between operators. Lombardi et al. [5] calculates the load of an operator based on the input load of the DSPA and the average selectivity of its upstream operators, and then analyzes the CPU utilization of the operator under the load to determine whether to scale elastically. Wei et al. [6] calculates the load of an operator according to the DSPA's load, average selectivity, and average network bandwidth of its upstream operators, which is then compared with the performance to adjust the parallelism. However, these methods ignore the dynamic characteristic of correlation between operators. This will result in lagging and inaccurate resource allocation, which increases processing latency in DSPSs.

In this paper, we propose an elastic resource allocation method based on the dynamic perception of operator influence domain to adjust the operator parallelism dynamically and in advance. The contributions of our work are as follows:

- To the best of our knowledge, our work is the first to utilize the dynamic correlation between operators for resource allocation. We use the static selectivity metrics and dynamic selectivity statistic metrics to evaluate the influence domain of upstream operators. Accordingly, we divide the DSPA into partitions adaptively, and plan the parallelism of the operators in units of partitions.
- We use the random forest regression (RFR) [7] to model the correlation between operators within each partition online and update it dynamically. For the input load of each partition, we compute the optimal parallelism of each operator in this partition.
- We use the meta-learning method to predict the load of each partition online. To the best of our knowledge, this is the first to combine the strong expressive long short term memory networks (LSTM) meta-learner and the efficient multi-layer perceptron (MLP) base-learner to catch the fluctuations features of the data stream in real-time.
- The experimental results show that our work ensures that the end-to-end latency meets QoS requirements while improving resource utilization efficiency.

The rest of this paper is organized as follows. Section 2 introduces the motivation of our work. Section 3 describes the design and implementation. Section 4 shows the experimental results. Finally, Sect. 5 concludes our paper.

2 Motivation

2.1 Dynamic Correlation Between Upstream and Downstream Operators

In DSPAs, the data stream is processed by the upstream operators and sent to the downstream operators. As mentioned above, when the load or parallelism of the upstream operator changes, the downstream operator will change accordingly. We take the word count application as an example. The directed acyclic graph (DAG) of the application is shown in Fig. 1. The operator *parser* parses each received article, *filtter* filters duplicate articles, *splitter* splits each article into words, and *counter* counts the frequency of each word. We illustrate our motivation with the operators *parser*, *filtter* and *splitter*, whose input loads are expressed in the number of articles received per second.

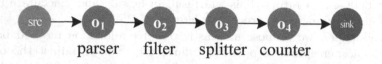

$$src \longrightarrow o_1 \longrightarrow o_2 \longrightarrow o_3 \longrightarrow o_4 \longrightarrow sink$$
parser   filter   splitter counter

Fig. 1. The DAG of word count application.

In Fig. 2(a), as the load of *parser* increases or decreases, the load of *filter* also increases or decreases. To handle the increased load, we increase the parallelism of *parser*, and subsequently the parallelism of *filter* also increases, as shown in Fig. 2(b). Thus, we get that the changes of load and parallelism of the upstream operator affect that of the downstream operator. Besides, we compute the Pearson Correlation Coefficient of load and parallelism between *parser* and *filter*, as shown in Table 1. We get that the correlation between the upstream and downstream operator is time-varying.

Table 1. Pearson Correlation Coefficient of load and parallelism between upstream and downstream operators.

Epoch	CORR (*parser,filter*)		CORR (*parser,splitter*)	
	Load	Parallelism	Load	Parallelism
Epoch 0–9	0.9973	0.9489	−0.0626	0.01856
Epoch 10–19	0.9993	0.9836	0.9993	0.9972

2.2 Dynamic Influence Domain of Upstream Operators

The operators in the DAG are not globally correlated but show the characteristics of local correlation. We find that there is little or no correlation between the upstream operator *parser* and downstream operator *splitter*. As shown in Fig. 2(a), from *epoch* 0 to 9, the load of *parser* increases first and then decreases, while that of *splitter* fluctuates. The reason is that *parser* receives a lot of duplicate articles during that period, which are filtered by *filter*.

So the correlation of load between *parser* and *splitter* is not obvious, as shown in Table 1. From *epoch* 10 to 19, the load of *parser* increases and that of *splitter* also increases. The load between *parser* and *splitter* shows a strong correlation. In Fig. 2(b) we get the same results for the parallelism of *parser* and *splitter*. Thus, we get that the influence of upstream operators on downstream operators has a range, which is named the operator influence domain. Besides, the influence domain of operators is time-varying, as shown in Table 1.

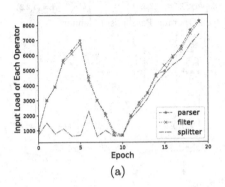
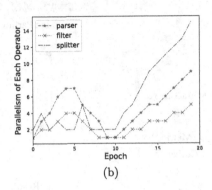

(a)   (b)

Fig. 2. The load and parallelism of operators in the DAG.

2.3 Importance of Dynamic Perception of Operator Influence Domain

QoS Guarantee. If we can accurately analyze the influence domain of upstream operators, we can adjust the parallelism of downstream operators within the domain in advance to satisfy the end-to-end latency requirements better.

System Stability. If we can refer to the upstream operators to adjust the parallelism of downstream operators in the influence domain, we can avoid adjustment jitter to improve system stability.

System Overhead. If we can model each influence domain instead of each operator, we will reduce the computational overhead and resource overhead.

3   Design and Implementation

3.1 Overview

We describe our model in Fig. 3. It contains three core modules: the online load predictor (OLPredictor), the adaptive operator partitioner (AOPartitioner), and the online partition-based operator parallelism planner (OPPlanner). We use the AOPartitioner to dynamically divide the DAG into partitions. The division is based on the influence domain of the upstream operators. Then we refer to the partition results and use the OPPlanner to determine the parallelism of operators in each partition. Besides, we use the OLPredictor to predict the load of each partition online to achieve proactive elastic scaling.

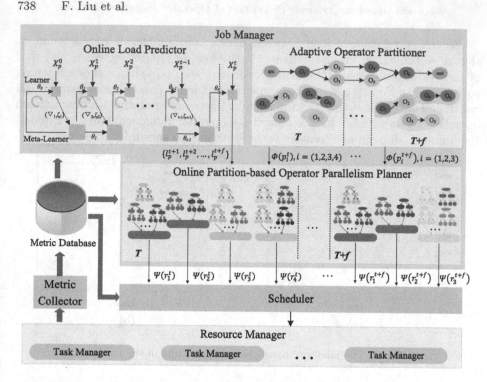

Fig. 3. Architecture.

The work process of our model mainly contains five stages. Firstly, the Metric Collector collects the dynamic selectivity statistic metrics of theoretically unstable operators, the load metrics of each partition, the parallelism metrics of operators in each partition covering the load and the resource metrics of each node, and stores them in MetricDatabase [8]. Secondly, the AOPartitioner refers to the static selectivity metrics and dynamic selectivity statistic metrics to evaluate the influence domain of upstream operators and partitions the DAG adaptively. Thirdly, the OPPlanner uses the load and operator parallelism metrics to build the RFR to plan the optimal parallelism of operators in units of partitions. Fourthly, the OLPredictor uses the load metrics to accurately predict the load of each partition in the future online. And we input the predicted load into the OPPlanner to get the optimal operator parallelism of each partition in the future. Finally, we refer to the scheduler [9] in our previous work to place the instances of operators in each partition on appropriate nodes.

3.2   AOPartitioner: Adaptive Operator Partitioner

Operator Classification. In DSPSs, a DSPA is usually modeled as a directed acyclic graph (DAG), expressed as $G = (O, D)$. In a DAG, a vertex represents an operator o_i ($o_i \in O$) for data processing and an edge represents the data stream d_{ij} ($d_{ij} \in D$), which flows from operator o_i to o_j. While the DSPA is running, the total input rate of the operator o_u^+ is $r_{in}(u)$. Then it sends the processed data to its downstream operators. The data output rate to each downstream operator is $r_{out}(u, v)$. We use O_u^- to denote the set of downstream operators.

There are various operators in the DAG, such as transformation, union, filtering operators, and so on. The selectivity of an operator is defined as the ratio between the data output rate and the total input rate. We divide the operators into two categories: stable operators and unstable operators, based on their selectivity as follows.

$$\begin{cases} o_u^+ \ is \ stable & \forall o_v^- \in O_u^-, \ r_{out}(u,v)/r_{in}(u) \ is \ constant \\ o_u^+ \ is \ unstable & \exists o_v^- \in O_u^-, \ r_{out}(u,v)/r_{in}(u) \ is \ variable \end{cases}$$

When the operator is deployed to DSPSs, we can infer its theoretical selectivity based on the operator's data processing logic, called the static selectivity. For the theoretically unstable operator, we can get the actual selectivity when the operator is running, called the dynamic selectivity.

Input and Output. We use $s_s(o_i)$ to denote the static selectivity metrics of the operator o_i, and $s_d(o_i)$ to represent the dynamic selectivity statistic metrics of unstable operators from MetricDatabase. So we use the dataset $D_{s_s} = \{s_s(o_1), s_s(o_2), ...\}$ and $D_{s_d} = \{s_d(o_j), s_d(o_k), ...\}$ as the input. We use the operator partition results $\phi(p_i^t) = \{o_j, o_k, ...\}$ as the output, where p_i^t denotes the ith partition at time t.

Partitioning Module. We propose an adaptive operator partitioning algorithm, called AOPA, to partition the DAG during runtime. Our AOPA consists of two phases: the startup and running phase, as shown in Algorithm 1. In the startup phase, we use the static selectivity metrics of operators to solve the cold-start problem of partitioning. At first, we divide the operators into stable and unstable sets according to the static selectivity. Then, we cut off all the input edges of their downstream operators for the operators in the unstable set. At last, we find all connected sub-graphs based on the result of the second step and aggregate them into a partition. In the running phase, we analyze the dynamic selectivity statistic metrics of unstable operators during runtime and update stable and unstable sets to re-partition the DAG.

In AOPA, we use the trend of online statistics to judge the dynamic selectivity of operators and partition the DAG more accurately. We also present a solution for the cold-start problem.

3.3   OLPredictor: Online Load Predictor

Input and Output. We predict the multi-step load of partition in the future with the past multi-step load. We use the dataset $X_p^t = \{L_p^{t-W+1}, L_p^{t-W+2}, ..., L_p^t\}$ as the input, in which $L_p^t = (l_p^{t-h+1}, l_p^{t-h+2}, ..., l_p^t)$. In particular, we use t to represent the current time, h to represent the length of historical time window and W to represent the number of continuous load sequences. So we use l_p^t as the load of partition p at time t, L_p^t as the load sequence of partition p over the past h time period and X_p^t as continuous multi-step sequences of load with size of W. Besides, we use the dataset $Y_p^t = \{l_p^{t+1}, l_p^{t+2}, ..., l_p^{t+f}\}$ as the output, which denotes the load of partition p in

---

Algorithm 1. AOPA.

Step 1

1: **for** *operator* $\in$ *operatorSet* **do**
2:    **if** *operator* is theoretically stable **then**
3:       *stableSet*.put(*operator*)
4:    **else**
5:       *unstableSet*.put(*operator*)
6:    **end if**
7: **end for**

Step 2

1: **for** *operator* $\in$ *unstableSet* **do**
2:    **for** *downOperator* $\in$ *operator*.allDownOperators() **do**
3:       *downOperator*.inactivateAllInputs()
4:    **end for**
5:    *operator*.inactivateAllOutputs()
6: **end for**

Step 3

1: *minimalClusters* = NewEmptyClusterCollection
2: **for** *operator* $\in$ *operatorSet* **do**
3:    *cluster* = *minimalClusters*.findOrBuildNewRelatedCluster(*operator*)
4:    *cluster*.add(*operator*.allActivatedDownOperators ())
5:    *cluster*.add(*operator*.allActivatedUpOperators ())
6:    **for** *cOperator* $\in$ *cluster* **do**
7:       **if** *cOperator* $\in$ *minimalCluster* **then**
8:          combine(*minimalClusters*.getBeforeCluster(*cOperator*),*cluster*)
9:       **end if**
10:    **end for**
11: **end for**

Step 4

1: **while** the DSPA is running **do**
2:    **for** *operator* is theoretically unstable **do**
3:       **if** *operator* $\in$ *unstableSet* AND *operator* is stable in the recent period **then**
4:          *unstableSet*.remove(*operator*), *stableSet*.put(*operator*)
5:       **end if**
6:       **if** *operator* $\in$ *stableSet* AND *operator* is unstable in the recent period **then**
7:          *stableSet*.remove(*operator*), *unstableSet*.put(*operator*)
8:       **end if**
9:    **end for**
10:    **if** *unstableSet* is changed **then**
11:       repeat *Step 2* and *Step 3*
12:    **end if**
13: **end while**

---

the future f time period. In addition, we use the Min-Max scaler to normalize all the load metrics to the range $[0,1]$.

Prediction Networks. The load of each partition is a time series that exhibits long-term trends and short-term fluctuation. In order to capture the short-term fluctuation characteristics and realize accurate prediction of future load online, the prediction method is required to update the model quickly and accurately to give the latest inference results after receiving new data at each moment.

Compared with the existing methods, the meta-learning method [10], combining the meta-learner and base-learner, can not only learn the long-term regular characteristics of data but also capture the short-term unique characteristics, showing powerful nonlinear generalization ability. The Meta-LSTM method [11] uses the LSTM model as the meta-learner to guide the convolutional neural network (CNN) base-learner for classification training and achieves good performance. In this paper, we use the Meta-LSTM method for online load prediction. However, the complexity of the CNN model is very high, and it is difficult to quickly update the model after receiving new data. So we use a simple and efficient MLP network as the base-learner. We propose a model combining the LSTM meta-learner with the MLP base-learner.

During the training process, the MLP base-learner trains the arriving small sample of data to learn the short-term fluctuation characteristics. The LSTM meta-learner summarizes the training results in the base learner and then provides the base learner with better initial values on the new data. Specifically, after receiving new data, we calculate the loss function value and loss function gradient value of the MLP base-learner and input them into the LSTM meta-learner to update the cell state. The meta-learner provides updated parameters to the base-learner.

3.4 OPPlanner: Online Partition-based Operator Parallelism Planner

Input and Output. We use the output of the OLPredictor $Y_p^t = \{l_p^{t+1}, l_p^{t+2}, ..., l_p^{t+f}\}$ and the output of the AOPartitioner $\phi(p_i^t) = \{o_j, o_k, ...\}$ as the input of our OPPlanner. We use $\psi(r_i^t) = \{(o_j, n_j), (o_k, n_k), ...\}$ to denote the optimal operator parallelism as the output, in which o_j is the operator in the partition p_i and n_j is the optimal number of operator instances.

Planning Module. For each partition in the DAG, we collect the metrics of load and the optimal parallelism of each operator through experiments. The optimal parallelism of an operator is the minimum number of instances that can handle the load while meeting the QoS requirements. We spend a long time collecting data and the dataset is relatively small. Then we build an operator parallelism planner for each partition. Because the partitions of the DAG change dynamically over time, the planners are also time-varying. Besides, we need to update our planning model as we collect more data.

The relationship between the load and the operator parallelism in a partition is complex and nonlinear. And the dataset we collected is relatively small. In order to improve the accuracy of modeling and avoid overfitting problems, we use the ensemble learning method. The ensemble learning method integrates many weak models to improve the accuracy and robustness, which is effective for small sample learning.

Compared with the boosting models, such as the Adaptive Boosting (Adaboost) and the gradient boosting decision tree (GBDT) [12], the random forest regression (RFR) [7] adopts the bootstrap strategy and has strong anti-overfitting ability, which performs better. So in this paper, we use the RFR to build the relationship between the load and operator parallelism in each partition.

3.5   Scheduler

We use the scheduler from our previous work [9] to place the operators to the appropriate nodes. We build a cost model to evaluate the total cost of all elastic-scaling actions for all operators from the start time t_s to end time t_e. The cost is defined as:

$$W = \sum_{t \in [t_s, t_e]} w_t, \quad w_t = \sum_{o \in V} (c_o^r n_{ot}^r + c_o^u n_{ot}^u + c_o^d n_{ot}^d) \tag{1}$$

where c_o^r denotes the cost of running an instance of operator o per unit time, n_{ot}^r denotes the number of instances of operator o running at time t, c_o^u denotes the startup cost of an instance of operator o, n_{ot}^u denotes the number of instances of operator o that started at time t, c_o^d denotes the stop cost of an instance of operator o, and n_{ot}^d denotes the number of instances of operator o that stopped at time t.

4   Experiments

4.1   Settings and Datasets

Settings. Our experiments run on a cluster with eight servers. There are two GPU servers and six CPU servers in the cluster. Both GPU servers are comprised of 36 cores Intel Xeon CPU E5-2697 v4 2.30 GHz, 256GB memory, two NVIDIA GeForce GTX 1060ti cards, and 500GB disks. All CPU servers are comprised of 36 cores Intel Xeon CPU E5-2697 v4 2.30 GHz, 256GB memory, and 500GB disks. One GPU server is used to run Job Manager and MetricDatabase, and the other is used to train and evaluate our proposed model. Six CPU servers are used as Task Manager to run the instances of operators on Kubernetes. The version of our Kubernetes is v1.22.0.

Datasets. We build four databases from our online DataDock system [8]. The first database collects the dynamic selectivity statistic metrics of unstable operators. The second dataset collects the load and the optimal operator parallelism metrics of each partition to build the OPPlanner. The third dataset collects the load of each partition in 30 d to build the OLPredictor. We divide the dataset into two sets: a training set (from the beginning to the 20th day), and the test set (from the 21st day to the last day). The fourth dataset collects the resource metrics of each node.

We use a real preprocessing application to evaluate the performance of our algorithm. The DAG of the application is shown in Fig. 4, where the blue operators are unstable and the green are stable. The operators and edges cover the operator classification in Sect. 3.2.

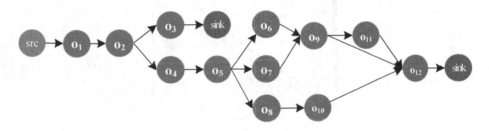

Fig. 4. The DAG of a real preprocessing application.

4.2 Evaluation

To evaluate the performance of resource allocation, we compare our method with the BriskStream [1], ELYSIUM [5] and Pec [6]. The goal of resource allocation is to ensure that the latency meets the QoS requirements while reducing resource utilization. So we use the end-to-end latency guarantee and the total cost to evaluate the performance of our method.

End-to-End Latency Guarantee. We run the preprocessing application on the DataDock [8] and record the end-to-end latency of four models. The results are shown in Fig. 5(a). Compared with other methods, the end-to-end latency of our model is smaller and always stays stable. As a result of considering the dynamic correlation of operators in the DAG, we can accurately adjust the parallelism of operators in units of partitions before the load changes. So we can process data stream load in time. While the BriskStream adjusts the parallelism of an operator after its load changes, which is lagging and increases data processing latency. The ELYSIUM and Pec adjust the parallelism of operators in units of DAG before the load changes, but they refer to the average selectivity to compute the load of operators for parallelism adjustment, which is inaccurate and the error is always larger for the more downstream operators.

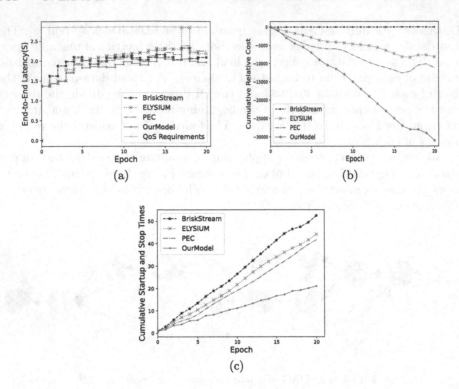

Fig. 5. The overall results of different methods.

Total Cost. We calculate the cost of each operator at each epoch and get the total cost of all operators in a period of epoch. In our experiment, we set $c_o^r = 8000$, $c_o^u = 3000$ and $c_o^d = 3000$ and collect the times of startup and stop action of operators.

We take the baseline of BriskStream to get the relative cost. Figure 5(b) represents the cumulative relative cost. Figure 5(c) represents the cumulative startup and stop times of all operators.

Compared with other methods, our method costs and adjusts less. The reason is that we can accurately adjust the parallelism of operators and reduce the number of startup and stop. Besides, we take the cost of startup or stop into consideration to improve the cost-effectiveness of adjustment.

4.3   AOPartitioner Evaluation

The AOPartitioner refers to the static selectivity metrics to get the initial partitions, which address the cold-start problem of partitioning. The initial partition result is depicted in Fig. 6(b). Then we analyze the collected dynamic selectivity statistic metrics of theoretically unstable operators and find the selectivity of operator o_9 is constant from *epoch* 332 to 450, as shown in the red part of Fig. 6(a). So we re-partition the DAG and the result is depicted in Fig. 6(c).

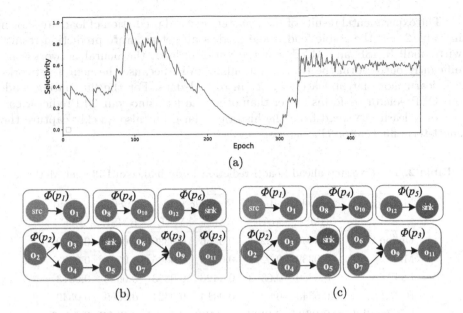

Fig. 6. The adaptive partition results.

4.4   OLPredictor Evaluation

To enhance the overall performance, we select the better prediction method in the OLPredictor. So we compare the prediction performance of our Meta-learning based OLPredictor with different methods, including separate LSTM, separate MLP, online support vector regression (SVR), and bidirectional gated recurrent units (BiGRU) [13]. We use the RMSE and the MAE to evaluate the performance.

The main parameters in the compared methods are depicted as follows. In our Meta-learning, we use *adam* as the optimizer, 0.001 as the learning rate and 16 as the size of hidden layer for the meta-learner, and use *relu* as the activation function, *mse* as the loss, and 128 as the size of hidden layer for the base-learner. In the LSTM, MLP, and BIGRU model, we all use *relu* as the activation function, *adam* as the optimizer, *mse* as the loss, and 128 as the size of hidden layer. In the online SVR model, we use 200 as the window size, *rbf* as the kernel, *scale* as the gamma, and *c* is set to 100.0.

The experimental results of the five-step forward prediction of load are shown in Table 2. For the stable load, these models all get accurate prediction results with small RMSE and MAE. For the periodic load, the neural networks significantly outperform the traditional online SVR, because the neural networks can learn more latent features from historical data. For the fluctuating load, our OLPredictor performs better than other models, since our OLPredictor can not only learn the trend from the historical data, but also quickly capture the short-term fluctuation characteristics.

Table 2. The Five-step ahead Load Prediction Performance of Different Methods.

Method	Stable load		Periodic load		Fluctuating load	
	RMSE	MAE	RMSE	MAE	RMSE	MAE
Online SVR	0.0348	0.0287	0.0456	0.0330	0.0541	0.0469
MLP	0.0339	0.0273	0.0439	0.0385	0.0515	0.0473
LSTM	0.0389	0.0302	0.0426	0.0353	0.0625	0.0513
BiGRU	0.0343	0.0272	0.0414	0.0324	0.0669	0.0549
OLPredictor	**0.0327**	**0.0258**	**0.0408**	**0.0312**	**0.0502**	**0.0406**

4.5   OPPlanner Evaluation

To enhance the overall performance, we select the better operator parallelism planning method in the OPPlanner. So we compare the planning performance of different methods, including the RFR model, the SVR model, the Adaboost model, and GBDT model.

The main parameters are depicted as follows. In our RFR based OPPlanner, we set $estimators = 100$. In the SVR model, we set $gamma = 'scale'$, $c = 1.0$, $kernel = 'rbf'$. In the Adaboost model, we set $estimators = 150$, $learning\ rate = 1.0$. In the GBDT model, we set $estimators = 100$, $loss = 'ls'$. Other parameters in the models are set to default values.

The performance of the above methods for partitions in Fig. 6(b) is shown in Table 3. The ensemble learning models, including the RFR, Adaboost, and GBDT, significantly outperforms the SVR model. The reason is that the ensemble learning models integrate many weak models to learn from a small sample effectively. Besides, our OPPlanner performs better than other ensemble learning models, since the bootstrap strategy of RFR avoids overfitting and improves robustness.

Table 3. The Planning Performance of Each Partition of Different Methods.

Partitions		SVR	Adaboost	GBDT	**OPPlanner**
p_1	RMSE	0.0649	0.0642	0.0641	**0.0467**
	MAE	0.0396	0.0370	0.0370	**0.0338**
p_2	RMSE	0.0764	0.0674	0.0634	**0.0591**
	MAE	0.0676	0.0362	0.0327	**0.0312**
p_3	RMSE	0.0758	0.0714	0.0628	**0.0602**
	MAE	0.0572	0.0357	0.0326	**0.0306**
p_4	RMSE	0.0562	0.0553	0.0548	**0.0523**
	MAE	0.0451	0.0214	0.0208	**0.0217**
p_5	RMSE	0.0559	0.0452	0.0541	**0.0437**
	MAE	0.0483	0.0143	0.0214	**0.0210**
p_6	RMSE	0.0794	0.0782	0.0714	**0.0549**
	MAE	0.0452	0.0426	0.0357	**0.0345**

5 Conclusion

In this paper, we present an elastic resource allocation method based on the dynamic perception of operator influence domain. It contains three core modules: the AOPartitioner, the OLPredictor, and the OPPlanner. Firstly, we use the AOPartitioner to adaptively partition the DAG based on the dynamic influence domain of upstream operators. Then we use the OLPredictor based on Meta-learning to get the online multi-step prediction result of load for each partition. At last, we use the OPPlanner to model the load and the optimal parallelism of operators in each partition with RFR. The experimental results illustrate our method is better than the state-of-the-art methods on the real-world datasets. We can ensure that the end-to-end latency meets the QoS requirements while reducing resource utilization.

References

1. Zhang, S., He, J., Zhou, A.C., He, B.: Briskstream: scaling data stream processing on shared-memory multicore architectures. In: Boncz, P.A., Manegold, S., Ailamaki, A., Deshpande, A., Kraska, T. (eds.) Proceedings of the 2019 International Conference on Management of Data, SIGMOD Conference 2019, Amsterdam, The Netherlands, June 30–5 July 2019, pp. 705–722. ACM (2019)
2. Mu, W., Jin, Z., Zhu, W., Liu, F., Li, Z., Zhu, Z., Wang, W.: QEScalor: quantitative elastic scaling framework in distributed streaming processing. In: Krzhizhanovskaya, V.V., Závodszky, G., Lees, M.H., Dongarra, J.J., Sloot, P.M.A., Brissos, S., Teixeira, J. (eds.) ICCS 2020. LNCS, vol. 12137, pp. 147–160. Springer, Cham (2020). https://doi.org/10.1007/978-3-030-50371-0_11
3. Borkowski, M., Hochreiner, C., Schulte, S.: Minimizing cost by reducing scaling operations in distributed stream processing. Proc. VLDB Endow. **12**(7), 724–737 (2019)

4. Hung, B.D.T., Omori, T., Ohnishi, T.: Ripple effect analysis of data flow requirements. In: van Sinderen, M., Maciaszek, L.A. (eds.) Proceedings of the 14th International Conference on Software Technologies, ICSOFT 2019, Prague, Czech Republic, 26–28 July, 2019, pp. 262–269. SciTePress (2019)
5. Lombardi, F., Aniello, L., Bonomi, S., Querzoni, L.: Elastic symbiotic scaling of operators and resources in stream processing systems. IEEE Trans. Parallel Distributed Syst. **29**(3), 572–585 (2018)
6. Wei, X., Li, L., Li, X., Wang, X., Gao, S., Li, H.: Pec: proactive elastic collaborative resource scheduling in data stream processing. IEEE Trans. Parallel Distributed Syst. **30**(7), 1628–1642 (2019)
7. Breiman, L.: Random forests. Mach. Learn. **45**(1), 5–32 (2001)
8. Mu, W., Jin, Z., Wang, J., Zhu, W., Wang, W.: BGElasor: elastic-scaling framework for distributed streaming processing with deep neural network. In: Tang, X., Chen, Q., Bose, P., Zheng, W., Gaudiot, J.-L. (eds.) NPC 2019. LNCS, vol. 11783, pp. 120–131. Springer, Cham (2019). https://doi.org/10.1007/978-3-030-30709-7_10
9. Liu, F., Jin, Z., Mu, W., Zhu, W., Zhang, Y., Wang, W.: DROAllocator: a dynamic resource-aware operator allocation framework in distributed streaming processing. In: He, X., Shao, E., Tan, G. (eds.) NPC 2020. LNCS, vol. 12639, pp. 349–360. Springer, Cham (2021). https://doi.org/10.1007/978-3-030-79478-1_30
10. Thrun, S.: Lifelong Learning Algorithms. In: Thrun, S., Pratt, L. (eds.) Learning to Learn, pp. 181–209. Springer, Boston (1998). https://doi.org/10.1007/978-1-4615-5529-2_8
11. Ravi, S., Larochelle, H.: Optimization as a model for few-shot learning. In: 5th International Conference on Learning Representations, ICLR 2017, Toulon, France, April 24–26, 2017, Conference Track Proceedings, OpenReview.net (2017)
12. Friedman, J.H.: Greedy function approximation: a gradient boosting machine. Ann. Stat. **29**(5), 1189–1232 (2001)
13. Qin, Y., Song, D., Chen, H., Cheng, W., Jiang, G., Cottrell, G.W.: A dual-stage attention-based recurrent neural network for time series prediction. In: Proceedings of the Twenty-Sixth International Joint Conference on Artificial Intelligence, IJCAI 2017, Melbourne, Australia, 19–25 August, 2017, pp. 2627–2633 (2017)

PRISM: Principal Image Sections Mapping

Tomasz Szandała[(✉)] [iD] and Henryk Maciejewski [iD]

Wroclaw University of Science and Technology, Wroclaw, Poland
{Tomasz.Szandala,Henryk.Maciejewski}@pwr.edu.pl

Abstract. Rapid progress in machine learning (ML) and artificial intelligence (AI) has brought increased attention to the potential vulnerability and reliability of AI technologies. To counter this issue a multitude of methods has been proposed. Most of them rely on Class Activation Maps (CAMs), which highlight the most important areas in the analyzed image according to the given model. In this paper we propose another look into the problem. Instead of detecting salient areas we aim to identify features that were recognized by the model and compare this insight with other images. Thus giving us information: which parts of the picture were common, which were unique for a given class. Proposed method has been implemented using PyTorch and is publicly available on GitHub: https://github.com/szandala/TorchPRISM.

Keywords: Deep learning · Xai · Convolutional neural networks

1 Introduction

Deep neural networks show excellent prediction accuracy, but the reasoning for their predictions is often difficult to understand [1]. Moreover it has been proven that classification might be based on latent factors unbeknown to the user. Visualization techniques aim to inspect the model and to prevent reliance on incorrect features. Their task is to identify parts on an input that contributed the most to the output.

Most attribution methods are based on backpropagation of the network's activations from the output back to the input. They are usually a modification of the backpropagation algorithm [2–6] and, for computer vision models, take the form of a saliency map that highlights the decisive regions on the input image.

We are introducing a new method that relies on Principal Component Analysis of features detected by neural network models and interpolates them on the initial image. We call it the Principal Image Sections Mapping (PRISM). The result of the formula is an RGB colored image mask that assigns one color to each feature identified by the model. Moreover the same color will be used to highlight the same feature across all pictures processed in the same batch, of course only if it is present in other samples. This allows exposure of a comparative set of features between images processed in the same batch, and thus facilitates the Explanation by Example technique.

In this paper we aim to provide a solution to identify features that contributed to the given classification. During our research, we have performed a deduction experiment

D. Groen et al. (Eds.): ICCS 2022, LNCS 13350, pp. 749–760, 2022.
https://doi.org/10.1007/978-3-031-08751-6_54

based on VGG-16 pretrained model [7]. We take two very similar classes and try to deduce discriminative features between them by clipping the input pictures. Finally we try to generate an adversal attack on the model by swapping identified features between images [8].

2 State-of-the-Art

One of the main concerns with the deep networks is that they provide no visual output by themselves, therefore it is not possible to know which part of the image influenced the decision the most [12]. As a consequence, there is a demand for methods that can help visualize or explain the decision-making process of such networks and make them understandable for humans.

Many methods have been devised for this purpose. The first noteworthy group consists of methods that rely on gradient backpropagation. They generate maps of pixels that influence the output of the network.

Other methods can be grouped under that term Class Activation Maps. Here we can find original CAM, Grad-CAM [2] as well several of its derivatives like Grad-CAM++ [6], full Grad-CAM [5] and Excitation Backpropagation [4]. The output of these techniques is a heatmap, where the regions that contributed the most for a certain class are highlighted at most. The setback for them is that they are computed for a single chosen convolutional layer. The deeper the layer is, the reliable given saliency map is. However to visualize the obtained map on a referenced image it has to be extrapolated. Since the deepest layers consist of the smallest convolutional masks. the output map might be in a low resolution.

To overcome this obstacle another method has been introduced: Guided Grad-CAM [3]. In it we just multiply the result of Guided Backpropagation and Grad-CAM methods. This gives us pixels that contribute the most to the given class. Likewise any other CAM method can be combined with Guided Backpropagation to determine the most significant pixels.

Apart from these methods several more have been proposed, but our attention caught Explanation by Example. Introduced by Chen et al. in 2019 [9] and rated as the most preferred explanation style by the average non-technical end-users [10]. In input domains spanning visual, audio, and sensory data, explanation by nearest training examples offers users an opportunity to compare features across a test input and similarly mapped ground-truth examples. In short: among the training set it points to a set of images that results in a similar response to the examined one. As stated by Jeyakumar et al. it was the most obvious for non-technical consumers, but it is not grounded enough to make deterministic conclusions.

3 Problem Description

Our goal was to provide a more faithful method to compare features among similar images. As seen on a picture below, features that determine both wolves and coyotes are their body and head (Fig. 1).

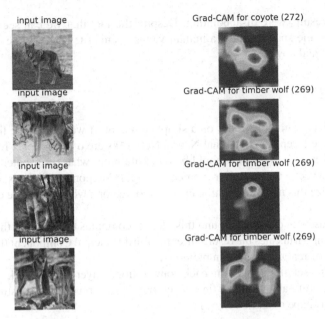

Fig. 1. Grad-CAM output for several pictures

We considered this as insufficient conclusion, therefore we have applied the Guided Backpropagation and combined it with Grad-CAM result which can be seen in Fig. 2. Results bring us closer to the features that distinguish these two species.

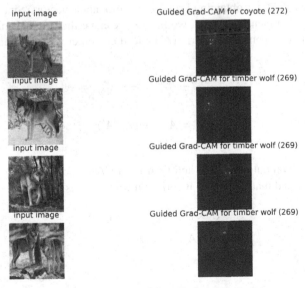

Fig. 2. Guided Grad-CAM result for several pictures

Still the results were disappointing. Despite the fact that we can see features that contributed to the specific class, a human viewer cannot tell which one differentiates between wolf and coyote.

4 Method

We proposed a method that relies on a simple concept: if we look at the final convolutional layer, the Deep Convolutional Neural Networks are only complex representation generators, therefore our focus should be on explaining: what contributed to the given representation vector the most. Here comes Principal Component Analysis, which allows us to reorganize this map into significance-sorted vectors, which could be consequently analyzed further.

PRISM has been designed around this idea. It computes the PCA for the last convolutional layer and truncates the results after the third Principal Component thus receiving an RGB map of features as seen in picture 1.

Assume is a set of outputs from each convolutional layer in a network. We choose a single output, preferable from the final layer, but there is no obstacle to study any other layer.. It has a shape of $n \times c \times h \times w$:

n - number of images in a batch
c - number of channels in layer
h,w - height and width of each mask in this layer.
v - multiplication of n*h*w.

Instead of straightforward PCA for dimensionality reduction we use Singular Value Decomposition (SVD) for obtaining Principal Components vectors. Since SVD is applicable only to two dimensional data, we have to reshape the original four dimensional batch into the two dimensional matrix (A') and then center it by subtracting the mean. (Eq. 1).

$$A^{n \times c \times h \times w}_{|\mathbb{A}|} \to^{reshape} \to A^{v \times c}_{|\mathbb{A}|} = A'$$

$$A'' = A' - mean(A') \tag{1}$$

Now we can compute the SVD and then the PCA outcome (Eq. 2). Last step is to reshape the obtained matrix back to its original four-dimensional form.

$$U \times S \times V^T = svd(A'') \tag{2}$$
$$A_{PCA} = U \times S$$

Fig. 3. PRISM raw output for final convolutional VGG-16 depicting coyote and timber wolf class representatives

This procedure results in a checkered representation of the processed images (Fig. 3). This Could be further processed using Gradual Extrapolation [11] (Fig. 4) producing a human-recognizable picture with colored features found by the network as seen on Fig. 5.

5 Experiments

In order to prove usability of our method we have performed and analyzed 3 experiments.

The first experiment was meant to use PRISM to find and make human-identifiable features that distinguish 2 potentially similar classes. We have started from coyotes and timber-wolves. We have compared these two classes and prepared an example that proves the differentiating features can be identified using PRISM.

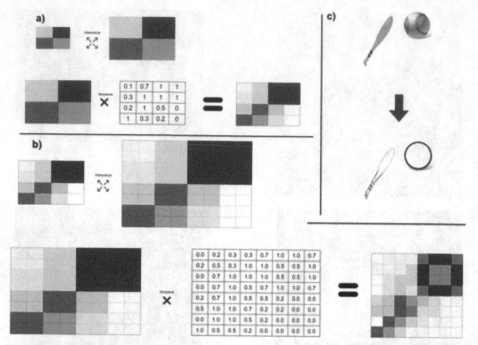

Fig. 4. Schematic concept of gradual extrapolation. Section a) and b) represents Gradual Extrapolation transition from an image of size 2 × 2 to 8 × 8. Section c) shows the concept of transition between image and its restoration using gradual extrapolation

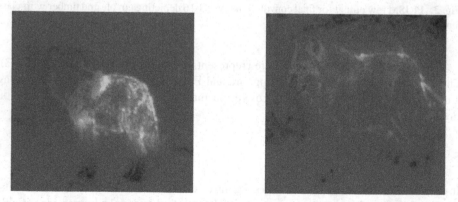

Fig. 5. PRISM output mapped to input image using Gradual Extrapolation technique

In Fig. 6 we see PRISM results for 4 pictures. First we started with full bodies of coyote and timber wolf (columns 2nd and 4th). Pawns, tail and back appear to be the similar feature vectors due to similar colors, but the difference appears to be in the animal's heads. So are taking a closer look at them by clipping the original images. Again features identified by PRISM seem mostly equal, but the differences appear in the eyes and ears areas.

Fig. 6. PRISM output for coyote and its zoomed face as well wolf and its zoomed head

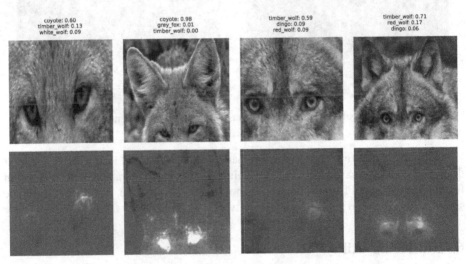

Fig. 7. PRISM output for coyote's and wolf's top parts of the heads and finally eyes

Figure 7 depicts further clipping. First we clip to the top parts of the heads - eyes and ears. This displays similar colors in regards to ears, but different in eyes area, so we perform the final clipping. This leads us to the conclusion that the feature which contributes the most to the distinction between timber wolf and coyote is in their eyes.

In order to check our assumption we have merged coyote eyes into the wolf and other way around. As we see on Fig. 8. the classification changed significantly. Although the most probable class is still a respective animal, the second guess is the one that we tried to induce. Also note that confidence dropped significantly in favor of the second class. Perhaps better results could be achieved with more sophisticated blending instead of simple paste.

timber_wolf: 0.52
coyote: 0.35
red_wolf: 0.12

coyote: 0.58
timber_wolf: 0.33
red_wolf: 0.05

Fig. 8. Classification confidence after swapping eyes between two animals

Of course a question comes, whether basic GradCAM could lead us with a similar deduction path? We have processed the original images and generated GradCAM output for them. As seen in Fig. 9 the outcome is not easily interpretable, therefore it would be significantly harder to generate a similar adversal attack.

6   Clustering Utility

In aforementioned form PRISM is usable mainly for manual inspection of suspicious classes, but it could be used for detecting ambivalent classes if combined with clustering technique, e.g. Self Organizing Maps (SOM). In picture 10 we have presented a first draft of PRISM's clustering utility. We have taken 5 canine classes (color from cluster map in bracket):

- coyote (orange)
- gray fox (red)
- timber wolf (green)
- samoyed (purple)
- border collie (blue)

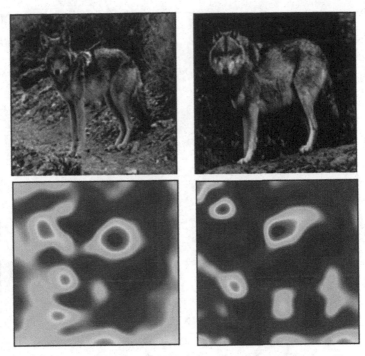

Fig. 9. GradCAM output for original image of coyote and wolf

Instead of drawing their PRISM-generated representation we have used them as feature vectors for SOM clustering. Note that PRISM is generating real domain values for coloring, therefore we had to quantize the colors to reduce the amount to a finite number of possible tints.

From Fig. 10 we can conclude that coyotes could be easily confused with timber wolves and gray foxes. On the other hand the Samoyed and Border collie specimens are well distinguishable from the rest.

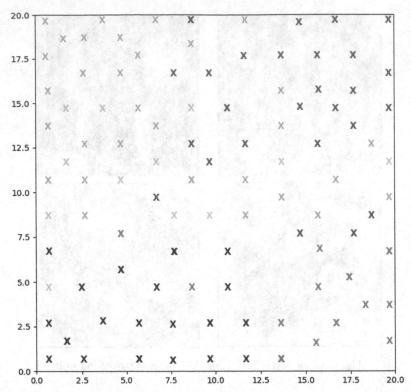

Fig. 10. SOM clustering outcome for 5 arbitrary chosen classes whose representation were generated by PRISM (Color figure online)

7 Conclusions

The proposed method can be a useful tool for CNN examination and can aid in how deep neural networks learn by specifying filters to detect single features, which combined together in the final fully connected layer, provide a classification decision.

Currently the method is dedicated for manual investigation. Our next step will be to improve PRISM's clustering utility, thus giving us an automated detector of problematic classes and possible outliers.

To sum up: PRISM distinguishes from other visualization techniques as it focuses on singular features present on the object. This could give us a multitude of applications, starting from better understanding of the classification process, through automated validation, to pruning of the network.

8 Future Works

The biggest setback of the proposed method is lack of actual saliency indicator. In case of an image displaying 2 or more entirely different classes it will still highlight all detected features as seen in Fig. 11. Left image depicts a dog alongside a mushroom.

Fig. 11. PRISM output displaying detecting fungus features – a minor issue concerning the proposed method

Lack of saliency factor may suggest that both dog and mushroom are equally important for the network, still if we look at the classification scores we see that model mentions only canine classes to be identified on this image. The countermeasure for this could be blending a saliency method into PRISM's output.

References

1. Nguyen, A., Yosinski, J., Clune, J.: Deep neural networks are easily fooled: high confidence predictions for unrecognizable images. In: Proceedings of the IEEE Conference on Computer Vision and Pattern Recognition (2015)
2. Selvaraju, Ramprasaath, R., et al.: Grad-cam: visual explanations from deep networks via gradient-based localization. In: Proceedings of the IEEE International Conference on Computer Vision (2017)
3. Bargal, S.A., et al.: Excitation backprop for RNNs. In: Proceedings of the IEEE Conference on Computer Vision and Pattern Recognition (2018)

4. Zhang, J., et al.: Top-down neural attention by excitation backprop. Int. J. Comput. Vision **126**(10), 1084–1102 (2018)
5. Morbidelli, P., et al.: Augmented grad-CAM: heat-maps super resolution through augmentation. In: ICASSP 2020–2020 IEEE International Conference on Acoustics, Speech and Signal Processing (ICASSP). IEEE (2020)
6. Chattopadhay, A., et al.: Grad-cam++: generalized gradient-based visual explanations for deep convolutional networks. In: 2018 IEEE Winter Conference on Applications of Computer Vision (WACV). IEEE (2018)
7. Simonyan, K., Zisserman, A.: Very deep convolutional networks for large-scale image recognition. arXiv preprint arXiv:1409.1556 (2014)
8. Dinh, V., Ho, L.S.T.: Consistent feature selection for analytic deep neural networks. arXiv preprint arXiv:2010.08097 (2020)
9. Chen, C., et al.: This looks like that: deep learning for interpretable image recognition. Advances in Neural Information Processing Systems (2019)
10. Tomsett, R., et al.: Sanity checks for saliency metrics. In: Proceedings of the AAAI Conference on Artificial Intelligence, vol. 34. No. 04 (2020)
11. Szandała, T.: Enhancing deep neural network saliency visualizations with gradual extrapolation. IEEE Access **9**, 95155–95161 (2021)
12. Behzadi-Khormouji, H., Rostami, H.: Fast multi-resolution occlusion: a method for explaining and understanding deep neural networks. Applied Intelligence, pp. 1–25 (2020)

The New UPC++ DepSpawn High Performance Library for Data-Flow Computing with Hybrid Parallelism

Basilio B. Fraguela(✉)🆔 and Diego Andrade🆔

Universidade da Coruña, CITIC, Grupo de Arquitectura de Computadores. Facultade de Informática, Campus de Elviña, S/N. 15071, A Coruña, Spain
{basilio.fraguela,diego.andrade}@udc.es

Abstract. Data-flow computing is a natural and convenient paradigm for expressing parallelism. This is particularly true for tools that automatically extract the data dependencies among the tasks while allowing to exploit both distributed and shared memory parallelism. This is the case of UPC++ DepSpawn, a new task-based library developed on UPC++ (Unified Parallel C++), a library for parallel computing on a Partitioned Global Address Space (PGAS) environment, and the well-known Intel TBB (Threading Building Blocks) library for multithreading. In this paper we present and evaluate the evolution of this library after changing its engine for shared memory parallelism and adapting it to the newest version of UPC++, which differs very strongly from the original version on which UPC++ DepSpawn was developed. As we will see, while keeping the same high level of programmability, the new version is on average 9.3% faster than the old one, the maximum speedup being 66.3%.

Keywords: Data-flow computing · Hybrid parallelism · PGAS · Runtimes · High-performance computing · Task-based parallelism

1 Introduction

The data-flow computing paradigm is very attractive for parallel computing because it does not impose any particular order of execution among tasks, only relying on the necessary satisfaction of the dependencies inherent to them. This approach is particularly interesting in the presence of complex and irregular patterns of dependencies, which require the fine grained synchronization among tasks that it provides in order to achieve good performance. Furthermore, its application can also simplify the development and maintenance of applications if the paradigm is applied through tools that can automatically extract the tasks dependencies from simple information such as the inputs and the outputs of each task. This is the case of the `depend` clause introduced in OpenMP 4.0, which is the most popular approach in this field. In fact, due to the higher complexity of distributed memory programming, with its need to distribute and

D. Groen et al. (Eds.): ICCS 2022, LNCS 13350, pp. 761–774, 2022.
https://doi.org/10.1007/978-3-031-08751-6_55

communicate data among processes, most proposals in this field are restricted to shared memory systems.

In recent years there has been increasing interest in bringing the advantages of data-flow runtimes to distributed memory environments [2,10,11,14,19,21], not only because they allow working on larger problems, but also because their benefits can be even larger in these systems. Indeed, since a data-flow runtime has all the information on the placement of the data and their relation with the tasks and the dependencies, it can also automate the transmission of data among the distributed memories on which the different processors work, a task particularly cumbersome for programmers in the presence of the complex irregular patterns of dependences for which these tools excel.

This paper focuses on one of the most recent proposals in this area, the UPC++ DepSpawn library [14], which allows exploiting this strategy in clusters of multi-core processors. One of the most original features of this library is that it operates on a Partitioned Global Address Space (PGAS) environment. In this model each process has a private space that is exclusive to it as well as a portion of a common shared space that all the processes can access. As expected, this shared space, which is manipulated by means of one-sided accesses, offers much lower latency and larger bandwidth in the portion that has affinity to the accessing process or to other processes that are located in the same node in the cluster compared to the portions located in other nodes. The PGAS model arguably simplifies the development of distributed memory applications compared to the traditional MPI-based paradigm based on totally separate memory spaces.

The name UPC++ DepSpawn stems from the fact that its PGAS environment is provided by UPC++ [23], a library that supports in C++ the abstractions of the UPC (Unified Parallel C) language [12], while most of the runtime for data-flow computing is derived from a shared-memory library called DepSpawn [17]. Both DepSpawn libraries provide native shared memory parallelism on top of the well-known Intel TBB library.

In the past few years the two external components on which UPC++ DepSpawn relies experienced very strong changes, which called for its redesign. First, while UPC++ had not received meaningful updates since its inception in 2014, a radically new version called UPC++ 1.0 [4] was released in 2019. This version both provided new functionalities and removed features provided by the original library, now called v0.1. Although some of the features lost were intensively used by UPC++ DepSpawn, UPC++ v0.1 had lacked support for a long time, while UPC++ 1.0 has a very active community and is strongly supported. This, together with the advantages of the new version validated in [4] called for changing this component of UPC++ DepSpawn. The second very relevant change was the decision by Intel of replacing Intel TBB by a new library called oneTBB [1]. Similar to the case of UPC++, the new version lacks components intensively used by UPC++ DepSpawn, and since Intel TBB is no longer distributed, an update in this area was also required.

This paper describes the modifications performed in the library due to the changes in UPC++ and TBB. This includes a brief description of our experience with the new limitations in UPC++ 1.0, and more particularly in the aspects related to its interaction with multithreading, which have not been discussed in the literature as far as we know. The changes performed are evaluated showing a positive impact on performance in both cases. The final contribution of the manuscript is the public release of UPC++ DepSpawn under a permissive open source license at https://github.com/UDC-GAC/upcxx_depspawn.

The rest of the paper is organized as follows. The related bibliography is discussed in Sect. 2, which is followed by a description of UPC++ DepSpawn in Sect. 3. The changes performed in the runtime are described in Sect. 4, while Sect. 5 is devoted to the evaluation. Finally, Sect. 6 is devoted to our conclusions and future work.

2   Related Work

There are two main approaches for providing data-flow computing in distributed memory applications, a static one and a dynamic one. The first one requires statically providing the set of tasks and dependencies among them in some domain-specific language (DSL). This is fed to a tool that generates a code that supports the parallel execution of the tasks described respecting the dependencies specified and performing the data communications required. Since the discovery of the dependencies is performed off-line and the code generator has all the information on the whole program, this approach should lead to minimal overheads during the execution and the best possible planning and discovery of potential optimizations. On the other hand, there are two shortcomings to this strategy. One is the impossibility of dealing with dependencies that can be only known at run-time, and sometimes, depending on the tool, with irregular dependencies, even if they can be known statically. The second problem is the need to identify and correctly state all the dependences in the code, which can be cumbersome. The static strategy is followed by the Parameterized Task Graph (PTG) model [13], on which the DAGuE/PaRSEC framework [10] is based. The most notable software developed on this framework is DPLASMA [9], the leading implementation for dense linear algebra algorithms in distributed memory systems.

As for the dynamic approaches, which can deal with dynamic and irregular dependencies, we can break them in two families. The first one relies on a wide array of mechanisms to define and enforce dependencies among tasks thanks to their manipulation during the execution of the program. This is the case of locks, full/empty bits, synchronized blocks, and futures, among others.

The last alternative consists of writing an apparently sequential version of the algorithm in which the parallel tasks are identified, and their inputs and outputs are labelled. Then, during the execution of this instrumented code in a parallel environment, a runtime finds the tasks to run and their dependencies, and it schedules the executions and performs the necessary data transfers in such a way that all the dependencies are met. In our opinion this is the most attractive strategy from the point of view of simplicity and elegance, but it is also the one with the higher expected overheads, as it is the most demanding on the runtime. We identify two main alternatives for implementing this approach. The first one, tested in [11,20], consists of running this code that provides the tasks and dependencies in a single process, which acts as master or client, while all the other processes involved act as slaves or servers, executing the tasks under its control. SYCL [18], which can manage multiple distinct memories and devices under this model from a single host application can also be classified in this group. The second strategy, followed by UPC++ DepSpawn and the MPI-based StarPU [2,3], executes the code in parallel in all the processes involved, so that they all have a view of the task graph and they agree on which process runs what and when. Another key difference of UPC++ DepSpawn with respect to StarPU in addition to the PGAS approach is the transparent exploitation of thread-level parallelism within each process.

It deserves to be mentioned that there are projects that involve both complex runtimes and advanced compilers. For example, a project that started following a library-based dynamic approach but later developed a new language and compiler in order to be able to apply more optimizations and remove runtime overhead was [5], the new alternative being presented in [19]. Another alternative based on a new language that achieves parallelism through data-flow is [21], which is characterized by focusing on workflows and being purely functional.

Finally, although [22] deals with a superscalar scheduler of PLASMA restricted to shared memory, it is strongly related to our work, since it describes the change of this scheduler from a proprietary solution to the OpenMP standard. Our work, however involves changes both in the shared and the distributed memory aspects of our library.

3   Data-flow Computing with UPC++ DepSpawn

Since UPC++ DepSpawn integrates in applications written using UPC++ [23], this section first presents the basics of the latter before introducing UPC++ DepSpawn, so that its syntax and semantics can be understood.

3.1   UPC++

UPC++ [23] supports the development in C++ of parallel programs following the PGAS paradigm and more concretely the concepts of the UPC language [12] but without the need for a new language and related compiler. In UPC++ programs the standard datatypes define data that is located in the local private memory of the process, while the UPC++ types `shared_array<T,B>` and `shared_var<T>` define, respectively, unidimensional arrays of elements and scalars of type `T` located in the shared memory that all the processors can access. By default, when the optional parameter `B` is not provided, the distribution of the elements of the arrays is purely cyclic, so that if there are P processes, element i has affinity to process $i \bmod P$, which means that it is placed in the portion of the shared space associated to that process. The template argument `B` allows changing the distribution to block cyclic, the size of the block being `B`. Relatedly, just as C++ has pointer and reference types, UPC++ provides `global_ptr<T>` and `global_ref<T>` that play the same role, respectively, for the data item placed in the global shared space.

Just as UPC, UPC++ follows a SPMD style, which is enabled by functions that provide a unique identifier to each process as well as the number of processes involved. The library also provides other interesting features, but they are not required to write UPC++ DepSpawn programs.

3.2   UPC++ DepSpawn

UPC++ DepSpawn provides data-flow computing on top of UPC++ in an elegant way that requires very little effort. This procedure can be summarized in three steps. The first one involves writing a sequential version of the code to parallelize where each task is encapsulated as a function that expresses all its dependencies with other tasks through its parameters. This library does not require to label each function parameter in order to inform the runtime on its usage by the function. Rather, the metaprogramming capabilities of C++ are exploited in order to analyze the data type of each formal parameter and infer from it its nature. This way, data passed by value or constant reference are assumed to be exclusively inputs, as the function can read but not modify them, while data passed by non-constant reference is assumed to be both an input and an output. In the case of UPC++ DepSpawn, the references will be objects of the template class `global_ref<T>`, where T will be a `const` type in the case of constant references. Then, the second step involves rewriting the function invocations from the usual `f(a, b, ...)` notation to `upcxx_spawn(f, a, b, ...)`. Finally, an invocation to `upcxx_wait_for_all()` must be inserted at the point where the algorithm finishes and we want to wait for the results.

Listing 1.1 exemplifies the usage of UPC++ DepSpawn for the implementation of a Cholesky decomposition on a shared array A of N×N tiles. The code uses the macro _ to map from a natural bidimensional indexing to the linear indexing required by the unidimensional `shared_array` class provided by UPC++. In addition to the main algorithm, the code includes the definition of the function `dgemm`, responsible for making a matrix product between tiles. From the data types of its parameters UPC++ DepSpawn can infer that `dest` is both read and written by the function, while a and b are only inputs.

As we can see, an algorithm written with UPC++ DepSpawn looks like a sequential implementation, making the development and the maintenance very easy. The code, however, must be run by all the UPC++ processes. This allows each process to learn the set of tasks and dependencies in the code and to schedule the executions making sure that all the dependencies are fulfilled. Each process also independently infers who will be responsible for running each task based on the location of its arguments following a strategy explained in [15]. The main topic introduced in this latter publication is however a variation of `upcxx_spawn` called `upcxx_cond_spawn`. This function takes as first argument a boolean that informs the runtime on whether the task is involved in the part of the global task dependency graph (TDG) that pertains to the tasks executed

```
shared_array<Tile> A(N * N);
#define _(i, j) ((i) * N + (j))
...

void dgemm(global_ref<Tile> dest, Tile a, Tile b) {
  // dest = dest + a x b
}

for(i = 0; i < N; i++) {
  upcxx_spawn(potrf, A[_(i,i)]);
  for(r = i+1; r < N; r++) {
    upcxx_spawn(trsm, A[_(i,i)], A[_(r,i)]);
  }
  for(j = i+1; j < N; j++) {
    upcxx_spawn(dsyrk, A[_(j,i)], A[_(j,j)]);
    for(r = j+1; r < N; r++) {
      upcxx_spawn(dgemm, A[_(r,j)], A[_(r,i)], A[_(j,i)]));
    }
  }
}

upcxx_wait_for_all();
```

Listing 1.1. Cholesky factorization in UPC++ DepSpawn

in this process or not. This can reduce the overhead of the runtime by quickly dismissing tasks that the process can safely ignore.

4 An Improved Runtime

As explained in Sect. 1, UPC++ DepSpawn was deeply changed due to the important recent novelties in the Intel TBB and UPC++, affecting its mechanisms to exploit both shared and distributed memory parallelism. This section discusses the decisions taken and the modifications performed in both fields.

4.1 Shared Memory Parallelism Migration

DepSpawn [17] and UPC++ DepSpawn [14] relied on the Intel TBB not only for the creation of threads and parallel tasks, but also for the scheduling of the ready tasks and the definition of many synchronization objects. At this point we must bear in mind that when DepSpawn [17] was written the C++11 standard, with its support for threading and synchronization, was not fully available everywhere, and relying on the Intel TBB solved this. Nowadays it no longer makes sense to not rely on the C++11 facilities whenever possible.

In 2020 Intel dropped the Intel TBB library as it had been known, and it integrated it into its oneAPI initiative, which goes well beyond the capabilities we need. The new version is called oneTBB [1], and while the high level functionalities are mostly the same, the low level tasking API was removed. Since the DepSpawn libraries relied on this API, a change was needed. One possibility was to adapt our libraries to the higher level API of oneTBB. Another option was to move the threading and tasking on top of C++11, designing ourselves the thread and task pools, task packaging and scheduling. This latter alternative was attractive because (a) removing the oneTBB dependency simplifies the deployment of the library, (b) being provided by a company, the portability of oneTBB is much more compromised than that of the C++11 standard, and (c) we suspected that while TBB/oneTBB provide very good performance, we could implement a runtime suited to our needs with even better performance for two reasons. The first one is that the generic nature of the TBB leads to generic APIs, with inherent encapsulation costs, and to a runtime that may be more complex than what we actually need. As a result, by writing our own implementation we could avoid TBB-related overheads. The second and most important reason is the lack of control on the order of execution of the ready tasks in TBB. This is in general an advantage of TBB that allows it to apply complex work-stealing strategies in order to balance the workload among its threads. However, a data-flow computing engine is a specific context in which we have a subset of tasks that are part of a general graph that are ready to be executed. Our intuition is that in this situation, if we have different ready tasks that can be run at a given point, it is better to run the oldest tasks before the newest ones. One reason is that this tends to fulfill older rather than newer dependencies in the TDG, resulting in a more balanced progress through the TDG. This results in more locality in the triggering of dependencies, so that older tasks on which newer tasks depend tend to complete their execution before.

As a result of this analysis, the threading, tasking and scheduling in DepSpawn and UPC++ DepSpawn were rewritten to be based on C++11 instead of the TBB. Our task objects are designed for maximum simplicity and minimum overhead, and just as other objects frequently built and destroyed, they have their own memory pools. Contention was minimized by basing almost every synchronization on atomic operations, resulting in a nearly lockless design. As for our new threading runtime, it is based on a pool of C++11 threads that extracts ready tasks from a thread-safe FIFO queue. The queue has a fixed maximum size in order to reduce memory management overheads and minimize insertion time. When this maximum is reached, the pushing thread runs tasks from the head until there is space for the new task. This policy also avoids storing too many ready tasks without actually participating in their execution.

The queue ensures that tasks are executed in the order in which they are ready, which is something we could not achieve directly with TBB. This helps follow the policy of executing older tasks before newer ones because, as a general trend, the older a task the earlier it tends to be ready for execution. However, our queue does not further enforce the execution of older tasks before newer tasks by acting as a priority queue. That is, it does not reorder its contents by sorting the ready tasks it holds according to their id, which is an integer that is smaller the older the task is. A critical reason for this is that DepSpawn, which provides the new threading system to UPC++ DepSpawn, does not support this id in its tasks, as its original purpose in UPC++ DepSpawn is to uniquely identify a task when its completion is notified to another process. A second reason is that with the current design, push and pop operations are extremely fast thanks to being based on very cheap non-blocking atomic operations and our decision to use a fixed size queue. Enforcing an order among the tasks stored would have implied locks, explorations of the structure upon every push and/or pop, and in general a more complex data structure with a much more expensive management and potential to become a bottleneck for the participating threads.

Two extra steps with low cost were taken to promote the ordered execution of the tasks in each process. First, whenever a task finishes, it releases its dependencies in order, i.e., from the oldest to the newest dependent task, so that tasks tend to enter in order the queue of ready tasks. Second, whenever our runtime receives notifications of termination of remote tasks, they are sorted according to their id so that older tasks release their dependencies before newer tasks.

4.2 Distributed Memory Parallelism Migration

The new UPC++ 1.0 [4] largely differs from the version 0.1 of 2014 on which UPC++ DepSpawn was originally written. In fact the new UPC++ is not an update to the old one, but a complete rewrite that is totally different both at the API and implementation levels. They do not even rely on the same low level libraries, as [23] used GASNet [6] for communications, while UPC++ 1.0 uses the new GASNet-EX [7] communication library for exascale.

Unfortunately, UPC++ 1.0 dropped elements that were used in UPC++ DepSpawn programs such as the distributed arrays or the global references. While UPC++ DepSpawn does not use itself global arrays, as seen in Listing 1.1 this abstraction is convenient for writing global data-flow programs. Global references were more related to UPC++ DepSpawn because they mimicked exactly the same semantics as regular references provide to DepSpawn in shared memory.

The good news is that UPC++ 1.0 offers all the elements needed to build and manipulate globally shared data, and thus any desired components can be built on top of it. We felt that global arrays are a nice abstraction that has been very successful, being widely implemented both by languages [12] and libraries [16], and thus we wrote our own global array class. Our class improves upon the one provided by [23] in several ways, mostly by being bidimensional, supporting not only full but also upper and lower triangular matrices to save space, and providing generic 2D block cyclic distributions and well as simple row and column cyclic distributions or even placement in a single process. Let us notice however that this type is provided for convenience and users are free to obtain the pointers to the global data they want to manipulate from any data structure they wish.

Contrary to global arrays, global references played a direct role in UPC++ DepSpawn. As explained in Sect. 3.2, the library analyzes the type of the formal parameters of the task functions and it infers from it the kind of usage that the function can make of the associated argument. Namely, arguments passed by value or constant reference i.e., parameters of type T or global_ref<const T>, are read-only inputs, while values passed by non-constant reference, that is, with parameters of type `global_ref<T>`, can be both inputs and outputs.

In this case we felt that forcing programmers to use a class written by us that replaced the now unsupported `global_ref` class was not the best option, particularly when we could just assign its meaning to global pointers, which do exist in UPC++ 1.0. This way we decided that when a formal parameter to a spawned function is a global pointer to a constant type, i.e. of the form `global_ptr<const T>`, the implication is that the function will read the associated data through the pointer, but it will not change it, and thus it will be a read-only input. Similarly, global pointer parameters that point to a non-constant type indicate that the function can both read and write the data they point to.

Besides these external changes, the migration to UPC++ 1.0 also implied internal changes in the runtime. Replacing the communication and synchronization mechanisms of [23] by those of [4] was relatively straightforward except for three details. The first one is related to thread safety, which is critical to our runtime given not only its multi-threaded nature, but also the aggressive level of asynchrony and optimization applied. Both versions of UPC++ can be compiled in a thread-safe mode, but while this suffices to ensure thread safety in UPC++ v0.1 this is not the case with the new UPC++. Indeed, as indicated in the UPC++ 1.0 specification [8], important elements such as futures and promises are not thread-safe. As a result, proper measures must be taken to safely manage these objects in a runtime like ours. Our strategy relied on minimizing as much as possible the use of these elements and trying to restrict the use of the ones used to a single thread. Finally, the few futures for which it is beneficial to allow several threads to operate on them are protected by mutexes.

The second issue is related to a new limitation on the activities that user code is allowed to perform when it is executed during calls to UPC++. This user code are the RPCs and the callbacks, as they are only executed when the UPC++ runtime makes what is termed as user-level progress, as opposed to the internal level progress, which cannot be observed by the application. The main way in which user-level progress takes place in UPC++ 1.0 is by invoking the new function `progress`. However, there are other very relevant and typical ways of entering this state such as waiting for a future to be ready. In fact this concept is so important that the UPC++ 1.0 specification [8] informs on the kind of progress that every single function can perform.

The new limitation that constitutes the second problem is that attempts to enter user-level progress in user code that is already run within a user-level progress context may result in a no-op every time. This way, for example, if during a callback or RPC our code tries to wait for a future to be ready, since this implies trying to make user-level progress in code already executed in that mode, the application can, and in fact will, very probably, hang. This situation happened in the UPC++ DepSpawn runtime because this limitation did not exist in [23]. As a result, the runtime was rewritten to move every attempt to enter UPC++ user-level progress out of RPCs and callbacks. This implied the creation of new task queues where tasks that required this kind of progress found during user-level progress were stored, so that they can be safely executed after leaving this state.

A third particularity of the new UPC++ that required special consideration were the new *persona* objects and the notification affinity for futures [8]. This latter concept is derived from the fact that in UPC++ 1.0 each future is associated to a single persona, each persona can only be associated to a single thread at a time, and only that thread is able to signal the completion of the futures associated to that persona. Thus additional care must be taken in the management of futures shared by several threads. In addition, only the thread that owns a special persona called *master persona* can perform important operations, such as executing RPCs. In our runtime any thread must be able to run these tasks if needed. Thus, the new UPC++ DepSpawn runtime also ensures the properly synchronized management of the master persona among all its threads.

5   Evaluation

The experiments have been run in a cluster with 32 nodes, each one consisting of 2 Intel Xeon E5-2680 v3 at 2.5 GHz and 128GB of memory. Since each processor has 12 cores, the experiments use up to 768 cores, which are configured with hyperthreading disabled, so that at most one thread can be run per core. Codes were compiled with g++ 6.4 and optimization level O3, release 2021.3.0 of UPC++ 1.0 being the one used.

The evaluation relies on the right-looking Cholesky factorization in Listing 1.1, the LU decomposition, and the Gauss-Seidel stencil. The BLAS computational kernels benchmarks rely on the OpenBLAS library version 0.3.1 using a single thread. As a result, all the shared memory parallelism comes from the exploitation of our runtime, which is configured to use a single process per node with 24 threads, one per core. Also, all the UPC++ DepSpawn codes apply the optimization presented in [15], which is thus the baseline performance.

Figures 1, 2 and 3 present the performance achieved by the old and the new version of UPC++ DepSpawn for each one of the three benchmarks. Each graph is a strong scaling study that considers two problem sizes and represents the performance as a function of the number of nodes used. In addition, the Cholesky and LU graphs include the performance obtained by the current state-of-the-art implementation, provided by DPLASMA [9] and relying on the same OpenBLAS library for the computational kernels. As explained in Sect. 2, DPLASMA follows a data-flow approach that relies on the Parameterized Task Graph (PTG) model [13] based on a static description of the algorithm. Another difference is that DPLASMA uses MPI for the communications, OpenMPI 2.1.1 being used in the experiments. Every performance point, both for the UPC++ DepSpawn and the DPLASMA measurements, corresponds to the best combination of tile size and matrix mapping on the number of nodes considered following a block cyclic distribution.

As we can see, the improvement is very noticeable in Cholesky. The new version always outperforms the old one, and it allows scaling to 32 nodes the

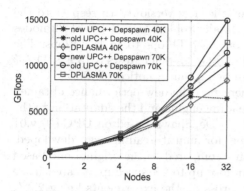

Fig. 1. Performance of the Cholesky descomposition benchmark

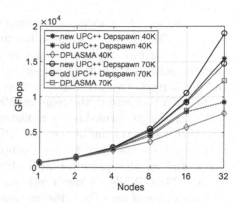

Fig. 2. Performance of the LU descomposition benchmark

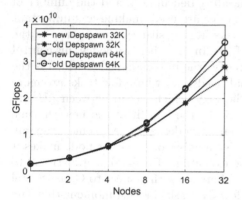

Fig. 3. Performance of the Gauss-Seidel benchmark

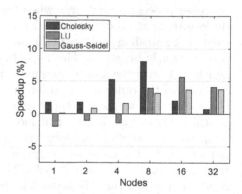

Fig. 4. Speedup of the new multi-threading runtime with respect to the old one

small problem size, which did not scale above 16 nodes with the old version. The improvement also allows UPC++ DepSpawn to clearly outperform DPLASMA for both problem sizes, the difference growing as the number of nodes used increases. In LU, the old UPC++ DepSpawn always scaled reasonably and outperformed DPLASMA, although we can see that the performance growth was slowed down for the small problem size when 32 nodes were used in the evaluation. The new version also systematically increases the performance, particularly as the number of nodes grows, and solves this scalability issue. Finally, Gauss-Seidel is a very lightweight computation with a pattern much simpler than Cholesky and LU. The performance of both versions is basically the same up to 16 nodes, but the tendency of the new version to provide better performance as the number of nodes grows is finally seen when 32 nodes are used.

Altogether, the maximum speedup for the new version with respect to the old one is 66.3%, which happens for the small problem size of LU at 32 nodes thanks to the increase in performance from 9.27 TFlops to 15.41 TFlops. The average speedup across the Cholesky, LU and Gauss-Seidel tests is 13.9%, 11.5% and 2.4%, respectively, the global average being thus a noticeable 9.3%.

In order to measure the relative influence on the new performance obtained of the changes performed in the multithreading engine and the adaptation to the new UPC++, a version of the old UPC++ DepSpawn based on UPC++ v.01 that relied on the new C++11 runtime for multithreading was developed. Figure 4 shows the speedup that it obtains compared to the original one based on Intel TBB. While the new runtime can lose up to 2% speedup, it can achieve improvements of up to 8%, the average across all the experiments being 2.3%. As a result, roughly one fourth of the improvement of the new runtime comes from the changes in the multithreading environment.

LU is the most computationally demanding benchmark, and the number of tasks per process is very large when few nodes are used, making less important the control of the order of execution of tasks in that situation compared to the smart work-stealing facilities of the Intel TBB. In the other two codes this control is more clearly beneficial. This is particularly true in Gauss-Seidel, which is the benchmark with the fewest computational needs and whose few tasks are more sensitive to the order of execution. In Cholesky, which sits in between LU and Gauss-Seidel in terms of computational needs, the benefit of the new runtime is still positive but drops when more than 8 nodes are used. This probably happens because after this point the smaller number of tasks per node makes it more unlikely that there are delays in the execution of tasks that are critical to trigger subsequent dependent tasks. The difference with respect to Gauss-Seidel resides in the fact that the latter has so few live tasks in each moment that any delay in the execution of tasks in the critical path will surely lead to idle cores, while Cholesky has a reasonable number of tasks to keep busy the cores used.

6   Conclusions

UPC++ DepSpawn is a recently introduced library for exploiting task parallelism in distributed memory systems. The tasks are executed under an an efficient scheduler that respects the implicit dependencies among them extracted by the library from their arguments and formal parameter types. In this paper the library largely evolved, changing both its shared and distributed memory components. This way, multithreading was moved from Intel TBB to a manually managed thread pool and related task queue, while in the PGAS component the obsolete and no longer supported UPC++ v0.1 library that relied on GAS-Net was replaced by the new UPC++ 1.0 based on GASNet-EX. This critical redesign of the implementation only involved minor changes in the API, not impacting the ease of use. Performance, however, noticeably improved, reaching maximum speedups of 66.3% when using 768 cores, the average speedup across varying number of cores, benchmarks and problem sizes tested being 9.3%.

The manuscript also describes the challenges that the new limitations imposed by UPC++ 1.0 on multithreading pose for our library, which did not exist in the older version of UPC++, and how we addressed them.

Furthermore, the library has been now made publicly available under an open source license at https://github.com/UDC-GAC/upcxx_depspawn.

As future work we want to evaluate UPC++ Depspawn in conjunction with heterogeneous devices typically found in clusters such as GPUs. Developing distributed BLAS and/or LAPACK libraries on top of UPC++ Depspawn also seems a useful contribution given the results observed.

Acknowledgements. This research was supported by the Ministry of Science and Innovation of Spain (PID2019-104184RB-I00/AEI/10.13039/501100011033), and by the Xunta de Galicia co-founded by the European Regional Development Fund (ERDF) under the Consolidation Programme of Competitive Reference Groups (ED431C 2021/30). We acknowledge also the support from the Centro Singular de Investigación de Galicia "CITIC", funded by Xunta de Galicia and the European Union (European Regional Development Fund- Galicia 2014–2020 Program), by grant ED431G 2019/01. Finally, we acknowledge the Centro de Supercomputación de Galicia (CESGA) for the use of their computers.

References

1. oneAPI Threading Building Blocks (oneTBB). https://github.com/oneapi-src/oneTBB. Accessed 26 Mar 2022
2. Agullo, E., Aumage, O., Faverge, M., Furmento, N., Pruvost, F., Sergent, M., Thibault, S.: Harnessing clusters of hybrid nodes with a sequential task-based programming model. In: Intl. Workshop on Parallel Matrix Algorithms and Applications (PMAA 2014), July 2014
3. Augonnet, C., Thibault, S., Namyst, R., Wacrenier, P.: StarPU: a unified platform for task scheduling on heterogeneous multicore architectures. Concurrency Comput. Practice Exp. **23**(2), 187–198 (2011)
4. Bachan, J., et al.: UPC++: a high-performance communication framework for asynchronous computation. In: 2019 IEEE Intl. Parallel and Distributed Processing Symposium (IPDPS), pp. 963–973, May 2019
5. Bauer, M., Treichler, S., Slaughter, E., Aiken, A.: Legion: expressing locality and independence with logical regions. In: International Conference on High Performance Computing, Networking, Storage and Analysis, SC 2012, pp. 66:1–66:11 (2012)
6. Bonachea, D.: Gasnet specification. Technical report CSD-02-1207, University of California at Berkeley, Berkeley, CA, USA, October 2002
7. Bonachea, D., Hargrove, P.H.: GASNet-EX: a high-performance, portable communication library for exascale. In: Languages and Compilers for Parallel Computing, LCPC 2019, pp. 138–158 (2019)
8. Bonachea, D., Kamil, A.: UPC++ v1.0 Specification, Revision 2021.3.0. Technical report LBNL-2001388, Lawrence Berkeley National Laboratory, March 2021
9. Bosilca, G., et al.: Flexible development of dense linear algebra algorithms on massively parallel architectures with DPLASMA. In: 2011 IEEE International Symposium on Parallel and Distributed Processing Workshops and Phd Forum, pp. 1432–1441, May 2011

10. Bosilca, G., Bouteiller, A., Danalis, A., Hérault, T., Lemarinier, P., Dongarra, J.: DAGuE: a generic distributed DAG engine for high performance computing. Parallel Comput. **38**(1–2), 37–51 (2012)
11. Bueno, J., Martorell, X., Badia, R.M., Ayguadé, E., Labarta, J.: Implementing OmpSs support for regions of data in architectures with multiple address spaces. In: 27th International Conference on Supercomputing, ICS 2013, pp. 359–368 (2013)
12. Burke, M.G., Knobe, K., Newton, R., Sarkar, V.: UPC language specifications, v1.2. Technical report LBNL-59208, Lawrence Berkeley National Lab (2005)
13. Cosnard, M., Loi, M.: Automatic task graph generation techniques. In: 28th Annual Hawaii International Conference on System Sciences, HICSS'28, vol. 2, pp. 113–122, January 1995
14. Fraguela, B.B., Andrade, D.: Easy dataflow programming in clusters with UPC++ DepSpawn. IEEE Trans. Parallel Distrib. Syst. **30**(6), 1267–1282 (2019)
15. Fraguela, B.B., Andrade, D.: High-performance dataflow computing in hybrid memory systems with UPC++ DepSpawn. J. Supercomput. **77**(7), 7676–7689 (2021). https://doi.org/10.1007/s11227-020-03607-1
16. Fraguela, B.B., Bikshandi, G., Guo, J., Garzarán, M.J., Padua, D., von Praun, C.: Optimization techniques for efficient HTA programs. Parallel Comput. **38**(9), 465–484 (2012)
17. González, C.H., Fraguela, B.B.: A framework for argument-based task synchronization with automatic detection of dependencies. Parallel Comput. **39**(9), 475–489 (2013)
18. Reyes, R., Brown, G., Burns, R., Wong, M.: Sycl 2020: more than meets the eye. In: International Workshop on OpenCL, IWOCL 2020 (2020)
19. Slaughter, E., Lee, W., Treichler, S., Bauer, M., Aiken, A.: Regent: a high-productivity programming language for HPC with logical regions. In: International Conference for High Performance Computing, Networking, Storage and Analysis, SC 2015, pp. 81:1–81:12 (2015)
20. Tejedor, E., Farreras, M., Grove, D., Badia, R.M., Almasi, G., Labarta, J.: A high-productivity task-based programming model for clusters. Concurrency Comput. Practice Exp. **24**(18), 2421–2448 (2012)
21. Wozniak, J.M., Armstrong, T.G., Wilde, M., Katz, D.S., Lusk, E., Foster, I.T.: Swift/T: Large-scale application composition via distributed-memory dataflow processing. In: 13th IEEE/ACM International Symposium on Cluster, Cloud, and Grid Computing, pp. 95–102, May 2013
22. YarKhan, A., Kurzak, J., Luszczek, P., Dongarra, J.: Porting the PLASMA numerical library to the OpenMP standard. Int. J. Parallel Program. **45**(3), 612–633 (2017)
23. Zheng, Y., Kamil, A., Driscoll, M.B., Shan, H., Yelick, K.: UPC++: a PGAS extension for C++. In: IEEE 28th International Parallel and Distributed Processing Symposium (IPDPS 2014), pp. 1105–1114, May 2014

Author Index

Abdelfattah, Ahmad 60
Ahn, Kwangwon 682
Al-Ghosoun, Alia 291
Alibasa, Muhammad Johan 276
Altıntaş, Ilkay 611
Anaissi, Ali 276
Andrade, Diego 761
Arzoglou, Ektor 651

Bamha, Mostafa 569
Basaj, Dominika 331
Belhamadia, Youssef 305
Benecki, Pawel 599
Bezbochina, Alexandra 502
Bielak, Piotr 178
Bifet, Albert 460
Bochenina, Klavdiya 164
Borowa, Adriana 318

Cerny, Matej 720
Chandola, Varun 75, 133
Chard, Kyle 403, 417
Chard, Ryan 403
Chen, Kun 233
Chen, Lin 584
Chen, Matthew 417
Chen, Yaowei 584
Chen, Zhipeng 554
Cheng, Zhenyu 345, 431
Choi, Gahyun 682
Cholodowicz, Ewelina 46
Chunaev, Petr 502
Clancy, Richard J. 445
Czarnul, Pawel 667

de Callafon, Raymond A. 611
Deighan, Dwyer 75
Deshmukh, Aadesh 118
DesJardin, Paul E. 75
Ding, Yu 584
Dongarra, Jack 60
Draeger, Erik W. 89

Dudek, Grzegorz 360
Dudziński, Marcin 31
Dziewa-Dawidczyk, Diana 31

Eghbal, Nooshin 488
El-Amrani, Mofdi 305
Elo, Tommi 651

Fan, Li 345
Fang, Fang 473
Foster, Ian 403, 417
Fraguela, Basilio B. 761
Furmańczyk, Konrad 31

Gawron, Karol 692
Gdawiec, Krzysztof 623
Geng, Jinbu 345, 431
Gjini, Rebecca 445
Gościniak, Ireneusz 623
Grindeanu, Iulian 150
Guggilam, Sreelekha 133
Gurhem, Jérôme 389
Guyet, Thomas 460

Hayashi, Akihiro 233
He, Jiayu 541
Heisler, Eric 118
Hong, Yi 584
Hovland, Paul 445
Hsu, Erica 417
Hu, Kai 192
Hu, Wei 584
Huang, Wentao 554
Hückelheim, Jan 445

Jeong, Jongwook 682
Jin, Zhicheng 584
Jo, Kihwan 682

Kajdanowicz, Tomasz 178
Kanclerz, Kamil 248
Khushi, Matloob 541
Kocoń, Jan 692
Kolingerová, Ivana 219

Kortesniemi, Yki   651
Kostrzewa, Daniel   599
Kovantsev, Anton   502
Koza, Blazej   599
Kruczek, Szczepan   318
Krzak, Mateusz   706
Krzywaniak, Adam   667

Lenz, David   150
Levental, Maksim   403
Li, Hao   473
Li, Mingyang   734
Li, Shuhao   345, 431
Li, Yangchun   473
Li, Zeyu   554
Limet, Sébastien   569
Liu, Fan   734
Liu, Qingyun   473
Liu, Zhicheng   431
Lu, Paul   488
Lytaev, Mikhail S.   205

Ma, Shijun   375
Maciejewski, Henryk   262, 749
Mahadevan, Vijay   150
Maheshwari, Ketan   516
Małecki, Krzysztof   706
Mańdziuk, Jacek   17, 103
Menickelly, Matt   445
Moutahir, Fatima-Ezzahrae   305
Mu, Weimin   734
Mycka, Jan   17

Nalluri, Prani   445
Nie, Yufeng   637
Nowosielski, Adam   706

Oleszkiewicz, Witold   331
Orlowski, Przemyslaw   46
Osman, Ashraf S.   291

Paczutkowski, Kacper   31
Pan, Shengli   554
Patra, Abani K.   133
Paul, Sri Raj   233
Peterka, Tom   150
Petiton, Serge   389
Piasecki, Maciej   248
Pogoda, Michał   692
Proficz, Jerzy   667

Puchalska, Daria   178
Putney, Sarah   3

Qin, Rui   345, 431

Randles, Amanda   3, 89
Rivault, Sébastien   569
Robert, Sophie   569
Ropiak, Norbert   692
Roychowdhury, Sayan   89
Ruutu, Sampsa   651

Saleh, Josef Yassin   720
Salunkhe, Amol   75
Sarkar, Vivek   233
Sato, Mitsuhisa   389
Seaid, Mohammed   291, 305
Sheraton, Vivek M.   375
Shi, Jinqiao   554
Silva, Rafael Ferreira da   516
Skala, Vaclav   720
Skluzacek, Tyler J.   417
Stavinova, Elizaveta   502
Suleiman, Basem   276
Sundar, Hari   118
Swędrowski, Michał   692
Szandała, Tomasz   749
Szkandera, Jakub   219
Szyc, Kamil   262

Tabor, Jacek   318
Tan, Li   611
Tanade, Cyrus   3
Tang, Lin   530
Tomov, Stan   60
Tran, Nguyen H.   541
Truong, Harrison   276
Trzciński, Tomasz   331
Tsuji, Miwako   389

Vandromme, Maxence   389
Volokha, Valery   164

Walkowiak, Tomasz   262
Wang, Hailong   473
Wang, Jingya   584
Wang, Meiqi   554
Wang, Weiping   734
Wang, Xuebin   554
Wang, Yan   530
Wildenberg, Gregg   403
Wilkinson, Sean R.   516

Xie, Jie   192
Xing, Jian   584

Yeh, Raine   150
Yue, Ruifeng   530
Yun, Xiaochun   345

Zhang, Jinyu   530
Zhang, Mingyu   473
Zhang, Shangyuan   637

Zhang, Wenbin   460
Zhang, Xiaoyu   584
Zhang, Yaru   584
Zhang, Yun   734
Zhou, Delun   530
Zhu, Mingying   192
Zhu, Weilin   734
Zhu, Ziyuan   734
Zieliński, Bartosz   318, 331
Żychowski, Adam   17, 103

Printed in the United States
by Baker & Taylor Publisher Services